4M2006

Second International Conference on
Multi-Material Micro Manufacture

4M2006

Second International Conference on Multi-Material Micro Manufacture

20–22 September 2006
Grenoble, France

Edited by:

Wolfgang Menz, Germany
Stefan Dimov, Cardiff University, UK
and
Bertrand Fillon, France

Organised by: FP6 4M Network of Excellence
Sponsored by: The European Commission

ELSEVIER

Amsterdam Boston Heidelberg London New York Oxford
Paris San Diego San Francisco Singapore Sydney Tokyo

Elsevier
The Boulevard, Langford Lane, Kidlington, Oxford OX5 1GB, UK
Radarweg 29, PO Box 211, 1000 AE Amsterdam, The Netherlands

First edition 2006

British Library Cataloguing in Publication Data
A catalogue record for this book is available from the British Library

Library of Congress Cataloging-in-Publication Data
A catalog record for this book is available from the Library of Congress

ISBN–13: 978-0-08-045263-9
ISBN–10: 0-08-045263-0

For information on all Elsevier publications
visit our web site at books.elsevier.com

Printed and bound in Great Britain
06 07 08 09 10 10 9 8 7 6 5 4 3 2 1

Working together to grow
libraries in developing countries

www.elsevier.com | www.bookaid.org | www.sabre.org

ELSEVIER   BOOK AID
International   Sabre Foundation

It is a pleasure to welcome you to the Second Multi-Material Micro Manufacture (4M) Conference in Grenoble, France. 4M2006 is an excellent opportunity to celebrate together the opening of MINATEC, the largest centre for Micro and Nano Technologies in Europe. MINATEC is one of few places in the world where research laboratories, training centres, and high tech companies share their know-how and facilities in order to rise to the challenge of extreme product miniaturisation and thus jointly excel in micro-nanotechnologies. The establishment of this Centre is in perfect alignment with the 4M vision and it is an excellent environment for multidisciplinary discussion in micro and nano technologies and the development of new production platforms for multi-material micro products.

Another year has passed by for the 4M community, with its challenges and discoveries. The second 4M conference is organised just before the official launch of the EC 7th Research Framework Programme, one of the priorities of which is to establish a new industry in Europe for the manufacture of products based on emerging micro and nano technologies. In this context, we believe that 4M2006 will be an excellent opportunity for experts from industry and academia to share the latest results of their in-depth investigations and to engage in interdisciplinary discussions about the creation of manufacturing capabilities that will underpin this new industry. In addition, the 2nd annual conference is the occasion to debate the latest technology and application advances, their implications for the future, and also the new directions for research that they open.

Faithful to its goal, the 2nd annual conference offers four complementary forums to engage in interdisciplinary discussions. In particular, the conference programme includes three plenary sessions in which six keynote speakers will provide overviews of important developments in micro and nano technologies. Eight technical sessions will cover important research and development issues in:

- Novel materials: characterisation and processing;
- Process modelling and simulation;
- Process characterisation including process chains;
- Metrology: inspection and characterisation methods;
- Components: fabrication and assembly technologies;
- Systems: novel product and system designs.

Two invited sessions will present the latest results from four major European projects and market studies on 4M Technologies and Applications. Finally, the programme includes two panel discussions to debate industry and academia priorities in developing and implementing micro and nano technologies together with different issues associated with their commercialisation.

It has been a great challenge for the partners in the 4M Network of Excellence to organise the Conference. We hope that the success will be a testament to the hard work of the Programme Committee, all the staff of the Network Office, and especially the Network Liaison Officer, Marika Takala. We would also like to thank the Local Organising Group at CEA and MINATEC, most notably Dr Bertrand Fillon and Pascal Conche, for their valuable advice and timely assistance in making the conference possible. We would like to acknowledge the support of all researchers in the Network without whom it would have been impossible to implement a policy of full peer review for all papers before acceptance for presentation at the conference.

Finally, and most importantly, we would like to thank the authors of all the papers, theme chairs, reviewers, speakers and all attendees for participating in the conference. Once more, we wish you all a very enjoyable and stimulating conference.

Bertrand FILLON

Conference Chair

Wolfgang Menz & Stefan Dimov

Conference Co-Chairs

4M2006 Conference Committee

4M2006 Programme Committee

B. Fillon, CEA/LITEN, France **(Conference Chair)**
W. Menz, Germany **(Conference Co-Chair)**
S. Dimov, Cardiff University **(Conference Co-Chair)**
A. Schoth, IMTEK, University of Freiburg, Germany
L. Mattsson, Royal Institute of Technology, Stockholm, Sweden
E. Jung, The Fraunhofer Institute for Reliability and Microintegration, Berlin, Germany
U. Engel, University of Erlangen-Nuremberg, Germany
P. Johander, IVF Industrial Research and Development Corporation, Sweden
S. Lange, Fraunhofer Institute of Production Technology, Aachen, Germany
M. Richter, The Fraunhofer Institute for Reliability and Microintegration, Munich, Germany
P. Kirby, University of Cranfield, U.K.

From left to right: S. Dimov, W. Menz and B. Fillon.

4M Scientific Committee

S. Faticow, University of Oldenburg, Germany
A. Dietzel, TU Eindhoven, The Netherlands
N. Mohri, University of Tokyo, Japan
G. Andrieux, Yole, France
F. Sauter, CEA-LETI, France
P. Coates, School of Engineering, Design & Technology, University of Bradford, UK.

4M2006 Industrial Advisory Board

C. Hanisch, Festo, Germany
R. Wimberger-Friedl, Phillips Research Laboratories, The Netherlands
D. Ulieru, Romes – SA, Romania
J. Anjeby, Saab Ericsson Space AB, Sweden
L. Cuccaro, AVIO, Italy
D. Lawrence, IDTechEx, U.K.
M. Knowles, Oxford Lasers, U.K.

Sponsors

IMRC, U.K.

The Manufacturing Engineering Centre, MEC, U.K.

SARIX SA, France

Minatec, European Centre for Micro and
Nano Technology, France

Welsh Assembly Government, WAG, U.K.

CEA, France

LITEN, France

Ville De Grenoble, France

Grenoble Alpes Métropole, France

Conseil Général de l'Isére, France

Grenoble Alpes Métropole, France

PLASTIPOLIS, France

IBS Precision Engineering, France

Oxford Lasers, U.K.

EU Sixth Framework Programme

A roadmapping study in Multi-Material Micro Manufacture

S.S. Dimov[a], C.W. Matthews[a], A. Glanfield[b], P. Dorrington[b]

[a] *Manufacturing Engineering Centre, Cardiff University, Cardiff CF10 3XQ, UK*
[b] *Cardiff Business School, Cardiff University, Cardiff CF10 3XQ, UK*

Abstract

This paper reports findings from a roadmapping study conducted by the FP6 Network of Excellence in Multi-Material Micro Manufacture (4M). The main aim of this study is to help inform European research and industry about current trends and application requirements in the development of Micro- and Nano-manufacturing Technologies (MNT) for the batch-manufacture of micro- components and devices. The results are based on a roadmapping workshop attended by 30 senior researchers and parallel questionnaires administered to 38 associated industrialists. Primary application areas addressed were micro-fluidics, micro-sensors & actuators, and micro-optics, while the technologies covered were surface modification and structuring processes, energy assisted and mechanical processes, and replication processes. Reported results include the main market sectors for these three application areas as perceived by the industrialists together with the main application requirements for developing/manufacturing micro products and the important current and future micro manufacturing technologies. Generic and specific conclusions are derived about the existing trends in developing micro technologies and their applications.

Keywords: micro manufacture, multi-material, roadmap

1. Introduction

Microsystems-based products are envisaged to be an important contributor to Europe's industrial and economic future, as a key value-adding element for many sectors of industry - and the predicted nanotechnology future will also be largely delivered by microtechnologies. While the late 20th century has seen a silicon-based microelectronics revolution, the 21st century looks forward to the adoption of micro- and nano- manufacturing technologies (MNT) making use of a variety of materials, components and knowledge-based technologies that provide functionality and intelligence to highly miniaturised systems. Markets for microsystems are predicted to double over 5 years from 2004 to 2009 [1].

According to a recently conducted study of research and development in micromanufacturing [2], the technologies that underpin the creation of production capabilities for innovative microsystems-based products have four main characteristics:
- They are *enabling* the widespread exploitation of nanoscience and nanotechnology developments;
- They are *disruptive* having the potential to change drastically the existing manufacturing paradigms;
- They are *transforming* the existing manufacturing systems and the way they are designed;
- They are *strategic* for the industrial competitiveness.

In this context, it is very important to study carefully the existing trends in developing micro technologies and their applications, and define research agenda and knowledge transfer provisions that reflect correctly the significance of this field for European industry. Such studies represent a background research that informs the EC 7th Research Framework Programme about the research and development priorities in establishing a new industry in Europe for the manufacture of products based on emerging micro- and nano-technologies.

This paper reports a roadmapping study conducted by the 4M Network of Excellence, one of the European knowledge communities in MNT. The main aim of this study is to help inform European research and industry about current trends and application requirements in the development of MNT for the batch-manufacture of micro-components and devices. According to its original scope shown in Fig. 1 [3], the approach adopted by the 4M Network is to link technologies, business drivers, and application requirements by establishing capabilities for Multi-Material Micro Manufacture. This generic approach is applied in the Network's roadmapping activities to examine the business and application drivers, and the corresponding trends in MNT development to meet demands for:
- Product miniaturisation through innovative integration and development of knowledge-based technologies and production concepts (especially micro and nano) for processing of non-silicon materials;
- Prediction of product and process performance to reduce/manage the risk during product development and production, and reduce time to market for the next generation of microsystems-based products;
- Future product platforms to meet the requirements of the next generation of microsystems-based products, and of more stringent regulations and environmental legislation;
- Production scale-up to ensure an effective and efficient transfer of product and technology ideas from laboratories to serial production.

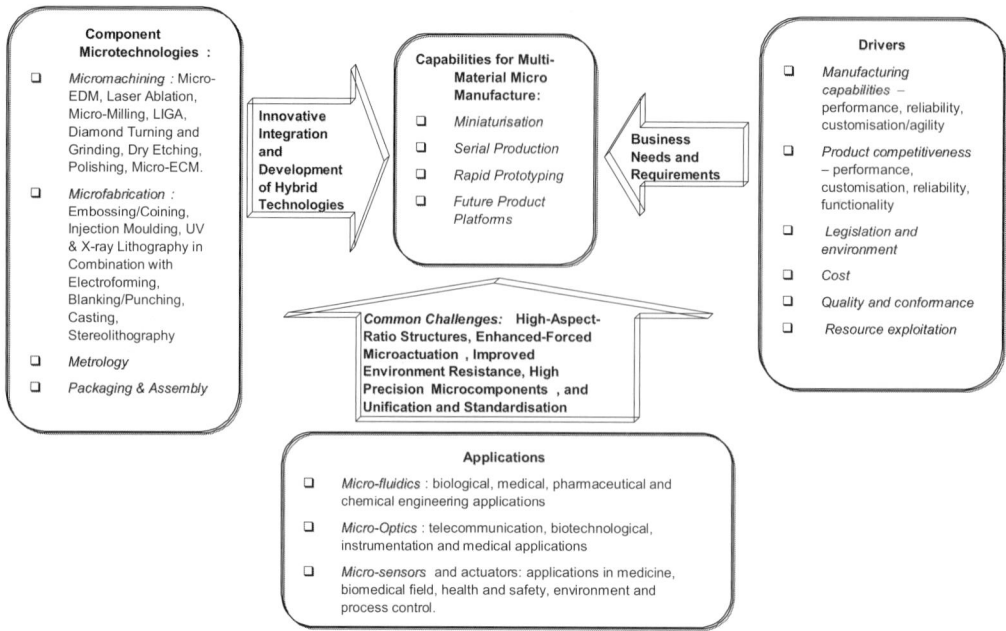

Fig. 1. 4M Scope: technologies, business drivers, and technical application requirements lead to required or provided capabilities.

In considering the above demands, this paper discusses the key factors affecting technology and application developments and tries to answer the following four main questions:

- What are the application requirements and the corresponding trends in technology/process development?
- Is there a potential mismatch between future application requirements and 4M capabilities?
- What is the current level of maturity and future potential of the 4M manufacturing technology areas?
- Are there 4M manufacturing technologies requiring further investment to meet perceived future application requirements?

2. Study Scope

Fig. 2 depicts the relationship between top-down and bottom-up manufacturing in developing technologies for future microsystems-based products [4]. The 'top-down' approach represents the technology developments in reducing the feature sizes towards nanoscale including continuous improvements of their accuracy and surface finish. At the same time, the figure presents the significant advances in chemical technologies for 'bottom-up' processing and control of ever-larger structures. As a result of this technology convergence the structures that can be produced by either approaches are of a similar order and leads to development of new hybrid methods of manufacture.

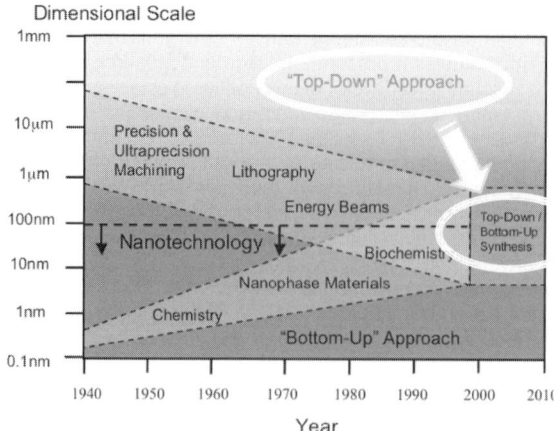

Fig. 2. The convergence of top-down and bottom-up production techniques [5]

Taking this into account, the roadmapping studies being conducted by the 4M Network address issues related to:

- Processing technologies for non-silicon materials;
- Development of top-down technologies towards the creation of top-down/bottom-up hybrid production platforms;
- Applications that will be underpinned by the establishment of such micro/nano manufacturing capabilities.

Fig. 3 represents the overall scope of the roadmapping studies being conducted by the the 4M Network of Excellence. This paper reports only the generic findings concerning application requirements and current and future technological trends. Other studies initiated by the 4M Network are focused on creating specialised application and technology roadmaps.

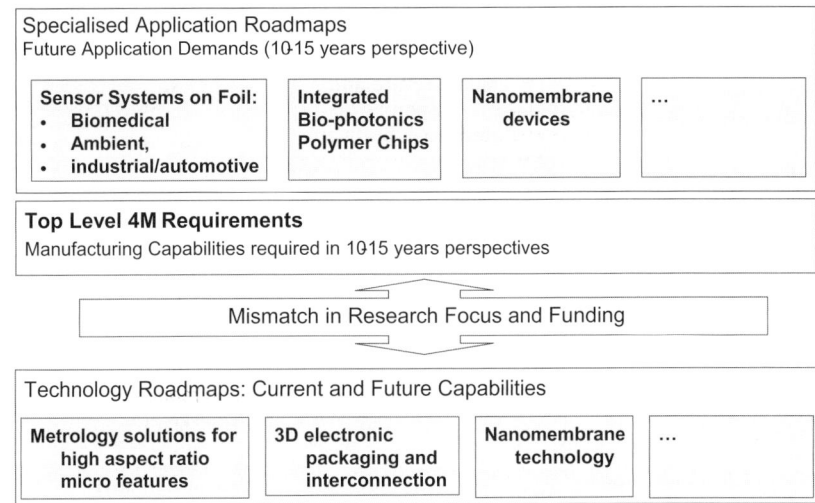

Fig. 3. The overall scope of the roadmapping studies conducted by 4M

In particular, the following roadmapping studies undertaken by Application and Technology Divisions of the 4M Network are currently on-going:
Applications:

- Sensor systems on foils;
- Integrated bio-photonics polymer chips;
- Nanomembrane devices;
- Micro fluidics on foil.

Technologies:

- Metrology solutions for high aspect ratio micro features;
- 3D electronic packaging and interconnection;
- Nanomembrane technology;
- Ceramic hybrid gravure printing process and microPIM for active µ-systems;
- Micro Scale Batch Self-Assembly of Microparts;
- Next generation Micro/Nano-metrology equipment.

The results of these specialised application and technology roadmaps will be reported separately from the findings of this "top-level" study.

The detail of the original 4M approach to mapping application requirements and technology capabilities is shown in Fig. 4 [3]. Common requirements from three application areas in the form of design/functional features are mapped onto manufacturing features produced by a range of micro manufacturing technologies to provide a set of 'Design for Manufacture and Assembly' rules. The objective of carrying out such analysis is to assist designers of new multi-material microsystems-based products and also to identify discrepancies between technological requirements of a range of applications and current and future manufacturing capabilities. In line with this approach, this study focuses on three application areas specifically addressed by the 4M knowledge community:

- A1: Micro-fluidics;
- A2: Micro-sensors and actuators;
- A3: Micro-optics.

It also covers three technology areas representing the range of the component technologies that constitute the 4M capabilities:

- T1: Surface modification and structuring processes;
- T2: Energy assisted and mechanical processes;
- T3: Replication processes.

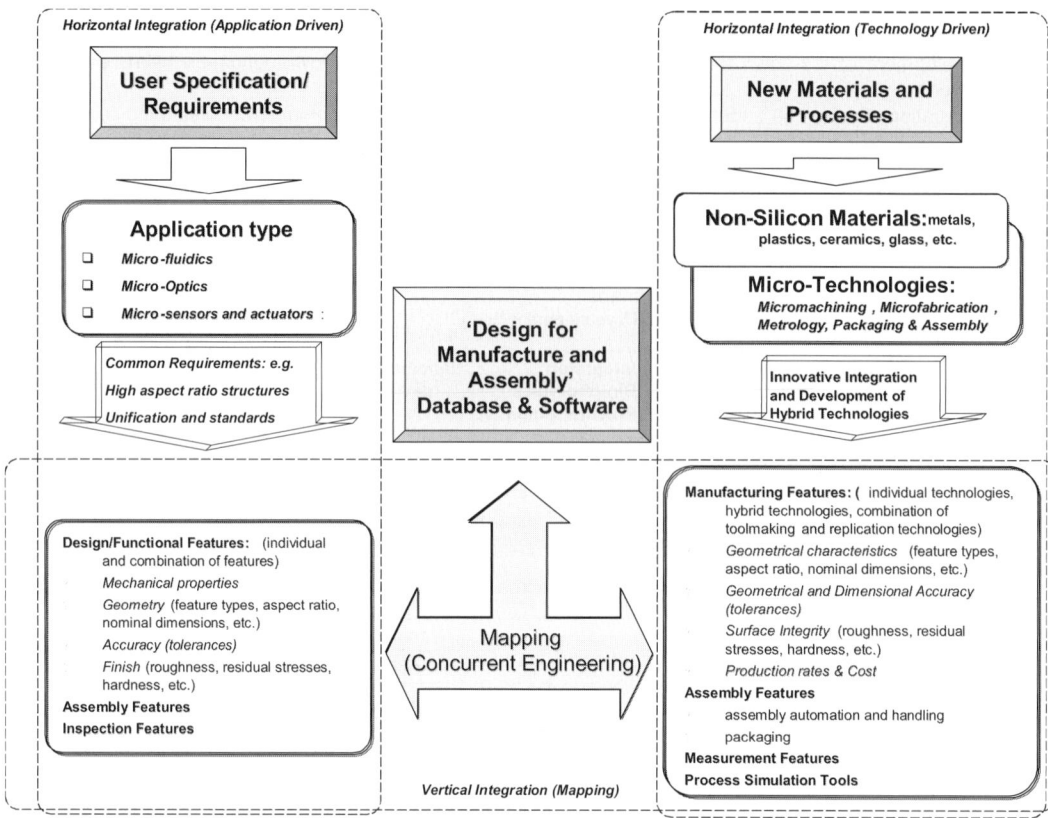

Fig. 4. 4M Approach: detail of mapping between application requirements and technology capabilities towards the creation of a new 'Design for MNT Manufacture' system

The technologies covered during this study are represented pictorially in Fig. 5, according to their material removal/addition/forming capabilities [6]. The first group of technologies performs material removal and deposition employing a 1D processing, e.g. structuring by a milling cutter or a laser beam. The second group includes technologies that utilise multiple 1D processing, e.g. 3D printing. Next, the technologies falling in the group of 2D processes perform structuring by employing masks, e.g. photo lithography. Finally, 3D processing can be carried out using technologies for surface modification and deposition, e.g. PVD, CVD and electroplating, or technologies for volume structuring such as injection moulding and embossing. Generally, the manufacturing flexibility of the technologies increases from right to left, e.g. less time and effort are required to setup the process, while the production speed increases from left to right. Usually, a combination of technologies in the form of a process chain is required for economic production of micro components and devices, incorporating micro/nano features. In Fig. 5, the numbers in brackets show the frequency of occurrence of technologies in the process chains suggested at the roadmapping workshop as described later.

3. Methodology

3.1 Researcher workshop

A one day workshop was held in February 2006 at Cranfield UK, facilitated by Cardiff University's Cardiff Business School and Manufacturing Engineering Centre. The 30 participants comprised the 17-strong 4M Executive Board and invited experts from among the 4M partners. A methodology was devised to answer application requirement and technology capability questions, based on the initial discussion round from a Delphi type study, supplemented by SWOT and timeline analyses.

After an introductory session, participants were split into 3 applications groups (A1, A2, A3 above) according to their area of expertise. During 3 parallel sessions, each group addressed 3 questions:

- What are the main drivers for developing micro-fluidic* products? (*or -optic or -sensor & actuator)
- What are the main barriers to the successful manufacture of micro-fluidic* products?
- What are the main 4M technical capabilities that are important for the successful manufacture of these micro-fluidic* products?

Dimension:	1D Processing 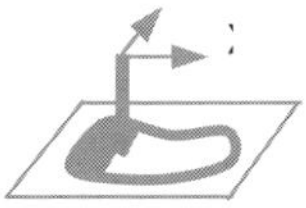	Multiple 1D Processing 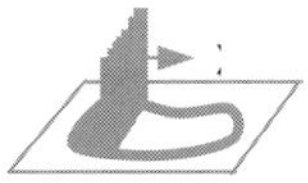	2D Processing 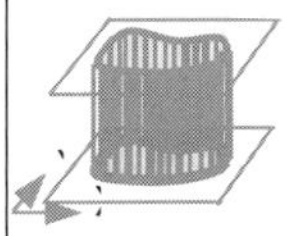	3D Processing (Surface) 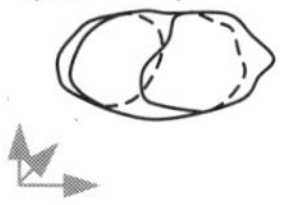	3D Processing (Volume) 
Metals	LH, EDM (4), ECM, Grinding (6),	MF (6), Grinding (6)	Lap, Pol (5), MF (6)	Lap, Pol (5), ECP, EF (3), EP (4)	EDM, MF (6)
Polymers	3DL	3DP (4)	EBL, IBL (2), LL, PUL (8), XL (1)		HUE (4), NIL (2), NI (2), R2RE (3), IM (17)
Ceramics	3DL, Grinding (6)	3DP (4)	IBL (2), LL,	Lap, Pol (5)	NIL (2), NI (2), R2RE (3
Any material	EBM (2), FIB (7), LA (9), PM, AWJ, Drilling (1), Milling (11), Turning (4), SLS		Etch (1), PMLP (2), SP (4)	PVD (9), CVD (2), SC (1), SA	Casting (2), MCIM, PIM (1)

Key:

3DL	3D Lithography	Lap	Lapping
3DP	3D Printing	LH	Laser hardening
AWJ	Abrasive water jet	LL	Laser lithography
Casting	Casting	MCIM	Multi-component injection moulding
CVD	Chemical vapour deposition	MF	Metal Forming
DL	Direct LIGA	Milling	Milling
Drilling	Drilling	NI	Nano-imprinting
EBM	Electron beam machining	NIL	Nano-imprint lithography
EBL	Electron beam lithography	PIM	Powder injection moulding
ECM	Electrochemical machining (ECM)	PUL	Photo / UV lithography
EDM	Electrical discharge machining (EDM)	PM	Plasma machining
EF	Electroforming	PMLP	Projection mask-less nanopatterning
ECP	Electro-chemical polishing	Pol	Polishing
EP	Electroplating	PVD	Physical vapour deposition
Etch	Etching	R2RE	Reel to reel embossing
FIB	Focused ion beam	SA	Self assembly
Grinding	Grinding	SC	Spin coating
HUE	Hot/UV embossing	SLS	Selective laser sintering
IBL	Ion beam lithography	SP	Screen printing
IM	Injection moulding	Turning	Turning / Diamond turning
LA	Laser ablation	XL	X-ray lithography

Fig. 5 Map of technologies according to process 'dimension' and material relevance.

In each session above the question was first "brainstormed" by the group - members were free to call out their ideas and proposals. Handouts were provided with ideas and prompts. After this, group members were asked to write down three answers each on sticky notes, which were then placed on posters. The group facilitator then brought similar notes together under common headings – in agreement with, and under the guidance of, the group members. Once under these headings, group members were asked to allocate their "budget" of 5 stickers to what they considered the most important headings. For drivers and technical capabilities (technical requirements), each group jointly placed the identified headings along a timeline, according to when they would have impact. For barriers and technical requirements, participants considered collectively what were considered to be the Strengths, Weaknesses, Opportunities, and Threats (SWOT) for Europe (barriers) or 4M (technical requirements) in relation to the rest of the world.

Participants were likewise split into 3 technology groups (T1, T2, T3 above) according to their area of expertise. During parallel sessions, each group addressed 3 questions:

- What are the 4M technologies/processes with the highest impact on the market currently?
- What are the key capabilities that can be used to characterise the 4M technologies in your group?
- What are the 4M technologies/processes that are currently under development and promise to become economically viable in 10 to 15 years?

Brainstorming, note writing, and budget allocation occurred as in the previous sessions. This resulted in ranked lists of current and future technologies and a ranked list of capabilities. The groups then ordered the top 10 current technologies within their group, according to their ability to satisfy each of the top three scoring capabilities.

An additional session combined applications and technologies by asking participants in each application group individually to write down promising process chains for specified products in their application area. Approximately 40 process chains were identified across the 3 application areas.

After the workshop, additional information was collected on the breakdown of research funding among 4M partner projects integrated within 4M. Project funding relevant to multi-material micro manufacture from institutional, national/regional, EC, and industry projects was broken down according to the technology areas shown in Fig. 5.

3.2 Industry questionnaire

A supporting questionnaire was prepared following, and based on the results from, the workshop, to canvas the views of industrialists through face-to-face interviews between 4M partners and associated industrialists.

For the application questions, interviewees were asked to select which of micro-optics/-fluidics/-sensors & actuators (or other ....) they would answer questions on. This was followed by additional questions on market shares and on market sectors, before moving to the 3 main application questions. The questions were altered slightly to try to obtain an 'industry-wide' viewpoint, and the third question mentioned 'technical requirements' and 'batch manufacture' explicitly, these being implicit in the workshop. Initial brainstorming was mimicked by first asking interviewees to answer questions before allowing them to see a list of answers (including an 'other' category) to vote on (invest a budget in), again with 5 votes. Common lists of answers for each application session were devised from the results of the workshop by combining answers from the 3 application areas, as shown in Table 2. This involved some interpretation in order to group similar answers across and sometimes within application areas. Timelines were mimicked by asking interviewees to choose a short (0-3 years) medium (3-8 years) or long-term (8-15 years) timescale for those answers voted on. SWOT analyses were mimicked by asking interviewees whether Europe was in front, comparable, or behind the rest of the world in for those answers voted on. An additional question asked which manufacturing capabilities and materials were most important for the application group chosen.

For the technology questions, interviewees were free to choose technologies from a composite list of technologies (T1, T2, and T3 combined), as shown in Table 4, when answering the first and third questions on current and future technologies. This list was based on an existing 4M technology classification augmented by the results from the workshop. The second question was not asked directly, but rather interviewees were invited to say how well their top two current and future technologies met a composite list of 8 capabilities, derived from the workshop. There were additional questions on which materials and application groups the technologies were most suitable for.

4M partners were free in their choice of industrialists, only aiming to find individuals with good technical and market knowledge. In total 38 industrialists from 38 companies (in 1 case a trade association) responded, normally by interview. The following indicates the profile of the companies involved:

- **Business descriptions**: tooling, injection moulding, etc. [5]; research & development, e.g. in MEMS [7]; machine tool manufacture [3]; micro- including electronic assembly & packaging [4]; sensor production [2]; heath service products [3]; coatings [2]; other product manufacture [6]; not specified [4]
- **Country**: Austria [2], Bulgaria [2], France [2], Germany [10], Greece [1], Hungary [2], Italy [1], Netherlands [1], Spain [3], Sweden [4], Switzerland [2], UK [6], International [2]
- **Technology provider/user**: technology provider [25], technology user [29]
- **Number of employees**: 1-10 [8] 10-50 [15] 50-250 [4], 250-2500 [4] >2500 [5], not specified [2]
- **Application area** (on which questions were answered): micro-fluidics [9], micro-optics [8], micro-sensors & actuators [10], other [7], although in 14 cases more than one category was chosen and in 2 cases no category was chosen. 'Others' were dental / medical / shaving products, display coatings, micro-tooling, metal processing, and electronics and semiconductors

No attempt was made to account for bias in those participating in the exercise and not all results from the workshop and questionnaires are reported in this paper.

4. Results: Applications

4.1 Market sectors

The market sectors perceived by industry as being most significant especially for micro-fluidic, micro-optic, and micro-sensor & actuator applications are shown in Table 1. Top sectors overall in order of their perceived importance are medical/surgical, automotive and transport, biotechnology, and consumer products. Second level sectors are information and communication technology, energy/chemical, academic/scientific, and pharmaceutical.

Table 1. Perceived market sector importance for micro products

Sector	Micro-fluidics	Micro-optics	Micro-sensors & actuators	Overall (inc. other)
Medical/Surgical	▪	▪	▪	▬▬▬▬
Automotive and Transport	▪	▪▪	▪▪	▬▬▬▬
Biotechnology	▪▪	▪	▪	▬▬▬
Consumer Products (Electrical, Games, ...)	▪	▪▪	▪	▬▬▬
Information and Communication	▪	▪	▪	▬▬
Energy/Chemical	▪	▪	▪	▬
Scientific/Academic Community		▪	▪	▬
Pharmaceutical	▪▪	▪		▬
Construction		▪	▪	▪
Domestic Products (Clothing, Furnishings, ...)	▪		▪	▪
Food	▪			▪
Aerospace/Space Science			▪	▪

4.2 Application requirements

Table 2 contains results from the three application questions regarding the perceived drivers, barriers and technical requirements for developing/manufacturing micro products. 'Other' answers from industry were '(total) flexibility' and 'automation' (both drivers), however their score was small compared to those of the pre-defined categories.

It is posited that the results from these three application questions and from the three application areas might be combined to form a common set of application requirements for micro products for these application areas. Table 3 shows the results of such a combination, where the results have been grouped into 11 categories, indicating both research and industry results. The entries are ordered according to their perceived importance, using a combined research plus industry score, though this ordering is clearly dependent on how the requirements have been grouped, which is in turn a matter of judgment. Market and product functionality / performance related answers have been excluded as not being application requirements. 'Tolerance & accuracy' and 'Metrology' answers were also excluded due to their very low score.

The top application requirement is for low cost / volume production. The second is for knowledge / training, including knowledge transfer to industry. The remaining categories are of roughly comparable importance. The two industry led requirements are low cost / volume production and quality. This might be interpreted as addressing short-term cost and quality imperatives. The research led requirements include integration, at the product, component, and process level, indicating a longer term viewpoint, though perhaps still aiming for low cost quality output in the end. The other research led requirements are for materials and for assembly & packaging.

Multidisciplinarity, process/technology knowledge (sharing), integrated process chains, and assembly & packaging, all address recommendations from the 4M Industrial and Scientific Advisory Boards at the time of the 4M2005 Conference, though the Boards' recommendation for metrology is not highlihgted here. The concentration of industry on cost and quality and of researchers on longer-term issues is also in line with the Advisory Boards' recommendation to recognise the different functions of industry and the research community. Design (for manufacture rules) corresponds to the ultimate 4M goal in Figure 2 of this paper as well as helping to fulfill the Advisory Boards' recommendation to provide 'toolboxes, characterising a broad range of technologies, that can be selected/ adopted in the design'. The importance of process chains may suggest the need to characterise processes not only for their individual capability but also for their suitability for integration into process chains. Novel combinations of processes may provide required capabilities not available through the conventional use of individual technologies.

Table 2. Results from application questions

Category[1]	Research				Industry
	Fluidics	Sensors	Optics	Overall	Overall inc. other
DRIVERS					
A Lower cost	■	■	■	■	■
A Volume production					■
M New markets/applications	■	■		■	■
P Improved product functionality		■		■	■
E Product function integration			■	■	■
E Physical product integration		■	■	■	
P Sensitivity + precision		■			■
H Integrated production (process chains)			■	■	
F Quality, efficiency + reliability	■			■	■
M Safety and environment			■	■	
M Biological epidemic risk	■				
BARRIERS					
A Cost	■				■
M Conservative markets	■			■	■
M Competition from silicon					
M No successful exemplar product					
G Material availability		■		■	
B Design knowledge	■				■
B Interdisciplinary knowledge	■		■	■	■
B Lack of trained people	■				
B Knowledge about processes/technology[2]	■		■	■	■
D Immaturity/lack of technology	■	■		■	■
H Integrated process chains		■	■	■	
D No prototyping		■		■	
C Process complexity for 3D		■			
D Technology diversity / lack of standards		■			
J Packaging & assembly			■	■	■
-- Metrology			■		
TECHNICAL REQUIREMENTS					
A Scale-up/replication	■			■	■
E Functional/system integration	■	■		■	
K Nano-micro or micro-meso integration	■	■	■	■	■
C 3D freeform structuring			■	■	
C Surface properties	■		■	■	■
-- Tolerance & accuracy		■			■
G New/improved/multi materials	■	■	■	■	■
B Design for manufacture		■			
I Process/machine technologies		■	■	■	■
B Modelling					
F Quality/reproducibility/reliability			■		■
D Standards	■				
J Assembly & packaging	■	■		■	

[1] A-J - common application requirements (Table 3); M - markets; P - product functionality / performance
[2] In the workshop this answer included the idea of knowledge sharing

Table 3. Common application requirements

Category	Importance	Time ordering	
		Research	Industry
A Low cost / volume production		5	4
B Interdisciplinary, design, and process/technology knowledge		-	-
C 3D features, surface properties		7	3
D Technology / maturity / standards		10	10
E Function and physical integration		4	7
F Quality/reproducibility/reliability		2	5
G New/improved/multi materials		9	6
H Integrated process chains		2	2
I Process / machine technologies		1	9
J Assembly & packaging		5	8
K Nano-micro or micro-meso integration		8	1

Key: ▮▮▮▮ Research ▨▨▨ Industry

Also indicated in Table 3 is a possible time ordering for when these requirements will have impact. Due to the groupings of results, timelines only being available for drivers and technical requirements, and the limited data for the lower scoring industry results, this must be considered only as a tentative ordering. Given this proviso, notable differences between research and industry results are the much earlier position given to process / machine technologies by researchers compared to industry and conversely the much later position given to nano-micro or micro-meso integration by researchers compared to industry.

The list of application requirements in Table 3 is a significant output of this study and might be considered potential research areas for multi-material micro manufacture.

4.3 SWOT analysis

Fig. 6 shows composite results from SWOT analyses carried out for the three application groups (see A1, A2, A3 above) with respect to barriers to the successful manufacture of micro products, for Europe versus the rest of the world. This represents the view only of researchers. Possible interpretations from the diagram are as follows:

- The strength of Europe is in the availability of trained people, well established machine tool industry and multidisciplinary product development expertise. These are the most important ingredients for the establishment of knowledge-intensive production capabilities in Europe.
- The main weaknesses that may slow down the take up of MNT in Europe are the costs required to adopt these technologies, the conservatism of some industrial sectors, the lack of knowledge sharing between R&D actors and industry, and lack of standards and metrology support.
- The possible opportunities for Europe in regard to the rest of the world are in integration of process chains, creation of capabilities for 3D processing, establishment of prototyping capabilities, effective implementation of design for manufacture, and improved research efficiency and focus.
- The indicated threats for Europe come from standards and packaging solutions developed outside Europe, and the dominance of Far East in key consumer markets.

5. Results: Technologies

5.1 Current and future technologies

Results from the two technology questions regarding the micro manufacturing processes with highest perceived current or future impact on the market are shown in Table 4. Technologies are divided according to the three workshop application groups (see T1, T2, T3 above) and further into the seven categories within the 4M technology classification. 'Other' answers from industry, one respondent for each, were glass moulding and soft embossing. Note that the scores within each of the three workshop groups were nominally constrained to be approximately equal; however there was some 'leakage' between groups, i.e. groups would sometimes choose technologies from one of the other two groups. Note also that in some cases scores were allocated to a category of technologies rather than to individual technologies.

Strengths	Weaknesses
A1 Trained people (linked to Opportunity)	A1 Lack of technology (linked to Opportunity)
A1 Multidisciplinarity	A1 No successful exemplar products
A1 Technical expertise	A1 Lack of awareness (linked to Opportunity)
A2 Technology diversity and many new ideas	A2 Infrastructure costs (ie funded by tax) (also Threat)
A3 Deep expertise in fundamental technologies	A3 Lack of shared knowledge with industry
A2 Technology maturity	A3 Labour cost
A3 Machine tool industry	A3 Need critical mass of research in design for manufacture
A3 Leaders in precision engineering	A3 Metrology is imported (Nano)
A3 Capability to adapt existing technology to new products/technologies	A3 Duplication of R&D efforts
A3 Political & economic stability	A3 Lack of students/education
	A3 Bureaucracy in funding R&D
Opportunities	A3 State aid issues
A1 Design knowledge (linked to Weakness)	A3 Spread versus Focus of funding allocation
A2 Infrastructure	
A2 Competition from silicon (also Threat)	**Threats**
A2 3D Capability	A1 Conservative markets (Linked to Weakness)
A2 No prototyping	A1 Risk/investment (Linked to Weakness)
A2 Integration (processes) EU strong in this	A2 Materials – stringent regulations in EU
A3 Integration of process chains	A2 Packaging
A3 Exploit development of integrated lines	A3 Materials development is occurring outside Europe[1]
A3 Established machine tool industry	A3 Lack of shared knowledge …. leading to increased development time
A3 Standardisation & unification of metrology stnds	A3 USA sets metrology standards
A3 Links to product development	A3 Reliance on adapted solutions
A3 Increase efficiency of collaboration	A3 Dominance of Far East in key consumer markets
A3 Funding agency "top of bill" research (military)	
A3 Deep vs Broad	
A3 Centres of Excellence for field development	[1]This threat is from the technical requirements SWOT

Key: Workshop groups: A1 = micro-fluidics, A2 = micro-sensors & actuators, A3 = micro-optics

Fig. 6. Barriers to the successful manufacture of micro products SWOT

The first analysis of current and future technologies may be made using the summary shown in Table 5. The research and industry results show reasonable agreement as regards the importance of current technology areas. Changes in the future, according to the research results, indicate an increase in the importance of beam-based processes, a large decrease in the importance of electrical / chemical processes, and a very large increase in the importance of prototype / layer-based processes, in particular 3D Printing and to a lesser extent SLS. These latter trends were not seen at all in the industry results, where EDM for example continues to be important. There is broad agreement on the trends for the remaining technology groups, which mostly retain the same relative importance, with the exception of coating that shows a decrease in importance.

With respect to the importance of individual technologies within the technology groups, there is some agreement as well as some noticeable differences. In addition to the prototype / layer-based processes differences noted above, there is particular discrepancy with regard to replication processes as to which technologies are important both now and in the future. While researchers highlighted metal forming, SLS, and particularly reel to reel embossing for the future, industry highlighted multi-component injection moulding, nanoimprinting, and powder injection moulding. Finally Table 6 lists individual technologies according to their perceived importance for the future. A consistent pattern to account for the differences between research and industry results has not been identified.

Table 4. Results from technology questions

	Technology	Importance				Process chain[1]	4M partner funding[2]
		Research		Industry			
		Current	Future	Current	Future		
Energy assisted and mechanical processes	**Beam-based**						
	E beam				▪	■	▪
	Focussed Ion Beam (FIB)					■■	■■
	Laser ablation	■	■	■	■	■■	■■
	Laser hardening			▪			
	Plasma machining		■	■	■		■
	Projection Mask-Less Patterning (PMLP)		■			▪	
	Electrical / Chemical			▪			▪
	Etching	■		■	▪	▪	■
	Electrical discharge machining (EDM)	■		■	■	■	■
	Electrochemical machining (ECM)				■		▪
	Electrochemical polishing	▪		▪			
	Electroforming			▪	■	■	■
	Mechanical machining	■	■	■			
	Abrasive water jet						
	Drilling			■	■	▪	▪
	Milling	■	■	■	■	■■	■■
	Grinding	■		■	▪	■	■
	Lapping				▪		▪
	Polishing			▪	▪	■	■
	Turning / Diamond turning	■		■■	■	■	■
	Prototype / layer-based manufacture			▪			▪
	3D Printing	■	■■	▪	▪	■	▪
	3D Lithography			▪	■		
	Selective laser sintering		■		▪		
Surface modification and structuring processes	**Lithography**						
	E beam lithography	▪	■	▪	■		
	Ion beam lithography		▪			■	▪
	Laser lithography	■■			▪		▪
	Nanoimprint lithography (NIL)	▪	■	■	■	■	■
	Photo / UV lithography	■■		▪	▪	■■	
	X-ray lithography				■	▪	
	Coating	■■		▪			▪
	Physical vapour deposition (PVD)			■	▪	■■	■
	Chemical vapour deposition (CVD)	▪		■	▪	■	■
	Electroplating			■	■	■	■
	Spin coating	■		▪		▪	▪
Replication processes	**Replication**		■■				■
	Casting	■		▪		■	
	Direct LIGA		■				
	Hot/UV Embossing			■	■	■	■■
	Injection moulding	■■■	■	■■■	■	■■	■■
	Metal Forming	■	■	■	▪	■	■■■
	Multi-component injection moulding			■	■		▪
	Nanoimprinting	■			■	■	
	Screen printing	■■	■			■	■
	Powder injection moulding		■	■	▪	▪	
	Reel to reel embossing		■■	▪	■		
	Self assembly		▪	▪	■		

[1]The relative frequency of occurrence of technologies in the process chains
[2]The relative R&D funding of 4M NoE partners categorised according to the technology classification.

Table 5. Summary results from technology questions

Technology	Importance				Process chain[1]	4M partner funding[2]
	Research		Industry			
	Current	Future	Current	Future		
Beam-based	▬	▬	▬	▬	▬▬	▬▬
Electrical / Chemical	▬		▬	▬	▬	▬
Mechanical machining	▬▬	▬▬	▬▬	▬▬	▬▬	▬▬▬
Prototype / layer-based manufacture	▪	▬▬	▪	▬	▪	▪
Lithography	▬	▬	▬	▬▬	▬	▬
Coating	▬	▪	▬	▪	▬	▬
Replication	▬▬▬▬	▬▬▬	▬▬▬	▬▬▬	▬▬▬	▬▬▬▬

[1]The relative frequency of occurrence of technologies in the process chains
[2]The relative R&D funding of 4M NoE partners categorised according to the technology classification.

Table 6. Perceived important future technologies

Technology	Perceived importance	Technology	Perceived importance
3D Printing	▬▬▬▬▬		
Milling	▬▬▬▬	Hot/UV Embossing	▪▪
Powder injection moulding	▬▬▬	Self assembly	▪▪
Reel to reel embossing	▬▬▬▬	Projection Mask-Less Patterning (PMLP)	▪
Nanoimprint lithography (NIL)	▬▬▬	Electrochemical machining (ECM)	▪
Injection moulding	▬▬▬	Electroforming	▪
Nanoimprinting	▬▬▬	Electroplating	▪
Multi-component injection moulding	▬▬▬	Direct LIGA	▪
Laser ablation	▬▬▬	Lapping	▪
Selective laser sintering	▬▬	Photo / UV lithography	▪
Metal Forming	▬▬	Physical vapour deposition (PVD)	▪
Plasma machining	▬	Etching	▪
E beam lithography	▬	Polishing	▪
Electrical discharge machining (EDM)	▪▪	E beam	▪
Screen printing	▬▬	Grinding	▪
Drilling	▪▪	Ion beam lithography	▪
X-ray lithography	▪	Laser lithography	▪
Turning / Diamond turning	▪	Chemical vapour deposition (CVD)	▪
3D Lithography	▪▪	Casting	▪

Key: ▬ Research ▪ Industry

5.2 Process chains

Technologies selected as important for the future may also be compared to the frequencies with which technologies are mentioned in the process chain exercise. There is partial overlap in the results in Table 4. There may be technologies in process chains that are not considered important in their own right but could become important because of their presence in an important process chain. The frequency of occurrence of technologies in the process chains is also indicated in Fig. 5. Examples of processes within the process chains that are not otherwise individually identified as important are:

- E beam
- Focused ion beam
- Polishing
- Ion beam lithography
- PVD
- Casting

5.3 4M technology funding

56MEuro of funding for partner projects integrated into the 4M Network and currently ongoing or recently completed (within the last year) were categorised according to the technology classification in Table 4. As an indication, the split by funding source for 113MEuro of partner projects integrated into 4M to date is approximately: institutional 14%, national/regional 53%, EC 18%, industrial 15%.

Analysing Table 5, in terms of groups of technologies, shows a reasonable fit between perceived importance of future technologies (research / industry) and 4M funding, with the exception of prototype / layer-based technologies ('deficit' of funding), and to a lesser extent electrical / chemical and coating processes ('excess' of funding). An examination of individual technologies (Table 4) likewise shows a relative excess or deficit of 4M project funding over the perceived level of future importance of the technologies. Technologies where the relative 4M project funding is significantly greater than the perceived future importance of the technology are:

- Focused ion beam
- UV/hot embossing
- Metal forming

Technologies where the relative 4M project funding is significantly less than the perceived future importance of the technology are:

- 3D printing
- Multi-component injection moulding
- Nanoimprinting
- Powder injection moulding
- Reel to reel embossing

This could indicate in which direction future research funding might move, although such an interpretation might be misleading since (a) full consideration of the technologies necessary for a complete process chain is not included, (b) funding decisions for some current 4M projects will have been made some while ago, (c) these 4M funded projects only cover part of the total research funding within these technology areas, and (d) 30MEuro of the 56MEuro funding is accounted for by only two 4M partners, with funding for some technologies, such as FIB, metal forming, laser ablation, plasma machining, and NIL, mostly accounted for by a single partner.

6. Conclusions

This paper reports the finding of the top-level roadmapping study conducted by the 4M Network of Excellence. The breadth and depth of technologies for multi-material micro manufacture are studied to identify mismatches between current and future application requirements and manufacturing capabilities. The following generic conclusions could be derived from the response of the research community and industry:

- There is no one technology that will prevail – the "breakthroughs" if any will come from an innovative integration of complementary technologies and their implementation in new manufacturing platforms.
- A "tool box" of technologies exists to support the move from designing MST-based products for specific materials and processes to designing processes/process chains to satisfy specific functional and technical requirements of new emerging multi-material products
- There are signs of bridging the "gap" between "mechanical" ultra-precision engineering and "MEMS/IC based" technologies. Many promising process chains suggested by the research community integrate component technologies from both groups.
- The breadth of micro-manufacturing technologies makes it difficult to select the most appropriate manufacturing route. Therefore, it is very important to advance the existing design for manufacture knowledge.
- The strength of Europe is in the availability of trained people, well established machine tool industry and multidisciplinary product development expertise. These are the most important ingredients for the establishment of knowledge-intensive production capabilities in Europe.
- The main weaknesses that may slow down the take up of MNT in Europe are the investment required to adopt these technologies, the conservatism of some industrial sectors, the lack of knowledge sharing between R&D actors and industry, and lack of standards and metrology support.

In addition, the following specific conclusions could be made based on the conducted roadmapping study:

1. The **market sectors** perceived by industry as most significant especially for micro-fluidic, micro-optic, and micro-sensor & actuator applications in order of importance were: medical/surgical, automotive and transport, biotechnology, consumer products, information and communication, energy/chemical, scientific/academic community, and pharmaceutical.

2. The **application requirements** for developing/manufacturing micro-fluidic, micro-optic, and micro-sensor & actuator products identified by the study in order of importance were: low cost / volume production; interdisciplinary, design, and process/technology knowledge; 3D features / surface properties; technology / maturity / standards; function and physical integration; quality / reproducibility / reliability; new / improved / multi materials; integrated process chains; process / machine technologies; assembly & packaging; and nano-micro or micro-meso integration.

3   The **manufacturing technologies** identified by the study as most promising for future batch-manufacture of micro products in order of importance were: 3D printing, milling, powder injection moulding, reel to reel embossing, nanoimprint lithography, injection moulding, nanoimprinting, multi-component injection moulding, laser ablation, selective laser sintering, metal forming, plasma machining, and e-beam lithography

Finally, Fig. 7 shows the interdependence of technologies and applications in developing new manufacturing platforms.

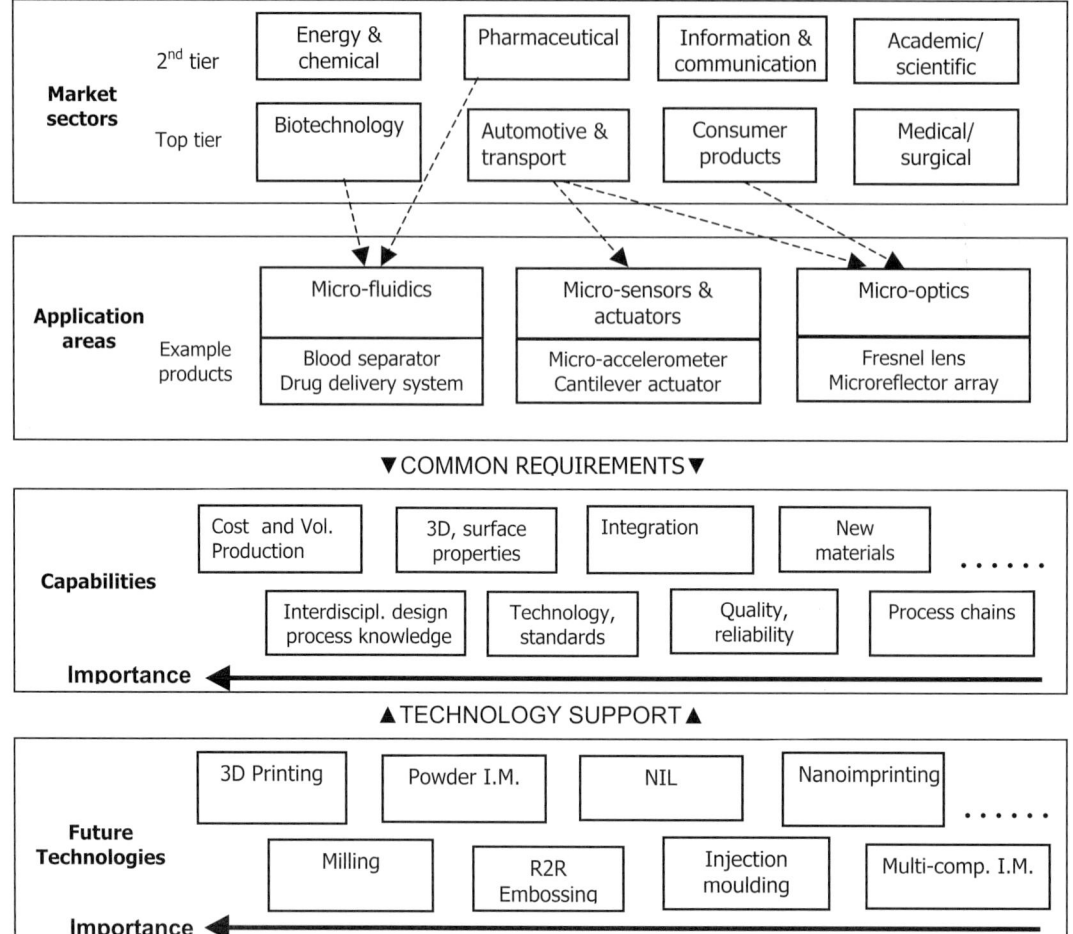

Fig. 7. Common application requirements and technology support

Acknowledgements

The help of 4M partners and associated industrialists, and the Cardiff University Innovative Manufacturing Research Centre funded by the UK Engineering and Physical Sciences Research Council are gratefully acknowledged in producing this study. The work was supported by the EC under the 4M Network of Excellence: Multi-Material Micro Manufacture: Technologies and Applications.

References

[1] NEXUS Market Analysis for MEMS and Microsystems III, 2005-2009, NEXUS, 2005.

[2] International Assessment of Research and Development in Micromanufacturing, Final Report, World Technology Evaluation Centre (WTEC), Inc., Baltimore, 2005.

[3] Dimov SS. 4M Network of Excellence: An Instrument for Integration of European Research in Multi-Material Micro Manufacture. 4M2005 - First International Conference on Multi-Material Micro Manufacture, Elsevier, UK, 2005.

[4] Nanoscience and nanotechnologies: opportunities and uncertainties, Final Report, The Royal Society & The Royal Academy of Engineering, UK, 2004.

[5] Whatmore RW. Nanotechnology: big prospects for small engineering. Ingenia, Issue 9, 2001, pp. 28-34. www.ingenia.org.uk.

[6] Pham DT and Dimov SS. Rapid prototyping and rapid tooling – the key enablers for rapid manufacturing. Proc. Instn. Mech. Engrs. Vol. 217, Part C, 2003.

CONTENTS

Process Modelling and Simulation

Process Characterisation including Process Chains

Systems : Novel Product and System Designs

Keynote Papers

Multi-Material Micro Manufacture
W. Menz, S. Dimov and B. Fillon (Eds.)

Lab on a Chip : advances in packaging for MEMS and Lab on a Chip

Dr. Fabien Sauter-Starace[a], C. Pudda[a], C. Delattre[a], H. Jeanson[a], C. Gillot[b], N. Sarrut[a], O. Constantin[a], R. Blanc[a]

[a]: CEA, LETI, Department Microtechnology for Biology and Health-care, France
[b]: CEA-LETI, Department Heterogeneous Silicon Integration, France

Abstract

In the field of bioMems or Lab on a Chip, one may have to handle fluid exchanges even at high pressure, moving parts, temperature sensitive probes. This specificity triggers huge packaging issues because most of the packaging techniques inherited from the microelectronic require thermal cycles at temperature above 350°C. In this paper, we first present the requirement of representative lab on chip projects carried out at the LETI in order to introduce two specific solutions of this field. Afterwards we present a brief review of the packaging processes used in the MEMS facilities eventually and introduce low temperature packaging processes developed at the CEA Leti to process lab on chip with MEMS and BioMEMS. We intend to develop robust and collective solutions (wafer level packaging). The first solution relies on a thin film approach develop for MEMS using silicon nitride cap on a sacrificial layer. The second one was especially designed for temperature sensitive lab on chips hence can be carried out at room temperature.

Keywords: packaging, lab on chip, thin film, in line screen-printing

Introduction

The fabrication processes of MEMS (micro electro mechanical system) are largely based on silicon technology derived from the microelectronic industry. However several parameters remains quite different from the processes of microelectronics. First as they are structural elements made or derived of semi conductor materials, the thickness of theses layers whether they are dielectric or conductive can be in the range of tens of microns or hundreds of microns, whence specific etching procedures and longer runs. Second interactions between the macro world and these devices are not restricted to electrons or photons exchanges. In our department we may have to handle fluid exchanges even high pressure fluids, moving parts. This specificity triggers huge packaging issues. In this paper, we first present the requirement of representative biochip projects carried out at the LETI in order to introduce two specific solutions of this field afterwards we present a brief review of the packaging processes used in the MEMS facilities eventually we discuss low temperature packaging processes developed at the CEA Leti to process lab on chip for MEMS and BioMEMS. We intend to develop robust and collective solutions such as wafer level packaging.

The first solution is based on an innovative in line glue process to package temperature sensitive biochips. A semi-automatic equipment is used to deposit through a mesh an accurate volume of UV curable glue of the top of a wafer with deep etched structures. Then the wafer is aligned with the top wafer (polycarbonate, SiO_2 glass or Pyrex) using dedicated alignment features and then bonded and cured using insulation in vacuum contact.

The second solution is an above IC process, this solution is a thin film packaging process and is well suited for ASIC or MEMS that can bare a 300°C process.

1. Several representative Constraints in projects using microfluidic

1.1 Biocompatibility

Since our applications deals with lipid bilayer, DNA strands and even living cells we have to restrict ourselves to biocompatible materials. Hopefully silicon technologies are well-suited for these fields since silicon is naturally oxidized in air or in water to silicon dioxide which is the major

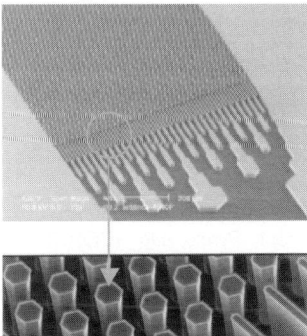
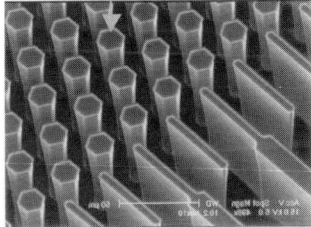

Fig. 1: Enzymatic protein digestion microreactor.

component of glass. Based on a huge amount of data, we can use silicon for in-vitro and even in-vivo devices. Obviously glass substrate, synthetic silica or borosilicate such as Pyrex can also be used. Consequently the strait forward materials of microelectronic that are silicon, silicon dioxide and silicon nitride fit perfectly with the biocompatibility constraints with a small limitation for long term stability of silicon in saline solution. This problem is said to be reduced by diffusing a boron etch stop[1].

1.2   Biochiplab- (high pressure)

This project is based on a liquid chromatography reactor. This biological sample is introduced in the reactor at high pressure commonly 20 bar and is forced to flow through a forest of pillars made by silicon deep etching. In order to obtain a high density reactor, micropillars of 10 μm diagonal, 15μm pitch and 50μm or 27μm height were etched in silicon[2]. Controlled thermal oxidation decreases the space between micropillars and allows electrical isolation of the channel essential for electroosmotic pumping. For digestion reactors, the micro-pillars have a hexagonal section (see in Fig. 1). For LC micro-columns, the micro-pillars have a square section and the pillar to pillar space is 1,5 μm (see in Fig. 2).

Fig. 2: LC micro-columns

A transparent Pyrex cap (CORNING 7740) containing inlet and outlet holes was then sealed on the microreactor by molecular bonding. Standard microfluidic connectors (Nanoport™ assemblies) were glued on the top of the cap opposite to the etched holes; they can be used either for electroosmotic or hydrodynamic flows. For electroosmosis, platinum electrodes were introduced directly into the connectors while for hydrodynamic pumping, another connecting part allows the fixation of a silica capillary of 360μm external diameter.

1.3   GoodFood

The GoodFood project is an integrated project of the sixth framework program (information society technologies). This project deals with the security and the tracking for quality assurance for the agrofood industry. GoodFood aims at increasing awareness at different levels:

- To the citizens: To show that Ambient Intelligence (AmI) and ubiquitous sensing can help on increasing food safety and quality.

- To the Agro-food Community: To demonstrate advantages of the MST, AmI and wireless solutions for new and/or current analytical tools and test methodologies, as well as their impact on improved farming.
- To the Microsystems industrial community: To show that Agrofood is a good niche market for MST solutions.
- To the SME and foundries: To develop demonstrators and market plans that show viability and lower the risks of future market access- To the MST scientific community: To give a path for take-ups, start-ups. A set of detection targets of vital importance has been also identified by the same relevant food industry representatives.

In this direction, the presence of chemical substances (antibiotics, pesticides and mycotoxins) and of life organisms (pathogens) has been considered the most relevant issues to be addressed from the safety point of view. The project restricts itself to a realistic limited number of applications that are nevertheless representative of a broad range of problematic involving solid and liquid products: milk, dairy products, fruits and fruit juices, wine, and fish[3].

Our lab is involved in the detection of antibiotics in milk which can trigger allergies or reduce the efficiency of human treatments based on antibiotics. Note the farmers are allowed to use antibiotics but the milk of these cows must be discarded for a strict period. A monitoring of the collected milk enable the control of this period of time.

The analytical concept is based on heterogeneous sandwich immunoassay and planar optical Wave guides (see in Fig. 3):

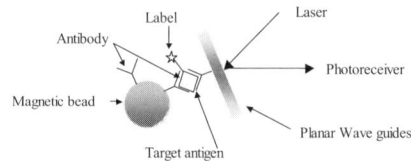

Fig. 3: Micro- Analytical concept of sandwich immunoassay and planar optical Wave guides.

An example of magnetic bead based immunoassay concept is shown in Fig. 4. Antibody coated beads are introduced to the micro component (1) and separated by applying a magnetic field (2). While holding the antibody coated beads, antibiotics are injected into the microchannel (3). Only target antibiotics are immobilized due to antibody/antibiotic reaction. Other antibiotic get washed out with the flow (4). Next, the antibody coated beads with the antibiotics are transported to the detection chamber (5). The antibody coated beads with the antibiotics bound at the secondary antibodies fix on planar optical Wave guides. (6) The evanescent field-based fluorescence detection is performed.

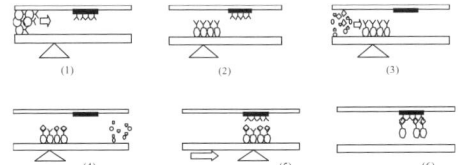

Fig. 4: Conceptual illustration of magnetic bead based immunoassay procedure using magnetic bead approach

The paradigm of the detection of the following: the milk sample is introduced in the microsystem, agitation or simple diffusion (depending on the authorized time) occurs so that the complementary antibody capture the antibiotics. The complementary antibody is a sandwich made of the complementary antigen with a fluorescent label and eventually the sandwich is grafted on a magnetic beads using an interaction such as biotin / streptavidine. Using a focused magnet the beads are gathered and driven to a smaller chamber which amounts to a 100 concentration factor. Then an elution based on thermal actuation or a change of the ionic force in the chamber break the link between the beads and the target labeled. Eventually the targets react with their complementary grafted on a wave guide. The evanescent wave excite the fluorophore and enable the detection of the antibiotics.

The last sequence of capture is based on antigen grafted on the wave guides, which triggers a major constraint: the sealing of the silicon wafer and the glass substrate including the wave guide must be very friendly to the antigen, i.e. the sealing temperature remain as low as possible.

2 Standards wafer level Solutions

Hereafter we describe very briefly three well-known bonding principles by decreasing temperature. Pros and cons of these techniques are highlighted and account for the development of new low temperature bonding process.

2.1 Silicon Direct Bonding, Pros and cons

Silicon direct bonding is by far the more strait forward way to obtain a bond between two silicon wafers. This process is well know and is described in Microfabrication handbook such as Microfabrication fundamentals of Dr. Marc Madou[4]. The principle is to create a hydrogen bond between two SiO_2 surfaces treated with a Brown solution or an oxygen Plasma (silanol Si-OH), which after a high temperature stage (1000 °C) under secondary vacuum (10^{-6} Pa) turns to a Si-O-Si bond. The fracture strength of the obtained stack is in the range of 20 MPa.

This process is front end compatible but very demanding on the cleaning processes to obtain a good bonding and there are obvious limits due to the bonding temperature.

2.2 Eutectic, Pros and cons

Eutectic bonding is based on the creation of a Si-Au layer to bond two silicon wafers (with a gold interface) or silica with a gold deposit. The mean fracture is as high as 150 MPa according to Tiensuu5 which is much higher than the silicon direct bonding.

There are important drawbacks with this technique. First the process is not compatible with most of the microelectronic fab since gold contamination is a major issue in Front-end clean rooms. Second there are large thermal stress associated to the thermal cycle whence failure and reliability issues.

2.3 Anodic, Pros and cons

Anodic bonding requires the contact of the silicon and a borosilicate glass substrate. Under 350 °C degree, and a voltage between 300 and 500V (at the Léti), one create a very robust bond. Glass such as Pyrex Corning 7740 are perfect for this technique since they are borosilicate glass and their coefficient of thermal expansion match the one of silicon (round $3.4E-6$ °K^{-1}). Consequently there are no thermal stress in the wafers. Note that the electrochemical, electrostatic and thermal mechanism could be combined to explain the anodic bonding. The exact phenomena is not perfectly understood, but migration of the sodium ions of the borosilicate glass in the silicon creates depletion which results in a high electrostatic force.

These three techniques require a wafer level packaging for sake of uniformity of the load, or the electrical field. This can be a drawback if to reduce costs a strategy of Multi Project Wafers (MPW) is chosen.

2.4 Polymer bonding using a laminated photoresist (Ordyl)

Laminated UV curable polymer such as Ordyl (SY 330 supplied by Elga Europe) gave good results when used to thermally bond above the glass transition a layer of Ordyl and to Indium Tin Oxide. It is also possible to bond two layers of Ordyl layers using thermo compression (load 6 kg.cm^{-2} at 70°C during half an hour and afterwards without load at 150 °C during two hours (see in

Fig. 5[6]). This technique allows watertight packaging and relatively low bonding temperature and gives excellent results compared to a SU8 photo-resist in terms of flatness a control on the whole wafer with an Altisurf 500 (supplied by Cotec) give 0.7 μm for the Ordyl vs. 5 μm for the SU8[7]. However this requires a photography, i.e. the use of a photomask and possible misalignment and the post bake is still too high for some our bio-labels.

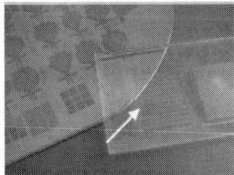

Fig. 5: a) Double bonded wafer and double bonded microscope glass. The wafer consists of a silicon substrate, two layers of SY550 dry resist in between and a glass wafer on top. The microscope slide had one layer of SY550 resist on top and was covered by a second microscope slide. The white arrow points to an 'unbonded' site, which can be recognized by a change in refractive index. b). Three-lane flow profile in a fluidic network (three layers of SY330)

3 Specific solutions for low temperature sealing at the CEA Leti

3.1 Thin film packaging

Silicon nitride has long been used for protection of ICs against moisture. All tests to date indicate that silicon nitride is an excellent thin-film material for hermetic encapsulation even in very thin layers so long as the films are pinhole free[8].

Another way to encapsulate MEMS at wafer level have been developed by MEMS engineers, called thin film packaging[9] see in Fig. 6. Closed cavities are formed above the devices with surface-micromachining techniques. As thin film packaging uses standard IC technologies and consumes less die area, it should offer a lower system cost than wafer bonding packaging. Moreover, it does not require wafer to wafer alignment and backside process technologies, which are not common in IC fabs.

The cavity above to device is formed with a sacrificial layer recovered by a cap. Holes are opened in the cap by a standard photolithography / dry etching process. Then, the sacrificial layer is removed through the holes. Finally, a film is deposited on the cap to seal the cap holes. Sealing can be performed by solder bumping, metal evaporation or dielectric deposit. In the technology developed at Leti, we use a polymer as the sacrificial layer. It is removed by dry etching (oxygen plasma) to avoid problems induced by wet etching such as sticking or pollution. The cap is formed by a silicon oxide deposition. Hermetic sealing under vacuum is obtained by a silicon nitride deposition.

The main interest of our technology is the compatibility with IC and MEMS as it is a low temperature process (<350°C).

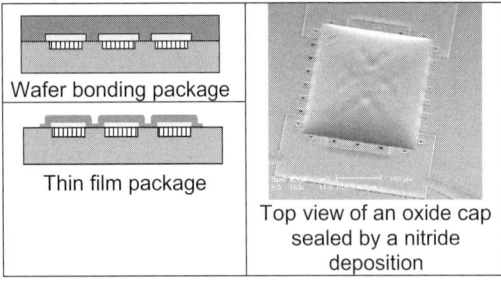

Wafer bonding package	
Thin film package	Top view of an oxide cap sealed by a nitride deposition

Fig. 6: silicon nitride deposition

3.2 In-line screen-printing solution

The in line glue process was especially developed to package chips including bio-label (antigen, antibodies or DNA strand). These species are dramatically sensitive to temperature. Packaging temperature must remain below 80°C unfortunately none of the classical packaging technique meets this requirement.

The in-line screen-printing process is based on the transfer of an amount of glue through a stencil. This process is based on an in-line screen printer supplied by Ekra© (model E5 STS High-Precision semi-automatic screen printer). The stencil is a mesh of Polyester as shown on the sketch in Fig. 7. Where the mesh is blank the glue is transferred through the stencil on the upper part of the substrate (hereafter so-called bottom substrate). Note that feature can be drawn on the mesh to avoid glue punctually on the wafer or to locate alignment marks. The volume transferred on the substrate depends on the parameter of the mesh and the characteristic of the squeegee such as angle, pressure, speed. The amount of glue is of secondary importance. A common formula (Eq. 1) is given to assess the amount of transferred glue:

$$v_{th} = \left(\frac{W}{W+d}\right)^{2} \cdot D \qquad (1)$$

, where d is the diameter of the fiber of the mesh, W the aperture between the fibers and D the thickness of the tissue.

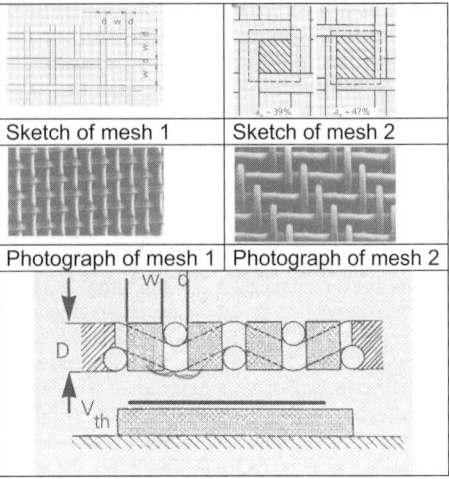

Sketch of mesh 1	Sketch of mesh 2
Photograph of mesh 1	Photograph of mesh 2

Fig. 7: mesh designs

In our group, a design of experiment (DOE) was carried out to minimize and control the thickness of the glue. The DOE was split in two studies for a 36 experiments trials. The study highlights the effect of the distance between the squeegee and the mesh. An optimized set of parameters allowed us to obtain a thickness in the range of 2.5 μm on 100 mm wafers with a standard deviation close to 0.5 μm. This value can be compared to the 8 μm obtained with a stamping process on 8'' wafers.

Note that the glue is an UV curable epoxy supplied by Delo and chosen to be biocompatible (with respect to USP class VI standard). After the deposit of the glue on the top of the bottom wafer, a mask aligner is used to align the top wafer with the bottom one using spacer similarly to a bond alignment process like Silicon Direct Bonding. Once the wafers are aligned, a pressure is homogeneously applied to the wafers by putting the whole wafers assembly under vacuum using either a flexible membrane or a blank mask. This amounts to a 80 daN force on the Z-axis for a 4'' wafers set. Moreover the vacuum procedure prevents the trapping of air bubble during the stacking. Once the wafers are held in contact the UV curing begins. Afterwards, the polymerization is completed in a oven under vacuum, i.e. a soft bake of several hours at 50°C.

3.2.1 Application to the GoodFood project

The GoodFood project requires a very low temperature sealing procedure, consequently it is perfectly well suited for the in-line glue process described above. In Fig. 8, we present two results of a deposition of the glue. The feature of the mesh is still very clear on the left picture and one can guess that there are unglued spots. Fortunately when the substrates are in contact and put under vacuum contact the glue spots merge and tend to create a continuous layer as shown on the right. Note that the width of the channel in the thinnest zone is 100 µm.

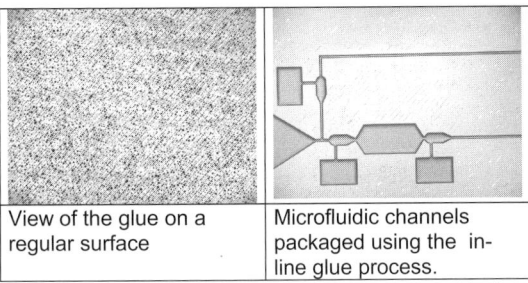

| View of the glue on a regular surface | Microfluidic channels packaged using the in-line glue process. |

Fig. 8: Results of glue deposit on structured and bulk silicon.

4   Conclusions

In this paper after an brief presentation of the usual bonding process we focused on low temperature and above IC packaging techniques. The first solution is based on an innovative in line glue process to package temperature sensitive biochips. The second solution is an above IC process, this solution is a thin film packaging process and is well suited for ASIC or MEMS that can bare a 300°C process. These techniques may become standards in the field of BioMEMS and MEMS respectively since they address a recurrent need low temperature and versatile packaging.

Acknowledgements

These works are developed with the help of the Nanobio program.

Reference:

[1] K.D. Wise, D.J. Anderson, J.F. Hetke, D.R. Kipke and K. Najafi, "Wireless implantable Microsystems, high density electronic interfaces to the nervous system", PROCEEDINGS OF THE IEEE, VOL. 92, NO. 1, JANUARY 2004

[2] Bing He, Niall Tait and Fred Regnier, Fabrication of nanocolumns for liquid chromatography, Anal. Chem. 70, 1998, 3790-3797.

[3] GoodFood: Food Safety and Quality Monitoring with Microsystems IST2002-508774

[4] M. J. Madou, Fundamentals of Microfabrication, CRC Press, Boca Raton, 2000.

[5] A.L. Tiensuu et al, "in situ investigation of precise high strength micro assembliy using Au-Si eutectic bonding", in 8th international conference on solid-state sensors and actuators (Transducers 95). pp 236-39.

[6] P.Vulto, N. Glade, L. Altomare, J. Bablet, L. Del Tin, G. Medoro, I. Chartier, N. Manaresi, M. Tartagni, R.Guerrieri, Microfluidic channel fabrication in dry film resist for production and prototyping of hybrid chips, www.rsc.org/loc | Lab on a Chip.

[7] I. Chartier, C. Bory, A. Fuchs, D. Freida, N. Manaresi, M. Ruty, J. Bablet, K. Gilbert, N. Sarrut, F. Baleras, C. L. Villiers and L. Fulbert, Fabrication of hybrid plastic-silicon micro-fluidic devices for individual cell manipulation by dielectrophoresis, Proc. SPIE-Int. Soc. Opt. Eng., 2003, 5345, 7–16.

[8] K. Najafi, "Micropackaging technologies for integrated microsystems: Applications to MEMS and MOEMS," presented at the SPIE. Micromachining and Microfabrication Symp., San Jose, CA, 2003.

[9] B.H. Starck, K. Najafi, "A Low-Temperature Thin-Film Electroplated Metal Vacuum Package", Journal of Microelectromechanical Systems, Vol. 13, No. 2, pp. 147-157, April, 2004.

Multi-Material Micro Manufacture
W. Menz, S. Dimov and B. Fillon (Eds.)

9

Micro- and nano- structuring using ion beams

A. Dietzel[a] , W.H. Bruenger[b], H. Loeschner[c], E. Platzgummer[c]

[a] *Eindhoven University of Technology, 5600 MB Eindhoven, the Netherlands*
[b] *Fraunhofer Institute for Silicon Technology (ISiT), Fraunhoferstr. 1, D-25524 Itzehoe, Germany*
[c] *IMS Nanofabrication GmbH, Schreygasse 3, A-1210 Wien, Austria*

Abstract

Ion beam techniques such as ion projection can be used to structure materials on micrometer and nanometer scales. It can complement current optical or e-beam lithography techniques for micro- or nanodevice fabrication. Besides ion beam lithography also direct ion beam structuring methods that do not rely on resist materials for the pattern transfer have been investigated. Local sputtering, magnetic nanopatterning and inducement of selective electroplating are examples for such resistless ion beam patterning techniques. A new type of instrument based on a parallel multibeam concept will allow combining high resolution with high throughput.

Keywords: ion beam lithography, direct ion beam structuring, nanopatterning

1. Introduction

Ion beams can be used to locally modify materials on the micrometer and nanometer scale. In principle there are two types of ion beam systems. One type is using a focused ion beam (FIB) [1] which can be scanned over the surface to be structured. Such instruments use a liquid metal ion source and can achieve a resolution below 10 nm. The other is a projection system in which a stencil mask is illuminated by a broad ion beam and projection optics is used to project a demagnified image of the mask onto the surface of the sample [2,3]. In both cases ions with energies of tens of keV up to hundred keV penetrate and interact with the material. The ion beam can be used to expose a resist layer which in the following steps allows all kinds of pattern transfer in a similar way as any other lithography technique. Besides ion beam lithography also direct interaction with the material can be used to locally induce material modification such as sputtering, ion implantation, ion beam induced intermixing, ion assisted etching and ion induced deposition. In comparison to optical techniques the resolution is not diffraction limited because the ions have an extremely small particle wavelength. In practice, features of 50 nm size were demonstrated using ion projection techniques [4]. Thereby patterns may be directly transferred onto materials without major forward and backscattering. The feature size is given by the beam shape and is not influenced by the proximity effect of backscattered electrons. The lateral exposure in an ion beam is very low and very narrow shapes can be written. Such properties make ion beams feasible for the direct fabrication of nanostructures. Hard materials like metals and also soft materials like polymers can be directly modified by ion beams on a nanometer scale.

2. Ion-beam interactions with materials

An energetic ion colliding with a solid target experiences elastic and inelastic collisions with the atoms and electrons in the solid. This leads to electronic and atomic interactions. Depending on the energy of the incident ion numerous interaction events such as backscattering, sputtering, implantation occurs. Nuclear reactions can only occur at much higher energies and will not be discussed here.

At relatively low energies the incident ion can be backscattered by the target atoms. This results in a deflection of the ion from its incident path. If momentum transfer to the target is sufficient it can lead to an atomic dislocation or even an ejection from the solid which is called sputtering or ion milling. There is threshold energy for the occurrence of sputtering. Above the threshold the sputter yield rises to a maximum and eventually decreases at even higher energies because the ion penetrates into the solid and dislocated atoms cannot reach the surface. The ion looses energy to the atoms and becomes trapped. As a consequence the ion is implanted and atoms around the ion path are dislocated from their original lattice sites. The implanted ions and displaced atoms can alter the properties of the solid. The depth of penetration decreases as the ion mass or the solid density increases. In summary, this type of elastic interaction leads to implantation, atom displacement, sputtering and formation of defects. Sputtering, implantation and displacement events can be simulated by Monte-Carlo methods and concentration profiles can be obtained [5].

The incident ion can also interact with the target electrons. The momentum transfer is very small and does not cause any appreciable scattering of the ion. This interaction creates electron excitation and ionization and is termed as inelastic. In addition to elastic and inelastic interaction with the target material the energy of the incident ions can also lead to chemical reactions which can be used for ion-assisted etching or ion-induced deposition and also to a change of the surface potential.

3. Ion projection technique

In an ion projection tool (see Figure 1) ions are extracted from a source resulting in a divergent ion beam. The collimated beam of ions passes an open stencil mask at energies around 5 keV. Extracted ions are H^+, He^+, N^+, Ar^+ or Xe^+. After they have passed the stencil mask the ions are further accelerated by electrostatic lenses to create a demagnified image of

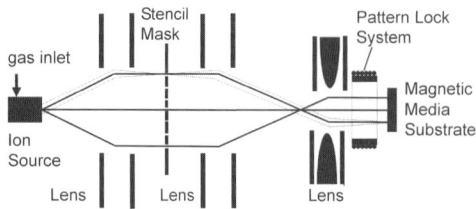

Fig. 1. Sketch of an ion projection tool from [6]. Ions are generated in a gas ion source and a broad ion beam illuminates a stencil mask. A demagnified image of the stencil mask is projected onto the substrate (here a magnetic media sample). The ion energy at the stencil mask is lower than the ion energy at the substrate. A pattern lock can compensate for a drift of the image.

the mask on the sample. The ions hit the surface at energies around 100 keV. The demagnification factor is between 4 and 10 and depending on the type of machine exposure areas between 4 mm^2 and approx 1.5 cm^2 can be achieved. The first practical ion projection tool was developed at IMS - Ionen Mikrofabrikations Systeme GmbH (IMS) in Vienna and is still operated by the Fraunhofer Institute for Silicon Technology. In 1997, a European MEDEA project was formed to build an process development tool (PDT) working at 4 x image reduction from mask to substrate and offering an exposure area of up to 17 mm in diameter.

A major challenge of the ion projection technique is the stencil mask. In contrast to the absorption and transmission of photons which is determined by the band structure of the mask material virtually any material can cause absorption or scattering of ion beams. Therefore, ion projection systems cannot use transparent mask blanks but rely on open stencil masks for pattern transfer. Letzkus et al. [7] developed a stencil mask process allowing a membrane thickness of 3 µm and mask sizes of up to 126 mm in diameter. The mask is fabricated on a silicon-on-insulator (SOI) wafer where the insulator layer is used as an etch stop. A deep RIE technique is used to etch the mask features that are written by e-beam lithography. The deep RIE process provides the steepness of the vertical sidewalls. The silicon membrane is coated with a thin carbon layer to protect the mask from ion damage

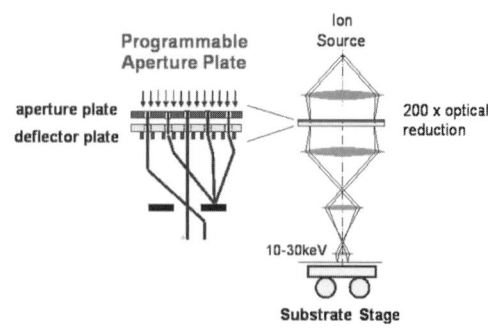

Fig. 2. Sketch of the CHARPAN (Charged Particle Nanopatterning) tool concept.

thereby extending the lifetime of the mask.

4. Ion projection maskless direct nanopatterning

There has been a tremendous development of focused ion beam (FIB) systems. Systems are available commercially which can provide resolutions below 10 nm [1]. They are used to create all kinds of micro- and nanostructures for research and industrial applications. In comparison to parallel beam systems such as ion projection machines they are inherently slower. Creating dense nanostructures over centimeter sized areas is practically impossible. Recently, a multi ion beam concept has been proposed [2]. The system combines the high resolution capabilities of a focused ion beam approach with the throughput advantage of a highly parallel system. In contrast to ion projection approach such a system can be programmed to deliver a specific pattern and does therefore not require individual masks for each application. An integrated European sixth framework project (FP6) named CHARPAN (Charged Particle Nanopatterning) has been started in 2005 in which such a tool will be developed and applications will be investigated. The innovative core of the system is an ion optical column comprising an ion source, a condenser system to form a parallel broad charged particle beam, a programmable aperture plate to structure the beam, and a 200x reduction ion beam optics to reduce the shaped beam to form a high resolution high intensity charged particle beam (see Figure 2). This versatile tool shall offer the flexibility to process a wide range of materials (metals, silicon, glass, ceramics, polymers) to sub - 25 nm feature range. The ultimate resolution limit is envisaged to be below 10 nm. One of the future applications can be the generation of templates for the nanoimprint lithography.

5. Ion projection maskless direct nanopatterning

Two different ways of producing micro- or nanopatterns by ion beams can be distinguished. The first is a lithographic approach similar to optical lithography using photoresists. In this case the ions are used to generate a latent image in a polymeric resist. The developed resist is then used to mask the subsequent etching or deposition in order to transfer the pattern in the substrate underneath. Features in

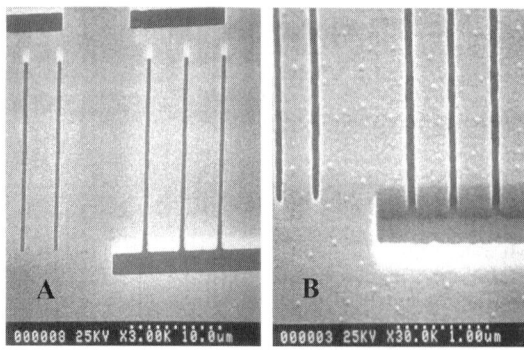

Fig. 3. Mask pattern (A) and corresponding 8.7 times demagnified wafer pattern showing ion beam projected 50 nm lines and spaces in chemically amplified resist (CAR) (B) [4].

developed resist of sizes down to 50 nm could be demonstrated by ion projection techniques (see Figure 3).

In the second case the ions are used to directly pattern the substrate. The direct interaction of ion beams with material can lead to sputtering, implantation or other material modification. Ion beams do directly modify a wide range of materials. In this case no resist is needed as a pattern transfer medium. Sometimes it is desirable to avoid the resist application because the surface may be contaminated by the resist. With ion projection techniques a surface can be treated in a parallel way. In addition the direct modification does in many cases not require subsequent etching or deposition und can therefore be a single-step process. Ion projection can been used to locally sputter materials using a stencil mask [8]. An array of lines is milled with a line width of 130 nm could be achieved.

5.1. Magnetic nanopatterning

Ion beam direct patterning is also very attractive for creating magnetic nanodots that can be used for future ultra high density magnetic recording. It is assumed that the conventional continuous magnetic media used in hard disk drives may be replaced in the future by patterned magnetic media. Those patterned media will consist of an array of well separated magnetic nano-islands. Resist based techniques are in this case not preferable because they will not keep the surface topography unchanged. An unchanged smooth surface topography is essential in a near contact recording scheme where the roughness of the disk surface shall not exceed a few nanometers. The ion projection technique was used to locally modify the magnetic

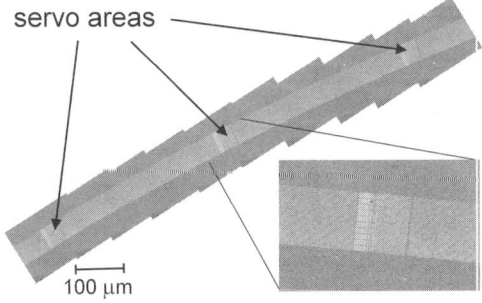

Fig. 4. Kerr images of a band of 50 adjacent tracks including single-bit data islands, as well as track positioning servo structures defined by ion projection modification of magnetic Co/Pt multilayer structures from [6].

properties of Co/Pt multilayers [6,9]. The advantage of this approach is that the magnetic contrast is a result of local ion beam induced intermixing of the thin Co layers

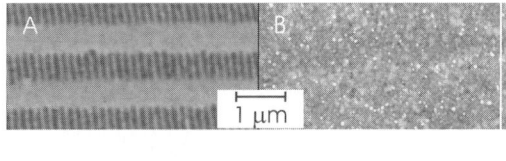

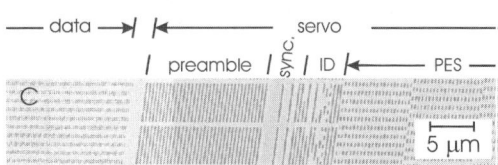

Fig. 5. A) Magnetic force microscopy (MFM) image taken after ion beam projection exposure of hard disk substrate coated with a Co/Pt multilayer. Three adjacent tracks with data islands, B) Atomic Force Image (AFM) of the same section shows a roughness of approx. 1.6 nm which does not differ from unirradiated regions. C) MFM images obtained from adjacent areas were stitched together. Regions typical for data areas as well as servo structures are indicated. From [9].

with the Pt layers. This intermixing is already efficient at doses which practically do not lead to any roughness induced by sputtering effects. Complete recording tracks consisting of well separated magnetic nano-islands were produced (see Figure 4).

The pattern was transferred using Ar^+ or He^+ ions with an energy of 45 keV using the PDT tool. Previously it was shown already that Xe^+ or Ar^+ ions are more effective for the magnetic modification by two orders of magnitude compared to He^+ ions [6].

The magnetic patterns obtained after an ion projection dose of 4×10^{13} Ar^+ cm^{-2} were investigated by Kerr Microscopy and by magnetic force microscopy (MFM) (fig. 5). It was confirmed that magnetic patterns with smallest dimensions of 70 nm could be created arranged in circular recording tracks. Moreover, atomic force microscopy (AFM) measurements indicated that the surface roughness of the magnetic media which was below 2 nm remained unchanged by the patterning process. It could also be shown that the nano-patterns can be magnetically switched between two directions perpendicular to the media surface without magnetically affecting the surrounding unirradiated thin film material.

5.2. Inducement of selective electroplating

It could also be demonstrated that ion projection systems can be used to locally create surface damage that allows for subsequent selective electroplating [10]. The surface damage locally induced by Ar^+ ion irradiation can significantly reduce the Schottky barrier breakdown potential of a semiconductor. Electro plating reactions can be selectively triggered on surfaces previously treated by ion beam projection at a dose of 2×10^{12} Ar^+ cm^{-2}. Figure 6 shows copper structures electroplated on p-type Si(100) surfaces after ion projection.

12

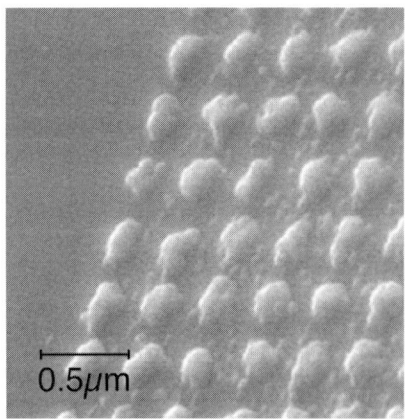

Fig. 6. Electron microscopy image (from [10]) of Cu structures electroplated on a p-Si(100) surface which was locally sensitized ion projection. The Cu structures resolved are in the order of 200 nm.

7. Conclusion and Outlook

Ion projection techniques are capable of producing nanostructures down at least to 50 nm in resist. They can be seen as an alternative to highly advanced optical or e-beam lithography. Moreover, ion projection techniques can in addition be used to do a direct, i.e. resistless patterning. A wide range of ion species and ion energies can be selected to adjust the beam properties, like penetration depth and sputter yield to the specific application. Sputtering, magnetic patterning and sensitizing substrates for selective electroplating have been demonstrated and produce structures on the nanometer scale. In the future direct interaction with a variety of materials such as polymers or metals shall be studied. With the development of a new ion beam tool for maskless nanostructuring (CHARPAN) a technique will be available that combines high resolution with high throughput. It will stimulate new ion beam nanostructuring applications.

References

[1] www.feicompany.com
[2] H. Loeschner,H. Buschbeck, A. Chalupka, G. Lammer, E. Platzgummer, H. Vonach, P.W.H. De Jager, R. Kaesmaier, A. Ehrmann, S. Hirscher, A. Wolter, A. Dietzel, R. Berger, H. Grimm, B.D. Terris, W.H. Bruenger, D. Adam, M. Boehm, H. Eichhorn, R. Springer, J. Butschke, F. Letzkus, P. Ruchhoeft, J.C. Wolfe, Proceedings of the SPIE, 4688, 595-606, 2002
[3] A.A. Tseng, small 2005, 1, No.6, 594-608.
[4] W.H. Bruenger et al. Appl Phys. Lett. 1999, 15, 403.
[5] TRIM computer simulation software, Biersack GbR,Bergstr. 18, 10115 Berlin, Germany, e-mail biersack@alaya.org
[6] A. Dietzel, R. Berger, H. Grimm, W.H. Brunger, C. Dzionk, F. Letzkus, R. Springer, H. Loeschner, E. Platzgummer, G. Stengl, Z.Z. Bandic, B.D. Terris, IEEE Trans. Magn. 2002, 38, 1952-1954
[7] R.F. Letzkus, J. Butschke, B. Hoefflinger, M. Irmscher, C. Reuter, R. Springer, A. Ehrmann, J. Mathuni, Microelect. Eng. 2000, 53, 609-612.
[8] W.H. Brunger, M. Torkler, C. Dzionk , B.D. Terris L. Folks, D. Weller, H. Rothuizen, P. Vettiger, G. Stangl, W. Fallmann, Micoelectron. Eng. 2000, 53, 605-608
[9] A. Dietzel, R. Berger, H. Loeschner, G. Stangl, W.H. Bruenger, F. Letzkus, Adv. Mater. 2003, 15, 1152-1155.
[10]A. Spiegel, W.H. Bruenger, C. Dzionk, P. Schmucki, J. Vac. Sci. Technol. B 2002, 65, 153-161.

Multi-Material Micro Manufacture
W. Menz, S. Dimov and B. Fillon (Eds.)

13

Micromoulding – precision processing for controlled products

P. D. Coates, B. R. Whiteside, M. T. Martyn, R. Spares, T. Gough

Polymer IRC/ Polymer CIC, School of Engineering, Design & Technology, University of Bradford, Bradford BD7 1DP, UK

Abstract

Micromoulding is an emerging field which has seen the evolution of conventional polymer injection moulding techniques for the manufacture of 3-dimensional components of sub-milligramme masses and/or microscale surface features. Micromoulding incurs extremely high strain rates and temperature gradients, particularly at the gate of the cavity. A range of issues associated with this, including high strain rate rheology of polymers (including experimental observation of in-cavity flows) and effects on orientation, residual stress and property developments, including surface replications have been explored in our laboratories and with collaborators. A major aim is to achieve controlled, repeatable dimensions and properties in micromoulded products, and to model this process – including molecular-feature models. The key factors determining control of precision products are discussed.

Keywords: micromoulding, miniature, precision moulding

1. Introduction

Micromoulding [1-3] is an emerging field, the evolution of conventional polymer injection moulding techniques, for the manufacture of 3-dimensional components of sub-milligramme masses (Figure 1) and/or microscale surface features. The process offers a route for mass production of complex microscale components using a range of materials including polymers, nanocomposites, metals and ceramics and has very low marginal production costs in comparison with alternatives such as self assembly of polymer structures and batch processes such as etching and lithographic techniques. Research in the field is now translating towards commercial applications in a wide range of areas including healthcare, optoelectronics, and micro-electro mechanical systems (MEMS) where demand for such components is rising exponentially each year. Barriers currently exist for companies wishing to transfer proof of concept designs into commercial manufacturing processes where variability in machine control and environmental conditions can have a significant impact on the yield of components meeting the desired specifications.

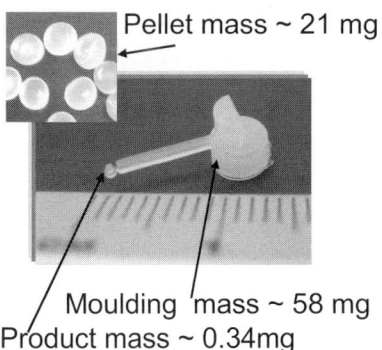

Pellet mass ~ 21 mg

Moulding mass ~ 58 mg
Product mass ~ 0.34mg

Figure 1: example of micromoulded product, with feed system in place

2. Extreme processing conditions

Conditions within a micromoulding process are typically much different from the conventional injection moulding process from which it was derived [4-7]. Product surface areas are large compared with product volumes, ensuring rapid heat loss from the molten material as the cavity fills, requiring high injection speeds to avoid premature solidification and incomplete products. The high speed injection can cause stresses and strain rates in the material which are orders of magnitude higher than those found in conventional injection moulding resulting in unusual material integrity and morphology. These factors ensure that the operating envelope is consequently small and process variations [8], temperature fluctuations, material inconsistency and ingress of impurities can result in sub-standard products which will not function in the manner for which they were designed (see also Section 5).

3. In-process measurements

A range of in-process monitoring techniques may be employed for quality assurance/ process evaluation. This is a particularly desirable feature for any high value-added manufacturing process, but is of course, rather challenging at micromoulding length scales. For example, the most sensitive of these techniques, measurement of melt pressure in the mould cavity using pressure sensors, can be difficult to employ due to the physical size of the cavity and/or the complexity of the mould tooling. Current research activity [9-12] includes development of novel sensor techniques, including sol-gel ultrasound sensors, which are optimised for the study of the micromoulding process and relevant materials.

Machine vision techniques are also attracting significant interest as they can offer a non-contact full field, multiple feature dimensional assessment of products which are difficult to see with the naked eye. These systems typically employ a number of cameras linked to pattern recognition software to allow evaluation of critical dimensions and feature sizes in the moulding process for every component in the manufacturing run. However, true micromoulded products with dimensions smaller than 1mm require evaluation using microscope lenses which usually have small working distances and shallow depths-of field, posing problems when attempting to employ them in machine vision applications. A novel solution to this technique uses the shallow depth of field advantageously by traversing the focus plane through the depth of the sample and recording a stack of images which can be subsequently processed using a computer to provide both a fully focused image and a 3-dimensional representation (Figure 2) of the product using a single camera system.

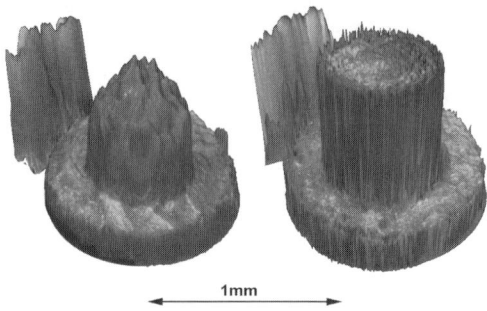

1mm

Figure 2: Short shot and full shot products, from the 3-d in-process imaging technique, for ~0.34mg product

4. Product evaluations

Products may be assessed, normally off-line, by a variety of techniques, all of which can be time consuming and difficult due to the small physical size of the products – hence the attraction of precision optical metrology in-process, to at least validate the dimensions of the product. Off line techniques currently include: Atomic Force Microscopy, with nanoindentation for morphology, structure and physical property determination [5,7]; white light interferometry/ surface profilometry for surface topology measurement [4]; birefringence (implies molecular orientation); x-ray scattering (wide and small angle) to evaluate crystal structures and orientation [13]; specialised micro-morphological optical measurements [5,7]; small angle neutron scattering (of deuterated polymer chains) for chain configuration measurements [14]; and polarised Raman spectroscopy for bond orientation measurements.

5. Materials

The supply of materials suitable for micromolding places very stringent demands on manufacturers. Polymer properties must be very tightly controlled with low rheological variation and minimal contamination. Previous studies in conventional polymer processing have highlighted the existence of batch-to-batch variations in material supply, but in the micromolding process variations within the same batch and even pellet-to-pellet variation can cause unacceptable fluctuations in process conditions and lead to poor repeatability. Contaminants introduced during the

compounding process may also adversely affect the rheology of a single shot of polymer and even damage the fine structures found in the mold cavities. The recent increase in awareness of the potential of the micromolding process has realized the opportunity for small-scale material suppliers to produce tailor-made products with rheological behavior, material properties and purity meeting the tight specifications required. The associated high cost of such materials is unlikely to perturb the end user, whose material costs form a small fraction of the total production costs for the high value–adding, but low mass, components.

The majority of micromolded products today are purely functional with tightly defined mechanical or optical criteria and little importance given to aesthetic or tactile properties. Therefore, typical adopted resins are engineering thermoplastics, such as liquid crystal polymers (LCP), polyoxymethylenes (POM), cyclo-olefin polymers (COP) and polyetheretherketones (PEEK), which exhibit desirable mechanical and optical properties and good processability.

The use of filler materials can improve the mechanical performance of the resins, but the small feature dimensions present in micromold cavities negate the use of conventional fillers, such as glass or carbon fibers. Nanofillers such as exfoliated clay platelets, polyhedral oligosilsesquioxanes (POSS), and carbon nanotubes show potential for use in the micromolding environment.

Micromoulding incurs extremely high strain rates and temperature gradients, particularly at the gate of the cavity, where shear strain rates well in excess of 10^6 s^{-1} are encountered. We have been exploring a range of issues associated with this, including characterisation of the high strain rate rheology of polymers (including experimental observation of in-cavity flows) and effects on orientation, residual stress and property developments, including surface replications in our laboratories, and with collaborators. A major aim is to achieve controlled, repeatable properties in micromoulded products, and to model this process – including the use of molecular-feature models. Modelling itself requires the experimental rheology data generated in our studies, for accurate material descriptions.

6. Process technology

Figure 3: View of part of the Centre for Micro & Nano moulding; 5-tonne Battenfeld microsystem machines on the left; Fanuc injection moulding on the right.

Precision process technology [3, 8, 9] is clearly a

vital part of the requirements for controlled precision products. At our Bradford Centre for Micro and Nano Moulding (Figure 3), we have an extensively equipped laboratory with five micromoulding machines, two mini moulding machines and one large scale thin-walled moulding machine, plus the novel in line compounding facility. This is part of a programme associated with miniaturisation of processing, which also includes very small scale extrusion, and development of sensors and in-process measurements, and characterisation techniques for products. A range of research programmes are in place, including exploration of bioresorbable materials for biomedical uses, medical technology devices, novel materials development, molecular feature rheology, and precision tooling developments.

6. Concluding comments

Controlled production of precision micromoulded products is challenging – especially if it is required to control the structure of the products, not simply the dimensions, to ensure a fit for purpose product. A range of scientific measurements and techniques are being used in our laboratories, in collaboration with suppliers of the technology (processing machinery, sensors, precision cavities) and raw materials, and end users, to develop confidence in such control – a vital step in the exploitation of micromoulding.

Acknowledgements

The authors gratefully acknowledge the EPSRC, Yorkshire Forward, DTI, and the support of the UK Micromoulding Interest Group (www.ukmig.com).

7. References

1. M Ganz, M., *Polymer Process Engineering,* ed. P.D. Coates, IOM Communications Ltd.: London (1999) pp. 8–17
2. L. Weber and W. Ehrfeld, Kunst. Plast. Eur, **89**, 10, 64-67/192-202, (1999)
3. C. Kukla, H. Loibi, H. Petter, W. Hannenbeim, *Kunstoffe Plast. Europe,* **6**, *p1331, (1998)*
4. BR Whiteside, R Spares, K Howell, MT Martyn and PD Coates (2005). *Micromoulding: extreme process monitoring and in-line product assessment.* In Polymer Process Engineering 05, University of Bradford, ISBN 1-85143-226-4, pp. 10-27
5. B.R. Whiteside, M.T.Martyn, P.D.Coates, G. Greenway; P. Allen; P. Hornsby, Plastics, Rubber and Composites (Maney Publishing), **32**, no. 6, pp. 231-239, 2003.
6. Zhao, J., Mayes, R.H., Chen, G., Chan, P.S., and Xiong, Z.J., *Plastics, Rubber and Composites,* (2003) 32(6), pp. 240–247
7. B.R. Whiteside.; M.T Martyn.; P.D. Coates ; G. Greenway; P. Allen; P. Hornsby. Plastics, Rubber and Composites, , **33**, no. 1, pp. 11-17, 2004.
8. Whiteside BR; Martyn MT; Coates PD, In-process monitoring of micromoulding-assessment of process variation, *International Polymer Processing* **20** (2), 162-169 (2005)
9. BR Whiteside, EC Brown, Y Ono, C-K Jen and PD Coates (2005). Polymer degradation and filling incompletion monitoring for micromoulding using ultrasound. *SPE Tech Paps, Vol. LI, Society of Plastics Engineers,* pp. 624-629
10. M. Kobayashi and C.-K. Jen, *"Piezoelectric thick bismuth titanate/lead zirconate titanate composite film transducers for smart NDE of metals,"* Smart Materials and Structures, **13**, pp. 951-956, 2004.
11. Y Ono et al. *ANTEC 2004 Proceedings of the 62nd Annual Technical Conference & Exhibiti*on, Vol. L, Chicago, IL,May 16---20. Society of Plastics Engineers, pp. 556-560
12. Y Ono, M Kobayashi, C-K Jen, C-C Cheng, BR Whiteside, EC Brown, L Mulvaney-Johnson and PD Coates (2005). *Real time diagnostics of injection moulding using ultrasound.* In Polymer Process Engineering 05, University of Bradford, ISBN 1-85143-226-4, pp. 52-79
13. T Gough, EL Heeley, W Bras, AJ Gleeson, PD Coates and AJ Ryan (2005). Using synchrotron radiation to follow structure development in commercial and novel polymeric materials. *SPE Tech Paps, Vol. LI, Society of Plastics Engineers*, pp. 2180-2185
14. Heeley, E.L., Gough, T., Bras, W., Gleeson, A.J., Coates, P.D., and Ryan, A.J., *Polymer Processing: using synchrotron radiation to follow structure development in commercial and novel polymer materials.* J Nucl Inst and Methods: Beam Interactions with Materials and Atoms, in press.

Multi-Material Micro Manufacture
W. Menz, S. Dimov and B. Fillon (Eds.)

17

Automatic nanohandling station inside a scanning electron microscope

S. Fatikow, Th. Wich, St. Kray, H. Hülsen, T. Sievers, M. Jähnisch, V. Eichhorn

Division Microrobotics and Control Engineering, University of Oldenburg
D26111 Oldenburg, Germany

Abstract

Current research work on the development of an automated nanohandling cell in a scanning electron microscope (SEM) is presented. Two experimental setups are shown, in which nanohandling robots ooperate in the vacuum chamber of an SEM. A client-server control system that can integrate various microrobots and sensors has been developed and evaluated by automatic handling of TEM (transmission electron microscope) lamellae by two nanorobots. The robots are controlled in a closed-loop way by using images from several CCD cameras and from the SEM. Algorithms for real-time processing of noisy SEM images have been implemented and tested. The experiment on automatic handling of TEM lamellae inside an SEM is described. The other setup contains a nanohandling robot using the probe of an atomic force microscope (AFM) as an end-effector. The manipulation of individual multiwall carbon nanotubes (MW-CNTs) and the characterization of nanofilms by nanoindentation are the applications being investigated. The experimental setup includesf a nanopositioning piezo sample stage with three degrees of freedom (DoF) and a three-axes nanomanipulator. Piezoresistive AFM probes are applied as an end-effector. This way the acting forces can be detected allowing force feedback for the station's control system. First investigations have been carried out by bending of MWNTs and calculating their elastic modulus.

Keywords: microrobotics, nanohandling automation, SEM image processing, nanocharacterization

1. Introduction

Micro- and nanohandling covers the field of handling objects with sizes in the range of μm, sub-μm and even a few nm. Important applications today are micro-assembly, semiconductor technology, nanotechnology, material research, medicine and biology. Within the last years a trend towards the automation of nanohandling processes emerged [1-2]. The nanohandling station that is currently being developed [3 4] aims at performing various nanohandling tasks in the vacuum chamber of a scanning electron microscope (SEM) by a microrobot system. Several only a few cm³ small mobile microrobots were developed [5]. The robots automatically operate with a precision better than 100 nm and can move at speeds of up to a few cm/s, using the slip-stick drive.

In the next sections an experimental set-up as well as the control system architecture for the nanohandling station is presented. The microrobot system has been evaluated by an automatic handling of TEM lamellae (TEM – transmission electron microscope). This task is of considerable interest for the semiconductor industry and for this reason a good bench mark for validation of the station's capabilities.

2. Setup of the Nanohandling Cell

The hardware setup of the nanohandling cell consists of an SEM, two multi DOF microrobots and the sensors for pose and contact detection.

For the nanohandling setup a LEO 1450 SEM had been used. It is equipped with a tungsten filament for the generation of the electron beam, which is sufficient for magnifications in the order of 50x to 1000x. In all the experiments a standard secondary electron detector is used for imaging. Fig. 1 shows the mechanical setup of the system. Every single setup for a specific application is built on its own SEM door. The door change takes

just a few minutes, which leads to high flexibility of the nanohandling cell and to reduction in vacuum pumping time. This approach also allows easy access to all major parts of the robot system.

Fig. 1. Set-up of the cell:
(1) mobile manipulator with tools, (2) robot plane, (3) stage robot with specimen holder, (4) SEM-door, (5) connector board. For training purposes on air, an optical microscope (6) is used

The robot plane is a glass plate, on which the mobile manipulator is driving around the stage robot. The stage robot can be moved in z-direction via a linear axis, which is also fixed to the SEM door. On top of this axis, the stage robot is driving on another glass plate. Additionally, a special connector board is implemented to distribute the power and signal lines separately to the corresponding sensors and actuators.

The stage robot places the specimen exactly under the electron gun of the SEM, and the mobile manipulator carries application-specific tools and positions them against the specimen. Both robots make use of the stick-slip drive [5], enabling a step size of better than 100 nm. However, the axes of the robot platforms

18

are not separated by 90 degrees, but by 120 degrees. Thus, the control system has to map the platform's coordinate system to Cartesian coordinates. The mobile manipulator carrying a touchdown-sensor and a gripper for handling TEM lamellae is shown in Fig. 2.

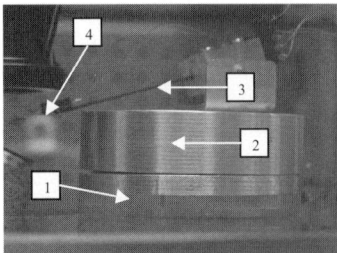

Fig. 2. The mobile manipulator, consisting of platform (1) with x-y-table (2) carrying the "touchdown"-sensor (3) and the gripper (4) on its top.

On top of the manipulator platform an x-y-table is mounted. Each axis of this table is a piezo stack actuator, which drives a leverage arm fixed by flexible hinges. The maximum stroke of the table is 100 µm. It takes care of fine positioning of the tool in the range of a few microns, after the coarse positioning via the slip-stick platform is done. The stage robot has a similar mobile platform with two translational and one rotational axis. The platform size has been reduced to a diameter of 30 mm in order to implement a more compact setup for the vacuum chamber. On top of the stage platform anoher platform is mounted upside down, to operate the ball-shaped specimen holder (Fig. 3). The upper platform of this "sandwich" can rotate the specimen holder around three axes.

Fig. 3. The upper platform of the stage robot carrying the specimen holder patterned for the position control. Also visible is the gripper of the mobile manipulator

For gripping the specimen, a gripper made by [6] has been used. It has two gripping jaws driven by electrostatic comb actuators. By applying a voltage of about 40 – 50 V, the jaws can be completely closed. The jaw-tips have a cross sectional area of 5 x 20 µm². The touchdown sensor consists of bimorph piezo bender, from which the first piezo ceramic is used as an actuator, while the second is used as the sensor. The first ceramic layer is driven by an AC-voltage with a low amplitude (appr. 10 mV), and the frequency is chosen at the set-up's resonance frequency of about 100 Hz. At the tip of the touchdown sensor, a connector bay for the gripper has been glued. When the touchdown sensor oscillates close to its resonance frequency, an AC-voltage of the same frequentcy and nearly the same amplitude can be measured on the second piezo ceramic, with a phase shift.

3. Control System of the Robot Cell

The control system has the client-server architectture with communication over TCP/IP and Ethernet or localhost to allow for a flexible use of the control and sensor modules in very different applications. Below a part of the cell control system, the low-level positioning control of the robot platform, is described. The aim is to move the platform from a given starting pose $\mathbf{g}_s^{(w)}$ to a desired pose $\mathbf{g}_d^{(w)}$, both given in a world coordinate frame. In Fig. 4 the low-level control structure is shown.

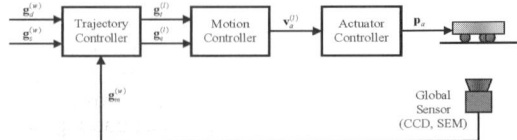

Fig. 4. Closed-loop control for mobile platform

The *trajectory controller* (Fig. 5) first calculates intermediate values on a given trajectory between the starting pose and the desired pose. The subsequent motion controller has two distinct objectives. The first objective is to move the platform along the trajectory towards the desired pose and the second objective is to keep the platform on the trajectory. Therefore, using the measured pose $\mathbf{g}_m^{(w)}$ the trajectory controller determines a local deviation $\mathbf{g}_e^{(l)}$ from the trajectory and an orthogonal desired local pose $\mathbf{g}_t^{(l)}$ along the trajectory. The *motion controller* then calculates corresponding desired local velocity vectors with help of a characteristic curve controller, which are then combined to a local actuation velocity $\mathbf{v}_a^{(l)}$ of the platform. The *actuator controller* finally tries to achieve the velocity $\mathbf{v}_a^{(l)}$ by setting parameters $\mathbf{p}_a$ of the actuator signals correspondingly. In the case of stick-slip-driven piezo-actuators of the mobile microrobot platform these parameters are amplitudes of six saw tooth signals.

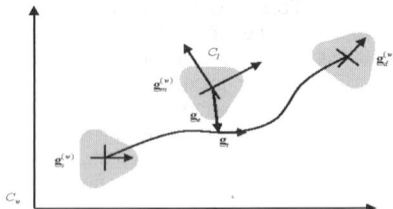

Fig. 5. Poses during trajectory control

For the current implementation actuation parameters for velocity vectors were found by trial-and-error. Six amplitudes of the sawtooth signals driving the microrobot must be determined for different combinations of absolute velocities and moving directions. Between these support vectors an interpolation algorithm [7] completes the mapping.

The control system has been tested by handling of TEM-lamellae (Section 5). The manipulator's positioning accuracy of 1 µm in combination with the sensor was sufficient for this application. It is important to note that the accuracy is only confined by the sensor, which is used for feedback. Depending on the SEM settings and on the robustness of the image processing, accuracies of even better than 100 nm were achieved in vari-

4. Real-Time Visual Feedback in SEM

Regarding resolution, image acquisition time and depth of focus of an SEM, it is a powerful sensor for nanohandling. However, the use of noisy real-time SEM images is a challenge for image processing.

Promising approach for object detection and position estimation in extremely noisy images are correlation-based algorithms [8]. The application of correlation-based template matching with SEM images is described in [9]. For correlation-based tracking, an image of the target object is defined as filter mask (matched filter). The disadvantage of this approach is that orienttation and scale estimation increases the computation time as additional patterns are needed. This drawback can be overcome with Active Contours approach.

The main difference between object tracking with cross correlation and with active contours is the representation of the target object. The parameterization of the target by its contour enables us multidimensional tracking by using a simple mathematical framework [10]. The contour of the target object is mathematically defined over the interval [0..L] by two dimensional spline curves $\mathbf{r}$ with (N_B -1) basis polynomials B of degree 2. Each basis polynomial is multiplied with a weight q. A one dimensional B-spline is defined as:

$$x(s) = \sum_{n=0}^{N_B-1} q_n B_n(s) \quad \text{for} \quad 0 \le s \le L \tag{1}$$

The shape of the B-Spline can be changed by a variation of the weight vector of the basis polynomials, therefore the weights are control parameters for tracking. A 2-dimensional spline curve $\mathbf{r}(s) = (x(s), y(s))$ is built up from two B-splines with the same basis polynomials. Usually just the shapes which can be adapted by the target object are of interest. Therefore, possible variations of the B-Spline are confined to a shape space of dimension n representing degrees of freedom of the target object. For tracking an object in the plane, a three-dimensional shape space is required, with the DOF x, y and φ. To work with different magnifications, a shape space with one more DOF for scaling is necessary. For this reason, the shape space of Euclidean similarities is the best choice.

To fit the contour of the target object with the B-Spline, an appropriate control vector has to be determined. A coarse control vector is initially defined, which is adapted to the target shape iteratively. The deviation between spline curve and object contour is estimated by measurement lines which are arranged orthogonal (with normal vector $\overline{\mathbf{n}}(s)$) to the B-Spline with equidistant space. Each measurement line is searched for intersections with the target object. The intersection points $\mathbf{r}_f(s_i)$ can be found by an edge detector applied to the lines. Then, a distance between intersection point and B-Spline is calculated. Finally, the control vector is adapted the target shape.

The crucial task to solve is the estimation of the deviation between initial curve $\overline{\mathbf{r}}(s)$ and measured intersection $\mathbf{r}_f(s_i)$ (feature curve) for N sample points:

$$\left\| \overline{\mathbf{r}} - \mathbf{r}_f \right\|_{\overline{\mathbf{n}}}^2 \approx \frac{1}{N} \sum_{i=1}^{N} \left[\left(\mathbf{r}_f(s_i) - \overline{\mathbf{r}}(s_i) \right) \cdot \overline{\mathbf{n}}(s_i) \right]^2 \tag{2}$$

The challenge of this approach is robust edge detection. A feature detector, which is robust against noise, is vital for successful tracking.

In the experiment described below a feed-forward backpropagation neural network has been used to determine a feature in a one-dimensional gray level array. The network has been trained by noisy data generated by propagating an edge through the input vector with additive noise. This approach has proven to satisfy the real-time requirements. However, a drawback of the neural network edge detector is that changing the length of a measurement line makes a new training of the network necessary, with a down- or up-scaling of the input vector. The algorithm has been implemented in Matlab™ to test its performance with SEM images and image sequences. In Fig. 6 an AFM tip is tracked by a spline curve.

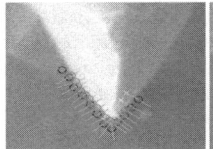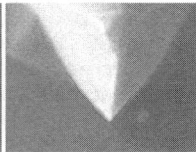

Fig. 6. Fitting a spline curve to the contour of an AFM tip (20 um). Left: start curve with measurement lines and edges. Right: adapted curve (one iteration)

The spline curve is fitted to a part of the AFM tip contour. Edge detection has been carried out both with a gradient filter and with the backpropagation network. The edge on the left side of the tip can be properly detected with the gradient filter, but the low contrast edge on the right side of the tip has not been detected for all measurement lines. The backpropagation network has made successful detection for both edges. The computational time for the active contours algorithm is less than for cross correlation as only the pixels sampled from the measurement line are needed for tracking. For the estimation of the motion vector, only simple matrix operations have to be carried out. The quantitative comparison of the performance is to carry out in the next step after the C++ implementation of the active contours algorithm.

5. Automatic Handling of TEM Lamellae

The performance of the nanohandling cell inside an SEM has been demonstrated by automatic handling of TEM-lamellae. Latter are tiny slices of a silicon wafer cut out of the wafer surface by Focused Ion Beam (FIB) etching (Fig. 7). The size of lamellae is usually in the range of 10µm x 2µm x 100 nm and below.

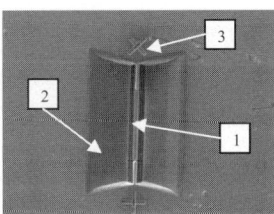

Fig. 7. SEM-image of a TEM-lamella (top-view): (1) lamella, (2) FIB-structured groove, (3) marker for closed-loop control during the lamella handling

The first step is coarse positioning of the stage robot, which is carried out automatically by using two CCD-cameras as position sensors, delivering bottom and lateral images of the stage robot to the positioning control system. The stage platform is driven to a defined position, at which the specimen with the lamellae can be scanned by the electron beam of the SEM. After that, the control system chooses the SEM as a position sensor, and, at a magnification of 50x, the wafer is automatically searched for TEM-lamellae. The image focus is adjusted using the auto-focus function of the SEM. As soon as one or more lamellae are detected, the user has to choose one for gripping via a graphical user interface.

The lamella chosen by the user is automatically centered at a magnification of 50x. A region of interest (ROI) is defined around the lamella. The SEM scans just the ROI and not the entire frame, so SEM images are acquired in real-time. They are sent to the closed-loop control system of the stage robot which automatically moves the chosen lamellae exactly to the middle of the SEM scan field. After that, the SEM magnification is switched to 300x, and the lamella's position is adjusted more precisely by using a new, smaller ROI. Finally, the SEM magnifycation is switched to 1000x, and a ROI is drawn around a cross-shaped marker on the wafer surface, allowing the final adjustment of the lamella to the scan field center with the highest possible accuracy.

After that, the mobile manipulator takes care of positioning the gripper. The first step to accomplish is to place the gripper above the specimen stage with no risk of a crash between gripper and stage. Knowing the geometry of the working scene, the control system of the stage robot ensures that there is always a "safety gap" between specimen and gripper. The gripper is then automatically moved by the mobile manipulaor to a pre-defined x-y position, using the coarse positioning closed-loop control via the third CCD-camera placed underneath the mobile manipulator. When the gripper tip reaches the aspired position, it is detected by the electron beam at a low magnification of the SEM. Finally, the gripper is adjusted to the lamellae at magnifications of 50x and 300x by using ROI as described in previous sub-section.

To reach the working position around the lamella, the specimen stage is lifted up by the stage robot. This operation is automatically controlled by using SEM images at a magnification of 300x. As soon as the touchdown sensor detects the contact between gripper and wafer, the lifting stops.

In the current series of experiments, the gripping and transporting a lamella still has to be done by teleoperation. The user closes the gripper by teleoperation control and breaks the lamella out of the groove. Fig. 8 shows a typical gripping sequence.

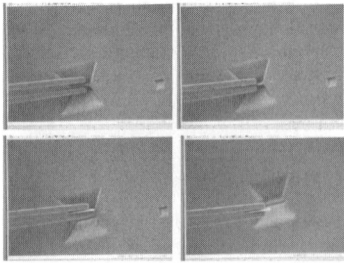

Fig. 8. Gripping sequence of a TEM-lamella

Finally, the mobile manipulator automatically takes the lamella to the TEM grid on the specimen stage. Firstly, the gripper moves into its waiting position outside the SEM scan field. Then the mobile platform of the stage robot is rotated, and the TEM grid is centered in the scan field, using CCD images for coarse positioning and SEM images for fine positioning of the stage robot. As soon as the TEM grid is centered, the gripper with the lamella is moved into the middle of scan field, keeping the distance of a few microns above the grid surface.

The user eventually places the lamella on the grid by teleoperation, monitoring the step on the SEM display. As expected, adhesive forces make the lamella stick to the gripper. The electrostatic forces can be overcome by using conductive and grounded grippers. To cope with the atomic forces, it is sensible to reduce the contact surface between TEM lamella and gripper jaw, which requires a special structuring of jaw tips.

6. Stereoscopic Depth Detection in an SEM

One restriction in the handling of objects inside an SEM is lack of 3D information. One way to overcome this problem is stereoscopic depth detection using two images from two different angles. In an SEM, these images can be generated by concentric tilting of the specimen table. It is advantageous to tilt the stage only a few degrees, meaning a trade-off between the angle and the possibility of depth calculation.

For the calculation of depth the relative displacements (disparity) between the two images has to be detected, which is hard to accomplish in noisy SEM images. A promising class of algorithms was developed from research into the visual cortex of mammals [11]. A formalisation of this approach was introduced in [11-13]. This algorithm has been used to detect depth information in SEM images. The result is that small disparities can be detected quite well by this algorithm [12] so that the angle can be quite small. In the experiment, the specimen table with the robot was tilted about 1°. The images show a tip between some objects with a low magnification of 200 times (Fig. 9).

Fig. 9. Two stereoscopic images tilted by about 1°

The result of the stereo algorithm is a high-density disparity map (Fig. 10, left) in sub-pixel accuracy of 1/16 pixel, which shows dark regions for high distances and bright regions for low distances with reference to the observer.

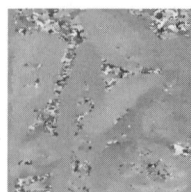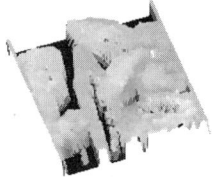

Fig. 10. High density disparity map (left) and the 3D-plot of the filtered disparity map (right)

In the disparity map, the tip is darker than the objects close to the tip i.e. the tip is lower than the objects.

There are some regions in the map, which show a heavy noise. The reason is the black background in the input images, which has a low texture. By this algorithm, it is not possible to estimate a robust disparity at these regions. Therefore, a special texture filter has been used [12]. The result is shown in Fig. 10, right, as 3D-plot. The calculation of the disparity map (at the resolution of 256x192) takes several seconds. With the disparity map, it is possible to detect the relative z-position of the gripper in sub-pixel accuracy for further automation processes.

7. Setup of the Nanocharacterization station

CNTs exhibit unique electrical and physical properties due to their specific atomic structure, and they have a huge application potential in nanotechnology. To learn more about CNT properties, the nanomanipulation and characterization of individual CNTs is required. Significant progress in nanohandling can be achieved by the use of an AFM [14]. However, during the manipulation the cantilever-shaped probe is not available for imaging the surface topography, so that AFM images can only be acquired before and/or after the manipulation. The integration of an AFM probe based nanohandling station into the vacuum chamber of an SEM is a promising way to deal with this problem.

The setup (Fig. 11) contains a three-axes micromanipulator MM3A (Kleindiek Nanotechnik, Germany) that is equipped with an AFM probe holder. The micromanipulator offers a resolution of 5 nm in x- and y-direction and a resolution of 0.5 nm in z-direction. Piezoresistive AFM cantilevers have been provided by Nascatec, Germany. The sample is located on the nanopositioning piezo stage with three DoF. The resolution of the nanopositioning stage (PI, Germany) is limited by the 16-bit D/A converter of the control module and amounts to 1.55 nm in closed-loop mode.

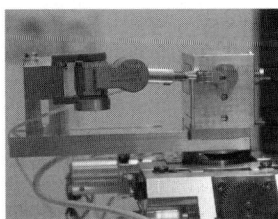

Fig. 11. Setup of the AFM probe based nanohandling robot station integrated within the SEM LEO 1450

For nanohandling in SEM the setup has been installed onto the positioning stage of a LEO 1450 SEM (Carl Zeiss). By operating the SEM stage the nanohandling station is adjusted within the vacuum chamber. This process is monitored by a CCD-camera mounted inside the SEM. To observe the sample and the AFM cantilever with the electron beam of the SEM, the stage can be tilted by the SEM stage. The automatic CNT characterization is implementted by pushing the AFM probe against a nanotube via the station's stage controller. The measurement process is implemented in LabView.

8. Nanoindentation and CNT Characterization

Nanoindentation is an important tool for characterizing films and parts on the micro- and nanoscale. A classical setup is shown in Fig. 12. An actuator generates a force for driving a diamond tip into a specimen. The generated pressing force is measured simul-

taneously. As a result, the tip makes an indentation into the material and its displacement is measured.

A standard procedure to determine hardness and elastic modulus from the load-displacement data was presented e.g. in [15]. Every indenter has a finite load-frame stiffness S, which is indicated in Fig. 12 as machine compliance $C=1/S$. If an indenter displacement occurs, the total measured displacement h is:

$$h = h_{lf} + h_{sp} = \frac{F}{S} + h_{sp} \qquad (3)$$

with the load-frame deflection h_{lf} and the true indentation depth h_{sp}. Eq. (3) shows that the device and the contact are modeled as springs in series. The stiffness-corrected load-force diagram can be evaluated to obtain hardness and elastic modulus.

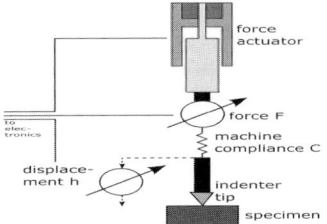

Fig. 12. Structure of a nanoindentation device

The nanohandling setup presented in Fig. 11 can also be seen as an implementation of the approach in Fig. 12. The three-axes manipulator is used to position the AFM cantilever – that is a combination of force sensor and indenter tip – above the sample. The nanopositioning stage is used as force actuator and – due to its excellent capacitive displacement readout – as indentation depth sensor. As stated above, load-frame stiffness that is equal to the cantilever stiffness in the nanohandling setup has to be subtracted from the total displacement. The resulting force is measured by the integrated piezoresistive Wheatstone bridge.

The basic setup in Fig. 11 can be used to perform sensor calibration. The first step is to measure the geometric parameters of the cantilever beam at different points. The average of these values can be used for the stiffness calculation. The geometric parameters (length l, width w, thickness t) of the piezoresistive cantilever used for the experiments are:

$$l - 507.5\,\mu m, \ w - 157\,\mu m, t = 7\,\mu m \qquad (4)$$

The stiffness is calculated according to Eq. (5):

$$S_{lf} = \frac{Ewt^3}{4l^3} \qquad (5)$$

where E is the elastic modulus of silicon and w, t, l are the geometric parameters of the AFM probe. The calculated stiffness in this setup is S=17.08 N/m.

Calibration has been done by applying the backside of the cantilever beam to a reference material. A hard silicon specimen was used to deflect the cantilever at its backside. By recording sensor output voltage and displacement, the calibration factor connecting voltage drop and force was calculated (Fig. 13). The linear slope between voltage and displacement in combination with the stiffness gives us the calibration factor as a voltage-displacement factor V of 3.76 mV/μm. By translating the displacement to the force via the stiffness it is possible to obtain the voltage-force factor K:

$$K = \frac{S}{V} \qquad (6)$$

22

With the calculated stiffness of S=17.08 N/m, this leads to a factor K=4.537E-6 N/mV. By multiplying the – offset corrected – voltage signal with this factor K, the total applied pressing force F can be obtained.

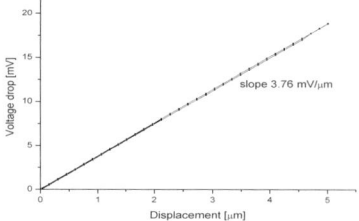

Fig. 13. Measured voltage drop-displacement ratio

First experiments on CNT characterization were performed to prove the feasibility of the new nanohandling cell. The setup has been tested by performing deflection measurements on single carbon nanotubes. The three-axes manipulator is used to bring the cantilever close to the nanotube. Fine positioning is done with the three-DoF nanopositioning stage of the station. The nanotubes had a length of 8.8 μm and a diameter of 370 nm at the bottom and 160 nm at the top. By moving the nanopositioning stage, the nanotube counteracts the probe tip and bends afterwards. The resulting deflection is measured by using the SEM images. The results of these measurements have shown good agreement with the position sensor of the nanopositioning stage in closed-loop mode. The bending force is measured during deflection on the basis of the cantilever calibration. Fig. 14 shows the measured force in dependence of CNT deflection.

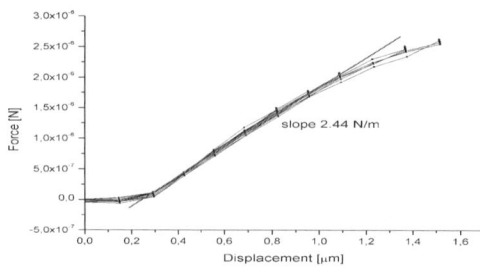

Fig. 14. Force-displacement measurements of a CNT

Maximum CNT deflection is approximately 1.5 μm, with the corresponding force of 2.5 uN. In contrast to a nanoindentation curve shown e.g. in [16], loading and unloading behaviour is identical. The correspondence between loading and unloading emphasizes the elastic behavior of CNTs. The slope of the linear region has been evaluated as m=2.44 N/m. Analysis of the bending moment and the elastic curve leads to an expression for the force-displacement relation. The elastic curve can be represented in dependence of the force for a solid cylinder, which represents a MWNT.

$$h = \frac{1}{3} \frac{F(h)l^3}{EI} ; I = \frac{\pi d^4}{64} \qquad (7)$$

The force $F(h)$ depends on the displacement h; l is the position of the nanotube having contact with the cantilever; E is the elastic modulus and d the diameter of the MWNT. By comparing coefficients, the elastic modulus E can be determined by the slope m of the measured data (see Fig. 14):

$$E = \frac{64ml^3}{3\pi d^4} \qquad (8)$$

The nanotube has different diameters at the bottom and at the top of d_b=370nm and d_t=160nm. The elastic modulus has been calculated for both values and the mean value leads to E=0.95 TPa, with l=4.17μm and m=2.44N/m. The result is in good conformity with theoretical calculations [17].

Acknowledgments

Parts of this research work have been supported by the EU (project NANORAC, NMP4-CT-2005-013680, and project ROBOSEM, GRD1-2001-41864).

References

[1] G. Yang, J. A. Gaines, B. J. Nelson, "A supervisory wafer-level 3D microassembly system for hybrid MEMS fabrication," Journal of Intelligent and Robotic Systems, 37, pp. 43-68, 2003

[2] S. Martel, et al., "Integrating a complex electronic system in a small scale autonomous instrumented robot: NanoWalker," Microrobotics and Microassembly, vol. 3834, Boston, pp. 63-74, 1999

[3] S. Fatikow, et al., "A Flexible Microrobot-Based Microassembly Station," J. of Intelligent and Robotic Systems, 27, pp. 135-169, 2000

[4] S. Fatikow, et al., "Development of a Versatile Nanohandling Station in a SEM," Int. Workshop on Microfactories, Minneapolis, pp. 93-96, 2002

[5] A. Kortschack, et al., "Driving principles of Mobile Microrobots for the Micro- and Nanohandling," Int. Conference of Intelligent Robots and Systems, Las Vegas, USA, pp. 1895-1900, 2003

[6] http://www.nascatec.com

[7] H. Hülsen, "Design of a fuzzy-logic-based bidirectional mapping for Kohonen networks," Int. Symp. on Intelligent Control, Taipei, pp. 425-430, 2004

[8] F. Goudail, P. Réfrégier, Statistical Image Processing Techniques for Noisy Images – An Application-Oriented Approach, Kluwer, 2003

[9] T. Sievers, S. Fatikow, "Visual Servoing of a Mobile Microrobot inside a SEM," IROS, Edmonton, pp. 1682-1686, 2005

[10] A. Blake, M. Isard, Active Contours, Springer, 2000

[11] R. Henkel, "Fast Stereovision by Coherence Detection", LCNS 1296, eds. G. Sommer, K. Daniilidis, J. Pauli, pp. 297, Springer, Heidelberg

[12] M. Jähnisch, M. Schiffner, "Stereoscopic Depth-Detection for Handling and Manipulation Tasks in a SEM," IEEE ICRA, Orlando, USA, 2006

[13] N. Qian, "Binocular Disparity and the Perception of Depth", Neuron, Vol. 18, 359-368, 1997

[14] A.A.G. Requicha, et al., "Manipulation of Nanoscale Components with the AFM: Principles and Applications, " IEEE Int.Conf. on Nanotechnology, Maui, Oct 28-30, 2001

[15] W.C. Oliver, G.M. Pharr, "An improved technique for determining hardness and elastic modulus using load and displacement sensing indentation experiments", J. Mater. Research, vol. 7, no.6, pp. 1564-1583, 1992

[16] M. R. VanLandingham, "Review of Instrumented Indentation", J. Res. Natl. Inst. Stand. Technol., vol. 108, pp. 249-265, 2003

[17] X.Y. Wang, X. Wang, "Numerical simulation for bending modulus of carbon nanotubes and some explanations for experiment", Composites. part B, vol. 35, pp.79-86, 2004

Recent evolution of electrical discharge machining

N. Mohri[a], T. Tani[b]

[a] *Department of Precision Machinery Engineering, The University of Tokyo, Tokyo, Japan*
[b] *Tsukuba University of Technology, Tsukuba, Japan*

Abstract

Electrical Discharge machining (EDM) is a comparatively new machining method which has several decades of history since it has been invented. At its beginning, it was developed as a precision machining method for hard materials. In recent years, several researches and methods based on the electrical discharge phenomena have been proposed. Mirror like finish machining, surface modification of mold die, machining of insulating materials and micro products manufacturing are noted among these researches and methods in the EDM field. These methods are particularly concerned with control of surface characteristics of the work piece and with control of electrode motion. In this paper, we introduce and discuss surface machining methods, insulating ceramics machining by EDM and micro EDM with which we have been associated.

Keywords: Electrical discharge machining, Finishing, Surface modification, Insulating ceramics, Micro EDM

1. Introduction

In EDM pulsative voltage is applied to the gap between two electrodes in working oil where one is a machining tool and the other is a work piece being machined by electrical discharge energy.

We can observe in the gap region some machining conditions as normal discharge, shortage and discharge concentration. Normal discharge frequency corresponds to cutting sharpness in traditional machining. The machining system of EDM consists of an electrical power source controlled by a gap condition observer and a servo mechanism of tool electrode motion. **Fig.1** shows an electrical discharge machining system. Many kinds of improvements on the gap condition have been achieved in EDM history.

New technological factors are summarized in **Table 1** comparing with traditional ones. Remarkable innovations have been found in each constituent technology, i.e. electrode materials and structure, target products, gap medium between electrode and work piece, control system for discharge pulse, and gap servo control.

Taking four examples of new technologies, developments and future trends in EDM are presented in the following sections.

2. Finish machining by EDM

In finishing by EDM, it is needed to make pulse discharge small and each discharge position dispersive in order to get uniform fine surface. **Fig.2** shows the effect of working area on the roughness of finished surface by EDM. Machined roughness increases according to machining area.

Graph A in Fig.2 shows the surface roughness machined by conventional EDM [1]. It can be observed that surface roughness reaches up to 15mm Ry for areas of several tens of cm^2. Under these conditions, hand finishing after finish EDM is sometimes required.

2.1 Silicon electrode

For the finish measurements, silicon was used as

Table 1 Constituent technologies in EDM

	Conventional technologies	New technologies
Electrode material	Cu, Graphite, W, Brass, Mo	Ti, Si, Conductive ceramics, Conductive diamond, Zinc alloy
Electrode structure	Solid electrode, Wire electrode	Micro-pin electrode, Flame electrode, Mesh electrode, Semi-sintered electrode, Assisting electrode
Workpiece material	Steel, WC-Co, W, Conductive ceramics	Silicon, Insulating ceramics, Boron doped Diamond
Working fluid	Kerosene, Deionization water	Powder suspended fluid, Dry EDM (Oxygen,Air, Ar)
Discharge power source	Condenser discharge, Iso-pulse discharge, Slope control, Super-posed voltage	Micro energy discharge, AC discharge in high frequency, Fuzzy control, Adaptive current waveform control
Driving motor	Voice coil motor, DC motor, Hydraulic motor, AC motor	Pulse motor, Linear motor, piezoelectric motor, Inch worm motor, Piezoelectric motor
Controlled object / motion strategy	Z axis control, Jump motion control, C axis control, Planetary motion, Scanning control	Hollow machining control, Mole machining control, Ultra-sonic vibration, Magnetic bearing control, Tracking machining control

a resistive electrode which divides the capacitance and enables the discharge dispersion [2]. With the silicon electrode, uniform fine surface can be achieved in very short time.

Graph B in Fig.2 shows the resultant roughness compared to conventional results. To our surprise, we noted that in this case, roughness does not depend on the area of the electrode. The whole electrode capacitance divides itself into smaller capacitances.

2.2 Powder mixed working fluid

The **graph C in Fig.2** shows the tendency of the roughness curve against machining area in the case where conventional electrodes are used with silicon powder mixed oil.

Since a discharge tends to occur at the point where powder is present, generated debris by each discharge should be the trigger of the next coming discharge. Then, if many debris remain in the narrow gap region, discharges concentrate in the region thereby making the machined surface coarse. When, a large amount of powder, more than the debris generated by discharges, is mixed in advance, discharges spread on whole area of the electrode surface.

3. Metal surface modification

Silicon electrode was used for finish machining of a steel plate. In the process of finishing a stainless steel by silicon electrode, it's been revealed that the machined surface has high corrosion resistance [3]. Specifically, the stainless steel surface (ANSI304) machined with the silicon electrode never corrodes in dilute hydrochloric acid or nitrohydrochloric acid (aqua regia). The top surface is covered with a silicon rich amorphous layer. This initiated the development of surface modification of metal.

3.1 Product of thermal conductivity and melting point.

The machining rate or electrode wear under certain conditions depends on thermal properties peculiar to each material. Experimental data reveal that in EDM, materials with large product of thermal conductivity and melting temperature have small wear.

Fig.3 shows the relation between the removal rate and the product of thermal conductivity and melting point [4]. Because this product is large in copper and tungsten, electrodes of these materials wear only slightly. Therefore, it seems difficult to use these materials as solids for surface modification.

Suddenly, it occured to us that composites or semi-sintered electrodes of the materials in question could be used. The apparent thermal conductivity of these composite and semi-sintered electrodes decrease sufficiently to allow wear. Substances of the electrode materials would leave the surface owing to discharges beyond the regard of the restriction in **Fig.3**.

3.2 Practical use of composite electrode and semi-sintered electrode

Surface modification of carbon steel was performed with TiC semi-sintered electrode [5]. The resultant surface hardness reaches up to 2 thousand

and several tens hundred HV as shown in **Fig.4** [6].

This method has been applied for modification of the cutting edge of punching dies in order to extend their tool lives.

The adhesive force in the modification by EDM is higher than that in other methods. The modification process can be performed on the same machine for shape generation.

In near future, combination of sinking EDM and surface modification will be performed by a machine tool for dies, cutting tools, or other parts manufacturing.

Surface modification by EDM has overtaken the up-to-now established assessments that EDM is a removal machining.

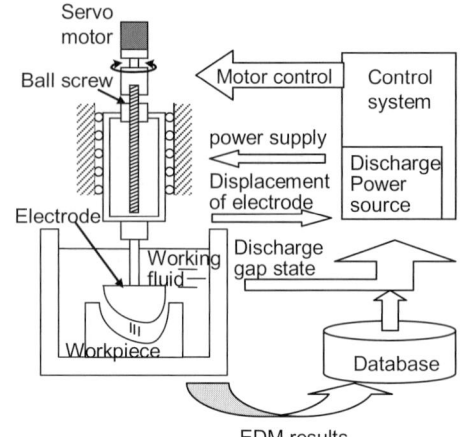

EDM results

Fig.1 Machining control system in EDM

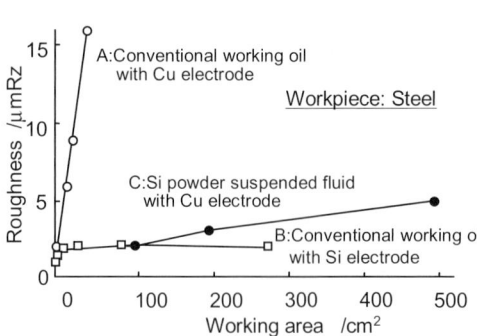

Fig.2 Relation between surface roughness and working area.

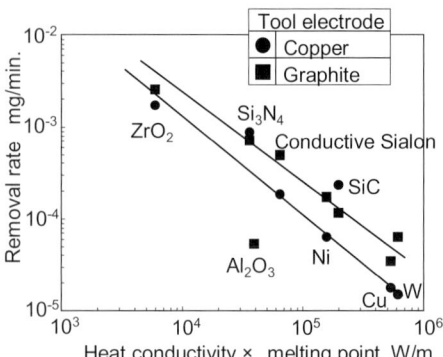

Fig.3 Relation between removal rate and product of heat conductivity and melting (sublimation) point.

4. EDM of insulating materials

Some of the functional materials such as dielectric materials or magnetic materials are made of ceramics. They are machined into complex shapes with high accuracy. As they are hard and brittle, EDM has always been applied, at first, if they are conductive materials. The machining rate of these ceramics does not depend on their hardness, which are determined by the product of thermal conductivity and melting point (or sublimate) as plotted in **Fig.3**.

4.1 Machining phenomena and their mechanism

In the machining of insulators, the surface of the work piece is covered in advance with a conductor which is called an assisting electrode. Metal plates, metal mesh or TiN coat by PVD process have been used as assisting electrodes. On the machined surface of the insulating ceramics, a carbonic layer is observed to form. As this carbonic layer is electrically conductive, it leads discharges to ceramics surface [7]. **Fig.5** shows the process of machining insulating ceramics. The insulator surface is covered with cracked carbon from the working oil, and renders it conductive. The specific resistance of the carbonic layer is measured to about $0.1\Omega cm$. The thickness of the layer, depending on machining conditions, varies around several micro meters.

4.2 Example of insulator machined by WEDM

Wire EDM is also applied to insulator machining. **Fig.6** shows an example machined by wire EDM. The work piece is an insulating silicon nitride plate (Si3N4: Specific resistance = 10^{14} Ωcm). In this case, the work piece is covered before hand with a TiN layer. Therefore, it can be safely assumed that there is no material that cannot be machined by EDM. The method described here can be applied to machining of a diamond.

5. Micro EDM process

Electrical discharge machining as a non contact machining is effective for micromachining with a thin electrode because of its small machining reaction force. But generally, less than several tens micrometers thread is difficult for handling in the machining with it. Forming a thin electrode and its handling is important in micro EDM.

5.1 Some methods for manufacturing thin electrode by EDM

There have been developed some methods for manufacturing thin electrode by EDM as shown in **Fig.7**[8][9][10]. An electrode formed by EDM with a block side as a counter electrode has been used on the same machine for successive machining onto a work-piece. Wire electrical discharge grinding (WEDG) invented by Prof. Masuzawa is well known as an innovative method for precision machining of thin tools [8]. It has advantages of keeping high precision through the whole processes to use feeding wire in making a thin electrode tool, and not to change the position of the tool electrode during its forming process and machining process with it. Many kinds of micro mechanical parts have been produced

by the WEDG method. This WEDG method has stimulated us into micro EDM including micro tool manufacturing. There are thin tool formations by using wear phenomena of counter electrode, by using perforated plate and self formation with single discharge.

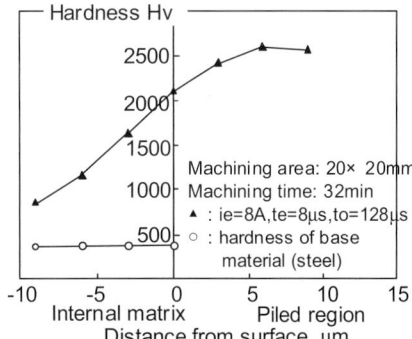

Fig.4 Surface hardness of carbon steel coated by TiC.

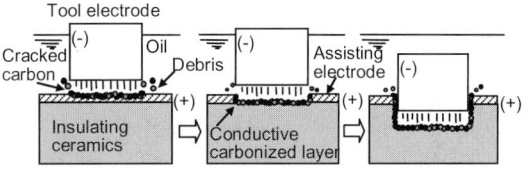

Fig.5 Machining process of insulating ceramics.

Fig.6 Machined Silicon Nitride (Si3N4) by wire EDM.

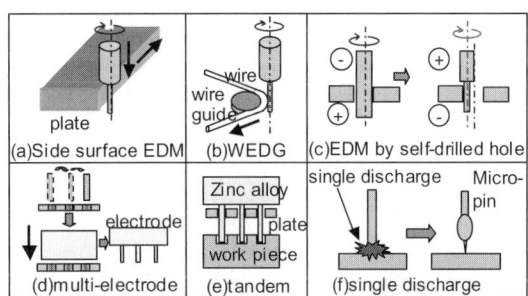

Fig.7 Micro machining of tool electrode by EDM.

5.2 Micro machining by scanning EDM process

Our previous study on EDM by scanning electrode method led to a novel idea of micro machining by EDM with a scanning rotary electrode in which a thin tool electrode could easily be formed by the scanning process. In this section, the story of the development of a micro machining device is described and obtained results are discussed.

The change of electrode diameter with respect to

machining time can be estimated by the following equation.

$$r = \sqrt{r_0^2 - \frac{\pi d}{V_V} t} \qquad (1)$$

Where, r: radius of micropin, r_0: radius of electrode, l: length of micropin, V_V: removal volume per unit time (=constant). This equation reveals that the diameter changes rapidly against time, and this is especially true for small diameters; therefore, it is difficult to control small diameter.

In order to solve the problems mentioned above, we set in position two electrically isolated plates as shown in **Fig.8**. Power supplying cables are connected to both of the plates with current transformers to measure discharge current. The movement of the electrode is controlled to keep its position in the center of the clearance by measuring currents to the isolated plates. Namely, during the machining, y-axis in **Fig.8** is controlled to the direction of B if the frequency of discharge or shortage at the A plate is higher than that at the B plate. This system is referred to as center tracking by frequency proportional control. With this control system, the position of the electrode being machined can very easily be controlled on the center line of the clearance and would not depend on initial position or inclination of the slit.

With the method described above, some experiments for electrode sharpening were carried out with working plates of brass and zinc alloy. **Fig.9** shows the result of machined electrode of WC material by condenser discharge machining. We cannot observe the wear at the edge portion in the

figure. This is attributed to the easy debris disposal owing to the clearance.

6. Conclusions

In this paper, we describe a series of surface machining and micro machining methods by EDM based on our experience.

The research appears to be a summary of the history of surface and micro machining by EDM. Including machining of manufacturing dies, these machining fields will widely expand in the future; for instance, machining of parts with complex structures and their surface modification seeking to enhance their functional use as mechanical parts.

Acknowledgments

The authors would like to thank Emeritus Prof. N. Saito of Toyota Technological Institute, Prof. T. Masuzawa of the University of Tokyo, Prof. Y. Fukuzawa of Nagaoka University of Technology for their suggestions in this study.

Reference

[1] N.Mohri, N.Saito, M.Higashi, A New Process of Finish Machining on Free Surface by EDM Method, Annals of the CIRP, Vol.40, No.1, pp.207-210, 1991.

[2] N.Mohri, N.Saito, M.Suzuki, T.Takawashi, K.Kobayashi, Surface Modification by EDM –An Innovation in EDM with Semi-Conductive Electrode, Research and Technological Developments in Nontraditional Machining-PED, Vol.34, pp.21-30, 1988.

[3] N.Mohri, H.Momiyama, N.Saito, Y.Tsunekawa, Surface Modification by EDM – Composite Electrode Method-, Proceeding of International Symposium for Electro-Machining (ISEM X), pp.587-593, 1992.

[4] N.Mohri, Y.fukuzawa, T.Tani, T.Sata, Some Considerations to Machining Characteristics of Insulating Ceramics –Towards Practical Use in Industry-, Annals of the CIRP, Vol.54, No.1 pp.161-164, 2002.

[5] N.Mohri, N.Saito, Y.Tsunekawa, Metal Surface Modification by Electrical Discharge Machining with Composite Electrode, Annals of the CIRP, Vol.42, No.1 pp.161-164, 1993.

[6] A.Goto, T.Moro, K.Matsukawa, M.Akiyoshi, N. Saito, N.Mohri, Development of Electrical Discharge Coating Method, Proceeding of International Symposium for Electro-Machining (ISEM XIII), Vol.2, pp.581-588, 2001.

[7] T.Tani, Y.Fukuzawa, N.Mohri, M.Okada, Machin-ing Phenomena in EDM Insulating Ceramics using Powder Suspended Working Oil, Proceeding of International Symposium for Electro-Machining XIII,Vol.1, pp.359-370, 2001.

[8] T.Masuzawa, M.Fujino, K.Kobayashi, T.Suzuki, Wire ElectroDischarge Grinding for Microm-achining, Annals of CIRP, 34/1,pp.431-434, 1985,.

[9] M.Yamazaki, T. Suzuki, N.Mori, M. Kunieda, EDM of micro-rods by self-drilled holes, Journal of Materials Processing Technology, 149, pp.134-138, 2004.

[10] N.Mohri, H.Takazawa, K.Furutani, Y.Ito, T.Sata, A New Process of Additive and Removal Machining by EDM with a Thin Electrode, Annals of CIRP, 49, 1, pp.123-126,2000.

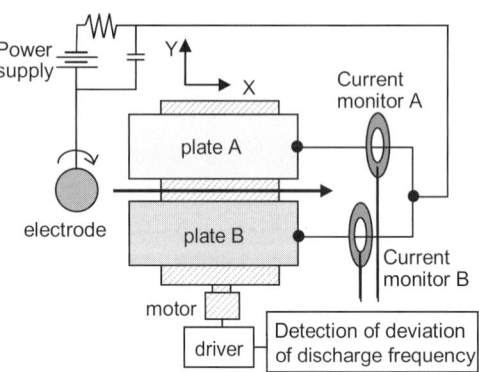

Fig.8 Experimental apparatus for slit tracking EDM.

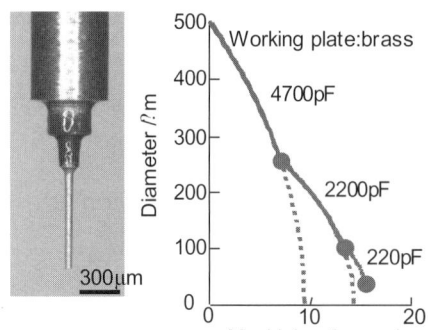

Thinnest diameter:41μm
Total machining time:16min. Change of diameter

Fig.9 Stepped micropin of WC by condenser discharge machining with variable capacitance.

Major FP6 Projects

Multi-Material Micro Manufacture
W. Menz, S. Dimov and B. Fillon (Eds.)

A dynamics-driven approach to precision machines design for micro-manufacturing and its implementation perspectives

D.H. Huo, K. Cheng

AMTRG, School of Technology, Leeds Metropolitan University, Leeds LS1 3HE, UK

Abstract

Precision machines are essential elements in fabricating high quality micro product or micro feature and directly affect the machining accuracy, repeatability and efficiency. Some design issues of precision machines are presented in this paper, and emphasis is placed on the mechanical and structural design of precision machine and relevant design methodology. In the discussion of mechanical and structural design of precision machine, particular emphasis is placed on the machining dynamics issue to identify the major structural factors affecting the performance of the machining. This paper begins with a brief review of the design principles of precision machine tools, including tool-workpiece loops and machine tool vibration, and stiffness, mass and damping issues with emphasis in the machining dynamics; Then design processes of precision machine are discussed; And finally two case studies about design of a fast tool servo system and a 5-axis micro milling machine tool are given.

Keywords: precision machines, machining dynamics, design, modelling

1. Introduction

Precision machines are essential in modern industry and directly affect machining accuracy, repeatability and efficiency. Generally speaking a design of precision machine mainly includes design of key machine elements (spindles, bearings, cutters and machine body, etc), control system and measuring system and their software. In this paper, emphasis is placed on the mechanical and structural design of precision machine and relevant design methodology for micro manufacturing purposes. In discussion of mechanical and structural design of precision machine, particular emphasis is placed on the modelling of machining dynamics issue to identify the major structural factors affecting the performance of the machining process with a view to improve machining accuracy, repeatability and efficiency.

Recently, new demands in the fabrication of miniature/micro products and micro features have appeared, such as the manufacture of microstructures and components with 3D complex shapes or free-form surfaces. Although many efforts have been put in developing IC-based fabrication method, mechanical ultraprecision machining has its unique advantage in the fabrication of real 3D miniaturized structures and free-form surface. Therefore precision machine design method and its machining dynamics should be researched to meet the requirement of fabrication of micro products.

This paper contains the following topics: tool-workpiece loops and machine tool vibration, and stiffness, mass and damping issues. The material presented here is a brief review with emphasis on machining dynamics; and precision machines design methodology, including a design process of a precision machine. Finally two case studies about the design of a fast tool servo system and a 5-axis bench-type micro milling machine tool are presented.

2. Tool-workpiece loops and machine tool vibration

From a machining viewpoint, the main function of a machine tool is to accurately and repeatedly control the point of contact between the cutting tool and the uncut material - the 'machining interface'. This interface is normally better defined as tool-workpiece loops. Fig. 1 shows a typical machine tool-workpiece loop. The position loop - the relative position between the workpiece and the cutting tools which directly contributes to the precision of a machine tool and directly lead to the machining errors.

On the other hand, deformations introduced by stiffness and thermal loop are two important aspects in tool-workpiece loops. The stiffness loop in a machine tool is a sophisticated system. The stiffness loop of the machine includes the cutting tool, the tool holder, the slideways and stages used to move the tool or the workpiece, the spindle holding the workpiece or the tool, fixtures, and internal vibration, and other dynamic effects. The physical quantities in the stiffness loop are force and displacement. During machining, the cutting forces at the machining point will be transmitted to the machine tool via the stiffness loop and return to the original point thus closing the loop. Influences outside of the structural loop, which still influence the loop and cause errors, include floor vibration, temperature changes, and cutting fluids. Thermal dynamic loop is similar to the stiffness loop and contains all the joints and structure elements that position the cutting tool and workpiece.

Machine tool vibrations play an important role in determining structural deformations and dynamic performance. Furthermore, excessive vibrations accelerate tool wear and chipping, cause poor surface, and damage to the machine tool component.

Vibrations can be classified as a number of ways according to possible factors. Vibrations can be classified in free vibrations, forced vibrations and self-excited vibrations based on external energy sources. It is useful to identify vibrations types in machine tools during the design stage and then control the vibrations from the machine tool design side.

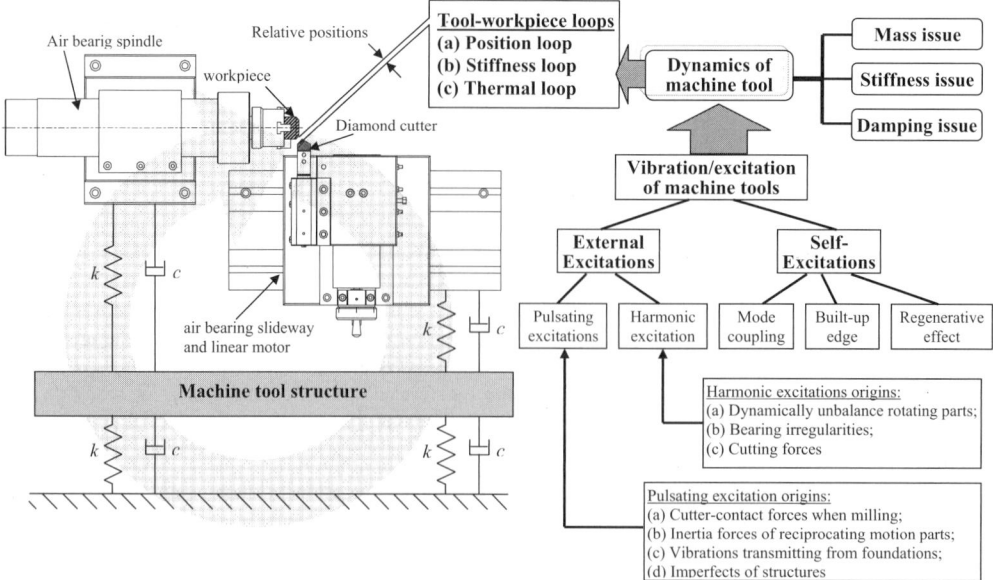

Figure 1 Machine tool loops and dynamics

If an external energy source is applied to initiate vibrations and then removed, the resulting vibrations are free vibrations. In the absence of non-conservative forces, free vibrations sustain themselves and are periodic.

The vibrations of machine tools under Pulsating excitations can be regarded as free vibration. Origins of pulsating excitations in machine tools include:

- Cutter-contact forces when milling;
- Inertia forces of reciprocating motion parts;
- Vibrations transmitting from foundations;
- Imperfects of machine tool structure material

If vibrations occur during the presence of an external energy source, the vibrations are called forced vibrations. The behavior of a system undergoing forced vibrations is dependent on the type of excitation. There are kinds of external forces, such as harmonic, periodic but not harmonic, step, impulse and arbitrary, etc. If the excitation is periodic, the forced vibrations of a linear system are also periodic.

Forced vibrations in machine tools can be generated from two kinds of energy sources, which are internal vibration sources and external vibration sources. External vibration sources, such as seismic wave, usually transfer vibration to machine tool structure via the machine tool base. Designing a good vibration isolator will eliminate or minimize forced vibrations caused by outside vibration sources.

There are many internal vibration sources which cause forced vibrations. For instance, an unbalanced high speed spindle, impact force in machining processes, and inertia force caused by a reciprocal motion component, such as slideways, etc.

Self-excited vibration is a kind of vibration in which vibration resource lie inside of the system. In machining self-excited vibration usually results in machine tool chatter vibration. Chatter occurs mainly because one of the structural modes of the machine tool-workpiece system is initially excited by cutting forces. Chatter is a problem of instability in the metal cutting process, characterized by violent vibrations, loud noise and poor quality of surface finish. Chatter reduces the life of the tool and the productivity of the

manufacturing process by interfering with the normal functioning of the machine. The problem has affected the manufacturing community for quite some time and it is a popular topic for academic and industrial research. Therefore it is very important to identify structural dynamic performance in the design stage.

3. Stiffness, Mass and damping

The dynamic behaviors of structures results from the exchange and dissipation of energies, dynamic loads transfer their energy to the structure, which then respond via several mechanisms, such as bending or extension [1]. During the process of energies transfer, stiffness, mass and damping dominate. Here stiffness, mass and damping issues in machine tool dynamics are discussed.

3.1 Stiffness issues

Stiffness is usually defined as the capability of the structure to resist deformation or hold position under applied loads. Static stiffness in machine tools refers to the performance of structures under the static or quasi-static loads. Static loads in machine tools normally come from gravity, cutting force etc. Apart from the static loads, machine tools are subjected to constantly changing dynamic forces, and the machine tool structure will deform according to the amplitude and frequency of the dynamic excitation loads, which is termed dynamic stiffness. Dynamic stiffness of the system can be measured using an excitation load with a frequency equal to the damped natural frequency of the structure.

The structural stiffness of a machine tool is one of the main criteria in the design of a precision machine tool. Generally speaking, both the static and dynamic stiffness of a machine tool structure determine the machine's accuracy and productivity. Therefore, high stiffness is required both statically and dynamically each affecting different aspects of the machining process.

Static stiffness affects displacement between

cutting tools and workpieces during cutting. High static stiffness is required to produce parts to a desired size and shape and although finish machining in precision machines often takes place with depth of cut of a few micron or even smaller and correspondingly light cutting forces, the resulting deflections can still be excessively large if the machine has inadequate static stiffness. The resulting deflection can thus exceed the tolerance of workpieces.

The need for high dynamic stiffness in precision machines results mainly from two separate aspects of the machining process. In the first case inadequate dynamic stiffness will result in poor quality surface finish of the machined parts due to relatively low levels of vibration occurring during machining processes. In the second case low dynamic stiffness can affects chatter and even damage to the cutting tool and machine structure.

Normally static stiffness is relatively easy to predict during the design stage, and hence can be optimized to obtain a machining accuracy. Dynamic stiffness is difficult to predict during the design stage, however with some modeling and analysis method, such as finite element method (FEA), dynamic stiffness can still be analyzed and identified, which is also important to determine control system parameters and obtain the desired machining accuracy. Section 5 in this paper provides a case study in analyzing static and dynamic stiffness for a 5-axis milling machine tool using FEA.

3.2 Mass issues

Mass or inertia is an issue that influences precision machine performance and cost. Generally speaking, reducing mass is an everlasting objective of machine design and highly desirable in precision machines design. A lightweight structure can benefit the precision machines in number of ways, including:
● less mass can enhance the ability of the machine to respond to high frequency input
● less mass can offer a higher natural frequency
● less mass can improve the damping ratio
● less mass can reduce cost of manufacture

There are two approaches to reduce mass. One is selecting lighter materials with higher performance, the other is structural optimal design in which mass is reduced and the requirement is met at the same time.

There are numerous structural materials available, some with very high stiffness-to-mass ratio, such as aluminum-magnesium alloy, titanium alloy etc. Unlike the aircraft and aerospace industries in which most advanced material has been widely used, precision machine design has to take the cost into account. Up to now only a few practical materials have been chosen to build machine tool structures.

Lightweight design or structural optimization design, on the other hand, became the main approach to reduce mass in precision design. Many structural optimization methods from parameter optimization to shape optimization and even topology optimization have been proposed. Numerical methods, such as finite element method, have been providing practical solving approaches for these structural optimization methods.

It also should be noted that higher mass helps reduce high frequency noise and reduce vibrations. Therefore, the machine bed in some precision machines is built using high mass materials like solid granite table to reduce vibrations.

3.3 Damping issues

Damping defines the ability of a system or structure to dissipate energy. Among stiffness, mass and damping that affect the dynamics of a machine tool, the importance of damping probably is the easiest issue to neglect. However, damping in fact plays a very important role in the dynamic performance of precision machines. Some of most important includes:
● to absorb energy from the process to reduce vibrations
● to prevent chatter and damage to the machined surface
● to absorb energy from structural modes excited by the servos

Sources of damping can be classified into internal damping and external damping: the internal damping includes material damping and joint damping. Material damping is a property of every material. Compared to other dampers material damping is very low and sometimes can be ignored. Joint damping exists between joint surfaces. The more joint surfaces a machine has, the more damping a machine has. However the stiffness will be reduced because of joint surfaces.

Damping can be from external sources, and it usually is termed dampers. There are kinds of dampers for machine tools include tuned mass dampers, constrained layer dampers, velocity feedback in servos and actively controlled damped masses attached to the structure.

4. Design processes of the precision machine

As illustrated in Figure 2 the design of a precision machine tool requires some basic steps: customers requirements and system functional requirements, conceptual design, analysis and simulation, experimental analysis, detailed design, design follow up, albeit the full design process is always iterative, parallel, nonlinear, multidisciplinary and open-ended to any innovative and rational ideas and improvements. The functional requirements of a precision machine may address the considerations in geometric, kinematics, dynamics, power requirement, materials, sensor and control, safety, ergonomics, production, assembly, quality control, transport, maintenances, cost and schedule, etc [2]. In this stage assessment of the state-of-the-art technology is needed to make the design more competitive and the cost economically. The final specifications will be determined after several specification iterations. The resultant conceptual design is important for the innovation of the precision machine design.

Brainstorming is a method most often thought of for generating conceptual design [2]. In this stage, selection of key components in precision has to be considered. These key components include machine structure and materials, main spindle and slides, feed drive, control units, inspection unit, tool and fixtures etc. The advantages and disadvantages of these components should be compared and evaluated with respect to system functional requirements and other factors such as cost. Some of these key components in precision machines have been briefly reviewed in

32

the previous section. Several design schemes may be proposed in this stage, which are followed by analysis and simulation processes and experimental analysis processes. From the dynamics point of view, the vibration should be avoided during this stage right first time, by the use of various integrated analysis and testing disciplines, from the component level to the final assembly.

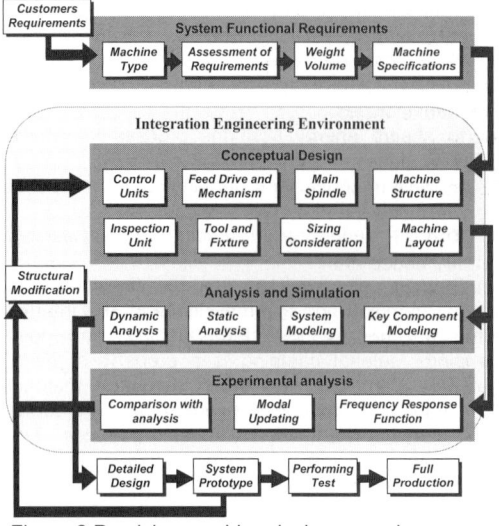

Figure 2 Precision machine design procedures

Analysis and simulation includes key component modeling, system modeling, static analysis and dynamic analysis etc. Analysis and simulation method that has been widely used is finite element analysis. The analysis results, together with errors budget and cost estimation, will be used to check the conformance to the machine's specifications. The analysis results also help to identify some weakest parts in machine tool structure and then provide data for structural modification hence speeding up the decision-making process.

Experimental dynamic analysis in precision machine involves in the selection of testing methods, frequency response function analysis, modal updating and comparisons with simulation results, etc. Through experimental dynamic analysis, some important dynamic characteristics of key machine components or even assembly such as modes and shapes, natural frequencies and damping ratio, will be obtained. Experimental analysis results can also be used to structural modification and verification of simulation model.

It should be stressed that the processes of structural design, structural dynamic analysis and tests are not necessary linear but interactive each other. Design of precision machines should involve structural design, analysis and experiments in an integrated engineering environment. First of all, experimental dynamic tests will be in support of dynamic simulation, since many unknowns prevail in a pure analysis and simulation process, especially when dealing with a fully assembled new design configuration. Insufficient understanding of the various simulation procedures, the characterization of new materials, or the use of different construction methods for structures, all generate unknowns and can lead to an inefficient use of simulation, and therefore more iterations. A principle role for

structural dynamics tests will provide the necessary feedback data to support the design and analysis process. Data feedback can often be understood in a broad sense. It is usually unnecessary to perform the dynamic testing for the whole assembled precision machine, or it will fall into traditional trial-and-error methods. For instance, data from dynamic testing of aerostatic slideways can be feedback to finite element analysis to establish an accurate model; data from nanoindentation tests will benefit the development of simulation criteria of nano/micro machining processes modeling. The experimental database should be integrated into the overall analysis and modeling processes to help update or correct the existing analytical model such as FEA model or to build new models based on experimental data.

The need for this integration is driven by the increasing demand of high precision machines. Fortunately advances in hardware and software have been contributing to this integration. The hardware and software available for executing structural testing and analysis has evolved from standalone instruments to computer based system and usually PC-based systems. Various successful commercial CAE software available in the market have been of benefit to designers in dynamic analysis. The data acquisition card and analyzers used for data acquisition and signal processing have become flexible, powerful and customizable to user requirements.

Once conceptual design, dynamic analysis and test has been finished a design plan can be formulated for detailed design. In the detailed design stage, all subsystem including mechanical structure, spindle and feed drive systems, tooling and fixture systems, control and sensor systems, and metrology and inspection systems will be completed, and if necessary more detailed analysis and simulation need to perform based on the detailed design.

After the detailed design is completed there is still much work that needs to be done in order to make the design successful, including development of test and user support programmes, update of design and document, etc.

5. Applications

5.1 Design case study 1: A piezo-actuator based Fast Tool Servo system

The piezoelectric actuator is a kind of short stroke actuator. It is very promising for application in the rotary table drive and slideways drive because of its high motion accuracy and wide response bandwidth [3]. Currently, piezoelectrics have been applied in the design of the fine tool-positioner in order to obtain high precision motion of the cutting tool. The piezoelectric actuator combined with mechanical flexure hinges is used for positioning control of the diamond cutting tool. More recently, Fast Tool Servo (FTS) system has been introduced for diamond turning components and products with structured and non-rationally symmetric surfaces such as laser mirrors, ophthalmic lenses molds, etc[4].

The piezo-actuator based fast tool servo (FTS) system is designed to perform precision positioning of the tool during short stroke turning operation which has been implemented in a test turning machine tool set up at Leeds Metropolitan University.

The FTS is designed to perform turning operations and hold diamond tools. Static deformation of the FTS structure caused by cutting forces during rough and finish machining, must be minimized to reduce form and dimensional error of the workpiece at nanometer scales. Therefore, high stiffness, particularly in the feed direction which affects the machined surface directly is required. A high first natural frequency is required in FTS structure to prevent resonance vibrations of FTS structure due to the cutting forces.

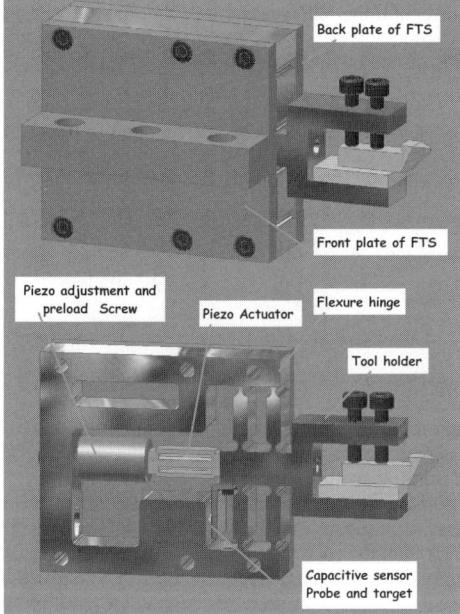

Figure 3. The schematic of the FTS

On the other hand, the high stiffness of FTS structure will reduce effective stroke of the piezo actuator to some extent. For this reason a compromise is made between high stiffness and actuator stroke reduction.

Figure 3 shows the schematic of FTS that comprises of a piezoelectric actuator (Cedrat PPA10M, PPA20M, or PPA40M), a flexure hinge, two cover plates, a tool holder, a capacitive sensor, and piezoelectric adjustment screw. The piezo actuator is housed under preload within the flexure hinge made from spring steel. Three piezo actuators, 18mm (PPA10M), 28mm(PPA20M) and 48mm(PPA40M) length respectively are used. Three different adjustment and preload screws with corresponding lengths are designed to house the three actuators in the same flexure hinge. Actuator displacement in actuated/feed direction is transmitted to the tool holder via four symmetric solid flexure hinges which have a circular hinge profile as shown in Figure 3. The tool holder is designed to be integrated with flexure hinge in a single part. The overall envelope of FTS is 90mm x 80mm x 55mm, with the tool holder extending 35mm. the design is compact and self-contained for easy of mounting on slideways and tool posts in other machine tools.

To determine the flexure hinge dimension, Both static and modal finite element analysis were conducted for the flexure, in which static stiffness and

natural frequencies were obtained. In both analyses, hinge radius r, hinge thickness t and hinge length were chosen as optimization variables, and optimization FEA results were obtained based on the requirement of stiffness, natural frequencies and strokes reduction.

In static FE optimization analysis, a force of 216Newtons was applied in piezo actuator feed direction and a maximum deformation of 0.223mm was obtained, corresponding to a static stiffness of 9.7 N/μm. Stroke reductions in the three cases are below the design requirement and it should be noted that stroke reduction will be a little higher if taking piezo-actuator preload effect into account.

The maximum Von Mises stresses predicted by FE analysis under maximum load are 45MPa, which are well below the strength value of spring steel.

The FE modal analysis was used to predict natural modes of the flexure structure. The first three natural frequences are 1262, 2086, and 2791 Hz. The lowest frequency mode is the translational motion along the actuated/feed direction with a natural frequency of 1262 Hz, which is above design requirement of 1000 Hz. Figure 4 shows a mirror surface with Ra<20nm obtain from preliminary cutting trails using this FTS.

Figure 4. Cutting trail results using the FTS

5.2 Design case study 2: A 5-axis micro-milling/grinding machine tools

This case study describes the conceptual design process of a bench-type 5-axis micro milling machine, which is currently being developed by the authors with the support of the EU MASMICRO project [5].

The machine aims at manufacturing the miniature and micro components in various engineering materials, potential applications include MEMS, optical components, medical components, mechanical components and moulds etc.

According to the machine specifications, the machining envelope, the 'bench-top' required dimensions, the types of components/materials to be machined, and the overall accuracy to reach and maintain, the open frame machine tool layout which greatly facilitates machining area access for fixturing and part handling as shown in Figure 5 was selected as the preliminary design for analysis.

The structural static and dynamic behaviours of the machine tool were simulated using ANSYS software. Because there are many air bearings in this machine tool configuration, it is difficult to simulate the compressed air in FEA. An equivalent method was proposed to simulate the air bearing. In this method spring elements in ANSYS was used to simulate the stiffness in different directions. The stiffness of the spring element is based on the stiffness data provided by experiments.

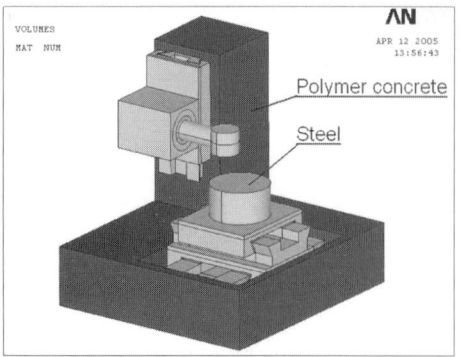

Figure 5. Preliminary configuration

First 10 natural frequencies were extracted in modal analysis using block Lanczos method. It should be noted that all the interfaces in machine tool are assumed to be rigid thus the natural frequencies are estimated higher than the real values.

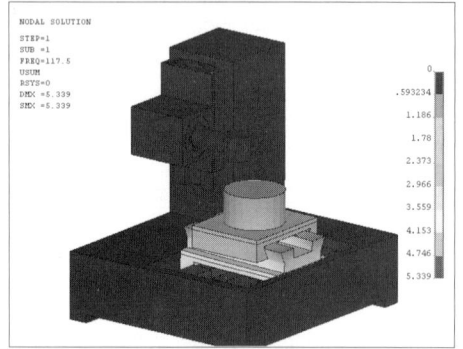

Figure 6. 1st natural frequency and its mode of origin configuration

After running a series of FEA simulations on the structure, we identified the important sensitive components on the machine structure, which can be seen from the mode shape of the first natural frequency (117 Hz). Due to the stack of equipment assembled on top of each other, the X slideway and the Z slideway are subject to important "tilting" effects, which are likely to affect the machine accuracy (illustrated in the Figure 6). Therefore, from the structural modification of view, improving the stiffness of the slideway is the most efficient method.

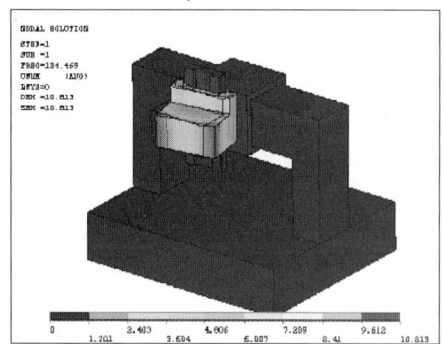

Figure 7. 1st natural frequency and its mode of gantry configuration

Following the information gained from the sensitivity analysis based on preliminary FEA results, a gantry type of machine configuration was proposed.

This gantry type of machine configuration will enable a much better overall machine stiffness, in static as well as in dynamic mode and overcome the problems identified in the original machine design. In order to improve the overall stiffness of the machine tool, machine structure material was changed from polymer concrete to granite. Modal analysis was conducted for the new configuration in which the horizontal slideway was neglected to increase computational speed because there is no tilting slideway. Figure 7 shows the first natural frequency and its vibration shape. The first natural frequency was increased from 117 Hz in preliminary design to 134 Hz, and the following natural frequencies are also improved. The final gantry 5-axis machine configuration is shown in Figure 8.

Figure 8. The gantry type of machine configuration

5. Conclusions

Precision machine tools with mechanical micro cutting capability have unique advantages in the fabrication of micro products and micro feature. Efforts will need to be made in the design methods of precision machine tools. Machining dynamics including machine structure dynamics and machining processes dynamics will determine the performance of precision machines and hence need to be taken into account during design stages to obtain the necessary high product accuracy. Dynamics-driven modelling and design method for precision machines can offer a better understanding of the fabrication of micro-products.

Acknowledgements

This work was funded by the EU 6th Framework IP MASMICRO consortium/project (contract No. NMP2-CT-2004-500095-2). Thanks are also extended for collaborative partners in its RTD 5 sub-group.

References

[1] Benaroya H Mechanical vibration — Analysis, Uncertainties, and Control. Marcel Dekler, 2004
[2] Slocum A H, Precision Machine Design, Englewood Cliffs, Prentice Hall, 1992
[3] Luo X, Cheng K, Webb D, Wardle F. Design of ultraprecision machine tools with applications to manufacture of miniature and micro components, Journal of Materials Processing Technology. 2005. 167(2-3): 515-528
[4] Weck M, Fischer S, Vos M. Fabrication of microcomponents using ultraprecision machine tools, Nanotechnology 8(1997) 145-148
[5] http://www.masmicro.net accessed in Feb. 2006

Multi-Material Micro Manufacture
W. Menz, S. Dimov and B. Fillon (Eds.)
© 2006 Elsevier Ltd. All rights reserved

A new technology platform for fully integrated polymer based micro optical fluidic systems

J. Nestler[a], K. Hiller[a], T. Gessner[a], L. Buergi[b], J. Soechtig[b], R. Stanley[b], G. Voirin[b], S. Bigot[c], J. Gavillet[d], S. Getin[d], B. Fillon[d], M. Ehrat[e], A. Lieb[e], M.-C. Beckers[f], D. Dresse[g]

[a] Center for Microtechnologies, Chemnitz University of Technology, Chemnitz, Germany
[b] CSEM Centre Suisse d'Electronique et de Microtechnique SA, Zurich, Switzerland
[c] Cardiff University, Cardiff, United Kingdom
[d] Comissariat à l'énergie atomique (CEA), Grenoble, France
[e] Zeptosens – a division of Bayer (Schweiz) AG, Witterswil, Switzerland
[f] Eurogentec S.A.,Seraing, Belgium
[g] Centre Hospitalier Regional de la Citadele, Liège, Belgium

Abstract

This paper describes a new technology platform for polymeric micro optical fluidic systems. The platform consists of active and passive optical and fluidic elements for a surface plasmon resonance (SPR) biosensor for the detection of proteins. The platform includes the integration of polymer light emitting diodes, polymer photodiodes as well as polymer based fluidic valves and pressure generation elements. Surface functionalization for micro optical and micro fluidic parts as well as advanced manufacturing methods are other important parts of the presented technology platform.

Keywords: Lab-on-Chip, Point-of-Care, Micro Optics, Micro Fluidics, Surface Plasmon Resonance, polymer LED, Biosensor, Hydrogels, Microspheres, Micro Actuators, Polymer MEMS

1. Introduction

The development of lab-on-chip (LoC) devices for bio analytical applications has been tremendously increasing during the past decade. Most of these devices are being fabricated by use of silicon or glass because processes for these materials are well known and surface properties of these materials are easy to match the requirements of biotechnology. However, since one core application of LoC devices is point of care applications, a strong demand on low-cost solutions for disposable LoCs has developed.

The use of expensive materials inhibit a high level of integration for point-of-care devices. Thus, most of the functionality is currently transferred to the "reader" – the controlling unit – instead of being integrated into the chip. Since the micro-macro interfaces between the chip and the reader are always difficult to handle, reader costs increase and sensitivity decreases. Thus, low-cost approaches for Lab-on-Chips have to be investigated which allow a high level of integration at an affordable price. Polymer materials show a high potential for cost efficient production of Lab-on-Chips not only as substrate materials but also as active components such as sensors and actuators.

To reach this goal, the development of a polymer based technology platform has been started within the European project "SEMOFS", which aims to integrate active optical sensors as well as micro fluidic actuators in a polymer cartridge. Fig. 1 shows the different aspects of the platform.

Miniaturized Surface Plasmon Resonance (SPR) has been chosen as optical detection method for the label-free detection of proteins. As active optical components, the integration of polymer light-emitting diodes (PLEDs) and polymer photodiodes (PPDs) for such a sensor application is investigated. Polymer

micro fluidic actuators will be part of the system as well as a special surface modification to avoid protein adsorption on the walls of the fluidic channels.

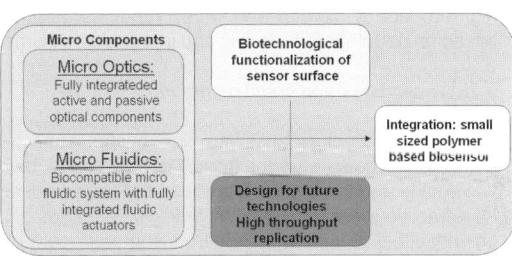

Fig. 1: Different aspects of the SEMOFS biosensor platform

2. Micro Optical Concepts

2.1. Polymer Light Emitting Diodes and Photodetectors

PLEDs and polymer photodiodes PPDs are potentially very attractive light sources and detectors for (disposable) lab-on-chip applications for several reasons. First, organic diodes are very efficient: As light sources their efficiency is comparable to fluorescent lamps (~ 100 lm/W) with our PLEDs emitting a power per unit area of up to 20 mW/cm^2 and our photodiodes show an external quantum efficiency of more than 70 % (see Fig. 2). Second, the peak wavelength can be tuned over the visible range by chemical modification of the emitting molecule, and thus can be matched to the specific requirements of the application. Third, these devices are ultra-flat (< 500 nm) and compatible with bendable substrates. Last and very importantly, PLEDs and PPDs can be produced by purely additive print processes at room-temperature and can thus be monolithically integrated into almost any system [1].

We currently investigate the near-field coupling of PLEDs and PPDs to waveguides. PLEDs have been successfully implemented directly on Ta_2O_5 planar waveguides.

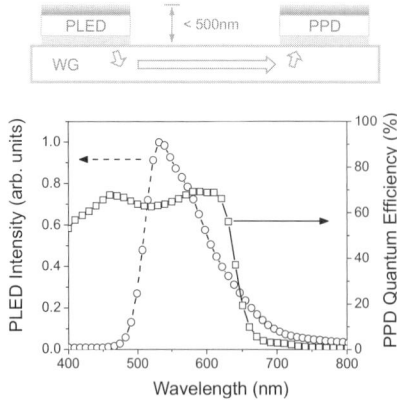

Fig. 2: Top: Schematic diagram of light-coupling from PLED into waveguide (WG) and detection by the PPD. Bottom: Electroluminescence spectrum of a polyfluorene-based PLED (circles) as well as external quantum efficiency of a typical bulk heterojunction C60-polythiophene PPD (squares).

The probably biggest challenge of using PLEDs and PPDs in a lab-on-chip environment is the severe requirements of these devices in terms of encapsulation and protection against oxygen and moisture. We are currently addressing this issue by developing ultra high-barriers to oxygen and water vapor permeation based on nanolayered organic and inorganic thin films.

2.2. Waveguides and coupling gratings

Grating couplers built into waveguides are very effective for coupling light from the free space to a waveguide and vice versa. Moreover, they can be used as refractive index sensor. High index waveguides can be very effective light collectors from sources located in the evanescent wave. They can also be used to perform optical processing inside a waveguide: wavelength filtration like in telecom or for polarisation processing. Amongst the different principles to couple light into and out of a waveguide (direct butt-coupling, prism coupling, grating coupling), the grating coupler is one of the most attractive and effective: it is fully planar and, therefore, compatible with technologies of the microelectronic industry for mass production. Grating couplers are also versatile because their efficiency can be controlled by the grating groove depth and profile shape and they can be adjusted to the free space beam size.

Waveguides with gratings have been used for refractive index sensing, the basic principle is to monitor the propagation constant of the waveguide mode using the coupling properties of the waveguide grating. The propagation constant of the waveguide mode is affected by the change of refractive index occurring in the medium located in the vicinity of the waveguiding layer (high refractive index layer). The propagation constant can be monitored by the coupling angle [2], the grating period [3] or the wavelength of the

light [4]. This was applied in biochemical sensing: in this case, the grating waveguide surface is functionalized with a recognition molecule such as an antibody, when molecules (antigens) bind to the functionalized surface, the refractive index and thickness of a thin layer on top of the grating change and modify the propagation constant that can be moni-tored in real time. The grating of the waveguide can be optimized for the different types of sensing such as thin layer sensing (case of antibody antigen interaction), matrix like sensing (case of a sensitive hydrogel) and particle sensing (case of cell sensing) [5].

When light sources are located in the vicinity of a waveguide mode (the so-called evanescent wave), the coupling efficiency to the waveguide mode is very high. This was applied for a fluorescent biosensor: the waveguide mode excited the fluorescent molecules and the coupled fluorescent light was analysed as a function of the analyte concentration [6]. This principle will also be applied in the case of the coupling of the PLED to the waveguide mode: the PLED layer system can be designed in such a way that the generated photons can efficiently couple to the evanescent mode of the waveguide.

Waveguide grating have also interesting properties for intra-waveguide optical processing: when guided light is incident at an angle onto the grating there will be only one wavelength under which light will be reflected (the Bragg wavelength). This reflection will also have polarization properties that can be used for polarization interferometry [7]. If we now consider a grating with a variable periodicity along the propagation direction - a chirped grating - , the Bragg wavelength will vary with the position and an integrated spectrometer can be realized this way (see Fig. 3).

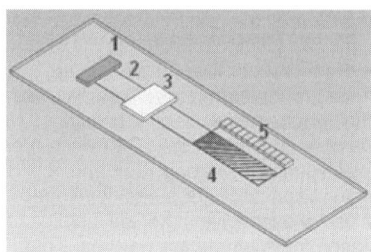

Fig. 3: Schematic diagram of a waveguide-plasmon sensor comprised of a PLED (1) coupled to a planar waveguide (2) region, a plasmon resonance section (3) and an integrated spectrometer (4). A slanted and chirped grating redirects the light to the detector line array (5).

A simplified version with less potential for integration is a simple straight grating without constant period. It also has dispersing properties but needs propagation through the free space and needs additional optics for light collection. It is, however, a good alternative in such cases where the detection part can be realized at an external readout system.

3.3. Replication technology for waveguides and gratings

Disposable sensor concepts also demand for low-cost fabrication technologies related to waveguides and gratings.

A few years ago, a sensor platform based upon an

integrated optical sensor chip was already developed for applications in biochemical and chemical sensing. The fabrication technology involves replication of a microstructured polymer chip followed by the vacuum deposition (DC magnetron sputtering) of a TiO_2 waveguide film [8,9]. The original grating micro-structure is fabricated by e-beam writing and, after processing, electroformed to a nickel shim for the replication process which can be hot embossing or injection moulding. Fig. 4 shows a replicated grating microstructure (periodicity 455 nm, depth 10 nm) in polycarbonate. Results of biosensing experiments based upon such platforms with on-chip referencing are described in [10].

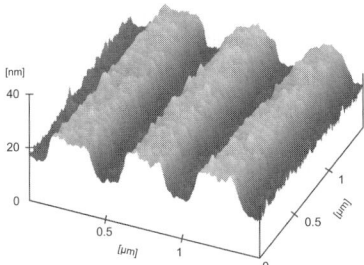

Fig. 4: Replicated grating microstructure relief measured with an AFM: 1.5 µm x 1.5 µm area.

As an alternative to pure polymeric chips we also developed waveguide grating sensor devices based on uv-replicated sol-gel materials that offer superior temperature resistance as compared to transparent polymers such as polycarbonate or PMMA. Planar waveguides combined with fine-period in- and out-coupling chirped gratings have been replicated into ORMOCER-materials [11]. The application aimed for data/telecom spectral monitoring related to dense wavelength-division multiplexing in the 1500 nm wavelength regime. For this application the uv-replication technology was combined with sputter deposition techniques. A coating of either TiO_2 or Ta_2O_5 films to establish the high index waveguides on top of chirped grating couplers has been developed.

3. Micro Fluidic Concepts

To avoid the usage of external interfaces for micro fluidics except of the sample inlets/outlet ports, integrated micro fluidic actuators for valves and pumps are desired for PoC devices. The integration of these actuators has to be realized within the technological platform and therefore be based on polymers. The actuators should further not influence or depend on the fluid properties. Therefore, methods for pressure generation in combination with the integration of polymeric valves will be investigated. Some examples are the usage of hydrogels, microspheres, bubble generation in a "working fluid" and the "simple" thermal expansion of a gas.

3.1. Hydrogels

Hydrogels are cross-linked, three dimensional networks of hydrophilic homo-polymers or co-polymers which undergo a swelling behaviour in water, resulting in an enormous increase of their volume. The ability of

a hydrogel to swell in water may change due to change of external parameters, such as temperature, pH, light, etc. Such stimuli-responsive hydrogels are well suited to serve as actuators especially for microfluidic applications. Fig. 5 (left) shows a possible working principle of an integrated hydrogel actuator.

There are different ways of preparing hydrogels. For micro actuators with the possibility for integration in LoCs, the preparation as a macro gel is not suitable due to its larger size and its slow response time. Thus, the gel has to be synthesized either as a micro gel (with gel particle sizes in the nanometer or micrometer range) or as a micro structured film. For integration, the second one might be even more feasible. In Fig. 5 (right) a lithographically structured thermo-responsive hydrogel (PNIPAAm) is shown.

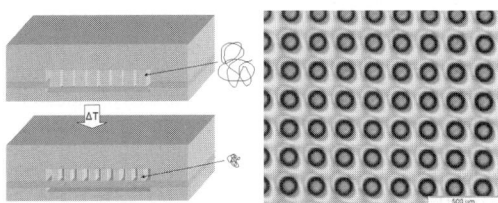

Fig. 5: Possible working principle of an integrated hydrogel based microfluidic actuator (left) and Lithographically structured hydrogel (right). The "circles" are PNIPAAm gel stamps with a height of approximately 15µm.

3.2. Expandable Microspheres

Another way of polymer based fluid actuation is based on expandable microspheres. Expandable microspheres are commercially available and consist of a thermoplastic shell with an expandable medium (gas or liquid) as the core. Heating such a microsphere results in the softening of the shell and its expansion due to the expansion of the core medium. The process is irreversible. Microspheres can be incorporated into a elastomeric matrix like PDMS which allows them to expand as a "solid" material [12]. This principle will here be especially investigated with respect to the possibility of its integration into the proposed technology platform and driven towards production possibilities.

3.3. Bubble Generation

The generation of bubbles offers a further possibility to generate a pressure in microfluidic systems. However, more "classical" principles like Joule's heating or resistive heating of a liquid are difficult to control, need high temperatures and high electric power. Especially for the targeted protein based application, high temperatures should be avoided. Thus, bubble generation by electrolysis offers an interesting alternative. This method is therefore currently being investigated especially with respect to the integration into a polymeric PoC device and its combination with hydrogels and super absorbing polymers (SAPs).

3.4 Surface Modification

While active fluid control is very important for LoCs, the control of the channel sidewalls is of the same importance for a microfluidic system of practical use. Especially protein adsorption to channel walls has to be avoided to allow the fabrication of LoCs with high sensitivity. New types of polymeric coatings with tuneable surface energy are investigated for the channel walls as an alternative to the widely used hydrophilic PEG coatings [13,14]. Besides matching PEG performance in reducing protein adsorption in its hydrophilic state a similar performance of the polymer in its hydrophobic state has also been observed by fluorescent tracking of labelled-proteins. This result is suggesting that low surface energy (hydrophobic) surfaces can compete with highly hydrated surfaces (hydrophilic) as a solution to non-fouling applications. Furthermore the tuneable wettability allows pre-defined conditions from a continuous range of hydrophilic-to-hydrophobic conditions on the same surface but at different areas that can help controlling the fluid flow passively.

5. Fabrication Technologies

A crucial step towards the development of a low cost lab-on-chip product is the fabrication of the components, which should facilitate mass production. In the last few years, micro-technologies have been the focus of numerous research initiatives agreeing on the fact that "mass production of micro-parts must be based on replication technologies" [15]. Thus, the use of a polymer platform becomes an obvious choice.

Using advanced Injection moulding machines the mass production of replicated micro parts is now possible. In addition, a number of studies in hot embossing, currently mainly used for prototyping and small batch production, are now focusing on mass production.

However, the main difficulty is the fabrication of the micro-mould, which should contain highly accurate 3D features while surviving enough replication in order to reduce the costs.

Silicon based technologies are well established and can produce features sizes below 5 μm. However, only 2½D features are usually achievable and the materials used cannot compete with typical mould material such as steel in term of tool life if used as such. So to overcome this limitation coating solutions are being investigated with duplex structures where a low friction polymeric film is deposited onto a wear resistant buffer coating.

Another type of technologies coming from the downscaling of processes used in precision engineering, such as micro milling, micro turning, laser machining and electro discharge machining, offers a good alternative. With these technologies features size down to 10 μm, and this boundary is constantly being pushed down, can be achieved on a wide range of materials including typical mould materials such as harden steel.

A combination of these technologies allows the production of most features required for replication of microfluidic systems.

6. Conclusion

A new polymer based technology platform for cost-effective Lab-on-Chips has been presented. Technologies for micro optics, micro fluidics, surface modification and biotechnology are being combined to develop a new generation of fully-integrated biosensors. Further work will concentrate on the selection of proper working principles, the integration and combination of different processes and upscaling of the technologies currently still in research stage.

Acknowledgements

The research work within SEMOFS has been started in 9'2005 and is being founded by the European Commission within the 6th framework program under the contract number IST-FP6-016768. Main tasks of the research partners include μOptics (CSEM), surface modification (CEA), μFluidics (Chemnitz University), fabrication technologies (Cardiff University), industrial application (Zeptosens), bio markers (Eurogentec), and proof of concept (Hospital Liège).

References

[1] L. Buergi et al. Optical proximity and touch sensors based on monolithically integrated polymer photodiodes and polymer LEDs. Organic Electronics 7. in print

[2] Ph. M. Nellen, K. Tiefenthaler and W. Lukosz. Integrated optical input grating couplers as biochemical sensors. Sensors and Actuators, Volume 15, Issue 3, November 1988, pp 285-295

[3] J. Dübendorfer and R. E. Kunz. Compact integrated optical immunosensor using replicated chirped grating couler sensor chips. Appl. Opt. 37:1890-1894, 1998

[4] K. Cottier, M. Wiki, G. Voirin, H. Gao, R.E. Kunz. Label-free highly sensitive detection of (small) molecules by wavelength interrogation of integrated optical chips. Sensors and Actuators B 91 (2003) 241–251

[5] K. Cottier, R. E. Kunz. Optimizing integrated optical chips for label-free (bio-)chemical sensing. Anal. Bioanal. Chem (2006) 384: 180-190

[6] P. N. Zeller, G. Voirin and R. E. Kunz. Single-pad scheme for integrated optical fluorescence sensing. Biosensors and Bioelectronics, Volume 15, Issues 11-12, December 2000, pp 591-595

[7] G. Voirin et al. Waveguide Grating for Polarization Preprocessing Circuits. Proceeding SPIE Vol. 1126, Electro-Optic and Magneto-Optic Materials and Applications, Paris, 1989, p. 50

[8] M.T. Gale et al. Replicated Optical Chips for Biosensor Applications, Lab Chips and Micro-arrays for biotechnological Applications, Jan. 13-15, 1999, Zurich, Switzerland

[9] J. Soechtig. Replicated plastical elements for sensors and optical Microsystems. 3rd International Conference on MOEMS, Aug. 30 – Sept.1, 1999, Mainz, Germany

[10] J. Dübendorfer and R.E. Kunz, Compact Integrated Optical Immunosensor using Replicated Chirped Grating Coupler Sensor Chips, Appl. Opt. 37, 1890-1894 (1998)

[11] H. Thiele et al. Tandem chirped grating couplers for data/telecom monitoring applications, J. Microlith., Microfab., Microsyst. Vol 2 No. 4, 283-291, October 2003

[12] B. Samel, P. Griss, and G. Stemme: Expandable Microspheres incorporated in a PDMS matrix, Transducers '03

[13] U.S. Pat. No. 5,002,794

[14] Darrel J. Bell et al., Using poly(ethylene glycol) silane to prevent protein adsorption in microfabricated silicon channels, SPIE Vol. 3258-0277-786X/98, pp. 134- 140

[15] L. Uriarte et al. A Comparison between Micro-fabrication Technologies for Metal Tooling. 4M Conf. on Multi Material Micro Manufacture, 2005

Market Studies on 4M Technologies and Applications

Multi-Material Micro Manufacture
W. Menz, S. Dimov and B. Fillon (Eds.)
© 2006 Elsevier Ltd. All rights reserved

Market forecasts for roll-to-roll electronics manufacturing

D.P. Lawrence[c]

IDTechEx, Ltd. Far Field House, Albert Road, Stow-cum-Quy, Cambridge, UK

Abstract

The significant developments in printed organic and molecular electronics will have a profound effect on the market for consumer and industrial products. By allowing the continuous production of functional electronic devices on roll-to-roll assembly lines, two fundamental changes are taking place in manufacturing. First, existing, costly processes may be replaced by more efficient production methods enabling widespread development of lower cost electronics. Second, entirely new markets may emerge that were not feasible prior to these developments. Several of these markets and applications are described along with their significance in reinforcing knowledge driven manufacturing bases within the European Community.

Keywords: Roll-to-Roll Electronics, Market Forecasts

1. Introduction

Economic progress has always been preceded by new ideas. Some evolutionary, others revolutionary, the world has been shaped by new technology. In manufacturing one of the most lasting improvements has been that of continuous production. Examples of which include continuous steel and paper production, printing, and the moving assembly line. Whenever a product can be produced in a continuous, rather than batch, process there is great opportunity for greater production efficiencies and affordability.

The vision of continuously manufacturing electronics using roll to roll printing presses is coming very close to reality. By using a materials set of printable conductors, dielectrics, and semiconductors, it is increasingly possible to create a new generation of fully printed devices. This has broad implications for the global electronics market. Though in a way disruptive, printed electronics will be far more enabling of new markets than disrupting of existing ones. By reducing the cost and time of assembly and eliminating limitations such as the scale of electronics entirely new markets will appear in the following general categories:

1. Low cost identification
2. Widespread energy collection
3. Novelty consumer electronics
4. Large area displays & sensing

2. Silicon versus Roll-to-Roll Electronics

The progress made in lowering cost per transistor of solid state devices over the last fifty years is profound. This progress has been predicated on making each transistor smaller with each new generation of manufacture, fitting more transistors in an increasingly small area and thereby reducing cost as wafer costs are generally fixed. Unit device cost is determined by the number of transistors required, size of each transistor and size of the wafer on which the device is patterned.

Current ICs are made using 200 mm (diameter), and more recently 300 mm, wafers. Depending on the modernity of the equipment and wafer size, wafers cost between €400 and €4000 in high volume. Much of this cost is to defray the enormous capital costs of creating an IC production facility. In fact, silicon based integrated circuits have reached a crucial point at which the current capital investment for production, now roughly €4B may increase by a factor of 10 to achieve to continue the advancement described by Moore's Law. While these devices will continue to operate at gigahertz frequencies, very few companies will likely be able to afford this type of investment. Contrast this with estimates for roll-to-roll 'organic fabs', which may require investments of less than €50M. Granted, the products produced in these facilities will be a small fraction of the performance of their silicon predecessors, but the economics for new applications will be very attractive.

Figure 1. depicts the trade-off between less capable, but more affordable, roll-to-roll electronics. Though this is certainly a classic example of an engineering tradeoff, it enables the creation on entirely new markets. Another differentiator between the left and right sides of figure 1 is that the high cost production methods rely on high temperature and/or vacuum process steps that are inherently slow and costly. As materials are developed that are solution processable, or suspendable in the case of inks, increasingly high speed processes such as offset (graphic arts) lithography, flexography, rotogravure and screen printing.

3. New Markets for Roll to Roll Electronics Manufacturing

A drastic decrease in the cost of manufacturing electronics provides entry into markets and applications that were not considered practical previously. While it is possible to attach a flat panel display to a box of cereal, using conventional technology this is cost prohibitive. With electronics printed directly on the package, simple displays and other functionality are nearing possibility.

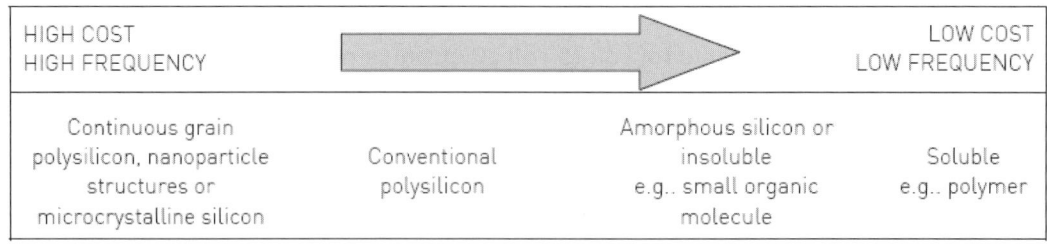

HIGH COST HIGH FREQUENCY			LOW COST LOW FREQUENCY
Continuous grain polysilicon, nanoparticle structures or microcrystalline silicon	Conventional polysilicon	Amorphous silicon or insoluble e.g.. small organic molecule	Soluble e.g.. polymer

Source: IDTechEx

Figure 1. Technical Span of Capability

Figure 2. includes the application types and market size forecasts for printed electronics in 2015 assuming the unit costs as shown. Though a thorough description of each of these technologies and participating companies is available elsewhere, we will describe the development of several of these technologies herein.

Radio frequency identification (RFID) is an attractive technology for logistics and contactless commerce applications. It has been estimated that hundreds of billions of Euros could be saved every year by more accurate tracking of goods in the supply chain.

Chip based RFID tag costs have dropped dramatically and have recent announcements place the cost per tag to under €0.10. While this price point creates an increased opportunity for the technology on lower cost products, true item level tagging of trillions of objects will likely require tags costing €0.01 or less. This will require technologies such as printed electronics. By using organic semiconductors, ideally printed upon the item itself, extremely inexpensive and still unimagined products become possible.

Photovoltaic cell manufacturers purchase the same silicon that IC manufactures use and there is constant uncertainty of cost and supply. Additionally, the return on an investment in solar cells is often greater than 7 years and systems are typically limited in surface area, and therefore energy collection, due to economic factors. While still in early stages of development, efforts to create large area, printed solar cells may eliminate the need for other forms of electrical generation in the next two decades. While they will likely be of much lower efficiency (perhaps a few percent) the ability to cover large areas, such as rooftops, could ultimately drive more economically justifiable investment in solar power. Additionally, small printed electronic devices may also include their own power source drawing from ambient light.

The blinking box: In retailing, packaging designers always strive to create a more eye catching graphic or design to promote a product at the point of sale. Imagine a store shelf comprised of plain brown packaging with black lettering. Across the aisle, imagine packages with vibrant yellows, oranges, and blues. It is unlikely that the plain packaging would attract greater consumer interest. Color and branding, though always in flux, are at a state of equilibrium that is going to change. As color packaging replaced the bland, ultra low cost printed displays and sensors will begin differentiating consumer products. Though a marketing promotion can occasionally tolerate a €0.30 toy or other gimmick, these are usually single run promotional tie-ins with a movie or new product launch. The possibility of an area of a package acting as a display for under €0.10 opens up increasing

opportunity for creative marketing departments to get their brand selected. Incorporate a game or other novel feature and this cost could easily be recouped by a brand price premium.

Another scenario involves the ability to produce a thin sheet of sensors that is a few meters wide and several meters long. This could be fitted to an airplane wing or spacecraft to gauge stress, damage, efficiency, or other metrics. It is unlikely the aerospace engineer

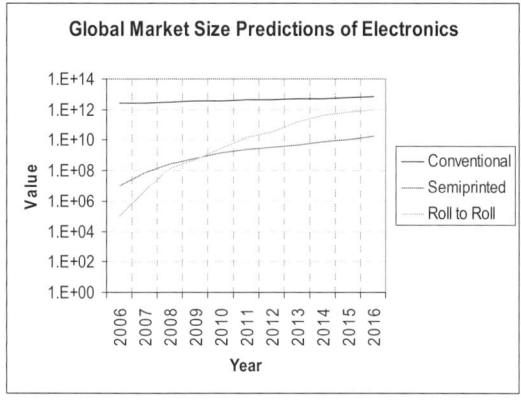

Figure 2. Market Growth of Electronics (value in Euros)

considers it even an option to conduct testing in this manner and that is part of the unpredictable markets made possible by electronics unconstrained by area or inflexibility.

This leads to an interesting challenge in market prediction. Breakthroughs in technology or application development could result in massive market growth over the next ten years. Figure 3 shows a gradient of scenarios where printed electronics bridges the gap in market size with that of the current electronics market. Here, semi-printed refers to devices which share both printed and conventional components. While these hybrids are not generally considered full realizations of the advantages of roll to roll manufacture, it is likely that they will be an interim approach, eventually eclipsed by printed devices once the technology is fully developed. This can be seen as the crossover point shown in Figure 3. It is assumed that the full market size of printed electronics approaches that of conventional electronics (trillions of Euros) due to the low cost, but high volume devices imagined in the near future. The figures for roll-to-roll differ from those described in Figure 2 because here we are assuming breakthrough applications not presently imagined and some erosion of existing (low end) electronics applications.

TYPE	NUMBER SOLD BILLION	UNIT COST OF DEVICE CENTS	GROSS VALUE $ BN
RFID	1000	1	10
Self-adjusting use by dates/sell by dates.	50	2	1.0
Electronic smart skin patches, diagnostic patches.	20	10	2.0
Moving colour advertisements, instructions, prompts on CPG and other goods.	10	10	1.0
Speaking and/or moving colour instructions and graphics on CPG, medicines etc	10	5	0.5
Timers on hair dye, curler, noodles etc	10	3	0.3
Electronic tearoffs on packages such as solar torches, games, radios, calculators, PDAs.	10	3	0.3
Electronically controlled chemically active packaging preserving food or emitting aromas	10	5	0.5
Patient compliance monitoring packages	5	5	0.25
EAS anti theft tags LC array type.	5	2	0.1
Electronic posters, wallpaper and billboard paper	0.2	1000	2
Membrane keyboards	0.1	5	0.005
Electronic books	0.1	10000	10
Wearable electronics	0.1	5000	5
Electronic gift cards and tags	0.5	10	0.05

Figure 3. Market estimates by application in 2016

Source: IDTechEx

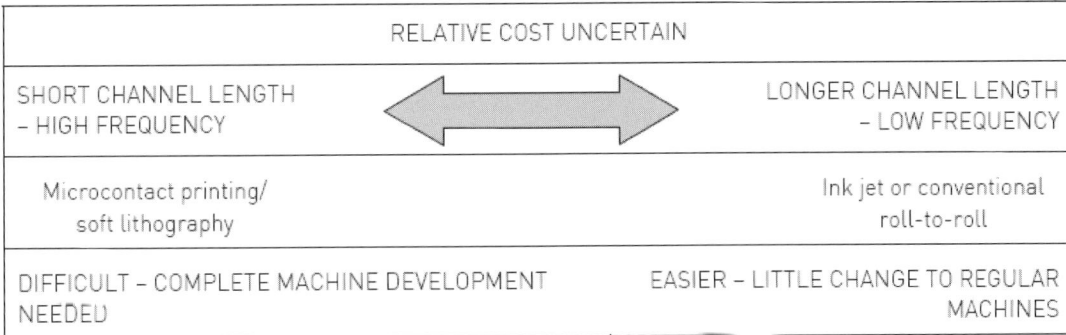

Figure 4. Manufacturing trends as a function of application complexity

Source: IDTechEx

4. Technical Path to Roll to Roll Manufacturing

To bridge the performance gap between conventional electronics and printed electronics much fundamental research is needed. Current semiconductors benefit from both high semiconductor electron mobility and very small feature sizes (<100nm). Both of these factors provide the GHz processor frequencies of modern CPUs. Printed electronics presently lack both of these advantages and feature sizes are two orders of magnitude larger. Thus, there is much research and development needed in several areas of existing technology.

Graphic printing has been traditionally limited to the resolution (line size) and registration (color alignment) by the resolving power of the human eye. This is more of a practical issue than a technical one. It makes little economic sense to build a conventional press with nanopositioning control elements. In the future, printed electronics platforms should be able to bridge much of the gap towards photolithographic processes.

Currently available semiconductor materials have electron mobility rivaling the amorphous silicon used in liquid crystal displays (~1 cm^2/V·s). This suggests that they should already be able to handle low speed switching such as displays, basic logic, and sensing function. As mobility improves, frequencies approaching those used in RFID (13.56 MHz +) will become increasingly possible. These types of developments will require industry/academic/governmental collaboration, which is a specific advantage of knowledge based economies such as the European Union.

5. Maintaining European Union Industry and Growth

Perhaps the most important facet of the vast potential of roll-to-roll electronics is that it shifts the advantages of manufacturing from ample, low cost labor, to knowledge intensive disciplines. An electronics printing press, or organic fabrication facility, with a team of specialized researchers, will be able to produce vast volumes of increasingly valuable electronics devices. They will be supported by technicians and skilled trades necessary to advance and maintain the equipment. Additionally, this value will provide employment in sales, administration, accounting, and all of the other important economic functions. And, as certain trillion Euro industries retract or shift geographically, printed electronics has the potential to replace both jobs and revenue in the developed world.

Acknowledgements

The author wishes to thank the staff of IDTechEx for compiling the market data and for energetic debate about the future of printed electronics.

References

[1] P. Harrop, R. Das, Printed Electronics

Multi-Material Micro Manufacture
W. Menz, S. Dimov and B. Fillon (Eds.)

SME's Perspectives: 4M current capabilities and future requirements

C. Neuy

IVAM Microtechnology Network, Dortmund, Germany

Abstract

The participation of small and medium sized enterprises (SME) in research and development is essential given their role in promoting innovation in this field. This is especially true in the field of microtechnology. The paper looks at different surveys and studies to analyse the potential for growth of SME in Europe especially in the 4M technologies. The bottom up survey within IVAM membership gives an overview on the technologies and application fields of European SME. These results are reflected in relation to the market reports of Yole Developpement and the NEXUS III report.

Keywords: SME, Europe, global competitiveness

1. SME in Europe

1.1. The role of SME in Europe for innovation

The participation of small and medium sized enterprises (SME) in research and development is essential given their role in promoting innovation in this field. SME play a vital role in the development and nurturing of new visions and transforming them into business assets. A significant involvement of SME is needed both as suppliers and as users of knowledge and technologies. This is especially true in the field of microtechnology.

1.2. Structural data of SME in Europe

A survey conducted by IVAM in 2005 [1] showed that in microtechnology SME are rather very small. Almost 52,3 % of the member companies in Europe have less than 20 employees. This definitely proves the predominance of small companies in this technology field.

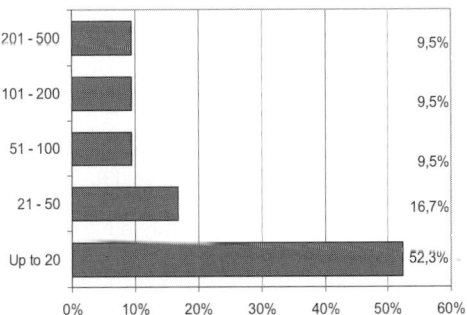

Fig. 1. IVAM Survey 2005:
How many employees do you have?
Here you see the replies of the IVAM members.
In percent.

This result was confirmed by a former survey in Europe within the European FP 5 project EMINENT in the beginning of 2004 [2]. About 2000 persons were interviewed with an online tool. 209 of them replied. There results were analysed. 27 % of the group has up to 9 employees, another 27 % 10 to 49.

1.3. Export rate

In spite of the small number of employees microtechnology companies show a very high export rate.

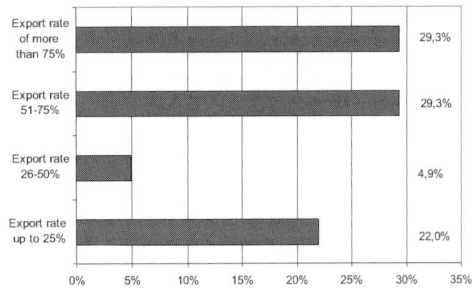

Fig. 2. IVAM Survey 2005:
What is your export rate?
Here you see the replies of the IVAM members.
In percent.

"Immediately global" is the reality of microtechnology start-ups. In addition, it is well known that most of the companies in micro- and nanotechnologies are at the very beginning of the value added chain. Meaning that the profit will probably made abroad, e.g. in the U.S: being export country No. 1. The most relevant export countries are listed here:

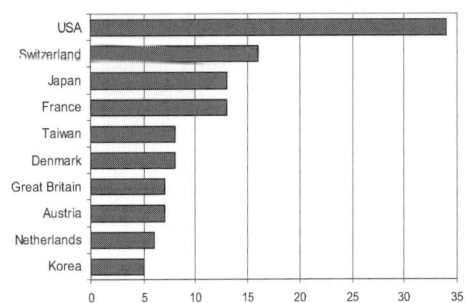

Fig. 3. IVAM Survey 2005:
Most relevant export countries of IVAM members
Weighted results.

2. 4M Technologies

2.1. Introduction to 4M Technologies

4M stands for Multi-material micro manufacture. 4M technologies are focusing on technologies apart from IC tools and materials. The work within the Network of Excellence 4M is divided into the technology fields:

- Polymers
- Metrology
- Assembly and packaging
- Metals
- Ceramics

as well as into the application divisions:

- Micro-optics
- Micro-fluidics
- Micro-sensors and actuators

2.2. Technological focus of European SME

Within the IVAM Survey 2005 the most relevant technology fields of European SME were identified:

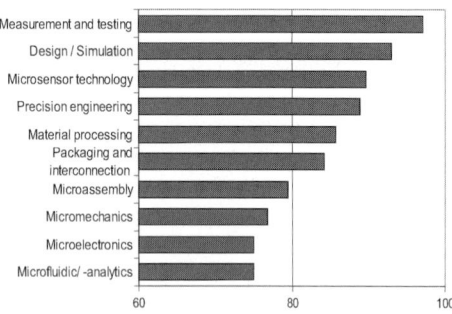

Fig. 4. IVAM Survey 2005.
Most relevant technologies of IVAM members.
Multiple answers were possible.

The results show that basic microtechnologies are used widely, e.g. sensor technologies, microfluidics, micromechanical engineering, precision engineering as well as different machining technologies for different materials. In addition, technologies before the machining and after the machining play a relevant role – which is not surprising: design and simulation, micro assembly as well as measurement and test.

2.3. Application focus of European SME

Since many of the IVAM members are in the components or equipment business, the most relevant "application field" microsystems technology is not surprising. The other application fields are in very good accordance with other surveys or studies. Medical technologies as well as automotive play an important role. In addition sensor, measurement and control technologies as well as automation are significant. The semiconductor industry but also biotechnology and genetic engineering are within the top ten application fields. The graph shows the most important application fields of the IVAM members in Europe:

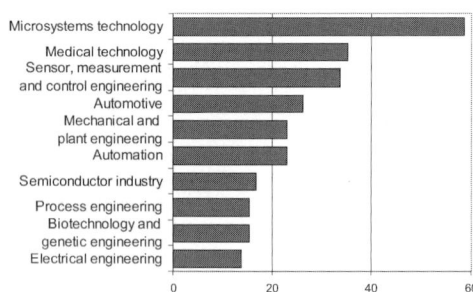

Fig. 5. IVAM Survey 2005.
Most relevant application fields of IVAM members.
Multiple answers were possible.

3. Market Forecasts

3.1. Study of Yole Développement

In addition to the IVAM survey, the studies from Yole and Nexus provide a good overview on the markets and products in microtechnology. The most important products were analysed by Yole:

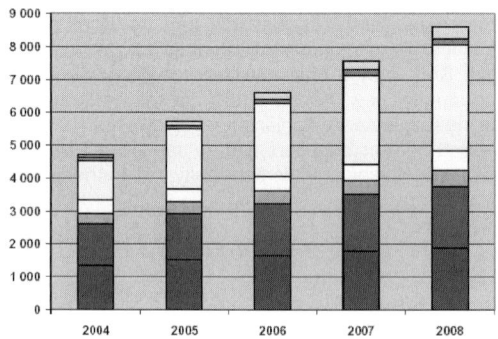

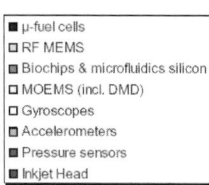

Fig. 6. MEMS Components Markets M$.
Source: Yole Développement [3].

If you look at the products analysed in the Yole study it is obvious that most of the products have a silicon based technology. The consequences of these findings for 4M technologies will be discussed in chapter 3.

3.2. NEXUS III Report

The NEXUS III report gives an outline of the growing application fields in microtechnology.

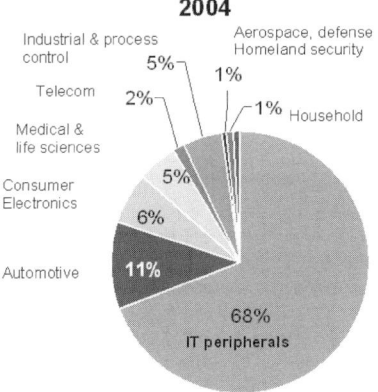

(total $12 billions)

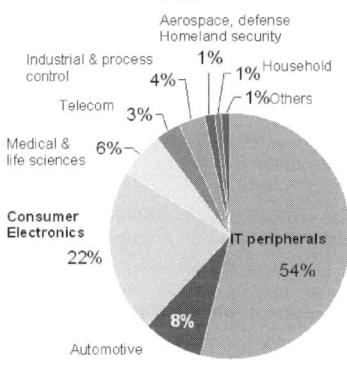

(total $25 billions)

Fig. 7. Market volume of microtechnologies.
Source: NEXUS III report, WTC [4].

The implications of report is that especially IT peripherals as well as consumer electronics become increasingly important.

The NEXUS III report looks into single products as shown in the following graph:

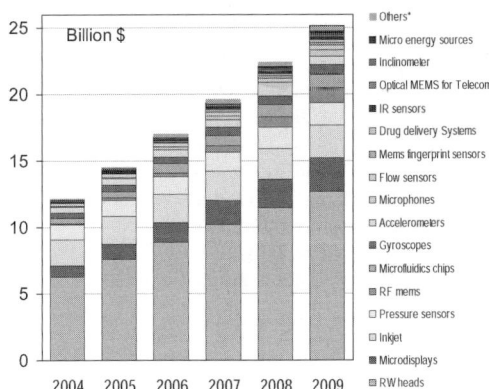

Fig. 8. Market volume of microtechnologies.
Source: NEXUS III report, WTC [4].

To make fig. 8 more clearly, the biggest market share products were selected in the following list, where No. 1 represents the highest market share:

1. Read/Write heads
2. Microdisplays
3. Inkjet
4. Pressure sensors
5. RF MEMS
6. Microfluidic chips
7. Gyroscopes
8. accelerometers
9. microphones
10. flow sensors

Again, it is obvious that mainly silicon based products are within the top ten products. Only a smaller number may also be made by 4M technologies.

3.3. Regional implications for European SME from the NEXUS III Report

Another perspective on the role of European SME in 4M technologies is given by the regional approach in the NEXUS III report.

Looking at North America it is the world's largest single manufacturing and application area for microsystem technologies. The industries that are the leading users of MST/MEMS devices are: IT peripherals, aerospace, automotive, biopharmaceutical, the aerospace and defense industry and telecommunications.

Table 1: Top 5 North American companies and their products [4].

Manu-facturers	main micro-system products	Turnover 2004 in MST
Seagate	magnetic heads	1700 Mio. US$
Western Digital	magnetic heads	930 Mio. US$
Texas Instruments	micro mirror array	840 Mio. US$
Hewlett Packard	inkjet heads	550 Mio. US$
Lexmark	Inkjet heads	225 Mio. US$

European industry is strong in the automotive, telecommunications as well as the industrial process control sectors. In the context of equipment industry, Europe is especially strong in the area of MST/MEMS systems, were EVgroup, Suss, AML, ASML, Adixen, STS, Trikon are the major suppliers worldwide for processing such as lithography, DRIE and bonding equipment.

Table 2: Top 5 European companies and their products [4].

Manu-facturers	main micro-system products	Turnover 2004 in MST
Robert Bosch (D)	accelerometers, gyroscopes, flow and pressure sensors	> 200 Mio. US$
STMicroelec-tronics (I/F)	inkjet heads, accelerometers for CE, micro-fluidic chips	250 Mio. US$
Infineon (D/N)	accelerometers, pressure sensors, RF MEMS	100 Mio. US$
VTI Techno-logies (FI)	accelerometers, pressure sensors, inclinometers	84 Mio. US$
Olivetti I-JET S.P.A. (I)	Inkjet heads, pressure sensors	80 Mio. US$

Traditionally, IT peripherals have been the main drivers for microsystems technologies in Asia with a strong commercial presence in hard disc drives and printers. Of late, the mass markets of consumer electronics and cellular /mobile phone handsets are establishing the new drivers for commercialization.

Table 3: Top 5 Asian companies and their products [4].

Manu-facturers	main microsystem products	Turnover 2004 in MST
Hitachi Global Storage Technologies (J)	magnetic heads	1100 Mio. US$
TDK (J)	magnetic heads	800 Mio. US$
Toshiba (J)	magnetic heads	680 Mio. US$
Fujitsu (J)	magnetic heads, RF MEMS, MOEMS, microfluidics	520 Mio. US$
Seiko Epson (J)	Inkjet heads, inertial sensors	210 Mio. US$

3. Relevance of 4M Technologies for SME

In automotive industry, silicon based technologies are the most important. But the meaning of 4M technologies in this area is growing. Looking at medical technologies, biotechnologies and genetic engineering, especially polymer technologies are widely used as disposables.

Looking for the competitive advantages of 4M technologies in respect to silicon technologies, process costs could play a role. Manufacturer of silicon based products often see a conflict between the investment in 6'' or 8'' processes. But on the other hand, customers pay less for a component. As a consequence the break-even elopes. This might open doors for substitute technologies.

New aspects include technologies such as moulded interconnect devices (MID) combining polymer and electronic properties. At the same time, "traditional" competences in mould making, material processing etc. – including 4M technologies – find new challenges to supply the polymer industry for microcomponents and systems. Even sensors in MID technology are under development, demonstrating the potential of the 4M technologies.

4. Outlook

McKinsey [5] reported about a future branch in distress referring to high tech companies in Europe. They draw a very critical picture of Europe, due to some boundary conditions, such as very fragmented markets. Their recommendation is the following:

CEOs of European high-tech companies must take action in 4 areas: speed-to-scale business models, developing top talent, capital structures designed for the long term but nevertheless focused on short-term profit, and actively shaping the environment.

Radically reversing this trend will require massively improving conditions in 4 areas:
- recognizing leadership
- Focusing on selected standards
- Concentrating public resources on developing top talent and on only a few clusters
- Supporting companies

Reversing the trend the European high-tech industry could create up to 4 million jobs in the medium term.

This trend has to be considered in the business strategies of high-tech SME. The innovative potential available in 4M technologies, especially for SME, has to be transferred into business successes. Activities such as the 4M network of excellence but also microtechnology networks such as IVAM provide support to SME.

References

[1] Sampling Description and Rate of Return

Delivered email-addresses	135
undeliverable email-addresses	3
Gross sampling	132
Return	65
Rate of return	49,2%

[2] www.eminent.ivam.de -> questionnaire 2004
[3] http://www.yole.fr/pagesAn/products/pdf/MIS.pdf, Seite 1, 30.11.2005.
[4] NEXUS Market Analysis for MEMS and Microsystems III 2005-2009.
[5] www.mckinsey.de/presse/ 051115_bb_hightech.htm

Components : Fabrication and Assembly Technologies

Multi-Material Micro Manufacture
W. Menz, S. Dimov and B. Fillon (Eds.)

51

Replication of micrometric pattern by hot embossing of silica based feedstock

L. Federzoni [a], O. Fourcade[a], P. Lourdin [b]

[a] CEA-Grenoble, DTNM//LCE, F38054 Grenoble cedex 9
[b] ECAM, 40 montée saint Barthélémy F69321 Lyon cedex

Abstract

The CEA is developing a new forming technique derived from the hot embossing of polymers. This technique consists in replacing the polymer by a feedstock composed on binders and inorganic nanopowders.
The process consists in three stages that are forming/pressing, debinding and sintering, similarly to the micro Powder Injection molding.
This process has been optimized for the replications of micrometric patterns. This work has led to the development of a feedstock composed of water-soluble binder and nanometric fumed silica. It was possible to replicate gratings of ~1 micron after sintering. The replication of sub-micronic plots is not possible at this stage. It is expected that the use of smaller particle size could allow reaching this objective.

Keywords: hot embossing, gratings, silica

1. Introduction

Hot embossing (or Nanoimprint lithography) [1–5] (including thermal and UV embossing) is a patterning method based on the mechanical structuring of a viscous material by pressing a stamp (mold) into a pre-defined layer of material, which is often thin in comparison to the lateral extension of the device. Its advantage in comparison to Powder Injection Molding is that a larger variety of materials can be used and higher aspect ratios can be achieved. Furthermore it is more appropriate for surface devices with different functional layers and for hybrid processing. Often the relatively low throughput is considered as its main disadvantage for mass-fabrication. Because the molding setup is simpler than that of injection molding and because process can be controlled more accurately, imprint is often used as prototyping method for PIM. However, imprint is easily scalable for high volumes by using wafer size processes and therefore also has a potential for mass fabrication. The physical properties of the material can be changed by thermal or UV induced chemical treatment. The processes, although all based on molding by imprint, have significant differences, particularly concerning the material used, the process temperatures and the post-processing.

Polymers currently used in nanoimprint are composed of relatively short linear chains. The ability to mold polymers is due to the increased mobility of the chains at higher temperatures, and, for thermoplastic materials, reversible. In most cases such polymers can be considered as amorphous, i.e. without any kind of orientation before and after molding. The radius of gyration of such polymer chains, which is a measure of the distance of two neighboring polymer coils, can be considered as a rough measure for the smallest sizes which can be molded by NIL. In case of the typically 100 nm long chains this is still below 5 nm.

When considering using powders in a polymer matrix, the smallest feature size is given by the grain size. The matrix enables these grains to flow into the cavities of the mold, or in the inverse case, to be displaced by protrusions of a stamp. The final shape is dependent on the density of grains achievable in a cavity or around a protrusion, and of shape modulations due to grain growth and densification during the sintering process.

This is more and more difficult if a high load of grains is used, typically of much higher than vol. 50%, as it is necessary for nanopowders. Therefore in contrast to experiments with filled polymers, e.g. for changing optical properties, the viscosity of the precursor is more dependent on the size and the amount of grains than of the initial viscosity of the matrix. The flow regime is expected to be more complicated than that seen in NIL using "standard" materials and de-mixing phenomena are expected at corners of small cavities. Strategies have to be found to avoid de-mixing and to assure sufficient flow. By using layers of materials with different properties (e.g. loads), it is expected that flow can be modulated.

Micro powder injection molding is an alternative technique which has been largely developed in the recent works [6-14]. This technique can be used to manufacture microparts for large-scale production. This process allows the replication of very fine details. However, the development of feedstocks for microPIM with nanopowders is difficult as the viscosity of the feestock increases drastically [7]. Then, the majority of feedstocks for microPIM are composed of micronic powders. They allow replicating details of a few 10th micrometers [6].

In this work, we have developed a new feedstock based on nanometric SiO_2 powders and polymers suitable for hot embossing (but not for microPIM) with high powder loading (75%) and tested for the replication of reference patterns.

This technique appears to be an alternative technique to microPIM for the manufacturing of surface textured ceramics or metals with improved replication performances.

2. Experimental

The precise feedstock composition is being patented and will be not exposed in this paper.

The feedstocks were prepared in a bi-screw extruder Thermo Electron Corporation. The feedstocks were composed a mixture of polymers and nanometric silica powders (Degussa OX50), representing a charge loading of 75%.

In this study, we have developed an experimental procedure for testing the feedstock. This procedure is described on Figure 1.

Hot embossing tests have been made on a press (METKON DIGIPRESS).

The feedstock is incorporated within the mold at room temperature then heated at a welding temperature of organic materials. The feedstock is then imprinted at 250 bars and demoulded at room temperature. Imprinted parts are then released from imprinting device for further steps.

The mould is a 200 mm silicon wafer, manufactured by a classical lithography process.

Debinding was performed in air atmosphere, in a furnace in which temperature was increasing at a rate of 1°C/min up to 550°C.

Sintering was performed in air atmosphere, in a furnace, at a constant temperature of 1000°C during two hours.

This sintering stage has been performed in order to provide sufficient mechanical properties to the part.

The surface of the sintered parts has been observed with a SEM.

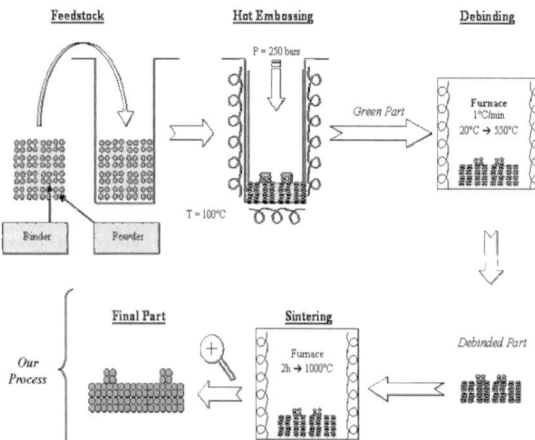

Fig. 1. Process description

3. Results

The silicon wafer tested for imprinting was made by several patterns among them gratings and plots. We have described below the replication obtained on them: tings and plots :in them.

3.1. Gratings replication

It can be seen that the replication of gratings of 1 micrometer is possible (fig 2 and 3). In this case, the aspect ratio was 1. With a higher magnification, it is possible to observe the nanopowder (fig. 3) and the residual porosity.

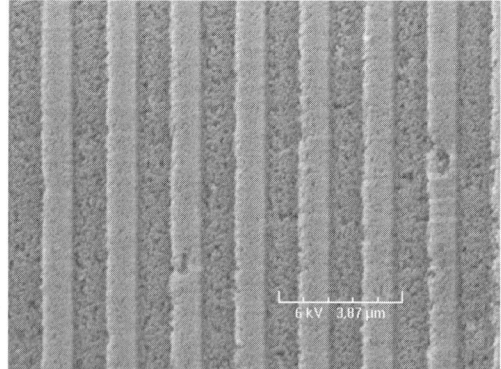

Fig. 2. gratings replication intermediate magnification

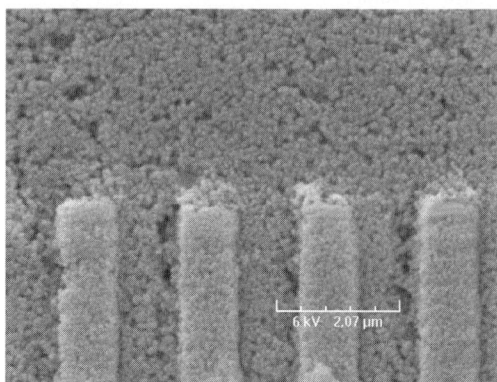

Fig. 3: gratings replication high magnification

Fig. 4. array of plot replicated - low magnification

The figure 4 displays the relatively good homogeneity of the initial feedstock through the good spatial repartition of the nanopowders on the part after sintering. One can also note that the porosity after sintering remains relatively high (measured to 15%) that means that the sintering treatment has also lead to a small volumic shrinkage of about 5%.

Fig. 5. plot replication – high magnification

3.2. Plots replication

Figures 4 and 5 represent the results of the replication of an array of plots of dimensions 0,7 μm x 0,7 μm. On the mould, the depth of the plots is 1 μm. One can see that the replication is only partial; this is particularly visible on figure 5. However, even if the replication is incomplete, it is possible to distinguish the location of the "expected" plots. For this case, we have observed that the weak replication of the surface is homogenous and reproducible on several trials.

4. Discussion

This preliminary work illustrates a new method to manufacture textured parts made of ceramics of metal. It is clearly to soon to conclude on the industrial interest of this method but it looks promising if considering that these results are really preliminary results.

We will comment below two specific aspects related to the replication: density of the final material and limits of replication.

4.1. Density of the final material

In this work, the samples have been sintered at relatively low temperature in order to consolidate the material. At 1000°C, the silica nanopowder has shrunk only of a few percents. This appeared to be the only way to preserve the shape of the gratings. When the samples are sintered at higher temperatures, the shrinkage related to the densification leads to an important distortion of the network of gratings and sometimes to the complete disappearance of the gratings by surface diffusion (example of a sintering at 1400°C for 3 hours). This distortion/disappearance will remain an important limitation of this process, when fully dense material are expected.

This distortion could be limited by the increase of the initial charge loading within the feedstock and by the increase of the packing density. The CEA has lunched a research program of that area. Preliminary results show the possibilities to increase the packing density by a modified granulometric distribution. Results will be published elsewhere.

4.2. Limits of replication

The comparison of the quality of replication between plots and gratings indicates that the particle size is probably the most important factor affecting the quality of replication. In the case of the plots, the surface is filled by a few grains.

It is expected that a particle size of 10 nm would improve the replication of the plots. Such tests will be performed in the next future by the authors.

In all cases, it is seen that the quality of replication has no relation with results obtained on pure polymers under 5nm).

5. Conclusion and outlook

In this paper, we have described preliminary results obtained on nanopowder feedstock hot embossing. This process has been set-up on a silica nanopowder feedstock with a relatively high charge loading (75%). With a good feedstock formulation (not described in the paper) it is possible to reach very good homogeneity and allow the replication of fine details.

However, these results do not hide the technical difficulties:
- to reach a full density material without geometrical distorsion
- to replicate submicronic plots.

The future work will concentrate on a new feedstock type with a very high packing density and high powder loading (above 85%) in order to limitate the shrinkage an preserve the gratings even on a full density material.

Applications will be developed in the field of textured transparent silica parts with modified optical properties low reflectivity, refractive index.

Acknowledgements

Authors would like to thank Mrs Oswald from Degussa who provided the powder and Dr Schiavone from the LTM CNRS in Grenoble who provided the moulds.

References

[1] H. Schift, C. David, M. Gabriel, J. Gobrecht, L.J. Heyderman, W. Kaiser, S. Köppel and L. Scandella, Nanoreplication in polymers using hot embossing and injection molding, Microelectronic Engineering 53, 171-174 (2000).

[2] Alternative Lithography Unleashing the Potential of Nanotechnology, Volume editor C. Sotomayor Torres, book series on Nanostructure Science and Technology in Kluwer Academic/Plenum Publishers, editor D.J. Lockwood. Hardbound, ISBN 0-306-47858-7, November 2003, 425 pp., 46-76 (2003).

[3] H. Schift, J. Gobrecht, B. Satilmis, J. Söchtig, F. Meier and W. Raupach, Nanoreplication in a Network, Kunststoffe plast europe 94 (6/2004), 1-4 (2004).

[4] L. J. Guo: Recent progress in nanoimprint technology and its applications, J. Phys. D: Appl. Phys. 37 (2004) R123–R141

[5] [5] M. Takahashi, Y. Murakoshi, K. Sugimoto, R. Maeda, Micro/nano hot embossing Pyrex glass with glassy carbon mold fabricated by focused-ion-beam-etching. DTIP of MEMS & MOEMS, Switzerland, 12-14 May 2004.

[6] Ruprecht R, Benzler T, Hanemann T, Müller K, Konys J, Piotter V, et al. Various replication techniques for manufacturing three-dimensional metal microstructures. Microsyst Technol.

54

1997;4:28– 31.

[7] Piotter V, Benzler T, Gietzelt T, Ruprecht R, Hausselt J. Micro powder injection molding. Adv Eng Mater 2000;2(10): 639–42.

[8] Piotter V, Bauer W, Benzler T, Emde A. Injection molding of components for microsystems. Microsyst Technol 2001;7: 99– 102.

[9] Liu ZY, Loh NH, Tor SB, Khor KA, Murakoshi Y, Maeda R. Binder system for micro powder injection molding. Mater Lett

[10] Murakoshi Y, Shimizu T, Maeda R, Sano T. High aspect ratio structuring by IPC etching and metal and ceramics forming. MicroMaterials Conf, Berlin, April 17–19; 2000.

[11] Merz L, Rath S, Piotter V, Ruprecht R, Kleissl JR, Hausselt J. Feedstock development for micro powder injection molding. Microsyst Technol 2002;8:129–32.

[12] Ruprecht R, Gietzelt T, Mqller K, Piotter V, Haugelt J. Injection molding of microstructured components from plastics, metals and ceramics. Microsyst Technol 2002;8:351– 8.

[13] Liu ZY, Loh NH, Tor SB, Khor KA, Murakoshi Y, Maeda R, et al. Micro-powder injection molding. J Mater Process Technol 2002;127(2):165–8.

[14] Rota A, Duong TV, Hartwig T. Micro powder metallurgy for the replicative production of metallic microstructures. Microsyst Technol 2002;8:323 –5.

[15] Rota A, Duong TV, Hartwig T. Wear resistant tools for reproduction technologies produced by micro powder metallurgy. Microsyst Technol 2002;7:225 – 8.

Multi-Material Micro Manufacture
W. Menz, S. Dimov and B. Fillon (Eds.)
© 2006 Elsevier Ltd. All rights reserved

Force fields with one stable equilibrium for micropart 2D manipulation

E.K. Xidias[a], N.A. Aspragathos[a]

[a] *Department of Mechanical Engineering & Aeronautics, University of Patras, Patras 26500, Greece*

Abstract

This paper proposes an approach for structuring a two dimensional (2D) force field with just one stable equilibrium at the desired location for the manipulation of microparts without sensing. The derived force field induces a unique stable equilibrium for any convex or non-convex polygonal micropart on the plane with arbitrary shape and size, where the equilibrium pose is the desired location of the micropart. Extensive simulations show the efficiency of the proposed method in the formulation of the force fields for 2D manipulation of microparts.

Keywords: stable equilibrium, force field, micropart manipulation

1. Introduction

Last years, the research on programmable force fields for microparts manipulation has received an increasing attention. The force fields can be used for the manipulation of a wide variety of microparts, since little or no sensing is required, for orienting and positioning microparts, towards assembly automation [1]. The force field is diffused on a planar surface where the micropart is placed, so the force and torque applied on the part translate and rotate it toward a stable equilibrium pose, which coincides with the desired target location.

A few types of potential fields proposed such as squeeze fields [2], elliptical fields [3], the combination of a radial and a gravity field [2] and a combination of a linear radial force and a constant force field [4] in which the work integral is zero along any circular path. When a micropart is placed into a squeeze field, it is translated and reoriented towards a stable equilibrium pose. The number of steps in the sequence of translational and reorientational motions depends on the complexity of the micropart. The elliptical fields [3] and the combination of radial and gravity fields (radial-gravity fields), can translate and rotate any micropart to equilibrium configurations. Sudsang and Kavraki [4] proposed a combination of a linear radial field and a constant field that induces a unique stable equilibrium for almost any micropart. The proposed field is defined by the parameters consisting of the magnitude of the constant force field and the coefficients defining the linear function associated with the linear radial force field.

The above fields can both translate and rotate a micropart to equilibrium configurations. These strategies typically consist of a sequence of force fields that cascade the micropart through multiple equilibria until the desired goal state is reached.

In this paper, we present an approach for determining a force field that induces a unique stable equilibrium for any convex or non-convex polygonal micropart. The proposed field is defined considering that the desired position and orientation of the micropart is the unique stable equilibrium.

2. Problem statement and equilibrium conditions

Assume a two-dimensional (2D) polygonal micropart $\mathbf{P}$ with a uniform mass distribution that is placed in a random location on the plane where a force field $\tilde{f}$ is activated. This micropart should be driven toward a stable equilibrium pose by the resultant force $\tilde{\mathbf{F}}$ and torque $\tilde{\mathbf{M}}$ induced by the field $\tilde{f}$ at the contact between the micropart and the plane. The motion of $\mathbf{P}$ under this field is a standard planar motion, which can be decomposed into a translation of its centre of mass (COM) and a rotation about its COM.

The micropart $\mathbf{P}$ reaches a global equilibrium (translation and orientation equilibrium) when the resultant force $\tilde{\mathbf{F}} = (\mathbf{F}_x, \mathbf{F}_y)$ and torque $\tilde{\mathbf{M}} = (\mathbf{M}_x, \mathbf{M}_y)$ applied to the COM of the micropart are zero,

$$\tilde{\mathbf{F}} = \iint_D \tilde{f}(x,y)\,dxdy = 0 \tag{1}$$

$$\tilde{\mathbf{M}} = \iint_D \binom{x}{y} \times \tilde{f}(x,y)\,dxdy = 0 \tag{2}$$

where, $\tilde{f} = (P(x,y), Q(x,y)): \mathbb{R}^2 \to \mathbb{R}^2$ is the force field and D is the plane region occupied by the micropart $\mathbf{P}$ in the stable equilibrium pose.

3. The formulation of the force field

In this section we present the proposed approach for the determination of the force field that can drive a micropart into a unique stable equilibrium.

The key idea is to construct a force field $\tilde{f}$ which is pulling a micropart $\mathbf{P}$ from a random location into the field toward the stable equilibrium pose D as it shown in fig. 1. The force field $\tilde{f}$ should be reduced to zero while approaching the region D and should be zero inside D.

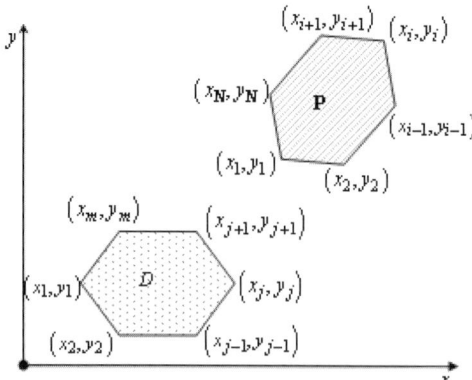

Fig. 1. The desired D location of a micropart $\mathbf{P}$, where $m = \mathbf{N}$.

The components P, Q of the force field $\tilde{\mathbf{f}}$ are given by

$$P(x,y) = -\frac{\partial u(x,y)}{\partial x} \text{ and } Q(x,y) = -\frac{\partial u(x,y)}{\partial y} \quad (3)$$

where $u(x,y)$ is the potential function. Since the field inside D should be zero, the polynomials scalar functions $P(x,y)$ and $Q(x,y)$ must become zero at all points (x_j, y_j) $j = 1,...,m \leq \mathbf{N}$, where $\mathbf{N}$ are the vertices of D and/or other characteristic points, inside D such as the COM. For simplicity, we consider that $P(x,y) = P(x)$ and $Q(x,y) = Q(y)$, thus we have,

$$P(x) = \prod_{j=1}^{m} (x - x_j) \quad (4)$$

and

$$Q(y) = \prod_{j=1}^{m} (y - y_j) \quad (5)$$

The polygonal region D, which is defined as:

$$D = \left\{ (x,y) \in \mathbb{R}^2 / P(x) = 0 \text{ and } Q(y) = 0 \right\} \quad (6)$$

Thus, the potential function $u(x,y)$ attains its minimum at (x_j, y_j), $j = 1,...,m \leq \mathbf{N}$, where $u(x_j, y_j) = 0$, and, it is differentiable everywhere in the plane. The number m of the points (x_j, y_j), where $P(x) = 0$ and $Q(y) = 0$ depends on the desired stability of the micropart $\mathbf{P}$ on the stable equilibrium pose. The higher m, the better approximation of the stable equilibrium pose. Thus it is possible to trade off the computation speed against the quality of the solution.

Experimental work showed that, an even degree polynomial potential function $u(x,y)$ offers higher

stability than odd-degree one. Therefore $P(x,y)$ and $Q(x,y)$ must be odd-degree polynomials functions. Figures 2 and 3, show the force field vectors for an even-degree $u(x,y)$ and for an odd-degree $u(x,y)$, respectively.

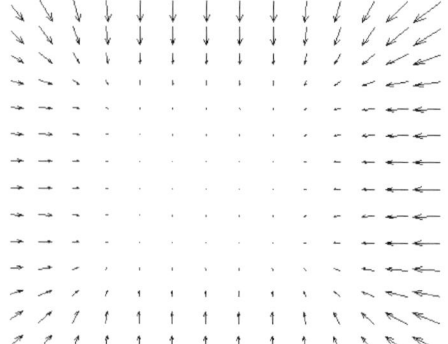

Fig. 2. The corresponding force field vectors for even-degree $u(x,y)$.

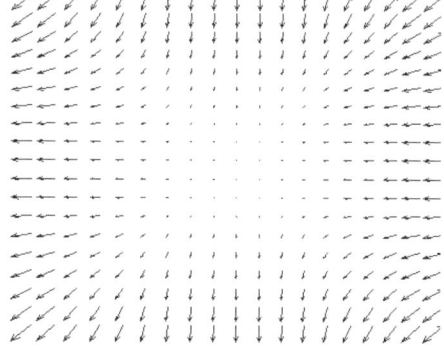

Fig. 3. The corresponding force field vectors for odd-degree potential function.

Fig.2 shows that all force field vectors are pointing towards the stable equilibrium pose. However, fig.3 shows that some areas of the force field vectors are pointing towards the outer limits of the considered region. Therefore, in such areas the field pushes the micropart away from the desired location.

4. Simulation results

The performance of the derived force fields are investigated for a variety of shapes of convex or non-convex polygonal microparts and the results show that the determined force field according to the proposed approach presents a unique stable equilibrium. Due the page limits, we present the results obtained by applying the determined field only for two polygonal microparts a convex and a non-convex.

As mentioned above (Section 2), any planar motion of a micropart $\mathbf{P}$ can be decomposed into a translational one of its COM and a rotational one about its COM. The translational motion is given by the solution of the following equation:

$$\tilde{\mathbf{F}} + \tilde{\mathbf{F}}_d = m_p \left[\ddot{x}_{COM}, \ddot{y}_{COM} \right]^{\mathrm{T}} \qquad (7)$$

where, $\tilde{\mathbf{F}}, \tilde{\mathbf{F}}_d$ are the resultant force and the damping force applied on $\mathbf{P}$ respectively, and m_p is the mass of the micropart. The rotation is given by the [3] of,

$$\tilde{\mathbf{M}} + \tilde{\mathbf{M}}_d = I_z \ddot{\theta} \qquad (8)$$

where, $\tilde{\mathbf{M}}, \tilde{\mathbf{M}}_d$ are the resultant torque and the damping torque applied on $\mathbf{P}$ respectively, and I_z is the moment of inertia of $\mathbf{P}$. The orientation of $\mathbf{P}$ is determined by the angle θ between the $x-$axis and the axis defined by the vertex (x_1, y_1) and the COM of the micropart $\mathbf{P}$.

In the following test cases, the number m is set equal to $\mathbf{N}$, the number of the vertices of the micropart.

In the first test case, the manipulation of a scalene triangle micropart is investigated. As aforementioned above, we choose $m = 3$ and $x_1 = 0, x_2 = 2.8284$, $x_{COM} = 1.8856$ and $y_1 = 2$, $y_2 = 0.5858$ and $y_{COM} = 1.0572$ are the coordinates of the vertices of the triangle and the COM of the triangle. Replacing in Eqn. (4) and Eqn. (5) we get,

$$P(x) = (x - 0)(x - 2.8284)(x - 1.8856) \qquad (9)$$

$$Q(y) = (y - 2)(y - 0.5858)(y - 1.0572) \qquad (10)$$

Figure 4, displays the obtained field $\tilde{\mathrm{f}}$ for the desired target location (grey area) of the micropart and the initial location of the triangle (black object).

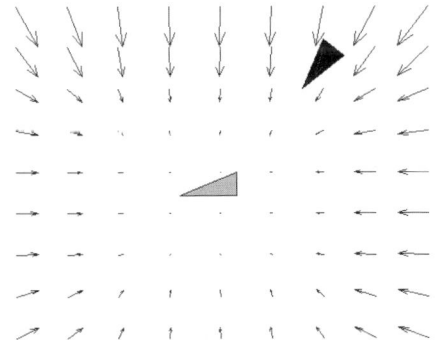

Fig. 4. The propose force field and the desired equilibrium pose D (grey area) and the initial location (black object).

Figure 5(a), shows the trajectory of the COM of the triangle reaching the stable equilibrium pose while, fig. 5(b) shows the variation of the COM position $x(t)$, the variation of the COM velocity $\dot{x}(t)$ and the variation of the cosine of the orientation angle θ of the triangular micropart.

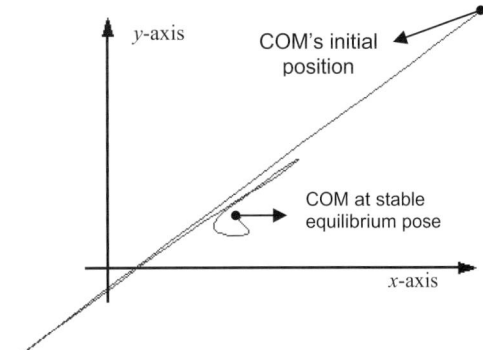

Fig. 5. (a) The path of the COM from the initial to stable equilibrium pose.

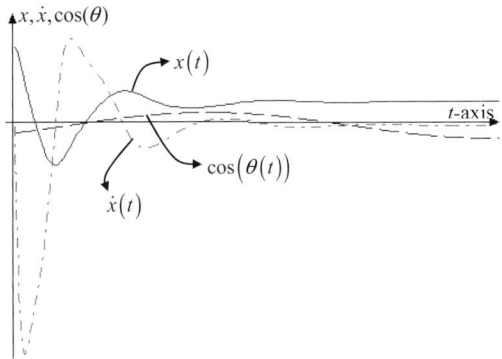

Fig. 5. (b) The variation of the position $x(t)$, velocity $\dot{x}(t)$ and the orientation of the micropart.

The desired pose of the COM is $(x_{COM}, y_{COM}, \theta) = (1.8856, 1.0572, 225°)$ and the final pose is $(x_{COM}, y_{COM}, \theta) = (1.88016, 1.05652, 224.45°)$ when the determined field is applied.

In the second test case, the manipulation of a non-convex polygonal micropart is investigated. We choose $m = 3$ and $x_1 = 2$, $x_2 = 3$, $x_3 = 4$ and $y_1 = 1$, $y_2 = 2$ and $y_3 = 3$ are the coordinates of the selected vertices of the micropart. Replacing in Eqn. (4) and (5) we get,

$$P(x) = (x - 2)(x - 3)(x - 4) \qquad (11)$$

$$Q(y) = (y - 1)(y - 2)(y - 3) \qquad (12)$$

Figure 6, displays the obtained field $\tilde{\mathrm{f}}$ for desired the target location (grey area) of the micropart and the initial location of the micropart (black object). Figure 7(a), shows the trajectory of the COM of the micropart reaching the stable equilibrium pose while, fig. 7(b) shows the variation of the COM position $x(t)$ the variation of the COM velocity $\dot{x}(t)$ and the variation of the cosine of the orientation angle θ of the micropart.

58

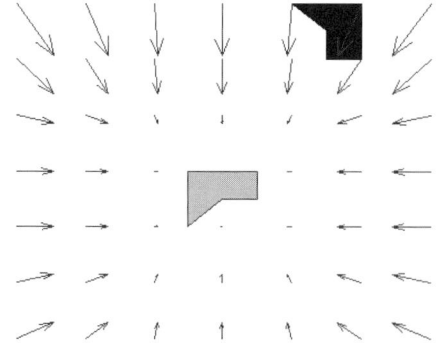

Fig. 6. The propose force field and the desired equilibrium pose D (grey area) and the initial location (black object).

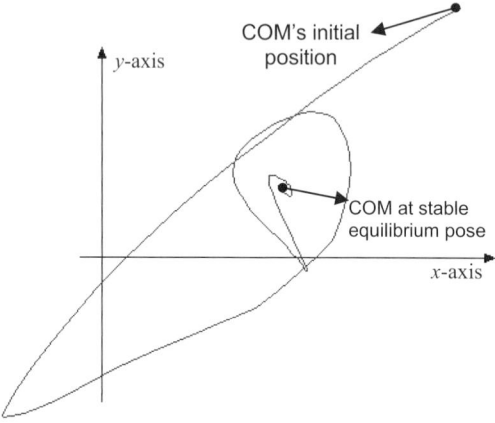

Fig. 7. (a) The path of the COM from the initial to stable equilibrium pose.

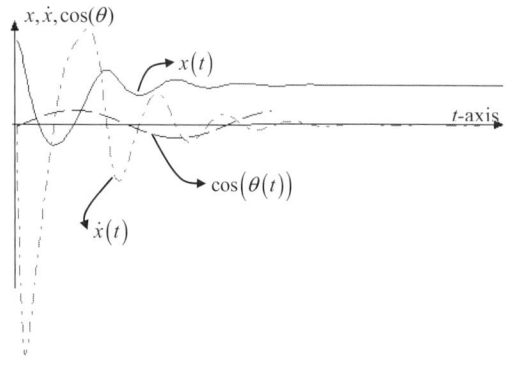

Fig. 7. (b) The variation of the position $x(t)$, velocity $\dot{x}(t)$ and the orientation of the micropart.

The desired pose of the COM of the micropart is $(x_{COM}, y_{COM}, \theta) = (3, 2.8, 221°)$ and the final pose is $(x_{COM}, y_{COM}, \theta) = (3.0124, 2.789, 220.015°)$ when the determined field is applied.

In both cases the microparts executes oscillations with small width while the micropart approaching the stable equilibrium target.

5. Conclusions

This paper introduces an approach for the formulation of a force field which is derived considering the target position of the polygonal micropart in order to induce a unique stable equilibrium for any micropart on the plane.

Simulations results show that the derived field manipulates efficiently any kind of micropart's shape. The future work will be focused on the criteria for automatic selection of the characteristic points of the desired pose

Acknowledgments

This work is financed by the Research Committee of the University of Patras as a part of the research project "An Optimal Motion Planning for a Robot based on Computational Geometry" under the K. Karatheodoris frame. University of Patras is a partner in the EC Network of Excellence "Multi-Material micro Manufacture: Technologies and Applications (4M)."

References

[1] Böhringer K.F., Donald B.R. and MacDonald N.C.. Upper and lower bounds for programmable vector fields with applications to MEMS and vibratory plate parts feeders. In Int. Workshop on Algorithmic Foundations of Robotics (WAFR), 1996.
[2] Böhringer K.F., Donald B.R. , Kavraki L. and Lamiraux F.. Part otientation with one or two stable equilibria using programmable vector fields. IEEE Trans. on Robotics and Automation, 1999.
[3] Kavraki L.. Part orientation with programmable vector fields: Two stable equilibria for most parts. In Proc. IEEE Int. Conf. on Robotics and Automation (ICRA), 1997.
[4] Sudsang, A. and Kavraki, L. E.. A geometric approach to designing a programmable force field with a unique stable equilibrium for parts in the plane. In Proc. IEEE Int. Conf. on Robotics and Automation (ICRA 2001), 2001, pp. 1079–1085.
[5] Shabana Ahmed. Computational dynamics. John Wiley & Sons, New York, 1994.

Multi-Material Micro Manufacture
W. Menz, S. Dimov and B. Fillon (Eds.)

59

Off-axis machining of NURBS freeform surfaces by Fast Tool Servo Systems

Prof. Dr.-Ing. C. Brecher, Dr. S. Lange, M. Merz, F. Niehaus, M. Winterschladen

*Fraunhofer Institute for Production Technology IPT,
Steinbachstraße 17, 52074 Aachen, Germany*

Abstract

The manufacturing of optical components with complex geometries like freeform surfaces is getting more and more important. Mass production of these optics is enabled by replication methods like injection moulding. For the machining of the required mould inserts with freeform geometries ultraprecision technologies like Fast Tool Servo turning can be applied to achieve the appropriate optical surface quality. For the mathematical description of freeform surfaces NURBS (Non Uniform Rational Basis Splines) are highly applicable. To avoid interpolation errors and to create set points with a continuous and smooth path of motion a trajectory generator directly based on NURBS data has been developed for Fast Tool Servo assisted turning [1], which evaluates the NURBS surface in each position control cycle. An approach of increasing the efficiency of manufacturing freeform surfaces is the off-axis machining of multiple workpieces at the same time. The paper presents the different process steps in simulation and off-axis machining of freeform surfaces using Fast Tool Servo assisted turning in conjunction with NURBS based data processing.

Keywords: Ultraprecision Machining, Fast Tool Servo System, Freeform Surfaces (NURBS), Off-Axis Turning

1. Introduction

The market for optical components is increasing strongly. Manufacturing technologies like grinding and polishing are established for the fabrication of spherical shapes but limited concerning the use for machining of complex optical components like freeform or micro-structured surfaces, which are today strongly penetrating into new areas of application. Examples for the use of freeform geometries are head-up displays in the automotive industry or customised contact lenses and glasses in the ophthalmologic sector. A field of application for microstructured surfaces in optical quality is the backlight for liquid crystal displays (LCD). Since the mentioned fabrication technologies are also not suitable for mass production, the development of methods for an economic and flexible large-scale production of complex optical components is of particular importance. The research activities of the Transregional Collaborative Research Center "Process Chains for the Replication of Complex Optical Elements" SFB/TR4 of the Universities of Aachen, Bremen and Stillwater (USA) have the objective to define the scientific basis for a deterministic and economic mass production of optical components with complex geometries, e.g. aspheric, non-rotational asymmetric or micro-structured surfaces potentially superimposed on freeform geometries.

In case of using the replication process injection moulding ultraprecision manufacturing of mould inserts is required and precision machining processes like diamond turning and fly-cutting are essential. To manufacture complex lenses with high accuracies the mould inserts must be manufactured with an adapted geometry to compensate for form deviation caused by the machining process or typical shrinking effects during the replication process. To determine the suitable geometry of the mould inserts several iterations loops are necessary.

For manufacturing an adapted mould geometry the trajectory generation is fundamental for ultraprecision machining. In a first step after design a Matlab routine analyses the NURBS encoded freeform surface, evaluating production parameters and prepares the data for

the manufacturing process. After this it is loaded to the control unit of the production machine. The calculation of the tool path of the diamond tool is processed online with an algorithm, described in detail in paragraph 4.

2. Optical design of freeform surfaces

The process chain for the manufacturing of complex optical components starts with the optical design. The main functions of the design process are the calculation of the optical path and the analysis of the optical components concerning applicability, producibility and manufacturing tolerances. Besides the homogeneity of the used materials, the achievable surface roughness and form tolerances of optical surfaces are responsible for the quality of the whole optical system.

In the following an optical system using a 3-D tailored freeform surface designed by the company OEC AG is presented. 3-D tailoring is a constructive method for the design of freeform illumination optics [2]. The freeform mirror or lens redirects the light (e.g. in fig. 1 from a point-like source) in such a way, that the prescribed illuminance distribution is exactly generated on a target surface. The freeform shape is found by solving a set of differential equations, which connect the continuity and the desired trimming of the surface as well as the redirection of the radiation defined by its local slope and curvature.

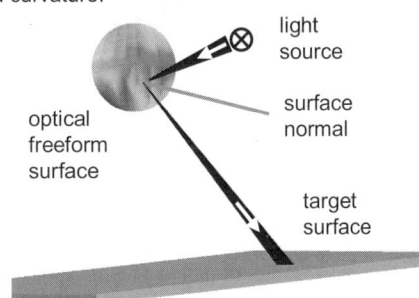

Fig. 1: Principle of 3-D tailoring [2]

The example in fig. 1 shows a 3-D tailored freeform mirror which redirects the light from a LED to form the Fraunhofer IPT logo as brighter lines in front of an evenly lit square. This illuminance distribution is produced by means of geometric optics with a single reflection at the freeform mirror surface. The mirror has a diameter of 50 mm. With these dimensions the scaleable illumination has a projection size of 90 mm x 90 mm.

The mathematical description of the given freeform mirror is based on NURBS. Until now a reproducible and competitive manufacturing has not been possible, since a convenient data processing for ultraprecision machine systems did not exist. Following the state of the art the mirror had to be manufactured by milling and hand polishing. The goal of the presented research activities was to solve those problems by building up a data interface based on NURBS to provide adequate surface information for the manufacturing of even complex optical surfaces by ultraprecision processes such as FTS-turning with a diamond cutting tool.

3. Pre-processing of off-axis machining

There are at least two alternatives to machine the above presented freeform mirror by FTS assisted turning. One is conventional FTS turning with the mirror in the centre of the spindle. Disadvantages of that scenario are highly varying cutting speeds from the outer radius of the workpiece towards the centre having a bad influence on the surface quality, which should be better than 20 nm Ra. The off-axis arrangement prevents a decrease of the cutting speed to zero and in that way enables better cutting conditions. In addition, avoiding artefacts in the surface centre depends heavily on an extremely accurate height adjustment of the diamond cutting tool. A remaining nub with a diameter of about 2 µm, see fig. 2, is not tolerable. Both problems are eliminated by off-axis turning. Also, multiple and even different workpieces can be machined simultaneously improving the efficiency. In contrast to one single part in the centre of rotation, where just the radius is processed, in the case of the off-axis arrangement the feeding length is the diameter of the part, which takes twice the time. However, if four mirrors are machined at the same time, the total gain of time, comparing the two arrangements, is roughly 50 %, which increases productivity considerably. Considering the setting-up time of the machines as well, the overall time saving is larger than 50 %.

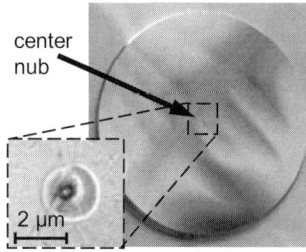

Fig. 2: Artefact in the surface centre caused by misalignment

The off-axis machining causes new demands on the NURBS data processing. The major challenge to be solved is the automated creation of surfaces that by interpolation fill the non-defined areas between the individual workpieces with the given functional optical geometries. The combination of the single workpieces is essential to simplify data processing in the machine control system on just one large surface.

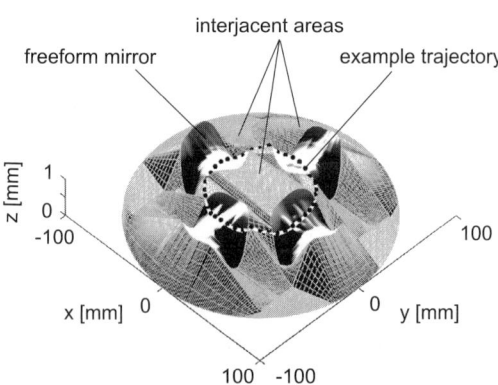

Fig. 3: Example geometry with four mirrors and interjacent areas

The arrangement of four separate freeform surfaces with the introduced optical function prepared for off-axis machining is shown in fig. 3. The freeform mirrors are positioned in a circle. The orientation of the surfaces is chosen in such a way, that the resulting inclinations of the tool path influencing the cutting tool clearance angle are best possible maintained for the FTS assisted turning process. In this situation the surfaces are rotated around the centre by 90°.

After arranging the single surfaces an interpolation algorithm fills interjacent areas by connecting opposite surface edges with linear interpolation, which is sufficient, since interjacent surface information is only needed for positioning the FTS axis without removing material from the workpiece. The functional surfaces, where accuracy of the geometry is strongly required, remain untouched by maintaining their original control polygons. The algorithm introduces knots with maximal multiplicity at the surface edges to inhibit surface errors. Aspects regarding harmonic tool path generation include acceleration-continuous generation of the tool path trajectory. The attended way of interpolation, maintaining the functional surface with the technique of maximising knot multiplicity, prohibits the usually inherent property of continuity with splines. In the first step of linear interpolation, the trajectory remains only position-continuous and therefore inharmonic at the edges of the surface, which would lead to transient effects in the actual tool movement. Acceleration-continuous extrapolation of the functional surface geometry into the filling areas is done by modifying the control points at the edge of the filling areas with a fitting algorithm minimising discontinuities at the surface edge.

Before the manufacturing is done the process is simulated using a model with the non-linear transfer behaviour of the FTS system to ensure machinability of the surface geometry, e.g. regarding the maximal clearance angle of the diamond tool. After fundamental analysis of the machinability, the machining parameters like feed rate and spindle speed are evaluated. The parameters are mainly constrained by the dynamics of the FTS system. The dynamic behaviour of the FTS system is mainly limited by the maximum drive force. Increasing the frequency requires a reduction of the stroke to prevent an intolerable following error of the cutting tool during the manufacturing process. Using a simulation tool, which emulates the FTS, the dynamic behaviour can be analysed and the following error can be determined by reviewing different machining parameters. The aforementioned techniques enable influence on the form accuracy by calculating the effects of parameter manipulations.

4. Fast algorithm for trajectory generation

After defining the desired freeform surface as NURBS data file and simulating the machining process a method has to be provided to process the NURBS data online on the FTS control system. Therefore a quick algorithm is needed to compute an adequate number of tool path supporting points. The position control clock of 8 kHz requires an according rate of set point generation. The frequency of 8 kHz means that every 125 µs the current computing and the information transfer from and to the physical I/O have to be completed. The transfer and control operations take at most 35 µs, so that 90 µs remain for the set point generation. The present processor of the FTS controller with 650 MHz CPU (Central Processing Unit) frequency enables approximately 8000 FLOPs (Floating Point Operations) during the remaining time for the set point generation. The set point z for the FTS system is given implicitly in the Cartesian surface $S(u,v)$. The lacking analytical inversion of the projection of the surface $S(u,v)$ to the X-Y-plane defined by the coordinates x,y, which are supplied by the control system of the base machine, to the NURBS parameters u,v proves to be the critical point.

$$(S_x, S_y) \xrightarrow{\text{Newton}} (u,v) \xrightarrow{\text{analytical}} (S_z) \qquad (1)$$

That inversion can only be done efficiently with a Newton optimisation [3]. Selecting convenient starting points, the Newton algorithm converges super-linear. Nevertheless, the numerical complexity is demanding compared to the available calculation time of 90 µs. The following abstract demonstrates how the algorithm calculates the inversion giving the coordinates x,y.

The objective function for the Newton algorithm is the squared distance vector in two dimensions

$$|r(u,v)|^2 = \left\| \begin{bmatrix} S_x(u,v) \\ S_y(u,v) \end{bmatrix} - \begin{bmatrix} x \\ y \end{bmatrix} \right\|^2 \qquad (2)$$

satisfying convexity. The objective function's negative gradient is set to zero for minimum condition.

$$\kappa_i = -\begin{bmatrix} f(u_i,v_i) \\ g(u_i,v_i) \end{bmatrix} = -\begin{bmatrix} r(u_i,v_i) \cdot \partial_u S(u_i,v_i) \\ r(u_i,v_i) \cdot \partial_v S(u_i,v_i) \end{bmatrix} \overset{!}{=} 0 \qquad (3)$$

It can be calculated from the tangent vectors of the NURBS surface $\partial_u S(u_i,v_i)$ and $\partial_v S(u_i,v_i)$, which are projected on the X-Y-plane. A new approximation u_{i+1}, v_{i+1} for a minimum distance r is calculated with the Newton iteration

$$\begin{bmatrix} u_{i+1} \\ v_{i+1} \end{bmatrix} = J_i^{-1} \kappa_i + \begin{bmatrix} u_i \\ v_i \end{bmatrix}; J_i = \begin{bmatrix} \partial_u f & \partial_v f \\ \partial_u g & \partial_v g \end{bmatrix} \qquad (4);(5)$$

using the matrix of second order derivatives, the Jacobi matrix J_i, which is computationally complex, requiring the explicit calculation of two first order derivatives ($\partial_u S$, $\partial_v S$) and four second order derivatives ($\partial_{uu}S$, $\partial_{vv}S$, $\partial_{uv}S$, $\partial_{vu}S$). The matrix operations of the 2x2-equation-system appear minor on the other hand. A NURBS surface evaluation of order two needs about 140 FLOPs for the calculation of one point. All in all, one Newton step requires 600 FLOPs. Depending on good start values for u,v the algorithm needs three to seven iterations leading to 1800 to 4200 FLOPs in total. Hence, even in worst case the chosen algorithm is fast enough to run on the available FTS controller during the machining of surfaces given in NURBS.

5. Aerostatic Fast Tool Servo System

During the work an aerostatically guided Fast Tool Servo System, developed by the Fraunhofer IPT, was used for manufacturing the freeform mirror mentioned above [1]. The supported slide of the FTS consists of CFRP (carbon fiber reinforced plastics), in which a permanent magnet is incorporated (Fig. 4). It is driven by a linear motor comprising of two coils, one above and one beneath the slide, which can be cooled by a pipeline system, if that is required. The monophase linear motor is powered by a current controlled analogue amplifier with a maximum current of ±8 A. The motor is capable of a maximum stroke of ±5 mm. The position feedback control system is PC-based with a control cycle of 8 kHz.

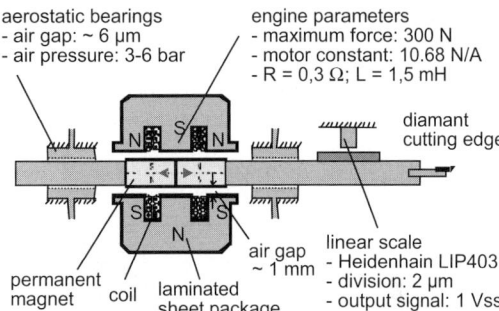

Fig. 4: Schematic drawing of Fast Tool Servo System

The software of the control system includes the device driver for the PCI I/O board and a server system, which facilitates the building of a control algorithm and a set point generator as well as elementary features like online representation of position data. To provide deterministic time response the realtime operating system RTLinux from FSMLabs is used. The Debian Linux distribution features a standard Unix environment and a graphical user interface. The control algorithm and a NURBS based set point generator are implemented in Matlab/Simulink. A massive reduction of latency was achieved by optimising the time-critical parts of the set point generator for the aforementioned server system. The applied control algorithm is a cascaded PI/P-controller. The not measured actual velocity is generated by differentiation of the actual position. In order to reduce the following error the inverse transfer function of the closed position loop is implemented as anticipatory control to level the frequency response. A couple of compensations of alignment and measurement errors are introduced to advance the performance of the control system. The magnetic attraction between linear motor and enclosure is compensated by a lookup table measured in an initialisation sequence. Another correction concerns the periodic interpolation errors of the open linear scale, which cannot be oriented ideally.

6. Off-axis machining of freeform surfaces

The machining of the freeform surfaces takes place on an ultraprecision lathe equipped with the above mentioned FTS system. The mechanical part of the FTS is mounted on the z-slide of the machine while the controller is connected to the position sensors of the x-slide and the spindle of the lathe. Using the above described algorithm the controller generates the trajectory depending on the current radial and angular position of the workpiece. The FTS is set-up with a mono-crystal diamond cutting tool with a radius of 0.5 mm and a rake angle of 0°. Concerning inclined surfaces a clearance angle of 15° is chosen. During the earlier simulation of the machining process the size of the clearance angle is verified as well as the machining parameters for the finishing. Due to the use of diamond cutting tools non-ferrous metals have to be used for the mirrors. The machined workpieces are made of an aluminium alloy and have a diameter of 50 mm each.

To enable the off-axis machining a special clamping device, capable of carrying four workpieces has been designed, see fig. 5. Additional references are needed, which permit an exact alignment and repositioning on different machine tools. The clamp consists of a round base plate thick enough to prevent deformation. The circumference of the plate is machined by diamond turning and serves as reference for centring the clamping device. The rotational position is defined by a flat surface at the circumference. Based on that references four counter bores with a tight fit have been machined on the top to set the position of the workpieces. Each mirror is mounted without introducing tensions in the surface, which will be machined, by screws from the backside.

Fig. 5: Clamping device and FTS on ultraprecision lathe

Additionally to the machining on the ultraprecision lathe, the manufacturing sequence of the freeform mirrors is extended by a pre-machining step on a precision milling machine. The pre-machining is required, because the manufacturing of the maximum surface depth of about 1.7 mm is due to low cutting depths extremely time consuming on an ultraprecision lathe. Hence, the pre-machining is performed on a milling machine using three axes and hard metal ball end mills. The surface information for generating the tool path for the milling process was provided with an IGES (Initial Graphics Exchange Specification) data file, which like the FTS system also uses NURBS for freeform encoding The trajectory generation of the milling preprocess offers a reasonable accuracy for this manufacturing step. The rough machining using a 6 mm tool diameter and a line feed of 2 mm is followed by the smoothing process using a smaller ball end mill with 3 mm diameter. The line feed is reduced to 0.1 mm. So, an allowance of 50 µm is left for the machining on the ultraprecision lathe. The whole pre-machining on the precision mill takes about 2.5 hours compared to estimated 35 hours the pre-machining by FTS turning would need. For the final machining the clamping device has to be aligned at the vacuum chuck on the ultraprecision lathe. The change between the two machines during the manufacturing of the mirrors necessitates the high demands on the references of the clamping device. After arrangement on the lathe the alignment of the clamp is so precise that after one roughing process with a depth of cut of 20 µm the freeform surface is already machined completely.

The following finishing is performed with a spindle speed of 50 rpm, a depth of cut of 4 µm and a feed rate of 1 mm/min. With these parameters the final cut takes about 50 minutes. All in all the machining on the ultra-precision lathe takes about 1.5 hours, including all roughing and finishing steps. The surface roughness of the mirrors is measured with a white light interferometer and is in a range of 15 to 20 nm Ra. The form deviation can not be measured directly yet, but the recording of the position signal of the FTS linear scale exemplifies a following error of the tool below ±0.4 µm. This value can also be estimated with the earlier mentioned simulation tool before the actual machining. Thus machining parameters like spindle speed are adjusted to realise a minimum following error and a high performance of the FTS.

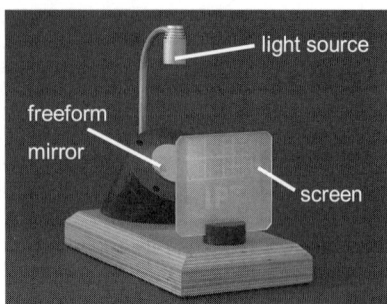

Fig. 6: Function test of the manufactured freeform mirror

7. Off Summary and outlook

In this paper a new approach for machining complex optical surfaces is presented, instancing a freeform mirror for the demo application illustrated in figure 6. The employed off-axis machining allows for an efficient manufacturing of optical freeform surfaces by avoiding typical issues of conventional turning like artefacts in the surface centre and highly varying cutting speeds. In that way the surface quality is increased as well as the production time is cut down significantly leading to a reduction in machining time of 50 % referring to the machining of a single workpiece. In addition, a solution for the data pre-processing of the off-axis machining is demonstrated. In this case four single freeform surfaces encoded in NURBS are combined automatically and manufactured in one machining step.

Acknowledgements

The research work presented in this paper is part of the SFB/TR4 "Process Chains for the Replication of Complex Optical Elements", funded by the German Science Foundation DFG.

The 3-D tailored freeform mirror was designed by OEC AG, Germany.

References

[1] Weck, M.; Winterschladen, M.; Pfeifer, T.; Dörner, D.; Brinksmeier, E.; Autschbach, L.; Riemer, O.: Manufacturing of Optical Molds Using an Integrated Simulation and Measurement Interface; Proceedings of SPIE, 5252 (2003).

[2] Ries, H.; Muschawek, J.: Tailored freeform optical surfaces, Journal Optical Society of America, 19 (2002) 3.

[3] Piegl, L.; Tiller, W.: The NURBS Book, Springer Verlag, 1997.

Multi-Material Micro Manufacture
W. Menz, S. Dimov and B. Fillon (Eds.)

63

Automated tool exchange for ultraprecision diamond milling and turning applications

Prof. Dr.-Ing. Christian Brecher, Dr.-Ing. Sven C. Lange, Frank Niehaus, Christian Wenzel

*Fraunhofer-Institute for Production Technology IPT,
Steinbachstrasse 17, 52074 Aachen, Germany*

Abstract

Active tool wear control is a crucial factor for fulfilling the high shape and surface requirements in the large area ultra-precision diamond surface structuring of optical components. The Fraunhofer IPT has developed opto-mechanical setups for a fully automated in situ tool characterisation and a reproducible tool exchange. The systems embody a submicron accuracy for both, the machining of planar masters for replication by means of planing and fly-cutting operation as well as for the turning of large rollers for embossing applications.

Keywords: diamond turning, ultra-precision machines, automated referencing and tool exchange, tool wear control

1. Introduction

The quality of microstructured optical foils for reflectors, display technology or even diffractive elements is strongly influenced by the moulds with ultraprecise surfaces used for the replication of the plastic elements [1]. Single crystal diamonds with defined cutting edges are employed for microstructuring these surfaces by means of milling, planing or turning operations. Machining time for these extreme ultraprecise operations lasts from several days up to weeks, while working plan or cylindrical areas of several square meters. Process stability as a guarantee for constant surface quality suffers from tool wear of the diamond tools, causing insufficient accuracy in the geometry of the structures and low surface quality on the flanks [2]. Currently, defective parts can only be identified after an entire machining run, causing enormous rejecting costs and accordingly endangering the companies that are active in these fields [3].

The Fraunhofer IPT has developed add on solutions for diamond turning and milling machines that enable a fully automated diamond tool exchange to ensure process stability and constant surface quality for up to a multi-day machining operation. In comparison with state of the art high precision passive mechanical tool exchange systems with an operating position repeatability of two to five microns the new solution reaches down to submicron lateral and angular accuracy. The equipped machines at the Fraunhofer IPT are able to visually characterise and judge the condition of the tools cutting blade down to submicron defects, to change worn tools if necessary, to subsequently measure the tool tips position with the same submicron accuracy, and to correct for possible position deviations in a fully automated manner.

2. Sample geometries and precision requirements

For the manufacturing of micro structured optical mould inserts diamond tools with a defined cutting geometry are used. Depending on the structure design, different tools have to be utilized. The precision achieved during the exchange of the tools is crucial for the overall optical performance. Accuracies below one

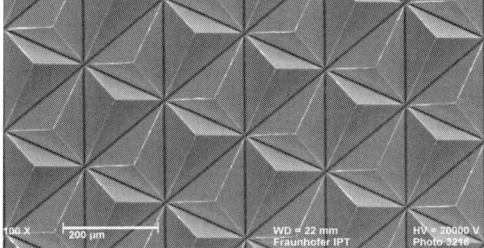

Fig. 1. Mould insert for an optical reflector

micron are requested to ensure the best optical performance. Fig. 1 shows a mould insert used for an optical reflector with three intersecting grooves and two different apex angles. With the new metrology system integrated into the ultra precision machining unit, a tool exchange precision of below 0.5 µm could be achieved in this example.

Fig. 2 shows a master tool with a size of the optically structured surface of 600 x 500 mm². The diamond radius tool has been used in fly-cutting operation for 300 h. The iterative optical control of the tools condition was inevitable in order to ensure a superb and continuous surface finishing of the entire master. A tool exchange had to be conducted due to tool wear after approximately 200 h.

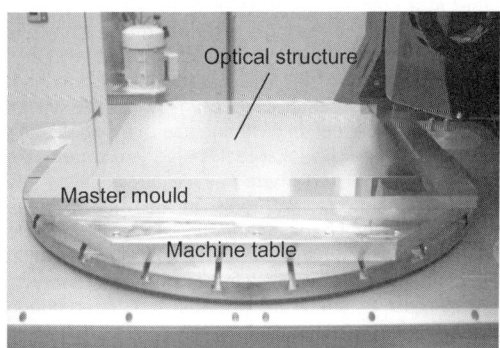

Fig. 2. Mould insert for 24"-TFT screen.

3. Ultra precision machining centre UHM

For the planar structuring of optical mould inserts, the Fraunhofer IPT has developed a large area ultra-precision-machine that can be used for planing, fly-cutting and turning operation of parts up to one square meter in size. The UP-machining system is equipped with an automated tool exchange unit consisting of a high resolution CCD Camera and a robot for the automated tool exchange (Fig. 3) [A-B].

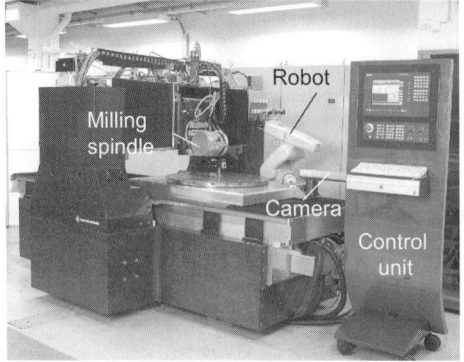

Fig. 3. Large area UP-machining centre UHM

3.1. Optical and mechanical setup of the exchange unit

With the use of a Sony HRC camera for the UHM planing machine 1024 x 768 pixels are available with an individual pixel size of 4.65 x 4.65 μm^2. A telecentric telescope is used for magnification (10x) in combination with the camera leading to an overall projection surface of 0.49 x 0.37 mm^2 in a focal distance of 48 mm. Due to the magnification of the telescope almost free of distortion a single pixel can resolve a theoretical spot size of 0.465 x 0.465 μm^2 reaching down for the minimal possible resolution of a visual system. The sensitive illumination of the projection surface is achieved with a telecentric red light LED mounted on the opposite side of the objective (Fig. 2). The entire unit is placed in a compact housing for protection during machining. Beyond, the housing unit has four degrees of freedom for the alignment of the optical path of the camera system during the initial set-up.

For the exchange unit a six DOF articulated robot equipped with a pneumatic gripper and an electric drive for a bolt fixation of a Posalux standard diamond tool holder is capable of reaching up to a magazine for a tool pick up. Subsequently, the tool can be placed either in a planing tool adapter or an air bearing fly-cutting spindle. The spindle is therefore mechanically fixed in a defined position for the characterisation and the exchange of the tool. The reproducible alignment of the spindle rotor is crucial for allowing the visual inspection of the tool tip. With the use of a self aligning adapter for the Posalux-tool holder a repeatable mechanical mounting in the spindle as well as a fine tuned balancing for all tools in use is possible. However, the mechanical mounting does not fulfil the requirements deriving from the ultra-precision application. A subsequent referencing is necessary.

In order to check for possible tool wear during diamond chipping the machine will position the used diamond tool in the CCD-Camera reference station to characterize the condition of the diamond blade and to save the lateral position of the tool tip relative to the optical system. On the base of predefined wear marks the analysis software of the camera is capable of autonomously deciding whether the current tool needs to be replaced or is still suited for the continuation of the structuring. In case of a tool exchange, the robot will unbolt the Posalux tool base from the holding fixture of the machine to mount it in the tool magazine. A new tool, also taken from the magazine, can be aligned and fixed in the holding fixture for the planing operation. Subsequent to the mechanical replacement the machine runs the tool into the reference station to compare the new tool tip position with the one of the previous tool. The lateral displacement is automatically transmitted to the CNC control unit of the machine for compensation purposes. With the corrected tool position the structuring is continued.

3.2. Analysis of the accuracy

The performance of the system strongly depends on the accuracy of the visual inspection system. In addition to the repeatability of the camera itself and the analysis software algorithm especially a misalignment of the optical path can lead to a distortion of the illustration and therefore to misinterpretations of the determined distances to be corrected. With the help of a self designed gauge block the camera system has been adjusted on the machine with a remaining angular misalignment smaller than 0.05°. Beyond the alignment of the camera system the mechanical fixation of the installed velocity controlled spindle causes an error in repeatability. The magnitude of this influence has been analysed and optimized by means of empirical tests with a highly sensitive inductive contact sensor. With a measured repeatability in the radial spindle rotor fixation of around 3 to 5 microns, the corresponding theoretically determined geometrical error due to the camera alignment is in the range of one nanometre being significantly below the camera resolution. The remaining uncertainty due to the camera system as well as the mechanical repeatability could be proved to be in the range of half a micron.

As a verification of the system sample work pieces have been structured with linear grooves. Within one single groove the diamond tool has been replaced and a second cut with the same infeed has been undertaken for half of the length of the groove. After the replacement of the worn tool a lateral offset of 2.3 μm in x-direction and of 4.8 μm in y-direction has been measured with the optical reference station as values to be compensated.

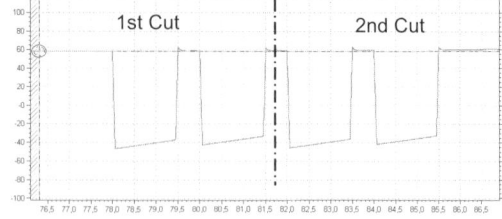

Position deviation between 1st and 2nd cut:
Max. deviation in Z: 0.1535 μm

Max. deviation in X: 0.0592 μm

Fig. 4. Lateral deviation after tool exchange.

These values can be used in the control system of the UP machine tool for a tool offset correction. At the interface of the two cuts on the work piece a lateral and a horizontal displacement caused by the tool exchange could be analysed to smaller than 0.153 µm by means of a Taylor Hobson Talysurf tactile system.

4. Ultra precise roller turning machine UPRTM

A second machine has been designed for microstructuring embossing rollers in cooperation with industrial partners in the scope of the project »MICROSTRUCT« [C]. Objective of the project is the development of a machine for direct and seam-less microstructuring of large embossing rollers that are used for the replicating mass production of plastic films in a continuous process. The dimensions of the rollers are diameters of up to 600 mm and lengths of up to 2000 mm, while the structure sizes are in the range of 5 to 150 µm and generated by means of ultra-precision machining with mono-crystalline diamond tools to ensure an optical surface quality.

The dimensions of the rollers require a new machine concept with respect to layout and process automation (Fig. 5). In order to reach high stiffness and damping characteristics, the guides of the linear axes as well as the bearings for the roller are carried out hydrostatically. Linear motors are used to drive the slides. Thus surface finishes below 15 nm Ra and position accuracies of 5 µm along the roller can be achieved. A dynamic axis to machine non-rotationally symmetric microstructures can supplement the machine system. The range of microstructures is wide and not all can be machined by turning operations. Especially reflecting functions use pyramidal structures, which require an additional machining process. Hence the new machine allows for turning as well as for planing processes along the surface line of the roller.

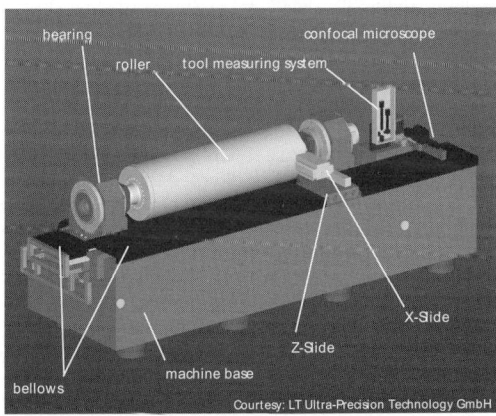

Fig. 5. Ultra precise roller turning machine UPRTM

The machining of a large area roller is time consuming and requires new concepts of process automation, therefore different measurement devices systems are integrated in the machine. A confocal microscope is used for 3D-surface measurement and a CCD camera system enables measuring of the tool position and characterisation of the tool quality. In combination with an automatic tool exchange device it is possible to replace a tool when it is worn.

The developed tool exchange device is different from the tool changer that is installed in the UHM machine for fly-cutting processes. A standardized HSK-25-E (hollow-shaft taper) interface with a standard automated clamping device is used for the exchange of diamond tools (Fig. 6). The HSK unit guarantees for a high position repeatability ensuring a high stiffness at the same time. With an additionally applied alignment pin, a rotational fixation of the HSK-25 is achieved. The alignment pin is also used as an index to be able to provide tools for turning and planing operations. For planing the tool has to be turned around its longitudinal axis by 90° to orientate the tool towards the cutting direction. The HSK-25 adapter enables the mounting of diamond tools with a standard 6 x 6 mm^2 shaft. Before the tool holder can be deposited in the machine, the angular orientation and position of the tool have to be pre-adjusted. The height is defined by a support while the other orientations are adjusted by set screws.

Fig. 6. HSK tool holder

To handle and store the tools inside the machine, a magazine was developed, which is capable of ten tool holders. The magazine with the revolver mounted on a slide moved by a pneumatic cylinder is shown in Fig. 7. The tool holders are arranged star-like in the rotating magazine to enable an easy pick-up. They are clamped by spring loaded levers that can be opened by pneumatic cylinders. The tool exchange works without additional handling devices by mounting the magazine on a slide that moves back and forth to hand over the tool to the x-slide.

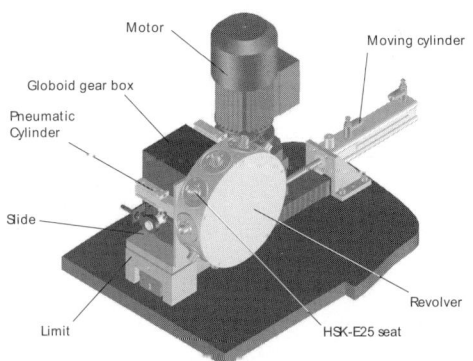

Fig. 7. Tool exchange device

After a tool exchange the new tool has to be checked with a camera system. The camera system is capable of an automatic detection of the tool position and an angular displacement, which works with radius as well as v-shaped tools. Further the system is able to determine the tool quality and recognises e.g. a broken tool tip. Therefore two CCD cameras, each with a telecentric lens, are integrated in the machine. Two cameras are needed to be able to measure tools for turning and also for planing processes in a 90° turned orientation. Hence there is a camera for measurements in the X/Z-plane for turning and a X/Y-plane for planing (Fig. 8).

The chip of the CCD camera offers 1400 x 1000 pixels and in combination with a telecentric lens (10x magnification) an overall lateral resolution of 0.5 µm is realised, while the focal distance is 48 mm. Like the measurement system in the planing machine at Fraunhofer IPT the illumination is achieved by a red-light LED mounted on the opposite side of the objective (Fig. 8). To protect the entire system a housing with an opening to place the tools in front of the camera is provided.

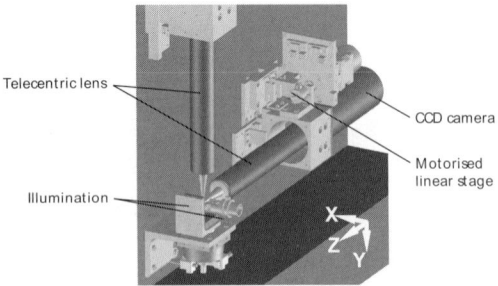

Telecentric lens

CCD camera

Motorised
linear stage

Illumination

Fig. 8. Camera system

To determine the exchange accuracy using a HSK interface a test bench has been designed that allows for an automatic tool exchange. The test bench is equipped with one of the above described cameras and an automatic HSK clamping device. Across from the clamping device a pneumatic cylinder, which is able to move the tool seat back and forth, is mounted to load and unload the tools.

At first the accuracy has been determined using the camera system. The image processing calculates the tool tip position of a v-shaped tool by intersecting the two cutting edges, which are determined by an averaging process. Due to this averaging process the accuracy of the tool position is higher than the resolution of the camera itself. The results of the test series were very satisfactory and the position deviations were below +/- 0.3 µm conducting 100 cycles, while the deviations in axial direction were less (Fig. 9).

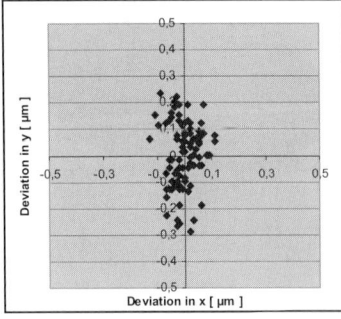

Fig. 9. Accuracy of tool exchange procedure

To verify the camera measurements additional tests have been conducted using capacitive sensors to measure the position deviation. Those measurements have been done in axial and radial direction concerning the HSK tool holder. By means of the sensors it is also possible to break down the deviations in linear and angular failures. The positive results of the first measurements have been confirmed by the new test series. An axial displacement of the tool has been determined by +/- 0.1 µm. Further it has been shown that the deviations in z-direction measured by the camera are consisting of radial deviation of +/- 0.1 µm and an angular failure of +/- 0.02' (angular minute). Further the function and the accuracy of the angular alignment by use of the pin has been tested and determined to +/- 0.004'.

The results are very positive, especially because of the angular accuracy is higher than the angular precision, which is guaranteed by a diamond tool manufacturer.

The camera system is the only way to determine the position of the cutting tool inside the machine and in that way the accuracy and the position of the camera system influence the overall structure accuracy of the workpiece. Hence a reference mark is required to calibrate the camera inside the machine. It has to be position stable and free of thermal drift. The mark has to be connected to the granite of the machine base and is positioned in the field of view of the camera without influencing the measurement.

5. Conclusion

With the innovative tool characterisation and exchange systems for UP-machining systems an analysis of the tool condition can be conducted during a machining break with an accuracy below 0.5 µm. The reproducible automated exchange of a diamond tool in a standard aerostatic fly-cutting spindle could be conducted with a position accuracy of 0.65 µm. The biggest potential for a future optimization could be identified within the mechanical fixation of the components. In slightly changed details the described systems are applicable to any high- and ultra precision machine tool.

Acknowledgements

[A] The work conducted has been funded by the AIF within the public project "Großmikro" – No. 13525 N/1. Special thanks are directed for the support.
[B] The analysis of machine tools and minimum structure resolution has been conducted within the European project "4M" - Multi-Material-Micro-Manufacturing. Special thanks for the support.
[C] The work conducted has been funded by the BMBF within the public project "MICROSTRUCT" – No. 02PW2092. Special thanks are directed for the support also to PTKA-PFT. In addition to Fraunhofer IPT six industrial partners are engaged in the work.

References

[1] De Chiffre, L. et al.: Surfaces in Precision Engineering, Microengineering and Nano-technology, Technical University of Denmark Annals of the CIRP, Vol. 52/2/2003, pp. 1-17
[2] Hesselbach, J. et al.: mikroPRO - Untersuchung zum internationalen Stand der Mikroproduktionstechnik, Springer-VDI-Verlag, 2003
[3] Ikawa, N. et al.: Ultraprecision Metal Cutting – the Past, the Present and the Future, Annals of the CIRP, Vol. 40/2/1991, pp. 587-594

Multi-Material Micro Manufacture
W. Menz, S. Dimov and B. Fillon (Eds.)
© 2006 Elsevier Ltd. All rights reserved

Polymer based multifunctional 3D-packages for microsystems

W. Eberhardt[a], D. Ahrendt[a], U. Keßler[a], D. Warkentin[a], H. Kück[b]

[a] *Hahn-Schickard-Institute for Microassembly Technology HSG-IMAT, Stuttgart, Germany*
[b] *University of Stuttgart, Institute for Micro and Precision Engineering, Germany*

Abstract

Multifunctional 3D-packages can be fabricated with nearly any geometry using injection moulding. 3D line patterning can be performed on injection moulded parts with very flexible laser techniques. Finest line pitches can be realized with subtractive patterning as well as with additive laser direct structuring (LDS). For mounting of SMD and bare dies different assembly techniques can be used.

Keywords: MID, laser patterning, assembly, bare die, SMD

1. Introduction

Polymer based multifunctional 3D-packages made by MID (Moulded Interconnect Devices) technology combine high functionality and miniaturization at low cost. MID not only provide the SMD (Surface Mounted Device) and IC package including the wiring, it also offer the possibility to integrate functional interfaces such as sensors, membranes, fluidic channels, optical windows, light guides, plug connectors etc. [1]. Assembly technologies for bare dies on MID substrates were developed following the general trend in miniaturization. Injection moulding of polymers is suitable for 3D applications made from thermoplastic. Different technologies are available for selective metallization of 3D thermoplastic parts [2]. Laser techniques are well suited for patterning fine lines for assembling SMD and bare die. Furthermore with laser techniques layout changes can be performed easily. With laser direct structuring (LDS) as well as the subtractive patterning process both free-forming surfaces and geometries with high angles can be processed.

In this paper the results of investigations on laser patterning technologies on MID and SMD and bare die assembling on different MID substrates is shown.

2. Laser patterning of multifunctional 3D-packages

2.1. Subtractive laser patterning

The process flow of the subtractive laser patterning process is shown in Fig. 1.

After one shot injection moulding the plastic part is completely electroless plated with copper, nickel and gold. A fine focussed ultraviolet laser beam is used for patterning the metal layer by an ablation process. The debris is removed from the surface by a subsequent cleaning step e.g. using aqueous tenside solution in an ultrasonic bath. Fig. 2 shows a subtractive patterned 3D-MID. Pads on different height levels are interconnected over ramps by conductor lines.

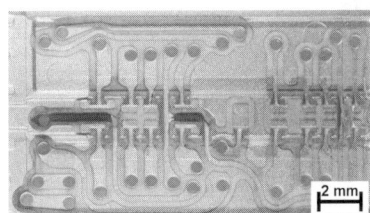

Source: MIDEE, EADS, HSG-IMAT

Fig. 2. Subtractive laser patterned 3D-MID

2.2. Semiadditive laser patterning

A further process including laser ablation is semiadditive laser patterning. After the ablation of a thin metal starting layer the remaining metal is reinforced by an electroless plating process. The process flow is shown in Fig. 3.

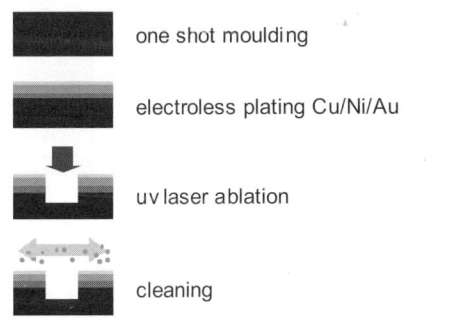

one shot moulding

electroless plating Cu/Ni/Au

uv laser ablation

cleaning

Fig. 1. Subtractive laser patterning of MID substrates

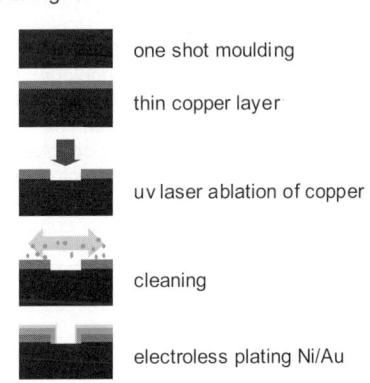

one shot moulding

thin copper layer

uv laser ablation of copper

cleaning

electroless plating Ni/Au

Fig. 3. Semiadditive laser patterning of MID substrates

68

Using the semiadditive process the device is injection moulded with a non catalysed polymer. A thin copper layer can be deposited on the surface by electroless plating or physical vapour deposition as well. An ultraviolet laser beam is used for ablation of the starting layer. After a cleaning step the remaining copper layer is reinforced by electroless nickel and immersion gold.

Compared to the subtractive laser process less processing time is required, because only a thin layer has to be removed. Another advantage is the coating of complete conductor line sides with nickel and gold.

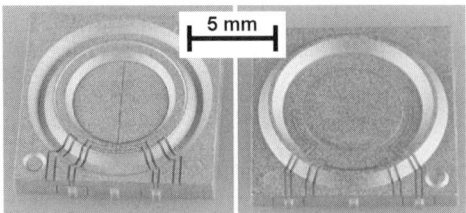

Fig. 4. Capacitive inclination sensor

Fig. 4 shows two halves of a capacitive inclination sensor with a sensor pattern including fine conductor lines produced by the semiadditive process. After assembling of the two halves a complete sensor is obtained [3]. Another MID fabricated by use of the semiadditive process is a Braille display module [4]. The sensor pattern used in this module is shown in Fig. 5.

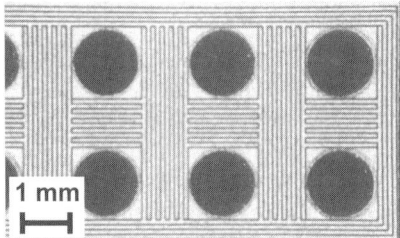

Fig. 5. Sensor pattern on touch sensor

2.3. Laser Direct Structuring (LDS)

Laser direct structuring (LDS) is a full additive process. The plastic part is made by injection moulding of a polymer with special additives. At present polymers based on LCP (Liquid Crystal Polymer), PBT (Polybutylenterephthalat) and PA6/6T (Polyamide) are commercial available for this technology. With an infrared laser beam the polymer surface is activated and roughened. The activated areas can be selectively metallized by electroless plating [5]. Electroless copper is deposited first and subsequently electroless plated with nickel and gold. A good adhesion of the metal layers results from anchorage of the copper to the roughened surface. The process flow is shown in Fig. 6.

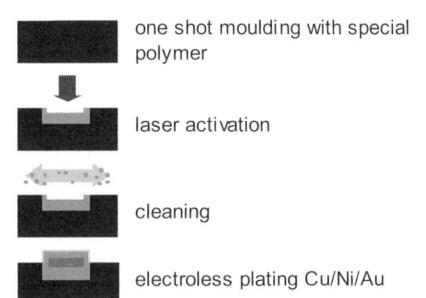

one shot moulding with special polymer

laser activation

cleaning

electroless plating Cu/Ni/Au

Fig. 6. Laser Direct Structuring of MID substrates

The complete sensor based on LDS shown in Fig. 7 is used to measure the partial CO_2 level in human tissue. First tests in animal experiments were successfully.

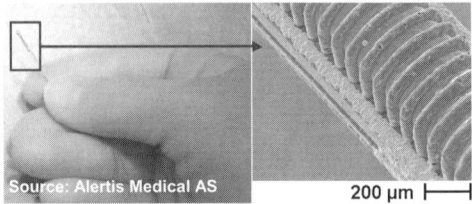

Source: Alertis Medical AS   200 μm ├──┤

Fig. 7. Sensor fabricated in LDS technology

LDS enables also the fabrication of micro vias with the same process [6]. The vias shown in the cross sections (Fig. 8) are made by laser patterning of injection moulded conic shaped holes in a LCP substrate. The side walls show angles of 70° respectively 80° and are laser activated.

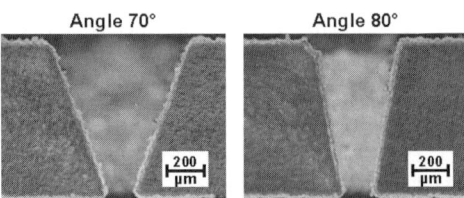

Angle 70°   Angle 80°

Fig. 8. LDS of injection moulded conic shaped holes

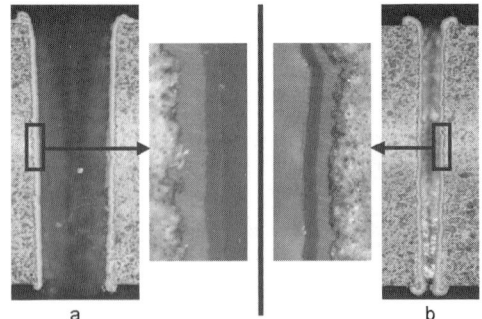

a   b

Fig. 9. Micro vias in LDS

The infrared laser beam can also be used for drilling the holes in the plastic part first and then for activating the conductive pattern. While drilling the holes, the side walls are activated. Vias of different diameters were metallized and the thickness of the metal layer in the middle of the via was measured.

Fig. 9 shows cross sections of two micro vias with a maximum diameter of 240 µm (a) and 50 µm (b) drilled in LCP substrates with a thickness of 800 µm. The copper layer in the via with smaller diameter is thinner but uniformly coated as well. At present further research is done to investigate the reliability of the micro vias under environmental stress like e.g. thermal shock.

3. Assembly of multifunctional 3D-packages

3.1. SMD assembly

SMD assembling on multifunctional 3D-packages can be performed by reflow soldering or by using electrically conductive adhesive. For reflow soldering thermoplastics with high thermal stability have to be chosen. However many thermoplastics with appropriate thermal stability are commercial available.

To investigate the reliability of lead free soldered devices, SMD resistors with size 0603 were mounted on LDS substrates made of PBT/PET Pocan TP710-004. Vapour phase soldering as well as reflow soldering were used. Both processes show comparable shear strength of the SMD directly after assembling and after temperature shock and humidity storage (Fig. 10).

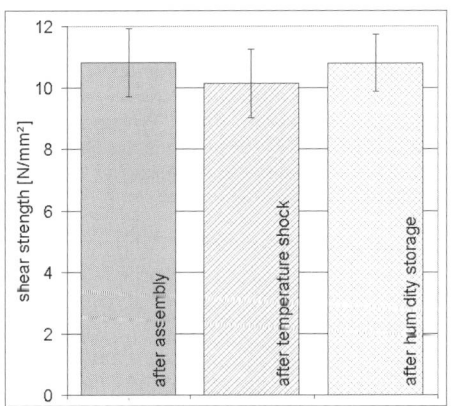

Fig. 10. Shear strength of lead free soldered SMD on LDS substrates

Fig. 11 shows a detail of the housing of a rotation angle sensor system. SMD have been soldered on three different levels [7]. The lead free SnAgCu soldering paste was applied by dispensing. Vapour phase soldering was chosen for smooth and homogeneous heating of the device.

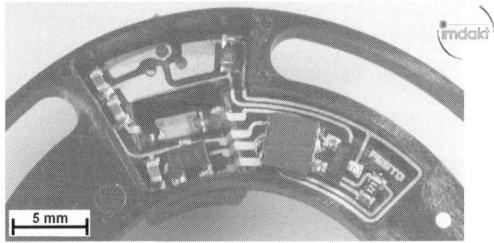

Fig. 11. Rotation angle sensor system in LDS technology with lead free soldered SMD

Assembling of SMD with electrically conductive adhesives perform a much lower thermal stress on the substrate. Therefore a wide range of substrate materials can be used. However humidity is disadvantageous for the properties of the adhesive. Therefore it is important, that both the substrate and the SMD terminals have suitable surfaces, e.g. immersion gold for the substrate and AgPt for the SMD pads, respectively.

3.2. Bare die assembly

For higher miniaturization and functionality not only SMD but also bare dies are mounted on multifunctional 3D-packages. One of the most important issues of bare die assembly on MID regarding reliability is the thermomechanical behaviour of the substrate. Suitable processes for bare die assembling are flip chip and wire bonding.

On MID ultrasonic wire bonding offers major advantage to thermosonic wire bonding due to its capability as a process at ambient temperature. So the thermoplastic with its poor thermal conductivity will not be heated. The surface roughness of electroless plated metal layers highly affects bonding on MID. In general adherent metal layers on thermoplastic surfaces are obtained if an adequate pre-treatment step is performed before plating. By this step the surface is roughened depending on the pre-treatment conditions. Reliable wire bonding results for e.g. LCP substrates are obtained using appropriate process conditions [8].

Fig. 12. Module of a multi axis inertial sensor system

Bare dies and bond wires are coated e.g. with silicone gel to protect the assembly against damage and environmental stress. Fig. 12 shows a module of a multi axis inertial sensor system. The assembly is made by subtractive patterning of an electroless plated LCP substrate and subsequently mounting of several bare dies by ultrasonic wire bonding and coating with silicone gel. Wire bonding on LDS substrates is also feasible if appropriate substrate materials, laser patterning parameters and plating conditions are used.

Flip chip assembly with non conductive adhesive (NCA) is a simple process which provides only low thermal stress [9]. After applying the NCA on the substrate the flip chip is attached. Then the adhesive is instantly cured by thermal snap curing within a few seconds. Especially when chips with low I/O-counts are used, flip chip assembling with serially manufactured Au stud bumps is a relatively cost-effective process. Due to the mechanical properties of the Au studs, they are suited for rough substrate surfaces as well. The Au studs will adjust perfectly to the rough substrate surface (Fig. 13).

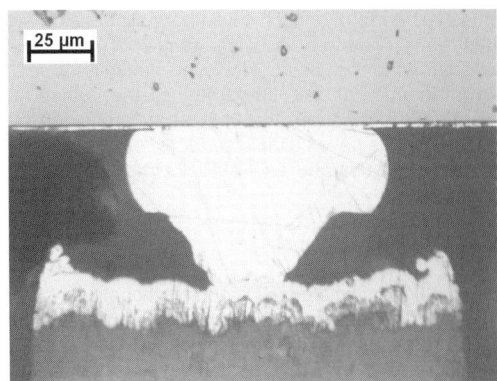

Fig. 13. Flip chip with Au stud bump on subtractive patterned substrate

Flip chip assembling on moulded polymer bumps is a unique technique having the capability to substitute stud bumping. By injection moulding stud bumps with height tolerances lower than 5 µm can be realized. It was found that the maximum bump deformation without bump damaging is sufficiently higher than the maximum height difference within a bump array of ten or less polymer bumps. Fig. 14 shows an LCP substrate with polymer bumps fabricated in LDS technology.

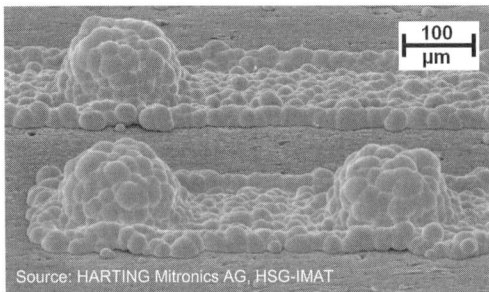

Source: HARTING Mitronics AG, HSG-IMAT

Fig. 14. Moulded polymer bumps in LDS technology electroless plated with Cu/Ni/Au

4. Conclusion

Injection moulding, laser patterning, electroless plating and SMD and bare die assembling techniques are well suited for realization of multifunctional 3D-packages. Thus high functional micro systems can be realized in a cost-effective way.

Acknowledgements

We gratefully acknowledge the support on the research project on micro vias (AiF-Vorhaben-Nr. 14282N: Untersuchungen zu Micro-Vias bei Laser-Feinst-Pitch-MID für die Mikrosystemtechnik) by BMWi by "Arbeitsgemeinschaft industrieller Forschungs-vereinigungen Otto von Guericke e.V." (AiF).

5. Literature

[1] Warkentin, D., Ashauer, M., Briegel, R., Eberhardt, W., Kück, H., Mohr, R. Münch, M., Schilling, P., Scholz, U.: Miniature Flow Sensor Systems and Accelerometers Based on MID, Proceedings 5. International Congress: Molded Interconnect Devices MID '02, 25.-26.9.02 Erlangen.

[2] 3-D MID e.V., 3D-MID Technologie - Räumliche elektronische Baugruppen, Hanser Fachbuchverlag, München, 2004 .

[3] Benz, D., Botzelmann, T., Kück, H., Warkentin, D.: On low cost inclination sensors made from selectively metallized polymer, Sensors and Actuators, A 123 – 124, 2005, p. 18 – 22.

[4] Grotz, U.: MID-Modul mit Touch-Sensorik für grafikfähige Braille-Displays, Workshop: Innovative Anwendungen der MID-Technik, Stuttgart, 5.10.2005.

[5] Schlüter, R., Rösener, B., Kickelhain, J., Naundorf, G.: Completely Additive Laser-Based Process for the Production of 3D MIDs – The LPKF LDS Process, 5th International Congress Molded Interconnect Devices, Erlangen, 2002.

[6] AiF-Vorhaben-Nr. 14282N: Untersuchungen zu Micro-Vias bei Laser-Feinst-Pitch-MID für die Mikrosystemtechnik.

[7] Eberhardt, W., Ahrendt, D., Keßler, U., Kück, H., Blassmann, L., Hanisch, C., Schauz, S., John, L., Heininger, N.: Laserbasierte Herstellung von multi-funktionalen 3D-Packages für innovative Mikro-drehgeber in der Automatisierungs- und Kraftfahr-zeugtechnik, PLUS VTE, 8/2005, S. 1452-1457

[8] Scholz, U.: Untersuchungen zur Eignung des Ultraschalldrahtbondens für die Chip-Montage auf MID-Substraten, Shaker Verlag, Aachen, 2004

[9] Keßler, U., Kück, H., Eberhardt, W.: Adhesive Technology for Flipchip Assembly on Moulded Interconnect Devices (MID); Proceedings of the Sixth IEEE CPMT Conference on High Density Microsystem Design and Packaging and Failure Analysis (HDP'04), Shanghai, June 30 – July 03 2004

Multi-Material Micro Manufacture
W. Menz, S. Dimov and B. Fillon (Eds.)

71

Laser transmission welding of micro plastics parts

E. Haberstroh, W.-M. Hoffmann

Institute of Plastics Processing (IKV), RWTH Aachen University, Aachen, Germany

Abstract

Only few welding processes for plastics meet the special demands of microtechnology. Laser transmission welding has distinctive advantages like low mechanical and thermal load of the joining parts. Thus the laser is a joining tool which is particularly suitable for the welding of micro plastics parts. Contour welding is a process variant of laser transmission welding enabling the welding of complex and even three-dimensional weld contours. In addition to that, this is a very flexible process which can be easily adapted to changing part geometries. So far it is only applied for welding plastics parts of macroscopic scale in the industrial practice. Recent research at the Institute of Plastics Processing in Aachen shows that it is also possible to use this process to weld filigrane micro parts. Thus another promising welding process has proven its suitability for the microtechnology. Welding trials with different thermoplastics show, however, that there are big differences in the material behaviour that make some platics more suitable than others for the contour welding in microtechnology.

Keywords: laser transmission welding, contour welding, tear-out force, weld seam morphology

1. Introduction

Micro systems technology is a strongly growing industrial sector in which plastics are applied more and more [1]. Apart from low material costs the huge design possibilities of plastics parts have to be pointed out. By means of micro injection moulding and micro hot embossing, micro parts can be produced in high numbers and short cycle times.

However, existing micro systems often consist of several components which have to be joined together to a working micro system. In order to realize welded joints for parts of macroscopic scale there are several welding processes available which use different mechanisms of energy input. However, the micro technology has special requirements to a suitable welding process. Therefore, only few welding processes can be applied for the welding of micro plastics parts.

2. Requirements to a welding process in micro technology

A welding process for plastics which is suitable for the application in micro technology has to meet the following demands [2]:

- precisely controllable energy input
- low mechanical load of the parts
- low thermal load of the parts
- one step process
- small flash and no abrasion
- high positioning accuracy during welding

Only the laser and ultrasonic welding are suitable for welding mirco parts. Laser radiation can be well focused, so that the energy can be inserted locally and exactly metered into the welding area. Furthermore, it is a contactless process. This means that the mechanical load of the plastics parts is very low. Additionally, there is no abrasion of the material during the welding process and flash can be mostly avoided.

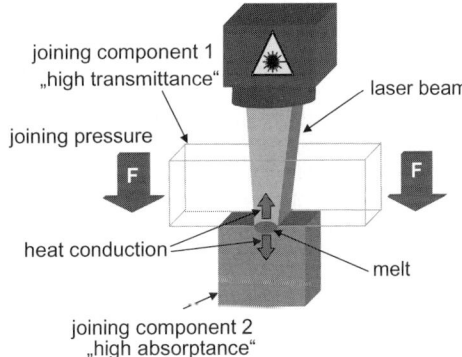

joining component 1 „high transmittance"

laser beam

joining pressure

F   F

heat conduction

melt

joining component 2 „high absorptance"

Fig. 1: Laser transmission welding

3. Laser transmission welding

Of the existing variants of laser welding only the laser transmission welding could prove itself applicable in the industrial practice.

It is a one step process, in which the heating of the plastics and the joining take place simultanously. One of the joining parts must have a high transmittance in the range of the laser wavelength and the other must show a high absorptance. Before welding both parts are positioned in the preferred final location to each other and the joining pressure is applied [2,3].

The process sequence is shown schematically in Figure 1. The transparent joining component is transmitted by the laser beam without any considerable heating. Only in the second component the laser beam is absorbed entirely in a thin near-surface layer, whereby the laser energy is transformed into heat and the plastic is melted. By heat conduction the transparent joining part becomes plastified, too [4]. A small amount of melt flow and interdiffusion processes enable the build up of intermolecular forces which sum up to the adhesive force. This happens under the influence of the outer as well as the inner joining

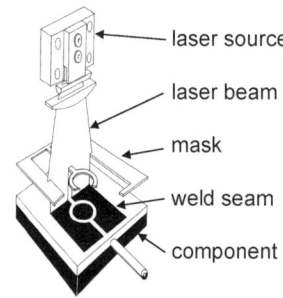

Fig. 2: Mask technology [Leister]

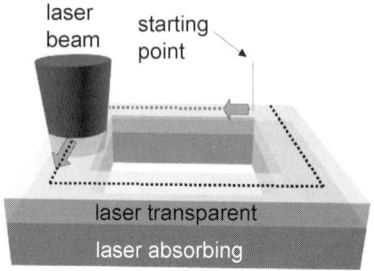

Fig. 3: Contour welding

pressure. The latter results from the expansion of the melt.

Laser sources which are usually applied on this process are solid state lasers (Nd:YAG, λ = 1064 nm) and high power diode lasers (λ = 800 – 1100 nm) [2,4].

Out of the different existing process variants, both the mask technology and the contour welding are being investigated at the Institute of Plastics Processing in Aachen.

Mask technology means that a mask is placed between the joining parts and the laser source [5,6]. This mask only transmits the laser radition in the regions where a joining should happen. The mask can consist of a laser cut steel sheet or a lithographically coated glass plate. The laser, which has a line-shaped focus, proceeds the mask and images the mask structured onto the sample parts (see Fig. 2). The dimension and precision of the weld are determined primarily by the mask and the quallity of the laser beam. Small weld seams with a width of less than 100 µm can be produced [7]. Additionally, complex shaped welds can be realised with a simple laser movement. Curved and straight lines can be welded as well as lines with varying width or even flat shaped weld seam geometries.

Contour welding means that the weld seam is irradiated sequentially by a focused laser beam and is melted locally (see Fig. 3). As a result of the geometric conditions, the melt volume remains small, so that the flash can be reduced and avoided, respectively. The advantages of this process variant are the large flexibility and the possibililty to weld complex three-dimensional weld geometries [8]. Since there is almost no relative movement of the two joining components towards each other, the gap between both parts should not exceed 50-100 µm [9]. Therefore, already during the production of the joining parts narrow tolerances have to be assured. The process time increases linearly with the length of the weld.

3. Results

In this paper results of recent research concerning the contour welding of micro parts are presented.

Figure 4 presents schematically the complex three-dimensional micro sample part which is welded by laser transmisson welding. As mentioned above the contour welding has the advantage that it is flexible concerning the part geometry. This means, three-dimensional weld lines can be realized. In order to show that this advantage is also suitable for the welding of micro parts, besides a weld line width of the catwalk-shaped structure of 300 µm this sample part has a step of 500 µm in the welding plane. Two of these parts are welded so that the surfaces of the catwalks are in close contact. For the accurate positioning of both parts to each other positioning structures are implemented.

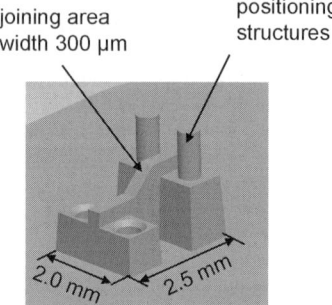

Figure 4: Complex micro part welded with contour welding

Both a semicrystalline (polyoxymethylen) and a amorphous (polycarbonate) thermoplastic material are welded. The content of carbon black in the absorbing component is 0.5 weight-%. The applied laser source is a high-power diode laser with a wavelength of λ = 940 nm a focus diameter of d_F = 370 µm and d_F = 230 µm respectively. The laser power and the velocity of the welding movement are the parameters to be varied.

For each material the range of welding parameters is examined in which welding is possible. The laser power P_L is varied in three and the velocity v_L of the laser source scanning the weld seam in five steps, which yields the experimental design shown in Table 1 exemplarily for the polycarbonate (PC).

Table 1
Experimental design for PC

line energy E [J/mm]		velocity v_L [mm/s]				
		15	20	25	30	35
power P_L [W]	25	1,66	1,25	1,00	0,83	0,71
	30	2,00	1,50	1,20	1,00	0,86
	35	2,33	1,75	1,40	1,16	1,00

The line energy E can be used to characterize the welding process. According to Eq. 1 the line energy E depends on laser power and velocity.

$$E = P_L / v_L \qquad (1)$$

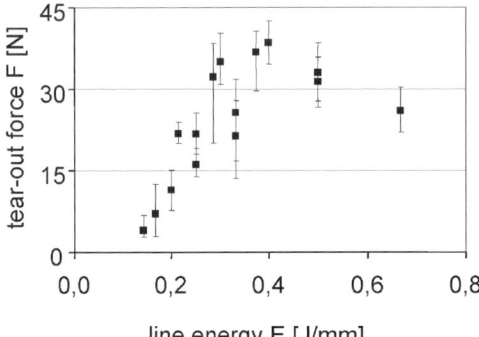

Fig. 5: Tear-out forces vs. line energy
(material: PC; d_F = 370 µm)

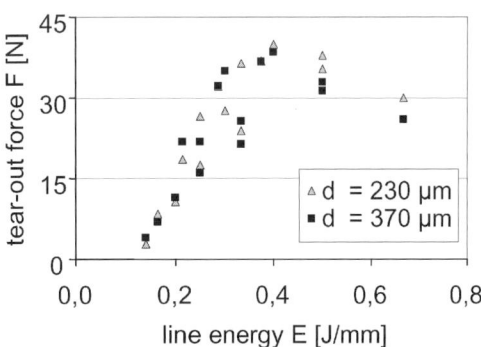

Fig. 7: Tear-out forces vs. line energy for
different laser focus diameters (material: PC)

In general, the tensile strength of the joint is taken to evaluate the quality of the weld. In this case, the weld has a complex threedimensional geometry so that the stress conditions are very complex. Therefore, the real uniaxial tensile strength cannot be determined. Instead, the tear-out forces are measured by tensile tests. Thereby, the tear-out force is the force at which the joint breaks under load.On the other hand the weld morphology is examined in detail by microscopic analysis.

3.1 Polycarbonate (PC)

Figure 5 shows the resulting tear out forces for PC achieved with a laser focus of d_F = 370 µm. The course of the tear-out force over the line energy can be observed which is characteristic for the laser transmission welding of thermoplastics. If the line energy is too low, the created amount of melt is too small to assure a tight connection between the joining parts. With increasing energy the tear-out forces increase as well, but only to a certain extent. High line energies lead to a reduction of the weld seam strength, because the plastics material disintegrates due to thermal overload. By this, products of this decompostion like pores and voids decrease the possible tear-out force.

The analysis of the weld seam morphology shows an increasing flash at a high line energy, which is correlated to a increased formation of melt (see Fig. 6). The more melt is created, the better the joint strength between the components and accordingly the higher is the tear-out force. However the flash deteriorates the

appearance of the weld, so a compromise between weld strength and flash size has to be accepted.

A flash is formed when the welding zone is entirely plastified and the melt is pressed out of the weld into the flash by the joining pressure. Thus, a relative movement of the welding components can be observed. Normally, a flash is no problem regarding contour welding, since the weld zone is melted only locally next to the laser beam. Most of the weld zone remains solid and prevents a joining movement and the formation of a flash, respectively. Due to the small dimensions of this micro part a relatively big amount of melt is generated related to the part size. Thus, the plastified area is large enough to enable a joining movement which results in a flash.

The welding tests with a laser focus d_F = 230 µm yield comparable results. Initially, the tear-out forces increase and drop to high energy input. Although the focus diameter is smaller, i.e. although the laser power is distributed among a smaller area, which means the energy density is higher, no differences of the tear-out forces are observed. Solely, towards higher energies a smaller laser focus seems to yield better tensile strengths (see Fig. 7).

Regarding the weld seam morphology the same dependencies on the energy input can be observed, i.e. the higher the line energy the more melt is created and thus the larger is the flash. A distinctive difference between both laser focusses is not clearly obvious. However it can be stated that the use of a smaller focus diameter tends to result in smaller flashs (see Fig. 8).

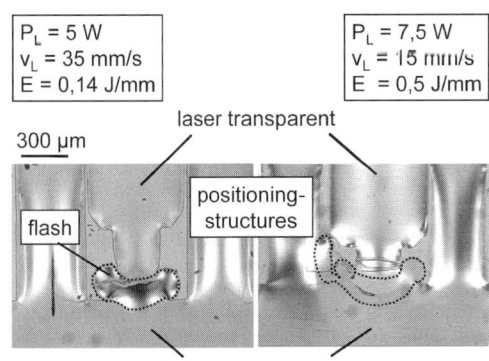

Fig. 6: Weld seam morphologies at different line energies (material: PC; d_F = 370 µm)

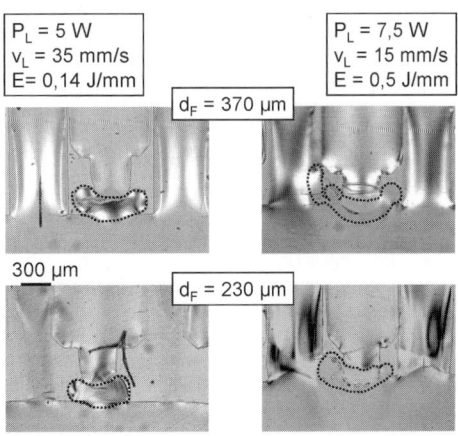

Fig. 8: Weld seam morphologies at different line energies E and laser focus diameters d_F (material: PC)

74

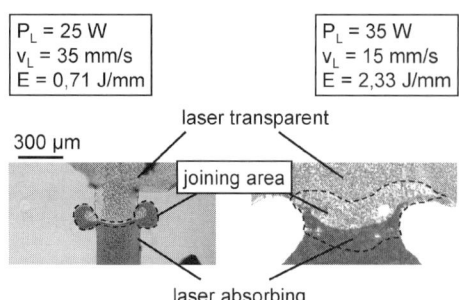

$$P_L = 25\ W$$
$$v_L = 35\ mm/s$$
$$E = 0.71\ J/mm$$

$$P_L = 35\ W$$
$$v_L = 15\ mm/s$$
$$E = 2.33\ J/mm$$

300 µm

laser transparent

joining area

laser absorbing

Fig. 9: Weld seam morphologies at different line energies (material: POM; d_F = 370 µm)

3.2 Polyoxymethylen (POM)

The welding of Polyoxymethylen (POM) has shown itself to be more difficult. On the one hand, the transmittance of semicrystalline thermoplastics is smaller compared to amorphous plastics. So higher laser power is needed to supply the joining area with enough energy to create sufficient melt. Besides, the laser beam is widened by effects like scattering at the spherulites of the semicrystalline structure. Thus, the energy density is reduced.

Due to the beam widening the focus diameter in the joining interface exceeds by far the width of the weld structure. For the welding of tiny, filigrane structure semi-crystalline thermoplastics are not suitable.

At high line energies so much material is melted due to the hihgly increased focus diameter that the weld seam width is noticeably larger than the original 300 µm. The catwalk structure is entirely melted, so that the joining parts are joined on a large area. The energy input is too high und the microstructure is destroyed (see Fig. 9).

4. Conclusion and prospect

It can be concluded that the contour welding is a capable process for the welding of complex micro parts. However, semicrystalline thermoplastics are not that suitable for the welding of filigrane micro parts due to their scattering effect on laser radiation. The use of alternative laser sources like Nd:YAG-lasers, which can be focused more strongly and have a better laser beam quality may yield better results. Furthermore, concepts to minimize the flash should be developed and implemented. One possibility could be the limitation of the relative joining movement and by this a reduction of the melt flow into the flash.

5. Comparison to classical approaches

Besides welding, also glueing and hot stamping can be used to join plastics microparts.

Hot stamping is only suitable for joining thin foils with solid substrates. For example, microfluidic structures in a flat substrate can be sealed with a thin film [10]. The advantages are the low costs for the hot stamping equipment and the fact that there are no restrictions concerning the optical properties of the joining components. Regarding the laser welding process, an absorbing additive has to be incorporated in one of the joining parts in order to yield the required absorptance. This is not necessary for hot stamping. However, the heat input cannot be controlled exactly, so temperature-sensitive components might be damaged. Moreover, solid complex parts like the micro geometry discussed in this paper cannot be joined with hot stamping.

Glueing is also an assembly method for micro parts [10]. The advantage is the possibility to join different materials which cannot be welded and - like hot stamping - the independence from the optical properties of the material. However, a further process step is necessary in order to apply the glue. Additionally, it takes some time for the curing process. Besides, for certain applications glue might not be acceptable for example due to medical reasons.

Acknowledgements

The authors gratefully acknowledge the financial support of the Deutsche Forschungsgemeinschaft (DFG) within the Collaborative Research Centre SFB 440 "Assembly of Hybrid Microsystems".

References

[1] Gächter, P. An den Grenzen des Machbaren mit herkömmlicher Spritzgießtechnik. Fachtagung Mikrospritzgießen – Mikroformenbau, Würzburg: Süddeutsches Kunststoff Zentrum, May 2000

[2] Klein, H. Laserschweißen von Kunststoffen in der Mikrotechnik. RWTH Aachen, dissertation, 2001

[3] Rotheiser, J. Joining of Plastics: handbook for designers and engineers . Munich: Carl Hanser Verlag, 2004, ISBN 3-446-22454-8

[4] Schulz, J. Werkstoff-, Prozess- und Bauteiluntersuchungen zum Laserdurchstrahl-schweißen von Kunststoffen. RWTH Aachen, dissertation, 2002

[5] Chen, J.-W., Zybko, J. Laser Assembly Technology for Planar Microfluidic Devices. Society of Plastics Engineers (SPE), Proceedings of Annual Technical Conference (ANTEC), San Francisco, USA, 2002

[6] Leister, C. Modulare Diodenlasersysteme zum Fügen von empfindlichen Bauteilen in der Elektronik und Mikrotechnik. Aachener Kolloquium für Lasertechnik, Aachen, 2000, pp. 303-313

[7] Chen, J.-W. Mit der Maske in die Mikrowelt. TAE aktuell (2000) 12, pp. 2-4

[8] Haberstroh, E., Lützeler, R. 3D Laser Transmission Welding. Society of Plastics Engineers (SPE), Proceedings of Annual Technical Conference (ANTEC), Nashville (TN), USA, 2003

[9] Russek, U.A. Innovative Trends in Laser Beam Welding of Thermoplastics. Proceedings of the 2nd International WLT-Conference on Lasers in Manufacturing 2003, Munich, Germany, 2003

[10] Eberhardt, W., Kück, H., Münch, M., Sandmaier, H., Spritzendorfer, M., Steger, R., Willmann, M., Zengerle, R. Low Cost Fabrication Technology for Microfluidic Devices Based on Micro Injection Moulding. Proc. Micro.tec 2003, Munich, Germany, 2003, pp. 129-134

Multi-Material Micro Manufacture
W. Menz, S. Dimov and B. Fillon (Eds.)

Prototyping of multilayer waveguides with V-grooves in COC/Topas®

F. Bundgaard[a], J. Ducrée[b], R. Zengerle[c], O. Geschke[a]

[a] MIC – Department of Micro and Nanotechnology, Technical University of Denmark,
DTU Building 345 East, DK-2800 Kgs. Lyngby, Denmark
[b] HSG-IMIT, Institute for Micromachining and Information Technology, Villingen-Schwenningen, Germany
[c] IMTEK – University of Freiburg, Laboratory for MEMS Applications,
Georges-Köhler-Allee 106, 79110 Freiburg, Germany

Abstract

A number of fast and low-cost methods are used for prototyping of optical waveguides in the cyclic olefin copolymer Topas®. This polymer has a number of advantages compared to polymers traditionally used for micro systems. The advantages include good chemical resistance, good optical properties, and low water absorption, making it well-suited for lab-on-a-chip systems. Using micro milling, spin coating, and thermal bonding, and exploiting the different refractive indices of the different grades of Topas®, waveguides with a width and heigth of 130 μm are created. 45° V-grooves at both ends of the waveguides are used for coupling light in and out of the waveguides, using the principle of total internal reflection. In this manner light can be coupled into the waveguides perpendicularly from above or below the structure, depending on the groove orientation. This coupling scheme has a large potential for rotating microfluidic system, the so-called lab-on-a-disk systems.

Keywords: COC/Topas®, micro milling, waveguides, prototyping, V-grooves

1. Introduction

In the chain from concept to finished product, rapid and cost-effective ways of prototyping are important. In recent years a number of new methods for prototyping of polymer microsystems have emerged. Many of the methods are relatively low-cost compared to traditional cleanroom processing methods such as photo lithography. Typically, the desktop equipment used for the experiments described below are in the 10,000-20,000 € price range, while typical cleanroom equipment is an order of magnitude more expensive. Also, the desktop methods are often more flexible, since design changes can be applied more rapidly than when using lithographic methods.

The prototyping methods for direct structuring of polymer microsystems include laser ablation, using for instance a CO_2 laser micro, and micro milling, using traditional desktop milling machines. The machines are, however, used with tools down to 5 μm in diameter. The prototyping methods are also applicable for creation of moulds for use in injection moulding or hot embossing processes [1, 2]. Either the moulds can be machined directly, or a master can be made, from which a negative is then casted in, e.g., an elastomer like polydimethylsiloxane (PDMS) or in some epoxy resin. This mould is then used to manufacture the final structures [3].

Bonding of polymer microfluidic systems, used to effectively seal the channels, is another important issue. Traditional gluing can be difficult when working with structure sizes below 100 μm. Thermal bonding is a useful way of sealing large areas, but care must be taken, since too large a pressure or too high temperatures can deform the structures. Also, the strength of the bonding depends very much on the surface quality of the parts to be joined. Protrusions on the interface will prevent lower lying parts of the interface to bond properly, thus yielding a lower bonding strength, and possibly causing leaks in the system.

Laser bonding, or laser welding, is used when only selected areas on a surface are to be bonded. An absorbent layer is required on the interface between the parts. This layer then absorbs the energy from an infrared laser beam moved over the surface, thus melting the surrounding material and bonding the parts together. The fact that the interface is only heated if absorbing material is present can be used to avoid clogging of microfluidic channels. The absorbent material is removed by the milling process, leaving the channels open after the bonding [4].

As mentioned, all of the processes above are available using relatively low-cost desktop equipment, and can be carried out in short time (typically minutes), thus providing an efficient and versatile toolbox for micro system manufacturing.

1.1. Waveguides in COC/Topas®

In this paper, we present different prototyping methods mentioned above used for manufacturing optical waveguides in the polymer Topas®. Joining two different grades of Topas® with slightly different refractive indices, light can propagate in the milled waveguides. Instead of coupling the light in and out of the structure using fibres lying in the plane of the structures, so-called V-grooves are used. Exploiting total internal reflection above a critical angle, light can be sent perpendicularly into the structure, and reflected into the waveguide. This coupling scheme is less sensitive to alignment and positioning than butt-coupled fibre-waveguide interconnections, and can be used, e.g., with rotating platforms, so-called lab-on-a-disk systems. The V-groove principle has previously been used for glucose measurements on this platform [5].

76

2. Cyclic Olefin Copolymer Topas®

Traditionally, polymer-based microfluidic systems and lab-on-a-chip (LOC) systems have been manufactured using standard polymer materials like poly(methylmethacrylate) (PMMA), polycarbonate (PC) or PDMS. The relatively new polymer type Topas® [6] has a number of interesting features making it well-suited for microfluidic applications.

Topas® is a copolymer of the two monomers norbonene and ethylene. The glass transition temperature T_g can be varied from 65°C to 170°C by altering the ratio between the two monomers.

The refractive index is around 1.53, and changes slightly with T_g, so that a low glass transition temperature yields a higher refractive index [7]. Generally, the optical properties of Topas® are superior to those of the above mentioned polymers. For instance, the transparency of Topas® to ultraviolet light extends below 250 nm.

Topas® is chemically resistant to acids, hydrolysis, acetone, ethanol, and polar solvents, and is resistant to standard photolithographic processes [8]. It is soluble in non-polar solvents like toluene and heptane, a fact which is exploited when spin coating Topas® solutions to obtain thin layers.

Topas® has a low water absorption of 0.01%, which is 30 times lower than that of PMMA. This makes Topas® well-suited for microfluidic systems, since swelling is limited, and so-called memory effects, where the system can interfere with the concentration of analytes etc., largely are avoided.

Also, the chemical resistance of Topas® makes it well-suited for harsh environment or long term measurements, such as waste water analysis.

3. Waveguides and V-grooves working principle

The function of both the waveguides and the V-grooves relies on the principle of total internal reflection. As shown in Fig. 1, the light is coupled vertically into the structure. When hitting one of the sides of the V-groove at an angle higher than the critical angle θ_c, the light is reflected at the boundary, rather than being refracted out of the structure. The critical angle θ_c is given by

$$\sin \theta_c = n_2/n_1 \qquad (1)$$

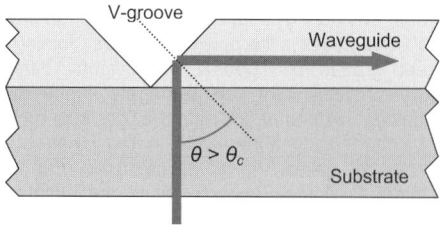

Fig. 1. Principle of total internal reflection, seen from the side. The light beam is not refracted out into the air, but is totally reflected into the waveguide layer, when hitting the right interface of the V-groove at an angle θ above the critical angle θ_c. For the Topas®/air interface, the angle is around 41°.

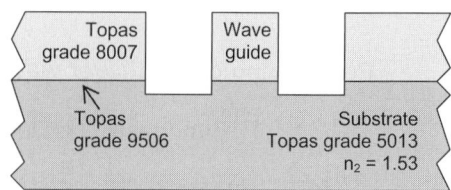

Fig. 2. Cross sectional view of the waveguides. Two parallel grooves are milled, creating a waveguide in between. The grooves are slightly deeper than the grade 8007 core layer to avoid light leakage. The extruded top layer is thermally bonded to the substrate using a thin layer of spin coated low-T_g Topas® grade 9506.

where n_1 is the refractive index of the Topas®, and n_2 is the refractive index of air. Hence, for an air/Topas® interface the critical angle is 41°. The angle between the two sides of the V-groove is set to 90°, since this is a standard angle for engraving tools and mills, but the V-groove could be altered by using engraving tools with other angles, in order to accomodate other refractive indices and geometries. Also, strictly speaking, only one side of the V-groove is needed to couple light into adjacent waveguides, but the symmetry is given by the engraver tool.

The difference in refractive index between the two grades of Topas® used for the substrate and core layers, is quite small, in the order of 0.005. Therefore the critical angle is 85.4° in the bottom of the waveguide on the interface with the substrate. The three other sides of the waveguide are surrounded by air, meaning that the critical angle is 41°. In the cross sectional view in Fig. 2 it is shown how the sides of the waveguide are milled deeper than the thickness of the foil layer. This is done to ensure that no light 'leaks' out at the bottom of the waveguides. Also, thickness variations caused by the bonding are taken into account in this manner.

4. Manufacturing processes

Injection moulded Topas® grade 5013 with a T_g of 135°C was used as substrate. A 130 µm thick extruded Topas® grade 8007 film with a T_g of 75°C was used as core layer.

4.1. Bonding

In order to thermally bond the extruded film, acting as core layer, to the substrate, several processes have been investigated. First, simple thermal bonding was attempted. Using a roll lamination device the two layers were pressed together at a temperature of 140°C and a pressure of 5 Bar. The bonding was, however, very weak. A stronger bonding was obtained by treating both substrate and the extruded foil in an oxygen plasma, thus changing the surface energy of the parts. The bonding was clearly stronger, but the foil would still peel of the substrate in simple pull-experiments. Finally, it was decided to add an intermediate layer of Topas grade 9506 with a T_g of 65°C. This layer acts as glue between the two other layers due to its low glass transition temperature, and has roughly the same refractive index as the grade 8007, thereby not spoiling the optical properties of the sandwich structure.

Pellets of grade 9506 were dissolved in toluene using ultrasonic aggregation and heat to accelerate the process, producing a 3 weight% solution. This was spin coated onto the substrate, which was subsequently annealed for several hours at a temperature slightly below the glass transition temperature of the substrate, in order to remove the toluene.

The roll lamination was then repeated at the same temperature and pressure as previously used. Including the grade 9506 intermediate layer, a bonding strength comparable to that of the material itself was obtained, meaning that the extruded film would tear before splitting of the interface between the substrate and the extruded film. Also, the long term behavior of the bonding seems to be good – after several months, no degradation of the bonding has been observed.

4.2. Milling

The waveguides were designed using standard CAD/CAM software, and were milled using a high precision desktop CNC milling machine (Mini-Mill/3 Pro from Minitech, Norcross, Georgia, USA) with a resolution of around 1 µm. The milling machine is equipped with a brushless, electronic spindle (Astro-E500Z from NSK, Schaumburg, Illinois, USA) with a run-out of less than 1 µm and a maximum rotational speed of 50,000 rpm.

The most significant parameters in obtaining high quality milled surfaces are the tool and spindle used. The quality and sharpness of the tool is vital for achieving smooth surfaces. Also the run-out, or deviation from perfect axially symmetric rotation, of the spindle can become important, if too large. In normal, macro scale milling the run-out is negligible compared to the diameter of the tool, but for tools with a diameter much below a millimetre the run-out can become comparable to the diameter of the tool itself. This leads to geometrical changes in the structure (e.g. making channels wider than intended), and to an increased stress and wear on the tool. Also, the quality of the milled surface decreases. Different tool sizes between 200 µm and 600 µm were used in order to find the best compromise between miniaturisation and structural

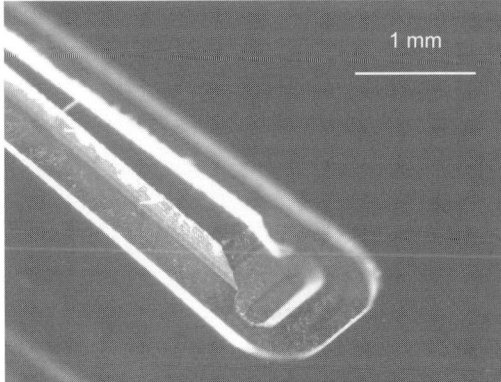

1 mm

Fig. 3. The end of the waveguide with half a V-groove; the other half has been removed by the other milling steps. The 45° angle of the V-groove is machined using an engraver tool, while the rest of the waveguide is machined using a cylindrical tool. Light coupled in perpendicularly from below will be reflected in the waveguide. To enlarge the V-groove area which reflects the incoming beam, the waveguide is tapered at the ends. The surface roughness Ra of the V-grooves has been measured to 650 nm.

stability and cutting speed of the tool. Since 130 µm wide waveguides theoretically can be milled with tools several centimetres in diameter, the advantage of using a small tool is the possibility of diminishing the distance between adjacent structures. Also, using a sufficiently small tool, microfluidic channels in the same structure can be milled in one pass, using the same tool.

The optimal tool diameter is also an important parameter if the milled structure is used as master for embossing. A negative is casted from the master, and the milled grooves become protrusions with the same dimensions. The width of the negative structure's protrusions ultimately determines the limits of the pressure which is applied, for a given force, and thereby the quality of the embossing.

For making 130 µm wide and deep waveguides, the best compromise between surface quality, tool durability and structure miniaturisation was to use a tool with a diameter of 400 µm. Typical process parameters are a rotational speed of 40,000 rpm and a feed speed of 50 mm/min. No clear connection has been found between the surface roughness and the processing parameters, but generally the average surface roughness Ra for the channels milled with micro end mills lies from 50 to 450 nm. Ra of the V-groove surfaces lies somewhat higher, around 650 nm.

The low glass transition temperature of Topas® grade 8007 makes the material soft when energy is deposited during machining. This leads to the creation of burrs on the edge of the milled structures, which are difficult to remove due to their size and softness. Since the burrs prevalently occur on one side of the milled path, the burrs can be removed by repeating the path with a slight offset and in reverse direction. Ultrasonic treatment does also have a certain effect, especially removing smaller burrs.

The waveguides have been designed with tapered ends, so that they broaden from 130 µm width in the middle to 500 µm at the ends. In this manner a larger area is available for coupling light into the waveguide. A tapered end is seen in Fig. 3.

Previous experiments [9] have shown a propagation loss in this type of waveguides, without tapered and V-grooved ends, below 1 dB/cm in the visible range. This value is expected to be similar for the waveguides presented here. The total loss of this type of waveguides, however, will presumably be lower, since the coupling loss will diminish compared with traditional planar butt-coupling of optical fibres and waveguides.

5. Summary and discussion

We have shown a number of different rapid and low-cost prototyping methods used for manufacturing a micro milled waveguide with tapered ends and V-grooves for coupling in and out light. The prototyping methods include micro milling, roll lamination, thermal bonding, and spin coating.

Compared with traditional cleanroom processing of waveguides, the process shown in this paper could be improved in one important field. The surface roughness Ra of the milled surfaces is still, despite optimisation, not entirely adequate for optical applications such as waveguiding. To improve the uniformity of the surface, subsequent treatments should be performed. These treatments can be chemical, such as using a solvent to smoothen and 'smother' the sub-micrometer surface features. Physical treatments include polishing using abrasive particles, plasma treatment or thermal treatment. The latter should be carried out at very high temperature and for short time, thus preserving the large-scale structures, like the waveguides, while smoothing the milled surfaces. Preliminary experiments using oxygen plasma and CO_2 laser annealing have, however, not significantly improved the surface quality.

Acknowledgements

The authors would like to thank the Danish Ministry of Science, Technology and Innovation (VTU, Centre Contract µKAP) for financial support. FB would like thank IMTEK for hospitality during his stay in Freiburg.

References

[1] Jensen M.F. *et al.* Rapid prototyping of polymer microsystems via excimer laser ablation on polymeric moulds. Lab-on-a-chip **4** (2004) pp. 391-395.

[2] Dahms, S., Bundgaard, F., and Geschke, O. A new approach in polymer waveguide fabrication. Proceedings of the 4M2005, 29 June – 1 July 2005, Karlsruhe, Germany, pp.99-101.

[3] Reinecke, H. et al. Hot Embossing for the Rapid Prototyping of Microstructured Polymers. Proceedings of the 4M2005, 29 June – 1 July 2005, Karlsruhe, Germany, pp.3-8.

[4] Bundgaard, F., Perozziello, G., and Geschke, O. Rapid prototyping of all-COC/Topas® fluidic microsystems. Proceedings of the 4M2005, 29 June – 1 July 2005, Karlsruhe, Germany, pp.405-407.

[5] Grumann, M. *et al.* Optical beam guidance in monolithic polymer chips for miniaturized colorimetric assays. Proceedings of the 18th IEEE International Conference on Micro Electro Mechanical Systems, MEMS 2005 Miami, pp. 108-111.

[6] Topas Advanced Polymers GmbH, *www.topas.com*

[7] Khanarian G. Opt. Eng. **40**(6) (2001) pp. 1024-1029.

[8] Nielsen T. *et al.* Nanoimprint lithography in Topas, a highly UV-transparent and chemically resistant thermopast. J. Vac. Sci. Technol. B **22** (2004) pp. 1770-1775.

[9] Bundgaard, F., Perozziello, G., and Geschke, O. Rapid prototyping methods for all-COC/Topas® waveguides and microfluidic systems. Proceedings of the µTAS, October 9-13, 2005, Boston, Massachusetts, pp. 1200-1202.

Multi-Material Micro Manufacture
W. Menz, S. Dimov and B. Fillon (Eds.)

An integrated solenoid-type inductor with Fe-based soft magnetic core

Chong Lei[a], Yong Zhou[a], Wen Ding[a], Xiao-Yu Gao[a], Ying Cao[a], Hyung Choi[b],Jonghwa Won[b]

[a] National Key Laboratory of Nano/Micro Fabrication Technology, Key Laboratory for Thin Film and Microfabrication of Ministry of Education, Institute of Micro and Nano Science & Technology, Shanghai Jiao Tong University, Shanghai 200030, China (corresponding author: yzhou@sjtu.edu.cn)
[b] Samsung Advanced Institute of Technology (SAIT), Samsung Electronics Co., Ltd., 416 Maetan-Dong, Yeongtong-Gu, Suwon, Kyungki-Do 442-742, Korea

Abstract

This paper reports on a technological process that combines copper as conductor, soft FeCuNbCrSiB magnetic thin film prepared by magnetron sputtering as magnetic core and polyimide as the insulation material to complete a solenoid-type inductor with high inductance and high quality factor. The shape of the magnetic core scheme is rectangular, and the size of the inductor is 4 mm × 4 mm × 0.084 mm. The results show that the fabricated solenoid-type inductor has high inductance and high quality factor in the frequency range of 1-20 MHz, the inductance is 1.19 µH at a frequency of 1 MHz and the maximum quality factor is around 4.5 with an inductance value of 1.1 µH at a frequency of 12 MHz.

Keywords: high inductance, high quality factor, Fe-based magnetic thin film, solenoid-type inductor

1. Introduction

Recently there has been a great demand on micro DC/DC converter for the applications of the portable electronic products such as mobile communication product CDMA, notebook computer, microprocessor, digital camcorder and so on [1]-[5]. Magnetic components such as inductors and transformers are the essential components in constructing the micro DC/DC converter. But there are several key factors that must be solved in order to produce an inductor with adequate inductance value (1 µH or more), high quality factor (Q-factor) (at least greater than unity), small area and low temperature fabrication processes [2,6,7,8]. However, the fabrication of the above-mentioned inductor is an extremely difficult task at present, and the size reduction of the magnetic inductor has been much slower than that of the other passive devices such as resistors and capacitors. This is due to the conventional semiconductor technology has great limitations in realization of the integrated 2D or 3D electromagnetic devices with high performance and very small size. In recent years, with the development of MEMS (Microelectromechanical Syetems), the non-silicon fabrication technology of UV-LIGA (UV-Lithografie, Galvanoformung, Abformung) [9] has especially become one of the most advanced technologies to fabricate the 3D structures and RF-MEMS [10]-[12]. These micromachining techniques provide the advantages for miniaturization of microdevices and microstructures, but particular attentions have been paid to achieve a high performance inductor. This paper describes that the following key issues that must be addressed to implement the microfabrication processes of the solenoid-type inductor: 1) removal of seed layer by dry etching instead of wet etching in avoiding of the erosion of conductor; 2) fine polishing of polyimide instead of O_2 plasma etched in avoiding of copper oxidation; 3) planarity of multilevel metallization; 4) strong adhesion between different layers.

2. Design

Fig. 1 shows the solenoid-type inductor that is composed of a magnetic thin film core and multilevel metal conductors. It is designed to have a complete closed magnetic circuit with rectangular shape core as to minimize the flux leakage. Electroplating technique is used to fabricate the conductor lines and the conducting vias. Polyimide is used for the insulation material between the coils and the magnetic thin film core. For a solenoid-type inductor, the inductance at low frequency is calculated by the following equation [6, 13]:

$$L = \frac{\mu_0 \mu_r A_c N^2}{l_c} \qquad (1)$$

where A_c is the cross-sectional area of the magnetic core, l_c is the total length of the closed magnetic core, N is the number of coil turns, μ_0 and μ_r are the vacuum and relative permeability of the core material, respectively.

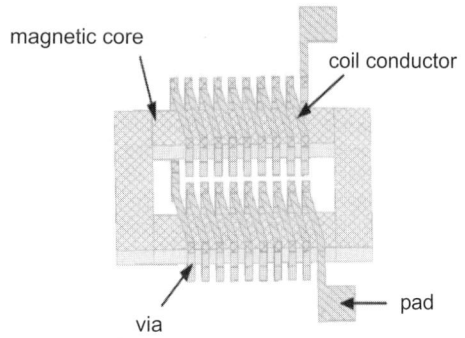

Fig. 1. Schematic diagram of a fully integrated micromachined inductor.

The Q-factor can be defined as follows:

$$Q = \frac{\omega L}{R} = \frac{\omega \mu_0 \mu_r N A_c A_w}{2(w+l)\rho l_c} \qquad (2)$$

where A_w is the cross-sectional area of the conductor, ω the angular frequency, $2(w+l)$ the length of coil per turn, ρ the resistivity of the conductor material. These two equations may not be so available for high frequencies when the eddy current and skin effect can't be neglected, but it still has the relevance in designing the solenoid-type inductor for frequencies below 10 MHz.

From Eq. (1) one can find that there are three ways to improve the inductance, (1) using the magnetic core with high-relative permeability, (2) increasing the cross-sectional area of the magnetic core through adding the magnetic core thickness, or (3) increasing the number of coil turns, which means decreasing the line width and space between the adjacent coils in a certain surface area, and also increasing the parasitic capacitance between the windings. In our experiment, the solenoid-type inductor has a profile of 4 mm × 4 mm × 0.084 mm with 80 turns of coils, line width of 20 μm, space of 35 μm and aspect ratio of 4:1. In order to achieve high inductance and miniaturize the size of the inductor, the shape of the magnetic core scheme is closed rectangular, of which the width of the long side and short side are 1.4 mm and 0.6 mm, respectively.

3. Fabrication process

A brief fabrication process for the complete microinductor was shown in Fig. 2. The process started with a clean glass substrate. First, the double-side alignment marks were formed on the other side of the glass substrate in order to improve the precision of alignment photolithography (Fig. 2(a)). Then the chromium/copper layer (100nm) was sputtered as a seed layer for electroplating and then covered by thick photoresist. The patterns of the bottom conductor traces, vias and pads were transferred photolithographically, followed by the selective electroplating of the bottom conductor lines, vias and pads (Fig. 2(a)). After that, all the photoresists were removed by using the acetone and the seed layer was dry etched in order to avoid the erosion of bottom copper conductor lines by wet etching. In order to isolate the conductor lines and the magnetic core, thick polyimide was used as the insulation materials and structure holder. After coating, polyimide was cured at 250°C for 2 hrs in argon atmosphere. Then polyimide was fine polished instead of RIE etching [11] until the vias and pads were exposed out (Fig. 2(b)). The procedure of fine polishing is done by a machine, which is similar to the CMP. When polishing the polyimide, the sample is put on the machine using grinding dust whose particle sizes can be in the range from several tens microns to below 1 micron, in avoiding leaving scratch on the sample. After fine polishing, the sample is well cleaned by acetone with supersound vibration and pre-sputtered by sputtering machine.

To fabricate the magnetic core, amorphous soft FeCuNbCrSiB film was deposited by magnetron sputtering and patterned by UV photolithography, and then the magnetic film was wet etched by the special chemical solutions. During the fabrication of the magnetic core, a special protective layer was sputtered onto or below the FeCuNbCrSiB film in order to avoid the etching of the magnetic core and the vias respectively during the wet etching of magnetic core (Fig. 2(c)). After the fabrication of the magnetic core, the vias and pads were electroplated with copper again. Next, another polyimide layer was spin coated, hard cured and fine polished to insulate the magnetic core from the top conductor lines (Fig. 2(d)). Finally, the top conductor lines were electroplated in completing the solenoid-type inductor (Fig. 2(d), (e)).

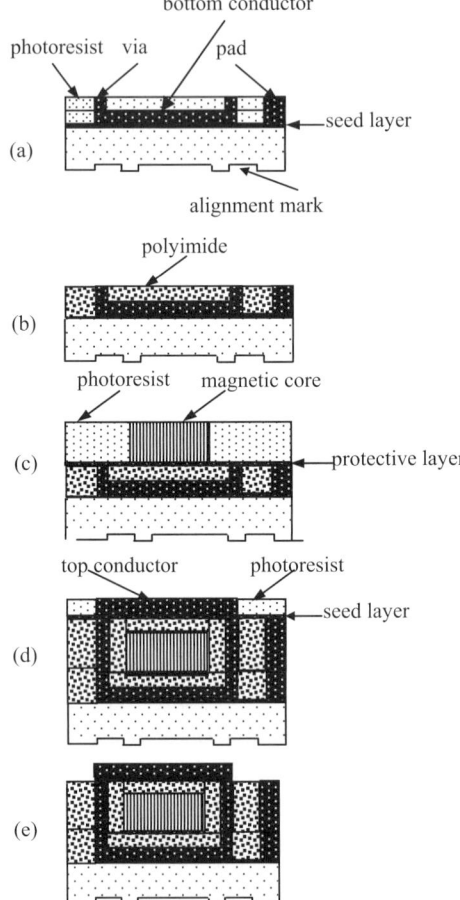

Fig. 2. Fabrication process of a solenoid-type microinductor: (a) fabrication of alignment mark, electroplating of bottom conductor lines, vias and pads; (b) spin coating, curing and polishing of polyimide; (c) fabrication of magnetic core; (d) electroplating the vias and pads; spin coating, curing and polishing of polyimide; electroplating the top conductor; (e) removal of seed layer.

Fig. 3 shows the photograph of the fabricated solenoid-type inductor, which consists of the electroplated copper conductor lines, amorphous soft FeCuNbCrSiB magnetic core and polyimide as an insulation materials. The width of the electroplated copper conductor lines is 20 μm, space of 35 μm and thickness of 30 μm. Amorphous soft FeCuNbCrSiB magnetic core is 1400 μm in width and 4 μm in thickness.

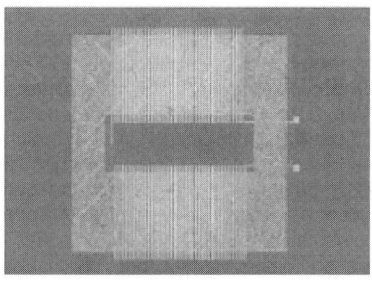

Fig. 3. Photograph of the fabricated solenoid-type inductor

4. Results and discussion

NiFe film is widely used as a magnetic material for microinductor and microtransformer since it has high relative permeability (~800) and high saturation flux density (B_s) (~1T). While, the typical disadvantage of low electrical resistance has limitations for their high frequency applications. It has been found that Finemet alloy showed superior soft magnetic properties beyong the conventional soft magnetic materials, this kind of materials has high relative permeability (~10^4) and high B_s (>1.3T) and high electrical resistivity (~130 µΩ·cm). However, there are rarely reports on the solenoid-type inductor with this new Fe-based soft magnetic thin film. Here, we have fabricated the solenoid-type inductor with FeCuNbCrSiB film as the core materials. Amorphous soft FeCuNbCrSiB films were prepared on glass substrate by magnetron sputtering. The background

pressure was 8×10^{-7} Torr, and during the deposition process, the sputtering power and the Ar pressure was 600 W and 4×10^{-4} Torr respectively. X-ray and AFM were used to analyze the structure of the magnetic thin film, and our experiments show that the as-deposited film is in amorphous state and still remained in amorphous after annealing at 300^0C for 30 minutes.

The magnetic properties of the FeCuNbCrSiB films were measured by VSM (vibrating sample magnetometer), and the saturation flux density (B_s) is 1.62 T and 1.39 T for the as-deposited film and after film annealed at 300^0C, respectively. According to $B_s = \mu_0 \mu_r H_k$, where H_k is the anisotropy field, using H_k =800 A/m as measured from the giant magneto-impedance effect, the estimated permeability μ_r is around 1600 for the as-deposited FeCuNbCrSiB film, which is higher than that of the electroplated NiFe film. Fig. 4 shows the magnetization curve of the as-deposited FeCuNbCrSiB film, and also the electroplated NiFe film is shown in order for comparison. The thickness of the as-deposited FeCuNbCrSiB film and the electroplated NiFe film is 4 and 20 µm, respectively.

For the fabricated microinductor, the thickness of the magnetic thin film is selected as 4 µm, one reason is that the thick in thickness will result in larger internal stress by sputtering; another reason is that the skin effect become stronger in thicker magnetic thin film at high frequencies due to the eddy current effect.

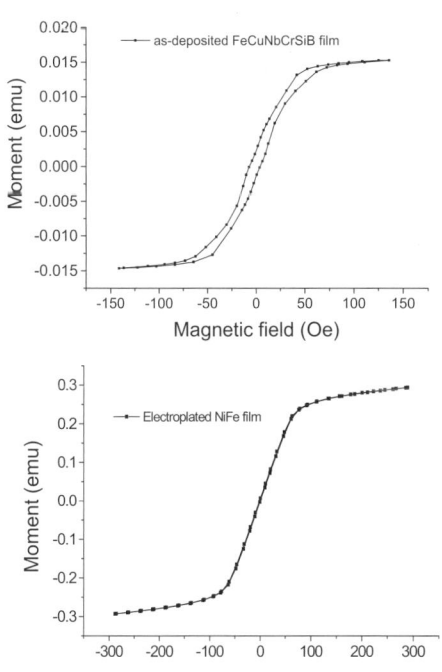

Fig. 4. Magnetization curves of the as-deposited FeCuNbCrSiB film and electroplated NiFe film.

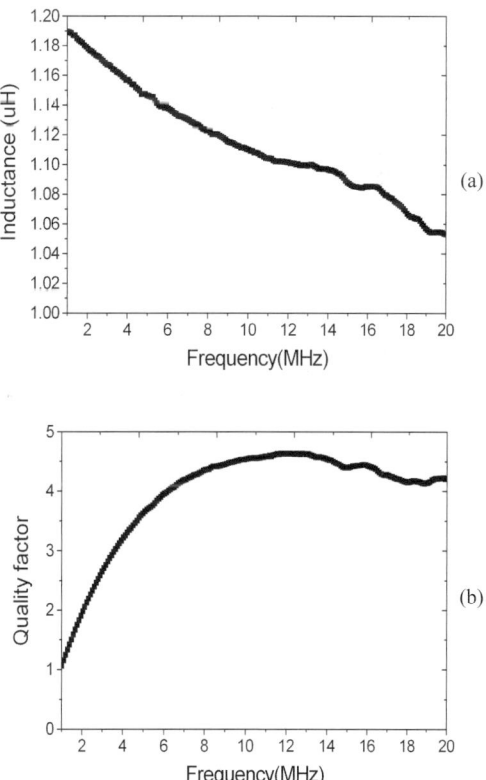

Fig. 5. The dependence of the measured inductance (a) and quality factor (b)
Measurements of the fabricated inductor were

done in the frequency range of 1-20 MHz by using the Agilent E4991A RF Impedance/Material Analyzer, combined with a microwave probe station produced by Cascade Microtech manufacturer. The Cascade Microtech ACP40-GS series probe was used as the probe head. Measurements were carried out after making the open/ short/ load calibration of the equipment. Fig. 5 shows the measured inductance, quality factor (Q-factor) and resistance as a function of frequencies. As shown in Fig. 5 (a), the inductance decreases slowly with the increase of frequency in the range of 1-20 MHz, and the inductance is 1.19 µH and 1.05 µH at frequencies of 1 MHz and 20 MHz, respectively. Park et al. [6] reported a microinductor in size of about 4 mm× 4 mm with electroplated magnetic core (35 µm), and the inductance of about 1 µH was achieved with Q-factor of 1.7 at 1 MHz. It shows that the FeCuNbCrSiB film has good frequency behavior in the frequency range of 1-20 MHz than that of the electroplated NiFe film. The curve of quality factor along with the frequency is shown in Fig. 5(b). The maximum Q-factor value is about 4.5 with an inductance value of 1.1 µH at a frequency of 12 MHz. The high Q-factor with high inductance at high frequencies is appreciable for the applications of micro DC/DC converter, further improvements in inductance will be done possibly by increasing the thickness of the FeCuNbCrSiB magnetic core.

5. Conclusion

A solenoid-type inductor was fabricated using copper as conductor, amorphous soft Fe-based magnetic film as magnetic core and polyimide as the insulation materials. The size of the inductor is 4 mm × 4 mm × 0.084 mm with magnetic core in rectangular shape. The experimental results show that the inductance is 1.19 µH at a frequency of 1 MHz, and the maximum Q-factor is 4.5 with an inductance value of 1.1 µH at a frequency of 12 MHz, which shows very attractive for the applications of micro DC/DC converter.

Acknowledgements

This work was supported by Samsung Advanced Institute of Technology (SAIT), Samsung Electronics Co., Ltd., Shanghai-Applied Materials Research and Development (No. 0515), the National High Technology Research and Development Program (No. 2004AA302042), and the Nanotechnology Program of Shanghai Science & Technology Committee (No. 0352nm014).

References

[1] Liang TJ and Tseng KC. IEE Proceedings: Electric Power Applications. 152 (2005) 217.
[2] Randon EJ, Wesseling E, White V, Ramsey C, Castillo LD and Lieneweg U. IEEE Trans. Magn. 39 (2003) 2049.
[3] Park JW and Allen MG. IEEE Trans. Magn. 39, (2003) 3184.
[4] Prieto MJ, Pernia AM, Lopera JM, Martin JA and Nuno F. IEEE Trans. Ind. Appl. 38 (2002) 543.
[5] Zhuang Y, Rejaei B, Boellaard E, Vroubel M and Burghartz JN. IEEE Electr. Device Lett. 24 (2003) 224.
[6] Park JY and Allen MG. IEEE Trans. Electron. Pack. Manuf. 23 (2000) 48.
[7] Kurata H, Shirakawa K, Nakazima O, Murakami K. IEEE Translation Journal on Magnetics in Japan 9 (1994) 90.
[8] Atta RMH. Sens. Actuators A 112 (2004) 61.
[9] Bryzek J. Sens. Actuators A 56 (1996) 1.
[10] Laurent D,Olivier G and Rooij NFD. Proc. SPIE, Santa Clara, CA, 2000, pp. 16-27.
[11] Ahn CH and Allen MG. IEEE Trans. Ind. Electron. 45 (1998) 866.
[12] Pisani MB, Hibert C, Bouvet D, eaud P and Ionescu AM. Microelectron. Eng. 73-74 (2004) 474.
[13] Soohoo RF. IEEE Trans. Magn. 15 (1979) 1803.

In-process assembly of micro metal inserts in a polymer matrix

G. Tosello, H. N. Hansen

*Department of Manufacturing Engineering and Management, Technical University of Denmark,
Produktionstorvet, Building 427S, DK-2800 Kgs. Lyngby, Denmark*

Abstract

New functionalities and smaller dimensions of micro products can be achieved by means of a higher degree of integration of both materials and components. Smart micro assembly techniques (such as on-the-machine assembly) together with hybrid structures (as metal inserts in polymer matrix) are suitable solutions to manufacture new micro products with several integrated functionalities, reduced number of components and assembly phases, as well as the possibility to be replicated in a high number of specimens. Innovative manufacturing systems, as well as new design rules and testing methodologies, have to be established in order to be able to develop new and more integrated micro products. In this paper a method for testing the bonding between micro thickened metal inserts and the polymer matrix they are moulded in is presented. A specific demonstrator has been manufactured by means of a hot embossing-like process which allows fast developing time and the possibility of batch process. Different levels of surface roughness and metal insert thickness were applied in a systematic design of experiments. The results show a strong influence of surface texture on bonding strength. The testing procedure assists the designer in giving data to be used when dimensioning micro products involving integrated metal parts in a polymer product.

Keywords: bond strength testing, micro metal inserts, polymer hybrid structures, in-process assembly

1. Introduction

Integration of different materials and components in micro systems production is at the same time an opportunity and a challenge. On one hand, a higher degree of integration brings to smaller dimension, lower number of components, and increased functionality. On the other hand, the production of small and highly integrated devices is complex due to miniaturized dimensions and presents issues on storing, handling, positioning, and joining the micro parts. Alternative techniques to the conventional pick-and-place approach have to be applied.

Miniaturized devices with increased functionality can be achieved by means of integration of polymers and metals in the same components. Polymers are relatively cheap materials; they exist in an enormous variety, in a way they are suitable for virtually any kind of application, and allow good replication capabilities at micro scale. Metals provide additional functions and properties such as stiffness, elasticity, conductivity. Hybrid structures are characterized by properties of both materials, for example providing an insulating matrix with conductive and stiff features. Typical micro devices exploiting such combination of properties in order to achieve higher integration and functionality are micro mechanical actuators, micro regulators, hearing aid components, and integrated sensors for e.g. medical analysis, etc. Because of the high amount of possible applications, it is clear the importance of being able to provide properties and characteristics to design micro products involving integrated metal parts in a polymer based product. This paper focuses on the manufacturing and testing of hybrid structures composed of micro metal inserts in a polymer matrix.

2. Manufacturing and assembly of hybrid polymer based micro systems

When medium batch or mass fabrication of polymer based products are required, replication processes such as hot embossing and micro injection moulding are suitable to be used. The former primarily is to be employed in laboratory and R&D applications, the latter mostly to produce final products in a larger scale. Depending on the chosen replication manufacturing process, special integrated assembly solutions can be applied:

- Smart assembly techniques as on-the-machine assembly: micro insert moulding and micro assembly injection moulding [1,2].
- Parallel assembly techniques: multi-cavity micro injection moulding, hot embossing and pre-adjusted assembly polymer magazine [3].
- Use of a metal frame connecting all the metal inserts to be overmoulded, to be applied either to multi-cavity micro injection moulding or to a water-like approach with hot embossing.
- Automation system in order to feed inserts, pick and place the hybrid (polymer and e.g. metal inserts) moulded or hot embossed parts by handling directly the sprue or the metal frame and not the single micro parts.

In this work the hot embossing process was chosen. It allows short developing time, quick set-up changes, rapid prototyping and fast material screening test [4]. The large hot embossing area was furthermore exploited in order to fabricate components with different characteristics at the same time.

3. Experimental

3.1. Demonstrator design and manufacturing

A test part was designed in order to evaluate the bonding strength between the micro metal insert and the polymer matrix where it is inserted in. The demonstrator had a cross section of 6x4 mm^2, and a length of 6 mm. The metal insert was positioned in the

84

parting line of the polymer part, at 2 mm from both planar surfaces.

The overmoulding was obtained by means of a steel tool composed by two plates (total thickness 4 mm) and a hydraulic hot press to perform the embossing process (see Fig. 1). The cavity of 6x6 mm^2 was manufactured by wire EDM. The housing pocket for the metal insert was then fabricated using micro EDM on the internal surface of the lower plate (which represents the parting surface of the moulding).

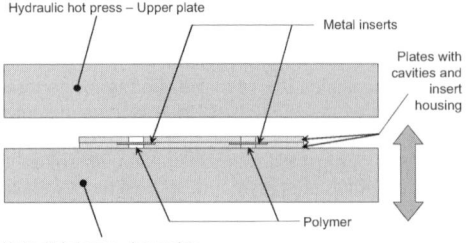

Fig. 1. Hydraulic press and hot embossing set-up.

Polystyrene (PS) and polyoxymethylene (POM) were selected as polymers for the hybrid structure to be produced. The transparency of polystyrene allows to check the inside structure integrity and insert placement. Hot embossing temperature was chosen at 250 °C. POM process temperature was chosen at 180 °C. Both materials are particularly suitable for micro moulding applications due to their very easy flowing. Metal inserts made of rolled steel were used for the experimental batch (see Fig. 2). Steel inserts covered by a thin layer of palladium were also used. This particular material combination is used in the production of miniaturized electro-mechanical systems (e.g. micro switches) because it presents both stiffness and high electrical conductivity. Asymmetry of shrinkage with respect to the insert might cause bending. However, due to the symmetry of the demonstrator and a reduced shrinkage because of the smooth temperature cycle, the metal inserts did not show any deformation (e.g. bending) or displacement from the position given by the housing in the cavity.

Fig. 2. Over moulded hybrid demonstrator composed of micro thickened steel insert in a polymer matrix (steel, 400µm / PS on the left; steel and Pd, 80µm / POM on the right).

3.2. Metal inserts: selection, surface treatment, and characterisation

During the hot embossing process the polymer flows the very short way from its solid state into the micro structures. High melt and cavity temperatures allow a high degree of replication of the micro surface texture of the insert. Therefore the filling of even the deepest grooves of the surface can be achieved. This brings the insert to be completely surrounded by the polymer, down to the micro dimensional level of the surface texture. Mechanical interlocking at the polymer/metal interface can therefore be obtained.

Moreover, higher bonding strength with the polymer matrix can be achieved when such interlocking between the polymer and the metal is increased. This can be obtained by means of a modification of the metal insert surface, e.g. by increasing its roughness. Hence, the influence of the surface texture of the metal inserts on the mechanical bonding strength of the hybrid structures was investigated.

A superficial treatment consisting in a glass blast jet was applied to the metal insert in order to obtain higher roughness of the metal insert's surface. Glass powder with granules from 40 up to 70 µm was used.

Table 1
Surface topography analysis of three metal insert configurations investigated. Measured area is 100x100 µm^2.

Material / Insert thickness: Steel / 40 µm			
Ra	0.35 µm	Ra	0.85 µm
Sa	0.33 µm	Sa	0.84 µm
Sdr	16.6 %	Sdr	84.7 %

Material / Insert thickness: Steel / 80 µm			
Ra	0.30 µm	Ra	0.98 µm
Sa	0.29 µm	Sa	1.00 µm
Sdr	14.2 %	Sdr	74.5 %

Material / Insert thickness: Steel with Pd layer / 80 µm			
Ra	0.44 µm	Ra	1.37 µm
Sa	0.46 µm	Sa	1.35 µm
Sdr	29.2 %	Sdr	119 %

The manufactured demonstrator reproduces the case of an actuator in which the two components do not have to change their mutual position. In this case, a surface parameter suitable to be used for surface characterization is the average surface roughness (Ra) [5]. Moreover, though three-dimensional surface parameters are not yet defined by standardisation bodies, 3D surface average roughness (Sa) can be obtained by means of calculations based on two-dimensional standards extended to three dimensions [6]. Hybrid three-dimensional parameter such as the surface area ratio is suitable to be employed as well. The surface area ratio (Sdr) expresses the ratio between the area of the actual measured surface (taking the Z height into account) and the projection of the actual surface on the X,Y plane. For a totally flat surface, the real surface area and the projected area of the X,Y plane are the same and Sdr = 0 %.

Investigation of the surface texture of the micro inserts was carried out with a three-dimensional laser focus detection instrument. Three-dimensional roughness parameters as well as surface mapping of the metal inserts were determined [7]. Results regarding smooth and treated surfaces of 40-80 μm insert thick made of steel and 80 μm covered by a palladium layer are presented (see Table 1). The same treatment was applied for all the insert configurations. Steel inserts covered by a layer of palladium showed a deeply modified surface texture, resulting in a rougher surface because of the softer material on the surface.

4. Bonding strength testing

The hot embossed hybrid structures were tested by using a pull test machine. A special device has been designed to hold the polymer matrix and to clamp the micro metal insert (see Fig. 3). Pull test speed was set to 1.0 mm/min. Different factors were varied and combined in order to evaluate their influence on the hybrid demonstrator's bonding strength:

- Polymer materials: PS and POM.
- Metal insert materials: steel and steel covered with a palladium layer.
- Metal insert thickness: 40 μm, 80 μm, 100 μm, 400 μm.
- Metal insert surface: treated and not-treated.

Variation of the pull test results for the same configuration was within a range of ±13 %. The indicated bonding strength is the highest load recorded during the tensile test.

As alternative, peel-tests could have been used. However, this test seems more suitable to determine mechanical properties at the interface between an injection moulded polymer part and a thin film on its surface.

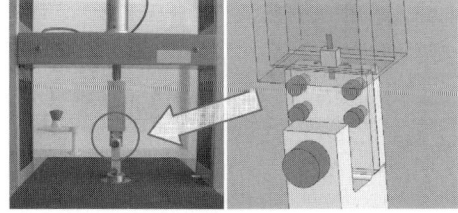

Fig. 3. Pull test (left), device for holding the polymer matrix and to clamp the metal insert (right).

4.1. Influence of the insert material and surface treatment

Steel micro inserts with thicknesses of 80 μm and different type of surfaces (glass treated and not treated, covered with palladium layer and uncovered) were hot embossed in a PS and POM polymer matrix. Bonding test shows increased strength for both polymers when the metal surface roughness increases, as well as when using the covered steel with palladium instead of the uncovered one (see Fig. 4). This is also correlated to the increased surface area ratio of the modified surfaces.

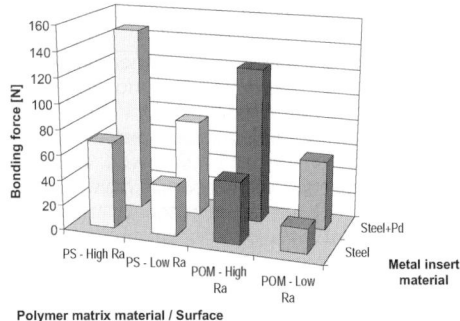

Fig. 4. Bonding strength test of hot embossed hybrid demonstrator composed of micro steel insert with different surface treatments in a POM and PS polymer matrix (metal insert thickness 80 μm).

4.2. Influence of the metal inserts thickness

PS polymer matrix was hot embossed with steel inserts of different thicknesses. A first experiment concerned inserts 40 μm and 80 μm thick, with both glass treated and not treated surfaces. Influence of the increased roughness Sa and surface area ratio Sdr (which were for the 40 μm thick insert, 0.33 μm and 17 % before treatment, and 0.84 μm and 85 % after treatment respectively) results to be higher than the increased surface due to the thicker insert (see Fig. 5). Variation of the insert thickness from 40 μm to 80 μm has no significant influence on the bonding strength.

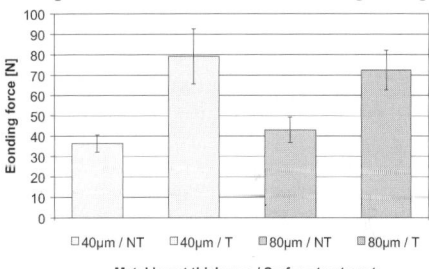

Fig. 5. Bonding strength test of hybrid demonstrator composed of micro steel insert with different surface treatment and thicknesses in a PS polymer matrix.

Thicker inserts of 100 μm and 400 μm made of steel with no surface treatment have also been tested to evaluate the influence the insert thickness in a larger range. Polymer matrix materials are POM and PS (see Fig. 6). The increase of the contact surface (due to the thicker inserts) results in a higher bonding strength of the hybrid demonstrator with the 400 μm thick insert.

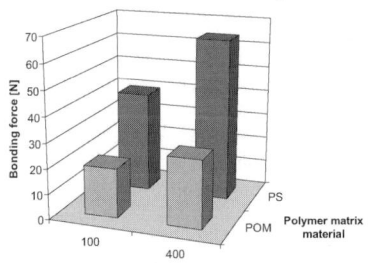

Fig. 6. Insert thickness influence on bonding strength of hybrid demonstrator (steel + PS, steel + POM).

4.3. Influence of pull test speed

A standard configuration (polymer matrix made of PS, metal insert 100μm thick with no treatment) was chosen and the pull test speed was varied. Three levels were defined: 0.1 mm/min, 1.0 mm/min, and 10 mm/min. Two effects can be observed. On one hand the bonding strength increases with higher pull speed. On the other hand a decrease of one order of magnitude from the standard speed has more effect on the measured bonding strength than when the speed is increased 10 times (see Fig. 7).

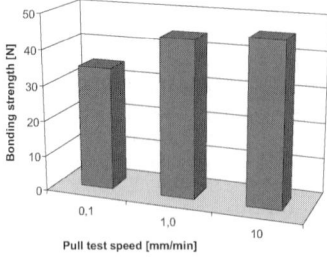

Fig. 7. Influence of pull test speed on the bonding strength of hybrid demonstrator (steel + PS).

5. Polymer and metal interface investigation

Bonding strength of the hybrid components and surface roughness as well as surface area ratio are proportionally related. Because of the almost zero adhesion between polymers and metal, bonding is provided by mechanical interlocking of the metal into the polymer matrix at micro range level.

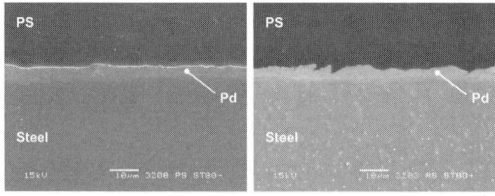

Fig. 8. Full penetration of PS in the metal insert surface texture: not treated (left) and glass treated (right).

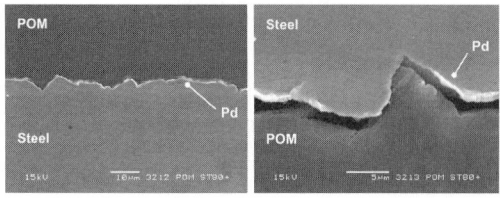

Fig. 9. Full penetration of POM in the metal insert surface texture of a glass treated metal insert.

Scanning electron microscopy (SEM) was employed to characterize such interlocking, i.e. penetration of the polymer in the valleys of the metal surface texture. Due to the favourable hot embossing process characteristics and the easy flowing of the material complete penetration is achieved for both POM and PS (see Fig. 8 and Fig. 9, left). Moreover, by means of the surface treatment, the depth and the quantity of polymer flowing into the surface texture has been increased. Ejection forces show to have pulled apart the polymer from the metal insert creating a gap of 1μm. Mechanical interlocking is still possible because of mutual embedding (see Fig. 9, right).

6. Summary and outlook

Manufacturing and testing of a hybrid demonstrator composed by a micro metal insert made of steel in a polymer matrix was presented in this paper. Two different polymers (PS, POM) were tested. Pull test results were reported showing bonding strength dependency on the materials involved, testing conditions, and on the insert surface treatment and thickness. In particular, bonding strength was observed to be strongly dependent on metal insert roughness surface. Scanning electron microscopy investigation at the polymer/metal interface showed that complete penetration of the polymer on the surface texture of the metal part was achieved creating mechanical interlocking between the two components.

Surface treatment was found to be a key point to enhance bonding strength of hybrid micro structures. The need to explore different surface modifications to increase integrated components is clear. Process downscaling is going to be investigated as well, in order to determine the effect of thinner polymer parts as well as ultra-thin metal insert on bonding strength. An increase of influence of the surface roughness is expected to occur. Further investigations will focus on the establishment of micro injection overmoulding technology to increase production and replication capabilities. Indeed, a larger variety of both polymeric and metallic materials is on its way to be investigated.

Acknowledgements

This research was carried out as a part of the innovationskonsortium "Selective micro metallization of polymers" (project no 61568) supported by The Danish Ministry of Science, Technology and Innovation. Thomas W. Juhl and Martin Bondo Jørgensen, Sonion Roskilde A/S, are thanked for their collaboration and support in the project.

References

[1] Michaeli W., Ziegmann C. Micro assembly injection moulding for the generation of hybrid microstructures. Microsystem Technologies. (2003) Vol. 9, pp 427-430.

[2] Michaeli W., Opfermann D. Micro assembly injection moulding – Potential application in medical science. Proceedings of 1st International Conference on Multi-Material Micro Manufacture (4M). (2005) pp 79-82.

[3] Ehrfeld W. et al. Highly parallel mass fabrication and assembly of microdevices. Microsystem Technologies. (2001) Vol. 7, pp 145-150.

[4] Heckele M. Hot embossing – a flexible and successful replication technology for polymer MEMS. Proceedings of SPIE. (2003) Vol. 5345, pp 108-117.

[5] De Chiffre L. et al. Quantitative Characterisation of Surface Texture. Annals of the CIRP. (2000) Vol. 49/2, pp 635-652.

[6] Barbato G. et al. Scanning tunnelling microscopy methods for roughness and micro hardness measurements - EU programme for applied metrology (CD-NA-16145 EN-C), Brussels (1995).

[7] Scanning Probe Image Processor (SPIP™). Image Metrology A/S, (2005).

Low temperature adhesive bonding in MEMS

D. Andrijasevic[a], W. Smetana[a], D. Esinenco[b], W. Brenners[a]

[a] Institute of Sensor and Actuator Systems, Vienna University of Technology, 1040 Vienna, AUT
[b] The National Institute for Research and Development in Microtechnologies, 077190 Bucharest, RO

Abstract

A new concept for adhesive bonding of components with dimensions less than 300 µm in complex 3D structures is presented in this paper. Two different kinds of adhesives - polyurethane adhesive foil and polyethylene hot melt glue - were applied to the basic substrate by different techniques. The focused stream of hot gas directed to the substrate softens glue which had been applied as a solid. Afterwards, micro-parts were embossed in the softened glue, or covered and shielded by it. After cooling down at room temperature, a rigid and compact bond was formed. The parameters which induce the heat transfer and the process performance generally are considered through various experiments. The advantages of this new approach are compact system which can be easily integrated in existing production lines, low capital costs, partial reversibility, applicability to many material combinations, localized heating with reduced thermal stresses, etc.

In order to confirm the advantages of proposed bonding technique, a single mode glass optical fiber (core diameter 9 µm; cladding diameter 125 µm) was positioned in V-grooves to achieve requested alignment and positioning precision. V- grooves were made by chemical etching of silicon in the "111" plane. Bonding glass to glass was also successfully demonstrated.

Keywords: adhesives, low temperature bonding, hot gas stream

1. Introduction

Assembling and packaging of microcomponents with dimensions less than 300 µm are very difficult due to the limitations in use of manipulating tools and external equipments. Conventional microbonding techniques, like anodic bonding, fusion bonding, eutectic bonding and other are followed with demanding process requirements (high voltage, high process temperature, specific materials to be used, quality of surface, etc) which can restrain or disable assembly process. This paper discusses a new approach to adhesive microbonding which can overcome some of drawbacks of the conventional MEMS packaging techniques. Some advantages of this technique are: low process temperature, multi material applicability, partial reversibility and partial bio-compatibility. Additionally, this technique can be performed on all levels of packages integration: single chip, wafer based single chip, wafer level packaging.

Using adhesives in micro-system techniques was not very popular due to its chemical instability and remarkable changes of mechanical properties during the life cycle. But with development of advanced dispersing techniques and improvement of properties of adhesives in general sense, they are getting more and more an important role in the field of microbonding. Some of the most favoured adhesives characteristics in comparison with other bonding techniques are:

- improved stress distribution - stresses are evenly distributed over the entire bonding area, thereby minimizing high localized stress concentrations,
- fatigue resistance of adhesive bonds to cyclic loading,
- high resistance to mechanical shocks and vibrations,
- ability to bond dissimilar as well as similar materials including metals, plastics, elastomers, glass, ceramics,
- outstanding humidity and corrosion resistance at both ambient and elevated temperature conditions,
- substantial weight and significant cost savings without decrease of relevant bonding strength parameters,
- unexcelled dimensional stability and long term durability [1] - [3].

In the research presented in this paper, the "hot air stream concept" is employed for joining micro-components into complex and sophisticate structures. The adhesive is deposited on the substrate and then the micro-component is carried and placed at the requested position. Afterwards, the stream of hot air is applied in order to soften the glue and to emboss the micropart into the glue. After cooling down to the room temperature, the glue hardens and final bond is achieved.

The quality of bonds was evaluated by measuring of the mechanical strength of bonds obtained for different process temperatures and varying softening times. The achieved results are comparable with maximal bonding strengths reported in the literature for conventional processes. This confirms the usability of the technique in several MEMS applications, for instance, in assembling of optical components, as presented in this work.

2. Experimental set-up

In order to prove the working principle of proposed technique including its advantages and to perform different experiments, a special setup was developed, and its schematic is shown in Fig. 1.

The main parts of the set-up are a gas reservoir (technically clean air), a compressor, a heating element on the metal tube, a nozzle, and additional pipes for connecting these elements. The heating element has been developed on the metal tube by "a thick film on the tube" process, a technology traditionally used in electrical circuits manufacturing. In this way, a small and extremely compact set-up was obtained. As a heating tube, an austenitic stainless steel tube (inner diameter of 2 mm, outer diameter of 4 mm, length 120 mm) was used. The heating element with a total length of 80 mm was formed on the tube in a spiral shape in the following way: dielectric glass ceramic paste type ESL D-4916 was applied first on the metal tube to provide insulation function. Low resistance ESL 29115 (sheet resistance: 100 mOhm) resistor paste formed the heating element. On the ends of the insulation layer, paste ESL 9695 formed contact pads, and both connectors have the total length of 1 cm. With this arrangement, a peak temperature of 600° C could be obtained on the heating tube within 5 seconds by applying heating power of 28.1 W.

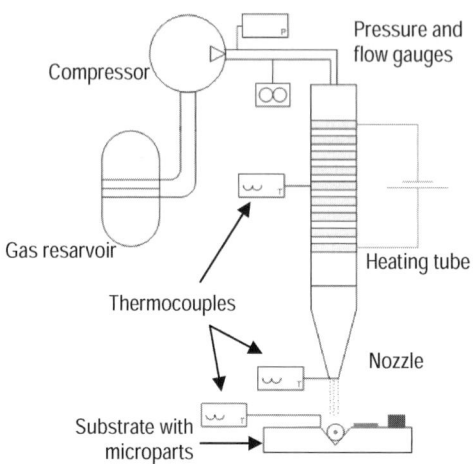

Fig.1. Schema of experimental set-up for micro-joining by hot gas stream

Metal nozzles (CrNiMo), with an inner diameter in the range of 150 μm and 300 μm and a capillary length of 11 mm, were mounted on the bottom end of the heating tube to focus the gas stream on and to direct it at the targeted area. The resulting temperature on the nozzle, on the substrate and on the heating tube were measured respectively with three K-type thermocouples made of Ni-CrNi wires with diameter of 80 μm, to provide negligible influence of the thermal mass of the thermocouples on the measurement. The flow meter reduces the velocity of the air to the value optimized through experiments and simulations.

3. Bonding process

3.1. Optimisation of the softening conditions

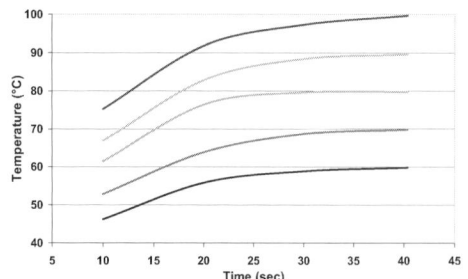

Fig.2. The temperature rise on the surface substrate versus exposing time

Due to the relatively big mass of the substrates (2 inch × 2 inch), the temperature on the substrate surface

requested for softening of adhesives was obtained after some period of time. This means that thermal losses must be considered and they influence directly the temperature rise on the substrate surface. In order to find out the minimal period of time within this temperature is achieved, the rise course was measured, and the results are shown in Fig. 2. It could be proved that the requested temperature on the surface of the ceramic substrate (for the starting room temperature) has been achieved app. after 40 seconds. According to this, the measurement of the curing time has started 40 s after the exposing of the glue to heat.

3.2. Adhesives

Two kinds of commercially available adhesives – Polyurethane foil and hot melt glue on the Polyethylene base were investigated. Their main properties are shown in Table 1.

Table1
Adhesive properties

Type of adhesive	State	Softening point (°C)	Thickness (μm)	Deposition method
Hot melt glue	solid	65	~ 12	spinning
Adhesive film	solid	72	50	lamination

The adhesive foil has been laminated to the basic substrate. The hot melt glue was firstly, dissolved in PGMA ((1-Methoxy-2-Propyl)- Acetat) in mass-proportion of 55% of PGMA and 45% of glue, and then applied by spinning (800 rpm, 40 s) on the basic substrate. Afterwards, the substrate was baked at 70 °C for 5 minutes in order to evaporate the solvent from the applied film, leaving the pure glue on the substrate. The average thickness of the adhesive layer after spinning was 12 μm.

3.3. Mechanical strength of adhesives

Mechanical strength of joins made with investigated adhesives was measured by destructive pull test on Accuforce Ametek Force Gage whereby the velocity of the moving part of tool was 5 cm/min. [4]. By increasing the curing temperature, the strength of bond made with adhesive foil increased significantly, while the strength of join made with hot melt glue dropped rapidly because of adhesive evaporation (the layer of adhesive was not thick enough to realize rigid join).

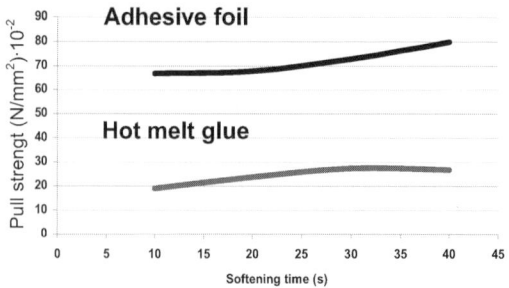

Fig.3. Maximal mechanical strength of bonds attained with hot melt glue and PU foil for different softening time

3.4. Bond realisation

3.4.1. V-grooves etching

In the first phase of experiments, the investigated technique was used for bonding single mode optical fibers (core diameter 9 μm; cladding diameter 125 μm) to the basic substrate and for bonding glass to glass. Those both enabled effective assessment of the bonding-technique quality.

Optical fibers are conventionally positioned in V-grooves made by anisotropic chemical etching of silicon <100> wafers, or for compact disposing of fibers, by anisotropic RIE.

One of the big advantages of using anisotropic etching for fiber alignment is the fact that due to the closure of etched <111> plans the depth of the V-grooves can be controlled very precisely. The wall roughness is high below 5nm as inspected and mentioned in [5], which is another critical factor in precise alignment, and the depth uniformity along the wafer doesn't depend on the chemical process conditions as long as the etching duration is longer than the maximum possible.

Wet chemical etching in Si <100> wafers in KOH and TMAH (25%) solutions at temperature of 80 °C with etching speed up to $R_{<100>}$ = 69 μm/h was used.

The etched v-grooves were characterized using SEM surface inspection, some results can be seen in Figure 4.

Fig. 4. SEM photo of etched V-groove

The sidewall angle of V-grooves is 54.7°. The depth was calculated and defined through the width of open windows in the mask using:

$$\tan 54.7 = \frac{h}{w/2} \qquad (1)$$

where:
w – width of etching windows
h – depth of v-grooves

V-grooves with depths in the range of 50 μm to 80 μm were defined for fiber positioning.

Rectangular channels etched in Si <110> were also used for fiber alignment using the same masks. A few disadvantages have been noticed: the depth is not totally uniform along the wafer and strongly depends on the initial planarity of the wafer also on the etching reaction uniformity along the wafer surface. The depth factor can be controlled changing the duration of the etching process, which is in many cases convenient. Due to the low roughness of the bottom wall and etch depth control precision not better than 1 μm this type of trenches are less used for fiber alignment.

In high density MOEMS precise alignment is a key factor. V-grooves for alignment of fibers with 125 μm diameter with SiN$_x$ waveguides of 5 μm to 15 μm width and 1 μm to 2 μm thicknesses were used. For this case a misalignment in vertical direction of more than 0.5 μm is critical.

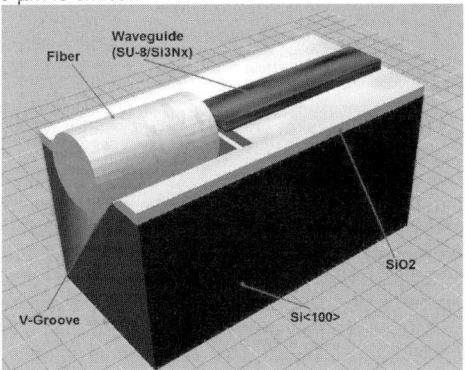

Fig. 5. 3D schematic view of an V-groove alignment principle.

3.4.2. Adhesive applying

For positioning and joining optical fibers in V-grooves, few different approaches were employed: The hot melt glue which hardens after baking was softened at low temperature of around 65 °C. Optical fiber (or any other micro component) was brought into requested position by means of magnetically actuated microgripper with visual and force feedback [8]. The stream of hot air melted the solid glue, and in parallel, pressed the fiber into the groove. In this way, the fiber was precisely positioned. Bonding with foil can be performed in two ways. First, the foil is laminated to the substrate, softened at the 90 °C and the fiber, previously positioned by microgripper, is embossed in the soft adhesive by means of hot air stream. Second, the fiber is positioned onto the V-grooves and covered with foil. Than, the foil is softened by hot air and it covered and shielded the fiber. After cooling down, the fiber adheres to the substrate. The quality of final, cured bonds was analyzed by optical inspection.

In both cases, for better self-alignment, one has to be sure that adhesion forces of the bonding liquid to the Si substrate and fiber cladding are high enough to "entrap" and keep the fiber inside the V-groove. Moreover, the viscosity/interfacial tension factor should be low enough for wetting the contact surfaces as uniform as possible and to form a thin film between fiber cladding surface and <111> planes of the groove.

The self alignment takes place during the softened state of the bonding polymer. Due to the normal component of the adhesion forces to both <111> planes of the grove, the fiber is automatically positioned in the center of the groove and tracked down by the liquid surface forces to the bottom of the groove. After the self alignment the bonding polymer can be hardened by cooling down at room temperature.

High viscosity bonding polymers should be avoided also due to the bubbles problem for tiny structures.

A bond realized by a foil applied onto the fiber is shown in Figure 6. The stage has been driven under the nozzle in direction corresponding to the fiber axis (velocity 2 mm/sec). The substrate-nozzle distance was 3 mm and temperature at the nozzle outlet was 120 °C.

The preform softened continuously, adhered to the substrate and fixed the fiber in the requested position.

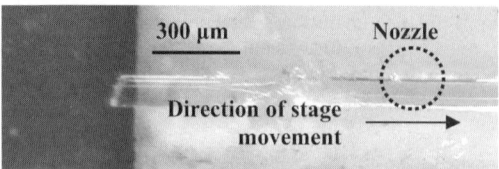

Fig.6. Top view of optical fiber (diameter 125 μm) joined with adhesive foil

In the second test, glass stamps with dimensions 10 mm × 10 mm are bonded to the glass substrate. The bond is made only over the stamp edge. The adhesive in the middle is not heat affected, and there are not any changes in the material or structure. This opens new possibilities for forming cavities or membranes using adhesives adequately. The appearance of single decompression bubbles was noticed. It is necessary to minimize them because they can cause performance degradation due to non-uniform stresses [6], [7]. This was tried to be solved by exposing the adhesive to the hot air stream for a longer period.

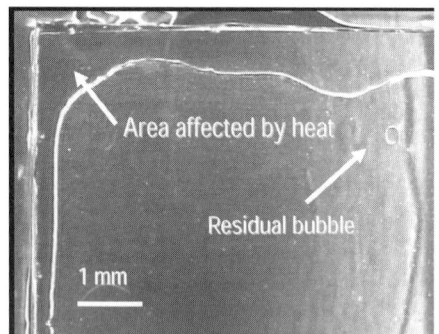

Fig.7. Top view glass stamp bonded to the glass substrate

4. Conclusion

As a new approach for MEMS packaging and MEMS assembly, the investigation on softening adhesives under hot gas stream is described. This innovative technique offers several advantages to meet recent requirements in MEMS manufacturing. The usability of the presented technique in view of existing technologies is also discussed. The experimental work on the real assembly of the optical fibre on pre-machined substrate, as well as glass to glass adhesive bonding on the wafer scale have been carried out. The results of experiments are presented and discussed here. In addition, it could be demonstrated that this technique allows local bonding.

Further investigation will be focused on the selection of appropriate adhesives which meet the requirements of bonding by hot air stream according to the melting point, viscosity, stress in adhesive joints, reliability and achievable mechanical strength. In order to reduce the curing time of the adhesive, the adhesives will be exposed to a pulsed hot stream with a temperature higher than the nominal softening point. Further work will concern the design and development of the special tool for pressing the part at the requested position (fibre to groove). The tool should integrate the nozzle for hot air flow and a nose for embossing the part.

Acknowledgment

This work was financially supported by FP6 Marie Curie Research Training Network "Advanced Methods and Tools for Handling and Assembly in Micro-technology - ASSEMIC", funded by the European Commission, Contract no. MRTN-CT-2003-504826, and it is a part of common work in the 4M Network of Excellence Multy-Material Micro Manufacture, Contract no. FP6 500274-2.

The authors thank to Mr. Zehetner from Vorarlberg University of Applied Science, for precious advice, fruitful discussions and cooperation.

References

[1]   W. Brenner. "Epoxies Pace Growth of High–Performance Bonding Technology", http://www.masterbond.com

[2]   F. Sarvar, D. A. Hutt and D. C. Whalley. "Application of Adhesives in MEMS and MOEMS Assembly: A Review", IEEE Polytronic 2002 Conference, Zalaegerszeg, Hungary

[3]   H.-s. Noh, K.-s. Moon, A. Cannon, P. J Hesketh and C P. Wong. "Wafer bonding using microwave heating of parylene intermediate layers" J. Micromech. Microeng. 14 (2004) 625–631

[4]   Force gauge data sheet: http://www.chatillon.com/Our%20Products/Spec%20Sheets/3003_DFE_Spec.pdf

[5]   C. Mihalcea, S. Khumpuang, S. Kuwahara M. Yang, Z. Maeda, R. Tominaga, J. and N. Atoda, „Ultra-Fast Anisotropic Silicon Etching with Resulting Mirror Surfaces in Ammonia Solutions", Transducers'01, Munich, Germany, June 10-14, 608-611

[6]   S. Behler. „The Formation of Decompression Bubbles in Polyimide Adhesive Tapes", Electronic Components and Technology Conference, San Diego, CA, USA, 1999, 727-732

[7]   Y. M. Cheung and Arthur C. M. Chong. "New Proposed Adhesive Tape Application Mechanism For Stacking Die Applications", 5th International Conference on Electronics Packaging Technology, Beijing, China, 1998, 309-315

[8]   I. Giouroudi, H. Hötzendorfer, J. Kosel, D. Andrijasevic, M. Ferros and W. Brenner. „Design of a Microgripping System with Visual and Force Feedback for MEMS Applications", MEMS Sensors and Actuators, London, UK, 2006

Multi-Material Micro Manufacture
W. Menz, S. Dimov and B. Fillon (Eds.)

Manufacture of recessed rotating microelectrodes for mass transport investigations in the LIGA process

M. Lisinenkova, L. Hahn, J. Schulz

Institut für Mikrostrukturtechnik, Forschungszentrum Karlsruhe GmbH, Postfach 3640, D-76021 Karlsruhe, Germany

Abstract

The enhancement of the knowledge of mass transport during the different manufacturing stages of high aspect ratio microstructures is important for further progress in the LIGA technology. The information about mass transport characteristics is essential in order to find the optimal processing parameters for the development and the electroforming.

A new construction and fabrication method for recessed rotating microelectrodes (RRME) for mass transport investigations in micro technology is presented. The manufacturing of these electrodes based on the LIGA process is described. This technology was tested in practice, its deficiencies were discovered and a new improved technology was presented. The improvement includes a substitution of electrochemical gold deposition to vapor deposition and using an alternative glue, which provides sufficient PMMA to gold adhesion.

The fabricated RRMEs were tested in electrochemical ferro/ferricyanide system and the first test results are reported. It has been shown that RRME, produced by improved technology, meet qualifying requirements and can be used for mass transfer investigations in micro technology

Keywords: fabrication technology, recessed rotating microelectrodes (RRME), LIGA process

1. Introduction

The LIGA process is used worldwide for the fabrication of high aspect ratio microstructures The main steps of LIGA process [1] are the generation of three-dimensional structure by means of deep X-ray lithography (DXRL) and electroforming of a complementary metal structure, which can be used as final product or as a mould insert for replication processes.

The enhancement of the knowledge of the mass transport during different manufacturing stages of the high aspect ratio microstructures is important for further progress in the LIGA technology. The information about the mass transport characteristics is essential in order to find the optimal processing parameters for the development and the electroforming.

The electrochemical methods, particularly current-voltage curves as obtained from a rotating disk electrode (RDE) and with the rotation speed being a central parameter, are frequently used for the investigation of mass transport rates. Owing to the proportionality of the electrical current to the flux of ions discharging at the electrode interface it is possible to conduct the measurements in situ. One more advantage of electrochemical methods is an ability to set and to control the potential (the driving force of electrochemical processes) and to measure the current (the material flux under the action of this force). [2,3]

For the investigation of mass transport in the high aspect ratio microstructures it is appropriate to use the recessed rotating microelectrodes (RRME). The first construction and manufacturing method of RRME are described in the literature [4]. The described manufacturing method is based on drilling of the required holes in the insulating layer, followed by the electrical deposition of gold into the holes. This way of microelectrode fabrication is moderately

expensive, but it is only possible to make circuar electrodes with the diameters down to 50µm.

The new construction and the manufacturing technology for RRME by means of LIGA technology allowing the fabrication of the microelectrodes of any 2D pattern and size up to 1µm is proposed in this work.

2. The construction of RRME

The recessed rotating microelectrodes need to meet several requirements: the geometry must fit the aspect ratio of interest, the current must be measurable, the interface area must be constant in order to obtain the same current with the identical conditions and, last but not least, the set up must be compatible with the rotating unit available.

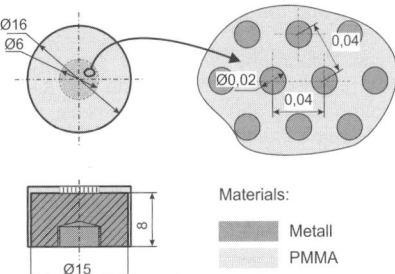

Fig. 1 Construction of the RRME

Based on these requirements the following construction was proposed (see Fig. 1). The basic part of it is a metal cylindrical blank, hereafter called substrate, with a diameter of 15mm and a height of 8mm, provided with a screw for mechanical and electrical contact with the rotating unit.

At the opposite side, in the centre, a microstructured insulating layer is placed. It is PMMA structured using the LIGA process and presents an array of round holes with the diameter of 20μm and a distance between the holes of 20μm. The whole array has a circular form with a diameter of 5.5mm and the total area of 5mm². The height of the microstructured layer can vary from 40 to 1000μm, the corresponding aspect ratio being 2 to 50. This design of microstructures was chosen as a convenient for measurement data processing, delivers currents in the order of 100μA and allows the manufacture of microelectrodes with various aspect ratios. The side face of the complete electrode is insulated with PMMA based glue.

In order to provide the constant interface area and to avoid concurrent electrode-electrolyte reactions the electrode surface is required to be made of chemically inert material. In this construction a gold coating is used.

3. The manufacture of RRME

3.1 The description of common techniques for the RRME manufacturing

The metal substrate with a diameter of 15mm was produced by eroding out a standard copper titanium coated (3μm) plate (84x54x8mm) normally used as mold insert base. For better PMMA adhesion and uniform start of electrochemical gold deposition wet oxidation of titanium was carried out for 4 minutes. These substrates were fastened in the custom-made holder made of copper plate for the exposure and the development.

The PMMA (GS-233, Röhm GMbH, average molecular weight $1.2 \cdot 10^6$g/mol) sheets used for the x-ray lithography were cut into circular pieces with a diameter of 15mm, polished, annealed and then glued to the substrate.

The application of resist was achieved using a JANOME gluing robot under argon atmosphere. Two kinds (A and B) of glues based on PMMA were used. The glues were prepared directly before use via addition to the glue base of starting initiators (15% BPO (Benzoyperoxide), 3% MEMO (Trimethil-methacrylate), 1%DMA (N,N-dimethylalanine), all weight %). The glue base A is a solution of PMMA in MMA. The glue type B base is a solution of PMMA in MMA containing GMA (Glycidylmethacrylate)[5]. After gluing and the glue polymerization all samples were annealed.

X-ray lithography was used for the pattern transfer. Exposures were made at the ANKA synchrotron using the Litho2 beamline (equipped with a Ni-coated Si-mirror, angle 8.65mrad, energy range: 2.5keV to 7.0keV).

The development was made in a conventional so-called GG-developer[6] with subsequent rinse in BDG-rinse and washing with distilled water. After development the samples were dried in air for 12 hours. It was followed by development quality control using an optical microscope for structure height up to 100μm. If this quality is satisfactory, the sample can be processed further. For the higher structures, quality control is only with electron microscope possible, but subsequent sample use becomes impossible. This means that quality control for such structures can be achieved only indirectly via electrochemical testing (see the part **4.**).

Electrochemical gold deposition for technology I was made in conventional gold sulfite electrolyte (concentration 25g/L) at a temperature of 55°C, a rotation rate of 200 rpm, a constant current density of 0.6A/dm², for 20 minutes. Gold layer thickness of 2.55μm was targeted.

The vapor deposition of gold for technology II was made in UNIVEX 450 (LEYBOLD-HEREAUS GMbH) e-beam evaporator.

3.2 Manufacturing Technology I

The manufacturing technology I, presented in the Fig. 2, was considered initially.

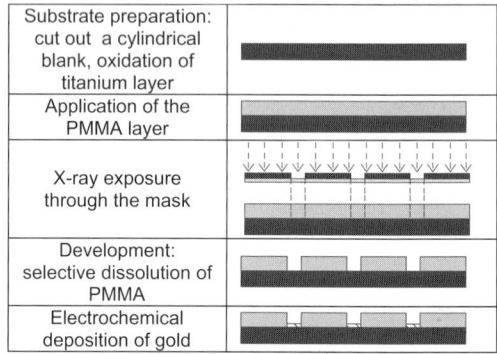

Substrate preparation: cut out a cylindrical blank, oxidation of titanium layer	
Application of the PMMA layer	
X-ray exposure through the mask	
Development: selective dissolution of PMMA	
Electrochemical deposition of gold	

Fig. 2. Manufacturing technology I

Its common stages are substrate preparation, gluing of the PMMA layer using the glue type A, x-ray exposure through the mask, development and electrochemical gold deposition.

Two deficiencies were found in the gold electroplating step. Firstly, plating at constant current started non-uniformly. Secondly, the adhesion of the insulating microstructured layer to the substrate became significantly worse. This lead to partial or complete detachment of this layer. The use of glue type B for better adhesion did not improve the situation. The reason for poor adhesion thought to be the large difference in thermal expansion coefficients of PMMA ($\alpha=67,3 \cdot 10^{-6}$/K) and copper substrate ($\alpha=17,0 \cdot 10^{-6}$/K) [7]. These complications were not reported for a conventional LIGA process, because gold is plated to the structure height, sometimes even overplated. In this case the non-uniformity of the plated layer is smoothed; the detached resist layer clings to the grown gold structures and is finally removed by stripping.

3.3. Manufacturing Technology II

Vapour deposition of gold was proposed in order to avoid the complications mentioned above. The steps of improved manufacturing technology II are shown in Fig. 3.

The main problem of this technology is insufficient adhesion of PMMA to gold. Its solution was found in the employment of the alternative glue type B and the deposition of chromium-gold layer (20nm of chromium and 300nm of gold) onto the rough titanium oxide layer improving the adhesion. The gold layer was deposited only in central part of the substrate in form of a circle with a diameter of 7mm in order to gain better adhesion and insulation at the substrate edge.

PMMA sheets were glued after the gold deposition using glue of type B. Further steps of this technology are fully identical to the technology I excluding the electroplating of gold.

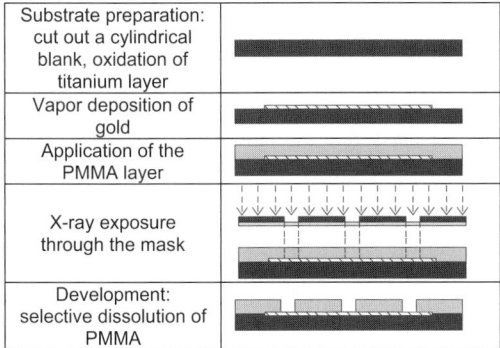

Substrate preparation: cut out a cylindrical blank, oxidation of titanium layer	
Vapor deposition of gold	
Application of the PMMA layer	
X-ray exposure through the mask	
Development: selective dissolution of PMMA	

Fig. 3. Manufacturing technology II

4. Electrochemical testing of fabricated RRME

4.1 Experimental

A conventional three electrode cell was used for electrochemical tests. As working electrodes the gold rotating disk electrode with a diameter of 2mm (RADIOMETER ANALYTICAL) and that fabricated by means of both technologies RRMEs were used. The Hg/HgSO₄-electrode (XR200, RADIOMETER ANALYTICAL) was used as reference electrode and platinum wire electrode (RADIOMETER ANALYTICAL) was used as the auxiliary electrode. The redox electrolyte (20/20mM $K_3[Fe(CN)_6]/K_4[Fc(CN)_6]$, in 1MKCl) was prepared, based on purified degassed water.

The testing of fabricated RRME is carried out by cyclic voltammogram recording for manufactured electrodes. The electrochemical system is the in well known ferro/ferricyanide redox couple:

$$Fe^{II}(CN)_6^{4-} \leftrightarrow Fe^{III}(CN)_6^{3-} + e^- . \quad (1)$$

All tests were performed at a temperature of 20±1°C, rotating speed of 500rpm and a scan rate of 2mV/s.

The cyclic voltammogram for gold RDE is used as the reference curve (Fig. 4).

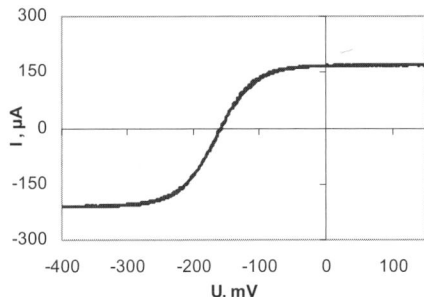

Fig. 4. The cyclic voltammogram for 2mm diameter gold RDE, scan rate 20mV/s

During the analysis of obtained curves, attention was paid to the following factors: the presence of limiting current plateau, value of limiting current, value of equilibrium potential (for the zero current), and the presence of hysteresis. The value of equilibrium potential in redox electrolyte is defined by the chemical composition of the couple electrolyte-electrode and of the electrolyte temperature. If the same electrode material at the same temperature is used, the equilibrium potential value stays constant. As it is shown in the Fig. 4. , this potential for the used system is equal to -167mV.

The presence of limiting current for U>0mV and for U<-200mV implies that, firstly, electrode is a good conductor, and secondly, there are no secondary reactions on its surface.

The magnitude of diffusion limiting current is defined, under the same other conditions, by electrode area (surface) and can be estimated by formula [8]:

$$I_{lim} \approx nFDA\frac{C}{h}, \quad (2)$$

where n=1 – number of electrons participating in reaction(1), F=96845C/mol – Faraday constant, D=(5.43·10⁻⁶ and 6.17·10⁻⁶)cm²/s – the diffusion coefficient of Fe^{II} and Fe^{III} obtained at RDE, C=2·10⁻⁵mol/cm³ – the concentration of Fe^{II} and Fe^{III}, h – height of the microstructured insulating layer varying from 20 to 1000µm, A=5mm² – surface area of the electrode. Calculated for Fe^{II} and Fe^{III} it gives the anodic and cathodic estimation shown at cyclic voltammograms for test results below as grey lines (Figs. 5 and 6).

Comparing the estimated value (2) with that obtained experimentally, one can assess the quality of fabricated electrode: if the experimental value is much lower than the estimation, this indicates insufficient development or wetting, if the experimental value much greater than the estimation, detachment of the microstructured insulating layer can be assumed.

4.2. Test results for RRME manufactured by technologies I and II

The obtained voltammograms for the RRME with aspect ratio of 5 manufactured by technologies I and II are shown in Fig. 5 and 6.

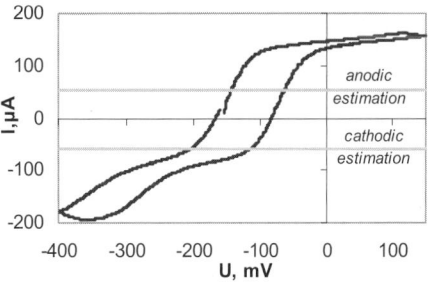

Fig. 5. Cyclic voltammogram for RRME I made by manufacturing technology I

As it can be seen in Fig. 5 the limiting current for RRME I has not been achieved, the current is larger as grey the line indicated estimated value (2) and a

94

wide hysteresis is presented. These facts signify the partial or complete detachment of the insulating microstructured layer.

The value of equilibrium potential for the RRME II (Fig.6) is similar to this for gold RDE. The limiting current plateau is present at the expected range of potential (U>0mV and U<-200mV). Thus the working surface of the electrode is identical to that of the reference (gold RDE), i.e. gold. The limiting current value does not exceed the estimated value and the hysteresis is not wide. This means, that the insulating microstructured layer is not detached.

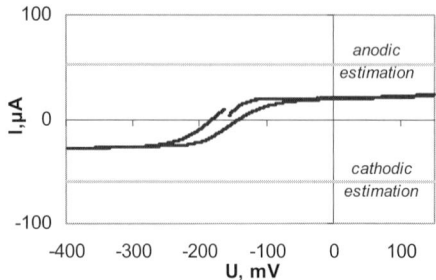

Fig. 6. Cyclic voltammogram for RRME II made by manufacturing technology II

Based on this observation is concluded, that that the manufacturing technology II can be used for the fabrication required RRME.

4. Conclusions

New construction and manufacturing technologies for RRME using LIGA process are outlined. Testing of this technology is carried out and its improvement has been shown through substitution of electrochemical gold deposition by vapor deposition and using an alternative glue providing sufficient PMMA to gold adhesion.

It has been shown that RRME, produced by improved technology II, meets qualifying requirements and can be used for mass transfer investigations in micro technology.

References

[1] Chung S.J., Hein H., Mohr J. et al. LIGA technology today and its industrial applications. Proc. SPIE 4194, 2000
[2] Landau U. Determination of laminar and turbulent mass transport rates in flow cells by the limiting current technique. AChE Symposium Series, No.204, Vol.77 (1981), 75-78.
[3] Selman J.R. Techniques of mass-transfer measurements in electrochemical reactors. AChE Symposium Series, No.204, Vol.77 (1981), 88-103
[4] Leyenecker K., Bacher W, Stark W., Thommes A. New Microelectrodes for the investigation of the electroforming of LIGA Microstructures. Electrochem. Acta, 39 (8/9) (1994), 1139
[5] Patent Nr. DE10135529 B4 2005.09.20
[6] Greeneich J.S. Developer characteristics of poly(methyl metacrylate) electron resist. J.Electrochem. Soc. 122 (1975), 970-975
[7] Achenbach S. Optimierung der Prozess-bedingungen zu Herstellung von Mikro-strukturen durch ultratiefe Röntgenithographie. PhD Thesis, Forschungszentrum Karlsruhe, Wissenschaftliche Berichte 3512, 2000
[8] BON: Bond A.M., Luscombe D. et al. A comparison of chronoamoerometric response at inlaid and recessed disk microelectrodes J.Anal. Chem., 249(1988);1

Multi-Material Micro Manufacture
W. Menz, S. Dimov and B. Fillon (Eds.)

95

Membrane-less mass flow micro sensor

T. Velten[a], H. Schuck[a], T. Knoll[a], N.Graf[a], W. Haberer[a]

[a] Fraunhofer Institute for Biomedical Engineering (IBMT), 66386 St. Ingbert, Germany

Abstract

Nowadays, most micro machined mass flow sensors which are based on the thermo transfer principle are made of silicon. Unfortunately, silicon is a rather good heat conductor. To avoid high thermal losses heaters and temperature sensing elements are commonly placed on thin membranes of silicon nitride or silicon dioxide. The fragile membranes make these sensors sensitive to any mechanical stress or to pressure peaks in the fluidic system. In this paper we report on the development of a membrane-less micro mass flow sensor. The sensor is optimized for an extremely low range of 50 µl/h which is a typical value in implantable drug delivery systems. The focus of this paper is on the fabrication process as well as on sensor characterization. The sensor mainly consists of two glass parts which are fabricated using typical micro machining techniques. The sensor dimensions are 5 x 3 x 1 mm^3. The measured sensor uncertainty is in the order of ±10% full scale output. Most noticeably, the sensor output seems to be rather independent of changes of the ambient temperature.

Keywords: microfluidics, micro flow sensor

1. Introduction

Mass flow sensors are widely-used to control the exact dosing of gases or liquids in many technical systems. The size of the sensor is adapted to the diameter of the pipe through which the fluid or gas is transported. Small diameter tubes make it necessary to use miniaturized sensors which are usually fabricated by exploiting silicon technology.

A rather new application field for mass flow sensors is the medical field of drug delivery. Here, small amounts of liquid drugs are handled and need to be metered. Only miniature sensors are suited for this application. The demand for small-sized sensors is even more severe if the sensor is part of an implantable drug delivery system. Another requirement in medical applications is reliability and safety, i.e. failure of the sensor has to be strictly avoided. The missing mechanical robustness of the membrane of silicon-based micro mass flow sensors is their main drawback which hampers the use of micro machined silicon mass flow sensors [1-4] in medical implants.

Here, we report on a membrane-less micro mass flow sensor which is based on glass substrates. Compared to silicon, glass has a lower specific heat conductivity as well as a higher chemical resistance against most chemicals.

2. Sensor design

The sensor presented in this paper is designed to be part of an intra-oral drug delivery system which has the size of two molar teeth. The concept of this drug delivery system and also the sensor requirements has already been reported in detail in [5].

The main requirements for the sensor is small size, mechanical robustness and low power consumption. The sensor range is 50 µl/h, the target resolution is 5 µl/h and the sensor uncertainty should not be worse than ±10% full scale output. As already mentioned above the mechanical robustness is guaranteed due the membrane-less design.

The sensor is working according to the heat transfer principle which is depicted in Fig. 1. The sensor core, comprising a heater and two temperature sensing elements, is located inside a fluidic channel. The main design parameters affecting the sensor resolution is the distance between heater and temperature sensing elements as well as the height and width of the fluidic channel. The sensor geometry furthermore affects the response time of the sensor. This is the minimum on-time of the heater which leads to a thermal steady state condition. A low response time is highly desirable because short heating times lead to a low power consumption which is a main sensor requirement.

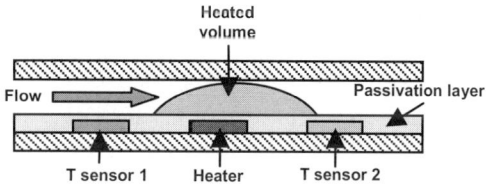

Fig. 1. Flow sensor (thermotransfer principle).

Details of the design optimisation process will be reported elsewhere. Here, we will only report the results of this process: The meandric resistor structures extend to 95 µm (heater) and 67.5 µm (temperature sensors) along the channel. The fluidic channel has a height of 15 µm and a width of 50 µm. The optimal distance between the centre of the heater and the centre of the temperature sensors is about 100 µm.

3. Fabrication process

The developed micro machined flow sensor consists of two parts, which are micro fabricated on 4" glass wafers (Pyrex 4470). The bottom part contains resistor structures made from platinum as well as conducting paths made from gold. Both metal layers are structured using lift-off processes. The metal layers are passivated by a combination of silicon nitride and silicon dioxide thin films. The channel walls are realised by the lithographic structuring of a negative photoresist (C-AR 4400-25, AllResist) with a thickness of about 15 µm. The width of the flow channel is 50 µm. The core of a patterned bottom part is depicted in Fig. 2.

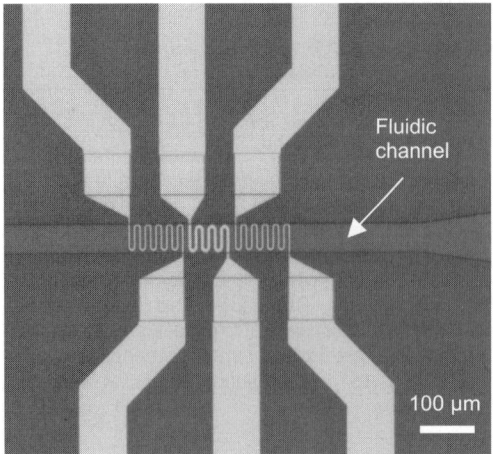

Fig. 2. Patterned bottom part of the sensor: meandric heater and temperature sensors inside the fluidic channel .

The top part of the chip contains the holes for the fluid inlet and outlet, which are realised by micro-powderblasting. Both chips are bonded to each other using a low-temperature adhesive bonding process at 373 K. At this temperature, the top part was manually aligned to the bottom part by means of an optical microscope. The resulting sensor chip has dimensions of 5 x 3 x 1 mm^3 (see Fig. 3). The fluidic connection of the sensor is established by tubings (PEEK) which are glued into the powder blasted holes of the top chip.

Fig. 3. Unpackaged flow sensor chip.

The assembled flow sensor is mounted on an interface ceramic substrate containing screen printed conducting pads. A connector is soldered to the ceramic substrate to allow to connect the packaged sensors to the sensor electronics easily. Fig. 4 shows a packaged flow sensor chip. The fluidic interconnects is not shown in this figure.

Fig. 4. Packaged flow sensor chip on ceramic substrate.

3. Sensor characterisation

For sensor characterisation a precision syringe pump (Type MDSP3f, MMT, Germany) generated different flows (5, 10, 20, 30, 40, 50 µl/h) which were pumped through the sensor under test. Additionally a commercial sensor (SLG1430-150 from Sensirion, Switzerland) was fluidically connected in series and both outputs were compared. To avoid any temperature influence from the environment, the sensor under test was placed in a climatic chamber at fixed temperature conditions. Pump and reference sensor were located outside this chamber. Temperature fluctuations inside the chamber were in the range of ± 1 K. Rather long tubes were necessary to fluidically connect pump, sensor under test and Sensirion reference sensor.

To determine the response time the sensor signal was measured at a constant flow of 5 µl/h for heater-on times ranging from 50 ms to 500 ms in steps of 50 ms. The results are shown in Fig. 5.

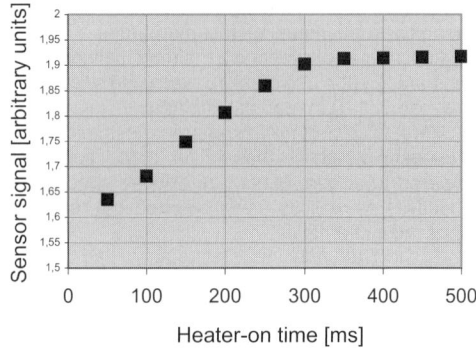

Fig. 5. Sensor signal for different heater-on times.

The sensor response to various flow values is depicted in Fig. 6. The sensor raw signals have been amplified by about 18000. This measurement has been made with deionised water at a temperature of 37°C.

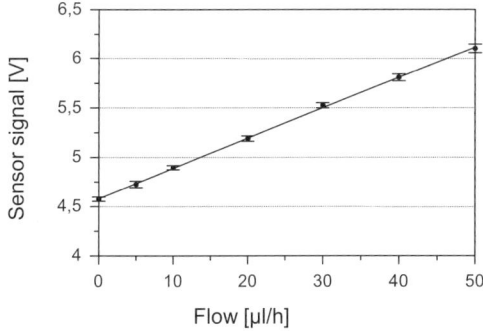

Fig. 6. Sensor signals for various flows. Amplification of sensor signals is about 18000.

Using the results of this measurements the measured flow [µl/h] can be calculated on basis of the sensor signal. This relation is shown in Fig. 7.

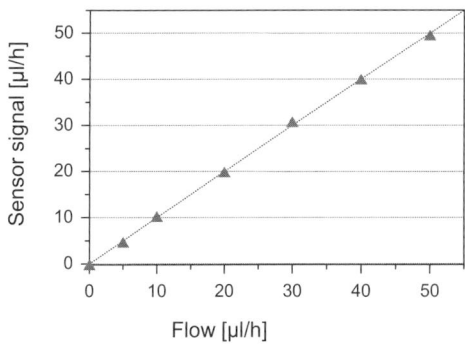

Fig. 7. Sensor signals [µl/h] for various flows at 37°C.

To determine which effect temperature variations have on the sensor signal the measurements were repeated for temperatures of 20, 30 and 50°C. The observed temperature effect on the sensor signals was rather small. In a type of diagram like shown in Fig. 7 the differences were almost not visible. Therefore, in Fig. 8 a different type of diagram is chosen. Fig. 8 shows the deviation (% full scale output) of the response curves measured at different temperatures from the response curve measured at 37°C. This demonstrates the behaviour of a flow sensor which is calibrated at a temperature of 37°C but which cannot compensate for any temperature effects.

Up to now, all measurements have been done with deionised water. But as described above the mass flow sensor has been developed for a drug delivery system which will deliver Naltrexone as drug. Therefore, additional measurements have been done with an almost saturated solution of Naltrexone. The result is shown in Fig. 9. Again, the sensor has been calibrated at 37°C.

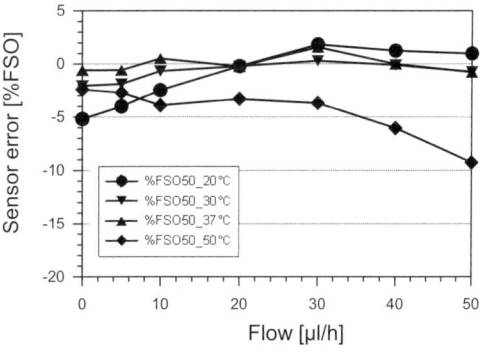

Fig. 8. Measurement with deionised water: sensor error [full scale output] for various flows at various temperatures for a sensor which was calibrated at 37°C.

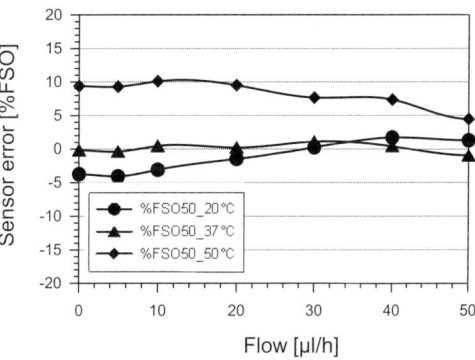

Fig. 9. Measurement with Naltrexone: sensor error [full scale output] for various flows at various temperatures for a sensor which was calibrated at 37°C.

4. Results and discussion

The measurements for different heater-on times showed that the sensor signals are higher for longer heater-on times. This is due to the fact that the amount of heat which is generated by the heater and which is transferred to the fluid inside the microfluidic channel is proportional to the on-time of the heater. On the other hand, there are heat losses because heat is conducted through the walls of the microfluidic channels. After some time a thermal steady state is reached. Fig. 5 shows that this steady state is reached after about 300 ms. This is the minimal heater on time which is required to achieve stable and defined measuring conditions.

Fig. 6 and Fig. 7 indicate that the sensor shows a linear behaviour for flows which are within the aimed at measuring range of 50 µl/h. The linearity error seems to be rather small. To be able to give quantitative figures for the linearity error and the errors due to changes of temperature a more appropriate diagram has been chosen for Fig. 8. This diagram indicates that temperature variations do not affect the sensor sensitivity nor the sensor offset. It seems that the sensor errors shown in Fig. 8 are caused by the reproducibility of measurements. No clear tendency of

98

temperature effects is observable. As the measurement at 37°C has been used to calibrate the sensor by applying linear regression the best linearity is achieved for a temperature of 37°C. All curves measured at other temperatures show similar linearity errors. Fig. 9 confirms these findings. This diagram also shows that sensor errors are almost the same for deionised water and for a solution of saturated Naltrexone.

The sensor uncertainty, including linearity errors and temperature effects, are in the range of ±10% full scale output. A temperature effect on the sensor signal is not clearly visible.

As mentioned above, the different components of the fluidic system have been connected by rather long and flexible tubes. The authors assume that this setup is very sensitive to any mechanical effects like e.g. vibrations. Keeping in mind that the measuring range of the sensor is extremely small (50 µl/h) it seems to be very challenging to perform measurements which have a reproducibility of better than 10% of the measuring range. Hence, it can be concluded that the sensor uncertainty is at least ±10% full scale output. Further and more accurate measurements would be necessary to determine if the fluidic setup or the mass flow sensor itself is responsible for the measured sensor errors.

5. Conclusions

A membrane-less micro mass flow sensor has been fabricated and characterized. The sensor consists of two glass plates and has a size of $5 \times 3 \times 1$ mm^3. The sensor fulfils the requirements of mechanical robustness and small size. Additionally it shows a very short response time which is in the range of about 300 ms. The sensor uncertainty is at least ±10% full scale output at thus it fulfils the requirements. The measurements indicate that the sensor is not sensitive to any temperature variations in the range of 20-50°C.

Acknowledgements

This work was supported by a European Grant under the Sixth Framework – Project IntelliDrug – IST-FP6 Contract No 002243.

References

[1] M. Ashauer et al. Thermal flow sensors for very small flow rate. Proc. 11th International Conference on Solid-State Sensors and Actuators (Transducers'01), Munich, Germany, June 10-14, (2001).

[2] Thorbjörn Ebefors, Edvard Kälvesten, and Göran Stemme. Thermal flow sensors for very small flow rate. IEEE Int. Workshop on Micro Electro Mechanical System (MEMS'98), Heidelberg, Germany, January 25-29, 1998.

[3] Takashi Yoshino, Yuji Suzuki, Nobuhide Kasagi and Shoji Kamiunten. Optimum design of micro thermal flow sensor and its evaluation in wall shear stress measurement. Proc. Int. Conf. MEMS'03, Jan. 2003, Kyoto, pp. 193-196.

[4] A. J. van der Wiel, C. Linder, N.F. de Rooij. A liquid velocity sensor based on the hot-wire principle. Sensors and Actuators A, 37-38, 1993, pp. 693-697.

[5] T. Velten, H. H. Ruf, T. Knoll, T. Koch, O. Scholz, A. Wolf, B. Z. Beiski, IntelliDrug Consortium, "Intelligent intraoral drug delivery microsystem". Proceedings of the 1st International Conference on Multi-Material Micro Manufacture (4M), 29th June – 1st July 2005, Karlsruhe, Germany, pp. 393-396.

Fabrication of a solenoid-type inductor with electroplated NiFe magnetic core by MEMS

Chong Lei[a], Wen Ding[a], Yong Zhou[a], Xiao-Yu Gao[a], Hyung Choi[b], Jonghwa Won[b]

[a] *National Key Laboratory of Nano/Micro Fabrication Technology, Key Laboratory for Thin Film and Microfabrication of Ministry of Education, Institute of Micro and Nano Science & Technology, Shanghai Jiao Tong University, Shanghai 200030, China (corresponding author: yzhou@sjtu.edu.cn)*
[b] *Samsung Advanced Institute of Technology (SAIT), Samsung Electronics Co., Ltd., 416 Maetan-Dong, Yeongtong-Gu, Suwon, Kyungki-Do 442-742, Korea*

Abstract

A solenoid-type inductor was fabricated by MEMS (Microelectromechanical Systems) technique. NiFe film was electroplated as the magnetic core, and polyimide with a low relative permittivity was used as the insulation material. In the fabrication process, UV-LIGA, dry etching, fine polishing and electroplating technique have been adopted to achieve high performance of the solenoid-type inductor. The inductor was in size of 3.9 mm × 2.66 mm × 80 μm with coil width of 20 μm, space of 35 μm and aspect ratio of approximately 4:1. The inductance was 1.973 μH at a frequency of 1 MHz, and the maximum quality factor was 1.52 at a frequency of 2.14 MHz with an inductance of 1.82 μH, which is attractive for the applications of micro DC/DC converter.

Keywords: Inductance, NiFe magnetic core, MEMS, solenoid-type inductors, quality factor

1. Introduction

Recently there has been a great demand on Micro DC/DC converter for the applications of the portable electronic products such as mobile communication product CDMA, portable notebook computer, microprocessor, digital camera, digital camcorder, flash memory, network product ADSL, PDA and so on [1]-[5]. Magnetic components such as inductors and transformers are the essential components of the electronic circuits, where the miniaturization of inductor and the integration of inductor with electronic circuit are the key to realize the electronic products with high performance, small size, light weight, high saturation current, high efficiency and low product lost. However, the fabrication of microinductor with high inductance and high quality factor (Q-factor) at high frequencies is an extremely difficult task at present, and it is difficult to fabricate the microinductor using the conventional semiconductor technology. In recent years, with the development of MEMS (Microelectromechanical Systems), especially the non-silicon fabrication technology of UV-LIGA has become one of the most advanced technologies in fabricating the 3D structures and RF MEMS [6-10]. Ahn [11] has reported a solenoid-type inductor with dimension of 4 mm × 1.0 mm × 120 μm, and the inductance value is 0.4 μH at 10 kHz with maximum Q-factor of 1.5 in the frequency range from 10 kHz to 1 MHz. In present work, microinductor with ultra-low profiles (3.9 mm × 2.66 mm × 80 μm) is fabricated using MEMS techniques, which has high inductance and high Q-factor in the frequency range of 1-10 MHz.

2. Design

The solenoid-type inductor is composed of a magnetic core and multilevel metal conductors. It is designed to have a completed closed magnetic circuit with rectangular shape as to minimize the flux leakage. Electroplating technique is used to fabricate the conductor lines and the vias, as electroplated metal contacts usually have relatively low metal contact resistance [11]. The inductance value is calculated by the following equation [5,12]:

$$L = \frac{\mu_0 \mu_r A_c N^2}{l_c} \qquad (1)$$

where A_c is the cross-sectional area of the magnetic core, l_c is the total length of the closed magnetic core, N is the number of coil turns, μ_0 and μ_r are the vacuum and relative permeability of the core material, respectively.

The Q-factor can be defined as follows:

$$Q = \frac{\omega L}{R} = \frac{\omega \mu_0 \mu_r N A_c A_w}{2(w+l)\rho l_c} \qquad (2)$$

where A_w is the cross-sectional area of the conductor, ω the angular frequency, $2(w+l)$ the length of coil per turn, ρ the resistivity of the conductor material.

From the above Eq. (1) and (2), it is seen that the inductance value and Q-factor are linearly proportional to μ_r and A_c in the microinductors. However, unlike the conventional inductors, the inductance value is not proportional to the square of the number of coil turns in the microinductor [13], since due to the fabrication constraints on the micromachined inductor, larger numbers of coil turns require a longer core length.

3. Fabrication process

Fig.1 shows a brief fabrication process for the complete inductor. First, the double-side alignment symbols were formed on the clean glass substrate [Fig. 1(a)], which was very important for double-side mask alignment photolithography. The chromium/copper layer was sputtered as a seed layer for electroplating and then covered by thick photoresist. The patterns of the bottom conductor traces were transferred photolithographically, followed by the selective electroplating of the bottom conductor lines and vias [Fig. 1(a)]. After that, the photoresist and chromium/ copper seed layer were removed with dry etching, in avoiding of the erosion of bottom conductor lines by using wet etching. In order to isolate the conductor lines and the magnetic core, thick polyimide was used as an insulation material and the structure holder. After coating, polyimide was cured at 250 C for 2 hrs in argon. Then polyimide was fine polished instead of RIE etched [12] until the vias were exposed [Fig. 1(b)]. Through this process, the bottom and top conductor lines can be interconnected together. For electroplating the magnetic core, the chromium/copper seed layer was sputtered again. Thick molds for electroplating the magnetic core were patterned and filled with electroplated nickel – iron [Fig. 1(c)] adopting the standard electroplating technique [14]. To fabricate the top conductor lines [Fig. 1(d)], the same process, which was used to fabricate the bottom conductor traces was used. The completed inductor was illustrated in Fig. 1(e). Fig. 2 shows the SEM (scanning electron microscopy) photograph of the fabricated microinductor, which consists of the electroplated copper conductor lines, the electroplated magnetic core and polyimides.

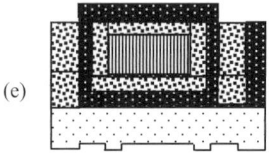

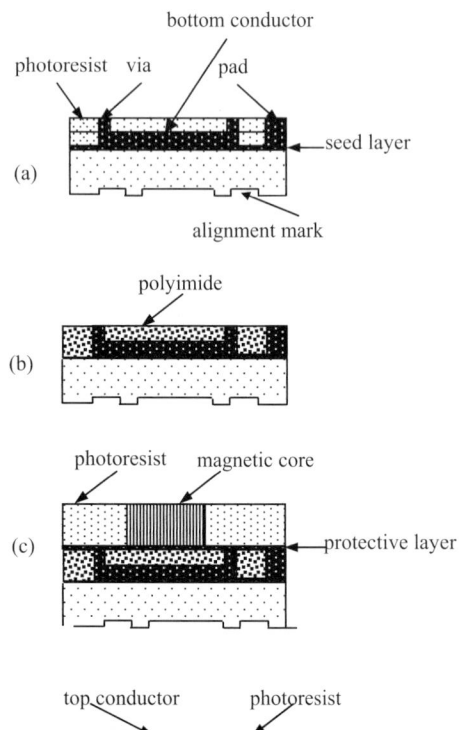

Fig. 1. Fabrication process of a solenoid-type microinductor: (a) fabrication of alignment marks, electroplating the bottom conductor lines, vias and pads; (b) spin coating, curing and polishing of polyimide; (c) fabrication of magnetic core; (d) electroplating the vias and pads; spin coating, curing and polishing of polyimide; electroplating the top conductor; (e) removal of seed layer.

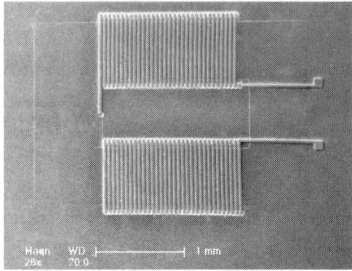

Fig. 2. Photograph of the fabricated solenoid-type inductor.

4. Results and discussion

According to Eqn. 1, there are three kinds of ways to improve the inductance values: 1) using magnetic core material with high relative permeability; or 2) increasing the cross-sectional area of the magnetic core material through adding magnetic core thickness; or 3) increasing coil turns, which means decreasing the coil width and pitch between the adjacent coils for a certain value of surface area. In this experiment, microinductor which has ultra-low profile of 3.9 mm × 2.66 mm × 80 μm, 64 turns, coil width of 20 μm separated in space of 35 μm and aspect radio of 4:1, was fabricated by MEMS techniques. At last it was measured in the frequency arrange of 1 MHz to 10 MHz by using the Agilent E4991A RF Impedance/Material Analyzer, which was connected to a probe station provided by Cascade Microtech manufacturer. The Cascade Microtech ACP40-GS Series were used as the probe head. After making the open/ short/ load calibrations, measurements were performed for the inductor. Fig. 3 shows the inductance, Q-factor and resistance as functions of frequencies. The inductance value is 1.973 μH at a frequency of 1 MHz, and the maximum Q-factor of 1.52 is achieved at a frequency of 2.14 MHz with an inductance value of 1.82 μH. In Ref. [8], it is reported that the fabricated inductor has inductance value of 0.4 μH at 10 kHz with maximum Q-factor of 1.5 at 1 MHz. As shown in Fig. 3(a), the inductance value is fall-off at higher frequencies, this is due to the dependence of the permeability of the nickel-iron core on frequencies. The resistance increases along with the frequency, this is because the proximity effect and skin effect become stronger in the frequency range from 1 to 10 MHz [Fig. 3 (c)].

To determine the circuit parameters of the integrated inductor, an equivalent circuit shown in Fig. 4 was assumed in which a stray capacitance is in parallel with the series connection of the inductance and the internal resistance. According to the equivalent circuit, impedance Z can be calculated by the following expression:

$$Z = \frac{\dfrac{(R + j\omega L)}{j\omega C}}{R + j\omega L + \dfrac{1}{j\omega C}} = \frac{R + j\omega L}{1 + j\omega C(R + j\omega L)} \qquad (3)$$

According to Eq. 3, the stray capacitance C was approximately equivalent to zero, where $\theta_z \approx 50°$, L=1.973 μH, f=1 MHz, R=10.389 Ω at 1 MHz as shown in Fig. 3. The same results were obtained from 1 to 10 MHz. From these data, it can be concluded that the winding capacitance and other parasitic capacitance do not appear to affect the inductor performance in the measured frequency ranges.

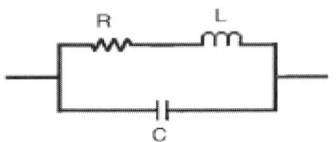

Fig. 4 Equivalent circuit of the inductor, where R, L and C is referred to resistance, inductance and capacitance, respectively.

5. Conclusion

Microinductor with ultra-low profile was fabricated by MEMS techniques. The total dimension of the inductor is 3.9 mm × 2.66 mm × 80 μm, having 64 turns, with coil width of 20 μm separated in space of 35 μm and high aspect radio of 4:1. The maximum quality factor of 1.52 at a frequency of 2.14 MHz with an inductance of 1.82 μH, is achieved.

Acknowledgements

This work was supported by Samsung Advanced Institute of Technology (SAIT), Samsung Electronics Co., Ltd., Shanghai-Applied Materials Research and Development (No. 0515), the National High Technology Research and Development Program (No. 2004AA302042), the Nanotechnology Program of Shanghai Science & Technology Committee (No. 0352nm014).

References

[1] Park JW and Allen MG. IEEE Trans. Magn. 39 (2003) 3184.
[2] Zhuang Y, Rejaei B, Boellaard E, Vroubel M and Burghartz JN. IEEE Electron device lett. 24 (2003) 224.
[3] Prieto MJ, Pernia AM, Lopera JM, Martin JA and Nuño F. IEEE trans. industry applications 38 (2002) 543-552.
[4] Korenivski V. J. Magn. Magn. Mater. 215-216 (2000) 800.
[5] Soohoo RF. IEEE Trans. Magn.15 (1979) 1803.
[6] Bryzekj. Sens. Actuators A56 (1996) 1.
[7] Laurent D, Olivier G and Rooij D. Proc. SPIE 4176(2000) 16-27.
[8] De Los Santos HJ, Fischer G, Tilmans HAC, and van Beek JTM. IEEE Microwave Magazine (2004) 50.
[9] Strohm KM, Sehmiickle FJ, Schauwecker B, Luy ,JF, and Heinrich W. IEEE MTTS Digest, IF-WE-16 (2002) 1209-1212.
[10] Becker B, Notarp DL, Vogel J, Kieselstein E, sommer JP, Bramer K, Großer V, benecke W, and Michel B. Microsystem Technologies 7 (2001) 196-202.
[11] Ahn CH and Allen MG. IEEE Trans. Industrial Electronics 45 (1998) 866.
[12] Park JY and Allen MG. IEEE Trans. Electronic Packaging Manuf. 23 (2000) 48.
[13] Liu C. Study of silicon integrated inductor and CMOS RFIC. Ph.D. dissertation, Chinese Academy of Science, Shanghai Institute of Microsystem and Information Technology, Shanghai, China, 2002.
[14] Anderson RL, Gangulee A and Romankiw LT. J. Electron. Mater.2 (1973) 161.

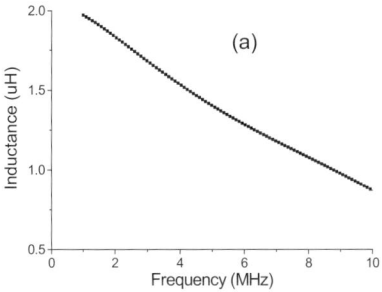

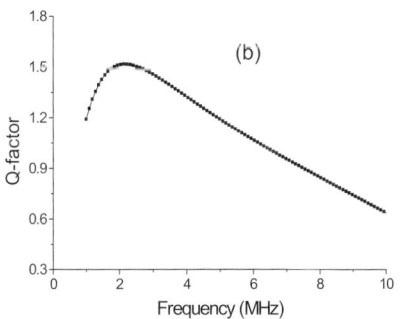

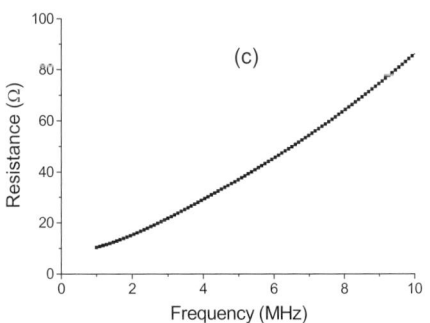

Fig. 3 Measured inductor parameters as functions of frequencies: (a) inductance vs frequency, (b) Q-factor vs frequency, (c) resistance vs frequency.

Multi-Material Micro Manufacture
W. Menz, S. Dimov and B. Fillon (Eds.)

Surface micromachined three-dimentional solenoid-type inductor

Dong-Ming Fang, Yong Zhou, Wen Ding, Xi-Ning Wang, Xiao-Lin Zhao

National Key Laboratory of Nano/Micro Fabrication Technology, Key Laboratory for Thin Film and Microfabrication of Ministry of Education, Institute of Micro and Nano Science & Technology, Shanghai Jiao Tong University, Shanghai 200030, China (corresponding author: yzhou@sjtu.edu.cn)

Abstract

Miniaturized integrated inductor of high performance is very important in mobile communications and microwave integrated circuits. Existing on-chip spiral inductors based on conventional planar integrated-circuit fabrication technology suffer from substrate loss and parasitics, and have relatively large area. In this paper, We present the design, fabrication, and performance of surface micromachined on-chip three-dimentional (3-D) air-core solenoid microinductors. The fabricated microinductors are characterized at high frequencies from S-parameter measurements. The resulting 7-turn air-core solenoid inductors have inductances between 1.03 to 1.81 nH, peak quality (Q) factors between 19.1 to 38.0 at peak-Q frequencies between 4.4 to 6.0 GHz, and self-resonant frequencies between 10.5 and 13.0 GHz.

Keywords: Inductor, RF MEMS, quality factor, 3-D

1. Introduction

In many on-chip radio frequency (RF) passive components, the inductor is an essential element and is critical for integrated wireless communication systems. Most microinductors to date are planar spiral inductors because the structural and fabrication simplicity is one of well-known advantages. However, in order to achieve high performance, a planar spiral inductor has a large area of metal traces in close proximity to the substrate and the eddy current losses become a dominant factor, which limits its quality factor. So in recent years, works have been done to realize three-dimensional (3-D) solenoid inductors as an alternative to improve inductor's performance. Yung et al. proposed a concise technology to enable the fabrication of good quality 3-D suspended RF microinductors on bulk silicon without utilizing the lithography process on sidewall and trench-bottom patterning [1]. A series of solenoid-type inductors with ferromagnetic cores were demonstrated and compared to control devices with air cores [2]. In order to achieve high performance passive components integrated with CMOS circuit, on-chip solenoid inductors on Si substrate embedded WL-CSP were fabricated [3]. A high-Q on-chip solenoid inductor was also fabricated by 0.18 μm CMOS technology with air-gap structure [4].

It is important to note that an optimal inductor solution would need to provide not only good performance, but also low cost, simple fabrication process to save time, and minimum to save the chip area. In this regards, we proposed a new method for fabricating air-core solenoid inductors using simple surface micromachining techniques.

2. Design and fabrication

The most important figures of merit of the microinductor are Q-factor, inductance and self-resonant frequency (f_{SRF}). The Q-factor of the inductor is defined as [5,6]:

$$Q = \frac{\omega L_s}{R_s} \tag{1}$$

where L_S is the series inductance and R_S is the series resistance of the inductor. The first factor in Eq. 1 is the intrinsic quality factor of the overall inductance. The middle factor models the substrate loss, and the final factor models the self-resonance loss due to the overall parasitic capacitances. To improve the Q-factor, the inductance should be increased and the series resistance should be reduced. A low resistivity of the conductor and a high resistivity of the substrate are desirable. Because the series inductance is equal to the sum of the self-inductance of the conductors and the mutual inductance between coils, which can be written as:

$$L_S = L_0 + \sum M = L_0 + \sum M_+ - \sum M_- \tag{2}$$

where L_0 is the sum of the self-inductance of all the straight conductors, $\sum M_+$ is the sum of the positive multual inductances, and $\sum M_-$ is the sum of the negative multual inductances. The self-inductance of a straight conductor and the mutual inductance between two parallel conductors are given by the Greenhouse theory [7]. The geometrical structural parameters dominate the inductance and Q-factor besides the type of the substrate and the metal material of the inductor.

The structure of solenoid inductor is shown in Fig. 1, supposing that N is the number of coil turns, l, w, and t is the length, width and thickness of each conductor line, s is the horizontal space between two conductor lines, and h is the height of the pillars. For N=7, l=350 μm, w=50 μm, t=5 μm, s=40 μm and h=45 μm, the calculated inductance is about 1.74 nH and the peak Q-factor is about 40.

104

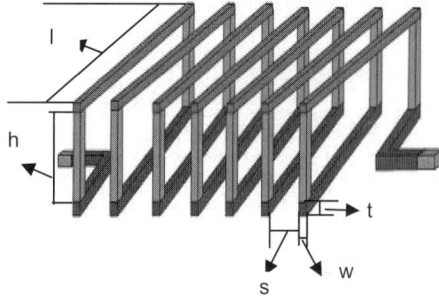

Fig. 1. Structure of air-core solenoid inductor.

Fig. 2 shows a brief description of the fabrication process. Inductors were all fabricated on 3-in glass substrates.

(1) A Cr/Cu (10nm/90nm) seed layer was sputtered onto the wafer and the AZ P4903 positive photoresist with a thickness of 5 μm was spin-coated, then the patterns were transferred and the bottom conductor lines, probe contact pads and ground planes were electroplated with copper [Fig. 2(a)].

(2) The 5 μm photoresist was removed by the acetone, then a thick layer photoresist was spin-coated and patterned, after that, the pillars were electroplated with copper [Fig. 2(b)].

(3) The thick layer photoresist was removed and the seed layer was dry etched, and then a very thicker photoresist was spin-coated [Fig. 2(c)].

(4) After that, the thicker photoresist was simply fine polished [Fig. 2(d)].

(5) The second seed layer was deposited onto the top of the fine polished photoresist, and a thickness of 5 μm photoresist was spin-coated onto the second seed layer, then the patterns were transferred and the top conductor lines were electroplated with copper [Fig. 2(e)].

(6) Finally, the 5 μm photoresist was removed by maskless exposure, the second seed layer was dry etched and the photoresist below was removed, then the solenoid microinductor was completely fabricated [Fig. 2(f)].

The fabrication process is simple, compatible with IC technology, and the cost is very low. We have fabricated four groups of solenoid-type inductors on glass substrates. Group A: l=350μm, w=50μm, h=45μm; Group B: l=350μm, w=100μm, h=15μm; Group C: l=250μm, w=100μm, h=15μm; Group D: l=350μm, w=50μm, h=15μm. The other geometrical structural parameters are N=7, s=40μm, t=5μm. Table 1 shows the geometrical structural parameters of the four group inductors.

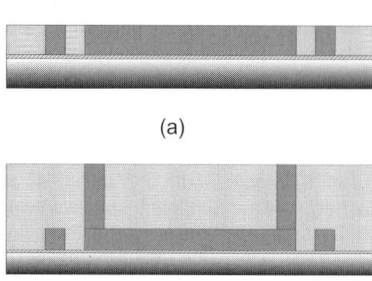

(a)

(b)

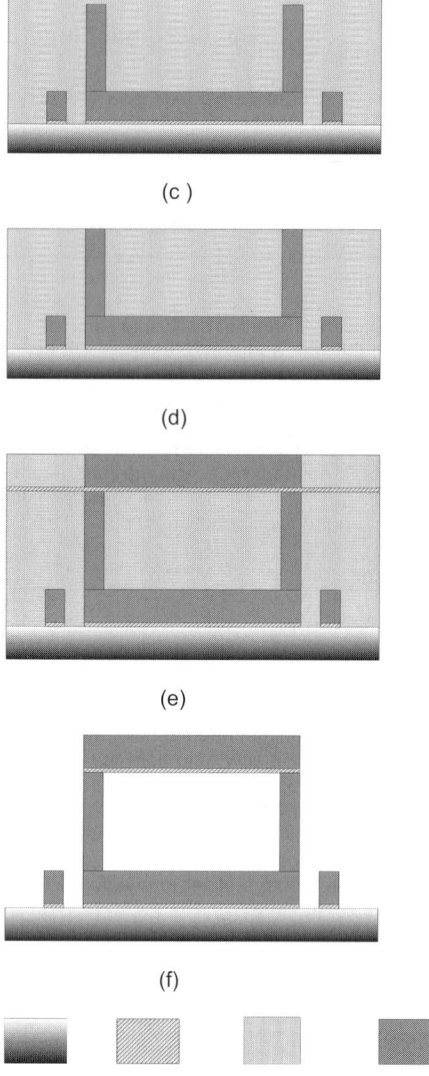

(c)

(d)

(e)

(f)

Substrate   Seed layer   Photoresist   Copper

Fig. 2. Schematic of the fabrication process flow.

Table 1
The geometrical structural parameters of four group inductors

Group	N	l (μm)	w (μm)	t (μm)	s (μm)	h (μm)
A	7	350	50	5	40	45
B	7	350	100	5	40	15
C	7	250	100	5	40	15
D	7	350	50	5	40	15

3. Results and discussion

Fig. 3 shows the SEM photographs of the fabricated air-core solenoid-type inductors (Group A and Group C) on glass substrates.

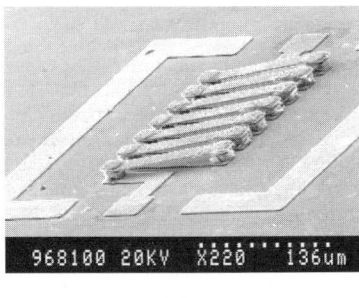

(a)

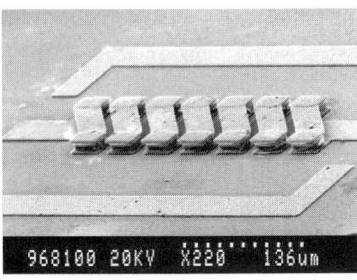

(b)

Fig. 3. SEM photographs of the fabricated solenoid inductors: (a) Group A and (b) Group C.

Two-port *S*-parameters of the inductors were measured using a HP 8722D Network Analyzer and a Cascade Micro-tech coplanar ground-signal-ground (G-S-G) probes with 250μm pitch size in the frequency range of 0.05 GHz to 10 GHz. The parasitic capacitances and the contact resistances are accurately subtracted from the measured data with calibrations on the open, short and 50-Ω resistor sites. The inductor's characteristics such as quality factor, inductance and self-resonant frequency were extracted from the measured data of S-parameter.

Fig. 4 shows the measured inductances of four group inductors as a function of frequency. For each group, the inductances increase gradually with the increase of the frequency. The inductances of Group A (higher than 1.70 nH) are much higher than those of Group B, C and D (between 1.0 nH and 1.3 nH). Compared Group A with Group D, one can see that the higher pillar of the inductor is better to improve the inductance. Compared Group B with Group C or Group B with Group D, one can find that if we increase the length (*l*) or decrease the width (*w*) of the conductor lines, the inductance will increase. These conclusions are also can be drawn from the Greenhouse theory. In addition, for a traditional solenoid-type inductor, neglecting the fringing effect, the inductance of the inductor can be approximately expressed as follows:

$$L = \frac{\mu_0 \mu_r N^2 A_c}{l_c} \qquad (3)$$

where μ_0 and μ_r are the vacuum permeability and the relative permeability of the core, respectively, A_c is the cross-sectional area of the core, l_c is the total length of the core. When *l* or *h* enlarges, A_c will increase, thus the inductance will be raised. If *w* increases, the inductance will decease.

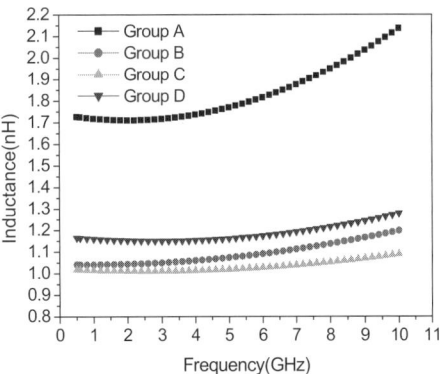

Fig. 4. Measured results of inductances of various inductors as a function of frequency.

Fig. 5 shows the measured Q-factors of four group inductors as a function of frequency. For each group inductor, it is observed that with increasing frequency, Q-factor grows initially, however, as the frequency continues to rise, the Q-factor reaches at its peak and then decreases. These inductors have high Q-factors over wide ranges of operating frequency. The inductances and Q-factors at peak-Q frequency (6, 4.4, 5.8 and 5.6 GHz) of group A, B, C, D are 1.81, 1.07, 1.03, 1.17 nH and 38, 19.1, 24.1, 21.9, respectively. These inductors have high f_{res} (self-resonance frequency) more than 10 GHz, from 10.5 GHz to 13.0 GHz, and the detail is given in Table 2.

Table 2
The inductance, Q-factor, self-resonance frequency of four group inductors

Group	Inductance L (nH)	Q-factor Q_{peak}	Peak-Q frequency f_0 (GHz)	Resonance frequency f_{res} (GHz)
A	1.81	38	6	>10
B	1.07	19.1	4.4	>10
C	1.03	24.1	5.8	>10
D	1.17	21.9	5.6	>10

According to Eq. 1, the Q-factor is synthetically dominated by the substrate, the metal material and the geometrical parameters of the inductor. Using non-conductive material such as glass or high resistivity silicon, or removing the part of silicon substrate underneath the inductor is the ways to increase the Q-factor. On the other hand, shielding the substrate with metal ground planes to decrease the resistance of the substrate is another way to improve the Q-factor. It is reported that the conductor layer composed of low resistivity metal is critical to maintain a high Q-factor for inductor at very high frequencies [8]. Since copper has low resistivity (ρ =1.724×10^{-8} Ω · m), it contributes to get high Q-factor. The reason that the Q-factor values arrive their peak-Q and then decrease as the increasing frequency is that the dissipation of energy in the substrate begin to increase faster than the inductive reactance with the increasing frequency [9].

At the same high frequency, the magnitude of the inductance values of the four groups is A>D>B>C, while the Q-factor is A>C>D>B. It is obviously that large gap between the bottom and top conductor lines, i.e. high pillar, will increase the inductance and decrease the parasitic capacitance, which determines the f_{res}, hence the inductor operation range and Q-factor. Although inductors in group B have higher inductances than those of group C, higher length (or width) of conductor lines will cause larger direct facing surface area among conductor-to-conductor and conductor-to-substrate, thus increases the parasitic capacitance, series resistance and make Q-factor lower. Moreover, higher length (or width) of the conductor lines will occupy more chip area.

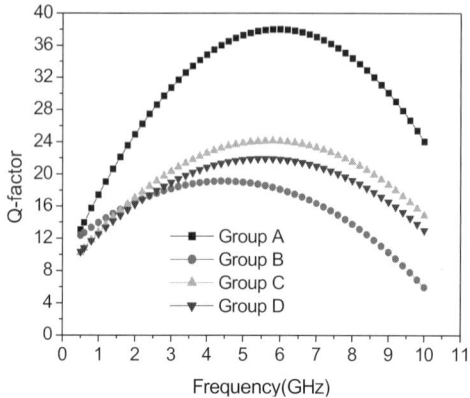

Fig. 5. Measured results of Q-factors of various inductors as a function of frequency.

4. Conclusion

Four groups of on-chip 3-D solenoid-type inductors have been successfully fabricated using surface micromachining fabrication techniques. The fine polishing of the positive photoresist is introduced to simplify the fabrication process. The inductor has an air core and an electroplated copper coil to reduce the series resistance, and the solenoid-type structure with laterally laid out structure saves the chip area significantly. The measured results show that the inductors exhibit high Q-factor, self-resonant frequency, and rather wide operation frequency range. Higher pillar will effectively improve both the inductance and the Q-factor of the inductors.

Acknowledgements

This work was supported by Key grant Project of Chinese Ministry of Education (No. 0307) and Shanghai-Applied Materials Research and Development (No. 0515). The authors would like to thank Zhi-Ping Ni, Hai-Ping Mao and Zhi-Ming Wang for experimental help, Rong Qian for technical assistance in RF measurements.

References

[1] Yung CL, Zeng WJ, Ong PH, Gao ZX, Cai J and Balasubramanian N. IEEE Electron Device Lett. 23 (2002) 700.

[2] Zhang Y, Rejaei B, Boellaard E, Vroubel M and Burghartz. IEEE Electron Device Lett. 24 (2003) 224.

[3] Itoi K, Sato M, Abe H, Ito H, Sugawara H, Okada K, Kazuya Masu and Ito T. Proceedings of HDP, 2004, pp.105-108.

[4] Lin CS, Fang YK, Chen SF. IEEE Electron Device Lett. 26 (2005) 160.

[5] Booker HG. Energy in Electromagnetism. London/New York: Peter Peregrinus (on behalf of the IEE), 1982.

[6] Yue CP and Wong SS. IEEE Journal of Solid-State Circuits 33 (1998) 743.

[7] Greenhouse HM. IEEE Transactions on Parts, Hybrids, and Packaging PHP-10 (1974)101.

[8] Kim YJ and Allen MG. IEEE Trans. Comp. Packag. Manufact. Technol 21 (1998) 26.

[9] Long JR and Copeland MA. IEEE Journal of Solid-State Circuits 32 (1997) 357.

Multi-Material Micro Manufacture
W. Menz, S. Dimov and B. Fillon (Eds.)

Development of a MID LED housing

R. M. de Zwart, R. A. Tacken, P.J. Bolt

TNO Science & Industry, De Rondom 1, P.O.Box 6235, 5600HE Eindhoven, The Netherlands

Abstract

The development of microdevices would be well served by integration of electronic connects into 3D packaging. A number of technologies have been proposed in which conductive paths are created on polymer structures, of which MID (Molded Interconnect Device) technology excels in 3-dimensional flexibility. In this study, the feasibility of manufacturing a three-colour light emitting diode housing with MID technology is demonstrated. Requirements with respect to connection to the housing included absence of wire bonding and SMT relflow solderability. Two technologies were investigated: (i) two-component molding using pre-catalysed LCP and (ii) SKW process with polyamide 4,6. In both processes successive electroless copper deposition was used followed by nickel and gold coating to obtain good solderability. In processing the pre-catalysed LCP the critical step appeared to be the initial electroless copper deposition which requires tight control to obtain the metallization selectivity required. In the SKW process proper handling of parts in the catalytic activation step between molding the first and second shot is the main factor influencing yield and quality. A fairly large number of products were made with both technologies on lab scale. The finished housings were assembled on a PCB in a reflow solder process and performance of the products was evaluated positively with respect to mechanical integrity as well as electrical behaviour.

Keywords: molded interconnect devices, metallization, integrated electronics, two-component molding, LED

1. Introduction

Molded Interconnect Devices (MID) is a generic term for technologies in which conductive paths and molded polymer parts are integrated into one component. MID technologies give new opportunities for miniaturization of complex devices, e.g. antennae in a cell phone. Integration of functionalities into one component reduces assembly costs and product volume.

The core of MID technology is the ability to realise (isolated) conductive paths on or in a polymer molded structure [1]. A number of approaches are used, some of which are suitable for 3D geometries. These include (i) insert molding with metal foil structures, (ii) physical or chemical vapor deposition with masks, and (iii) electroless plating of a surface that is locally perceptive for metallization [2]. The first is complicated because of the manufacturing and vulnerability of the foil structures, the second suffers from shadow effects and complicated mask design for 3-D projections. The last method is the most important technique in practice.

Two variations are used: (i) two-component injection molding where only one component is perceptive for subsequent metallization and (ii) local activation by laser of a plastic filled with an organometallic component ('LDS' process developed by German firm LPKF). Two-component molding is more suited for complex 3-dimensional structures but more suited for large production series since tooling costs are high; the laser method is more flexible and typically suits smaller production series, but geometrical design freedom is limited.

This paper describes the development of a MID version of a three-colour light emitting diode (LED) housing that was produced using two-component molding technology in combination with electroless plating. The targeted production volumes are very high, making two–component molding the technology of choice.

2. Existing two-component molding technologies for MID

2.1. Conventional

By selection of two molded compounds that are respectively inherently plateable and 'difficult-to-plate', a satisfying selectivity of metallization can be obtained for a very small number of combinations of plastics, which includes for example ABS (or ABS/PC blends) in combination with PC and PA66 with PA6/6T. The selectivity of metallization must be obtained by effective chemical pretreatment of the plastics that renders the first plastic receptive for activation with plating catalyst (mostly Pd) while the second remains free of catalyst.

2.2. Precatalysed polymers

The two-component molded component consists of a pre-catalysed and a non-precatalysed polymer. Because the former contains metal catalysts (nuclei) it can be metallized by an electroless plating process. The latter cannot and hence selective metallization of the surface results.

Over the last years, a number of dedicated precatalysed polymer granulates have been introduced on the market, e.g. LCP Vectra by Ticona and PBT Vestodur by Degussa. Disadvantages of these precatalysed systems are higher costs, limited number of suppliers and negative effects on the properties and processing of the polymers. Furthermore, the metallization process is slow because of a low surface concentration (which itself is influenced by the molding process) of active catalytic sites and requires very active electroless metallization formulas that require intensive process control [3].

2.3. SKW-process

In the SKW process, developed by Sankyo Kasei company, the first molding shot is taken from the mold and activated in a Pd catalyst containing solution, dried, put back in the mold and successively overmolded by the second component. The finished 2-component molded product is then selectively metallized by electroless plating on the catalysed surface of the first component. This procedure is quite elaborate and requires secure handling of parts during processing.

3. Development of a MID LED housing

3.1. Design aspects

A housing for a color controllable high power light emitting diode assembly was developed in MID-design. Fig. 1 shows the assembly of the LED schematically. The housing connects the PCB to the 3 LEDs (red, green, blue) under the yellow lens cap by 3x2 tracks. Requirements to the MID element include:
- use of robust technology, high yield, low cost
- high production volumes
- no wire bonds
- reflow solderable

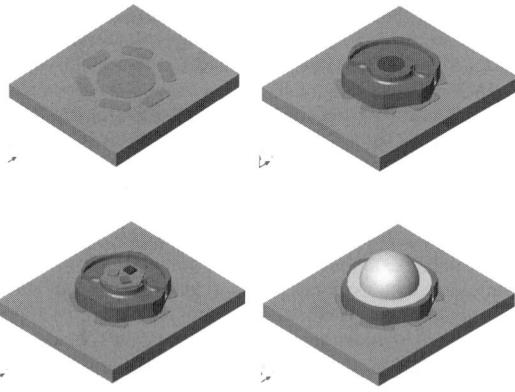

Fig. 1. LED assembly. *Up-left*: PCB with 6 solder pads and cooling slug base; *up-right*: housing with six connecting tracks placed on cooling slug; *down-left:*placement of chip with three solid state LEDs; *down-right:* with lens cover

3.2. Molding and material aspects

Choice of plastic materials was done on basis of flow characteristics, reflow solderability and above all plateability. Moreover, choice of material is directly linked to the processing route selected. To show the feasibility of the LED housing two approaches were taken:
- Two-component molding using precatalysed LCP Vectra E820i Pd in combination with non-catalysed LCP Vectra E820i (both supplied by Ticona)
- Two-shot molding using SKW process and PA4,6 Stanyl GF30 (supplied by DSM)

Large scale production of 2 shot molded housings would require a multiple cavity 2-shot mold with probably more than one injection point or film gate per shot. For this feasibility project a semi 2 shot method was used to manufacture the 2 shot housings. Two

mold cavities were fabricated by electrical discharge machining.

A first cavity produces the wheel like shaped product with 6 hammers in a cyclic proces. Part of the hammers constitute the surface to be metallized. This product is placed in a second cavity (using auxiliary tooling) which overmolds the first one and realizes the electrical isolation of the housing between the hammers and the inner circle. The second cavity has been fabricated after measuring the first molded product so shrinkage is compensated for. Fig. 2 depicts the first shot and final molded product.

 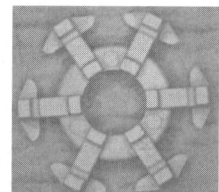
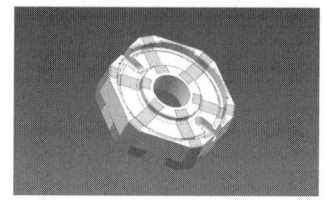 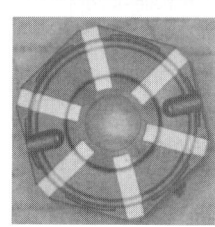

Fig. 2. Design (left) and actual (right) first-shot and final molded housing

The injection point strategy was limited by the use of a 2 plate generic test mold for the experiments. It proved that one melt entrance was sufficient to fill both cavities. It is well known that the weld-line strength of LCP is very poor. Using a single entrance for the first shot creates a mechanically weak weld line opposite the gate, see Fig. 3.

Fig. 3. Weak weld line formation in first shot with LCP Vectra 820i Pd

However in the second molding cycle this weldline is overmolded by the second plastic and reinforced. The weld-line forms at the same location when PA 4,6 molding, however the strength is much higher and cannot be broken by manual means.

Practical issues on the micro molding of LCP material include sticking of LCP in the cavity after molding. Because the cavities are so small it is difficult to use mechanical or thermal means to remove this material. In the project a solvent was developed which was able to remove LCP out of the cavity without corroding the steel mold. Details on the solvent are proprietary and cannot be presented here.

3.3. Metallization aspects

3.3.1. LCP route

In the case of the LCP route, the molded 2-shot product was etched and metallized using Enplate MID Select chemistry (supplied by Cookson Electronics): comprising pre-etching, alkaline etching, acid neutralization, and subsequent plating with Cu. Additionally, Ni-B and immersion Au coatings were applied to ensure good solderability of the finished part to the PCB (see Table 1).

Table 1. Applied coating stacks

Subsequent coating stack	LCP	PA4,6
Ni-P	-	3 µm
Cu	5-6 µm	8-10 µm
Ni-P	-	3 µm
Ni-B	3 µm	-
Au	0.2 µm	0.2 µm

There is actually no particular reason why Ni-P was used for PA4,6 and Ni-B for LCP other than reasons of practical availability at that time.

The etching step must be optimized to obtain not only a sufficient catalytically active surface but also a high adhesion strength by roughening the surface and providing anchoring sites for the metal.

Eventually a good adhesion was found that complies to ISO 9227 (cross-hatch test). Delamination or blistering was not observed. Unwanted metal deposition on the non-precatalysed LCP ('overplating') could be controlled by prevention of contact between parts in the plating baths (by racking the parts) and moreover by minimizing the plating time and precise control of the plating speed in the initial electroless Cu bath. Further, already in an early stage of plate-out of electroless Cu or Ni plating solution, the formation of very small metallic nuclei can lead to overplating ('plate-out' is the term for spontaneous decomposition of an electroless plating solution which are thermodynamically meta-stable by nature). Some examples of overplating are shown in Fig. 4.

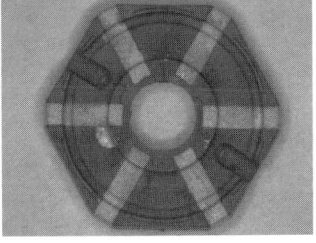

Fig. 4. Some examples of overplating due to product contacting and electroless bath instability

Initial metallization trials were performed on the first shot products, where it was observed (Fig. 5) that in a region close to the injection zone of the first shot, both etching and metallization proceeded much slower than on the remainder of the surface due to the aberrant surface condition. This phenomenon is known in literature [4] but not fully understood. In the LED design the injection point was chosen on a non-critical location which is overmolded by the second shot.

Fig. 5. injection point region on first shot LCP Vectra 820i after alkaline etching; white zones indicate slow etching

3.3.2. PA4,6 route

In the PA4,6 route, the molded first shot was taken out from the mold, etched, catalysed (nucleation+acceleration) using Noviganth AK chemistry (supplied by Atotech), dried with ethanol and hot air, and put back in the mold to inject the second shot. Then, metallization was performed according to table 1; an initial electroless Ni-P coating was applied since this showed to be superior to electroless Cu in terms of adhesion. Adhesion was tested using a pull-test method; pull strength was determined at 400 to 500 N cm^{-2}. It is clearly seen that, with proper handling of the intermediate first shot product, the catalytic activity of the surface is much higher compared to the pre-catalysed LCP. Surface coverage in the initial electroless plating step is reached within 10 to 20 seconds compared to several minutes in the case of precatalysed LCP. Therefore, a less active and more stable (higher activation energy) electroless Cu chemistry could be used in the case of PA 4,6 in the SKW route.

3.4. Assembly

Of the total of 1000 pieces produced in the laboratory environment, a smaller number was assembled in a SMT reflow process using a Heraeus oven at 260 °C, see figure 6. The Ni/Au coated tracks are wetted in an excellent way by the solder as can be seen by the solder creeping upwards along the tracks.

Fig. 6. Assembled LED with MID housing

Performance of the products was evaluated positively with respect to mechanical integrity as well as electrical behaviour.. No blistering was seen on either LCP or PA4,6 samples.

4. Summary and conclusions

Feasibility of production of a LED housing with MID technology was demonstrated using two different two-component molding techniques. In terms of process yield, the SKW process with PA 4,6 outperformed the pre-catalysed LCP route, the main reason being that the process window for metallization is larger with the SKW process. The SKW method however is more elaborate and requires more attention on aspects of logistics and timing. With both methods product requirements were met.

Acknowledgements

TNO would like to thank Lumileds, ITB Precisietechniek, DSM and the Brabantse Ontwikkelings Maatschappij (BOM) for their contribution to the work.

References

[1] "3D-MID Technologie", Carl Hanser Verlag, Munchen (2004)

[2] O'Mallory G, Hajdu J.B., Electroless Plating, AESF Orlando (1990)

[3] Prinz U, "Electroless Copper in MID", Proc. 6th MID Congress, Erlangen (2004) pp. 195-201

[4] "Kunststoff Metallisierung", Eugen G. Leuze Verlag, Saulgau (1991)

Multi-Material Micro Manufacture
W. Menz, S. Dimov and B. Fillon (Eds.)

111

Micropart manipulation by electrical fields for highly parallel batch assembly

P. Lazarou, N.A. Aspragathos
Erik Jung *

*Robotics Group, Department of Mechanical Engineering and Aeronautics, University of Patras, Patras T.K.
26500, Greece*
* Chip Interconnection Technologies, Fraunhofer IZM, 13355 Berlin, Germany

Abstract

This paper depicts the concept of micro part manipulation by electrical fields to achieve a high precision, highly parallel batch assembly of components required to put Microsystems together. The concept of orienting microparts in 3D as well as precision alignment on a temporary precision carrier substrate is described and simulations are performed to identify an optimal configuration of the electrical field. In order to move the microparts, they are included in a liquid droplet of high dielectric constant, allowing for fast movement and easy control.

The concept implementation idea is briefly described and the required electrode structure is depicted.

Keywords: MEMS, 3D manipulation, force fields, electrowetting

1. Introduction

Micropart manipulation is a very important step in the assembly of microsystems. Usually the manipulation is performed by physical contact using microgrippers in order to properly align and orient the parts to be assembled [1]. Unfortunately, micromanipulation with physical contact can be a difficult and often time-consuming task, due to the appearance of sticky effects in the microworld [2]. Thus, non-contact manipulation techniques have been proposed, such as aerostatic grippers, optical traps and laser tweezers [1].

One of the recent alternatives has been the introduction of static and dynamic force fields [3],[4], [5]. Radial, curl, squeeze and elliptical fields are used to manipulate a micropart in two dimensions, providing the necessary rotation and translation. Some implementations have been presented, such as MEMS actuator arrays in the microworld [6], arrays of directed air jets [7] and vibrating plates [8] in the macroworld.

Dr. Lydia Kavraki et al [3], presented a method to manipulate parts in 2D using force fields. Their goal was to find a sequence of force fields that would enable them to assemble two laminar parts lying on a plane. The introduced procedure includes three sequential stages: centring, rotation and translation. Their experiments showed that the microassembly of the two parts was successful.

This paper presents a concept for micropart manipulation and a potential technical implementation, overcoming the bottlenecks of a manipulation scheme based purely on field alternation. Using a combination of translational fields, electrowetting principle and structured surface properties, successful gripperless micro-assembly can be conducted. Flip operations, however, will be difficult to perform due to the large scale of movement and short time available for this, if a highly parallel, high speed batch assembly method is envisioned. For this, an alternative selection scheme for correctly face-aligned micro-parts is described, together with a high speed transportation mechanism based on electrowetting principle.

An initial description of force fields in 3D [9], their application to precision manipulation of microparts in

an optimized electrode configuration and the concept for a technical implementation is presented.

2. Electrode configuration and droplet polarization

The basic idea behind the design of the electrodes was to implement an electric field that centres a micro-object/droplet in the workspace, regardless of its original position. Kavraki [10] has suggested such a field in two dimensions. The problem of describing the motion of a sessile water droplet on a surface inside an electric field required a three-dimensional electrode configuration.

A configuration of a tiny central circular, negative electrode ("pin"), surrounded by a "ring" that is positively charged, as shown in Figure 1, is considered. Dimensions are of the order of a few µm for the radius of the central pin, the height of the electrodes and the "thickness" of the ring and a few hundred µm for the ring's radius. The objective of this layout is to create a proper radial electric field and to reduce the thickness of the electrodes so that the droplet's movement is free from obstacles. The central pin coincides with the position where self-assembly [11] takes place.

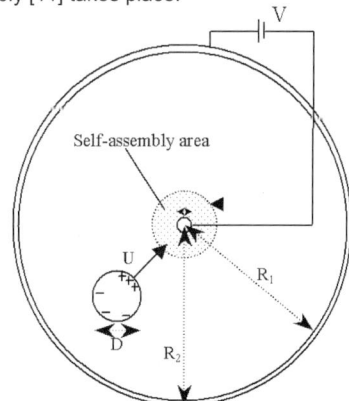

Figure 1: Electrode configuration, droplet polarization and movement towards the centre pin

Comsol Multiphysics [12], a finite element method program, is used for the electrode design and the calculation of the induced electric field, pointing from the ring to the pin (Figure 1). The intensity of the field is weak near the ring and strong near the central pin. A potential well that corresponds to this field is shown in Figure 2.

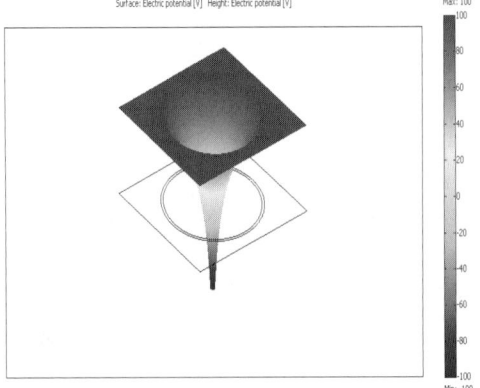

Surface: Electric potential [V]   Height: Electric potential [V]

Figure 2: Electric potential well

It is assumed that the droplet, which carries the micropart, is transported by e.g. electro-wetting from its original position inside the ring having a random position and carrying zero net charge. The voltage difference is subsequently applied on the two electrodes and, as a result, the electric field induces and re-orients the molecular dipoles of the droplet so that they become parallel to the field's intensity vectors at each point.

A negative charge –Q is accumulated in the part of the droplet close to the ring. The part that is closer to the pin acquires a positive charge +Q so the overall net charge of the droplet is zero. Because of the non-homogeneity of the field, the sum of the electrostatic forces acting on the charge +Q is greater than the sum of the forces acting on the charge –Q. This difference in electrostatic pressure (Maxwell stress) on the two sides of the droplet causes a motion towards the central pin along a straight line with speed U as shown in Figure 3. The advantages of the proposed system are the μm accuracy and the sensorless centering of the droplet, the final accuracy only depending on the precision of the structuring of the pin-ring electrode.

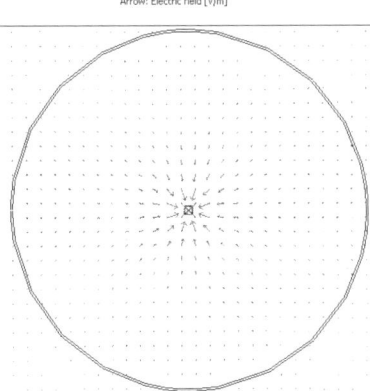

Arrow: Electric field [V/m]

Figure 3: Electric field's vectors in x-y plane

3. Droplet modelling

In this manipulation concept the considered droplets are originally hemispherical, with diameters of 50-100 μm, volumes of around 70-140 picolitres and masses of the order of 10^{-11}-10^{-10} kg.

In order to describe a droplet's motion on a surface under the influence of an electric field and to estimate its deformation (change of length), a 2D mechanical analogue of a bead-spring-bead is considered as a first approximation (Figure 4). This model has been used by Sumino et al. [13] in 1D to describe the motion of a self-propelling droplet due to surface tension forces on surfaces with different chemical properties.

In our approach, the 2 beads are considered to be almost point-masses. The bead that is closer to the ring electrode is named as "bead 1" and the one that is closer to the central pin "bead 2".

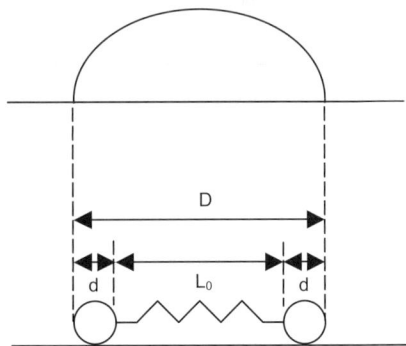

Figure 4: Droplet mechanical analogue. Lengths and distances are not indicative of the actual case

The droplet polarization inside the ring is described by assuming that the two beads carry charges

$$q1=-a*|Q| \text{ and } q2=+b*|Q|, \qquad (1)$$

where "a" and "b" are coefficients with positive values ranging from 0.1 to 0.9.

The reason for their introduction is that the bead-spring-bead model cannot describe the difference in electrostatic pressure nor the change of the charge distribution as the droplet moves closer to the centre, as shown in Figure 5 (a) and (b). When the droplet touches the central pin, the proposed mechanical analogue ceases to apply (Figure 5b). However, it is adequate for the purpose of self-assembly since it brings the system inside the self-assembly area. Therefore, these coefficients serve to approximate up to an extent the real phenomenon in our calculations, but do not represent any actual changes in the charges. The total charge of the system still remains zero. The coefficients a and b depend on the distance of the beads from the central electrode and it applies that

$$a+b=1 \qquad (2)$$

Bead 1 at the initial position is assigned a=0.4 and bead 2 b=0.6. As the beads move towards the central pin, a and b change linearly according to Eq. (2). When bead 2 reaches the central pin, a=0.1 and b=0.9.

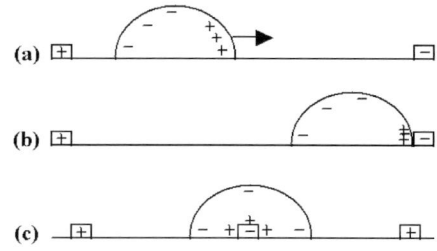

Figure 5: Change of the direction of polarization a,b) as the droplet moves towards the central pin c) when it surrounds the pin.

The forces acting on each bead are the electrostatic, the spring force and a friction force (Figure 6).

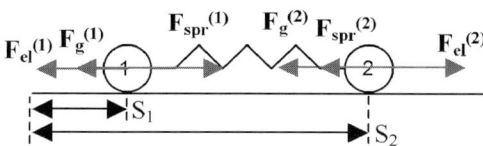

Figure 6: Forces acting on the bead-spring-bead model

Considering the direction of the motion as the positive direction, the kinematics equations for the beads 2 and 1 are respectively:

$$\frac{d^2 S_2}{dt^2} = -\frac{g}{m}\frac{dS_2}{dt} - \frac{K}{m}|S_2 - S_1| + \frac{EbQ}{m} \qquad (3)$$

$$\frac{d^2 S_1}{dt^2} = -\frac{g}{m}\frac{dS_1}{dt} + \frac{K}{m}|S_2 - S_1| - \frac{EaQ}{m} \qquad (4)$$

The first term of the second part of the equations represents the total friction force proportional to velocity, taking into account viscous drag, molecular displacement around the contact-line of droplet-planar surface (absorption-desorption phenomena) and surface tension forces acting on the solid-liquid-liquid interface. These phenomena are of secondary importance compared to the electrostatic driving of the system and mostly influence its motion velocity and shape. In this model approach, they are considered as damping effects with a coefficient g. This damping force can be reduced by properly selecting the material of the plane surface (Teflon or super-hydrophobic surfaces [14]) as well as the viscous medium that surrounds the system.

The driving force in this model is the electrostatic force on bead 2, moving the system towards the central pin. The electrostatic force acting on bead 1 retards the motion, however, due to the Maxwell stress, it will be weaker than the driving force. The change of the length of the spring $(S_2 - S_1)$ connecting the beads represents the deformation of the droplet along its axis of motion.

It has to be noted that the local deformation of the external electric field caused by the polarization of the dielectric droplet (internal electric field) is neglected in this model.

4. Simulation and results

A simulation is performed in Simulink. The characteristic dimensions and voltages of the

electrode configuration are the following: r=15 μm the radius of the central electrode, R_1=400 μm the inner radius of the ring electrode, R_2=405 μm the outer radius of the ring electrode, z= 5 μm the height of the electrodes, V=100 Volt the voltage of the ring (DC), V=-100 Volt the voltage of the central pin (DC).

The physical characteristics of the sessile water droplet are: diameter D=76 μm, volume V≈110 picolitres, mass M=1.1*10⁻¹⁰ kg, density ρ=998,23 kg/m³, total net charge Q_{tot}=0 C. In the bead-spring-bead model, each bead is assigned a mass of M/2 and a diameter of d=4μm. The length of the spring in its rest state is L_0=68 μm and its elastic constant K=10⁻⁶ N/m [15]. Charge on the beads is -a*|Q|, and +b*|Q|, with Q=8*10⁻¹⁹ C. The friction coefficient g is of the order of 10⁻⁸ kg sec⁻¹m⁻¹.

For the purposes of this simulation, when the system touches the central pin, the electrode voltage is shut off. The only forces acting on the system then are the friction and the spring restorative forces. Its motion continues till the resting state. Results showed that the motion towards the central pin is an approximately straight line. The change of the system's spring length over time is shown in Figure 7. The length is increasing as bead 2 is being attracted by the central pin and reaches its peak as the bead touches the pin. When the field is shut off, restorative spring forces act to bring the system into equilibrium. It has to be noted that this model can only provide an estimation of the droplet's change of length along the motion axis, but no other information about its shape, which is expected to be an ellipsoid of revolution along the motion axis-electric field line, similar to the findings in [16]. Anyway, the positioning of the droplet's central mass will be accurate to the central pin, providing the desired alignment accuracy.

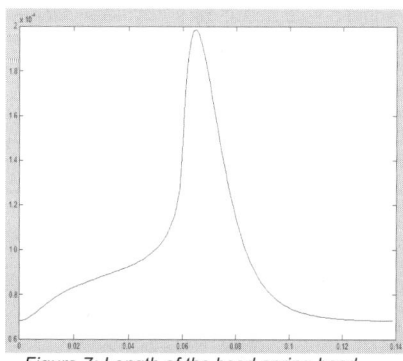

Figure 7: Length of the bead-spring-bead system over time

5. Technical implementation concept

The concept for a final, high precision batch assembly technique draws on a two stage approach, where in the first stage onto a precision temporary carrier the micro-components are aligned and fixed in position, and finally are transferred by a larger area bonding step to the final, receiving substrate. As the latter process is well understood and known from Wafer-to-Wafer bonding techniques [17], we need to concentrate on the alignment process on the precision carrier substrate only.

Using the precision alignment field approach described above, a threefold implementation to maneuver the parts to their final position is required.

a) dispensing of droplets containing exactly one microcomponent, verified by high speed camera

b) ensuring the correct face-up situation of the micropart, verified by camera

c) fast movement of the droplet into the vicinity of the precision alignment field

d) alignment according to the described procedure, passively controlled e.g. by capacitive feedback

e) submicron and rotational alignment to the final end position, performed by hydrophilic/hydrophobic structuring of the temporary carrier substrate surface

a) to d) can be summarized in the flow diagram (Figure 8).

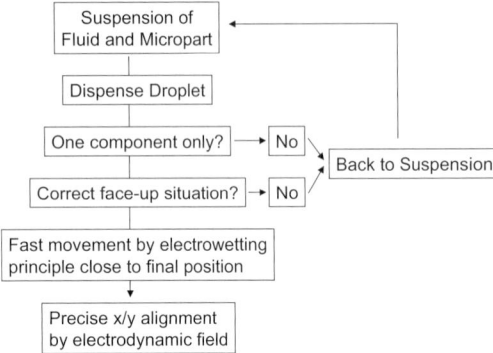

Figure 8: Flow Diagram of proposed implementation

As b) brings severe challenges with respect to identification of actual face-up situation in-time and the required field action for a 180° flip maneuver, for the technical implementation the discarding of false aligned parts back to the suspension reservoir is favored.

The fast movement described in c) is performed by utilizing electro-wetting principle, e.g. as published in [18], [19]. Then the field to a pin/ring electrode as described in the previous sections is applied to draw the microparts into their desired x/y position.

The final alignment, also with respect to rotational aspects, will require a surface modification on the temporary carrier substrate, facilitating the rotational aspect by minimizing the free surface energy of the system [20].

Thus, the final structure of the carrier substrate consists of three layers:

1) non structured, transparent top electrode (GND) layer for high speed movement by electrowetting (ITO)

2) structured bottom layer for high speed movement by electrowetting, according to [19]

3) dielectric separation layer between 2) and 4)

4) pin/ring electrode layer

5) dielectric cover (e.g. Parylene C) with structured hydrophilic and hydrophobic regions according to [20]

6) Spacer ring between 5) and 1)

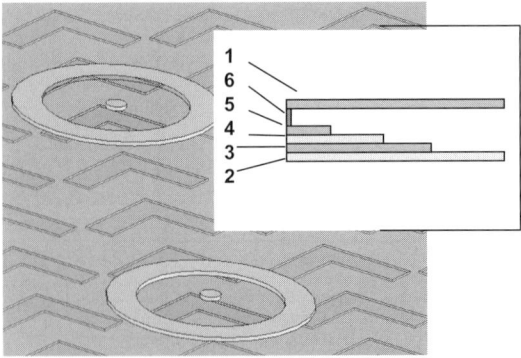

Figure 9: Representation of multilayered parallel batch assembly substrate with electrowetting structures and pin/ring configuration

6. Conclusion and Outlook

This paper presents an approximate model and technical implementation concept of a droplet that will be used to carry a micropart for self-assembly.

A pin/ring electrode configuration that centers the droplet in the workspace is designed; a simulation of the bead-spring-bead model for the droplet is conducted and the results show that the system moves towards the central pin with an increase in the spring length.

The proposed mechanical analog system is a first step towards modelling the sessile droplet and provides an estimation of the droplet's length along its motion axis; future work will include:

-a better and more thorough modelling of the droplet that takes into consideration its deformation, improvement in the design of the electrodes and thus in the induced electric field

-a study of the local deformation effect of the external electric field due to the polarization of the droplet and an investigation of the impact of the existence of a micropart on top of the sessile droplet with regards to droplet polarization, field local deformations and effect on droplet motion

The described concept and calculations require their actual technical implementation and verification. Surface roughness, viscosity drag and wetting behaviour are expected to have significant influence on the achievable placement precision. Subsequent electrical and mechanical interconnection techniques need to be developed, e.g. by novel reactive contacts [21].

A successful implementation will allow high speed, high volume placement of micro-components, performing the above mentioned process flow in parallel with a multitude of droplets and controlling these by switching electronic fields instead of moving heavy mechanical placement heads.

7. Acknowledgments

University of Patras and Fraunhofer IZM are partners in the EC Network of Excellence "Multi-Material Micro Manufacture: Technologies and Applications (4M)."

8. References

[1] H. Van Brussel, J. Peirs, D. Reynaerts, A. Delchambre, G. Reinhart, N. Roth, M. Weck, E. Zussmar, "Keynote Papers, Presented at the Opening Session of Assembly of Microsystems", Annals of the CIRP Vol. 49, 2000.

[2] Ronald S. Fearing, "Survey of sticking effects for microparts handling", Proc. Of IEEE/RSJ IROS 1995.

[3] J. Luo , L. Kavraki, "Part Assembly Using Static and Dynamic Force Fields". In Proceedings of The IEEE/RSJ International International Conference on Intelligent Robots and Systems, IEEE Press, 2000, pp. 1468–1474.

[4] K. Bohringer, M. Cohn, K. Goldberg, R. Howe, A. Pisano, "Parallel microassembly with electrostatic force fields", In Proc. IEEE Int. Conf. on Robotics and Automation (ICRA), Leuven, Belgium, 1998.

[5] K. Bohringer, B. R. Donald, N. C. MacDonald, "Programmable vector fields for distributed manipulation, with applications to MEMS actuator arrays and vibratory parts feeders", Int. Journal of Robotics Research, 1999.

[6] K. Bohringer, B. R. Donald, N. C. McDonald, G. Kovacs, J. Suh, "Computational methods for design and control of MEMS micromanipulator arrays", IEEE Computer Science and Engineering, 1997, pp. 17- 29.

[7] M. Yim, A. Berlin, "Contact and non-contact mechanisms for distributed manipulation", ICRA workshop at distributed manipulation, 1999.

[8] D. Reznik, J. Canny, "Universal part manipulation in the plane with a single horizontally vibrating plate", In P. Argarwal, L. Kavraki, M. Mason, editors, "Robotics: the algorithmic perspective", A. Peters, Ltd, Wellesley, MA, 1998.

[9] P. Lazarou, N.A. Aspragathos, "Three dimensional force fields for micropart manipulation", 4M 2005 conference, Karlsruhe, 29 June-1 July 2005.

[10] J. Luo, L. Kavraki, "Static and Dynamic Force Fields for Assembly Planning", Proceedings of the ITTE/RSJ International Conference on Intelligent Robots and Systems (IROS), pp.1468-1474, 2000.

[11] K. Bohringer, U. Srinivasan, R. Howe, "Modelling of capillary forces and binding sites for fluidic self-assembly", IEEE Conference on Micro Electro Mechanical Systems (MEMS), pp. 369-374, Interlaken, Switzerland, January 21-25, 2001.

[12] Comsol Multiphysics, http://www.femlab.com

[13] Y. Sumino et al., "Chemosensitive running droplet", Phys. Rev. Lett. 94, 068301, 2005.

[14] Lei Zhan, Fevzi Ç. Cebeci, Robert E. Cohen, and Michael F. Rubner, "Stable Superhydrophobic Coatings from Polyelectrolyte Multilayers". Nano Lett.; 2004; 4(7) pp 1349 – 1353.

[15] H. Sun, K. Takada, S. Kawata, "Elastic force analysis of functional polymer submicron oscillators", Applied Physics Letters, Vol. 79, Num. 19, pp. 3173-3175, 2001.

[16] S. Moriya, K. Adachi, T. Kotaka, "Deformation of Droplets Suspended in Viscous Media in an Electric Field", Langmuir, 2, pp. 155-160, 1986.

[17] F. Niklaus, G. Stemme, J. -Q. Lu, and R. J. Gutmann, "Adhesive wafer bonding", J. Appl. Phys. 99, 031101, 2006.

[18] Vamsee K. Pamula and Krishnendu Chakrabarty, "Cooling of integrated circuits using droplet-based microfluidics," Proc. ACM Great Lakes Symposium on VLSI, pp. 84-87, 2003.

[19] Vijay Srinivasan, Vamsee K.Pamula, and Richard B. Fair, "An Integrated Digital Microfluidic Lab-on-a-chip for Clinical Diagnostics on Human Physiological Fluids," Lab on a Chip , 2004.

[20] Jiandong Fang, Karl F. Böhringer, "Parallel micro component-to-substrate assembly with controlled poses and high surface coverage." IOP Journal of Micromechanics and Microengineering (JMM), accepted for publication February 15, 2006.

[21] K.-F. Becker, T. Löher, B. Pahl, O. Wittler, R. Jordan, J. Bauer, R. Aschenbrenner, H. Reichl, "Development of a Scalable Interconnection Technology for Nano Packaging", 2005 NSTI Bio Nano Conference, Boston, May 2006.

Multi-Material Micro Manufacture
W. Menz, S. Dimov and B. Fillon (Eds.)

117

Design and manufacturing of micro heaters for gas sensors

Per Johander[1], Igor Goenaga[2], David Gomez[2], Carmen Moldovan[3], Oana Nedelcu[3], Petko Petkov[4], Ulrike Kaufmann[5], Hans-Joachim Ritzhaupt-Kleissl[4], Robert Dorey[5], Katrin Persson[6]

[1] *IVF- Industrial Research and Development Corporation; Argongatan 30, S431 53 Molndal, Sweden,*
[2] *Fundacion Tekniker, Micro and Nanotechnology Department, Av. Otaola 20, 20600 Eibar, Spain*
[3]*National Institute for R&D in Microtechnologies, Erou Iancu Nicolae 32 B, Bucharest 077190, Romania*
[4]*Manufacturing Engineering and Multidisciplinary Technology Centre, Cardiff University;*
[5]*Forschungszentrum Karlsruhe, Institut für Materialforschung III, P.O. Box 3640, 76021 Karlsruhe, Germany ;*
[5] *Nanotechnology Group, Cranfield University, Cranfield, Bedfordshire, UK;*
[6]*IMEGO, Arvid Hedvalls Backe 4, SE 411 33 Goteborg, Sweden;*

Abstract

The paper presents the design and manufacturing steps of micro heaters, built on ceramic suspended membranes for gas sensor applications. The micro heaters are designed and fabricated by combining laser milling techniques, and conductive ceramic technology. Trenches are created in the ceramic substrate in order to define the geometry of the heater using laser processing of the substrate. The heater is completed by filling the trenches with conductive ceramic paste and then baking to remove the solvent from the paste.

The final step involves releasing the membrane by laser milling, enabling it to be suspended on four bridges, to minimise the dissipation of the heat in the substrate. The temperature of the heater element was measured with a heat camera from FLIR 40 system comparing the case of the heater positioned on top of a released membrane and that of the non-released membrane. The simulation of the heater build on top of a released membrane was compared with the heater measurements.

Keywords: Micro heaters, gas sensor

1. Introduction

Micro heaters are an essential part in chemoresistive gas sensors [1]. The heating element must produce accurate and uniform heating of the sensor surface. The heating element is normally made on silicon, LTCC or alumina substrates. The common methods to manufacture heating element are thin film deposition of metal and lithographic/ etch process [2] and thick film screen printing on LTCC and alumina [3]. The heat distribution in such 2D design concepts is not so good due to the spreading of heat and thus quite large power consumption. This could be improved by using a 3D design concept with a suspended heating element incorporating four bridges, so that the heating and sensing element are thermally isolated from the surrounding substrate. This has been demonstrated in silicon by bulk etching of silicon [4], [5]. This design gives a much more uniform heating of the sensor surface and much faster heating characteristics. In this paper we demonstrate a similar design concept produced by laser machining LTCC and alumina substrates.

1.1. Laser milling process overview

Laser milling involves applying laser energy to remove material through ablation in a layer-by-layer fashion. CNC programs for laser milling are obtained directly from a three-dimensional CAD model of the workpiece. Thus, apart from being a material removal rather than material accretion system, a laser-milling machine operates like any other layered technology manufacturing equipment.

1.2. Material removal

The laser milling process removes material as a result of interaction between the laser beam and the substrate or workpiece. Several removal mechanisms can take place, depending on a number of process parameters related to the beam and the workpiece material.

During interaction between the laser radiation and the material, electrons in the substrate are excited by the laser photons. As a result, the electron subsystem is heated to a high temperature and the absorbed energy is transferred to the atomic lattice [6]. Energy losses are caused by heat transport via electrons into the bulk of the substrate. According to classical linear theory, light absorption is described by the Beer-Lambert law, which states that the absorption of a particular wavelength of light transmitted through a material is a function of the material path length and is independent of the incident intensity. For the very high intensities, which can be achieved in laser processing, non-linear phenomena take place and cause stronger energy absorption. According to the linear absorption model, the electrons excited by photon absorption transfer the heat energy to the lattice and cause melting or vaporisation. In the case of extreme intensities, as with ultrashort pulse ablation, the bound electrons of the material can be directly freed. Effects such as multiphoton absorption and avalanche ionisation can be observed. Thermal conduction in the material draws energy away from the focal spot and leads to heat affected zones.

For metals, laser light absorption is not a major problem, because the band structure of the material allows the absorption of most low-to-moderate energy photons. Metals absorb laser radiation with their electrons in the conduction band and near to the Fermi level. For semiconductors and insulators, including ceramics and polymers, the Fermi level is between the valence and conduction bands. The requirement that an electron has to absorb enough energy to pass through the forbidden region between the valence and conduction bands sets a limit on the photon energies that can be absorbed by a linear process. There are two methods for non-linear absorption: an electron avalanche and a strong multiphoton absorption.

In the case of an electron avalanche, some electrons with intermediate energies are excited into the conduction band by single photon absorption. The electrons in the conduction band can absorb single photons of the incident light, which will increase their energy. These electrons collide with bound electrons, which lead to avalanche ionisation.

Multiphoton absorption is another mechanism for non-linear absorption in which an electron transfers from the valence band to the conduction band by absorbing several photons.

In general, as materials reach a critical density of electrons, they start to absorb sufficient photon energy to undergo ablation. It is important to note that there is a material dependent ablation threshold fluence (peak energy per unit area), below which the laser ablation cannot start and material removal is not feasible.

The response of the substrate to laser radiation is influenced by a number of material characteristics. Hence, for optimal machining results a proper match of laser source and material should be achieved. Generally, higher absorption efficiency leads to a more effective laser milling process. Thermal conductivity is another key material factor. This affects the dissipation of the absorbed energy into the bulk of the material, the energy losses, the material removal efficiency and the dimensions of the heat affected zone (HAZ) [7], [8].

2. Experimental set up

A commercial femtosecond Ti:sapphire laser system has been used to perform the micromachining of LTCC and briefly, the system is basically composed by:
(i) a femtosecond seed laser (Coherent, Vitesse LP SB) with an output average power of 250mW at a pulse repetition rate (PRR) of 80MHz.
(ii) a multipass amplifier (Quantronix, Odin-Compact DP 1.0) based on Chirped Pulse Amplification (CPA).
(iii) a LBO crystal for doubling the natural wavelength of Ti: sapphire beam from λ=800nm to λ=400nm by means of Second Harmonic Generation (SHG) techniques.

The resulting beam, used to perform all the experimental tests presented in this work, is gaussian (beam quality, M^2~1.2 (TEM$_{00}$)), circular (3mm in diameter), ultraviolet (λ =400nm), pulse length is 90fs and the maximum energy peak is 350μJ. Fluence on the sample has been controlled by the rotation of a half-wave plate that changes the polarization plane of the beam before the LBO crystal. Considering that the conversion efficiency strongly depends on the polarization direction, it was possible to control energy variations from 0μJ to 350μJ in steps of 1μJ.

The samples for the experimental tests were placed in a computer controlled XYZ-stage positioning system from Aerotech with resolution of 0.1μm and repeatability (3μm) of ±1μm in the case of the XY-stages and 0.5μm and ±2μm, respectively, for the Z-stage. A circular 2mm mask was used before focusing the beam on the sample in order to remove the tails of the gaussian profile. A fused silica plano-convex lens with focal length of 100mm (Newport, SPX022) was used as focusing optics. In all cases, focal point was focused on the surface of the sample and no image projection technique was used.

3. Design and manufacturing

The heating element is suspended by four supporting bridges that also serve as electrical connections to the heating and sensing elements (see Figure 1).

The heating element is produced by laser machining a serpentine groove directly into the substrate material and filling the cavity with AuPtPd paste. The heater element is released by laser milling. In addition, the back face of the heater element is thinned to reduce the thermal mass of the element (Figs. 2 and 3).

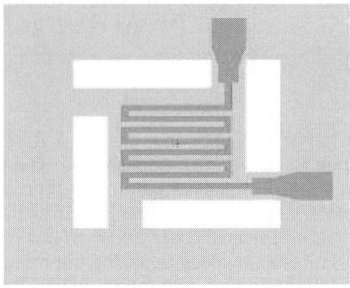

Figure 1. The design of the *electro ceramic heater* **(100μm width) on the ceramic membrane (2000x2000μm^2) suspended on four bridges (300x200μm^2 each)**

Laser machining was performed to:
(i) machine a groove structure into the substrate in order to embed a conductive ceramic paste to form the heating element.
(ii) reduce the thickness of the substrate under the heating element and thermally isolate the membrane from the substrate.

In the first case, the laser was slightly de-focussed on the sample in order to increase the spot size, leading to faster machining of the 100μm-wide grooves. The applied fluence was 5.5J/cm^2 at a feed rate of 0.5mm/s. The depth of cut obtained was 45μm. The same parameters were used to machine the electrode contacts. The machining of these elements was performed by means of concentric curves with an offset of 25μm.

Once the heater groove was machined and filled with AuPtPd paste, the next step was to proceed with the thinning and release of the heater. The releasing operation was performed first by machining the four bridges around the heater. This operation was completed by traversing the 0.5x2 mm rectangles several times until the full thickness of the wafer was machined, resulting in a rectangular hole. The laser was focused onto the sample through a 100FL lens and the spot size of the focused beam was 36 μm±1μm, the applied fluence was 1.8J/cm^2 at a feed rate of 0.5mm/s

Once the releasing step had been completed, the thinning of the back face was conducted. This was performed by "scanning" the 3x3 mm area with the laser beam as shown in Figure 3, where the lateral step size (**s**) was 6 μm. The fluence value used was 3.7J/cm^2 using the same feed rate as before. As a result of the thinning process, a total 175 μm of material was removed, resulting in a membrane thickness of about 225µm. The same focusing lens and spot size as used previously were used for thinning. The laser milling sequence is presented in Fig.4.

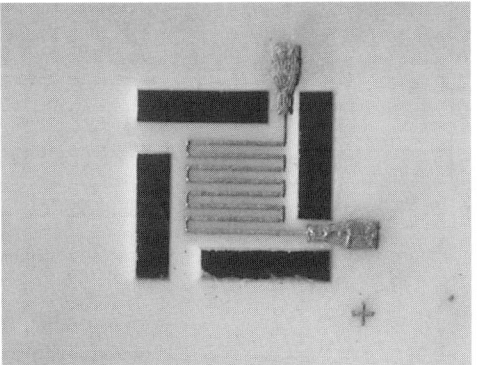

Figure 2. The suspended heating element

Figure 3. The heater element from the backside showing the thinning of the substrate

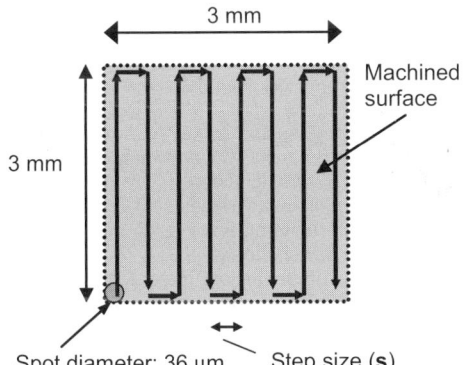

Spot diameter: 36 µm   Step size (**s**)

Figure 4. The laser milling sequence

4. Measurement of the thermal properties

The temperature of the heater element was measured using a thermal camera from FLIR 40 system. This system is capable of recording 50 images per sec allowing the dynamics of the system to be analysed. The heater positioned on the released membrane reached a temperature of about 490 °C in 5

seconds. The temperature versus time profile for the released heater element is presented in Figure 5. The heat distribution of the released heater, measured using FLIR 40, is presented in Figure 6. This should be compared with the results obtained for the non-released heater element (figure.7) shown in figure 8.

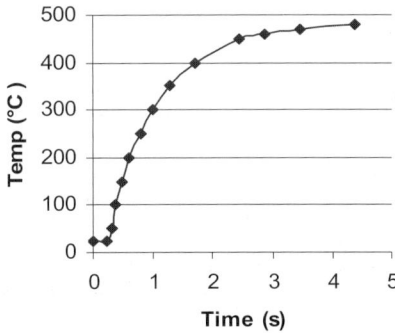

Figure 5. The heat distribution from the released heating element

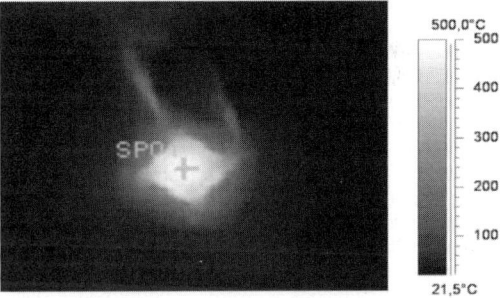

Figure 6. The heat distribution from the released heating element

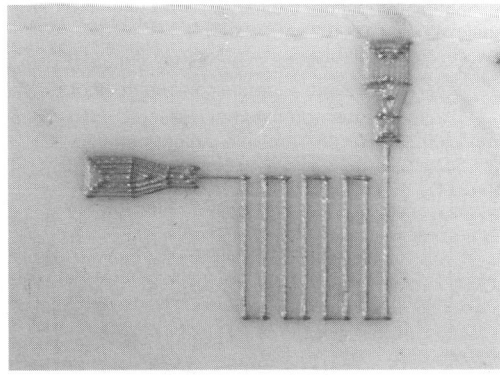

Figure 7. The Non-released heater element

Because heat is conducted in all directions away from the unmachined heater elements a circular temperature distribution around the element is observed compared to the angular temperature distribution observed for the released heater element. In figure 6 the temperature is uniform within the released membrane area. Outside of this area the temperature is at least 400 °C lower showing that the machined structure is effective at providing thermal isolation. The input power was 1.1 W and this should be compared with the input power required for the non-released heater element that requires 2.4 W of input power of to reach a temperature of 490 °C[1]. The heat distribution around the non machined heater element can be seen in figure 8 below.

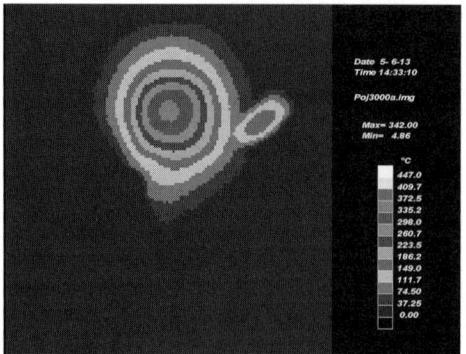

Figure 8. The heat distribution of a non- released heater element as seen in figure

5. Simulation of the thermal behaviour of the integrated heater on the ceramic membrane

The simulation of the heat dissipation was conducted to validate the chosen design and technology of the sensor. The simulations were made considering the case of the heater positioned on top of the released ceramic membrane. The comparison between the simulation results and measurements of systems, with identical layout and technological steps, allows validation of the heater design, processing and efficiency. In addition the simulations have the role to minimize the number of experiments required to optimise the design.

The simulations were performed with CoventorWare2004, MemMech module [1] and assumed a LTCC AuPdPt paste heater material.
The simulation conditions used were:

- 550^{0}C, Step time: 0.02 sec; Solver step: 0.01 sec; Analyzed time: 0 - 0.2 sec
- Constant temperature in heater volume (550^{0}C);
- Initial temperature in ceramic substrate: 293^{0}K
- Convection through external surfaces: (convection coefficient: 25 W/m^{2}K   ambient temperature: 293^{0}K)

Fig 9. Thermal distribution after 0.02s at 550^{0}C

The dimensions of the heater are similar to the Mask 1 (Fig.1). All material constants were considered for the conditions described above.
The results are presented in Fig. 9 and 10 showing the thermal distribution in the ceramic substrate after 0.02s (Fig.9) and after 0.2s (fig 10). A very uniform distribution of the heat can be observed after 0.2s.

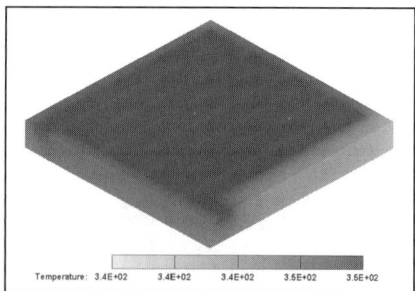

Fig 10. Thermal distribution after 0.2s at 550^{0}C

Figure 11 shows the evolution of the heating process as a function of time.

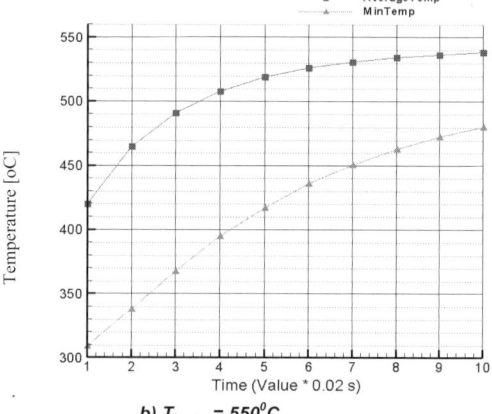

b) T_{heater} = 550^{0}C

Fig.11. Evolution of average and minimum temperature in alumina substrate

Comparison of figure 10 (showing the *simulation* of the thermal distribution) with figure 6 (showing the *measurements* of the heat distribution) shows that, for a released heater membrane, similar thermal behaviour, in terms of temperature achieved (500^{0}C) and area of temperature distribution (identical for both cases, concentrated on the released membrane), are observed

The membrane suspended by four microbridges is uniformly heated at 500^{0}C, providing the required temperature for a working sensor.

In the case of measurements of the heater, the time to achieve uniform heating of the membrane depends on the power supply.

In the case of simulation, there are no electrical parameters involved in the model, and the model assumes that the substrate will be kept at constant temperature 550°C. These assumptions led to the differences between simulation and measurements (Figures 5 and 11) observed for the time to achieve uniform heating of the substrate at 500 °C.

6. Conclusions

The purpose of the work was to establish a new technology for obtaining heater elements for use in gas sensors requiring a uniform heating. The technology combined different techniques including laser milling machining and conductive ceramic paste deposition.

Thermal simulations of heater and membranes were used to reduce the number of experiments required to optimise the heater design.

The heater design and manufacturing methodology described in the paper act as a demonstrator for such micro processing technology.

The main goal/overall objective was to obtain miniaturized, low cost devices on non-silicon substrates, with high sensitivity and low power consumption for use in portable devices.

The described technologies allow the generation of a new type of microsensor with existing tools (software, technological facilities, etc).

Acknowledgements

The authors express the acknowledgements to the 4M NoE for supporting the work described in this paper. All results have been obtained in the frame of 4M, WP8 (Ceramic Cluster). Anders Löfgren Ericsson MicroWave for kind helps with the IR measurements.

References

[1] C.Moldovan O. Nedelcu, U. Kaufmann, HJ Ritzhaupt-Kleissl, S. Dimov, P. Petkov, R.Dorey, K. Persson, D. Gomez, P. Johander "Mixed technologies for gas sensors microfabrication", Proceedings 4M Conference on Multi Material Micro Manufacture 29 june-1 July 2005 Karlsruhe pp 211-z1217

[2] J. Shuehl, R.E Cavicchi, M. Gaitan, S. Semancik, "Tin oxide Gas Sensor fabricated using CMOS micro-hotplates and in situ processing" IEEE Electron Devices Lett. 14 (1993) 118-120

[3] T. Zawada, A. Dziedzic and L.J Golonka, "Heat Sources for Thick-Film and LTCC Thermal Microsystems" 14th European Microelectronics and Packaging & Exhibition, Friedrichhafen Germany 23-25 June 2003

[4] S. Semancik, R.E Cavicchi, M.C Wheeler, J.E Tiffany, G.E Poirier, R.M Walton, J.S Suehle, B. Panchapakesan and D.L DeVoe, "Microhotplate platforms for Chemical Sensor Research" Sensor and Actuators p 579-591 B77 (2001)

[5] Welch, "Micro-Machined Thin Film Hydrogen Gas Sensors" Proceedings 2002 US DOE Hydrogen Program Review NRFI /CP-610-32405

[6] Shirk M D and Molian P A, 1998, "A review of ultrashort pulsed laser ablation of materials", Journal of Laser Applications, Vol. 10, No 1, pp 18-28

[7] Kautek W and Krüger J, 1994, "Femtosecond pulse laser ablation of metallic, semiconducting, ceramic and biological materials", Proceedings SPIE, Vol. 2207, pp 600-610

[8] Von der Linde D and Sokolowski-Tinten K, 2000, "The physical mechanisms of short-pulse laser ablation", Applied Surface Science, Vol. 154-155, pp 1-10

Developments in Micro Ultrasonic Machining (MUSM)

J.J. Boy*[+], M. Aiguillé**, A. Boulouize**, C. Khan-Malek***

* FEMTO-ST / LCEP – 26 chemin de l'Epitaphe – 25000 BESANCON - France
** TEMIS Innovation (Projet µUSM) – 18, rue A. Savary – 25000 BESANCON - France
*** FEMTO-ST / LPMO – 32, avenue de l'Observatoire – 25030 BESANCON Cedex
[+] Corresponding author : Tel : 33 (0) 3 81 40 28 23 ; E-mail : jjboy@ens2m.fr

Abstract

Ultrasonic machining (USM) presents a particular interest for the cutting of non-conductive, brittle materials such as ceramics, glasses or fused silica and quartz crystal. Unlike other non-traditional processes such Electrical Discharge Machining (EDM and micro-EDM, adapted to conductive materials), laser ablation or wet chemical etching, USM does not thermally damage the workpiece and does not create significant levels of stresses. Production of complex 3-D shapes with a volume of a few cubic millimeters is presented.

Keywords: Ultrasonic machining, micro technology, aspect ratio, tool wear, PZT.

1. Introduction

Quartz and other piezoelectric single crystals ($LiNbO_3$ or $LiTaO_3$, $GaPO_4$ and more recently crystals from the Langasite family) are used as raw materials to build resonators, filters sensors or other microsystems which, introduced in an electronic oscillator, work at their resonant frequency. Usually, the quality and the stability of the output frequency intrinsically depends, naturally, on the material quality but also on the possible damages induced by the mechanical operations. For example, the resolution of quartz accelerometers or gyrometers working at low frequencies (generally a few tens of KHz) is drastically linked to the stress distribution in the entire volume of the device (and not just in the vibrating part due to the role of the sismic mass) and so, it is important to prevent and avoid damages and particularly twins which can be generated by suitable stress and can propagate in the crystal [1].

In the field of MEMS production, glass or fused silica are also widely used as a structural and functional material in micro-total-analysis-systems (µTAS). If the device is produced by classical microtechnics developed for Silicon, the packaging is generally made in glass in which it is necessary to machine cavities or via holes.

Whereas machining technologies starting from bulk materials are well established for metals and alloy, the machining of brittle, hard, non-conductive materials such as silicon still poses considerable problems. Particular challenges are the generation of non-rotationally symmetrical 3D shapes, with high aspect ratio and processes inducing neither residual stress nor cracks. Traditional machining of ceramics is done with diamond-cutting tools for which the resolution is limited by the tool size to a few hundred microns. Pulsed-laser systems can produce enough energy which is focused onto a spot to ablate even the hardest materials. However, they often cause surface deterioration such as a heat-affected zone due to thermal effects and micro-cracks. If composite materials are ablated, a different chemical composition may be left on the surface behind. Techniques such as photolithography and etching (wet or dry) derived from semi-conductor industry are also sometimes used but they need special processes to be developed for each materials, adapted to its chemical reactivity. Moreover, in the case of wet etching, the geometry of the workpiece is limited by the crystallography of the materials. So such processes may be interesting for mass production but not efficient for prototyping or small series.

2. Ultrasonic Machining

The MUSM technique exploits the tool (or microtool) vibration at the ultrasonic frequency to force abrasive grains to hit the workpiece material and erode its surface. MUSM has proved successful in the precise machining of hard and brittle materials, in particular non-conductive substrates which are difficult to machine otherwise, due to the nature of the material, e.g. hard and brittle non-conducting materials.

The tool rests on a blanket of abrasive grains atop the workpiece. Vibrating generally at about 20 kHz, it needs to transmit a sufficient energy to the grains to remove material by a hammering process and cavitation.

We use water as the slurry medium due to its good property for transferring ultrasonic waves. The tool is excited at its ultrasonic resonant frequency by a power generator which applies high voltage to the piezoelectric transducer, composed by one or two pairs of PZT pre-stressed discs. By the inverse piezoelectric effect, the applied voltage (a few hundreds of volts) on their faces produces a thickness variation transmitted to the end of the sonotrode by means of conical and bi-cylindrical mechanical amplifiers (preferably in Titanium) linking rigidly the transducer to the machine (see Figure 1 and 2). The piezoelectric properties of the Z-cut PZT discs fed in phase opposition (U voltage) induce "breathing" of the discs along the Z axis, which produces a variation of thickness Δh as indicated in Fig. 1-a, and simultaneously bending on both sides of the Z plane (deflection noted as δ in Fig. 1-b).

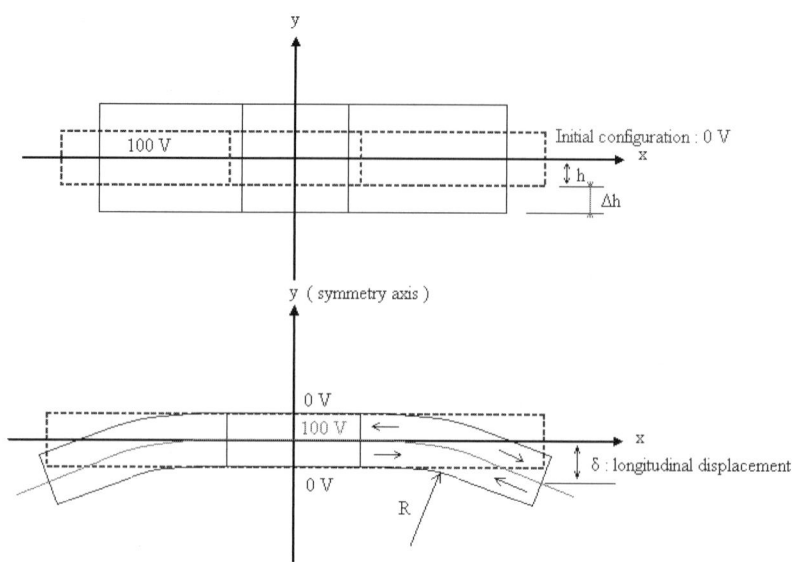

Figure 1-a and 1-b: Thickness variations $\Delta h = d_{33}.U$ and $\delta = \frac{3}{4}.d_{31}.U.(l/h)^2$, where d_{33} and d_{31} are the piezoelectric coefficients of the PZT disc and l and h its radius and thickness respectively.

To illustrate the comparative values of the sinusoidal thickness variation induced by the power supply (typically a few hundreds volts), we have calculated Δh and δ for various PZT materials used as circular transducers. As we can see in the Table 1, the bending effect δ is more efficient than the stretching Δh to create the longitudinal waves in the sonotrode (Figure 2). The amplification ratio at the end of the sonotrode is about 9 due to its bi-cylindrical shape resulting in a maximum amplitude of vibration of 30 microns with a power supply of 700 Watts.

Table 1: Numerical application for various PZT materials (U = 100V, l = 18.75 mm, h = 6 mm)

Mat	d_{31} (10^{-12} m/V)	d_{33} (10^{-12} m/V)	δ (µm)	Δh (µm)	$\delta / \Delta h$
PZT5A	171	374	0.125	0.037	3.35
PZT5 H	274	593	0.200	0.059	3.40
P1 94	305	640	0.223	0.064	3.45
P1 89	108	240	0.079	0.024	3.30
P7 62	130	300	0.095	0.030	3.15

Finally and up to now, the maximum area that can be machined in a single step is within a circle of about 1 inch diameter. But research in machine development is also conducted. It consists in developing first a new and smaller machine with a horn that will function at higher frequency, 35 KHz instead of 20 KHz and second a tool with a working surface reaching 3 inches diameter. The definition of a complete acoustical system with a piezoelectric transducer operating at new frequency and with a new horn was computed using Finite Element Analysis and validated with the existing ultrasonic machine. The following table (Table 2) gives examples of calculations made on the complete acoustical system presented in the Figure 2 with different Finite Elements Modellings and compares the amplification ratios and

resonant frequencies calculated with different elements and integrating the piezelectricity behavior for the last one. These results have been obtained with ANSYS software.

Figure 2: Longitudinal wave in the acoustical system

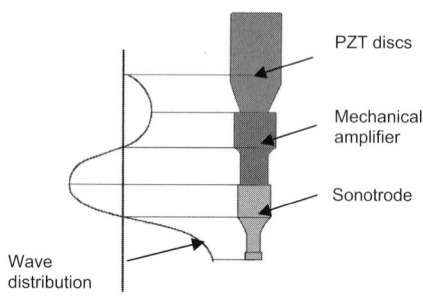

Table 2: Examples of results obtained with ANSYS software

	Vibration mode	F(Hz)	Ampli ratio
Theoretical study	-	-	9.27
FEM with 3D elements	3rd long	20156	9.24
2D axisym model	3rd long	20062	9.28
The same with piezoelectricity	3rd long	20207	9.40

3. Tool wear

The grains hit the end of the vibrating tool and tend to erode it. So, tool wear depends on several parameters such as:
- the tool (also called sonotrode) material,

- the workpiece material,
- the static load of the tool (linked to the acoustic unit) on the blanket of abrasive grains,
- the amplitude of vibration (which is a linear function of the electric power supplied to the transducer,
- and to a lesser extent the dimension of particles.

Here, we detail just the effect of the static load on the machining speed (for more informations, see [2]). Our example concerns tool wear machining glass. Similar experiments are conducted in our lab focusing on the wear tool and not on the machining speed.

Measurements of tool wear as a function of static load were performed on pyrex glass using a circular steel tool with a diameter of 6 mm. The realisation of 200 µm deep non-through holes induces a wear of approximately:

- 1 µm/hole when using a static load of 6N and 9 µm abrasive grains (mesh 600)
- 2 µm/hole when using a static load of 6N and 17 µm abrasive grains (mesh 400)
- 0.3 µm/hole when using a static load of 3N and 17 µm abrasive grains (mesh 400).

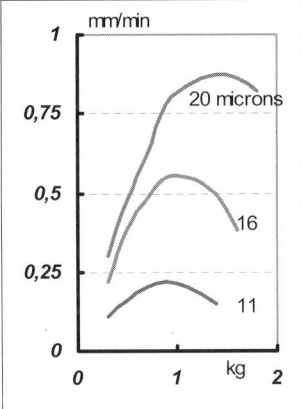

Figure 3: Machining of glass: erosion rate as a function of static load (after [3]).

The lower static load, the smaller is the wear, down to a value where the erosion becomes negligible. The influence of static load is dominant, which would require its tight in-situ control during machining. The dimension of abrasive grains, though less influential, also contributes to the wear, with an increase of erosion rate with bigger grain sizes (see Figure 3 presenting erosion rate with 3 different grain sizes).

The realisation of holes of much smaller diameter, in the hundred-micrometer range, which is our target, induces a much larger tool wear. However, micro-USM remains competitive because it allows the production of multiple microstructures in parallel using a tool matrix instead of a single structure at a time as it is performed in direct mechanical micromachining with carbide or diamond drills.

When tools of very small dimensions are used, the static load needs to be small to avoid breakage of the tool. For example a load between 50 to 100g for a tool with 65 µm diameter or a load of 10 grams for a 20 µm^2 square tool seem to be the best values [4].

4. Machinable materials and machining capabilities

Hard and brittle materials can be machined using this technique, in particular non-conductive substrates. The present machine was developed for producing electronic components based on piezoelectric quartz crystals. Indeed, this process is efficient because it generates neither a temperature gradient nor mechanical stress in the crystal, and it preserves the surface integrity of the specimen (3). It was extended easily to less hard materials such as glass or pyrex, silicon and polycrystalline piezoelectric ceramics. In contrast, it needed to be adapted for harder materials such as sapphire using harder abrasive grains with more cutting power (natural diamond instead of silicon carbide) at the expense of an important reduction in machining speed.

Whereas through-holes of high quality were easily produced, non-through-holes were also machined with a depth accuracy within 10 µm (anti-mesa) but their surface smoothness is generally not as good and not sufficient for applications using very high-frequency resonators (around a few hundreds MHz). Any 2-D pattern of 5 cm^2 maximum area can be transferred within the bulk material, and the pattern stepped and repeated on the substrate. At present, within a 500 µm thick substrate of hardness similar to that of quartz crystal, the minimum feature size of a through-hole was 120 µm, corresponding to an aspect ratio of 4, with a negligible conicity. An example of such machining in silicon is shown in Fig. 4. An aspect ratio of 10 can be obtained with larger holes with some conicity (12 µm/mm, i.e. less than 1 degree) that is still acceptable compared to other techniques such as LIGA known for producing very straight walls. Rotation of the tool would improve the roundness and surface quality of the circular holes.

All above-mentioned structures have been machined using tools produced by conventional techniques (EDM, drilling, etc.). These tools wear relatively slowly (as indicated in the previous paragraph). We are now exploring machining with tools produced by a number of lithography-based high-aspect-ratio micro-techniques (LIGA, DRIE).

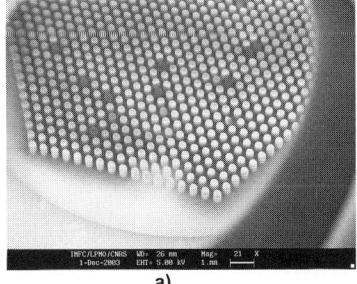

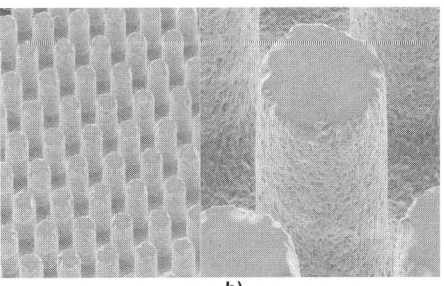

a)   b)

Figures 4: Example of ultrasonically-machined microstructures:
a): Si 2D matrix with cylinders of 280 µm diameter and 600 µm depth; b) close-up view.

126

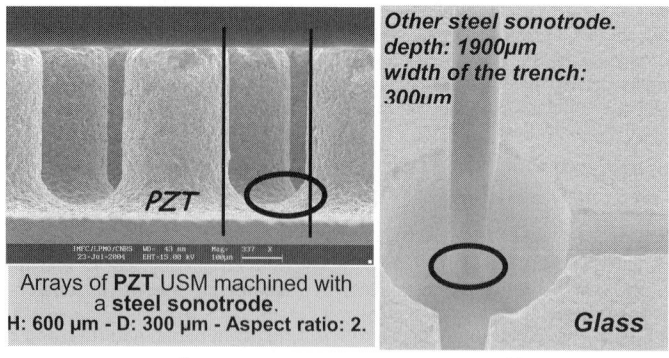

a)   b)

Figure 5-a and 5-b: Array of pillars in PZT (a) and hole and trenches in glass (b).

The previous figures (Fig. 4 and Fig. 5a) illustrate an array of pillars in PZT which was produced using a steel disk with a honeycomb of holes of 300 microns diameter. The granular aspect of the sidewalls is characteristic of material removal through the erosion of the workpiece due to abrasive grains. The curved bottom is due to the inhomogeneous wear of the sonotrode whose edges become rounded. Nevertheless, the sidewalls of pillars are vertical on most of the height. And we can ensure that a much deeper hole machining (more than 1 mm) will present also vertical sidewalls.

5. Conclusion

As we see here, this "old technique" can be used to realise microsystems developed in the frame of new applications born in the microtechnics field. We have proved that the definition of well adapted parameters such as static load value or grain size enable the realisation of microholes with acceptable roughness. Nevertheless, future developments concern:
- optimisation of the tool material to decrease drastically the wear (in particular, we study the efficiency of tool in polycarbonate easily realized by moulding),
- realisation of tool with a larger working surface defined by finite elements modelling,
- improvement of the roughness of the walls and of the bottom of the non through-hole,
- ...

One of the advantages of the Micro-Ultra-Sonic Machining comes from the simplicity of the pattern transfer into the substrate in one single step once the tool has been produced, even if it presents various heights for multi-level complex structures. Nevertheless, the wear tool can limit this application when the shape of the steel tool is difficult to pattern and so becomes too expensive.

Finally, the advantage of MUSM is that it is not material dependant and can be used, for example, to machine microchannels in glass for microfluidic applications or microsensors with output frequency in quartz crystal.

References

[1] Yamni K., Boy J.J., Yacoubi A.: "Temperature effect on the ferrobielastic behaviour of the quartz crystal" – Ann. Chim. Sci. Mat, 2001, pp. 19, 22.

[2] Andrey E., Boy J. J., Khan Malek C.: "Tool wear for micro-ultrasonic machining (MUSM)" - 5th Int. Conf. European Society for precision Engineering and Nanotechnology, Montpellier, 8-11 mai 2005, Proc. EUSPEN 05, pp. 345-348.

[3] Rozenberg et al: "Physical principles of ultrasonics technology. Ultrasonic cutting", Part 1, Consultants Bureau, New-York, 1973.

[4] Egashira K., Masuzawa T., Fujino M., Inter. Jour. of Electrical Machining n° 2, Jan. 1997.

The effects of material microstructure in micro-milling

K. Popov[a], S. Dimov[a], D. T. Pham[a], R. Minev[a], A. Rosochowski[b], L. Olejnik[c], M. Richert[d]

[a] Manufacturing Engineering Centre, Cardiff University, Cardiff, UK,
[b] Design, Manufacture and Engineering Management, University of Strathclyde, UK,
[c] Institute of Materials Processing, Warsaw University of Technology, Warsaw, Poland,
[d] Faculty of Non-Ferrous Metals, AGH - University of Science and Technology, Krakow, Poland

Abstract

Micro-milling is one of the technologies that is currently widely used for the production of micro-components and tooling inserts. To improve the quality and surface finish of machined microstructures the factors affecting the process dynamic stability should be studied systematically. This paper investigates the machining response of a metallurgically and mechanically modified material. The results of micro-milling workpieces of an Al 5000 series alloy with different grain microstructure are reported. In particular, the machining response of three Al 5083 workpieces whose microstructure was modified through a severe plastic deformation was studied when milling thin features in micro-components. The effects of the material microstructure on the resulting part quality and surface integrity are discussed and conclusions made about its importance in micro-milling. The investigation has shown that through a refinement of material microstructure it is possible to improve significantly the surface integrity of the micro-components and tooling cavities produced by micro-milling.

Keywords: micro-milling, material microstructure, grain size effects

1. Introduction

The current trend for product miniaturisation fuels the demand for new micro-engineering technologies that are suitable for manufacture of meso-scale components incorporating micro features. Micro-milling is one of these technologies that is currently widely used for the production of microstructures and tooling inserts for micro-injection moulding and hot embossing. In particular, important application areas of this technology are the manufacture of micro parts for watches, keyhole surgery, housings for micro-engines, tooling inserts for fabrication of micro-filters, and housings and packaging solutions for micro-optical and micro fluidics devices. A common challenge across all these applications is the machining of thin features [1]. To machine reliable micro-components with thin features and improve their surface finish, the factors affecting the dynamic stability of the process should be investigated systematically.

The typical cutting conditions in micro milling imply that the material removal process is governed by the interfacial interaction between the cutting edge and the workpiece material. Because of this, the microstructure of the workpiece can play a fundamental role in the cutting process [2].

The effects of material properties on the micro cutting mechanism are considered extremely important. The effects of the high frequency vibrations that could occur during the micro-milling process are investigated in [1]. A conclusion is made that the quality of the machined micro-structures is highly dependant on the milling strategies employed. This research could be considered as a continuation of the experimental study in [1]. In particular, the milling strategies that were optimised for machining thin features by Popov et al. were applied in this research to investigate the effects of the workpiece microstructure on part quality and surface integrity.

Other researches reported that the integrity of the machined surfaces was highly dependant on the number of passes and the depth of the cuts [3].

Especially, it was found that the depth of the cuts had increased the burnishing caused by the friction between the generated surface and the tool flank face. This contributed to the recovery of the elastic strains of the crystal grains and their subsequent stress relief. Also, it was suggested that a quantification of the resulting surface integrity after the machining operations might offer important information about the component performance. It was found that the depth of the damaged layer in non-ferrous metals, for example Al and Cu, machined with very sharp cutting tools could range from 1 up to 17 µm. However, the effects that a modification of the workpiece material, both metallurgically and mechanically, could have on the machining response in micro milling were not investigated.

The aim of the paper is to investigate the machining response of metallurgically and mechanically modified materials. The results of machining three workpieces of the Al-5000 series alloy with different grain sizes are reported. The effects of the material microstructure on the resulting surface integrity are discussed and conclusions made about their importance in machining micro-components.

2. Material microstructure and processing

2.1 Grain size effects

When performing metal cutting, it is important to study the chip formation in order to understand the surface generation mechanism. The crystalline texture of the material resulting from its processing could lead to variations of the chip thickness. In addition, such variations could be caused by changes in shear angle from grain-to-grain due to varying material properties such as elastic modulus (E). However, it should be said that the anisotropic cutting conditions resulting from these effects may be attenuated or even eliminated by refining the grain structure or strain hardening the material before machining.

128

The defects in the crystalline structure influence strongly the material properties and affect directly the metal cutting conditions. In micro-milling, during the cutting process micro-cracks along the grain boundaries develop and also dislocation slips occur in metal's crystal structure. The specific processing energy required to initiate the chip formation depends directly on the ability of metals to produce dislocation slips. By enhancing the mechanical strength of metals the mobility of the dislocations is reduced and higher cutting forces will be required to move sharp tool through the material [4]. Also, during the cutting, the dislocation density increases due to the formation of new dislocations and dislocation multiplication.

It is considered that refinements of the grain structure could lead to more "favourable" conditions during cutter/material interactions and thus result in a better machining response, especially at micro-scale. For example, when cutting simultaneously a larger number of grains, chatter vibrations due to crystallographic changes can be attenuated. The material refinement also increases grain boundaries that disrupt the motion of dislocations through a material. Therefore, reducing the crystallite size is a common way to improve strength, often without any sacrifice in toughness.

In this research the machining response of Ultra Fine Grained (UFG) metals is studied. The UFG metals are also known as nano-crystalline metals or nano-metals and have grain sizes of 0.1 to 1 μm. This is about 100 times smaller compared to the grain size of the material in its initial state before undergoing any refinement of its microstructure.

2.2 Material processing

A number of techniques exist for creating UFG metals, for example compaction of nano-powders, crystallisation of amorphous materials, electro-deposition, and severe plastic deformation (SPD). In this research the machining response of a metal created through SPD is employed. By applying this refinement technique dislocation bands are created that subdivide original coarse grains into much smaller grains.

The SPD technology enables processing of all kinds of metals using a new class of metal working processes that do not change the shape of the metal billet [5]. One of those new processes, Equal Channel Angular Pressing (ECAP), is employed in this study to create a UFG metal. In particular, this is achieved by forcing a billet material through an intersection of two channels having the same cross-section as shown in Figure 1. Simple shear occurring in the diagonal plane at such intersections leads to grain refinement. To achieve uniform and stable UFG structure, the billet is passed several times through an ECAP die and rotated about its axis, e.g. by 90°, between any two consecutive passes.

3. Experimental set-up

3.1 Material

In this research three different workpieces of Mg4.5Mn Aluminium alloy, with ISO designation Al 5083, were used to assess the effects of material microstructure on part quality when micro milling thin features. Al 5083 was selected due to the capabilities of the available ECAP machine set-up. The material is a non-heat treatable alloy which can only be hardened by

plastic deformation. Prior to micro-milling its microstructure was modified metallurgically and mechanically. In particular, the following three workpieces were used in this experimental study:

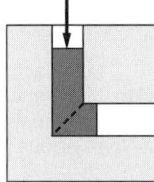

Figure 1 Equal channel angular pressing (ECAP)

1. *"As Received" (AR) Al 5083*. In its initial state the material was in the form of a 20 mm diameter bar that was annealed. The bar was produced through hot extrusion then annealed at 540°C. The microstructure of the annealed material was non-uniform across the bar and the smallest grains of approximately 200 μm were located close to its centre. Figure 2 shows a micrograph of the AR Al 5083.

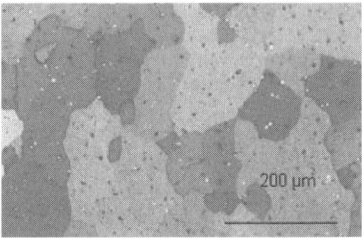

Figure 2 The micrograph of AR Al 5083

2. *Conventionally Processed (CP) Al 5083*. The second workpiece was strain hardened to ε=0.9 by reducing the diameter of the AR bar from 20 mm to 13 mm by forward extrusion at room temperature. After the extrusion process, the same area of the bar cross-section as in the AR sample was inspected. The micrograph showed that the size of the grains was approximately the same as for the AR bar and almost uniform across the core of the bar.

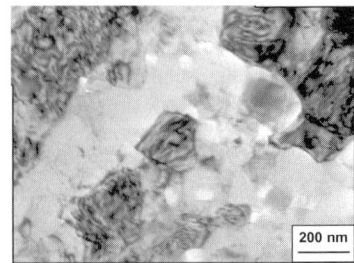

Figure 3 A TEM micrograph of UFG Al 5083

3. *UFG Al 5083*. To create the UFG workpiece, the material underwent the following processing. Square samples with dimensions 8×8×46 mm were cut out from the CP bar, lubricated and subjected to ECAP at 180°C. Plastic strain generated during one ECAP pass was approximately ε=1.15. The process was repeated four times, with samples being rotated by 90° between consecutive passes [6, 7]. After the four ECAP passes the size of the grains was reduced to approximately 200 nm, as

shown in Figure 3, which represented a size reduction in the order of 1000 times.

The Rockwell hardness of the three workpieces was measured using a 1/16" ball according to ISO 1024:1996.

The following results were obtained: H_R^{AR} = 28 HR30T, H_R^{CP} = 67 HR30T and H_R^{UFG} = 66 HR30T. Table 1 provides further information about the mechanical properties of the three samples used in this study. It can be seen that the material hardness reacts very quickly to the strain generated by plastic deformations while tensile properties change more gradually.

Table 1 Mechanical properties of the three samples

	Al5083	Hardness Rockwell 30T	Yield stress MPa	Ultimate strength MPa	Elongation %
1	Annealed	28	152	315	18.6
2	Forward extruded	67	228	379	12.9
3	After four ECAP passes	66	433	472	6.1

The AR sample was used in the carried out experiments as a reference in assessing the machining response of the CP and UFG workpieces.

3.2 Machining set-up

The test part in Figure 4 [1] that was proposed specially to investigate the machining response during micro milling of thin features was employed in this research. The part was designed taking into account dynamic flexibilities, natural frequencies, and mode shapes of common thin features and thus was representative of the cutting conditions that could occur when milling microstructures.

The machining of the test parts was carried out on a KERN HSPC 2216 micro-machining centre. Its

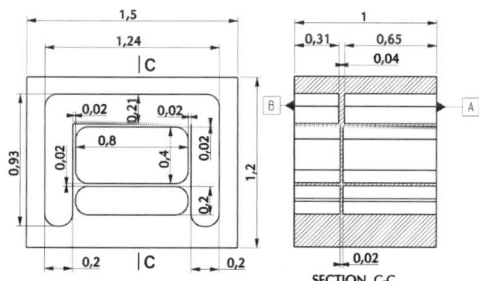

Figure 4 Test part

polymer concrete mono-block frame absorbs high frequency vibrations much better than cast iron frames which is very important in micro-milling. The factors affecting the performance of micro milling operations investigated in [8] were taken into account in selecting the machining parameters in this experimental study. This included the selection of cutting depth that would keep milling forces within predefined limits along the machining path. Spindle speeds and feed rates were chosen depending on the workpiece - tool material combination.

In addition, milling strategies that were optimised for micro machining of thin features were implemented in the experiments [1]. The machining of the workpieces in the three materials investigated included the following steps:

- All samples were face cut on a wire EDM machine;

- Side A of the test part was machined entirely using strategy described in [1];
- The U-type channel and two 800x200 μm pockets on Side B were then machined to their full depth of 310 μm by applying reciprocating plunge-cut cycles. Figure 5a shows the test part after these three machining steps.
- Finally, the 800x400 μm pocket located in the centre of the part was machined applying the milling strategy presented in Figure 5b. At this step the thin ribs of 20 μm around the pocket and the biggest web of the part were formed.

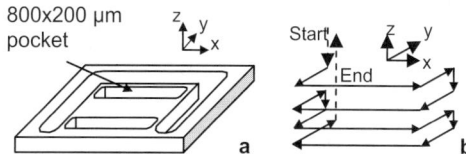

Figure 5 Machining strategy

A 150 μm diameter end-mill (DIXI 7242 tool) was used to mill the U-type channel and the 800x400 μm pocket. The machining parameters used to carry out the micro-milling operations are provided in Table 2.

Table 2 Micro-milling parameters

Cutting Speed (m/min)	Spindle speed (RPM)	Feed / Tooth (mm)	Step Depth (mm)	Step Over (mm)
15.51	33000	0.007	0.005	0.070

To generate 3D profiles and assess the resulting surface finish the relevant areas of the test parts were scanned using a surface mapping system, Micro-XAM.

4. Experimental results

To investigate the effects of the material microstructure on part quality during micro-milling a series of experiments was conducted. Two workpieces were prepared from each of the available AR, CP, and UFG Al 5083 bars, six in total. A test part with the design shown in Figure 4 was produced from each of the workpieces applying the machining sequence outlined in Section 3.2. Two test parts were machined for each material by milling workpieces in two different planes, A – perpendicular to the extrusion direction and B - along it. Thus, it was also possible to investigate the effects of the crystalline texture of the materials on the machining response. During the machining of the six workpieces there were not any noticeable differences in the cutting conditions.

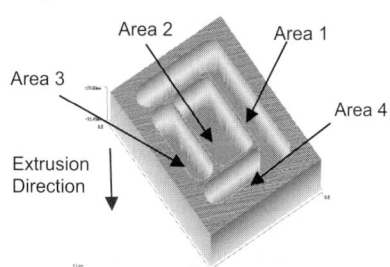

Figure 6 The profile of the UFG sample after milling in the A plane

The generated 3D profiles of the machined six parts did not show any bending of the ribs. One of these profiles is provided in Figure 6 however the others were almost identical. In addition, the profiles were analysed to verify if there were any variations of the ribs' thickness in the C-C cross-section (see Figure 4) between the samples. Again, no variations were found. This consistency in thin features' quality could be explained with the use of milling strategies that were optimised specially for machining micro ribs and webs.

The surface roughness of four different areas on the bottom on the machined samples, as shown in Figure 6, was measured over a sampling length of 0.16 mm and a cut-off length of 0.08 mm.

The size of the scanned areas was chosen according to ISO 4288:1996 and ISO 11562:1996 (http://www.predev.com/smg/standards.htm). The parameter used to evaluate the surface roughness was the arithmetic mean roughness (Ra) because relative heights in micro topographies are more representative, especially when measuring flat surfaces. The average results of all measurements and standard deviation (S.D.) are presented in Table 3.

Table 3. Roughness results

	Ra^{AR} (µm)		Ra^{CP} (µm)		Ra^{UFG} (µm)	
	A($\downarrow$)	B($\rightarrow$)	A($\downarrow$)	B($\rightarrow$)	A($\downarrow$)	B($\rightarrow$)
1	0.49	0.45	0.30	0.39	0.09	0.10
2	0.64	0.62	0.26	0.60	0.12	0.16
3	0.47	0.51	0.46	0.56	0.16	0.18
4	0.33	0.49	0.45	0.47	0.15	0.17
S.D.	0.127	0.073	0.102	0.094	0.032	0.036
Av.	0.48	0.52	0.37	0.51	0.13	0.15

As expected the surface roughness of the AR samples was the highest. At the same time the average roughness of the UFG samples was approximately 3 and 4 times better than that achieved on the CP and AR test parts, respectively. The experimental results show a strong relationship between material microstructure and attained surface finish. In addition, the grains' orientation in polycrystalline materials could affect their machining response albeit at much smaller scale. In particular, the increase of the average surface roughness of the CP and UFG samples, when the milling was carried out in the B plane (horizontal) instead of A (vertical), exceeded 35% and 15%, respectively.

5. Conclusions

This research investigates the effects of material microstructure on part quality in micro-milling. The machining response of mechanically and metallurgically modified Al alloy when milling thin features in micro-components was studied. The investigation has shown that through refinement of materials' microstructure it is possible to improve significantly the surface integrity of the machined micro features. In particular, the following conclusions can be made.

- The roughness of micro-features produced by micro-milling is highly dependant on the material grain size. For example, the surface roughness of thin features in micro-components improved more than three times as a result of the reduction of the grain sizes of the Al alloy used in the experiments from 100-200 µm to 200 nm.

- A "favourable" crystalline texture of the material in regard to the machining direction could lead to surface roughness improvements. However, such improvements are of a magnitude smaller than those achievable through a refinement of the material grain structure.

- The use of optimised micro-milling strategies results in manufacture of thin features with a consistent quality. For example, the ribs' thickness of the machined test parts was consistent throughout the carried out experiments and did not show any variations when using workpieces with different material microstructure.

Further research is required to benefit from the grain size effects in manufacturing micro components in a range of micro-engineering applications. For example, the use of UFG brass as a material for producing micro-optical components and micro-tooling inserts could improve significantly their surface finish and ultimately lead to significant improvements of their functional performance.

ACKNOWLEDGEMENTS

The authors would like to thank the European Commission, the Department of Trade and Industry, the Welsh Assembly Government and the UK Engineering and Physical Sciences Research Council for funding this research under the ERDF Programmes 53767 "Micro Tooling Centre" and the EPSRC Programme "The Cardiff Innovative Manufacturing Research Centre". Also, this work was carried out within the framework of the EC Networks of Excellence "Innovative Production Machines and Systems (I*PROMS)" and "Multi-Material Micro Manufacture: Technologies and Applications (4M)".

References

[1] K. Popov, S. Dimov, D.T. Pham, A. Ivanov, "Micro milling of thin features", Proceedings of the First International conference on Multi-Material Micro Manufacture (4M 2005), July 2005, pp.363-366

[2] Jasinevicius, R. G., Campos, G. P. de, Montinari, L. et al. "Influence of the mechanical and metallurgical state of an Al-Mg alloy on the surface integrity in ultra precision machining". J. Braz. Soc. Mech. Sci. & Eng., July/Sept. 2003, vol.25, no.3, p.222-228. ISSN 1678-5878.

[3] Duduch, J.G., Porto, A. J. V., Rubio, J. C. C. Jasinevicius, R. G., "La Influencia del Numero de Pasadas en la calidad de las Superficies Opticas de Refleccion", Revista Informacion Tecnologia del Chile, Vol. 11, n.4, pp.53 – 58, 2000.

[4] Rosochowski, W. Presz, L. Olejnik and M. Richert, "Micro-extrusion of ultra-fine grain aluminium", International conference on Multi-Material Micro Manufacture (4M 2005), July 2005, pp.161-164.

[5] A. Rosochowski, L. Olejnik, M. Richert, "Metal forming technology for producing bulk nanostructured metals", Journal of Steel and Related Materials - Steel GRIPS Vol. 2 Suppl. Metal Forming (2004) 35-44.

[6] V.M. Segal, V.I. Reznikov, V.I. Kopylov, D.A. Pavlik, V.F. Malyshev, "Processy Plasticheskogo Structyroobrazovania Metallov", Science Engineering. Minsk, 1994, 231 p. (in Russian).

[7] L. Olejnik, A. Rosochowski, "Methods of fabricating metals for nano-technology", Bulletin of The Polish Academy of Sciences, Technical Sciences, Vol. 53 No. 4 (2005) 413-423.

[8] Dimov S, Pham D T, Ivanov A, Popov K, Fansen K, "Micromilling strategies: optimization issues", Proceedings of the Institution of Mechanical Engineers, Volume 218 Part B, Engineering Manufacture, July 2004, pp.731-736

Multi-Material Micro Manufacture
W. Menz, S. Dimov and B. Fillon (Eds.)

Laser micro-milling of ceramics, dielectrics and metals using nanosecond and picosecond lasers

M.R.H Knowles[a], G. Rutterford[a], D. Karnakis[a], T. Dobrev[b], P.Petkov[b], S.Dimov[b]

[a]Oxford Lasers Ltd, Unit 8, Moorbrook Park, Didcot, OX11 7HP, UK
[b]MEC, Cardiff University, Newport Road, Cardiff, CF24 3AA, UK.

Abstract

Laser micro-milling of industrial materials like ceramics, dielectrics and metals is of significant commercial interest for fabrication of micro-moulds and other micro-system devices. 2.5D laser machined structures were generated in alumina, tungsten and steel substrates using a nanosecond copper vapour laser (511nm) at 10 kHz. Preliminary results in fused silica, alumina and steel are also presented from a high repetition rate amplified mode-locked picosecond Nd:vanadate laser. It is shown that high quality surface finish can be achieved with both laser types; for example, average surface roughness, Ra ~ 300nm has been demonstrated in steel. Fused silica could only be processed with picosecond laser pulses. Volume removal rates are analysed, which are especially high for difficult materials like tungsten (~0.1mm^3/min) and are greater compared to other milling technology like micro-EDM. Surface roughness measurements in these materials using white light interferometry are reported along with SEM analysis.

Keywords: laser, micro-milling, ceramic, dielectric, metal, ceramic, nanosecond, picosecond

1. Introduction

Laser micro-milling is emerging as an important technology for rapid prototyping [1], and serial production of microdevices using batch fabrication methods [2]. The attraction to pulsed lasers in particular, available with high repetition rates up to 500 kHz and short ns- or ps- pulse durations, stems from their ability for very finely controlled material removal at reasonable processing speeds. Ultrafast (pico and femto second) lasers can be optimised to process more than one class of material. Competing micromilling techniques such as micro electro-discharge machining (microEDM), ion milling or lithography are generally considered material dependent or slow. Although laser ablation suffers from the usual compromise between high ablation rates and good surface quality [3], we show that careful laser choice and process optimisation can result at a satisfactory compromise for both. Here we discuss the merits of laser micromilling using ns- and ps-multi kHz lasers emitting in the near IR (1064nm), visible (511nm) and UV (355nm). Metals (tungsten, steel), ceramics (alumina) and dielectrics (synthetic diamond, fused silica) have been investigated.

2. Experimental Set-up

The lasers used were a nanosecond copper vapour laser (CVL) and mode-locked Nd:YVO$_4$ picosecond lasers. They were focused on the target surface to spot sizes typically ranging between 5-20 μm either using achromat lenses (f=125, 100, 75mm) or multi-element flat field lenses (f=100mm) when galvanometer scanners were used. The surface morphology of the milled features was examined by SEM (JEOL JSM-5310) and a non-contact high precision 3D surface profiler with height resolution of 0.1nm (MicroXAM, ADE Phase Shift Corp). Loosely attached surface debris was removed by ultrasonic cleaning in alcohol.

Table 1 Copper Vapour Laser[a] characteristics

Laser Properties	CVL
Wavelength (nm)	511&578
Max.Average Power (W)	45
Repetition Rate (kHz)	10
Pulse Energy (mJ)	4.5
Beam Diameter, 1/e^2 (mm)	20
Pulse Duration (FWHM) (ns)	17
Beam Quality factor, M^2	<1.5
Polarisation Ratio	random

[a] manufacturer Oxford Lasers Ltd.

Table 2 Mode-locked Nd:YVO$_4$ Laser[b] characteristic

Laser Properties		
Wavelength (nm)	1064	355
Max.Average Power (W)	10	2
Repetition Rate (kHz)	0-500	0-500
Max Pulse Energy (mJ)	0.3	0.02
Beam Diameter, 1/e^2 (mm)	1	1
Pulse Duration (FWHM) (ps)	<15	<12
Beam Quality factor, M^2	<1.5	<1.2
Polarisation Ratio	100:1	1000:1

[b] manufacturer Lumera Laser GmbH

Different laser hatching patterns were used as shown in Figure1 in order to minimise surface roughness.

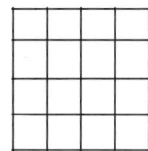
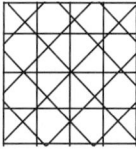

(a)   (b)

Fig.1. Examples of laser hatching patterns.

3. Laser micro-milling of metals

Laser milling of metals finds specific industrial applications but is often considered a complex operation. In many laser milling and engraving systems, long pulses (microsecond duration) are used. The long pulse duration produces relatively high ablation rates but the extensive thermal penetration depth, laser melting and recast dominate the process resulting in poor feature quality. More recently, progress in laser milling of metals has been achieved using shorter laser pulses at higher repetition rates.

Tungsten has a high melting and boiling temperature therefore requires very high peak power for efficient laser machining. Figure 2 shows a SEM image of a CVL machined feature used in biomedical tissue repair. The inverse pyramidal shape was machined in layers of 20μm thickness. Figure 3 shows removal rate versus laser fluence and shows that relatively high ~10^6 μm^3/sec rates could be achieved. A combination of high quality and attractive laser milling speed was achieved by careful optimisation.

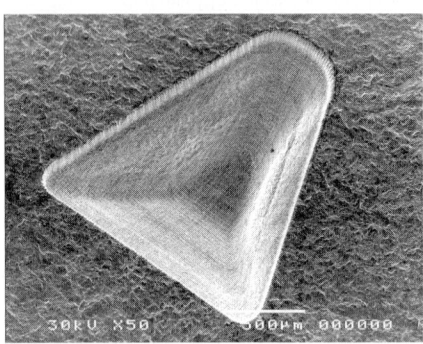

Fig.2 Laser micro-milled inverse pyramidal feature in tungsten.

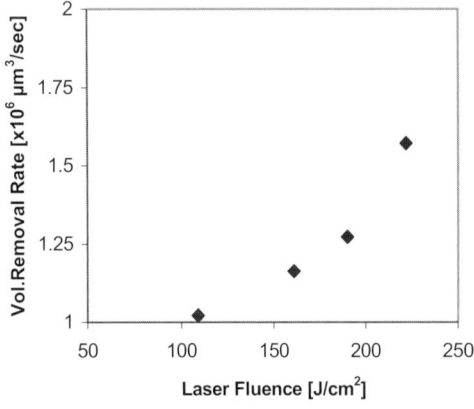

Fig. 3. Volume removal rate (x10^6 μm^3/pulse) versus laser fluence for tungsten micro-milling.

Initial tests have also been conducted using the 1064nm picosecond laser. Figure 4 shows a high magnification view of the bottom of picosecond laser milled feature in stainless steel. The surface roughness is Ra < 0.3 μm which is substantially better than that achieved with longer laser pulses. Figures 4, 5 and 6 show examples of laser milled bottom surfaces in stainless steel using pico, nano and micro-second laser

pulses respectively. It is clear that the surface roughness improves with decreasing pulse duration. Note that the nanosecond example is shown at a higher SEM magnification than the other examples. However, the picosecond milled features show higher taper than the nanosecond milled examples and so further process optimization is required.

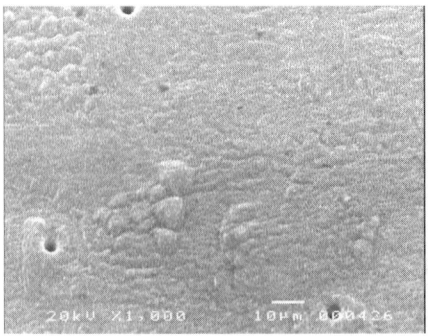

Fig. 4. Surface roughness in stainless steel achieved with picosecond laser (preliminary results). SEM image at X1000 magnification.

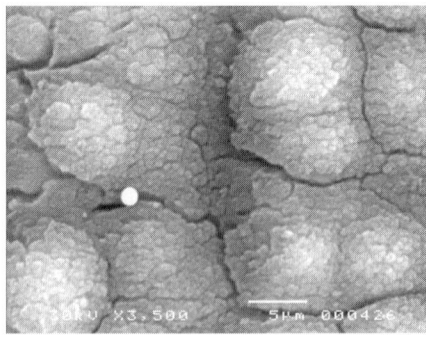

Fig. 5. Surface roughness in stainless steel achieved with nanosecond laser. SEM magnification X3500.

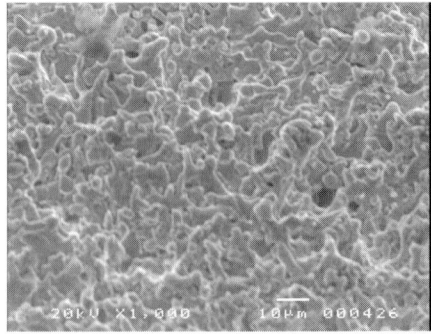

Fig. 6. Surface roughness in stainless steel achieved with microsecond laser. SEM magnification X1000.

4. Laser micro-milling of ceramics

The interaction of most hard ceramics with high intensity laser pulses is quite different to that with metals. The threshold for ablation is higher (factor of 2 – 10) and more obviously defined. In many ceramics such as alumina and silicon nitride there is no evidence of melting. One challenge for laser micromilling of ceramics is the large scattering exhibited at common laser wavelengths, which restrict energy absorption. A combination of short pulse and short wavelength usually shows best results.

Figure 7 illustrates high quality ps-laser milling of alumina using the modelocked laser at 1064 nm. The 1mm square was milled using rastering steps of 6μm, scanning speed of 300 mm/s and repetition rate of 50 kHz with a laser spot size ~17.5μm. The milled area appears flat with no evidence of laser-induced cracking or other collateral damage and given the already granular texture of alumina, the milled surface can be considered smooth.

Fig. 7. Laser milled pocket in alumina using 1064nm picosecond laser.

The floor surface roughness measured as Ra~0.9 μm, compares extremely well with the unirradiated parent surface (Ra~0.73μm) and is smaller than the irradiating wavelength. This, combined with the high volume removal rates of ~10^7 μm^3/sec and above, makes the laser very attractive for micromilling as compared to UV excimers with lower removal rates reported earlier (~10^4-10^5 μm^3/sec) [4]. The surface roughness increases gradually from Ra~1μm to 1.45μm with increasing rastering step from 4 μm to 10 μm. Similarly surface roughness increases with increasing volume removal rate as seen in Figure 8. For the highest recorded Ra value of 1.7μm, the rastering step used during milling was 10 μm.

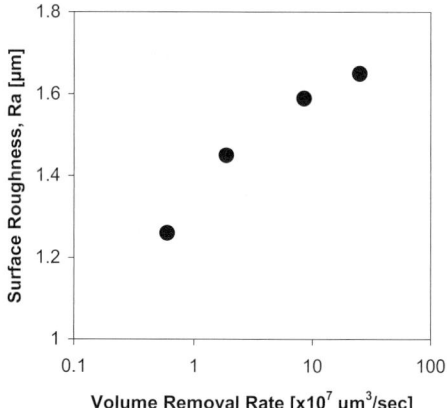

Fig. 8. Surface roughness versus material removal rate in μm^3/sec for alumina using a 1064nm picosecond laser.

The direct-write method of many laser machining processes can make it a very flexible tool for micro-machining. Figure 9 that shows an example of a component fabricated using drilling, cutting and milling in a single operation using the nanosecond CVL. The component was fabricated from a 0.3mm thick sheet of alumina. The 45° chamfers along the edge of the bar were created using controlled depth ablation. The depth being varied by locally adjusting the total number of pulses used.

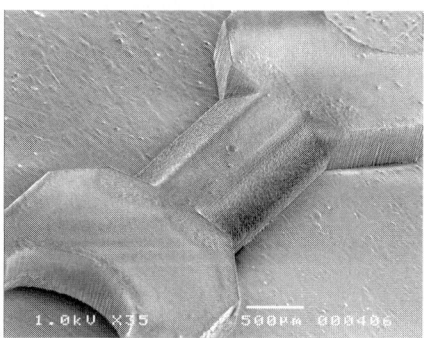

Fig.9. Example of drilling, cutting and milling in alumina ceramic.

5. Laser micro-milling of dielectrics

Laser machining of dielectrics is usually limited by their optical properties. Many dielectrics, such as fused silica, are transparent to most common laser wavelengths and controlled laser ablation can only take place at VUV wavelengths [4] or by using ultrashort laser pulses. Other materials such as synthetic CVD diamond are not totally transparent owing imperfections in the material and trace elements. As such it is possible to process CVD diamond with longer wavelength nanosecond lasers. Figure 10 illustrates an example of a laser milled sensor mold on synthetic diamond using high peak power from the CVL. The device consists of a meander-line ridge with 0.12mm height. The meander lines have a period of 170μm, wall thickness of 30μm and inner and outer radius of curvature of 25μm. A total area of 11.2 mm^2 was milled in 14 min. A layer of approximately 3-4 μm in depth was milled out for each laser pass of the rastered area. The overall volume removal rate in this case was 0.09 mm^3/min. The shot overlap used was maintained 50% and 55% in x- and y-axis respectively.

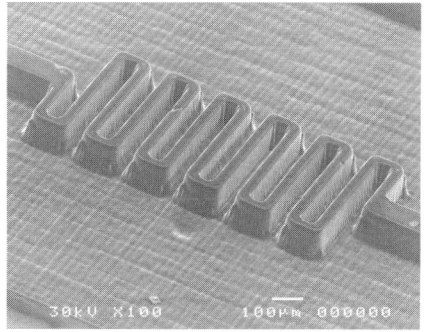

Fig. 10. CVL milled sensor mold in CVD diamond.

134

Fused silica is a wide band gap material, transparent at 511nm and so it was not possible to successfully micromill using the CVL. But a tightly focused 1064 nm ps-laser beam of average power 0.5W was capable of 3D micromilling. The laser was rastered on the surface to create 1mm square pockets using a rastering step of 1-2μm. The scanning speed ranged between 50-200 mm/s and the repetition rate was fixed at 50 kHz. By varying the number of repeat passes over the exposed area the milled depth ranged between 27-210 μm. Although ps laser ablation of fused silica has been shown before [6], it is interesting that high repetition rate laser ablation enables such high quality micromilling of a brittle material without evidence of microcracking or other collateral damage, see figure 11. Some edge chipping is present but generally the surface roughness on the floor of the milled area was kept low with measured values of Ra~0.9μm albeit higher than the unirradiated surface (Ra ~20nm).

Fig. 11. Picosecond laser (1064nm) milling of fused silica.

3D laser micromilling of fused silica was also attempted with the 355nm frequency upconverted mode locked laser. It was noted that the average floor surface roughness was reduced with increasing milled depth as shown in figure 12 and that for depths greater than 60 μm, the surface roughness was better than that achieved with 1064nm.

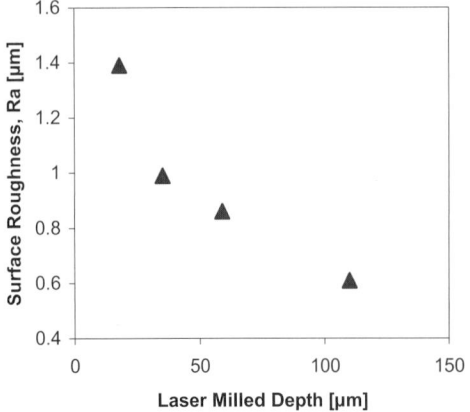

Fig. 12. Surface roughness Ra, versus laser milled depth in fused silica using 355nm picosecond laser.

The lowest Ra value measured was Ra~ 0.434 μm. Generally an uneven surface morphology on the milled floor is commonly encountered in ultrafast laser machining of glasses [7] and is attributed to the resolidification of the molten layers generated during the laser pulse overlap. Pressure gradients can be generated by the plasma formed above the molten surface, particularly large near the edges of the laser spot. During the melt lifetime, the thin layer of molten material will flow from the centre of the laser spot outwards. It is proposed that an averaging effect from a repetitive action of this process with increasing milled depth might produce the improving Ra observed.

6. Conclusions

We have shown that both nanosecond and picosecond lasers can be used for high quality laser micromilling of materials such as ceramics, dielectrics and metals difficult to machine with using conventional methods. Both laser types are capable of excellent surface finish with relatively high material removal rates. The picosecond laser can machine transparent dielectrics with low surface roughness with values comparable or lower than the irradiating wavelength.

Acknowledgements

The authors would like to acknowledge Lumera Laser for use of their application labs and Mark Cheverton for the SEM images.

References

[1] Pham, D. T., Dimov, S. S., Ji, C., Petkov, P. V., Dobrev, T. "Laser milling as a 'rapid' micromanufacturing process". Proc. Inst. Mechanical Engineers, Journal of Engineering Manufacture, **218** Part B:, p. 1 – 7, (2004)

[2] Fleischer J, Kotschenreuther J, "Manufacturing of Micro Molds by Conventional and Energy Assisted Processes", *4M 2005,* ed. W.Menz, S.Dimov, 1st Int. Conf. on MultiMaterial Micromanufacture, p.9, Elsevier, Karlsruhe Germany 2005

[3] Chryssolouris G, *Laser Machining, Theory and Practice*, Springer NY (1990)

[4] Duley WW, *UV lasers, effects and applications*, p.264, Cambridge University Press, Cambridge 1997

[5] Lenzner M et al, "Femtosecond optical breakdown in dielectrics", Phys.Rev.Lett., **80**, 4076 1998

[6] Herman P, Oettl A, Chen K, Marjoribanks R, "Laser micromachining of transparent fused silica with 1-ps pulses and pulse trains", SPIE Conf. Commercial and Biomedical Applications of ultrafast Lasers, **3616**, 148 San Jose, CA USA 1999

[7] Ben Yakar A, Byer R, Harkin A, Ashmore J, Stone H, Shen M, Mazur E, "Morphology of femtosecond-laser ablated borosilicate glass surfaces", Appl.Phys.Lett., **83**, 3030 2003

Multi-Material Micro Manufacture
W. Menz, S. Dimov and B. Fillon (Eds.)

Dimensional tolerances of micro precision parts made by ceramic injection moulding

M. Beck[a], V. Piotter[a], R. Ruprecht[a], J. Haußelt[a,b]

[a] Institute for Materials Research III, Forschungszentrum Karlsruhe GmbH,
P.O. Box 3640, D-76021 Karlsruhe, Germany, e-mail: martin.beck@imf.fzk.de
[b] Institute of Microsystem Technology (IMTEK), University of Freiburg,
Georges-Köhler-Allee 102, D-79110 Freiburg, Germany

Abstract

The aim of the study is to analyse the dimensional variation in ceramic injection moulding of micro precision parts made of zirconia ceramics in order to improve the quality of the injection-moulded parts and to reduce costly reworking of the sintered parts. Dimensional variation and surface quality of injection- moulded single-mode ferrules in the green and sintered state were measured with high precision in the sub-micrometre range. The influence of the process parameters on dimensional tolerances and surface quality of the parts was analysed. Furthermore, the dimensional tolerances and surface quality of the mould cavity used were measured and compared with those of the moulded parts. The results show that process parameters in injection moulding strongly influence the absolute dimensions as well as the dimensional scatter of the green compacts and sintered parts. The coefficient of variation of the outer diameter of the sintered ferrules (nominal value 2.5 mm) was between 0.2% and 0.3%. Furthermore, the injection-moulded green compacts showed shape deviations which could not be explained by the shape tolerance of the mould insert.

Keywords: PIM, CIM, micro precision ceramic injection moulding, optical ferrule, zirconia

1. Introduction

Structural ceramics are often used for high-end applications because of their outstanding properties, e.g. high wear resistance and high hardness. Traditionally, advanced ceramics have been machined from a fired or pre-fired blank to the final desired shape. However, machining of ceramics is very expensive and difficult, especially for complex shaped parts. Near-net-shape processing by Ceramic Injection Moulding (CIM) can remarkably reduce manufacturing costs for mass production of such parts. Nevertheless, for applications which require a very high surface quality and dimensional tolerances in the micrometre or even sub-micrometre range, costly reworking of injection-moulded ceramic parts cannot be avoided.

Therefore, the research presented in this article is focused on analysing and improving the geometrical quality of injection-moulded high-precision ceramic parts. An optical single-mode ferrule has been chosen as a demonstrator because of its extremely tight tolerances. This cylindrical part made of zirconia ceramic is used in optical fibre connectors for precisely positioning the glass fibres. Dimensional and geometrical tolerances of less than 1 micrometre and the bore of only 125 micrometres in diameter make it a real challenge for the manufacturing processes [1].

Dimensional variation of parts produced by powder injection moulding is influenced by many factors, thus the final dimensions reflect the effects of all process steps mixing, moulding, debinding and sintering, while, according to [2], sintering and differences between mixture lots of the feedstock are the most important factors. [3] showed that the dimensional variation of injection moulded ceramic parts increases substantially compared to the green mouldings. The variation of the thickness of those parts measured at different locations in the fired and unfired stage could be reduced by using a low mould temperature combined with a high hold pressure. The current study also confirms that for ultra

precision parts where dimensional scatter in the micrometre range is not negligible, different moulding parameters already have a significant influence on the dimensions of CIM parts in the fired and unfired state.

2. Experimental Work

2.1. Fabrication of the samples by CIM

Figure 1 shows the three steps of the ceramic injection moulding process: the moulded green compact, the brown part after debinding, and the sintered part.

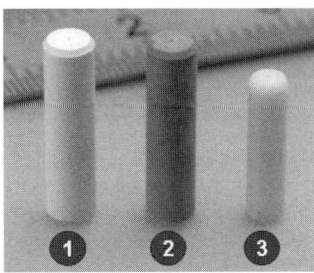

Fig. 1. Steps of the CIM process: green compact (1), brown part (2), and sintered part (3).

2.1.1. Equipment and feedstocks

For the experiments, the commercially available zirconia feedstock Inmafeed® K1011 (Inmatec Technologies Ltd.) based on the powder TZ-3YS-E (Tosoh Corp.) with an average particle size d_{50} of 0.51 μm and a BET surface area of 5.81 m^2/g was used.

The green compacts were fabricated on an injection moulding machine Arburg Allrounder 420C, purpose-made for micro powder injection moulding with a wear-resistant injection unit (screw diameter 15 mm). The high-precision ferrule mould used was made by Junghans Feinwerktechnik Ltd. (see also section 3.1).

2.1.2. Process parameters

To investigate the influence of different injection moulding parameters on the quality of the moulded parts, injection rate, mould temperature, hold pressure and cooling time were varied.

Debinding of the green compacts was done in two stages: at first, the parts were placed in distilled water (25 °C) for 40 hours. After drying, thermal debinding as well as sintering at 1500 °C took place in a chamber furnace Carbolite RHF 17/3 E with the temperature regime recommended by the feedstock producer [4].

2.2. Quality control

Dimensional tolerances and surface quality of the injection-moulded parts were measured in the green and sintered state. The outer diameter and roundness of the ferrules were determined in a contact-free manner with a laser scanner Z-Mike 1210 Gold LX (repeatability ± 0.15 µm) equipped with a high-precision rotary air-bearing table. The surface quality was determined with a MicroGlider$^®$ (Fries Research Technology Ltd.) using a CWL optical sensor CHR 150N.

3. Results and Discussion

3.1. Quality of the mould

The best possible green part quality is limited by the quality of the mould cavity. Thus, its geometry and surface quality were measured.

The geometry of the mould insert was controlled with a 3D coordinate measurement machine Werth Video Check$^®$. A spherical Werth Fiber micro sensor with a diameter of 136 µm was used. Over a depth of 13.5 mm, 96 points in six measuring planes were probed. For the 16 points in each measuring plane, a circular fit was calculated. The shape tolerance of the cylinder, into which the six circles were fitted, was 7.1 µm at a diameter of 3.198 mm.

The surface quality of the cylindrical mould insert was measured with a contact stylus instrument Perthometer$^®$ 6P (tip radius: 10 µm), while the core and the contour ejector were measured with the MicroGlider$^®$. The core and ejector showed a roughness of R_{max} = 1.2 up to 2.0 µm, while the surface roughness of the cylindrical die was R_{max} = 3 µm (R_a = 0.3 µm).

3.2. Analysis of the CIM process

Green compacts were injection-moulded using the commercial zirconia feedstock Inmafeed$^®$ K1011 (Inmatec Technologies Ltd.). The process parameters recommended by the producer [5] were used as a basis for the experimental design. Injection rate and mould temperature were varied on three levels.

In powder injection moulding, jetting is a common problem and the created weld lines can cause critical defects in the sintered parts [6]. A filling analysis revealed that jetting also occurs in the ferrules. The cavity is filled via two gates at the rear end of the ferrule. The moment when the molten zirconia feedstock has just passed the gates and enters the cavity is shown in figure 2. Two jets with the dimensions of the gate cross-section can be observed and the flowing feedstock is coiling while filling the cavity. The material on the right side is slightly hurrying ahead. As a result, the weld line created is asymmetric.

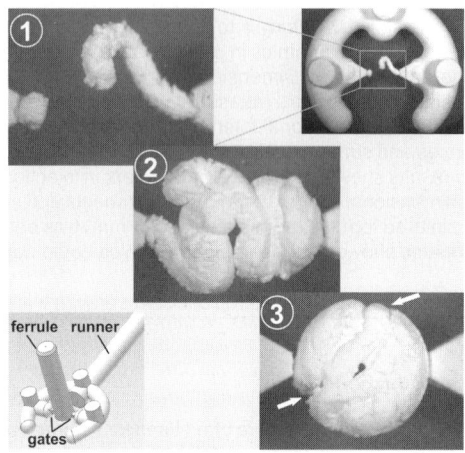

Fig. 2. Mould filling pattern with Inmafeed$^®$ K1011: jetting (1, 2) when the feedstock enters the die and weld line (3).

3.3. Quality of moulded parts

For each set of injection moulding parameters, three ferrules were measured in the green and sintered state. In 10 measuring planes, the ferrule was rotated from 0° to 180° and 17 outer diameter values (OD) were determined (see figure 3 on the left). Thereafter, in 5 angular positions from 0° to 180° (see figure 3 on the right), the ferrule was moved stepwise with a step width of 0.1 mm, so that 100 diameter values were determined for each angular position. Thus, for each of the ferrules, about 600 diameter values were determined. The variation of the part diameter was analysed with statistical methods.

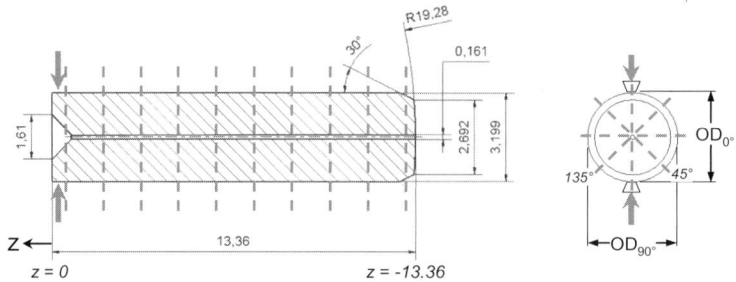

Fig. 3. Nominal dimensions of a ferrule green compact [mm] and positions of measuring planes (– – –) in relation to the gate locations (➤).

3.3.1. Geometrical tolerances of moulded ferrules

Figure 4 shows the dimensional variation of the sintered ferrules for different moulding conditions (injection rate v_i, nozzle temperature T_n and mould temperature T_m). The outlier box plots describe the statistical distribution of the measured values measured for each condition: the box shows the values in the interquartile range IQ (between the 25% percentile Q1 and the 75% percentile Q2), the whiskers represent the range of the data, whereas values beyond the inner fences LIF and UIF are considered outliers.

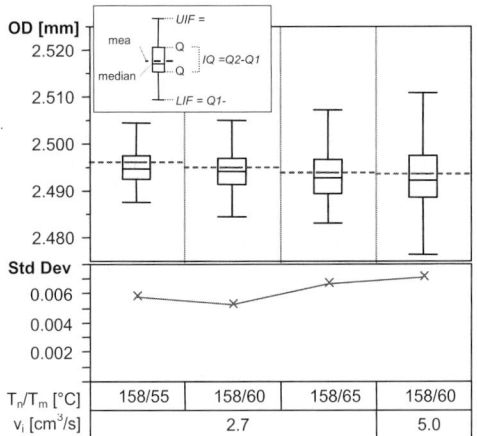

Fig. 4. Variation and standard deviation of outer diameter (OD) of sintered ferrules as a function of the injection rate v_i, nozzle temperature T_n and mould temperature T_m.

The parts with the smallest coefficient of variation (0.21%, calculated by dividing the standard deviation σ by the mean value) were achieved with a mould temperature T_m of 60 °C and an injection rate v_i of 2.7 cm³/s. the range of the diameters measured is 12 µm (σ = 5 µm). Increasing v_i or T_m induces a higher variation up to 18 µm (60 °C, 5 cm³/s) which corresponds to a coefficient of variation of 0.29%. The different tool temperatures have a slight but significant effect on the mean value (significance level 0.01); for higher mould temperatures, the temperature of the demoulded green compact is higher and thus the thermally induced shrinkage after demoulding is more important. The higher injection rate of 5 cm³/s causes a highly significant decrease of the mean value of about 1.5 µm compared to the condition 2.7 cm³/s, 60 °C (significance level 0.001).

All ferrules – both green and sintered parts – showed a characteristic increase of the outer diameter directly behind the 30° chamfer. The surface profile in figure 5, measured with the MicroGlider® surface measurement instrument, illustrates this tendency exemplarily for a sintered ferrule that has been injection-moulded at 55 °C mould temperature and 2.7 cm³/s. The shape deviation shown in figure 5 can also be observed in figure 6 which represents diameter values of the same part. Its diameter was measured over the whole length of the cylindrical surface up to the beginning of the chamfer in steps of 0.1 mm in two angular positions (0° and 90°). At the rear end close to the two gates (z = -0.5 mm), the diameter exceeds 2.520 mm and at the front of the ferrule between z = 9 mm and 10.2 mm, it increases up to 2.505 mm again. As no high diameter

differences were observed while measuring the mould insert surface, it is assumed that these diameter variations are caused by effects of mould filling, cooling or possibly ejecting. While the larger diameter at the rear end near the gates might be explained by the higher packing, the diameter increase behind the chamfer far away from the gates remains to be investigated in the following studies.

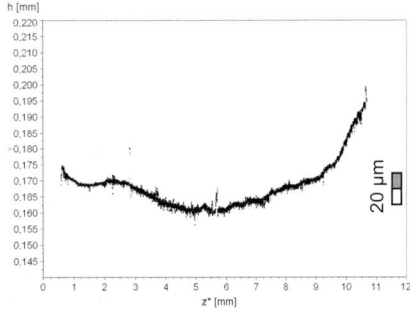

Fig. 5. Topography profile scan of the outer diameter surface of a sintered ferrule versus part length z* (gate positions on the right; T_m 55°C, v_i 2.7 cm³/s).

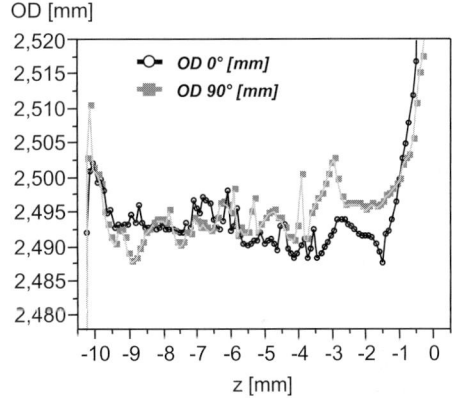

Fig. 6. Variation of the outer diameter (OD) of a sintered ferrule (at 0° and 90°) versus part length z (gate positions on the right; T_m 55°C, v_i 2.7 cm³/s).

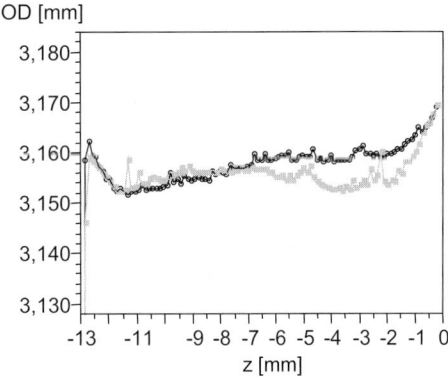

Fig. 7. Variation of the outer diameter (OD) of a green ferrule (at 0° and 90°) versus part length z (gate positions on the right; T_m 55°C, v_i 2.7 cm³/s).

As the dimensions of the sintered parts depend on the injection moulding parameters v_i and T_m, the shrinkage due to sintering (calculated with respect to the diameter of the mould insert) also depends on v_i and T_m and varies between 21.95 % (T_m 55 °C, v_i 2.7 cm^3/s) and 22.05 % (T_m 60 °C, v_i 5 cm^3/s) for the different sets of parameters.

3.3.2. Surface quality

The surface quality of $R_a = 0.3$ µm of the cylindrical surface of the green compacts is very similar to the roughness of the corresponding mould insert surface.

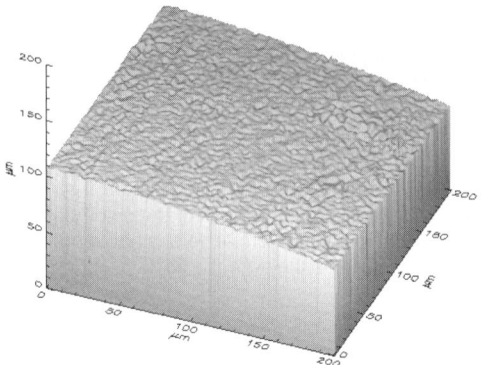

Fig. 8. Topography of the cylindrical surface of a sintered part (200 × 200 µm, FRT MicroGlider®).

The roughness of the sintered parts is much higher (up to $R_a = 0.8$ µm). Figure 8 shows the topography of the cylindrical surface of a sintered ferrule that has been injection-moulded at an injection rate of 2.7 cm³/s and a mould temperature of 55 °C. Figure 9 shows the surface of a sintered part, moulded under the same conditions, at a higher magnification. The structures on the surface have a depth of 0.5 to 2 µm.

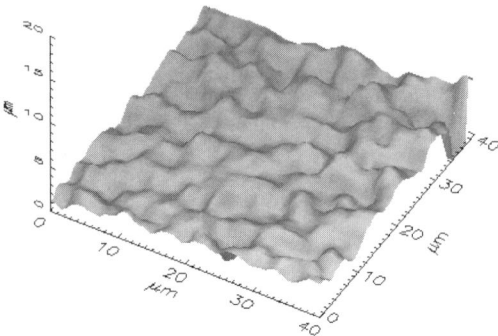

Fig. 9. Topography of the cylindrical surface of a sintered part (40 × 40 × 20 µm, FRT MicroGlider®).

4. Conclusions

Single-mode ferrules can be manufactured by ceramic injection moulding under laboratory conditions with an outer diameter of 2.496 mm ± 8 µm (± 0.31%). The results show that injection rate and mould temperature slightly influence the mean values and variation of the part dimensions.

However, although the cylindrical mould insert used had a very good shape tolerance of ± 7 µm, the green compacts showed significant shape deviations over part length and perimeter. The highest shape deviations were observed near the gate positions and at the end of the ferrules directly behind the 30° chamfer. Further experiments with different feedstocks and taking into account the effect of more process parameters will have to be carried out to understand these phenomena observed. A pressure sensor has been installed in the mould cavity in order to measure the pressure during the filling and packing stage. Furthermore, variation of debinding and sintering parameters will show the influence of these important process steps on the part dimensions.

The surface roughness of the green compacts is similar to the surface quality of the mould insert, thus it is assumed that a better surface quality of the injection-moulded parts could be achieved with a better mould surface. However, fabricating a mould with less roughness but maintaining the excellent shape tolerances is not possible at this time.

Acknowledgements

We thank the project partners Junghans Feinwerktechnik Ltd. and ADC Krone Ltd. for their cooperation. Moreover, we appreciate funding of the project 02PD2136 Micro-P-PIM by the Projektträger Forschungszentrum Karlsruhe and the German Federal Ministry of Education and Research BMBF. We also express our thanks to all our colleagues who have contributed to our work, especially to Heinz Walter for carrying out the injection moulding experiments and to Otto Jacobi for his advice and help while setting up the quality control equipment.

References

[1] Kato T, Molding systems for ferrule of optical connector. Purasuchikkusu 53 (2002) 36-41.
[2] German R M. Powder injection molding. Metal Powder Industries Federation, Princeton, NJ, 1990, pp 486-487.
[3] Tseng WJ. Statistical analysis of process parameters influencing dimensional control in ceramic injection molding. J. Materials Processing Technology 79 (1998), pp 242-250.
[4] Inmatec Technologies Ltd. INMAFEED K1011. Technical data sheet, 2002.
[5] von Witzleben M and Kollenberg W. Keramik-Spritzgießen in der Praxis. Kunststoffe 92 issue 6 (2002), pp 52-56.
[6] Krug S, Evans J R G and ter Maat J H H. Jetting and weld lines in ceramic injection moulding. British Ceramic Transactions 98 (1999), pp 178-181.

Metrology : Inspection and Characterisation Methods

Multi-Material Micro Manufacture
W. Menz, S. Dimov and B. Fillon (Eds.)
© 2006 Elsevier Ltd. All rights reserved

Fiber optic temperature sensor based on spectral transmitivity of CdTe

Z.V. Djinovic[a,b], M.C. Tomic[c], S. Ivankovic[c], A. Vujanic[b]

[a] Institute of Sensor and Actuator Systems, Vienna University of Technology, Vienna 1040, Austria
[b] Integrated Microsystems Austria, Wiener Neustadt 2700, Austria
[c] Institut Bezbednosti, Belgrade 11000, Serbia and Montenegro

Abstract

A fiber optic sensor for measurement of temperature in an environment with strong electromagnetic interferences is presented. The method is based on the change of energy gap of single crystal CdTe with temperature. The spectral transmission of sensing crystal is detected by a special spectrometric arrangement using a diffraction grating and a linear CCD sensor. The obtained measurement uncertainty is lower then ±1°C, in the measuring range from 15°C to 100°C, with the data rate of several hertz.

Keywords: temperature measurement, fiber optic sensor, CdTe, spectrometer

1. Introduction

Temperature measurement in harsh environment is always a challenge. Common approach based on measurement of change of an electrical property is very often connected with troubles mostly induced due to strong effect of environment such as high voltage and existence of EMI and RF radiation. According to this, sensing head, signal processing and transmitting have to be done in such a way to provide an accurate data to the receiving place. Sometimes it is of paramount importance to have a miniature sensing head enable to be placed in a confined room, such as biomedical application, smart structural monitoring, and nuclear environment [1, 2, 3]. Fiber optic temperature sensors could meet most of the above requirements utilizing different techniques [4, 5, 6]. However, main drawback of these techniques is complexity.

In this paper we present a relatively simple fiber optic temperature sensor based on the change of energy gap of single crystal of CdTe with temperature [7]. This is a well known phenomenon already used for temperature measurement based on fiber optic sensing configuration [8, 9]. Most of papers have been dealing with measurement of intensity of radiation transmitted through the CdTe or GaAs single crystal. In order to reach good measurement accuracy it is ultimate to include some referencing technique. However, it makes the sensing system to be rather complex and still not sufficiently accurate mostly due to misalignment of optical fibers into the sensing head. We overcome these problems by involving of spectrometric technique to detect and analyze of transmitted spectrum through the crystal. We developed of an algorithm for tracking and determination of threshold point along the transmitting curve of CdTe in dependence of temperature. Sensing head, connected with standard multimode fiber 62,5/125 µm, was made of CdTe dice with overall dimensions of 1x1x0,5 mm. We experimentally obtained measurement uncertainty lower then ±1°C in measuring range from 15°C to 100°C, with data rate of several hertz.

2. Theory

The presented fiber-optic sensor is based on the light absorption/transmission properties of Cadmium Telluride (CdTe). The sensor concept relies on the temperature dependence of the band gap of the CdTe crystal. Meaning that light transmission through a semiconductor (crystal) is a wavelength dependent phenomenon with a sharp rise in photon absorption occurring when photon energy exceeds the band gap energy $E_g(T)$. The transition wavelength above which light transmission increases substantially is given by $\lambda_g(T)=hc/E_g(T)$, where h is Planck's constant and c is the speed of light. The band gap energy, typically, drops monotonically with increasing temperature. As a result, the transition wavelength shifts to longer wavelengths as the temperature increases [7]

CdTe is a semiconductor with a direct energy gap E_g of around 1.5eV at 300K and with very broad transmission characteristic in the range of 0,83 µm till to 30 µm. Temperature coefficient of the CdTe band gap is $-dE_g/dT=3.610^{-4}$ eV/K, which indicates that gap is decreasing almost linearly with temperature increase. Expressed by wavelength the absorption threshold of CdTe is changing according to $d\lambda_g/dT=3.410^{-4}$ µm/K. When the light passes from a broad band GaAs LED diode (IRED) trough the CdTe crystal, the spectrum of IRED will be transformed almost totally if central wavelength of the light is beyond of the transmission threshold of CdTe. As the crystal's temperature increases, transmitted spectrum of CdTe shifts towards the higher wavelengths overlapping the emitted spectrum of the IRED. Higher crystal temperature means further shift of the transmission curve blocking the entire light spectrum of IRED to pass through the crystal. Hence, our sensor is based on measuring the transmitted spectrum and its threshold point in dependence on CdTe temperature. We define the threshold point as the start of the energy gap.

3. Experiment

The light travels down from the IRED light source via the fiber coupler to the temperature sensor head, where the light is partially transmitted through the CdTe crystal. A dielectric mirror reflects the transmitted light,

142

which returns to the fiber coupler and is directed to the spectrometer (see Figure 1). In the spectrometer we detect and analyze the spectrum of the transmitted light. The developed algorithm for determination of the threshold point position is applied to find actual temperature for each recorded spectrum.

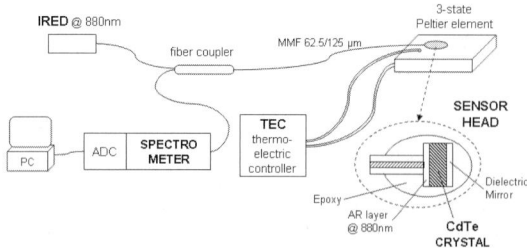

Fig. 1 Experimental set-up of the temperature measuring scheme

Spectroscopic Equipment: A fiber-optic-coupled miniature spectrometer was used in this study. It was Ocean Optics S2000 (wavelength range 400 -1000 nm) with 2048 pixel charge-couples device (CCD) detector Sony ILX511, 600 lines/mm grating, and 12-bit data acquisition.

Light source: The light source employed was GaAs LED Infrared Emitting Diode - IRED, emitting at λ=880 nm, with the spectrum width of 50nm

Temperature control: To control the temperature of the sensor and reference thermometer, we used the 3-stage Peltier element, as a thermoelectric cooler/heater, in the range of 15°C to 100°C. The temperature change at the Peltier element is controlled by the thermoelectric controller – TEC. Independent measurement was preformed by using a semiconductor thermometer Analogue Devices AD590KF, glued close to the CdTe sensor head by high temperature epoxy. In that way the simultaneous control measurement was made.

Fiber-optic waveguides: In this study we used standard multimode optical fibers 62.5/125 μm as shown in Figure 1 to connect IRED via coupler with the sensor in one way and the sensor via the coupler and over the spectrometer with the PC in other way. The optical fibers can be as long as 1000 m.

Sensor head: As can be seen in Figure 1 the sensing element is a CdTe prism with overall dimensions of 1 x 1 x 0.8 mm. The front side of the prism was covered by SiO_2 antireflection layer with a thickness of about 200 nm optimised for the wavelength of 880 μm. The back side of the prism was coated by a dielectric mirror. A fiber tip was in direct contact with the front side of the prism. The whole sensing head was glued in a thermal conductive epoxy. Small size was desirable to decrease the thermal mass and response time of the sensor.

Data collecting and processing: For collection of temperature data, the fiber-optic sensor head was put at the 3-stage Peltier element, together with the semiconductor thermometer AD590KF. Then, we have been changing the temperature in the range of 15°C to 100°C and simultaneously detected and analyzed the

transmitted spectrum. The number of sampling points was 300 in the range of 835 to 935 nm for different temperatures. In this way we obtained the transmitted spectra, with a threshold specific for each temperature as shown in Fig. 2.

The algorithm: The recorded transmitted spectra were analyzed using the MATLAB in order to obtain the calibration curve showing the threshold points against the temperature. The algorithm for determination of the threshold position is based on the shape recognition. The characteristic threshold shape includes the short horizontal segment of non-transmitting spectrum part, the arc and the short, about 72°-angled line corresponding to the CdTe gap. The searched shape was previously normalized using the non-affected spectrum part, beyond 910 nm. This algorithm, in fact, does not find the threshold "point" but the pixel position where starts the searched shape, having the minimum deviation from the examined spectrum.

4. Results and discussion

Fig. 2 presents the transmitted light spectrum in dependence on the temperature of CdTe crystal. We can see that the intensity of the transmitted light decreases with temperature in the same time with narrowing of the wavelength range of the spectrum. This is in accordance with theoretical explanation given ahead. The intensity change has previously been used commonly for manufacturing of these types of sensors [8, 9]. However, it was recognised that this technique requires having some referencing means in order to reach a reasonable accuracy. Instead to measure the light intensity we used to capture and analyse the spectrum of the transmitted light.

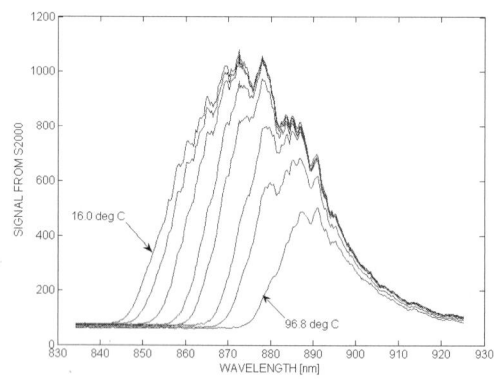

Fig. 2 Transmitted spectrums depending on the temperature acquired by spectrometer Ocean Optics S2000

The shape of the spectrum is actually the result of superposition of the initial bell shape radiation spectrum of the applied IRED light source (see spectrum part on the right side) and threshold position of the transmission curve of CdTe (see spectrum part on the left side). As we said earlier this part is changing part in function of the temperature of CdTe. In a case that temperature of CdTe is higher than 125°C the transmitted spectrum will disappear, i.e. the entire source radiation will be absorbed. In this case for this kind of light source we can say that this temperature is upper limit.

In order to extend the measuring range it is

possible to use new one light source with some another spectral characteristic, e.g. with central wavelength allocated toward larger values. It could be also possible to involve more than one IRED by multiplexing of light sources via 2x1 or 3x1 fiber optic couplers and optical switchers at the input light source arm (see Fig. 1). Ideally it would be good to have a white light source with a broad spectral range that can cover rather large spectral range. Such a source could be some bulb, e.g. a xenon lamp.

After collecting the transmitted spectrums for the whole temperature range we analyzed them further by using the MATLAB program. The main goal was to obtain the calibration curve of threshold points against the CdTe temperature. The final result is shown at Figure 3.

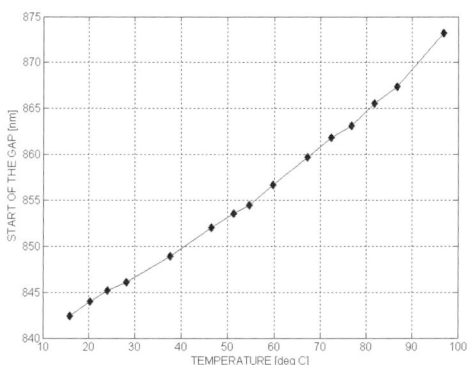

Fig. 3 The calibration curve of the threshold points in dependence on temperature

The curve is relatively smooth, but not linear. The several measurement cycles show no hysteresis and a good repeatability of the curve, yielding the measurement uncertainty lower than ± 1°C. Our measuring range was from 15°C to 100°C, because of limitation of the temperature control, but the real possible range can be estimated from Fig. 2, noticing the boundary temperatures where the threshold in the reduced spectrum still can be clearly resolved.

The maximum data rate is limited by the CCD reading speed, number of spectrum averaging, A/D conversion speed and the algorithm execution speed. Since the signal processing in these experiments was performed off-line, the later time could not be determined. However, since the CCD reading time is several milliseconds and the necessary averaging is 8 times, the data rate of several hertz can easily be achieved.

Measurement uncertainty may be affected by several reasons. First of them are spectrometer features, such as density of the grating lines, aberration of the optical system, number of pixels of the CCD as well as read-out electronics and A/D conversions. Second, S/N ratio depends on the light intensity as well as on the parasite reflections in the system. Also, complexity and precision of applied algorithm have a great importance, especially regarding the immunity to the variation of optical power through the fiber-optic link. The time caused deviation is still under the examination, but it is not expected that it would affect the measurement accuracy because of long term stability of spectrum transmission properties of optical fibers and CdTe crystal.

5. Conclusion

In this paper we presented a miniature fiber optic sensor that could be used for temperature measurement in specific and confined rooms. The principle of operation of the sensor is based on the change of energy gap of single crystal of CdTe with temperature. We included the spectrometric technique to detect and analyze of transmitted spectrum through the crystal. We developed of an algorithm for tracking and determination of threshold point along the transmitting curve of CdTe in dependence on temperature. In this way we achieved an accurate sensing system, with measurement uncertainty lower then ± 1°C in the measuring range of 15°C to 100°C.

Having in mind that this fiber optic sensor is based on the change of the optical characteristics of the crystal only in dependence on temperature and there is no metal part is included in the measuring technique, this sensor is in fully resistant to electromagnetic and radiofrequency radiation. Therefore it can be used for temperature measurement in harsh environment, like in high voltage transformers. This sensor can be installed directly in transformer winding to accurately monitor the hot spot temperatures. Accurate winding temperature measurement enable users to optimize transformer loading, maintain maximum capacity and prevent major transformer damages. Besides it is easy to install, it also should have long-term reliability and easy system integration into the existing infrastructures.

Acknowledgements

We are grateful to the Integrated Microsystems Austria, IMA that partially supported the research activities in this paper.

References

[1] Xu J., Wang X., Cooper K. and Wang A., Miniature all-silica fiber optic pressure and acoustic sensors, Opt. Lett., 30 (2005) 3269

[2] Berghmans F., Fernandez A. F., Brichard B., Vos F., Decreton M., Gusarov A., Deparis O., Megrer P., Blondel M., Caron S.and Morin A., Radiation Hardness of Fiber-Optic Sensors For Monitoring and Remote Handling Applications in Nuclear Environments Harsh Environmental Sensors, Proceedings of SPIE 3538, 1998

[3] Martin J. M. M., Esquer P. M., Lence F. R and Guemes J. A., Fiber Optic Sensors for Process Monitoring of Composite Aerospace Structures, Proceedings of SPIE 4694, 2002

[4] Burt M. C. and Dave B. C., An Optical Temperature Sensing System Based on Encapsulation of a Dye Molecule in Organosilica Sol-Gels, Sensors abd Actuators, B 107 (2005) 552

[5] Jung W. G., Kim S. W., Kim K. T., Kim E. S. and Kang S. W., High-Sensitivity Temperature Sensor Using a Side –Polished Single-Mode Fiber Covered with the Polymer Planar waveguide, 13 (2001) 209

[6] Ferreira L. A., Ribeiro A. B. L., Santos J. L. and Farahi F., Simultaneous Displacement and Temperature Sensing Using a White Light Interrogated Low Finesse Cavity in Line with a Fiber Bragg Grating, Smart. Mater. Struct., 7 (1998) 189

144

[7]   Zanio K., Cadmium Telluride in Semiconductor and Semimetals ed. by Willardson R. K. and Beer A. C., Academic Press, New York, 1978

[8]   Kyuma K., Tai S., Sawada T. and Nunoshita M., Fiber-Optic Instrument for Temperature Measurement, IEEE J. Quantum Electron., QE-18 (1982) 676

[9]   Sultan M. F. and O' Rourke M. J., Temperature Sensing by Band Gap Optical Absorption in Semiconductors, Proceedings of SPIE 2839, 1996

Multi-Material Micro Manufacture
W. Menz, S. Dimov and B. Fillon (Eds.)

Mechanical testing of laser welded micro seams

C.B. Nielsen[a], D.Buccoliero[a], T. Ussing[b]

[a]Centre for Microtechnology and Surface Analysis, Danish Technological Institute, 2630 Taastrup, DK
[b]Fluimedix ApS, Gregersensvej, 2630 Taastrup, DK

Abstract

In a previous work, a low power diode laser has been presented, which is capable of creating laser welded seams as narrow as 10 μm in polymers. The ability to produce narrow seams makes this low power diode laser suitable for perimeter welding of micro fluidic channels in Lab On a Chip (LOC) systems. Perimeter welding is useful when bonding polymeric LOC systems since clogging of the micro channels is prevented during the bonding procedure. In the present work a new mechanical test method is put forward that tests the mechanical strength and homogeneity of a laser welded micro seam, both of which is important when a micro seam is used to bond the lid in a LOC. The test is a pressure test, where the weld seam is subjected to the stress that arises from an isostatic pressure in a cavity that is defined by two pieces of polymer and the weld seam in question. The paper presents the experimental work together with mechanical finite element simulations, both of which show how the mechanical test results are affected by the width of the laser welded micro seam. Among the results the yield pressure-seam width relationship is shown to have a contra intuitive behavior. Instead of having an ever decreasing yield pressure for decreasing seam widths, a local minimum in this relationship is found.

Keywords: Mechanical testing, micro welding, Lab on a chip, micro fluidics

1. Introduction

When a polymeric microfluidic Lab On a Chip (LOC) system is fabricated, normally at least two polymeric pieces are required; one base piece that contains the microfluidic channels and the other piece which constitutes the lid [1]. There are more ways to bond the lid to the microfluidic substrate, one of which is thermal bonding. To do so, the two pieces of polymer are put together in a mechanical press and held just below the glass transition temperature T_g for an hour or so. Depending on the cross sectional dimension of the microchannel, the distance between channels, the ability to control the mechanical press, the temperature and the bonding time, there is a potential risk that the bonding procedure will clog the LOC system. Therefore laser transmission welding along the microchannels is an interesting alternative to the above. However, bonding microfluidic systems with micrometer distances between channels would require a laser bonder that has the capability to produce weld seams at least as narrow as said distances, see Fig. 1. In a previous work

[2] it has been demonstrated that the Fluimedix ApS low power diode laser is able to produce weld seams as narrow as 10 μm, which makes it useful for perimeter bonding of polymeric microsystems.

However, when optimizing the laser parameters for a specific weld task, it would be of interest to know the upper limit of the strength of such a weld seam. Furthermore it would be interesting to know how homogeneous a weld seam is, since local weaknesses could render a microsystem useless when fluids short circuits.

Traditional strength tests like a pull test or a peel test [3] will give data that corresponds to an average strength. However, it will not give any information on local weaknesses of a micro seam, and hence no information on the homogeneity of the seam strength. Furthermore using a peel- or a pull test does not emulate the complex stress state, a micro weld is subjected to when bonding a micro fluidic system.

In this paper we put forward a pressure test method that is more likely to test micro seams in accordance with their intended use.

2. Experimental

2.1. Test specimen

2.1.1. Basic geometry

A number of different test geometries have been examined by practical laboratory work and using Finite Element (FE) simulations. A suitable geometry that was found is presented in Fig. 2. The test specimen consists of a 50x32x1.5 mm³ black PolyMethyl MethAcrylate (PMMA) base piece, a 32x32x1.0 mm³ transparent top piece. The two pieces are bonded using a laser transmission welded circular seam, where the inner diameter of the seam is 28 mm. In the base piece, an Ø1 mm hole is drilled, where water under pressure is directed into the cavity that is defined by the two PMMA pieces and the weld seam.

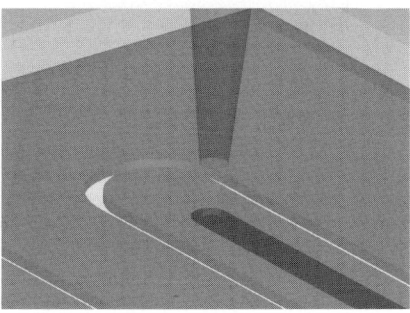

Fig. 1. Illustration of a laser transmission welding between micro channels

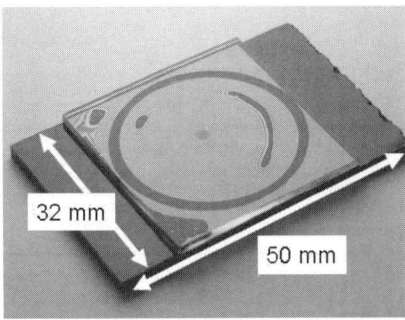

Fig. 2. Image of the pressure test geometry. The test specimen consists of two pieces of PMMA that are laser transmission bonded using a circular weld seam. The two pieces and the weld seam create a cavity that is subjected to a pressure during the test.

2.1.2. Seam width

In the present work it has been of interest to characterize the quality of micro seams. Especially to establish if a micro seam has the same yield stress as the bulk material. However, using a pressure test to find the yield stress is not straight forward, since the stress state in the presented geometry is complex. Therefore, pressure experiments with cavities that have different Seam Widths (SW) have been carried out to get an overall picture of the Yield Pressure (YP) versus SW.

The YP should be understood as the pressure at which there is a deformation of 5 % of the inner edge of the weld seam. Using this definition, the yield stress correlates to the YP through the equations of solid state mechanics with the Von Mises yield equation [4].

Since the Fluimedix laser in the experiment is used to make weld seams preferably below 200 μm, a Fisba Ironscan laser (λ= 808 nm) is used to create wider seams. Using both lasers, careful control of the laser parameters has made it possible to vary the SWs. Additionally FE simulations have been performed for comparison.

2.2. Pressure testing

In order to perform pressure tests, a test stand has been developed. The test sample is fixed into a suitable clamp that allows pressurized water to enter the cavity without influencing the strength of the cavity itself. In order to increase the pressure in the cavity, water is pumped into the cavity at a flow rate of 40 μL per sec, using a syringe pump from CAVRO (XL3000). The pressure is measured using an absolute pressure sensor from Sensor Technics (24PCGFA). The time-pressure data are obtained using a data acquisition card from National Instruments (DAQ 6036E).

Using this equipment, the Yield Pressure (YP) of the weld seam can be found. The YP is found using an optical microscope during the pressurization of the cavity and is identified as a change in color, going from transparent to whitish.

2.3. Finite element simulations

For comparison, FE simulations have been performed, using Mechanica/ProEngineer® for model definition and as solver. The FE model is build using the same geometrical values as the real samples. The

material properties and simulation conditions are listed in table 1.

In the table the top-base spacing is set to 20 μm. This spacing has been found experimentally using cross-sectional optical microscopy.

Table 1

FE simulation conditions

Elastic modulus	3300 MPa
Poisson's ratio	0.37
Yield strain	5%
Top-base spacing	20 μm
Strain criterion	von Mises

Table1: Above table contains the conditions that have been used in the FE model. The material properties represent common table values for bulk PMMA.

Similar to the physical experiments it is possible to define a YP criterion in the FE simulations. The results presented here, are found when an applied pressure in the model gives a von Mises strain of 5% in the modeled weld seam. In practice the YP is found by doing FE simulations with a fixed SW at different pressures. This experiment yields a pressure-strain curve for the specific SW where YP subsequently is found by interpolation between strain values below and above 5%.

In Fig. 3 an example from a simulation is shown. The color grades indicate the strain level at different locations on the sample where red has the highest strain level and blue has the lowest strain level. In the figure, only a quarter of the entire sample is shown. All simulations have been carried out using mirror conditions on the two cut surfaces and the lower circular edge on the base piece was fixed with respect to translation and rotation.

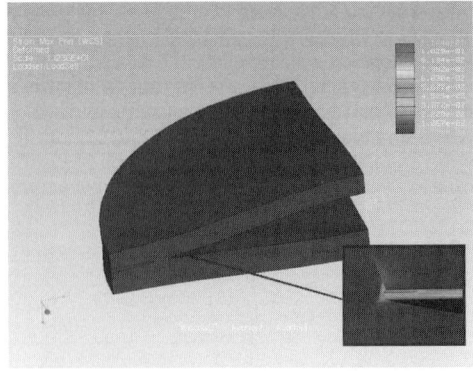

Fig. 3. Screen shot from a FE simulation. The colors indicate the strain level, where red has the highest strain level, and blue has the lowest. In the insert it can be seen that there is a strain concentration in the weld seam that points into the top piece of the sample. The used layer spacing is found experimentally.

3. Results

3.1. FE simulation results

From the FE simulations the YP data was extracted and transferred to the graph in Fig. 4. The graph shows at what pressure the seam will begin to yield at different weld SWs. The graph can be divided in three regimes. For SWs wider than 2000 μm, the YP is constant. For SWs below 50 μm the YP decreases rapidly, and for SWs in between there is a local minimum for the YP.

3.2. Pressure test results

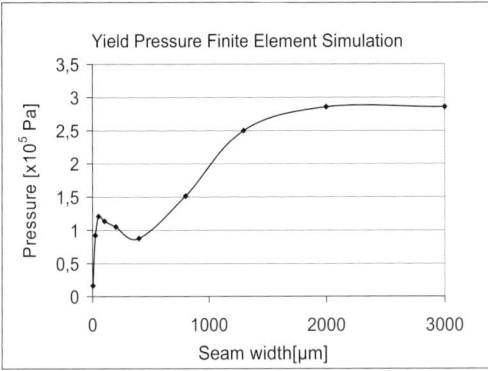

Fig. 4 Graph showing the YP curve found using FE simulations. In the graph three YP domains can be identified. For SWs above 2000 μm, the YP becomes constant. For SW below 50 μm the YP decreases rapidly for SW going towards zero. In the SW region between 50 μm and 2000 μm a local minimum is found.

In Fig. 5 the YP data found in the real experiments are shown in a graph (dots) together with the averaged YP data (dashed line) and the YP found by FE simulations (full line).

In general, the YP found by real experiments is a factor of two higher than that of the simulations.

On a qualitative level, it can be seen that for SWs above 1500 μm the YP goes towards a constant value. Furthermore, the local minimum present in the FE model is also present at a SW of about 600 μm. In the FE model, the YP goes rapidly towards zero for SW less than 50 μm. In the real experiments, YP data for SWs above 20 μm exist, which is why this behavior could not be verified. The YP data points for SWs between 20 μm and 200 μm, produced by the low power diode laser, can be seen to continue the trend from the Fisba laser, in the sense that the data substantiate the local minimum in the FE simulation.

4. Discussion

In Fig. 5 it can be seen that there is a certain amount of spread in the pressure test data set. This spread in YP data can be used as an indication on the homogeneity from weld to weld, i.e. the spread tells about local variations in seam width and in material-to-material bonding strength. To assess this problem further, the experimentalist would have to use for example optical microscopy.

Comparing the averaged pressure test data to the simulation results, there is a high degree of similarity

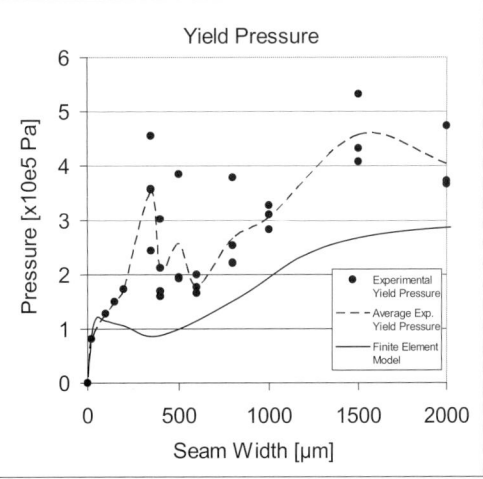

Fig. 5. Graph showing the found YP in the real pressure test together with averaged values and the YP results from the FE simulations. Although the experimental values in general are a factor of two higher than the modeled values there are qualitative similarities between the simulated values and experimental values.

between the two SW/YP curves. For SW larger than 1500 μm, the simulated YP goes towards a constant level. This behavior is not entirely seen for the experimental data that has a decrease in YP of approximately 12% for SW going from 1500 μm to 2000 μm. The flattening of the simulated YP/SW curve can be explained by the size of the strain field. The thickness of the weld seam is about 20 μm. Therefore it is highly likely that the strain field at a SW of 1500 μm does not extend all the way to the outer edge of the seam, and hence the YP becomes constant.

Assume now that the material involved is homogenous, isotropic, has an ideal stiffness and that displacements are small compared to the pressure test geometry. Assume further that the top and bottom part are infinitely stiff i.e. that the strain in the edge would only originate from stresses normal to the plane of the test specimen, the stress in the weld seam would then follow Eq. 1.

$$YP(\pi/4)D^2 = \sigma_{yield}A_{seam} \qquad (1)$$

Where D is the inner diameter of the weld geometry, σ_{yield} = 72 MPa (the yield stress of PMMA) and $A_{seam} = (\pi/4)((D+2SW)^2 - D^2)$. Eq. 1. reduces to:

$$YP = 4\sigma_{yield}(SW^2 + SWD^2)/D^2 \qquad (2)$$

Eq. 2 is shown in Fig. 6, where it is seen that YP goes towards zero for decreasing values of SW in a monotonical manner, i.e. there is no local minimum of YP.

However, as shown in Fig. 3, the strain at the inner edge of the seam is a combination of a normal tensile stress from the forces that will separate the top and base piece, and stresses induced by the bending moment at the inner weld edge. Therefore the local YP minimum at approximately 400 μm probably is due to a complex interaction between the bending moment and the lift off stress, which is influenced by the flexibility of

148

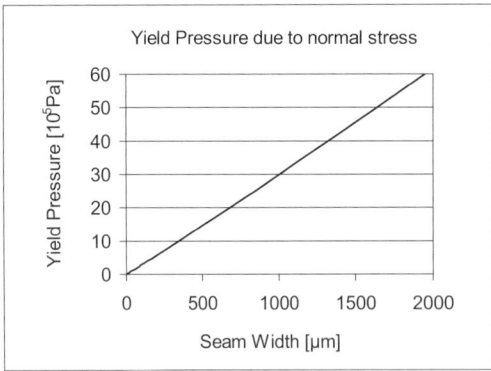

Fig. 6 The above graph shows how the YP-SW behavior would be, if only normal tensile stress was present in the weld seam.

the weld seam, which again is influenced by the overall geometry and the stiffness and Poissons ratio of the PMMA. For SWs below 50 μm the lift off force becomes the dominant strain factor and so the YP decreases monotonically towards zero described qualitatively by Eq. 2 and illustrated in Fig. 6.

In Fig. 5 the low power diode weld seams and the Fisba weld seams together give a SW-YP curve that is qualitatively similar to the FE simulated curve. In doing so, this points to the fact that the low power diode weld seams render a bonding, which has a quality that is at least as high as the one produced by the Fisba laser.

Since all experimental SW-YP data have higher values for YP than the simulated data, it must also be so, that the weld parameters were such that the bonding strength was comparable to the strength of the bulk material.

5. Conclusion

A new method for pressure testing of laser welded micro seams has been put forward. The methods ability to test micro seams has been evaluated using seam widths ranging from 20 μm to 2000 μm. This has been done using finite element simulations as well as real experiments. The two experiments show qualitatively the same yield pressure to seam width behavior. For high seam widths the simulated yield pressure is seen to be constant, which is also true within an error of 12% for the real experiments. Furthermore, both experiments have a decline in yield pressure at low seam widths and finally both have a local minimum in yield pressure at approximately 400 μm. However, knowing this relationship the presented method can be used to evaluate the strength and homogeneity of a micro weld at different seam widths.

Acknowledgements

This work was funded by the Danish Ministry of Science, Technology and Innovation (VTU), through the innovation consortium "Centre for Microsystems for Chemical and Biochemical Analysis Based on Polymers - μKAP".

References

[1] Beck W.A. IR Laser Welding of thin Polymer films as a fabrication Method for Polymer MEMS, Proceedings of SPIE 2002, Vol. 5067, pp167-178.

[2] Ussing T. Micro laser welding of polymers using low power laser diodes, 4M 2005 - First International Conference on Multi-Material Micro Manufacture, ISBN 0-08-044879-8. ELSEVIER, pp 291-293.

[3] Dosser L. Transmission Welding of carbon nano composites with direct diode and Nd:YAG solid state lasers. Proceedings of SPIE, 2004, Vol. 5339 Issue 1, pp 465-474.

[4] Alexander J.M., Strength of materials, Horwood 1991.

Multi-Material Micro Manufacture
W. Menz, S. Dimov and B. Fillon (Eds.)

149

Reactive magnetron sputtering deposited ITO thin films : influence of O$_2$ admixture on microstructure and optical properties

C. Secouard[a,b], C. Ducros[a], P. Roca i Cabarrocas[b], S. Noël[c], F. Sanchette[a]

[a]*CEA Grenoble, Laboratoire des Technologies des Surfaces, 17 rue des Martyrs 38054 Grenoble CEDEX, France*
[b]*Laboratoire de Physique des Interfaces et des Couches Minces (CNRS UMR 7647), Ecole Polytechnique, Route de Saclay, 91128 Palaiseau Cedex, France*
[c]*CEA Grenoble, Laboratoire des Composants Hybrides, 17 rue des Martyrs 38054 Grenoble CEDEX, France*

Abstract

Indium-tin oxide (ITO) thin films have been deposited onto glass substrates by reactive magnetron sputtering. Effects of O$_2$ content in Ar/O$_2$ gas mixture on coatings properties were investigated. The structure was analysed by X-ray diffraction. In the region between 0 and 8 O$_2$ vol.% in the gas mixture the crystallisation is improved and the lattice is expanded with increasing oxygen flow rates because of the incorporation of the overflowing oxygen. Above 8 O$_2$ vol.% in the gas mixture the lattice is shrunk and the crystallinity is lowered. The crystallites size decreases with increasing oxygen flow rate after reaching a maximum at 4 O$_2$ vol.% in the gas mixture. In fact, the excess oxygen is bound to be segregated to the grain boundaries preventing the grain growth and relaxing the lattice. The optical transmittance of the films was determined using a spectrophotometer. The transparency improves with increasing oxygen flow rates. The filling of oxygen vacancies and the deactivation of Sn donors both explain the red-shift in the UV region and the great increase of the transmittance in the near infra-red region. Good transmission rates (>80%) seem promising for photovoltaic application.

Keywords : indium tin oxide, reactive magnetron sputtering, X-ray diffraction, transmittance

1. Introduction

Tin-doped indium oxide (known as indium-tin oxide or ITO) is a wide band gap semiconductor. It has been widely investigated since it was found out about 40 years ago that good optical and electrical properties could be achieved in ITO thin films. They find applications in various domains such as humidity sensors, flat panel displays (LCD), OLED, electrochromic devices, and heatable glass [1,2]. They are also used in solar cells. In general, transparency and high electrical conductivity are mutually exclusive. In other words good electrical properties can be achieved but to the prejudice of optical transmission [3] so that a compromise between conductivity and transparency has to be found, since the efficiency of the cells depends on the amount of light available for the conversion and on the ability to collect the carriers.

ITO films can be deposited using a various set of methods such as sputtering, spray pyrolysis, evaporation, pulsed laser deposition, ion plating, screen printing and CVD [4,5]. Among these, sputtering — including DC, pulsed MF, RF magnetron sputtering and facing target sputtering — yields resistivity values as low as 115 µΩ.cm (DC power) [6], and is extensively used because of its cost effectiveness [1], high versatility [7] and controllability [5].

Electro-optical properties of ITO films are strongly sensitive to crystalline state and microstructure. In the deposition technique used in this work — reactive magnetron sputtering — different parameters can affect the crystallisation and the morphology of the coatings. For instance high process temperature and annealing post-treatment promote higher adatomic mobility supporting grain growth and changes in the films texture. Hence a beneficial effect on the conductivity [5,8]. Increasing the sputtering pressure reduces the energy of the impinging particles, changes the crystalline orientation and may even lead to the formation of amorphous coatings with poor electrical properties due to higher electron scattering [7,9]. Moreover,

the excitation mode and the power of the magnetron discharge influence significantly the structural and thus the electrical properties of tin-doped indium oxide films [10].

The purpose of this work is to study the influence of the relative amount of oxygen in the gas mixture on the structural and optical properties of ITO thin films deposited by pulsed DC reactive magnetron sputtering. The main application of the developed films is the use as transparent and conductive electrodes for solar cells.

2. Experimental details

Thin films of ITO were deposited on glass substrates using a pulsed DC reactive magnetron sputtering system. The vacuum chamber was pumped down to a base pressure ranging from 4 to 8x10^{-4} Pa prior to deposition. A 210x90 mm² rectangular-shaped ceramic plate of 90 wt.% In$_2$O$_3$ and 10 wt.% SnO$_2$ (purity 99.99%) was used as target. The target to substrate distance was kept constant at 80mm. Both substrates and target were separately pre-sputtered in pure Ar atmosphere respectively for 25 minutes at 5 Pa and for 10 minutes at 1 Pa. Then the deposition was carried out in argon and oxygen atmospheres at a total pressure of 0.5 Pa and a discharge power of 100 W. The admixture of oxygen to the argon sputtering gas was varied from 0 to 10 vol.% keeping the total gas flow constant (50sccm) through MKS mass flow controllers. During deposition, the substrates were not biased and heated up at 400°C. The process duration was adjusted so that the thickness of the coatings was 200nm±25nm.

Thickness was determined at the film edge with a stylus profiler. The crystalline structure of the films was analysed by X-ray diffraction using the Cu Kα radiation in a Siemens D 5000 diffractometer in θ/2θ mode. Ultra-violet/visible/near infra-red spectrophotometer was employed to measure the optical transmittance of the films.

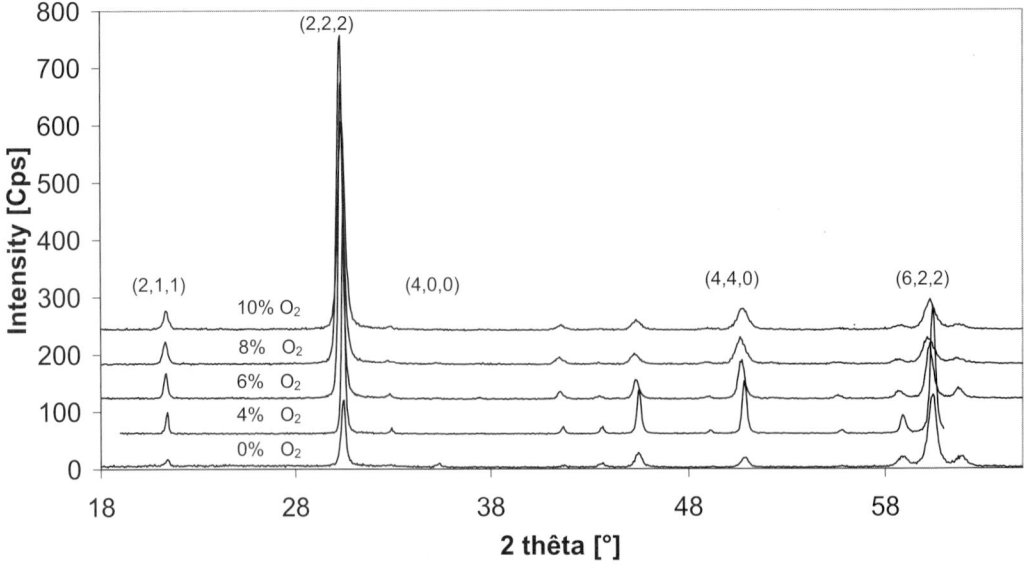

Fig. 1 Variation of the XRD spectra of the ITO thin films at various O_2 admixtures.

3. Results and discussion

In this work, the effect of the O_2 content in the gas mixture on the films properties (structure, transmittance) was studied. It was varied to 0, 4, 6, 8 and 10 vol.% while Ar was respectively 100, 96, 92, 91 and 90 vol.% keeping the total gas flow constant.

The XRD spectra of ITO films deposited at different O_2 admixtures are displayed on Fig. 1. All coatings show clear diffraction peaks and a preferred orientation with (2,2,2) planes parallel to the substrate surface. With increasing O_2 admixtures the peaks intensities are amplified except for the (4,0,0) peak which disappears. It has been found that crystallisation with (4,0,0) planes parallel to the substrate surface is due to reduced oxidation while (2,2,2) oriented films correspond to extended exposure to oxygen atmosphere [11]. As the oxygen flow rate is raised the reflections undergo a shift toward lower angles. The evolution of the (2,2,2) peak is plotted on Fig. 2(a). For O_2 admixtures ranging from 0 to 8 vol.% the intensity increases and the peak is shifted to lower angles. Then above 8% O_2, the intensity is weakened and the peak shifts back to a larger angle (see Fig. 2(b)). The coherence length D of the scattering material can be determined from the XRD spectra by means of the Scherrer formula. D has been calculated for (2,2,2) oriented planes thanks to the analysis software EVA. It is displayed on Fig. 3 as a function of the oxygen content in the gas mixture and is interpreted as the extension of the corresponding grains. After reaching a maximum value at 4% O_2 admixture the crystallite size decreases continuously with increasing O_2 rates. ITO exhibits an In_2O_3 bixbyite structure which is derived from the CaF_2 lattice (face-centered-cubic cation lattice with anion on tetrahedral sites) by introducing constitutional vacancies, so that every fourth anion position is not occupied [12].

During ITO thin films deposition the oxygen is incorporated in the crystals presumably by filling the constitutional vacancies and complexing the substitutional

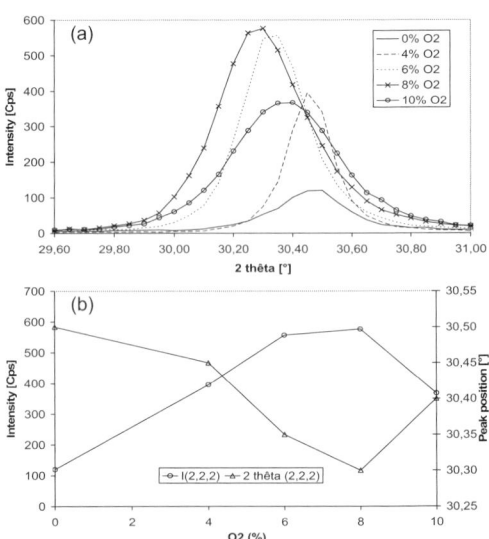

Fig. 2 Evolution of (a) the (2,2,2) peak and (b) the intensity and position of the (2,2,2) peak at various oxygen flow rates.

Sn atoms, therefore resulting in the formation of films with expanded lattices [12,13]. In addition, it has been reported that the excess oxygen atoms are also segregated to grain boundaries [12,13]. This leads to a reduction of the lattice constant, a partial amorphization of the films (see Fig. 2) and prevents the grain growth. Thus we can assume that in our experiments the excess oxygen is first incorporated in

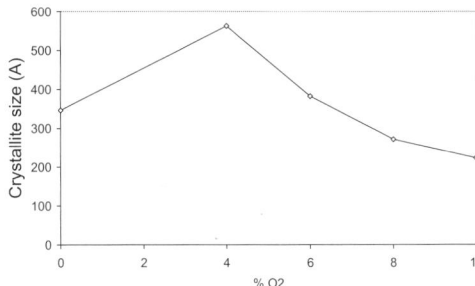

Fig. 3 Size of (2,2,2) oriented grains evaluated with the Scherrer formula as a function of %O₂.

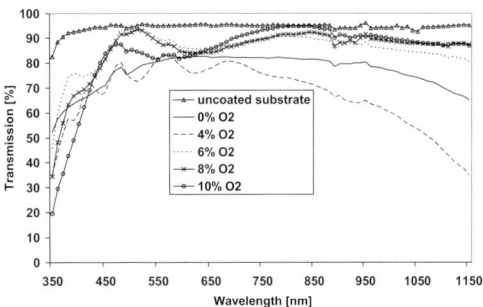

Fig. 4 The optical transmittance spectra of the ITO films deposited at different O_2 contents in the gas mixture.

the lattices and then segregation occurs above a certain oxygen flow rate.

Fig. 4 shows the optical transmittance of the ITO films grown with different O_2 admixtures for wavelength between 350 and 1150nm. For a comparison purpose, the spectrum of an uncoated glass substrate is also plotted. The transmittance is gradually improved as the O_2 content in the gas mixture is increased. Transmittance above 80% in the 450-1150 nm region is achieved with amounts of oxygen ranging between 6 and 10 vol.%. It has been reported that the contribution to the carriers concentration arises from both the oxygen vacancies and the activated Sn^{4+} on indium sites [6,12]. In the near infra-red region the variation of the transmittance is strongly related to the variation of the electron concentration, above 1000 nm the absorption by free electrons becoming critical [7,8,13]. Thus the continuous increase of the transmittance at the near IR region with the oxygen flow rate implies a gradual decrease in the carrier concentration, as the vacancies are filled and Sn donors deactivated. Furthermore, it is known that the Burstein-Moss effect can induce a bandgap shift. When the electron density is higher than a certain critical density, a blocking of the lowest states in the conduction band occurs, hence a widening of the gap. This leads to a shift of E_g toward shorter wavelengths i.e. a blue-shift in the transmittance onsets with increasing carrier concentration. This effect is counterbalanced by electron-electron and electron-impurity scattering which tends to narrow the band gap [7,14]. Consequently, the sequential red-shift in the transmittance onsets (UV region) with increasing O_2 admixture (see Fig. 4) indicates —in accordance with previous considerations on the near-IR transmittance— that the carrier concentration progressively decreases. As discussed previously, the excess oxygen in the ITO films favours the formation of grain boundaries and Sn^+O^- complexes resulting in the creation of negatively charged and neutral scattering centers [6,13] therefore contributing to the red-shift of the transmittance spectra.

4. Conclusion

In this work the influence of the oxygen flow rate on the structure and optical properties of ITO thin films deposited by pulsed DC reactive magnetron sputtering was studied. It has been found that the excess oxygen are incorporated in the crystallites by filling the constitutional vacancies and complexing the substitutional Sn. As a result, the lattice constant and the transmittance are increased with increasing oxygen rates. With a further increase in the O_2

admixture, the excess oxygen atoms are segregated to the grain boundaries in an amorphous phase. Consequently, the lattice is relaxed and the grain size is reduced. Lower carrier concentration and increased electron scattering explain the red-shift in the transmittance onsets.

These considerations on the variation of the electron concentration and scattering are to be confirmed with upcoming carrier concentration and mobility investigations.

The good transmission rate of the films in the visible/near infra-red region (>80% for oxygen flow rates between 6 and 10 vol.%) makes them suitable for photovoltaic applications. It should be corroborated by resistivity measurements.

References

[1] A. Rogozin, M. Vinnichenko, N. Sevchenko, A. Kolistsch, W. Möller, Thin Solid Films 496 (2006) 197-204.

[2] H.-C. Lee, J-Y. Seo, Y-W. Choi, D.-W. Lee, Vacuum 72 (2004) 269-276.

[3] A. S. A. C. Diniz, C. J. Kiely, Renewable Energy 29 (2004) 2037-2051.

[4] C.-H. Chung, Y.-W. Ko, Y.-H. Kim, C.-Y. Sohn, H. Y. Chu, S.-H. K. Park, J. H. Lee, Thin Solid Films 491 (2005) 294-297.

[5] L.-J. Meng, M. P. Dos Santos, Thin Solid Films 289 (1996) 65-69.

[6] D. Mergel, M. Schenkel, M. Ghebre, M. Sulkowski Thin Solid Films 392 (2001) 91-97.

[7] L.-J. Meng, M. P. Dos Santos, Thin Solid Films 303 (1997) 151-155.

[8] H.-C. Lee, O. Ok Park, Vacuum 75 (2004) 275-282.

[9] Y.-S. Kim, Y.-C. Park, S. G. Ansari, J.-Y. Lee, B.-S. Lee, H.-S. Shin, Surface and Coatings Technology 173 (2003) 299-308.

[10] R. Mientus, K. Ellmer, Surface and Coatings Technology 142-144 (2001) 748-754.

[11] P. Thilakan, J. Kumar, Vacuum 48 (1997) 463-466.

[12] D. Mergel, W. Stass, G. Ehl, D. Barthel, Journal of Applied Physics 88 (2000) 2437-2442.

[13] H.-C. Lee, O. Ok Park, Vacuum 77 (2004) 69-77.

[14] C. G. Granqvist, A. Hultaker, Thin Solid Films 411 (2002) 1-5.

Multi-Material Micro Manufacture
W. Menz, S. Dimov and B. Fillon (Eds.)

153

Comparison of the characterization of laser striation on lamellar cast iron with various techniques

C. Vincent[a, b], G. Monteil[a], T. Barrière[b], J.C. Gelin[b]

[a] *Laboratoire de Microanalyse des Surfaces, 26 chemin de l'épitaphe 25000 Besançon, France*
[b] *Institut FEMTO-ST, département mécanique appliquée, 24 chemin de l'épitaphe 25000 Besançon, France*

Abstract

After engraving specific grooves by means of a laser Nd:YAG, three measurement techniques have been used in order to characterize the shape and geometry of the grooves. Different types of grooves have been studied. They are defined by their profiles and their spatial repartition. The first characteristic is the profile. Four cross section geometries have been selected: semicircular, rectangular, trapezoidal and triangular. The second characteristic is the spatial distribution. The grooves are classified in two categories: continuous and discontinuous lines with a specific pattern for each one. The engraved material is lamellar cast iron which is a heterogeneous material. Consequently, the conditions of engraving are more difficult to define. Thus, an accurate metrology is the key point to adjust the laser machining in order to obtain the desired grooves.

Three measuring techniques have been used to measure one or more characteristics of the grooves like dimensions, distribution or profile. These techniques are: roughness profilometry, scanning electron microscope and non contact optical imaging. The combination of the information coming from those techniques leads to a better control of the geometry of the engraved surfaces.

Keywords: roughness profilometer, scanning electron microscope, non contacting measurement, laser machining.

1. Introduction

Laser machining is more and more employed in industry and research. The application fields are wide as in precision moulding, surface engraving [1, 2]... A precise knowledge of this tool allows realizing high performance pieces but it still remains a question: how controlling the final shape precisely.

The aim of this work is to present different techniques of control which allow measuring the result of the laser surface texturing in a heterogeneous material in this case lamellar cast iron.

In this work, the pieces are engraved by means of a laser Nd:YAG. The engraved patterns are lines with different specifications: spatial distribution and cross section geometries. Three techniques of control have been used: the roughness profilometry, the scanning electron microscope and the non contact measurements imaging. SEM is the most employed technique for that purpose. Sometimes, interferometric non contacting measurement is used [3]. In the present work an optical microscope combined with a camera was employed.

2. Experiences

2.1. Laser machining

2.1.1. Laser manufacturing machine

Laser manufacturing machine is a DML 40 SI of Deckel Maho (Germany). The laser is a Nd:YAG with a mean power of 100 Watts. Thanks to a galvanometric scanner, the laser beam can be piloted and tilted up to 20°. The laser beam has 2 focuses: Ø40 µm and Ø100 µm.

Different machining strategies can be selected during the CAM. The first engravement uses a perpendicular laser beam. The beam incidence is perpendicular to the surface. The second engravement uses a tilted laser beam. It is employed for the engravement up to 100 µm of depth. This inclination is possible with the galvanometric scanner (see Fig. 1). An other reason for the use of a tilted beam is the limitation of stairs in the flanks of the tilted surface which appears with the first machining strategy.

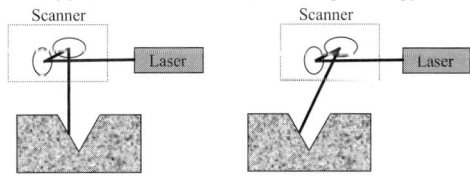

Fig. 1. Machining strategies. The first machining uses a perpendicular laser beam to engrave all kind of surfaces. The second machining is employed for the tilted surface with an important depth thanks to the tilted laser beam.

In addition to the machining strategies, other parameters concerning the laser machine can be adjusted. The most important parameter is the intensity of the current of the lamp which allows regulating the power of the laser and consequently adjusting the machined depth. Following the engraved material and the machined depth, the intensity of the current of the lamp must be adjusted. When the engraved depth is constant and in the same time the desired depth is reached, the parameter is regulated [4].

2.1.2. Laser striation

The engraved patterns are lines with various distribution and profile. For the spatial distribution, the grooves are classified in two categories: the continuous lines and the discontinuous ones. The first grooves are called C following the angle between the lines. The second ones are called D1 to D4. For each one, a single pattern allows to determine the characteristics of the distribution (see Fig. 2).

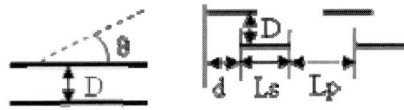

Fig. 2. Spatial distribution of grooves. A single pattern characterizes each groove.

For the profiles, four cross sections have been selected: semicircular, rectangular, trapezoidal and triangular (see Fig. 3).

Semicircular	Rectangular
Trapezoidal	Triangular

Fig. 3. Cross section geometries. Four profiles have been engraved.

All type of grooves has been engraved but with different dimensions: some micrometers up to millimetre. This modification is due to the problem of the characterization of the grooves but this point is developed in the next lines.

The engraved material is lamellar cast iron. It is an atypical material and there is a lack of available work. So, the first set of machined grooves has been made with roughly adjusted parameters. A second set of larger grooves has been engraved for a more accurate determination of the adjustment of the laser parameters (machined depth and profile shape accuracy). The aim of this final engraving process is to find out the proper laser adjustments by means of the first metrological feed back.

During the process, the material is pulverized by the laser. An extraction system aspired the dust but a part of metal cools at the edge of the grooves. These burrs must be eliminated by grinding before controlling the grooves.

2.2. Groove characterization

Several types of groove have been engraved: with a continuous or interrupted spatial distribution, with different dimensions. In proportion as the grooves were engraved, the control has to be changed in order to overcome the limitations itinerant to the apparatus and the measurement technique. Three measuring tools have been employed.

2.2.1. Roughness profilometry

The first method of characterization which has been used is based on a roughness profilometer. A square surface of 3 mm has been measured. Thanks to the specific software, each profile has been saved and fit together to obtain an image in three dimensions which can be processed (see Fig. 4).

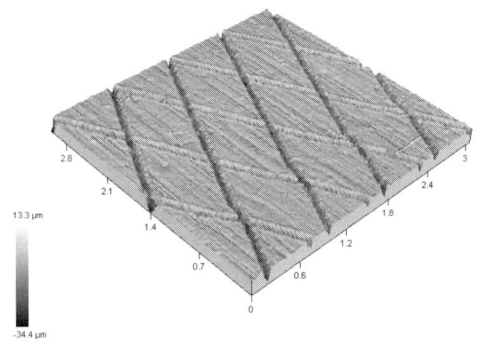

Fig. 4. Image from 3D roughness profilometry of C50. The image is built from all profiles acquired by the roughness profilometer.

With such an image, it is possible to measure the dimensions of the grooves and their distribution.

The surface of the flanks of the grooves is cleaner than the bottom (see Fig. 5). The problem comes from the sensor. In fact the sensor is a conical diamond one with a tip cone angle of 45°. As the angle of the flanks is less than 45°, the measure corresponds to the flank angle of the sensor. Nevertheless, this method allows controlling the repartition of the grooves but the cross section geometries cannot be obtained accurately.

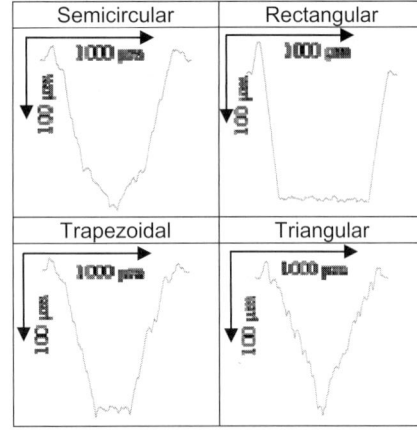

Fig. 5. Profiles extracted from 3D roughness images. The flanks of the rectangular groove correspond to the flanks of the sensor. Else, the profiles are good.

2.2.2. Scanning Electron Microscope

The SEM allows controlling the surface without contact. The grooves are not deteriorated. It needs only a cleaning of the samples.

When scanning the whole surface of the sample (see Fig. 6), the measurement is focussed on the spatial distribution and the roughness. The differences between the machining strategies can be explored too.

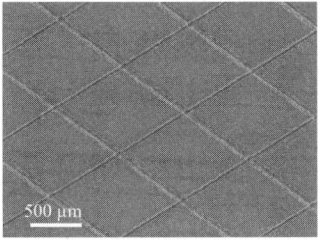

Fig. 6. Image from SEM of C75. This image allows only controlling the aspect of the engravement.

In addition, thanks to the enlargement of the SEM, the appearance of the surface after laser machining can be visualized. Scories appears and little trenches too (see Fig. 7). The engraved material is lamellar cast iron and it is a heterogeneous material with ferrite and graphite. The parameter of the intensity of the current of the lamp determines the engraved depth. With such a material, the adjustment of this parameter is delicate because the engraved depth is not stable. The differences are due to the presence of two different materials. The energy to pulverize the graphite is lower than the one for the ferrite. The consequence in terms of the intensity of the current of the lamp is about: 26% for the graphite and 45% for the steel.

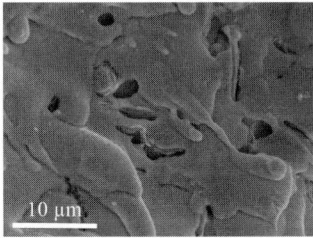

Fig. 7. Scories appear at the bottom of the groove due to the energy given by the laser.

In order to realize a precise observation of the machining quality of the groove geometry, their widths have been increased to the millimetre scale. For the SEM imaging a silicone groove replica has been made. Then in order to measure the cross section of the groove, this replica is cut and metallized (see Fig. 8).

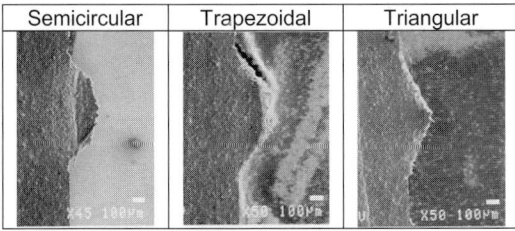

Semicircular	Trapezoidal	Triangular

Fig. 8. Profiles from SEM extracted thanks to a silicone mould of the grooves.

By means of the SEM photographs, dimensions of the grooves and their profiles can be measured. However the accuracy of this method is not sufficient to insure a close loop control of the machining conditions. In addition, this technique is rather difficult to operate on a wide scale. Consequently it has been avoided for the check of the precise control of the groove's shapes. It must be employed only for the metallographic and surface quality of the machined areas.

2.2.3. Non contacting optical measurements

For the non contacting optical measurements, the apparatus employed in the present work is an InfiniteFocus® from Alicona (Austria). This optical technique is founded upon an optical microscope coupled with a digital camera. The whole heights extend of the measured area is divided in a set of focus planes. The resulting set of images is analyzed by a software reconstruction to build a 3D map of focus points. The selection of the focussed pixels is made by a neighbouring comparison of the contrast. The image is the topography of the measured area. Joined area can also be measured leading to a greater accuracy in case of large fields. The resolution in heights rises up to 20 nm with an objective x100.

As a result three methods allow the dimensions and the shapes of the engraved patterns to be measured. These data come from the classical roughness profilometry, the SEM imaging and the optical measurements.

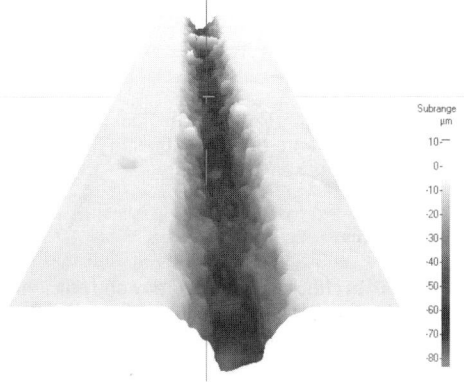

Fig. 9. Image of triangular groove.

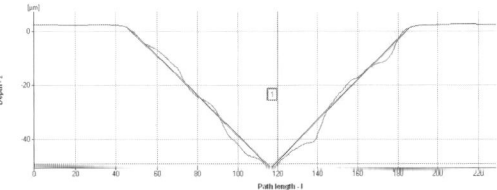

Fig. 10. Profile of the triangular groove.

Practically, the control has been realized on samples with only one groove. The purpose of these measurements was only to verify the accuracy of the machined shape of the grooves; profile and dimensions. Rectangular, trapezoidal and triangular grooves have been correctly engraved (see Fig. 9 and 10). For the semicircular groove, the profile seems to be closer to a trapezoidal shape than to a semicircular one. This discrepancy can be explained by the CAD synthesis of the machining process. The desired dimensions are micrometrics: 20 μm depth and 200 μm width. The software Catia© is not able to manage the calculation of a semicircular surface with these dimensions. By its discrete way of calculating the semicircular shape it subdivides this one in segments. Consequently the resulting simulated semicircular shape is closer to a trapezoidal groove (see Fig. 11).

156

Fig. 11. Calculated and desired profiles of semicircular groove.

On the contrary, sometimes the profile of the trapezoidal groove can look like a semicircular one (see Fig. 12). This area where this phenomena is visible is very fine along the engraved line. Elsewhere of the groove, the profile is trapezoidal. This remark confirms our supposition on the difference between the calculated and the engraved profile of the groove. So the respect of the profiles of the engraved grooves is delicate and the skill of CAD too.

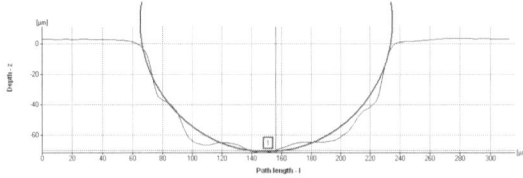

Fig. 12. Mean profile of trapezoidal groove. In addition, the software allows controlling the radius of the semicircular profile.

3. Discussion

In order to check out the accuracy and efficiency of these three methods to precisely control the machining process, the striation C75 has been compared using the three measuring techniques. The theorical dimensions are the dimensions of the groove in the CAD file. Some parameters are described in figure 2.

Table 1
Comparison of the dimensions of the C75 groove with the three techniques mentioned above.

	Depth (μm)	Width (μm)	Distance D (μm)	Crossing angle θ	Triangular profile sharpe	Profile angle
Theoretical	6	60	700	75°	yes	157°
Roughness profilometry	13	80	690	75°	yes	154°
SEM	X	40	655	70°	yes	X
Non contacting optical measurements	15	73	697	75°	yes	151°

The first remark concerns the SEM. This control allows checking the general look of the engraved grooves. The dimensions are measured directly on photographs. This micro graphs cannot be the image of the reality because the sample can be tilted and so there is an angle between the camera and the surface. Else, the roughness profilometer and the non contacting optical measurements data are the same.

4. Conclusion

The aim of the present work was to identify an accurate method of the characterization of the laser micrometric grooves on a surface. The laser machining process is a delicate technology and its control too. The management of the laser engraving machine tool is a crucial point in order to obtain a good accuracy of the engraved patterns on a surface. The classical method based upon contacting roughness profilometry suffers some physical limitations for this purpose. In order to overcome these limitations, non contacting methods have been employed. SEM and optical methods have been shown to be very promising techniques to check the dimensions and the shapes of the striated patterns. Unfortunately in practice, the SEM technique is requiring too long sample preparation to be employed. Nevertheless by means of a non contacting optical measurement the precise control of the engraved depth is possible.

Consequently as precise measured data are now available, the adjustment of the parameters of the laser and the realization of the desired grooves can be reached. As an example, for the machining of the triangular grooves in a lamellar cast iron the intensity of the current of the lamp has been adjusted because the grooves were found too deep, the value has been set at 25,8% instead of 32,3%. For the machining strategies, all engravement are realized with a perpendicular configuration to engrave faster and more accurately. To avoid the problems of the construction of the semicircular groove, the CAD is going to be changed. All these changes have been possible thanks to the use of the new techniques presented here.

Now, it is possible to engrave various profiles in a heterogeneous material thanks to these techniques of measurement. Consequently, the field of the applications becomes wider. In addition to the micromoulding area of the applications, this knowledge allows to use these striated surfaces in friction components generally made in heterogeneous materials like cast iron, or in the decorative marking.

Acknowledgements

The research reported here is the result of collaboration between ENSMM (Ecole Nationale Supérieure de Mécanique et des Microtechniques), l'école des Mines de Paris / ARMINES, Total, JPX, PSA (Peugeot – Citroën), Renault and ADEME (Agence de l'Environnement et de la Maîtrise de l'Energie). The authors gracefully thank the support of the partners.

References

[1] Ike H. Materials Processing Technology 60 (1996) 363-368.
[2] Makoto I. and al. Development of high image clarity steel sheet by the application of laser texturing (1987) IDDRG.
[3] Moon K and Microscale patterning of mesoscopic PZN-PT single crystal films by crystal ion slicing and laser induced etching, Ferroelec. Lett. 30(1-2) (2003).
[4] Documentation Lasersoft 3D, Pfronten GmbH, version 2.5.8.

Multi-Material Micro Manufacture
W. Menz, S. Dimov and B. Fillon (Eds.)

Measurement and characterisation of the microtopography in the contact area between workpiece and tool in microforming

S. Weidel, U.Engel

*Chair of Manufacturing Technology, University of Erlangen-Nuremberg,
Egerlandstr. 11, D-91058 Erlangen, Germany*

Abstract

In metal forming processes, the forming load is transmitted from the tool to the workpiece only by a certain fraction of the nominal contact area. According to the mechanical-rheological model these fractions are mainly the real contact area (RCA) and the so-called closed lubricant pockets (CLP). As the number of these contact areas (RCA and CLP) is reduced drastically in micro forming processes due to small part dimensions and approximately scale invariant surface topographies, their influence on the tribological conditions is increased significantly. Therefore, the knowledge about the real contact state in these areas in micro forming is much more important compared to conventional "macro" length scale. In the macro case, the real contact area is regarded as completely flattened during the forming process while the sub-topography which emerges on single asperities is not considered. In the present study, a test rig is introduced for the characterisation of the flattening behaviour of single asperities by in-situ observation and for the recording of the force-displacement characteristics. Additionally, a strategy for the evaluation of the sub-topography on single asperities is presented leading to an improved understanding of the tribological behaviour in micro forming processes. Thus, an improved simulative process design of micro forming processes is enabled.

Keywords: microforming, tribology, surface characterisation

1. Introduction

The ongoing miniaturisation of electronic and micro-mechanical components leads to an increasing demand for smallest metallic parts. When the part dimensions decrease, the influence of the surface on the component's functionality and its production process increases due to the changing surface to volume ratio. Therefore, all effects which are influenced by the surface, e.g. the tribological conditions, play a much more important role when the part dimensions are scaled down.

At macroscopic length scale, the desired tribological condition in forming processes characterised by low friction and low wear is mainly adjusted by the choice of the mating material or the lubricant. Besides this, there are also a few examples for a specific adjustment of the surface in order to reduce friction and wear. For instance, in sheet metal forming it is a common technique to apply a certain texture to the sheets by skinpass rolling with textured rolls in order to generate geometrically defined lubricant pockets [1]. These lubricant pockets improve the storage and the dispersion of the lubricant in the forming zone during the process leading to an improved quality and stability of the forming process [2,3]. In bulk metal forming it is state of research to apply textures on tool surfaces, e.g. cold forging punches, for purpose of wear reduction and tool life improvement [4].

In order to design surfaces according to the functional requirements, e.g. reduction of friction and improvement of wear resistance, exact models describing the interrelationship between topography and functional impact are necessary. This presupposes the possibility of an appropriate characterisation of the surface. Up to now, surfaces are mainly described by statistical parameters making it often difficult to predict the functional behaviour. The existing approaches for a functional surface characterisation are based on the evaluation of the contact state and the lubricant dispersion in order to create models for the interaction of surface and lubricant [5,6]. Starting from these models, the most important surface features can be identified and quantified by surface parameters.

For example, the mechanical-rheological model assumes that the forming load is transmitted from a forging tool to a lubricated workpiece by three different bearing ratios [7]. These are the real contact area (RCA), open (OLP) and closed lubricant pockets (CLP). As CLPs have no connection to the edge of the workpiece, the entrapped lubricant is pressurised while the forming load is applied and the asperities are deformed plastically. The developing hydrostatic pressure takes a part of the external forming load, thus reducing the normal pressure on the asperities leading to a decrease in friction. In case of lubricant pockets with a connection to the edge of the surface, the lubricant cannot be pressurised and is squeezed out with increasing normal pressure. These OLPs transmit only a negligible part of the forming load.

The characterisation of micro part surfaces regarding their tribological behaviour has been hardly considered so far although the influence of the surface increases as mentioned before. It is difficult to transfer the knowledge gained at conventional length scale to micro scale due to the following reasons. As the models used for macroscopic applications are based on the assumption that a multitude of asperities is in contact, differences in the behaviour of single asperities are averaged. As the surface features are approximately invariant to the dimensions of the workpiece, in microforming only a few asperities are in contact with the tool surface (fig. 1). Thus, the different behaviour of single asperities is not longer negligible [8,9]. The different behaviour compared to the macro case is provoked by the existence of only a few grains in the

158

a) ground surface on a conventional workpiece -
many asperities contacting the tool

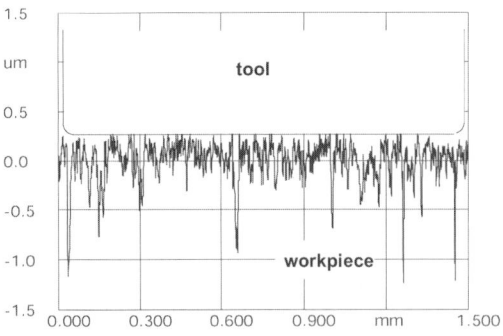

b) ground surface on a micro workpiece -
few asperities contacting the tool

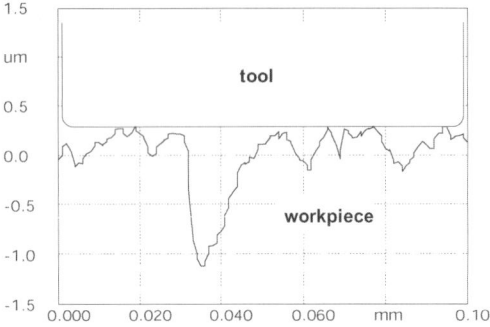

Fig. 1. Asperities in contact at conventional length scale and in the micro case

asperity resulting amongst others in a different plastic behaviour and a bigger influence of the surface layer.

Additionally, the consideration of the real contact area for the prediction of friction in the conventional length scale according to the generalised friction law developed by Wanheim/Bay [10] presumes that the solid contact area on top of a single asperity is flattened completely. This assumption does not consider a sub-topography emerging within these solid contact areas as already proposed by [11] describing the effectively existing contact state more precisely. In order to describe the influence of the surface on the tribological behaviour in micro forming processes with a small contact area and only a few asperities in contact, this sub-topography has to be taken into account [12].

The aim of this project is the characterisation of surface features relevant for the tribological behaviour of the surface in microforming processes. As the real contact area plays the significant role in mixed lubrication regimes [7] being predominant in forming processes, the formation of the contact area of a single asperity under load and the sub-topography has to be analysed in detail. This paper presents the experimental setup and the strategy used for the evaluation of the relevant surface features and their behaviour in forming processes. Additionally, preliminary results from conventional upsetting tests and the procedure for the assessment of the sub-topography under load are introduced.

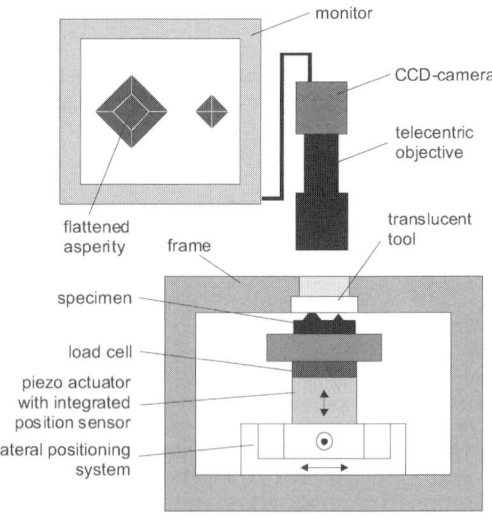

Fig. 2. Principle drawing of upsetting device

2. Experimental setup

For the derivation of surface parameters describing the tribological behaviour of micro parts, the flattening behaviour of the surface, in particular the flattening of a single asperity under load, is essential. Therefore, a micro upsetting test with a translucent tool for the in-situ observation of the asperity flattening has been designed. The experimental setup (fig. 2) consists of a piezo actuator with an integrated position sensor for the vertical movement of the specimen in 10 nm steps. The required force for flattening the asperities is measured by a load cell with a responsivity lower than 1 mN. The upper tool is designed as a frustum with a diameter in the nominal contact area of 1.2 mm and is made of quartz glass. For the in-situ observation of the surface flattening a telecentric objective with a reproduction scale of 5.0 and a CCD camera is used resulting in an image area on the specimen of 1.2 x 0.9 mm^2 and a resolution of about 1.0 µm. This test setup enables the direct observation of the formation of the real contact area during surface flattening.

In order to analyse the interaction of adjacent surface features and the lubricant behaviour under load and finally the tribological behaviour of the micro part surfaces in microforming processes, upsetting tests are conducted with cylindrical specimens with a diameter of 0.5 mm and a height of 3 mm. The face surfaces of the specimens have been produced by turning and grinding in order to produce surfaces with different topography features and a different arrangement of these features.

For a detailed analysis of the flattening behaviour of a single asperity, specimens have been designed as an array of pyramids with a base of 120 x 120 µm^2 and a height of 35 µm of each pyramid representing idealised asperities (fig. 3). The specimens are made out of OFHC (Oxygen-Free High Conductivity) copper. In order to ensure that there is no interaction between the topography elements, there is a distance of 1 mm between each other. The specimens have been produced by micromachining at the Laboratory for Precision Machining at the University of Bremen.

Thus, this micro upsetting test enables the exact measurement of the force displacement characteristics of single asperities leading to a better understanding of the behaviour of asperities under load and finally of the

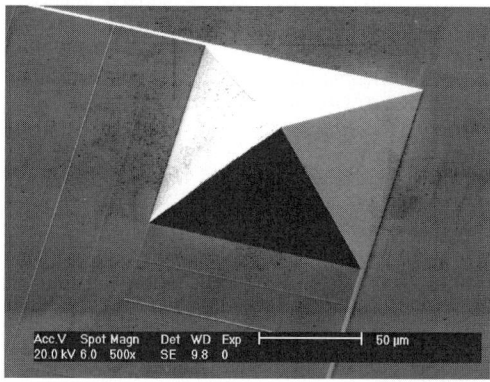

Fig. 3. SEM-image of a micro machined pyramid representing an idealised asperity

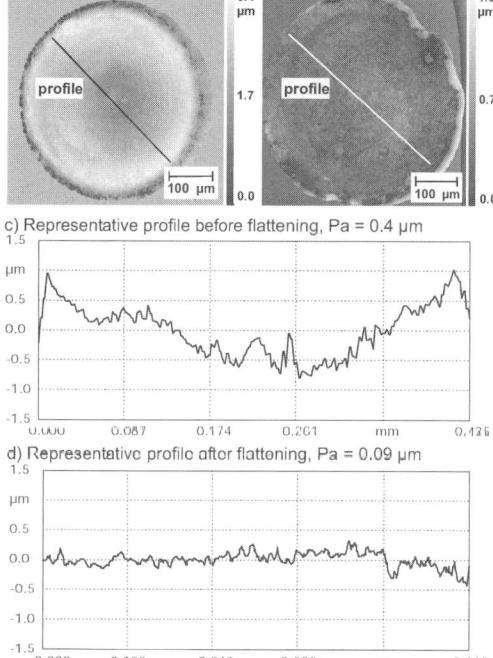

a) Topography before flattening   b) Topography after flattening

c) Representative profile before flattening, Pa = 0.4 µm

d) Representative profile after flattening, Pa = 0.09 µm

Fig. 4. Flattening of a turned surface without lubricant

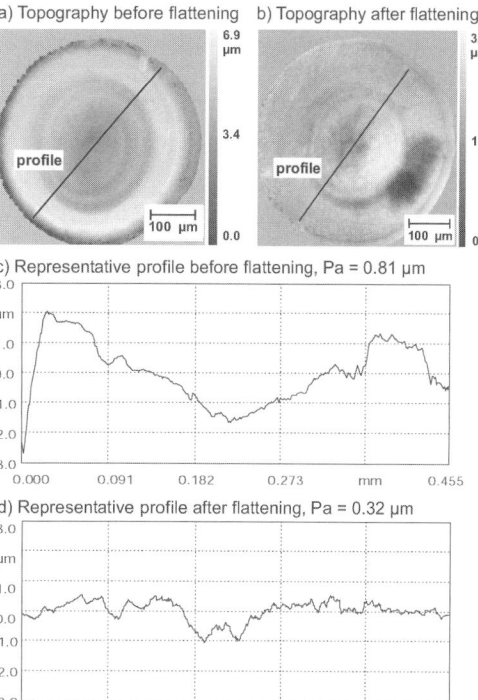

a) Topography before flattening   b) Topography after flattening

c) Representative profile before flattening, Pa = 0.81 µm

d) Representative profile after flattening, Pa = 0.32 µm

Fig. 5. Flattening of a turned surface with lubricant

tribological behaviour of micro part surfaces.

3. Preliminary upsetting tests

In order to identify topography features relevant for the tribological behaviour, cylindrical specimens with a diameter of 0.5 mm produced by turning and grinding have been upset. The experiments were carried out with a nominal surface pressure of 610 N/mm^2 and an upsetting speed of 1 mm/min. The specimens made from copper have been upset with and without lubrication.

Fig. 4 a,c and fig. 5 a,c show confocal microscope images and profiles of the topography of turned surfaces before flattening. Due to the concentric arrangement of the grooves, these features will act as closed lubricant pockets according to the mechanical-rheological model mentioned above. In contrast, the grooves of the ground specimens have a parallel arrangement provoking that the majority of lubricant will be squeezed out of the grooves (OLPs).

Therefore, the flattening behaviour of the turned surface is depending on the usage of lubricant as it can be seen in fig 4 b,d and fig. 5 b,d. There is a significant difference of the reduction of the arithmetic mean roughness S_a (3 dimensional equivalent to P_a) depending on whether lubricant has been used (reduction of 61 %) or not (reduction of 74 %).

Regarding the ground surface, there is no difference in the relative reduction of S_a, thus the flattening behaviour, visible (table 1). Hence, the formation of CLPs is essential for good tribological conditions also in microforming processes.

But due to the small surfaces present in microforming, many grooves are connected to the edge of the part surface. Hence, the incidence of a relevant amount of CLPs is not likely. Therefore, the locations where material contact occurs, thus asperities, are much more important. For the characterisation of the material contact within asperities and the formation of assumed micro-CLPs, a detailed analysis of this sub-topography is necessary.

Table 1

Relative reduction of the arithmetic mean roughness S_a (3D equivalent to R_a) after upsetting

	Turned surface	Ground surface
Without lubricant	74 %	73 %
With lubricant	61 %	75 %

160

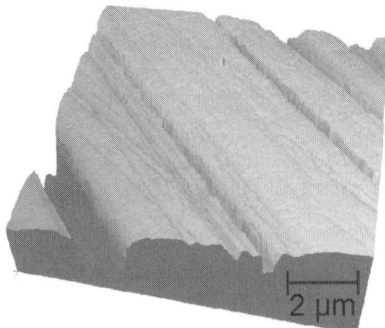

Fig. 6. SPM image of a ground surface

4. Evaluation of sub-topography

In order to perform a detailed evaluation of the sub-topography high-resolution measurements of the flattened area are necessary. This cannot be accomplished by the optical measurement system applied in the above mentioned test rig. Instead, a scanning probe microscope (SPM) is used for measuring sections of the flattened areas. Fig. 6 shows an SPM image of a ground surface upset by an almost ideally flat glass tool. There are still many non-flattened, submicron cavities present which cannot be explained completely by the elastic re-deformation of the material after unloading.

In order to trace back the topography under load from the unloaded surface the elastic flattening of the asperities is simulated numerically by FEM. Various input parameters for the FE model are not known exactly, e.g. the material properties may differ from the macroscopic case as only a few grains of the microstructure are involved in the flattening process. In order to determine these input parameters, the results of the micro upsetting test - the load displacement characteristics and the developing contact area - can be used for numerical identification. The simulation of the sub-topography under load enables a detailed characterisation concerning its impact on the tribological behaviour of the surface.

5. Conclusions

This contribution points out that the surface of micro parts plays a more significant role for the functional behaviour of the part than at conventional length scale due to increased surface to volume ratio. Therefore, the tribological behaviour of micro parts during manufacturing and usage can be influenced by an appropriate adjustment of the surface. Hence, the possibility of an accurate characterisation of the surface concerning its functional behaviour is necessary. The transfer of the knowledge for a functional description of the surface from the conventional length scale to the micro case is difficult: as only a small number of surface features, in particular asperities are in contact their flattening behaviour is much more important. Additionally, the asperities cannot be assumed to be flattened completely. Instead, the sub-topography on top of asperities has to be taken into account.

A test rig is introduced which enables the in-situ observation of the flattening behaviour of single asperities and the recording of the force-displacement characteristics. The sub-topography is measured by SPM and evaluated. Pyramidal specimens have been micro-machined representing idealised asperities.

The flattening behaviour of surface features in micro forming processes will be examined. This knowledge facilitates the functional characterisation of surfaces concerning their tribological behaviour, hence the derivation of functional surface parameters in micro forming applications.

Acknowledgements

This project is funded by the German Research Foundation (Deutsche Forschungsgemeinschaft DFG) and was carried out within the framework of the EC Network of Excellence "Multi-Material Micro Manufacture: Technologies and Applications (4M)".

References

[1] Popp U and Neudecker T and Engel U and Geiger M. Surface characterization with regard to the tribological behaviour of sheet metal in forming processes. Proc. of the 7th Int. Conf. on Sheet Metal, 1999, Erlangen, pp. 303 – 310.

[2] Steaves J Beurteilung der Topographie von Blech im Hinblick auf die Reibung bei der Umformung. Berichte aus Produktion und Umformtechnik, Shaker-Verlag, Aachen, 1998.

[3] Azushima A and Miyamoto J and Kudo H Effect of surface topography of workpiece on pressure dependance of coefficient of friction in metal forming. Annals of the CIRP, 47(1998)1, pp. 479-482.

[4] Geiger M and Popp U and Engel U Tool life improvement by surface laser texturing. Proc. of the 4th Int. Conf. THE Coatings in Manufacturing Engineering, Erlangen, Germany; 2004, pp. 57-68.

[5] Sørensen CG and Bech JI and Andreasen JL and Bay N and Engel U and Neudecker T. A basic study of the influence of the surface topography on mechanisms of liquid lubrication in metal forming. Annals of the CIRP, 44(1995)1, pp. 209-212

[6] Azushima A and Yoneyama S and Yamaguchi T and Kudo H. Direct observation of microcontact behaviour at the interface between tool and workpiece in lubricated upsetting. Annals of the CIRP 45(1996)1, 1996

[7] Sobis T and Engel U and Geiger M. A theoretical study of wear simulation in metal forming processes. J. Mater. Process. Technol. 34(1992) pp. 233-40

[8] Messner A. Kaltmassivumformung metallischer Kleinstteile – Werkstoffverhalten, Wirkflächenreibung, Prozessauslegung. In: Geiger M and Feldmann K (edtrs) Reihe Fertigungstechnik Erlangen, Meisenbach 1997

[9] Engel U and Messner A and Tiesler N. Cold forging of microparts - effect of miniaturisation on friction. In: Chenot, J.L. et al (edtrs.): Proc. of the 1st ESAFORM Conf. on Materials Forming. Sophia Antipolis France 1998, pp. 77-80

[10] Wanheim T and Bay N and Petersen AS. A theoretically determined model for friction in metal working processes. Wear 28(1974) pp. 251 – 258

[11] Steffensen H and Wanheim T. Asperities on asperities. Wear 43(1977) pp. 89 – 98

[12] Engel U. Tribology in Microforming. Wear 260(2006)3, pp. 265 - 273

Micro particle sizing by using of ray optics Monte Carlo code

M.D. Mikrenska[a], P.I. Koulev[a], J.-B. Renard[b]

[a] Institute of Mechanics, Bulgarian Academy of Sciences, Sofia, 1113, BG
[b] Laboratoire de Physique et Chimie de l'Environnement, CNRS, 45071 Orléans, FR

Abstract

Micro particle sizing is necessary for development, manufacturing and usage of particulate materials in biotechnology, micro and nanotechnology. This paper presents a ray optics Monte Carlo light scattering simulation software and its capabilities for sizing of semitransparent particles (the imaginary part of the refractive index is nonzero) suspended in a non absorbing fluid. Polarization phase functions, characterizing micro particles can be both calculated by means of ray optics Monte Carlo code and measured by an appropriate instrument (e.g. a polarization goniometer) in conditions ensuring single light scattering by randomly situated and oriented micro particles. Particle sizing is realized by identification of those polarization phase curve among the curves calculated by the ray optics Monte Carlo code which is most closed to the measured one. In this paper some results for the dependence of the polarization phase function on the particle size for different shapes (sphere, cube and rounded cube) are presented.

Keywords: micro particle sizing, light scattering, polarization phase function, ray optics, Monte Carlo simulation

1. Introduction

Micro particulate materials are frequently used in micro manufacture technologies. By this reason the particle size assessment (granulometry) as well as other parameter identification is an object of many developments. Early granulometric methods such as microscopy examination, consecutive sieving by sieves with different apertures, sedimentation (method of the gravitational settling) and method of electrical sensing zone (e. g. Coulter method) based on tracking out of the change of the electrolyte resistance at settlement of space between electrodes by particles, are reviewed in [1]. Other granulometric methods as chromatographic, acoustic analysis and streaming potential measurement (reviewed in [2]) are used also.

Laser light scattering (so called diffraction or Mie) methods [2], applicable to particles of all kind of materials, are widely used in granulometry at size range of 0.1 to 100 micrometers.

In general, the scattering of the incident light gives distinct patterns which are measured by a detector or calculated by means of a theoretical model. Each material, size and shape influences scattering and diffraction properties. Both laser scattering measurements and numerical simulations are useful for particle characterization technology and are suitable for particles with relatively simple shapes: spheres, cubes and mono crystals.

In the present work we propose a method for sizing of grains of particulate material. The method is based on the created ray optics Monte Carlo code for light scattering by randomly oriented semitransparent particles [3, 4, 5, 6] suspended in non absorbing fluid (vacuum, gaze, liquid). The effect of the particle size on the polarization phase function defined as the dependence of the degree of polarization

$$P = \frac{I_\perp - I_{II}}{I_\perp + I_{II}} \qquad (1)$$

on the angle between the sample illumination and the scattering light acquisition directions (phase angle) is used.

In (1) I_{II} and $I_\perp$ are the intensity components, polarized parallel and perpendicular to the scattering plane (defined by the directions of incidence and scattering), respectively.

A brief model description is given in Section 2. Some of the results of the numerical simulations performed are presented in Section 3. The computer model is verified by comparison with experimental data in Section 4.

2. Model

The present model of single light scattering by randomly oriented micro particles is based on direct simulation Monte Carlo ray tracing. The diffraction and interference effects are not considered. The scattering particles are supposed to be semitransparent, optically homogeneous and isotropic with sizes much larger then the wavelength so that the principles of geometric optics are valid. The imaginary part of the refractive index is supposed to be nonzero (absorbing particles). We consider spherical, cubic and rounded cubic particle shapes. The light source is randomly polarized laser.

The rounded cube is defined in [4] as the intersection of a cube with side a and a sphere with diameter d which centres coincide. The class of rounded cubes is characterized by the so called degree of roundness k:

$$k = \frac{a}{d} \qquad (2)$$

for $k \in [\frac{\sqrt{2}}{2}, 1]$. The smallest value of k ($k = \frac{\sqrt{2}}{2}$) corresponds to a perfect cube and the biggest one

($k = 1$) corresponds to a perfect sphere. The values of k which belong to the open interval ($\frac{\sqrt{2}}{2}$,1) define essentially rounded cubes.

2.1. Physical model

In geometric optics approximation for each ray – particle surface interaction Snell law determines the possible propagation direction and the changes of the intensities of light polarized parallel and orthogonal to the scattering plane are calculated by means of Fresnel law (see [7]):

$$T_{II} = \frac{2n_1 \cos(\theta_i)}{n_2 \cos(\theta_i) + n_1 \cos(\theta_t)} A_{II} \tag{3}$$

$$T_\perp = \frac{2n_1 \cos(\theta_i)}{n_1 \cos(\theta_i) + n_2 \cos(\theta_t)} A_\perp \tag{4}$$

$$R_{II} = \frac{n_2 \cos(\theta_i) - n_1 \cos(\theta_t)}{n_2 \cos(\theta_i) + n_1 \cos(\theta_t)} A_{II} \tag{5}$$

$$R_\perp = \frac{n_1 \cos(\theta_i) - n_2 \cos(\theta_t)}{n_1 \cos(\theta_i) + n_2 \cos(\theta_t)} A_\perp , \tag{6}$$

where A_{II} and $A_\perp$ represent the amplitudes of parallel and perpendicular components of the electric vector of the incident light with respect to the scattering plane. T_{II} (R_{II}) and $T_\perp$ ($R_\perp$) are the parallel and perpendicular components of the electric vector of the transmitted (reflected) wave respectively. n_1 (n_2) is the refractive index of the first (second) media. θ_i (θ_t) is the angle between incident (transmitted) ray and the normal to the surface at the point of hit. The coefficients in Eqs. (3, 4, 5, 6) are used as probabilities for the choice of the outcome of the photon – particle surface interaction: reflection or transmission (refraction).

In case of elliptic polarization (cubic and rounded cubic shapes) as a result of total internal reflection

($n_1 > n_2$ and $\theta_i > \theta_{cr}$; $\sin(\theta_{cr}) = \frac{n_2}{n_1} = n$) Fresnel

formulas give:

$$|R_{II}| = |A_{II}| \quad \text{and} \quad |R_\perp| = |A_\perp| \tag{7}$$

The phase jumps of the parallel and perpendicular component of the electrical vector and changes of ellipse orientation on the optical interface are described in [7].

2.2. Monte Carlo simulation. Software implementation

The main idea of Monte Carlo simulation is based on the tracing out of photon path for a great number of photons in correspondence with geometric optics laws. As a result of each photon-particle interaction both of

the following chains of elementary events are possible: an external reflection; a refraction (the photon gets into the particle), followed by one or more internal (or total internal) reflections and finally by a refraction (the photon gets out of the particle). For each of these photon-particle surface interactions Snell law determines the possible propagation direction and the changes of light intensities (parallel and orthogonal to the scattering plane) are calculated by means of Fresnel law.

The outcomes of tracing of all launched photons are summed and the final result for polarization phase function is calculated. The polarization is obtained by adding the calculated intensities. More detailed description of the algorithm used can be found in [6].

3. Numerical results

The developed computer code for Monte Carlo simulation of light scattering by partially absorbing particles is applied to study the dependence of the degree of polarization on the size of the scattering particles. Three series of numerical experiments were performed: for spherical, cubic and rounded cubic shapes. The number of launched photons varies from 3.10^7 to 5.10^7. The real part of the refractive index in the experiments presented below is taken to be 1.5567 that corresponds to the refractive index of KBr at wavelength of 0.632 µm. the refractive index of the surrounding media is taken to be 1 (vacuum).

Figs. 1-6 show the dependence of the polarization phase function on the particle size for different shapes. The numerical experiments provided show that for all of the examined shapes there exist a critical size values (corresponding to the upper curves in Figs. 1-6) depending on the shape and absorbing properties of the scattering particles for which the polarization curve of the optical mirror is obtained. The polarization curves tend to the mirror curve when the particle size increases. This numerical result has the following physical meaning: the scattered light energy has an unique component – the reflected by the particle surface energy. The remained energy is fully absorbed inside the particles. As it could be expected for a given particle shape the critical size increases when the value of imaginary part of the refractive index decreases.

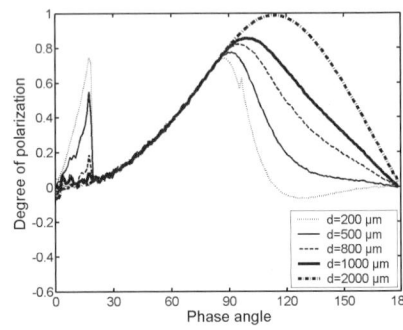

Fig. 1. Dependence of the degree of polarization on the size for spheres with n=1.5567+ix0.0001.

Fig. 1 and Fig. 2 show the obtained polarization phase curves for spherical particles with imaginary part of the refractive index 0.0001 and 0.001, respectively. It is important to note that the curves for different sizes of spherical particles are distinguishable each other for small phase angles and angles grater than 90^0. The observed peak at 18^0 becomes smaller and vanish when the size increases and the second peak for large phase angles becomes greater and moves to the forward direction.

As it can be seen in Fig. 1 the polarization phase curve for a sphere with n = 1.5567+ix0.0001 may be used for particle sizing in range of few microns (large enough so that the geometric optics to be applied) to 2000 µm.

Fig. 2 shows that the polarization phase curve for a sphere with n=1.5567+ix0.001 may be used for particle sizing in range of few microns to 200 µm.

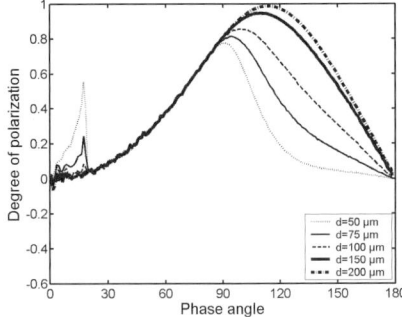

Fig. 2. Dependence of the degree of polarization on the size for spheres with n=1.5567+ix0.001.

The numerical results for particles with perfect cubic shape are presented in Fig. 3 (n= 1.5567+ix0.0001) and Fig. 4. (n=1.5567+ix0.001). Note that the negative backscattering peak (typical for cube) decreases and the positive peak increases when the particle size increases.

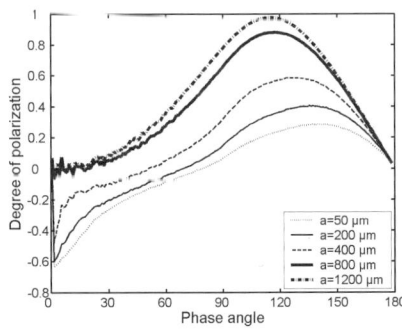

Fiq. 3. Dependence of the degree of polarization on the size for cubes with n=1.5567+ix0.0001.

Fig. 3 shows that the degree of polarization for a cube with n=1.5567+ix0.0001 may be used for particle sizing in range of few microns to 1200 µm.

The numerical results presented in Fig. 4 clearly show that the polarization phase curve for a cube with n=1.5567+ix0.001 may be used for particle sizing in range of few microns to 130 µm.

The obtained numerical results for polarization properties of particles with rounded cubic shape are plotted in Fig. 5 (n=1.5567+ix0.0001) and Fig. 6 (n=1.5567+ix0.001). As it could be expected the positive peak increases when the particle size increases.

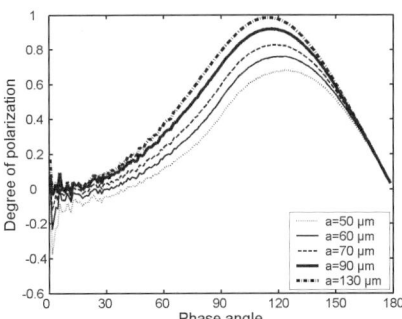

Fig. 4. Dependence of the degree of polarization on the size for cubes with n=1.5567+ix0.001.

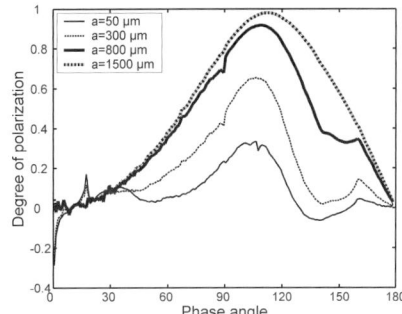

Fig. 5. Dependence of the degree of polarization on the size for rounded cubes with n=1.5567+ix0.0001 and degree of roundness k=0.85.

Fig. 5 shows that the degree of polarization for a rounded cube with n=1.5567+ix0.0001 and degree of roundness k=0.85 may be used for particle size characterization in range of few microns to 1500 µm.

From Fig. 6 it becomes clear that the polarization phase function for a rounded cube with n=1.5567+ix0.001 and degree of roundness k=0.85 may be used for particle size characterization in range of few microns to 150 µm.

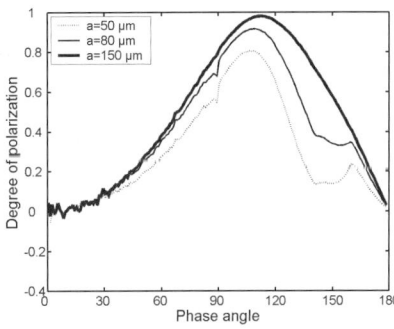

Fig. 6. Dependence of the degree of polarization on the size for rounded cubes with n=1.5567+ix0.001 and degree of roundness k=0.85.

The presented here numerical results show the strong qualitative and quantitative dependence of the polarization curve on the particle size (see Figs. 1-6) and shape (e.g. compare the polarization phase functions shown in Figs. 2,4,6 which correspond to particles with spherical, cubic and rounded cubic shape with size of 50µm).

4. Comparison between calculated and measured results

In order to verify the proposed technique a comparison with experimental data for polarization phase function of real crystals of KBr (see Fig. 7) with mean size of 300 µm is done. Measurements are conducted in microgravity with the new version of the PROGRA^2 instrument [8, 9].

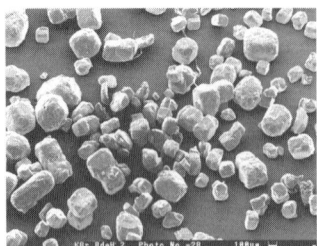

Fig. 7. Scanning Electron Microphotography of crystals of KBr with different sizes.

The shapes of the crystals of KBr with mean size of 300 µm are approximated by rounded cubes (60%) and parallelepipeds (40%) with uniform distribution of the roundness parameter in the interval $[\frac{\sqrt{2}}{2}, 0.72]$. The roughness of the particle surface is taken into account by randomly orientation of the scattering plane for 80% of the incident rays. The imaginary part of the refractive index is taken to be 0.000075 as it was found in a previous paper [5], where a comparison with experimental results for different sample is done. Fig. 8 shows good agreement between calculated and measured data. The mean value of the relative error is 3.8%.

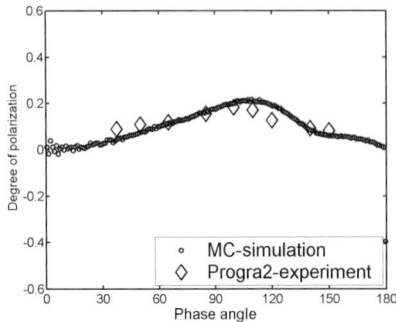

Fig. 8. Comparison between numerical and experimental results for KBr

The calculated curve on Fig. 8 is obtained for a priori known mean particle size. In case of unknown mean particle size it can be identified by the following way: By use of the ray optics Monte Carlo code several polarization curves for different particle sizes should be calculated; the particle size can be determined as the correspondent of the curve of best fitting with measured with polarization goniometer.

5. Conclusions

In the present paper the ray optics Monte Carlo code is applied for sizing of semitransparent particles with three types of shape: sphere, cube and rounded cube. A series of numerical simulations is performed and the critical size values depending on the shape and the imaginary part of the refractive index are determined. The proposed method for computer simulation is validated by comparison with measured in microgravity by the PROGRA^2 instrument data.

The proposed numerical techniques can be applied to create a methodology for particle sizing as well as for shape identification by creating a database of calculated polarization phase functions for particles with a variety of optical properties, sizes and shapes and determining the limits of applicability (size range) for each material. The database can be used to identify refractive index for known shape and size, or particle size for known shape and refractive index, or particle shape for known refractive index and size.

Acknowledgements

This research is supported by Bulgarian NSF grant TN-1521/2005.

References

[1] Riego Sintes, J.M.Guidance document on the determination of particle size distribution, fibber length and diameter distribution of chemical substances, ISBN 92-894-3704-9 - EUR 20268/EN (2002), European Commission.

[2] Xu R. Particle characterization: Light scattering methods, Kluwer Academic Publishers, 2000.

[3] Mikrenska M, Koulev P and Hadamcik E. Direct Simulation Monte Carlo of Light Scattering by Cube. C. R. Acad. Bulg. Sci. v. 57 n.11, (2004), 39-44.

[4] Mikrenska M, Koulev P, Renard J-B, Hadamcik E and Worms J-C. Direct Simulation Monte Carlo of Polarization at Single Light Scattering by Rounded Cubes. C. R. Acad. Bulg. Sci. v. 58 n. 3 (2005) 275-280.

[5] Mikrenska M, Koulev P, Renard J-B. Metrological characterization of micro particles by direct simulation Monte Carlo. In: Menz W, Dimov S (Ed.) Proc. Ist Int. Conf. 4M, 29 June-1 July, Karlsruhe, Elsevier, Amsterdam, 2005, pp 237-240.

[6] Mikrenska M, Koulev P, Renard J-B, Hadamcik E and Worms J-C. Direct simulation Monte Carlo ray tracing model of light scattering by a class of real particles and comparison with PROGRA2 experimental results. J. Quant. Spectrosc. Radiat. Transfer 100 (2006) 256-267.

[7] Born M, Wolf E, Principles of optics. Oxford: Pergamon Press, 1964. pp 25-96.

[8] Renard JB, Worms JC, Lemaire T, Hadamcik E, Brun-Huret N, Light scattering by dust particles in microgravity: polarization and brightness imaging with the new version of the PROGRA2 instrument, Appl. Opt. 41 (2002) 609-618.

[9] Renard JB, Hadamcik E, Lemaire T, Worms JC, Levasseur-Regourd AC, Polarization imaging of dust cloud particles: improvements and applications of the PROGRA2 instrument, Adv Space Res. 31 (2003) 2511-2518.

165

Repeatability analysis of two methods for height measurements in the micrometer range

C. Ferri[a], E. Brousseau[a], S. Dimov[a], L. Mattsson[b]

[a] The Manufacturing Engineering Centre, Cardiff University, Cardiff CF24 3AA, UK
[b] Department of Production Engineering, Royal Institute of Technology, SE-10044 Stockholm, Sweden

Abstract

A precision study of two height measuring methods is carried out. The first method is based on a White Light Interferometer (WLI) and the second on a Co-ordinate Measuring Machine (CMM) equipped with an optical probe. The height measurements considered are in the range [150; 250] μm. Point and interval estimates of repeatability are reported in the paper. This study presents experimental evidence that, under repeatability conditions, the precision of the WLI method is about five times higher than that of the optical CMM method. Furthermore, the precision of WLI is constant over the investigated height range whereas a dependency of the CMM precision on the nominal dimensions is identified. For both methods a linear relationship is detected between the random error and the sequence in which the measurements are taken.

Keywords: Repeatability, White light interferometry, CMM

1. Introduction

The current trend for product miniaturisation drives the demand for high precision manufacturing equipment [1]. Consequently, measuring and inspection methods with an even higher level of precision should be available to characterise both products and manufacturing processes in the micrometer range. The importance of quantifying the precision arises from the fact that performing measurements under seemingly identical circumstances do not generally produce identical results. In this paper precision refers to the ISO terminology, i.e. the closeness of agreement between test results of a measurement method [2]. This is one component in the assignment of the uncertainty of a measurement process.

Many methods for performing measurements in the micrometer range are already available, for example those currently used for inspecting surface textures [3] However, due to their relatively large number and different features in hardware and software, it is far from straightforward to compare them and thus define clearly their respective advantages and limitations. Therefore, investigations aiming at understanding and quantifying the sources of variability present in a measurement process are very important in determining the areas for their most appropriate applications.

The objective of this paper is to compare the precision of two measurement methods based on two widely used systems: a White-Light Interferometer (WLI) and a Coordinate Measuring Machine (CMM) provided with an optical head. Established methodologies for quantitative analysis were deployed. In particular, the precision of these two methods was compared when measuring steps in the range from 150 to 250 μm. A linear statistical model was used to describe the measurement results. Then, the precision was assessed by the ANalysis-Of-VAriance (ANOVA) method [4].

In this study, the optically provided CMM (OCMM) contained a Charged Couple Device (CCD), and a Computer Numerical Control (CNC) system for carrying out visual inspections. Thus, the machine is designed to perform contact-less 3D measurements.

To investigate the capabilities of WLI the tests were conducted on a white light profiling microscope for surface mapping. A CCD camera is used on the microscope to record the intensity of the bright and dark bands for each pixel in the field of view while the object is scanned in a direction perpendicular to the illuminated surface. Based on the intensities recorded, the topography of the inspected surface is determined by applying built-in algorithms.

The structure of the paper is as follows. Section 2 presents the experimental set up used in this research. Section 3 describes the applied statistical model for assessing the precision of WLI and OCMM. Section 4 discusses the test results together with some of the assumptions made in their modelling. Next, conclusions are made about the precision of both measurement methods.

2. Experimental set-up

2.1. Step heights

A procedure was developed to establish step heights in the micro-range by using certified gauge blocks of grade 1 [5] traceable in accordance with BS 4311-3 [6]. In particular, two blocks were used in preparing each step height. The gauge blocks were wrung side by side (see Fig. 1) onto a quartz optical parallel. The use of the transparent optical parallel allowed the quality of the wringing procedure to be assessed by detecting the presence of interference colour fringes and bright spots on the two wrung faces [7].

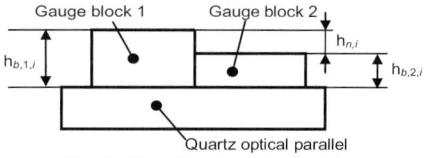

Fig. 1. Creating the step heights

The nominal height $h_{n,i}$ of the ith step is defined in this study by the following equation (cf. Fig.1):

$$h_{n,i} = h_{b,1,i} - h_{b,2,i} \quad \text{with } h_{b,1,i} > h_{b,2,i} \qquad (1)$$

where $h_{b,1,i}$ and $h_{b,2,i}$ are the nominal lengths of the two gauge blocks as defined in [5]. These nominal lengths are referred to as the heights of the blocks in Table 1. To prepare the required four step heights, eight blocks were used. The nominal dimensions of these steps and blocks are provided in Table 1.

2.2. Randomisation

In this research, the precision of WLI and OCMM in measuring steps in the interval of [150; 250] µm is investigated. The four selected steps, as shown in Table 1, span almost evenly the whole interval and are a representative selection of the infinite number of possible step heights belonging to it [4]. In fact, this sample has the same probability of being drawn from this interval as any other potential sample. Thus, the selected four heights qualify as a random sample [8-9].

For each of the four steps 16 measurements were taken at the same x, y position on the steps, in total 64 for the entire sample. Every single measurement started and ended in the origin of the instrument co-ordinate system. The measuring position was reached automatically by programming the instruments. Averages of the z co-ordinates, $\bar{z}_1$ and $\bar{z}_2$ for the top and bottom surfaces of each step respectively, were computed from subsets of points belonging to each surface. These subsets, although containing a different number of points for the two instruments, spanned a similar area of approximately 200 µm x 600 µm on each of the two step surfaces. Then, the jth measurement result of the ith step height was defined by $h_{ij} = (\bar{z}_1 - \bar{z}_2)_{ij}$, where $i = 1,...,i_{max}$ and $j = 1,...j_{max,i}$ with $i_{max} = 4$ and $j_{max,i} = 16$. The order in which the measurements were performed was established in the following way. To each measurement (of the four available steps) an integer number was assigned from 1 to 64. Then, the sequence in taking them was determined by selecting one of the possible 64! permutations [8]. This strategy usually allows the errors of random effect ANOVA models to be treated as independently distributed random variables. Furthermore, by selecting the test sequence in this way, it was possible to minimise the effects of potential disturbances not explicitly considered in the model, i.e. small unforeseen and unavoidable local oscillations of the room temperature. Variations of about ± 1.0 °C are expected around the 20 °C set point of the laboratory air temperature.

3. Statistical model

The deviation $\Delta h_{ij} = h_{ij} - \mu$ of a measurement

result h_{ij} from its general mean, μ, is modelled as the sum of two independent random variables a_i and e_{ij}.

The first variable, a_i, represents the contribution to Δh_{ij} due to the fact that h_{ij} is a measurement of the ith step height. This contribution is usually referred to as a random effect [8].

The second variable, e_{ij}, "is the random error occurring in every measurement under repeatability conditions" [3]. In particular, e_{ij} represents the contribution to the same deviation, Δh_{ij}, that is due to all those sources of variability that are difficult or expensive to identify and thus to control, (i.e. small vibrations in the measurement instrument). Consequently, this variable can be expressed as the difference between the measured step height and its expected value for a given level, namely:

$$e_{ij} = h_{ij} - E(h_{ij} \mid a_i) \qquad (2)$$

Accordingly, Δh_{ij} can be expressed as:

$$\Delta h_{ij} = a_i + e_{ij} \qquad (3)$$

It is assumed that a_i and e_{ij} are independent and identically distributed variables with mean value 0, and variance σ_a^2 and σ_e^2, respectively. In addition, a_i and e_{ij} are considered independent and thus:

$$cov(a_i, e_{ij}) = 0 \qquad (4)$$

From equations (3), and (4), it can be stated that:

$$var(h_{ij}) = \sigma_a^2 + \sigma_e^2 \qquad (5)$$

Based on this model, the estimate $\hat{\sigma}_e^2$ of σ_e^2 is obtained by the ANOVA method [10]. In particular,

$$\hat{\sigma}_e^2 = \left(\sum_{i=1}^{i_{max}} \sum_{j=1}^{j_{max,i}} j_{max,i}(h_{ij} - \bar{h}_{i.})^2\right) \bigg/ \left(\left(\sum_{i=1}^{i_{max}} j_{max,i}\right) - i_{max}\right) \qquad (6)$$

where $\bar{h}_{i.}$ is the average value of the measurements for level i. Instead, $\hat{\sigma}_i$ denotes the sample standard deviation obtained from the 16 measurement results of the ith step height.

In this study, the predicted realisations of the random errors, e_{ij}, are called residuals, $\hat{e}_{ij}$, and are defined from eq. (2) as:

$$\hat{e}_{ij} = h_{ij} - \hat{E}(h_{ij} \mid a_i) \qquad (7)$$

where, by applying the method of moments [12] to the ith sample, $h_{i1}, h_{i2},...,h_{ij_{max,i}}$, the mean $E(h_{ij} \mid a_i)$ is estimated by $\hat{E}(h_{ij} \mid a_i) = \bar{h}_i$.

4. Repeatability Results and Analysis

Based on the 64 measurements carried out for each of the two measurement methods investigated in this study, the point estimates, $\hat{\sigma}_e$ and $\hat{\sigma}_i$, were computed to assess the methods' repeatability. The results are summarised in Table 2.

The variations of the interquartile ranges shown in Fig. 2 lead to suspect that a dependency of the residuals' variability on the nominal step height may be present. Thus, a Levene's test was performed on the measurement data of both methods to reveal if the null hypothesis of homogeneous variances should be rejected (large value of the F-statistic) [8]. The result

Table 1
Specification intervals of the samples

	Nominal step height [µm]	Blocks' nominal heights [mm]
$h_{n,1}$	150	1.300 and 1.150
$h_{n,2}$	183	1.190 and 1.007
$h_{n,3}$	217	1.220 and 1.003
$h_{n,4}$	250	1.250 and 1.500

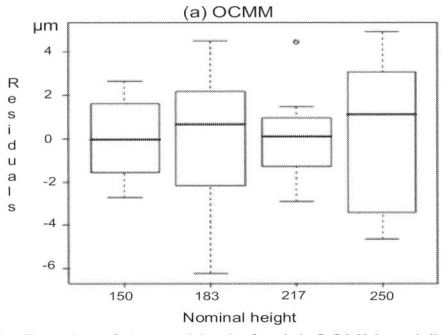
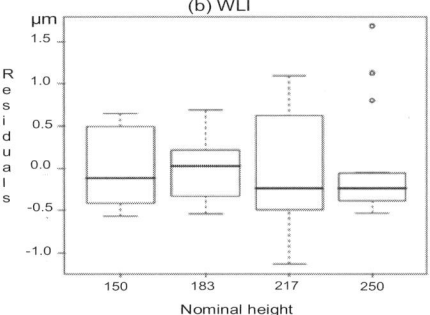

Fig. 2. Boxplot of the residuals for (a) OCMM and (b) WLI showing the median, upper and lower quartile and the max and min residuals

was $P_{value} = 0.0095$ and F_0 = 4.16 for OCMM and $P_{value} = 0.3345$ and F_0 = 1.15 for WLI. This signifies that the effect of the nominal step height on the precision under repeatability conditions is significant only for OCMM and not for WLI. In addition, this indicates that $\hat{\sigma}_e = 2.6\,\mu m$ obtained through the ANOVA estimation for OCMM is questionable. In fact, $\hat{\sigma}_i$ are used in order to avoid bringing together significantly different estimates of standard deviations into a common value as it is the case with $\hat{\sigma}_e$.

By performing an Anderson-Darling test, which tests if a sample of data came from a Gaussian population, a departure from normality of $\hat{e}_{ij}$ was detected only for the WLI. In particular, the P_{value} was $3.813 \cdot 10^{-4}$ for WLI while the result for OCMM showed a reasonable agreement with the null hypothesis of normality. In spite of this, to have consistency in this comparative study, the Jackknife procedure was adopted for computing the confidence intervals not only for σ_e in the case of WLI [11] but also for σ_i, in the case of OCMM.

Fig. 3 shows the residuals, $\hat{e}_{ij}$, of both measurement methods. A graphical examination of the plotted results highlights a possible violation of the assumption that e_{ij} are independent and identically distributed variables. Therefore, a first order regression model was fitted to the data to represent the dependence of $\hat{e}_{ij}$ on the test sequence:

$$\hat{e}_{ij} = \beta_0 + \beta_1 O_{ij} + e_{ij,new} \qquad (8)$$

where O_{ij} is the order in which the measurements are taken, β_0 and β_1 are the model parameters and $e_{ij,new}$ are normal independent and identically distributed variables with mean value 0, and variance σ^2.

The parameters in eq. (8) were estimated applying the ordinary least square method. P_{values} shown in Table 3 support the rejection of $\beta_1 = 0$. Analysing $e_{ij,new}$ did not show disagreement with the assumption that they are independently identically normally distributed (see $P_{values,A}$ in Table 3). The results also indicate two opposite tendencies, although moderate in values, descending in case of WLI and ascending for OCMM.

The existence of a linear dependence of $\hat{e}_{ij}$ on the test sequence could be attributed to the effects of one or more factors that are not explicitly included in the model. In particular, this could be due to a continuous shift in the environmental conditions or other factors directly inherent in the measurement process. For example, in the case of WLI, the motion control during the scanning might contribute to this drift of $\hat{e}_{ij}$ [13].

With an increased attention to the adopted measurement procedures, the causes of such a drift could be identified and in some cases even eliminated. Alternatively, these causes could be set under statistical control and thus, their effects on the measurement results could be compensated The regression models fitted in this experimental study suggest that in either way, improvements of about 28% and 25% in the repeatability can be achieved for WLI and OCMM, respectively (see R^2 in Table 3).

Table 2
Experimental results

	WLI		OCMM	
		Jackknife confidence interval (95%)		Jackknife confidence interval (95%)
$\hat{\sigma}_e$ (μm)	$5.5 \cdot 10^{-1}$	(0.33; 0.79)	2.6	
$\hat{\sigma}_1$ (μm)	$4.6 \cdot 10^{-1}$		1.8	(1.3; 2.3)
$\hat{\sigma}_2$ (μm)	$3.7 \cdot 10^{-1}$		3.0	(1.9; 4.1)
$\hat{\sigma}_3$ (μm)	$6.9 \cdot 10^{-1}$		1.7	(0.79; 2.8)
$\hat{\sigma}_4$ (μm)	$6.3 \cdot 10^{-1}$		3.3	(2.7; 4.0)

Table 3
A regression analysis of residuals, $\hat{e}_{ij}$

Instrument	WLI	OCMM
$\hat{\beta}_0$ (μm)	$4.95 \cdot 10^{-1}$	-2.18
$\hat{\beta}_1$ (μm)	$-1.53 \cdot 10^{-2}$	$6.70 \cdot 10^{-2}$
$\hat{\sigma}$ (μm)	$4.62 \cdot 10^{-1}$	2.18
F_0	23.8	20.6
P_{values}	$7.9 \cdot 10^{-6}$	$2.7 \cdot 10^{-5}$
R^2	0.277	0.249
A	0.533	0.162
$P_{values,A}$	0.166	0.944

168

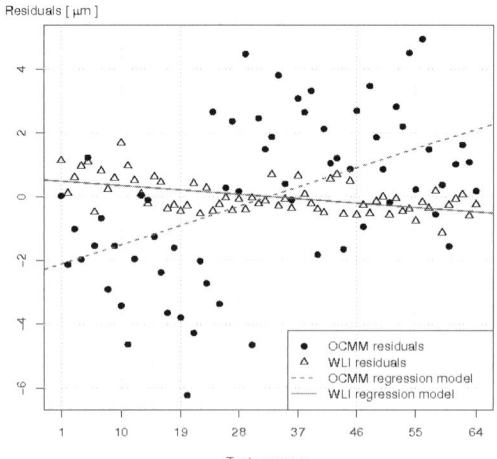

Fig. 3. Residual versus test sequence for the OCMM and the WLI

Once the dependence of $\hat{\sigma}_e$ on the nominal step height is established, it would be possible to characterise OCMM. However, the data available in this study did not reveal any particular pattern. In fact, an attempt to fit a regression line to the data resulted in F_0 = 0.78 and P_{value} = 0.47. Hence, the test failed to reject the assumption, $\beta_1\left(\sigma = f\left(h_{n,i}\right)\right) = 0$. A higher level model could have been fitted to the available data. However it was considered that this would give less insight into this relationship compared to extending the study to a larger number of levels. Hence, to characterise the OCMM process further research is required.

Some extreme values of $\hat{e}_{ij}$ in comparison to the majority of the data could be seen in Fig 3. They could be due either to randomness of the measurement process or to some unforeseen and unpredictable factors. In this study, these extreme values were considered to be caused by the randomness and thus they were regarded as typical of the phenomenon under investigation. Hence, they were not removed from the data set. As a result the estimates are either correct or constitute an upper limit of the repeatability standard deviation of both processes.

4. Conclusion

The precision of OCMM and WLI under repeatability conditions, while measuring step heights along the z direction and including lateral positioning in x, y, was investigated. For this purpose, measurements of step heights in the range [150; 250] μm were carried out. The main findings of this comparative study are:

- On average, WLI is about five times more precise than OCMM.
- The precision of WLI could be considered constant over the investigated range of step heights. However, this is not the case for OCMM. Furthermore, the dependency of its precision on the nominal dimensions of the step heights does not show any linear tendency.
- WLI and OCMM show a weak linear dependence of their random error on the sequence in which the

measurements are taken. Although this dependence is weak, its removal would lead to precision improvements of more than 20% for both methods, as the regression analysis has shown.

Acknowledgements

The authors would like to thank the European Commission, the Welsh Assembly Government and the UK Engineering and Physical Sciences Research Council for funding this research under the ERDF Programme 52718 "Support Innovative Product Engineering and Responsive Manufacture" and the EPSRC Programme "The Cardiff Innovative Manufacturing Research Centre". Also, this work was carried out within the framework of the EC Networks of Excellence "Innovative Production Machines and Systems (I*PROMS)" and "Multi-Material Micro Manufacture: Technologies and Applications (4M)".

References

[1] Masuzawa T. State of the art of micromachining. Annals of the CIRP. 49 (2), 2000, pp 473-488.
[2] BS ISO 5725-1. Accuracy (trueness and precision) of measurement methods and results - Part 1: General principles and definitions. BSI - British Standards Institution. 1994.
[3] Vorburger TV, Rhee H-G, Renegar TB, Song J-F and Zheng A. Comparison of optical and stylus methods for measurement of surface texture. International Conference on Multi-Material Micro Manufacture (4M). Karslruhe, 29th June – 1st July 2005, pp 35-42.
[4] Montgomery DC and Runger GC. Gauge capability and designed experiments. Part I: basic methods. Quality Engineering. 6 (1), 1993, pp 115-135.
[5] BS EN ISO 3650. Geometrical product specifications (GPS) - length standards - gauge blocks. BSI - British Standards Institution. 1999.
[6] BS 4311-3. Specification for gauge blocks and accessories - Part 3: gauge blocks in use. BSI - British Standards Institution. 1993.
[7] Decker JE and Pekelsky JR. Gauge block calibration by optical interferometry at the national research council of Canada. Measurement Science Conference. Pasadena, 23rd -24th January 1997.
[8] Montgomery DC. Design and analysis of experiments (5th edn). John Wiley & Sons, New York, 2001.
[9] Box EP, Hunter WG and Hunter JS. Statistics for Experimenters. John Wiley & Sons, New York, 1978.
[10] Searle SR, Casella G and McCulloch CE. Variance components. John Wiley & Sons, New York, 1992.
[11] Miller RG Jr. Beyond Anova, basics of applied statistics. Chapman & Hall, London, 1997.
[12] Mood AM, Graybill FA and Boes DC. Introduction to the theory of statistics (3rd edn). McGraw-Hill, New York, 1974.
[13] Schmit J and Olszak A. High-precision shape measurement by white-light interferometry with real-time scanner error correction. Applied Optics. 41 (28), 2002, pp 5943-5950.

Localizing micro-defects on rough metal surfaces.

A. Témun[a], L. Mattsson[a] and I. Heikkilä[b]

[a] Dep. of Production Engineering, KTH – Royal Institute of Technology, SE-100 44 Stockholm, Sweden
[b] KiMAB, Drottning Kristinas väg 48, 114 28 Stockholm, Sweden

Abstract

In this paper we present a study of reflectance measurements to a direct metrological problem. Our aim is to estimate progressive surface wear by detecting minute defect development in surface microtopography on Ra 0.7 μm rough stainless steel surfaces subject to mechanical wear from hard particle deposits on a metal bar. The investigation was carried out by using a standard CMOS digital camera with a macro objective. It demonstrates how surface scattering measurements can provide a viable alternative in circumstances where traditional roughness measurement techniques fail to deliver the expected results and, due to certain restraints, laboratory-type surface measurements are not applicable either.

Our work is based upon measuring variations in the amount and direction of scattered light, reflected from the inspected surfaces, featuring microirregularities. We will show that it is possible to identify the extent of surface flaws by statistically evaluating the recorded brightness information. The developments in this paper can provide basis for a future quality control system devoted to on-line surface measurements of micro components.

Keywords: surface roughness, defect assessment, reflectance, light scattering, digital camera

1. Introduction

Microscopic surface flaws and defects are easily detected and measured using white light interferometers, confocal microscopes or other area covering surface profilometers, provided the surrounding surface is distinctly different from the defect. For smooth surfaces with a micro-indent, there is no problem of finding it and measure it, provided the surface slopes of the defect are not too steep. However, when the surrounding roughness increases and comes close to the defect size itself it is practically impossible to distinguish the defects from random surface microstructures. Performing surface roughness measurements in the hope of finding a distinctly different feature for the defect is then very difficult, as the typical variation of roughness values from one place to the other can be about 15-20 %. The latter might well happen in a micro-manufacturing process, where milling or electrical discharges create a roughness of the order not far from the feature sizes being manufactured.

However, by taking advantage of the fact that light scattering is proportional to the square of submicron-height-steps, or for a larger area, to the standard deviation of the surface micro topography, usually referred to rms roughness, we might gain some sensitivity in locating minute defects in a "rough" surrounding. [1, 2] If we in addition also take advantage of angle dependent reflectance the chances of localizing critical defects increases even more. The drawback is that the technique is qualitative, as no theory cover the complex scattering - roughness relation when the surface roughness is in the vicinity of the wavelength of the incident light.

Anybody who has attempted at localising microdefects on a rough surface using a laser beam and a photodiode detector can realise that it is virtually impossible to distinguish the speckle intensity variations from the scatter variations caused by a micro defect. In this paper we therefore will report on a white light based technique operating in the vicinity of the specular, i.e. direct reflection for defect detection. Minute contrast changes caused by the micro-defects on a rough surface are detected by a standard digital camera.

2. Surface roughness measurement

The reasons for measuring surface roughness may vary, and so do the measurement methods used. For uni-directionally machined surfaces, where a similar surface structure, called lay, is present in one direction, we usually make use of contact stylus profilometry, and express surface roughness by the arithmetic mean deviation Ra or root mean square deviation Rq. [3, 4] If lateral distribution of the inspected surface structure is also of interest, as in tribological applications, we may use an optical profiler, or any other area topographic mapping instrument. When visual appearance and defect localisation are of primary interest, optical reflection and scattering techniques are the techniques to be preferred.

It is worth noting that local defects are not considered as part of the surface structure as defined in the criteria of standardised roughness measurement. Rather they are exceptions not to be measured. However for localising micro-defects we need to assess them by some means, and as was found in this case standard roughness measurements failed in determining the progress of a developing defect. The relatively rough general surface texture of Ra 0.6 – 0.7 μm "conceals" those defects. i.e. scratch marks which are obviously present on the surface, but very difficult to localize by mechanical or optical profiling methods.

3. Light scattering and reflectance

Reflection from a surface is a combination of both diffuse and specular reflection components. Specular reflection is the mirror-like reflection at the air-surface interface, and it occurs when the incident light is reflected from a smooth surface. A rough metal surface has multiple reflection lobes: a directional specular

170

beam with superimposed multi-directional lobes. The latter ones are the result of diffuse reflection, which can arise from multiple surface reflections in a rough, facetted surface as well as diffraction scattering from the microtopography of the top surface. [5, 6]

Variable surface roughness affects the relative ratio of specular to diffuse reflectance, and can hence be observed as contrast change in the appearance of the surface, if viewed in a particular direction. There are theoretical models predicting this fact. One of the first, and probably most popular models is the Torrance-Sparrow model [7], which provides a direct physical connection between a rough metal surface and its visual appearance. In this model the reflectance is based on geometrical optics, hence it is applicable only when the surface is rough (has sufficiently large facets) in comparison to the wavelength of the incident radiation. [8, 9, 10] One of the major drawbacks of the Torrance-Sparrow model, and similar physical-optics approaches of specular reflection from rough surfaces derives right from this fact, i.e. that they ignore the effect of the diffuse reflectance component from sub-micron structures. [11] More recently developed models [12] on the other hand show that it is possible to predict reflectance from isotropic rough surfaces, which have both specular and diffuse components.

For smooth surfaces there are well-developed methods for determining surface roughness by measuring the intensity of both specular and scattered radiation from an illuminated sample, this is the basis of e.g. total integrated scattering (TIS) and angle-resolved scattering (ARS) measurements.[2]

In our investigation however sensitivity to contrast changes, sufficient resolution, simplicity and speed were the main objectives.

4. Experimental work

4.1 The inspected specimens

Two series of stainless steel samples were evaluated in this investigation. The two series (marked as C and D, Fig. 2.) were parts of stainless steel sheets, being drawn and bent over a cylindrical tool within the frame of Bending Under Tension (BUT) testing. (Fig. 1.)

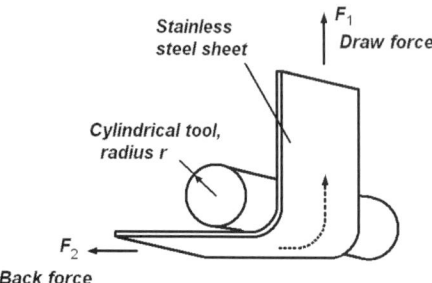

Fig. 1 Principle of the BUT tester

During the drawing process adhesion pick-ups, transferred from the sheet material, were formed on the tool pin and they grew in size during repeated testing. These protrusions induced scratching of the sheet material, and the size of the scratchmarks also grew during repeated use. The aim of our investigation was to verify the development of the defects, i.e. the micro-scratches in a designated area on the metal sheets.

Both series consisted of five specimens; starting from C1 and D1 to C5 and D5 they represented the impact effects of the protrusions, after 1, 500, 1000, 1500 and 2000 strokes at the bar respectively.

Fig. 2 Optimised images of the studied area on specimens C5 and D5

Previous to our study tests had been carried out on each of the specimens, where the surface had been scanned by an optical profilometer within an area of about 500x700 μm. These tests failed to show any correlation between the number of strokes and surface roughness of the specimens. The measured Ra and rms values from C1 to C5 showed only a modest variation of appr. 18%. A larger variation of Ra and rms values could be seen between samples from D1 to D5, but the measured roughness parameters were apparently independent of the number of strokes the specimens in both series had been subject to.

The optical profilometer therefore seemed to lack the capability to reveal surface characteristics of particular interest. The same is true for mechanical profiling with our Talystep (1 μm stylus radius) across the C series, which came out with Rq of 0.76 μm (Ra 0.62 μm) and with a standard deviation of 0.15 μm (0.13 μm). But for the D-series the most marked scratch of sample D5 appeared nicely, and had a double dip of close to 7 μm from the mean surface, with a width of 40-50 μm at half the depth for the two scratches, laterally separated by 100 μm.

4.2 Experimental method

To avoid laser speckles from the rough surfaces we designed our measurement setup using white light illumination from a fixed angle of incidence of 35°. The tungsten light-source had a nominal power of 20W, and the images were captured by a Canon 10D digital camera equipped with an F 50/2.8 macro objective in the setup shown on Fig. 3.

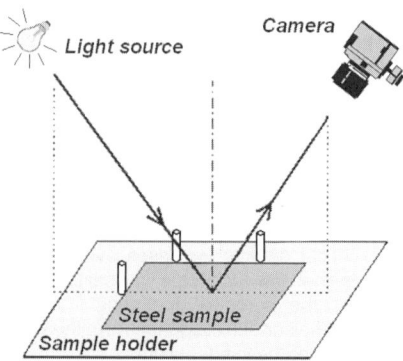

Fig. 3 Experimental setup

The camera axis was close to the specular angle, hence minor changes in the surface topography resulted in large variations in the detected brightness.

Linear conversion of the 12-bit images assured that the pixel-to-pixel brightness values, captured by the CMOS sensor, could be evaluated without any post-processing later on. The linear response also made it possible to do accurate photometry over the entire surface of the samples. To minimize compression artefacts the images were captured in raw mode.

The inspected area on each sample corresponded to 300 rows on the image sensor. This corresponds to an image resolution of 17 μm/pixel. To suppress the contrast variation caused by variations in microtopography at pixel level along the stripes, a mean intensity value was obtained by averaging the measured reflectance values in each row over a neighbourhood of 50 pixels. To bring out the major intensity variations as observed by the naked eye the image was furthermore lowpass-filtered, using a lowpass frequency of 30 pixels. Figure 4a. shows an example of variations in surface scattering on one of the samples after averaging and lowpass filtering.

The sample has been drawn around the metal bar in the direction of the horizontal stripes in the image.

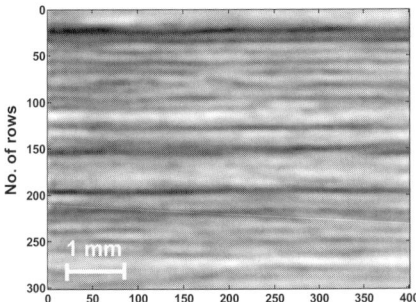

Fig. 4a Surface topography on specimen D3 after averaging and lowpass-filtering

4.3 Evaluation of the images

Image brightness values were averaged in each row, and by using a least squares method we determined the slope of the surface brightness. An average deviation from the calculated slope was then defined as

$$R_{average} = \frac{1}{N}\sum_{i=1}^{N}\left|\delta_i - \delta_{iav}\right| \qquad (1)$$

where N is the discrete number of rows in the image, δ_i is the mean brightness of a certain row, and δ_{iav} is the brightness value of the surface slope in the actual row. The rms deviation was also calculated, using the formula

$$RMS = \left[\frac{1}{N}\sum_{i=1}^{N}\left(\delta_i - \delta_{iav}\right)^2\right]^{1/2} \qquad (2)$$

To further enhance details of interest and to remove the overall surface texture from the manufacturing of the band, the mean grayscale values in each row of C2-C5 images were reduced by the RMS value calculated on sample C1. A similar subtraction of grayscale values was also made on the D2-D5 images (Fig. 4b). The reason for such a subtraction was that C1 and D1 specimens lacked those scratch marks that were present on subsequent

specimens, hence they represented scattering from intact stainless steel surfaces.

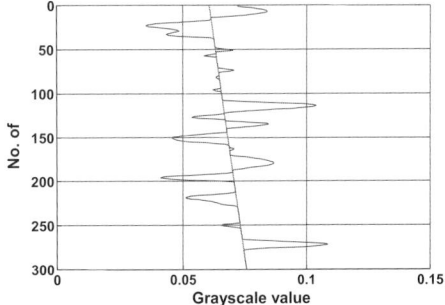

Fig. 4b Surface slope and RMS-reduced grayscale values on specimen D3

Once the grayscale values had been corrected by the "normal" brightness variation, the remaining "areas" on the histogram were also calculated as

$$ICV = \sum_{i=1}^{N}\left|\delta_{icorr} - \delta_{iav}\right| \qquad (3)$$

where δ_{icorr} is the RMS-reduced grayscale value of a given row. The extent of the ICV (Integrated contrast variation) value is a measure of the significant brightness deviations which likely reflects the variation in surface topography itself.

5. Results and Discussion

Two series of images were captured on both, C and D series. The setup and lighting conditions were similar during these series, however, there might have been minor differences between the positioning of the specimens. Most of the specimens featured a very apparent surface curvature, hence usually it was impossible to ensure that the scratch marks were perpendicular to the camera axis. Prior to further evaluations the captured images had therefore always been adjusted to the right angle.

$R_{average}$ values of the measurements are shown in Fig. 5. The first value in each series belongs to a specimen with an intact surface, while subsequent specimens are represented by an overall increasing $R_{average}$ value. In case of both C series the increase is very obvious, while in case of the D series we see a slight variation in $R_{average}$ values.

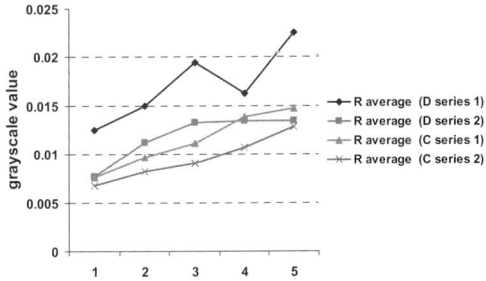

Fig. 5 $R_{average}$ values of two measurements on both series

It is worth to note that it is the relative differences

between the calculated values which are of particular interest rather than the absolute ones. As we have no obvious tendency in the roughness measurement of the surfaces, except for sample D5, no physical meaning can be derived from the absolute $R_{average}$ values. However, due to the linear response of the camera similar $R_{average}$ changes can directly be connected to similar changes in the surface structure, independent of which part of the grayscale these changes has been detected at.

ICV evaluations are presented in Fig. 6. These values also show a strong correlation with number of strokes the metal sheets has been subjected to. Once again it is the relative differences that are interesting, the two series show approximately the same increase in calculated area.

Image series taken on specimen D show a relatively large variation, which is either the result of some inaccurately positioned samples, or the waviness of the surface. This is because in the presence of any of these two circumstances brightness data within the same image row may comprise reflection information from both in- and outside of the scratch marks, which will result in smaller deviation of the actual row from the average surface reflection.

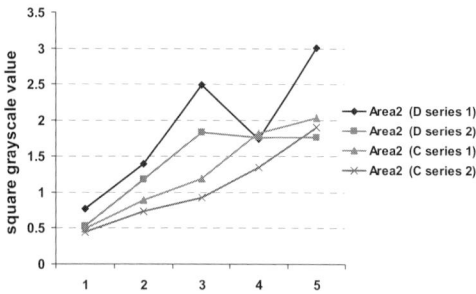

Fig. 6 Integrated contrast variation calculations of two measurements on both series

Despite of these variations we can see a clear evidence that the brightness measurements reflect the progressively increasing defect development on the stainless steel sheets.

6. Conclusion

In this paper we have studied if a white light based reflectance technique using a standard digital camera as detector, can reveal minute changes of defects in surface structure, not recognisable by optical and mechanical surface profilometers. The defect structure on stainless steel bands was developed from scratching by hard deposits on a cylindrical metal bar, subject to contact friction with the band.

The investigation has shown that the reflected intensity distribution from the surface has a strong correlation with the impact parameter. The number of strikes were well reflected by the $R_{average}$ and integrated contrast variation values, which were based upon the captured image brightness.

Despite the promising results, the relatively large variation in the calculated values for the the D series is not yet explained satisfactorily. Here, we will perform more thorough investigations using the Talystep profiler, to see if we will be able to resolve some particular structure of the D4 sample. We know that the

D5 sample has a pronounced double dip, giving a very distinct contrast variation. If we assume that the higher number of strikes implies larger scratch marks, then a possible explanation might be that even though the scratch marks should be isotropic in the drawing direction of the plates, their extent still vary. Therefore brightness measurements might very much depend on the lay of the specimen.

To improve the reliability of the measurements we have to ensure that the specimens are always laid in the same position on the sample holder. A reliable method must also be developed in order to achieve the desired illumination geometry, independent of the presence of any curvature on the surface.

A direct continuation of this work will be to carry out comparative Talystep measurements in order to try to verify the relation between the grayscale values of the evaluated images and the true Ra values. Besides further series of images are planned to be taken, focusing on increasing the dynamic range of the images.

Acknowledgements

This paper was compiled with support from EC FP6 NoE on 4M, Metrology division.

References

[1] D. Brune, R. Hellborg, H.J Whitlow and O. Hunderi, editors "Surface characterization", Wiley-VCH, 1997

[2] J.M. Bennett, L. Mattsson. Introduction to surface roughness and scattering, Optical Society of America, 2nd edition, 1999

[3] EN-ISO 4287:1998 "Geometrical Product Specifications (GPS) – Surface Texture: Profile method – Terms, definitions and surface texture parameters."

[4] ASME B46.1-1995 "Surface Texture, Surface Roughness Waviness and Lay", The American Society of Mechanical Engineers, 345 East 47th St., New York, N.Y.10017

[5] P. Beckmann and A. Spizzichino. The scattering of electromagnetic waves from rough surfaces. Artech House, 1987

[6] X.D. He, K.E. Torrance, F.X. Sillion and D.P. Greenberg. "A comprehensive physical model for light reflection" SIGGRAPH proceedings, p. 175-186, July 1991

[7] K. Torrance and E. Sparrow. "Theory for off-specular reflection from roughened surfaces". Journal of the Optical Society of America, 57:1105-1114, 1967.

[8] L.B. Wolff, "On the relative brightness of specular and diffuse reflection", CVPR, p. 369-376, 1994.

[9] J. Blinn. "Models of light reflection for computer synthesized pictures". Computer Graphics, 11(2):192-198, 1977

[10] B.K.P. Horn. "Obtaining shape from shading information". The Psychology of Computer Vision, p. 115-155, 1975

[11] A.J. Lundberg, L.B. Wolff, D.A. Socolinsky. "New perspectives on geometric reflection theory from rough surfaces". Computer Vision, 2001.

[12] B.V. Ginneken, M. Stavridi and J.J. Koenderink, "Diffuse and specular reflectance from rough surfaces", Applied Optics, Vol. 37, No.1, p. 130-139, 1998

Novel Materials : Characterisation and Processing

Forming and machining of the nano-crystalline alloys

Gy. Krállics[a], M. Horváth[b], J. Nyirő[b]

[a]Department of Materials Science and Engineering, Budapest University of Technology and Economics,
H-1521 Budapest, Bertalan L. u.7., Hungary
[b]Department of Manufacturing Engineering, Budapest University of Technology and Economics,
H-1111 Budapest, Egry József u. 1, Hungary

Abstract

Equal-Channel Angular Pressing (ECAP) is an effective tool for producing ultra fine grained materials. repeated application of ECAP, the rotation of the sample along the longitudinal axis of the billet allows to carry out different routes of deformation. The applied route has a strong influence on the texture, microstructure and mechanical behaviour of ECAP processed metals. In the present study ECAP was successfully applied to produce ultra fine-grained microstructure in a commercial Al-Mg-Si alloy (Al 6082). The mechanical investigations of the ECAP deformed specimens revealed that after 4 ECAP passes the material had a very high strength but a significantly decreased ductility. Further ECAP processing to 8 passes by route C increased ductility dramatically and strength slightly. This indicates that the anisotropy of the structure of ECAP deformed materials may play an important role in achieving good ductility. We made an ultra-precision manufacturing that we analyzed the changing of structure of nano-crystalline materials and the surface roughness for the effect of machining. After manufacturing we prepared an analysis with AFM (Atomic Force Microscope). The results provided that the quality of the ultra-precision finished surface of the processed sample improved dramatically after eight passes.

Keywords: Equal channel angular pressing, high strength and ductility, nano-sized microstructure, surface quality.

1. Introduction

Equal-channel angular pressing (ECAP) is a processing method in which a metal is subjected to intense plastic straining through simple shear without any corresponding change in the cross-sectional dimensions of the sample [1]-[2]. This procedure may be used to create ultra fine grain sizes in bulk polycrystalline materials. The principles of the ECAP process have been examined with reference to the distortions introduced into a sample as it passes through an ECAP die with special attention to the effect of rotating the sample between consecutive passes. Significant distortions of the grain structure occurred when a sample passed through a standard ECAP die, so when a sample is pressed repetitively through the die, it has been recognized that the overall shearing characteristics within the crystalline sample may be changed by a rotation of the sample between the individual passes [3]. The repetitive pressing of the same sample is generally carried out in order to attain very high imposed strains. At the same time there is an opportunity to rotate the sample between consecutive pressings in order to activate different shear planes and directions, thus enhancing the mechanical properties at room temperature by applying different routes [4].

It is well known that plastic deformation induced by conventional forming methods can significantly increase the strength of metals. However, this increase is usually accompanied by a loss of ductility. It has been found recently that materials processed by ECAP after certain number of passes show high ductility along with high strength [5]. Such unusual behaviour of materials, which is in contradiction with 'classic' tendencies to lost ductility with increased strength,

needs to be understood deeper. In this paper the effect of different routes of ECAP techniques on the strength and the ductility of an Al-based alloy is investigated. The mechanical behaviour is related to the characteristic features of the nano-crystalline microstructure formed during ECAP deformation 13]. [

2. Producing of ECAP Specimens

The material used in this study was a commercial Al-Mg-Si alloy (Al 6082). The main components of the alloy are Al (97%), Si (0.7-1.3%), Mg (0.6-1.2%) and Mn (0.4-1%). Before the ECAP deformation, the material was annealed at 420 °C for 40 minutes. Specimens in this condition were regarded as the as-received material. Cylindrical billets of 15 mm in diameter and 145 mm in length were pressed through the ECAP die with 90 °C intersecting channels [6]. Four and eight passes were completed by the following routes: B_C (rotation of the billet around its longitudinal axis after each pass by 90 °C clockwise), BA (rotation of the billet around its longitudinal axis after each pass by 90 °C clockwise and counterclockwise, alternatively) and C (rotation of the billet around its longitudinal axis after each pass by 180 °C, clockwise). The temperature of deformation was 293 K and the displacement rate of the billet was 8 mm/min.

3. Experimental Studies

3.1. Mechanical Testing

The mechanical attributes of the specimens processed by ECAP were studied by tensile and compressive tests. The specimens for the tensile and compression tests were cut from the pressed material along the longitudinal axis. The tensile tests were carried out at ambient temperature at 2 mm/min crosshead velocity in order to collect information about dependence of the yield stress, ultimate tensile stress, area reduction and elongation on different process routes and number of passes. The results of the tensile tests are indicated in *Figs.* 1 and 2. In *Fig.* 1 the area reduction and elongation vs. the number of passes are shown for each ECAP route. In *Fig.* 2 the yield stress and ultimate tensile stress as a function of the number of passes are depicted for each route. Some ECAP processed specimens were investigated also by compressive testing. During compression, the shape of the cross-section of the specimens tended to change from circular to oval. In order to investigate the influence of process routes on the developing anisotropy, the ratio of the largest and smallest diameters of the cross-section was measured for each process route after the fourth and eighth passes of ECAP *(Fig. 3)*.

profiles were measured on the cross-section of the specimens by a high-resolution double-crystal diffractometer (Nonius, FR 591) using Cu Kα1, radiation. The parameters of microstructure were determined from the peak profiles by the Multiple Whole Profile (MWP) fitting procedure described in detail in [7, 8].

It was found that nanosized microstructure (mean crystallite size ≈ 80 nm) with high dislocation density (3 x 10^{14} m^{-2}) was achieved even after the first pass. The microstructure was refined only slightly during further ECAP passes. At the same time the dislocation density increased with the increase of ECAP deformation up to 4 passes. The dimensionless dislocation arrangement parameter, M, has a value of 4.0 ± 0.4 for the as-received specimen and it decreased to 2.2 ± 0.3 after 8 ECAP passes. This indicates that the dipole character of the dislocation structure became stronger with increasing deformation.

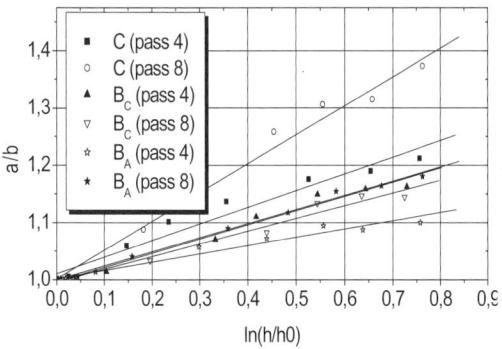

Fig. 3. Diameter ratio vs. compressive strain at ambient temperature for Al 6082 after various ECAP passes

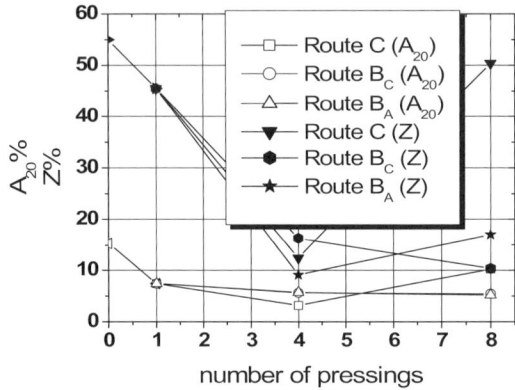

Fig.1. Elongation *(A$_{20}$)* and area reduction (Z) at ambient temperature vs. the number of ECAP passes for different routes

4. Ultraprecision manufacturing

4.1. Experimental setup

The experiments were machined with an ultra-precision machine (UP1, manufacturer: Csepel, license by Hembrug in Haarlem, the Netherlands). This machine has hydrostatic beard axes and spindle (Fig. 4.). The axes are moved by linear motors and the accuracy of rotation of spindle is under the 1 micron, we could measure this parameter. The resolution of the control system and the measuring system is 0.1 micron. But there is another important parameter, the temperature. The room temperature is controlled in the labour on T= 20.0 +/- 0.5 ºC. The processing was carried out with minimal liquid coolant and arising chips were blown away with compressed air from the field of action. [10]

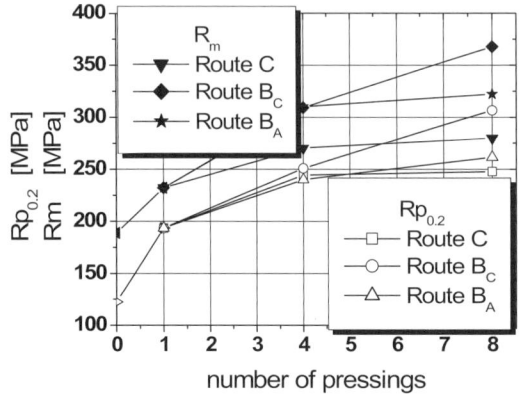

Fig.2. Yield strength *(R$_{p0.2}$)* and ultimate strength *(R$_m$)* at ambient temperature *vs.* the number of ECAP passes for different routes

3.2. *Microstructural Investigation*

The microstructure of the as-received and ECA pressed materials deformed by route C was studied by X-ray diffraction peak profile analysis. The peak

Fig. 4. Ultra Precision Machine (UPM)

During the manufacturing we used a mono-crystalline diamond tools that is made by Winter. It was worked with form lathe tools with a cutting radius of r_ε = 150 and 1000 micron and a relief angle α_o = 5°. The cutting edge radius was estimated in the dimension of r_n = 30...60 nm. Smaller cutting edge radius could not be reached by the tool manufacturer.

Main processing parameters:

- technology: front turning
- spindle velocity: 900...1100 rpm
- cutting speed: 55....65 m/s
- cutting depth: 5 µm
- feed rate: 2 µm/rev

4.2. *Surface Integrity Measurement*

The effect of crystal size and the ECAP process on the surface integrity was investigated by making mirror-like cylindrical and flat surfaces, using turning. [11], [12] The turning process has been applied to every specimen with the same cutting conditions, i.e. cutting speed was 62 m/min, while feed rate was 2 µm/rev. Some micro turned workpieces can see on the next picture (Fig. 5.).

Fig. 5. Workpieces after ultra-precision machining

Surface roughness was measured using an Atomic Force Microscope. The AFM picture you can see on the Fig, 6, The results are very interesting and very promising: surface finish of the cut raw part had the same value *(Ra - 55 nm)* for the specimens after 1 or 4 passes of ECAP process. At the same time after 8

passes Ra dramatically reduced to 10 nm. The results could see on the next table (Table 1.)

Table 1: Results of AFM analysis

Number of passes	Average Roughness (Ra, nm)	Roughness (Rz, nm)
0	49,6	283,1
1	54,8	402,9
4	53,3	272,7
8	9,5	70,87

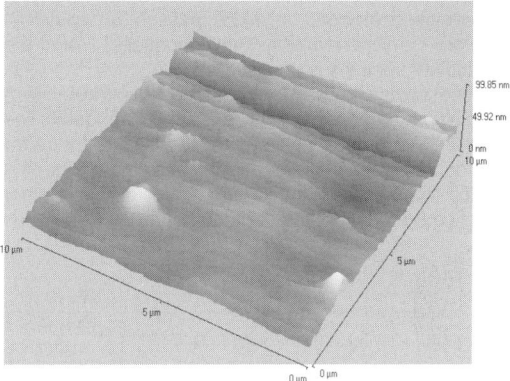

Fig 6. Surface roughness with AFM

5. Discussion

Yield stress *($R_{pO.2}$)* and ultimate tensile stress *(R_m)* grew with the increase of the number of ECAP passes for all the routes investigated. The maximum increment compared to the values before ECAP was reached after the first pass and after eight passes the highest absolute value for yield stress was obtained route B_C, the lowest value by route C. The area reduction (Z) for all routes shows a significant decrease after the fourth pass.

It is interesting to note that after eight passes the sample deformed by route C shows an increase in Z, up to almost the value of the as-received state. Route B_A results in a slight increase of the value of area reduction between four and eight passes. For route B_C, the tendency of decreasing value of Z can be seen even after eight passes but not so drastically as after the fourth pass. For the elongation, almost the same tendency can be observed with increasing number of passes: route C gives the increase of A_{20} after eight passes, while B_C and B_A have the tendency to decrease. These data show that high strength and high ductility exist together after eight ECAP passes applying route C.

The magnitude of the absorbed specific energy [9], which indicates toughness, was also investigated, considering samples produced by different ECAP routes. As shown in Fig. 7, it is clear that route C supplied the best results.

During the compressive tests, cross-section of the specimens tended to change from circular to oval, showing the presence of anisotropy caused by ECAP. The ratio of maximum and minimum diameters *(a/b)*

178

was measured and Fig. 3 shows the dependence of this ratio on different process routes and passes. It is worth noting that in some cases the cross-section of specimens has a 'pear-like' shape, thus the measurement of the diameter ratio is somewhat uncertain. It is also worth noting that the diameter ratio for the sample deformed by route C for eight passes is by far the highest among the specimens studied here. This indicates that anisotropy of the microstructure may play an important role in achieving the simultaneous existence of high strength and good ductility in nanocrystalline materials produced by severe plastic deformation. The study of the microstructure of the cross-section does not give any reasonable explanation for the outstanding ductility of the sample deformed in eight passes by route C, therefore microstructural investigations of the longitudinal sections is necessary.

The sample surfaces, prepared by ultraprecision machining, demonstrated interesting changes. The original surface quality was nearly identical to that after the fourth pass. After eight passes, however, the surface roughness dropped by a factor of five, further reflecting the effect of the structural changes.

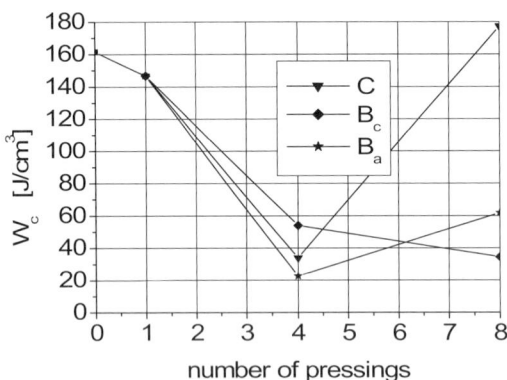

Fig. 7. Absorbed Specific Energy till fracture for Al 6082 after various ECAP passes

Conclusions

Completing the mechanical investigations of the produced specimens by the different routes revealed that route C showed both increasing of strength and ductility owing to the significant changes of the grain structure of the material. In turning experiments after eight steps of pressing - under equal cutting conditions - cut surface roughness reduced by a factor of five on average (from $Ra = 55$ to $Ra = 10$). It is another important impact of structural changes on engineering applications.

Acknowledgement

This work was supported by the „Multi-Material Micro Manufacturing: Technology and Applications (4M)" Network of Excellence, Contract Number NMP2-CT-2004-500274, by the "Application of nano-crystalline titanium for human implantation" Hungarian Government Fund, Contract Number GVOP 3.1.1-2004-05-0305/3.0 and by the Hungarian Scientific Research Fund, OTKA, Grant Nos. T-042714 and T-043247.

References

[1] Valiev, R. Z., Islamgaliev, R. K., Alexandrov, I. V., Progress in Material Science, 45 (2000), pp. 103-189.

[2] Segal, V. M., Materials Science and Engineering A271 (1999), pp. 322-333.

[31 Furukawa, M., Horita, Z., Nemoto, M., Langdon, T.G., Journal of Materials Science 36 (2001), pp. 2835-2843.

[4] Segal, V. M., Materials Science and Engineering, A197 (1995), pp. 157-164.

[5 Valiev, R. Z., Advanced Engineering. Materials, 5 No. 5 (2003), pp. 296-300.

[6] Krállics, Gy., Malgin, D., Raab, G. I., Alexandrov, I. V., Ultrafined Grained Materials III. Conference (2004), Charlotte.

[7] Ribárik, G., Ungár, T., Gubicza, L, MWP-fit: a Program for Multiple Whole Profile Fitting of Diffraction Profiles by Ab-Initio Theoretical Functions, J. Appl. Cryst., 34 (2001), pp.669-676.

[8] Zhilaev, A. P., Gubicza, J., Nurislamova, G., Révész, Á., Surinach, S., Baró, M. D., Ungár, T., Microstructural Characterization of Ultrafine-Grained Nickel, Phys. Stat. Sol. (a), 198 (2003), pp. 263-271.

[9] Czoboly, E., Havas, I., Gillemot, F., Proc. of Int. Symp. on Absorbed Spec. Energy/Strain Energy Density Criterion, Budapest (1980), pp. 107-129.

[10] U. Grimm, C. Müller, W. Menz, M. Wölfle, Fabrication of surfaces in optical quality on pretentious tool steels by ultra precision machining, proc. Euspen, Glasgow, 2004, 193.

[11]A. G. Mamalis, J. Prohászka, I. Mészáros: The Effect of the Anisotropy of the Material on the Surface Topography in Case of Ultraprecision Machining, 1st EUSPEN Topical Conference on Fabrication and Metrology in Nanotechnology, Copenhagen, May 28-30, 2000, pp. 440-446

[12]T. Moriwaki: Machinability of Copper in Ultraprecision Micro Diamond Cutting, Annals of the CIRP, Vol. 38/1/1989, pp. 115-118

[13]Gy. Krállics, M. Horváth and Á. Fodor: Influence of ECAP Routes on Mechanical Properties of a Nanocrystalline Aluminium Alloy, Periodica Polytechnica Ser. Mech. Eng. Vol. 48, No. 2, pp. 145-150 (2005)

Multi-Material Micro Manufacture
W. Menz, S. Dimov and B. Fillon (Eds.)

179

Modelling and design of GaN based piezoelectric MEMS

C.R. Bowen[a], D.W.E. Allsopp[b], R. Stevens[a], P.Shields[c], W.N. Wang[c]

[a] *Materials Research Centre, Department of Mechanical Engineering, University of Bath, Bath, UK, BA2 7AY*
[b] *Department of Electronic and Electrical Engineering, University of Bath, Bath, UK, BA2 7AY*
[c] *Department of Physics, University of Bath, Bath, UK, BA2 7AY*

Abstract

The conditions encountered in aerospace, in automotive and many industrial applications present a challenge for semiconductor based sensor technologies. High temperatures (>180°C) or ionising radiation inhibit the use of silicon transistors. This limits the scope for integrating silicon based sensors and MEMS with conventional electronics. The challenge of extreme environments requires a new approach. We examine here a solution based on GaN, a material with properties that offer wide ranging novel functionality and unexplored scope for integrating advanced sensor devices into single integrated systems for reliable operation in a wide range of extreme environments.

Keywords: GaN, piezoelectric, MEMS, sensor, actuator

1. Introduction

The conditions encountered in aerospace, in automotive and many industrial applications present a challenge for semiconductor based sensor technologies. High temperatures (>180°C) or ionising radiation inhibit the use of silicon transistors. This limits the scope for integrating silicon based sensors and MEMS with conventional electronics. The challenge of extreme environments requires a new approach. We propose here a solution based on GaN, a material with properties that offer wide ranging novel functionality and unexplored scope for integrating advanced sensor devices into single integrated systems for reliable operation in a wide range of extreme environments.

In extreme environments ideal sensor locations are often inaccessible or small, making sensor reliability of paramount importance and multifunctional operation a highly desirable attribute. Packaging and manufacturing requirements make a fabrication and device strategy attractive that offers many of characteristics of Si MEMS processing. However, for extreme environments this capability ideally should be combined with the novel properties of GaN to create a viable sensor technology.

The development of GaN heteroepitaxy has resulted in significant device advances such as blue-green diode lasers [1] and underpins the thrust towards high efficiency solid-state lighting. In addition, the wide energy band gap, high peak and saturation drift velocities and high breakdown field make Ga(Al,In) N the ideal material for fabricating both heterojunction field effect transistors (HFETs) and heterojunction

bipolar transistors (HBTs) capable of continuous operatiion at temperatures ≥300°C [2], [3].

Group III-Nitrides and related materials have further attractive properties that can be exploited for sensing in extreme environments. They demonstrate a large pyroelectric effect at temperatures up to 300°C [4], [5], well above the Curie temperatures of conventional materials. They have piezoelectric stress coefficients 6-7 times greater than those of GaAs [5], [6]. Owing to high bond energies, they have low chemical reactivity [7] and a higher threshold for radiation damage than Si, which is a desirable characteristic in aerospace applications. Furthermore a recent demonstration of a strain sensitive Schottky diode device provides evidence of the potential of GaN in sensor applications [8]. Semiconductor materials with the wurtzite crystal structure (GaN, ZnO, CdS, ZnS etc.), are often considered to have 'high' piezoelectric coefficients; however, it should be stressed that the coefficients are considerably lower than those of ferroelectric ceramics such as PZT. Table 1 presents a comparison of GaN properties with other materials, showing high piezoelectric e_{33} coefficients compared to GaAs, but smaller activity compared to ZnO and lead zirconate titanate type materials (PZT).

Rudimentary GaN thin film cantilevers grown by MBE on Si substrates, by selective etching of the Si, have been demonstrated recently [15]. However, the quality of the GaN was compromised by the mismatched crystal structures of adjacent layers. Recent advances in GaN epitaxial growth techniques offer opportunities for overcoming this limitation. Epitaxial lateral overgrowth (ELOG) over selectively

Table 1. Collated data of piezoelectric materials, relative permittivity, piezoelectric charge coefficients (d_{33}, d_{15}), piezoelectric stress coefficients (e_{33}, e_{31} e_{15}) and compliance (S_{33}) or stiffness (C_{33}).

Material	Relative Permittivity	d_{33} pC/N	d_{15} pC/N	e_{33} C/m	e_{31} C/m	e_{15} C/m	C_{33} (GPa) or S_{33} (pPa^{-1})
GaN	10[11]	3.7[6]	3.1[9]	0.67-1[10]	-0.33[10]	-0.37[10]	C_{33} 398 [11]
GaAs	12.5[12]	-	-	-0.185[10]	0.093 [10]	0.093[10]	S_{33} 12.64[12]
AlN	8.5[13]	5[6]	3.6[9]	1.55[10]	-0.58[10]	-0.48[10]	C_{33} 389 [14]
ZnO	9.15 [12]	11.67[12]	-11.34 [12]	1.32 [10]	-0.57 [10]	-0.48 [10]	S_{33} 6.94 [12]
PZT-5H	3400	593	741	23.3	-6.5	17	C_{33} 117

Figure 1. Self-supporting structures formed by ELOG of GaN and subsequent selective removal of the SiO_2 etch mask.

masked mismatched substrates is known to produce areas of GaN with low densities of dislocations over the masked regions of the substrate [16].

Our studies revealed that ELOG offers a route to fabricating self-supporting structures by selective removal of the growth mask. Figure 1 shows self-supporting structures formed by selective etching of the SiO_2 growth mask from beneath ELOG GaN grown by MOCVD. Control of the MOCVD growth conditions results in rectilinear cross sections with excellent morphology. Longer cantilevers may be formed from an asymmetric dogleg seed region, as shown in Figure 2. Subsequent inductively coupled plasma (ICP) etching along the long section of the dogleg and selective removal of the mask will release the cantilever. Greater clearances can be achieved by using thicker growth masks. The low reactivity of GaN in standard SiO_2 etches is a particular advantage of this approach, as it allows full device fabrication prior to release of the cantilevers, to form the basis of a practical manufacturing technology.

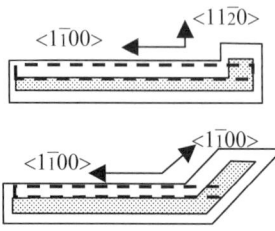

Figure 2. ELOG nucleated from dogleg-shaped growth islands (in-filled regions) pre-patterned on the substrate. The regions defined by heavy lines indicate where long cantilevers can be formed.

The aim of this paper is to report a study of GaN based cantilever and clamped membrane structures by finite element modelling to examine their electromechanical performance as sensors and nanoactuators. The output of the modelling is able to provide information on what structures can be viably manufactured by ELOG.

2. Coupled Field Modelling of GaN Actuators

2.1 Piezoelectric properties of GaN

The GaN piezoelectric [17] and stiffness [11] matrix used to undertake a coupled-field finite element

analysis of GaN bases MEMS are shown in Table 2. The properties were input into ANSYS finite element software using coupled field elements to model the electromechanical behaviour of the material.

TABLE 2. Material Parameters of GaN, [c] denotes the stiffness matrix, [e] is the piezoelectric matrix and $[\varepsilon]^s$ is the permittivity at constant strain matrix. The material is polarised in the c-axis, along the z-direction for this set of materials data.

$$[c] = \begin{bmatrix} 390 & 145 & 106 & 0 & 0 & 0 \\ & 390 & 106 & 0 & 0 & 0 \\ & & 398 & 0 & 0 & 0 \\ & & & 105 & 0 & 0 \\ & & & & 105 & 0 \\ & & & & & 105 \end{bmatrix} \text{GPa}$$

$$[e] = \begin{bmatrix} 0 & 0 & -0.37 \\ 0 & 0 & -0.37 \\ 0 & 0 & 0.67 \\ 0 & 0 & 0 \\ 0 & -0.33 & 0 \\ -0.33 & 0 & 0 \end{bmatrix} \text{C m}^{-2} \qquad \frac{[\varepsilon]^s}{[\varepsilon_0]} = \begin{bmatrix} 10 & 0 & 0 \\ 0 & 10 & 0 \\ 0 & 0 & 10 \end{bmatrix}$$

2.2 Modelling of the GaN cantilevers

The two underpinning assumptions of the modelling are (a) the starting materials is unstrained, as occurs in ELOG, and (b) the conductivity of epitaxial GaN thin films can be varied from semi-insulating behaviour through to strongly n-type conduction. In this respect, GaN layers with net free electron densities in the range $<1 \times 10^{15}$ cm^{-3} and $>1 \times 10^{18}$ cm^{-3} have been grown in our own laboratory by MOVPE, values compatible with most of the device designs considered here. Figure 3 shows a graph of the tip deflections of a GaN cantilever, which consists of a bottom conductive layer of n-type doped GaN which acts as a base electrode and an upper layer of undoped, insulating, and piezoelectric GaN. Each layer is 5μm thick and an electric potential of 100V is applied across the undoped piezoelectric layer. The length of the cantilever is varied

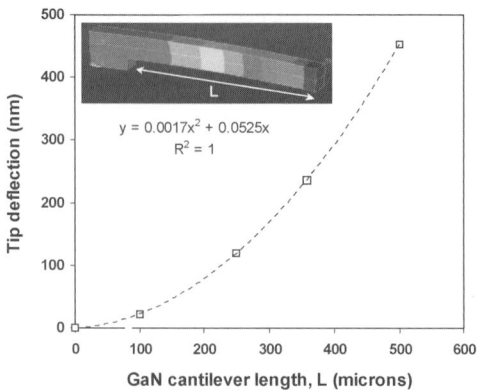

Figure 3. Cantilever (bottom region non-piezoelectric and conductive, top region piezoelectric). Each layer is 5μm thick, 100V potential difference applied across piezoelectric region. Cantilever length, L, varied from 100 to 500 μm. Influence of top electrode ignored.

from 100μm to 500μm, a range that potentially could be produced from an ELOG film. Longer cantilever lengths lead to larger tip deflections. While tip deflections of only ~20nm are predicted for 100 μm long cantilevers, a deflection of 450nm is obtained for a 500μm length structure.

The data presented in Figure 3, in essence,

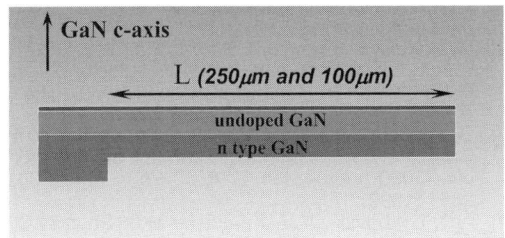

Figure 4(a) Upper layer is top electrode (1μm thick), middle layer is piezoelectric GaN (5μm thick), lower electrode layer is n-type GaN (1μm thick). c-axis is through thickness of cantilever

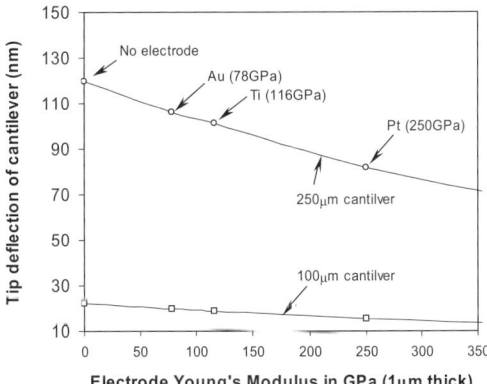

Figure 4(b) Model results for above cantilever for two lengths (L) 250μm and 100μm. Applied potential is 100V.

demonstrates the basic piezoelectric deformation of intrinsic/n-type GaN bi-layers, since the mechanical constraint associated the elastic modulus of the top electrode was ignored. With such thin structures the stiffness of the metal electrode deposited on top of the cantilever will constrain the piezoelectric induced deformation. Further simulations were undertaken to calculate the variation in the tip deflection of 100 μm and 250μm long metal/undoped GaN/ doped GaN cantilevers (Figure 4a). A potential of 100 V is applied across a 5 μm thick undoped GaN piezoelectric layer, sandwiched between a 1 μm thick upper metal layer and a conducting lower n-type GaN layer. The n-type GaN layer is also 5 μm thick. The Young's modulus of the electrode was varied from 78GPa (Au) and 250GPa (Pt). Figure 4b shows that increasing the Young's modulus of the electrode material leads to a decrease in the tip deflection since the overall mechanical stiffness of the structure is increased. For a 250μm cantilever a deflection of 106nm is predicted for a gold electrode compared to 84nm for a platinum electrode

Figure 5 shows the effect on the tip deflection of varying the thickness of the piezoelectric, undoped GaN

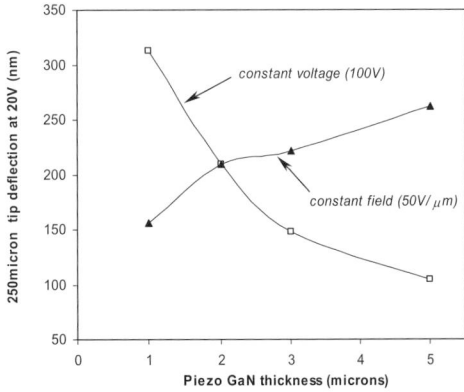

Figure 5. Variation in tip deflection with active layer thickness for a 250 μm long GaN cantilever formed by an undoped GaN layer sandwiched between a 5 μm thick lower n-type GaN layer with a 1μm gold top electrode. Applied voltage = 100 V.

layer. The data are for 100V applied between a 5 μm thick supporting n-type contact layer and 1 μm thick gold top electrode. Two cases have been considered (i) a reduction in undoped GaN thickness while maintaining a constant potential of 100V across the layer and (ii) a reduction in undoped GaN thickness while maintaining a constant electric field of 50V/μm. Considering the data for constant voltage, note how a reduction in the thickness increases the tip deflection of the bimorph cantilever, confirming the more extensive deflection of thin film structures. For the constant field condition a thicker active layer leads to improved actuation. Similar observations have been made by DeVoe and Pisano [18] on ZnO cantilevers, although to achieve maximum deflection under constant field an optimal ZnO thickness was observed. These authors also found that when designing for a voltage-limited drive signal, the optimal thickness for the piezoelectric layer was given by the thinnest achievable film, based on either breakdown field or fabrication constraints. This is consistent with the results shown in Figure 5.

2.3 Modelling a GaN shear actuator

In Group III-Nitrides, the piezoelectric shear coefficient, e_{15}, is comparable to the "e" values along the major axes [9]. In principle, this will give rise to a pseudo-rotational movement when an electric field is applied perpendicular to the c-axis. We have also examined a paddle-membrane device based on this unique property of wurzite crystal structures, that yields a degree of freedom capable of steering light beams. Fig 6(a) shows the novel concept schematically. Fig 6(b) shows the results of our detailed simulation of the rotation achieved by a 20×20 μm paddle when 50 V is applied to a 2×5 μm support beam perpendicular to the c-axis.

The structure utilises the piezoelectric shear coefficients and consists of a 20μm x 20μm main body attached to two 20μm x 5μm torsional arms, as shown in Fig 6a. The polarisation direction is through the thickness of the structure (z-axis) and an electric field is applied across the arms of the torsional mirror (y-axis); normal to the polarisation direction inducing a twist in

182

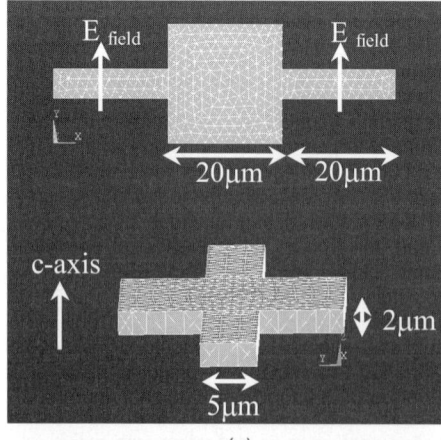

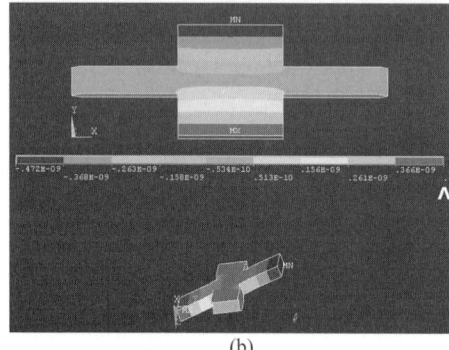

(b)

Figure 6. (a) geometry of GaN device (b) displacement profile, the colour contours indicate the displacement of the device in metres in the z-axis due to the induced twist.

the two arms.

The displacement of the device under the application of an electric field of $10V\mu m^{-1}$ is shown in Fig 6b. The results indicate a 0.5 nm vertical deflection, which equates to a rotation of 5×10^{-5} radians. An incident light beam undergoes twice this angular deflection, making the spatial deflection measured at a distance of 10 cm from a reflecting surface equal to $10\mu m$, i.e. one pixel on a CCD camera. Examination of a range of devices of different geometry indicates that the angular deflection increases on increasing the width of the arms or reducing the thickness. The arm length has a limited influence on the degree of angular rotation.

3. Discussion and Conclusions

The simulations are based on an assumption that the electromechanical behaviour of strained GaN epitaxial layers of wurtzite crystal structure is governed purely by piezoelectric properties described in Table 2. Under these conditions, it has been demonstrated that despite the relatively small piezoelectric coefficients, compared to ferroelectric materials, such as PZT, appreciable nanoactuation results from optimising the configuration of the GaN thin films and the device structure. This optimisation relies on the epitaxial growth of GaN layers with low residual strain and very low free carrier densities. The combination of ELOG growth and Mg (p-type) counter-doping to compensate residual donors that are usually found in nominally

undoped GaN offers a clear route for achieving these objectives. Furthermore, the micro-scale structures envisaged are compatible with semiconductor mass manufacturing processes.

References
[1] S. Nakamura, M. Seroh, N. Nagahama, N. Zwara, T.H. Uwento, S. Sano, K. Chocho, "InGaN/GaN/AlGaN-based laser diodes with modulation-doped strained-layer superlattices," *Jpn. J. Appl. Phys.* **36**, L1568 (1997).
[2] H. Morkoc, S. Strite, G.B. Gao, B, Sverdlov, and M. Burns, "Large band gap SiC, III-V Nitride and II-VI ZnSe based semiconductor device technologies," *J. Appl. Phys.* **76**, 1363 (1994).
[3] S. Yoshida, and J. Suzuki, "Characterisation of a GaN bipolar transistor after operation at 300°C for over 3000 hours," *Jpn. J. Appl. Phys.* part 2 **38**, L851 (1999).
[4] A.D. Bykhovski, V.V. Kaminski, M.S. Shur, Q.C. Chen and M.A. Khan, "Pyroelectricity in gallium nitride thin films," *Appl. Phys Lett.* **69**, 3254 (1996).
[5] M.S. Shur, A.D. Bykhovski, R. Gaska and M.A. Khan, "GaN based pyroelectronics and piezoelectrics," in *Semiconductor Homo- and Hetero-Device Structures*, Eds. M. Francome and C.E.C. Wood, Academic Press.
[6] I.L. Guy, S. Muensit and E.M. Goldys, "Extensional piezoelectric coefficients of gallium nitride and aluminium nitride," *Appl. Phys. Lett.* **75**, 4133 (1999).
[7] I. Adesida, C. Youtsey, A.T. Ping, F. Khan, L.T. Romano and G. Bulman, "Dry and wet etching for group III-nitrides," *MRS Internet J. Nitride Semicond. Res.* **4S1**, paper G1.4 (1999).
[8] R.P. Strittmatter, R.A. Beach, J. Brooke, E.J. Preisler, G.S. Picus and T.C. McGill, "GaN Schottky diodes for piezoelectric strain sensing," *J. Appl. Phys.* **93**, 5675 (2003).
[9] S.Muensit, E.M.Goldys and I.L.Guy, "Shear coefficients of GaN and AlN" App. Physics Letters, **75**, 3965-3967 (1999)
[10] S.Shur, A.D.Bykhovski and R.Gaska, "Pyroelectric and piezoelectric properties of GaN-based materials", MRS Internet J.Nitride Semicond. Res. 4S1, G1.6 (1999)
[11] A. Polian , M. Grimsditch , I. Grzegory, 'Elastic constants of gallium nitride' Journal of Applied Physics **79(6)** 3343 (1996).
[12] www.efunda.com/materials/piezo/material_data (accessed Feb 2006)
[13] Goldberg Yu. in Properties of Advanced SemiconductorMaterials GaN, AlN, InN, BN, SiC, SiGe . Eds. Levinshtein M.E., Rumyantsev S.L., Shur M.S., John Wiley & Sons, Inc., New York, 2001, 31-47.
[14] McNeil, L.E, Grimsditch M., French R.H., J. Am. Ceram. Soc. **76**, 5 1132-1136. (1993)
[15] S. Davies, T.S. Huang, M.H. Glass, A.J. Papworth, T.B. Joyce and P.R. Chalker, "Fabrication of GaN cantilevers on SI substrates for micromechanical devices," *Appl. Phys. Lett.* **84**, 2566 (2004).
[16] Z.R. Zytkiewicz, "Laterally overgrown structures as substrates for lattice mismatched epitaxy," Thin Solid Films **412**, 64 (2002).
[17] P.Tripathi and B.K.Ridley, 'Dynamics of hot-electron scattering in GaN heterostructures', Phys. Rev. B, **66** 195301 (2002)
[18] D.L. DeVoe and A.P. Pisano, "Modeling and optimal design of piezoelectric cantilever microactuators", Journal of Micromechanical Systems, **6**, 266 (1997)

Indirect tooling based on micromilling, electroforming and selective etching

P.T. Tang[a], J. Fugl[a], L. Uriarte[b], G. Bissacco[a] & H.N. Hansen[a]

[a] Dept. *of Manufacturing Engineering and Management (IPL), Technical University of Denmark, Produktionstorvet, bldg. 427S, DK-2800 Kgs. Lyngby*
[b] *Fundacion Tekniker, Avda. Otaola 20, 20600, Eibar, Spain*

Abstract

The tool inserts used for injection moulding or hot-embossing of polymer micro-components, are the most important and expensive and crucial part of this important mass-production process. In this paper a new fabrication scheme is introduce, consisting of a combination of micro-milling, electroforming and selective etching. The basic concept is to exploit the benefit of true 3D-machining in a soft substrate such as aluminium with the excellent replication capabilities of nickel electroforming. The term indirect machining covers the fact that the master that is produced by machining a positive structure, i.e. the opposite of what is needed for the actual mould insert.

Keywords: micromilling, electroforming, selective etching, tooling, injection moulding

1. Introduction

Fabrication of tools or tool inserts, for injection moulding or hot-embossing of polymer micro-components, is a cornerstone in micro- and nanotechnology today. Polymer components are in high demand, not because plastic is a superior material, but simply because it can be mass-produced.

Unique products based on micro- or nanotechnology, such as many lab-on-a-chip applications as well as sensors based on optical or magnetic properties, have yet to achieve a marked break-through - mainly because they are to expensive to produce.

This paper introduces the concept of indirect tooling into the field of micro-fabrication. The basic idea (see figure 1) is to machine the opposite of the tool in question – which turns out to be similar in shape to the final product. In particular for microfluidic components, it is far easier to machine the channels and reservoirs, than to machine the protruding walls and plateaus that are needed for a tool insert for injection moulding. By electroforming of a thick metal layer on top of the machined master, and subsequent selective etching of the master, the actual tool insert is produced.

The master can be produced using any type of micromachining, such as micromilling, laser machining, micro-electrodischarge machining (EDM), micro-electrochemical machining (ECM) or even micro machining of silicon (such as reactive ion etching, etc.). It is important to note that the master material is not selected for its hardness and durability as an injection moulding tool (which is usually the case), but for its machinability. The demand to the master is that selective etching is possible once the electroforming process is complete. Looking at micromilling and the examples of this paper, this means that the master is aluminium of the 6000 or 7000 series (such Al6063 or Al7075). These aluminium alloys are cheap, easy to machine, cause little tool wear and enable fast machining rates.

Using a newly developed pre-treatment process [1] prior to electroforming, an advanced electroforming process and an alkaline solution for selective etching of the aluminium master, a tool insert can be fabricated in 5-6 working days (depending on the thickness required for the tool insert).

Another important feature is the ability to fabricate complex 3D-inserts, by exploiting the full potential of micro-milling and the good replication of the master obtained by electroforming.

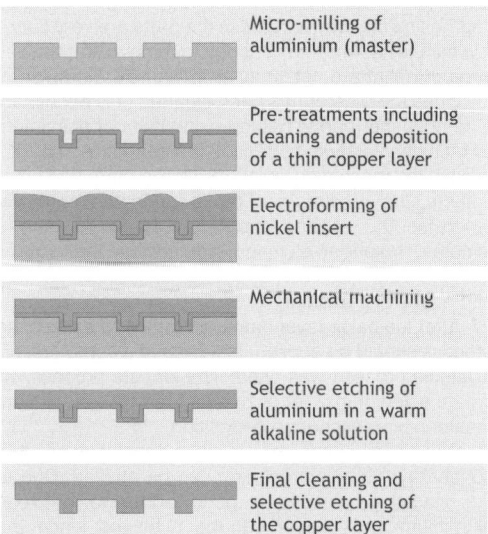

Micro-milling of aluminium (master)

Pre-treatments including cleaning and deposition of a thin copper layer

Electroforming of nickel insert

Mechanical machining

Selective etching of aluminium in a warm alkaline solution

Final cleaning and selective etching of the copper layer

Figure 1: Main principle of indirect tooling.

2. Experimental

The different processes listed in figure 1 will now briefly be discussed and typical parameters disclosed.

2.1 Micromilling of aluminium

Micromilling is still the most suitable technology when 3D complex micro structures are required [2]. Although this technique is similar to conventional scale milling from the operational point of view, the large reduction in dimensions means that cutting phenomena and mechanisms appear that are hardly ever encountered on a conventional scale. This scale reduction can be seen in some of the usual parameters of micromilling of aluminium: feed per tooth less than 1 µm, depth of cut 5 - 20 µm, spindle rotational speed more than 50000 rpm and tool diameter less than 0.5 mm.

The milling machine itself must also be specific to this application and designed and built to ultra precision requirements, with positioning accuracies on the order of 0.1 μm. Important efforts are being done to improve micromilling accuracy, productivity, and for downscaling the current capabilities, but actual limitations are still in the tool itself.

When comparing the machining of an aluminium master with the direct machining of an insert in hardened tool steel, clear advantages brought by the former are recognized in higher feed per tooth, depth of cut and longer tool life. These translate in a machining time up to five times shorter. Furthermore, if the focus is on microfluidic devices, much less material must be removed from the master and the machining time is reduced dramatically. Due to the lower cutting forces and tool wear, tool deflections are reduced and the dimensional accuracy of the master and thereby of the plastic parts is improved. Finally, machining of aluminium allows for higher cutting speeds than tool steel and therefore faster production. However, this potential advantage is rather unexploited due to the current limitations in rotational speed of commercially available spindles.

2.2 Pre-treatments

After the mechanical machining process, the aluminium master will be cleaned using a cathodic degreasing electrolyte. Unless the parts are very dirty a few minutes at 4 Volts is enough to remove oil, particles and burrs that are not firmly attached to the master.

Once the master has been cleaned, it is necessary to remove the (partly) natural oxide layer of aluminium by etching in sodium hydroxide (60 g pr. litre) at 60 °C. In case the machining introduced large burrs and other defects, the etching can be 3-4 minutes or more, otherwise the etching should be no more than 2 minutes. Prolonged etching will deteriorate the features fabricated by micromilling, by attacking sharp corners and edges in particular.

After the oxide layer has been removed a zincating process should be applied. In this case a commercially available process was used. The zincate process will create a very thin layer of zinc on top of the aluminium master by ion-exchange deposition.

With the zincating layer in place deposition of a thin copper layer by electroplating can be applied. Due to the zincating process a good adhesion between aluminium and copper can be achieved since the natural oxide layer of aluminium is etched away and replaced by a nanometer thick zinc coating. In order not to dissolve the zinc coating, the copper electroplating bath should have a mild pH-value. The most well suited type of copper bath is the pyrophosphate bath [3] which has a pH-value around 9. Furthermore is the pyrophosphate bath well-known for a relatively good material distribution.

2.3 Electroforming

During the electroforming process, a thick and mechanically durable nickel layer is deposited. Depending on the application, and the mould that the insert must fit into, the thickness of the this nickel layer can be anything from 300 μm to several millimetres. Typically a nickel sulphamate bath is used [3] because of the unique control of internal stresses that can be obtained using this type of plating electrolyte. In both examples described below the sulphamate bath was operated at 32°C with a relatively low current density of 1.0 A/dm^2 (corresponding to a deposition rate of 0.3 μm pr. minute).

In case of the microfluidic component (see section 4), the electroforming process consisted of a thin nickel layer (50 μm) for wear resistance and a thick (2.5 mm) copper layer [4] for good thermal conductivity. For true mass-production the thermal conductivity of the tool and the tool inserts is a crucial factor for achieving fast cycle times.

2.4 Final machining and selective etching

When the layer of metal deposited on the master is thick enough to provide the required mechanical strength, the block constituted by the master and the insert is removed from the electroforming bath. The block is then mechanically machined, typically by ordinary milling, in order to remove the material in excess, and to produce a flat reference surface on the back of the insert. Such surface is needed for the accurate positioning and alignment during the final assembly of the insert in the mould. Since this operation uses the back of the insert as a reference, it is important to maintain parallelism between such surface and the plane ideally dividing the master and the insert.

The next manufacturing step is constituted by selective etching of the master by immersion in an alkaline solution (e.g. sodium hydroxide). The duration of this step depends on the thickness of the aluminium master, but generally a 5-8 mm of aluminium can be dissolved in 24 hours. Since pure aluminium is very problematic to machine (burrs, continuous chips, etc.) the aluminium used is either the 7075 alloy containing zinc or the 6063 containing silicon and magnesium (in both cases also smaller amounts of other elements such as copper, chromium and manganese). Some of the alloying elements will form oxides or hydroxides, which eventually creates a thick black coating on the surface of the insert.

The last step is consequently to remove the intermediate copper layer (see section 2.2). By removing this copper layer the black precipitate will be lifted off (but not necessarily dissolved). The most gentle and selective way to etch copper without damaging the nickel underneath [2], is to use an alkaline mixture of ammonium, sodium chloride and carbonate, but solutions based on sodium, potassium or ammonium persulphate - or even cyanide – can also be used.

PINS	RIBS
Ø100 x 100-250 μm height	100 μm width x 100-250 μm height
Ø200 x 200-500 μm height	200 μm width x 200-500 μm height

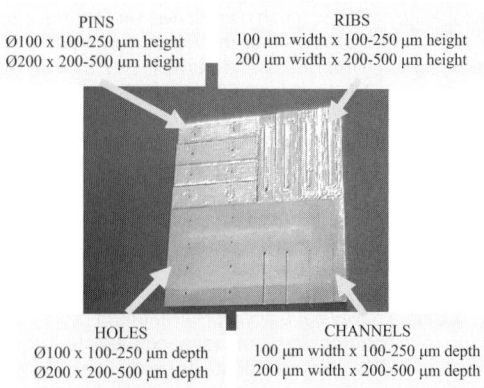

HOLES	CHANNELS
Ø100 x 100-250 μm depth	100 μm width x 100-250 μm depth
Ø200 x 200-500 μm depth	200 μm width x 200-500 μm depth

Figure 2: Overview of the test tool designed by the metal cluster of 4M

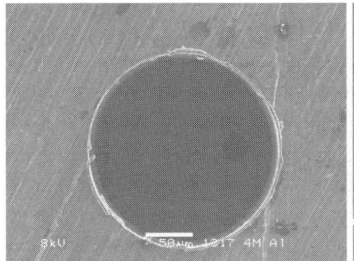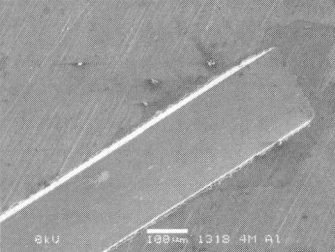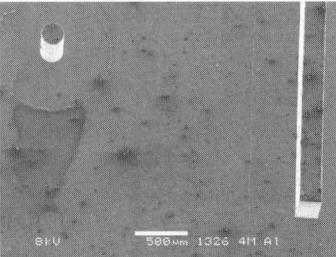

Figure 3: SEM micrographs of some of the features of the aluminium test specimen. The holes are made by drilling (left), the channels by cutting (middle) and the pins and ribs by milling (right). The scale bares are 50, 100 and 500 µm respectively (left to right).

3. 4M test part

Through the metals cluster of the 4M network, a test tool was designed containing both holes, pins, channels and ribs (see figure 2). For electroplating purposes, as mentioned above, the master was aluminium 7075. The following 3 figures show details of the surface of the nickel insert electroformed using the aluminium master (see figure 3).

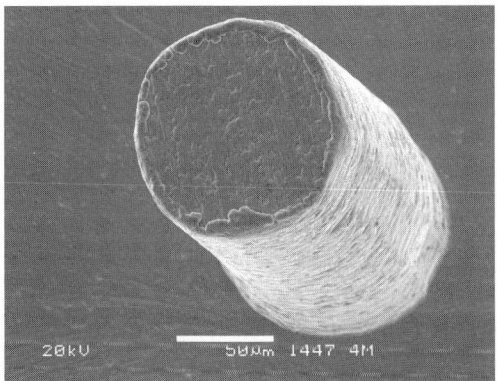

Figure 4: Holes in the master become pins in the electroformed insert (when the aluminium is dissolved). Scale bar is 50 µm.

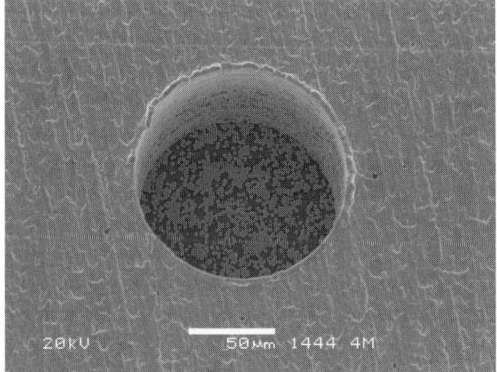

Figure 5: The pins of the aluminium master will become holes. Due to the excellent focus depth of scanning electron Microscopy it is possible to see the surface at the bottom of this hole. Scale bar is 50 µm.

Generally the test part was replicated meticulously, both for structures pointing out (poles and ribs) and structures pointing in (holes and channels). However, there will be a limit to how deep the structures can be without risking that they "grow together" at the opening without filling the hole or channel completely [4]. Such behaviour is a general problem with electroplating, especially when both the bottom and the side walls of the structure is conducting.

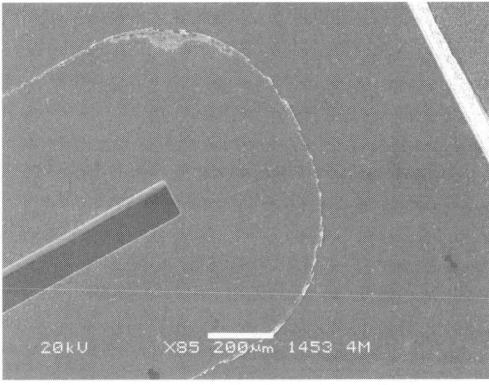

Figure 6: In order to machine the ribs the surroundings where removed. Even after electroforming and selective etching of the master, the machining path is clearly seen. At the bottom of the channel, some burrs from the machining can be seen to point "down". Scale bar is 200 µm.

4. Example of a microfluidic component

In order to test the abilities of indirect tooling, a test tool was made [4] containing various channels and reservoirs. The tool was designed to demonstrate the potential of indirect tooling by suggesting a number of features (such as the mixing chambers) that are very difficult – or even impossible – to fabricate using lithography based processed such as reactive ion etching of silicon or UV-LIGA (i.e. using the popular photoresist SU-8).

The tool also contained a number of chambers (see figure 7) that could function as inlet or outlet simply by penetrating the thin wall between the two chambers with a needle. In the following we will focus on the fabrication and characterisation of that particular feature, however information on the other features can be found in [4] and [5].

One tool insert was made using indirect tooling in a master made of Al6063 (AlMg0.5Si) - as described above. The other tool insert was fabricated directly by micromilling in a steel master ("Impax Supreme" which is a tool steel material made by Uddeholm with chrome, nickel and molybdenum as alloying elements).

186

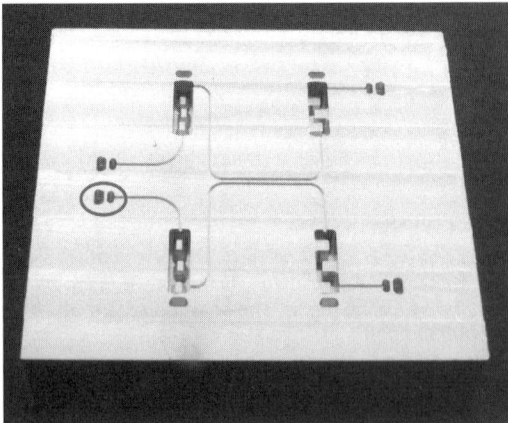

Figure 7: Indirect micromilling in Al6063. The channels between the reservoirs are 0.2 x 0.2 mm. Two of the reservoirs contain blocks with a height of 0.6, 0.9 and 1.2 mm. The two other reservoirs contain pyramids with a height of 0.6, 0.9 and 1.2 mm and with the corresponding angles of 53.1, 66.8 and 73.3 degrees. The feature in the red circle is studied more closely in the following.

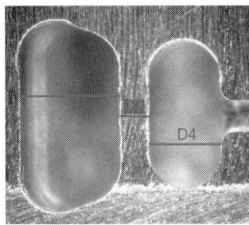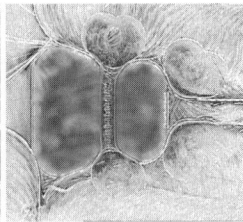

Figure 8: Stereo microscope pictures of the feature highlighted in figure 7. The left is indirect in aluminium the right is direct in tool steel.

Although polymer microfluidic parts were injection moulded successfully using both direct and indirect tooling concepts, the best accuracy was obtained by indirect tooling. The amount of burrs produced during the micro-milling of the aluminium master was a problem, but many were removed during the pre-treatment (especially the etching to remove the oxide layer described in section 2.2). Micro-milling of tool steel is possible, but the wear on the milling tool is severe and the machining time very long.

Table 1: Comparison of nominal and measured sizes of the high-lighted feature. D6, D5 and D4 refer to the distances illustrated in figure 8 (left). NiD5 is thus the distance between the two protrusions on the nickel insert (figure 9, left), while ImD5 is the same distance measured on the Impax insert (figure 8, right).

	NiD6	NiD5	NiD4	ImD6	ImD5	ImD4
Nominal	690	200	500	690	200	500
Measured	677	210	499	675	187	527
Deviation	-13	10	-1	-15	-13	27
Dev. (%)	-1.9	5.0	-0.2	-2.2	-6.5	5.4

5. Conclusion

The two examples have demonstrated that indirect tooling can be a viable fabrication route for metal inserts for injection moulding (or hot-embossing). The benefits of this route are that it is easier to machine the softer aluminium alloys and that – especially for microfluidic components – it is often faster to machine a "positive" structure than a "negative" one (easier to machine a channel than to machine a rib).

The introduction of electroforming and selective etching has also proven viable, yielding a good replication of the machined aluminium master.

6. Acknowledgements

The authors are grateful to the members of the metal cluster of the 4M network of excellence, for their contributions to the design and initiation of the 4M test part described in section 3.

7. References

[1] Pending patent application, PCT DK2005/000564
[2] L. Uriarte, A. Ivanov, H. Oosterling, L. Staemmler, P.T. Tang, D. Allen, "A Comparison between Microfabrication Technologies for Metal Tooling", 4M 2005 Proc., pp.351-354, ISBN 0.080-44879-8
[3] P.T. Tang, "Fabrication of Micro Components by Electrochemical Deposition", Ph.D.-thesis, Dept. of Manufacturing Engineering, KMT 980301-1 (1998)
[4] J. Fugl, "Værktøjsteknologier til mikrosprøjtestøbning" (in Danish), Master Thesis, Technical University of Denmark, Dept. of Manufacturing Engineering and Management, IPL-152-04 (2004)
[5] G. Bissacco, "Surface Generation and Optimization in Micromilling", Ph.D. Thesis Tech-nical University of Denmark, IPL-246-04 (2004)

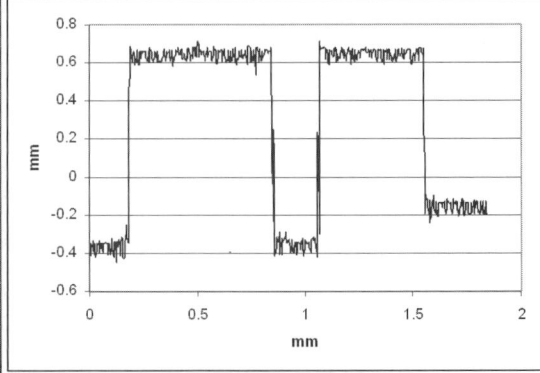

Figure 9: The highlighted feature of figure 7 examined closely after electroforming and selective etching. The nickel surface is seen in the SEM-image on the left and using an optical profilometer (right).

Multi-Material Micro Manufacture
W. Menz, S. Dimov and B. Fillon (Eds.)

Systematic interaction of sedimentation and electrical field in electrophoretic deposition

S. Bonnas[a,b], J. Tabellion[b], H.-J. Ritzhaupt-Kleissl[a], J. Haußelt[a,b]

[a] *Institute for Materials Research III, Forschungszentrum Karlsruhe, 76021 Karlsruhe, Germany*
[b] *Laboratory for Materials Processing, Department of Microsystems Engineering (IMTEK),University of Freiburg, 79110 Freiburg, Germany*

Abstract

The systematic interaction of sedimentation and electrical field in electrophoretic deposition allows the tailoring of specific properties of deposited green bodies. This technique permits a selective deposition of the nanosized fraction of conventional powders with broad or non-monomodal particle size distribution, thus making preceding classification obsolete. Potential applications are coatings with a very smooth surface or the replication of microstructures or moulds which are filled with nanosized particles and subsequently with coarser particles as support in one process step. Also graded structures can be fabricated with regard to particle size distribution, porosity and composition (e.g. zirconia toughened alumina). In this paper, the interaction of sedimentation and electrical field in electrophoretic deposition is described focussing on the characterisation of both processes, sedimentation and electrophoretic deposition. In addition the effectiveness of the combined process will be shown.

Keywords: electrophoretic deposition, sedimentation, particle size distribution, zirconia

1. Introduction

In micro system technology and micro chemical engineering, the application of submicron and nanosized ceramic particles is gaining importance because of the required smooth surfaces and dimensional accuracy. In conventional shaping techniques like axial pressing, slip casting or micro injection moulding nanosized powders are difficult to apply. For these powders, electrophoretic deposition (EPD) is a more suitable method [1, 2], because the deposition rate is independent of particle size as long as EPD is carried out perpendicular to sedimentation.

Electrophoretic deposition is a colloidal processing technique first observed in 1808 by the Russian scientist Reuss [3], the first description of the deposited yield was made in 1940 by Hamaker [4].

Electrophoretic deposition is achieved via motion of charged particles in a fluid medium under the influence of an electric field towards an oppositely charged electrode and the formation of a stable deposit on the electrode [1, 5-8] or on a membrane [6, 9]. Particle velocity is independent of particle size if EPD is carried out perpendicular to sedimentation. Homogen-eous deposits with a high mechanical strength and a low surface roughness can only be obtained by using well dispersed suspensions [10]. EPD is used for many applications, such as the application of coatings, the manufacturing of microstructures, laminated or graded materials or infiltration of porous materials [5-7, 9, 11]. A wide range of materials and combinations can be employed [1, 10, 12].

Simultaneously with electrophoretic deposition, sedimentation occurs depending on diameter and density of the particles and on the solids content of the suspension [13]. By using S-EPD, the particle size distribution of the deposited green body can be changed gradually. If, for example, the gravitational force acting on particles of a critical diameter is compensated by the force due to the applied electrical field, these particles

remain motionless [14], whereas smaller particles will be deposited.

S-EPD can be used for producing coatings or microstructures with a very smooth surface by depositing only the nanosized fraction of a conventional powder with broad or non-monomodal particle size distribution; thus, no preceding classifyca-tion is necessary and no expensive nanosized powders have to be used. After the micro structured moulds have been filled with the nanofraction of a powder a strong support of bigger particles is deposited subse-quently by increasing the applied electrical field strength. The process can be used for shaping com-ponents with continuously gradients in density or particle size across the layer thickness, as well as for compositional gradients.

For a better process control and the manufacturing of components with predictable gradients, kinetics of S-EPD have to be known, which is the main goal of this work. First of all a mathematical description of the kinetics of EPD is necessary.

In 1940 Hamaker proposed an equation to calculate the yield in electrophoretic deposition [4] which can be written as follows

$$\frac{dm}{dt} = a \cdot c \cdot \frac{\varepsilon_0 \cdot \varepsilon \cdot \zeta \cdot f}{\eta} \cdot E \cdot A \qquad (1)$$

In this equation a describes the quantity of particles reaching the electrode (in most cases a is assumed to be close to 1), $\varepsilon_0 \cdot \varepsilon_r$ is the permittivity, ζ is the zeta-potential f is the related correction factor f, η is the viscosity of the suspension medium, E is the applied electric field strength, A is the surface area of the electrodes A and t is the deposition time.

In 1962 Avgustinik developed a similar equation to calculate deposition yield for a cylindrical geometry [15] and upgraded it in 1967 by factoring the decreasing particle concentration within the suspension due to electrophoretic deposition into this equal [16].

Based on Hamaker's equation, Biesheuvel developed a correction term incorporating the volumetric cast concentration [10].

It turned out that Hamaker's description of the deposited yield in case of EPD was only valid for short deposition times [17]. The decrease in particle concentration of the suspension due to electrophoretic deposition was neglected. Another approach to include the decrease was set up by Zhang in 1994 [18]

$$m = m_0 \cdot (1 - e^{-k \cdot t}) \qquad (2)$$

where the parameter k is represented by

$$k = a \cdot \frac{A}{V} \cdot \frac{\varepsilon_0 \cdot \varepsilon_r \cdot \zeta \cdot f}{\eta} \cdot E \qquad (3)$$

The deposited mass m is as function of the initial mass m_0 in the electrophoretic cell, the parameter k and the deposition time t.

For a correct description of the interaction of sedimentation and electrical field in electrophoretic deposition the knowledge of the sedimentation behaviour of the dispersed particles is essential. The settling speed of an isolated particle in a diluted dispersion is described by the equation of Stokes [19]

$$v_S = \frac{2 \cdot g \cdot (\rho_P - \rho_F) \cdot r^2}{9 \cdot \eta} \qquad (4)$$

The settling velocity depends on the acceleration of gravity g, the density of powder ρ_P, and medium ρ_F, the viscosity of the medium η and the particle radius r.

In 1951 Kynch published a review of the theory of sedimentation, in which he specified an equation obtained by Einstein and Smoluchowski, where the density of the particles is small and their mutual distance is much bigger than their size [19]. Another review of sedimentation was given by Bürger in 2001. He presented amongst others the equation of Richardson and Zaki, where the batch-settling velocity of particles in real suspensions of small particles is described [20].

When sedimentation and electrical field interact during EPD, the critical radius of particles remaining motionless can be calculated in a simple form by using the equations of Hamaker and Stokes. The critical radius r_{crit} can be varied by adjusting the electrical field strength and can be calculated as follows

$$r_{crit} = \sqrt{\frac{9 \cdot \varepsilon_0 \cdot \varepsilon_r \cdot \zeta \cdot f}{2 \cdot g \cdot (\rho_P - \rho_F)} \cdot E} \qquad (5)$$

The goal of this work is to investigate experimentally the kinetics of S-EPD as a systematic interaction of sedimentation and electrophoretic deposition to allow the production of graded coatings and microstructures.

2. Experimental

2.1. Materials and suspensions

The following powder was used: ZrO_2 Unitec – 10 μm with 5 wt.% Y_2O_3 (Unitec Ceramics, UK), with a d_{50} of 1.42 μm. Aqueous slurries of the powder were prepared with a solids content of 10 wt.%. A commercial polyelectrolyte based on carbonic acid (Dolapix CE 64, Zschimmer & Schwarz, Germany), tetramethylammonium hydroxide (TMAH, Sigma-Aldrich, Germany) and hydrochloric acid (HCl, Merck, Germany) to adjust the pH were used to stabilize the suspensions. The slurries were prepared using a magnetic stirrer to suspend the powders in deionised water containing the dispersant. Dispersion of the particles and homogeni-sation of the suspension were carried out for five minutes in an ultrasonic bath and for one hour at 200 min^{-1} in a planetary ball mill using zirconia balls and a grinding beaker (PM 400/2, Retsch, Germany). After-wards, the pH value of the slurries was adjusted with TMAH or HCl.

2.2. Characterisation of the sedimentation

The particle size distribution was measured using laser light diffraction particle size analyzer (LS 230, Coulter-Beckman, Germany). This allows measuring particle sizes from 0.04 μm to 2000 μm by combiniation of laser light scattering for measuring the coarse particles and polarization intensity differential scattering (PIDS) for measuring the finer particles. To determine sedimentation kinetics, sedimentation of particles was investigated in cylinders and by in-situ measurements of the particle size distribution with the LS 230. This was executed by permitting particles to settle down in the sample chamber and continuously measuring the particle size distribution each 15 minutes with a measurement duration of 60 s.

2.3. Electrophoretic deposition

For conventional EPD a glass beaker was used as EPD-cell, in case of S-EPD a beaker of acrylic glass was used. The working electrode was made of brass, the counter electrode of stainless steel. The distance between the electrodes was 13 mm in case of EPD and 15 mm in case of S-EPD. For conventional EPD the electrodes were positioned vertically and for S-EPD horizontally. The applied electrical field strength varied between 30 V/m and 230 V/m for deposition times between 5 minutes and 60 minutes. The electrodes were weighed before and after electrophoretic deposition to determine the deposited mass.

3. Results and discussion

3.1. Sedimentation kinetics

To determine sedimentation kinetics, 10 ml of each slurry were filled in a cylinder. Slurries with a pH value of 10 were more stable than at pH 8. With addition of dispersant an apparent stabilising impact was revealed. The influence of the pH on sedimentation became less important with increasing dispersant concentration. Figure 1 shows the kinetics of sedimentation of Unitec – 10 μm powder suspensions with a solid concentration of 10 wt.% with dispersant contents of 0 wt.% and 0.2 wt.%, respectively at pH 8 and pH 10.

For the suspensions with 0.2 wt.% dispersant, a linear correlation of moving distance x of the particles and sedimentation time t was observed. Thus, the assumption of an isolated particle sedimentation is proved to be correct. Therefore, the description of sedimentation kinetics after Stokes (equation 4) is valid for the mathematical description of S-EPD for stabilised suspensions. For the suspension without dispersant at pH 8, an exponential correlation between x and t was found, indicating a hindered settling and a cooperative motion of the particles with 10 vol.% solids content. The suspension with 0 wt.% dispersant at pH 10 shows an approximately linear correlation, still the divergence from linear correlation shows little cooperative motion.

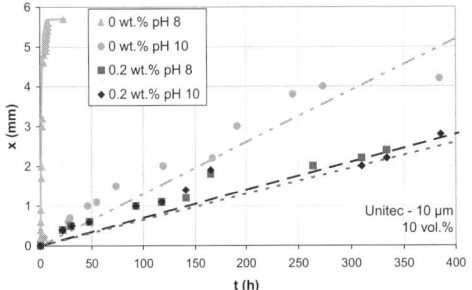

Fig. 1. Kinetics of sedimentation as function of time.

To allow determination of the sedimentation kinetics of the powders taking into account their real particle size distribution, in-situ measurements were carried out. In figure 2 the sedimentation kinetics of a zirconia suspension stabilised with 0.2 wt.% dispersant at a pH value of 10 is shown.

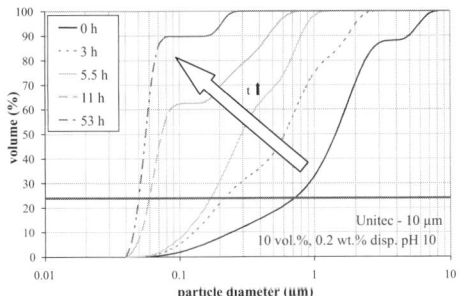

Fig. 2. In-situ measurement of the particle size distribution of a zirconia suspension with 0.2 wt.% dispersant at pH 10.

At the beginning of the measurement, the particle size distribution shows a bimodal size distribution with a d_{50} of 1.42 μm and a size variation from 0.04 μm to 9.82 μm. With increasing sedimentation time, the fraction of the coarser particles disappeared from the measurement field and the volume fraction of finer particles increases. After 53 hours of sedimentation, the size distribution is still bimodal with a d_{50} decreased to 0.06 μm and a size variation from 0.04 μm to 0.31 μm.

3.2. Electrophoretic deposition

For the mathematical description of S-EPD, kinetics of the electrophoretic deposition has to be analysed. For this, a ZrO_2 Unitec – 10 μm suspension with a dispersant concentration of 0.2 wt.% and a pH value of 10 was used for EPD. An electric field of 150 V/m was applied and deposition times were varied between 5 and 60 minutes. In figure 3 the deposited mass is shown as a function of time.

Up to 20 minutes a linear increase in deposition mass was observed resulting in a constant deposition rate of 23 mg/min*cm². With higher deposition times, the slope in m(t) decreases and so does the deposition rate. The Hamaker fit in figure 4 was achieved by solving

equation 1 with c = c(t) according to the approach of Zhang (equation 2). Using a = 1, ζ = 50 mV, $\varepsilon_0 \cdot \varepsilon_r$ = $7.08 \cdot 10^{-10}$ As/Vm, A/V = 77 1/m, η = 0.001 Pas and E = 150 V/m a value of k = 0.00042 1/s is obtained. The initial mass between the electrodes is calculated as 786 mg from the weighted sample. As can be seen in figure 3, the result does not fit the experimental data too well. With m = 750 mg and k = 0.0007 1/s, a very good fit is obtained based on Hamakers equation. The difference of m_0 can be explained by sedimentation, k differs by a factor 1.7. Possible explanations for k differing by a factor of 1,7 are the factor a, which was arbitrarily set to 1, the factor f 1.5 in the zeta potential and the fact, that an average applied field strength is used for the calculation, which may well differ from the true local field strength acting on each particle.

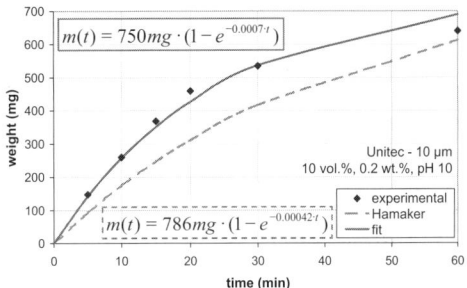

Fig. 3. Deposited mass as function of time for EPD of ZrO_2 Unitec – 10 μm.

3.3. Systematic interaction of sedimentation and electrical field in electrophoretic deposition

The results shown above indicate that equation 5 can be used for the calculation of the critical radius in S-EPD at least for stabilised suspensions with a solid content of 10 wt.%, because the kinetics of sedimentation and EPD can be described by the equations of Hamaker (equation 1) and Stokes (equation 4).

Similar suspensions to in 3.2 were used for S-EPD. An electric field of 130 V/m was applied and deposition times were varied between 5 and 30 minutes. In figure 4 the deposited mass as a function of time is shown for EPD and S-EPD.

Similar to EPD, the yield in S-EPD can be fitted by the Hamaker equation. In S-EPD only particles smaller than the critical radius are deposited onto the electrode. For the applied electrical field strength of 130 V/m, the critical radius was calculated to be 380 nm. Figure 2 shows that 24 % of the particles of the Unitec – 10 μm powder are smaller than this radius. Like in chapter 3.2., the factor k can be calculated as 0.00036 1/s and the initial mass m0 = 566 mg. Figure 4 demonstrates that with a factor k = 0.00053 1/s and an initial mass m0 = 200 mg a reasonably good fit can be achieved. Due to different EPD-cells with different surface areas and different sample volumes the factor k is different for EPD and S-EPD. The ratio of k1 determined for S-EPD and k2 for EPD is 1.3, which is exactly the ratio of the surface area of the electrode and the volume of the electrophoretic cell A1*V2/A2*V1.

The initial mass of the best fit to the experimental data is smaller than the calculated initial mass. A possible explanation is the loss of over potential at the electrodes, which inhibits EPD with voltages smaller than 1V. Thus the effective electric field strength acting on each particle is much smaller in S-EPD resulting in a significantly reduced critical radius of the deposited particles.

190

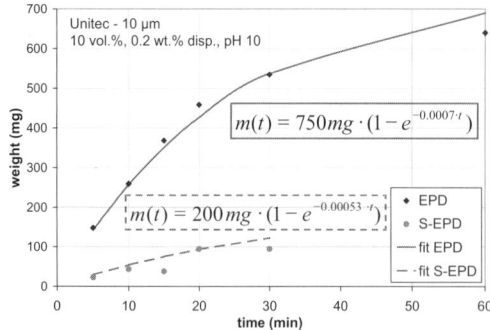

Fig. 4. Deposited mass as function of time for EPD and S-EPD.

4. Conclusions

It was shown that the experimentally determined kinetics of sedimentation can be described by Stokes equation. Also, the kinetics of electrophoretic deposition and S-EPD can both be very well described by Hamakers equation.

Acknowledgements

Our sincere thanks go to all colleagues for their support and encouragement for this work.

References

[1]   Boccaccini AR and Zhitomirsky I, "Application of electrophoretic and electrolytic deposition techniques in ceramics processing," *Curr. Opin. Solid. St. M.*, vol. 6, pp. 251-260, 2002.

[2]   Harbach F and Nienburg H, "Homogeneous Functional Ceramic Components through Electrophoretic Deposition from Stable Colloidal Suspensions - I. Basic Concepts and Applications to Zirconia," *J. Eur. Ceram. Soc.*, vol. 18, pp. 675-683, 1998.

[3]   Reuss FF, "Notice sur un nouvel effet de l'électricité galvanique," *Mémoires de la Societé Impériale des Naturalistes de Moscou*, pp. 327-337, 1808.

[4]   Hamaker HC, "Formation of deposit by electrophoresis," *Trans. Faraday Soc.*, vol. 36, pp. 279-287, 1940.

[5]   V. d. Biest OO and Vandeperre LJ, "Electrophoretic Deposition of Materials," *Annu. Rev. Mater. Sci.*, vol. 29, pp. 327-52, 1999.

[6]   Tabellion J and Clasen R, "Electrophoretic deposition from aqueous suspensions for near-shape manufacturing of advanced ceramics and glasses-applications," *J. Mater. Sci.*, vol. 39, pp. 803-811, 2004.

[7]   Sarkar P, Datta S, and Nicholson PS, "Functionally graded ceramic/ceramic and metal/ceramic composites by electrophoretic deposition," *Compos. Part B - Eng.*, vol. 28B, pp. 49-56, 1997.

[8]   Nagarajan N and Nicholson PS, "Nickel-Alumina Functionally Graded Materials by Electrophoretic Deposition," *J. Am. Cer. Soc.*, vol. 87, pp. 2053-2057, 2004.

[9]   Oetzel C, Clasen R, and Tabellion J, "Electric Field Assisted Processing of Ceramics," *cfi-Ceram. Forum Int.*, vol. 81, pp. 35-41, 2004.

[10]   Biesheuvel PM and Verweij H, "Theory of Cast Formation in Electrophoretic Deposition," *J. Am. Ceram. Soc.*, vol. 82, pp. 1451-55, 1999.

[11]   Simovic K, Miskovic-Stankovic VB, Kicevic D, and Jovanic P, "Electrophoretic deposition of thin alumina films from water suspensions," *Colloid. Surface. A*, vol. 209, pp. 47-55, 2002.

[12]   Sarkar P and Nicholson PS, "Electrophoretic Deposition (EPD): Mechanism, Kinetics and Application to Ceramics," *J. Am. Ceram. Soc.*, vol. 79, pp. 1987-2002, 1996.

[13]   Batchelor GK, "Sedimentation in a dilute dispersion of spheres," *J. Fluid Mech.*, vol. 52, pp. 245-268, 1972.

[14]   Dushkin C, Miwa T, and Nagayama K, "Gravity effect on the field deposition of two-dimensional particle arrays," *Chem. Phys. Lett.*, vol. 285, pp. 259-265, 1998.

[15]   A. I. Avgustinik, V. S. Vigdergauz, and G. I. Zhuravlev, "Electrophoretic Deposition of Ceramic Masses from Suspensions and Calculation of Deposit Yields," *Zh. Prikl. Khim.*, vol. 35, pp. 2090-2093, 1962.

[16]   Avgustinik AI, Zhuravlev GI, and Vigdergauz VS, "Calculation of the yield of deposit in electrophoretic deposition," *J. Appl. Chem. USSR*, vol. 35, pp. 379-383, 1966.

[17]   Tabellion J, "Herstellung von Kieselgläsern mittels elektrophoretischer Abscheidung und Sinterung," Dr.Ing. Thesis, Universität des Saarlandes, 2004, pp. 213.

[18]   Zhang Z, Huang Y, and Jiang Z, "Electrophoretic Deposition Forming of SiC-TZP Composites in a Nonaqueous Sol Media," *J. Am. Ceram. Soc.*, vol. 77, pp. 1946-49, 1994.

[19]   Kynch GJ, "A Theory of Sedimentation," *Trans. Faraday Soc.*, vol. 48, pp. 166-176, 1952.

[20]   Bürger R and Wendland WL, "Sedimentation and suspension flows: Historical perspective and some development," *J. Eng. Math.*, vol. 41, pp. 101-116, 2001.

Multi-Material Micro Manufacture
W. Menz, S. Dimov and B. Fillon (Eds.)

191

Particle size dependent viscosity of polymer-silica-composites

T. Hanemann[a,b], R. Heldele[a], J. Hausselt[a,b]

[a] Forschungszentrum Karlsruhe, Institut f. Materialforschung III, D-76021 Karlsruhe, Germany
[b] Albert-Ludwigs-Universität Freiburg, Institut f. Mikrosystemtechnik (IMTEK), D-79110 Freiburg, Germany

Abstract

In this paper the influence of micro- and nanosized particles on the flow behaviour of unsaturated polyester resin-silica-composites will be compared. Commercially available micro-sized quartz filler and different hydrophilic or hydrophobic nanosized Aerosils were investigated with respect to the change of the viscosity as well as the flow activation energy with filler load. Apart from particle size and specific surface area the polarity of the filler's surface has a strong impact on the resulting flow behaviour and the accessible maximum filler load. The dependence of the relative viscosity upon the filler load was described using different empirical approaches as established in thermoplastic or wax based feedstock systems containing porcelain, alumina or PZT.

Keywords: Polymer-ceramic-composites, rheological properties, nanosized particles

1. Introduction

With respect to the fabrication of ceramic micro-components via ceramic injection molding the whole process chain starting from composite or feedstock preparation, replication, debinding and finally sintering has to be evaluated very carefully. This is particularly important during the first process step, the feedstock preparation has a strong impact on the final part quality. Many parameters like filler amount, filler specific surface area, particle size distribution, surface chemistry and polarity as well as binder composition and additives such as surfactants or liquifiers affect the feedstock homogeneity and flow behaviour [1-3].

The change of the composite's relative viscosity with filler load (volume content Φ) was described firstly by Einstein (1) for diluted solutions or dispersions [4] introducing a coefficient k_E, which is 2.5 for hard particle spheres. The relative viscosity η_{rel} is defined as the quotient of the apparent viscosity of the composite (η_{comp}) and the pure binder (η_{binder}). A further modification of the Einstein approach considers geometric interactions between the particles in a power series approach (2) introducing additional parameters such as k_H and others. A large number of different empirical models has been developed for a correlation of the composite's relative viscosity with the ceramic load for highly filled composites. With respect to replication the estimation of the critical filler load Φ_{max} in organic binders is of particular interest [1]. Quite often the empirical approaches developed by Mooney (3), Quemada (4) or by Zhang & Evans (5) [5] among others are used. The latter one introduces a free parameter C which allows a more flexible approximation to experimental data.

In a liquid, dispersed particles are attracted by van der Waals forces and repelled mainly by electrostatic forces and steric interactions. Following the DLVO-theory the particle interaction energy between two particles can be described via the superposition of the attractive (E_{attr}) and repulsive (E_{repul}) energy. In particular nanosized particles tend extremely to agglomerate due to surface energy minimization.

$$\eta_{rel} = \left(\frac{\eta_{comp}}{\eta_{binder}}\right) = 1 + k_E \Phi = 1 + 2.5\Phi \tag{1}$$

$$\eta_{rel} = 1 + k_E \Phi + k_H \Phi^2 + \dots \tag{2}$$

$$\eta_{rel} = \exp\left(\frac{2.5\Phi\Phi_{max}}{\Phi_{max} - \Phi}\right) \tag{3}$$

$$\eta_{rel} = \left(1 - \frac{\Phi}{\Phi_{max}}\right)^{-2} \tag{4}$$

$$\eta_{rel} = \left(\frac{\Phi_{max} - C\Phi}{\Phi_{max} - \Phi}\right)^2 \tag{5}$$

$$\ln\frac{\eta_1(T_1)}{\eta_2(T_2)} = \frac{\Delta E_a}{R}\left(\frac{1}{T_1} - \frac{1}{T_2}\right) \tag{6}$$

The addition of fillers to a liquid changes the temperature influence on the composite's viscosity, which can be described with an Arrhenius-type approach [1, 6] (6), η_1 and η_2 are the apparent viscosities at the two different temperatures T_1 and T_2, R the gas constant and ΔE_a is the flow activation energy, which depends mainly on the composition of the investigated system.

Low viscous polymer-based reactive resins like polymethylmethacrylate solved in methylmethacrylate or unsaturated polyester solved in styrene can be used as model systems for high viscous polymer melts [5,7] and for a rapid prototyping of microstructured parts made of plastic, ceramic and metals [8]. The apparent viscosity of the pure reactive resins is below 5 Pa s under ambient conditions and enables a rapid composite processing using simple laboratory equipment. In this work the influence of the particle size and the specific surface of different micro and nanosized silica on the composite flow behaviour will be described.

2. Experimental

A commercially available low viscous mixture of an unsaturated polyester resin with a polymer content around 65 wt% and styrene as reactive thinner (Roth GmbH, Germany) was used as polymer binder. The average molecular weight of the polymer was measured

Table 1
Particle properties of investigated silica fillers

Material	Specific surface area (m²/g)	Primary particle size (nm)	Average particle size (μm)
Dorsilit 405	1.5[a]	-	6[a]
Aerosil 90	90[a]	20	-
Aerosil 200	189 [9]	12	-
Aerosil R7200	141	12	15
Aerosil R8200	142	12	15

[a] vendors data.

to be approximately 1700 g/mole [7]. Different micro- (Dorfner GmbH, Germany) and nanosized silica (Aerosil, Degussa AG, Germany) fillers were used for composite preparation; Table 1 lists all relevant filler details. Aerosil 90 and Aerosil 200 possess a hydrophilic surface, the R7200 and R8200 types are surface modified and therefore non-polar. All mixtures were processed using a laboratory dissolver stirrer at 1000 rpm under ambient conditions. A cone and plate rheometer (CVO50, Bohlin) was used for the rheological composite characterization. Due to experimental restrictions of the cone and plate rheometer composites showing large viscosities beyond 200 Pa s could not be investigated (sticking and gap emptying during measurement). All viscosity measurements were made at 20°C, 40°C and 60°C in the shear rate range between 1 and 100 1/s. The experimental uncertainty of the obtained data is in the range of ± 5%.

3. Results and Discussion

3.1. Filler dependent composite viscosity

The apparent viscosities of the pure unsaturated polyester resin as well as of the quartz filled composite with a load around 39 vol% at three different temperatures are shown in Figure 1. In general the pure resin follows a Newtonian flow behaviour at 20°C, at larger temperatures in the shear rate range between 2 and 20 1/s a pronounced flow anomaly can be observed, which is may be due to thermally activated relaxation processes like polymer chain untangling in the unsaturated polyester resin. The addition of the quartz filler introduces a slight pseudoplastic flow behaviour. In the following only the viscosities at a shear rate of 100 1/s will be compared. As expected two effects can be observed (see Fig. 1): The temperature increase yields a pronounced viscosity reduction in all systems, the filler addition causes a significant viscosity increase. Due to the large surface area only small amounts of the nanosized silica can be dispersed in the resin. The addition of only 0.5 vol% of the hydrophobic Aerosil R8200 polish the flow anomalies of the pure resin resulting in a Newtonian flow (see Fig. 2). The maximum measurable filler load was reached at 8.3 vol%, a pronounced pseudoplastic flow with a significant viscosity increase could be observed.

For a better comparison of the silica particle property influence on the composite's flow behaviour the change of the relative viscosity with the micro- and nanosized filler load was collected (see Fig. 3). The addition of the hydrophilic Aerosils causes a significant relative viscosity increase at small filler loads below 5 vol%. The hydrophobisation of the surface enables larger filler loads up to 8.3 vol% by a better compatibility

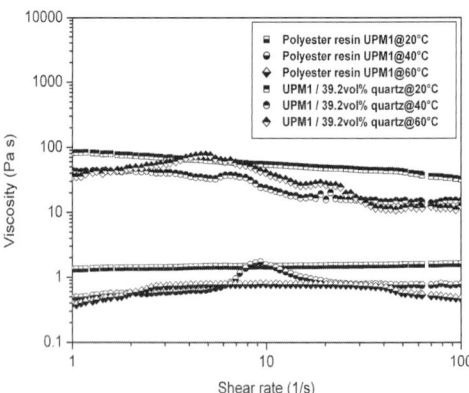

Fig. 1. Viscosities of the pure and the composite containing 39.2 vol% quartz filler.

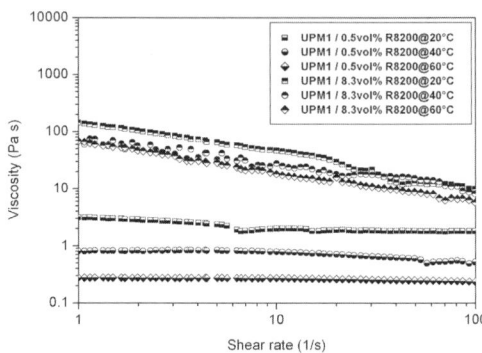

Fig. 2. Viscosity of the UPM1-R8200-composites at different temperatures and filler amounts.

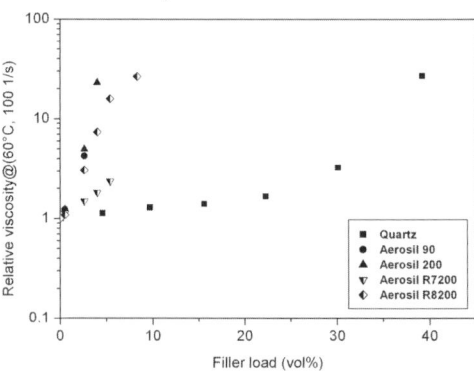

Fig. 3. Relative viscosity change of all investigated composites with filler load.

of the silica surface and the non-polar polymer resin. Due to the large Aerosil surface area the relative viscosity of a composite containing 8.3 vol% nanosized R8200 is quite similar to a system containing 39.2 vol% microsized quartz. The impact of the particle size or may be more important the large specific surface area (R8200: 142 m²/g, quartz: 1.5 m²/g) is obvious preventing large nanosized filler loads in polymer based binder systems due to large binder amount fixation.

3.2. Estimation of the flow activation energy

The addition of ceramic fillers should affect the influence of the temperature on the viscosity. For small temperature differences the use of an Arrhenius-type

Table 2
Flow activation energy and related fit stability index of investigated UPM1-quartz-composites

Resin UPM1, filled with	Flow activation energy (kJ/mole)	Stability index R^2
Pure resin	8.4	0.987
4.6 vol% quartz	9.1	0.981
9.7 vol% quartz	10.1	0.980
15.6 vol% quartz	13.9	0.999
22.3 vol% quartz	11.9	0.999
30.1 vol% quartz	10.0	0.989
39.2 vol% quartz	6.4	0.902

Table 3
Flow activation energy and related stability index of investigated UPM1-Aerosil R8200-composites

Resin UPM1 filled with	Flow activation energy (kJ/mole)	Stability index R^2
Pure resin	13.8	0.996
0.5 vol% R8200	12.4	0.989
2.6 vol% R8200	8.9	0.996
4.0 vol% R8200	7.5	0.999
5.4 vol% R8200	6.2	0.989
8.3 vol% R8200	3.5	0.955

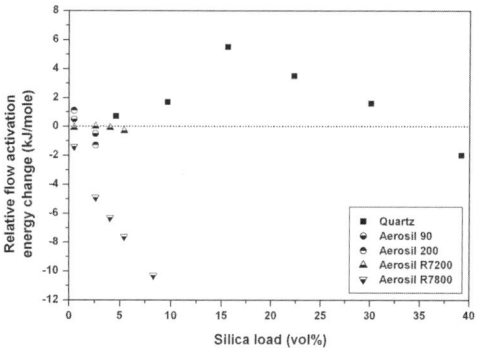

Fig. 4 Relative flow activation energy change with filler load of all investigated composites.

approach is reasonable. Table 2 lists the calculated flow activation energies ΔE_a and the fit stability index R^2 for all investigated UPM1-quartz composites. The filler addition raises the resulting flow activation energy slightly, the strong drop at the largest load is may be due to experimental uncertainties which can be deduced seen from the poor R^2-value. The influence of the hydrophobic Aerosil R8200 on the composite flow activation energy (see Table 3) is, however, quite different. With increasing nanosized filler amount a significant reduction of the flow activation energy can be observed. The numerical value of the flow activation energy of the pure resin is larger than the one listed in Table 2. This can be explained by the slow polymerization of the polymer based reactive resins and the resulting polymer chain growth. Therefore the related composites were prepared freshly using the listed pure resin to avoid experimental uncertainties. For a better comparison the relative change of the flow activation energy with filler load referred to the pure resin is summarized (see Fig. 4). The addition of the microsized quartz filler result in an increase up to a load of 15 vol%, a further addition reduces the change of the flow activation energy close to the initial value of the pure resin.

The hydrophilic nanosized Aerosils 90 and 200 can only be added to the resin up to a content of 2.6 vol%, a unique influence on the flow activation energy cannot be observed. The hydrophobic Aerosils R7200 and R8200 show a total different impact on the temperature influence on the flow behaviour. The surface of R7200 is modified with an organosilane which carries a methacrylate functionality additionally, the surface of R8200 is only covered with an organosilane moiety. Small R7200 amounts do not affect the composite's ΔE_a, with increasing load a slight decrease can be observed. The addition of R8200 results in a significant

drop of ΔE_a with increasing load. The stronger impact of R8200 in comparison to R7200 can be explained by a better compatibility of the organosilane surface to the polyester based resin.

In literature only a few papers dealing with the estimation of the flow activation energies of thermoplastic or wax based feedstock systems are published, systematic investigations on the dependence of ΔE_a with increasing load and particle properties have not been found. Trunec and coauthors investigated feedstock systems containing ethylene-vinylacetate-wax based and ZrO_2. At a filler load of 49 vol% a flow activation energy of 24 kJ/mol was estimated, at a load of 52.5 vol% the values increased up to 28.7 kJ/mole [10]. In case of reactive resin based binders Dufaud investigated 1,6-hexanediol-diacrylate (HDDA), filled with PZT; for the pure HDDA a ΔE_a of 32 kJ/mole was found, the values for the filled composites varied between 35 and 41 kJ/mole for loads between 40 and 83 wt%, a clear dependence could not be found [11].

In general a reduction of the flow activation energy corresponds with a reduced influence of the temperature on the viscosity e.g. by a pronounced interaction of the ceramic particles with the polymer chains e.g. due to an improved wetting and corresponding with a reduced free volume in the composite. This effect was described by Attarian and coworkers in a aluminum silicate-polyethylene wax composite doped with different amount of amphiphilic EVA-copolymers. The flow activation energy for a composite with 59 vol% could be lowered from 14.6 kJ/mole down to 6.8 kJ/mole by increasing the EVA-amount from 3 up to 8 wt% [12].

3.3. Empirical description of the relative viscosity

The experimentally obtained change of the relative viscosity with filler load can be described empirically using different established models which have been used e.g. in case of feedstocks for ceramic injection molding [1]. Quite recently the applicability of the models to composites containing reactive resins and alumina as filler has been published [5]. Composites containing the unsaturated polyester UPM1 as binder and the microsized quartz filler can be described in a similar manner (see Fig. 5). Using the Mooney or the more flexible Zhang-Evans approach the change of the relative viscosity with filler load as well as the critical filler load (Mooney: 0.59, Zhang-Evans: 0.45) can be estimated. The numerical value for the critical filler load obtained by the Mooney model seems to be more realistic because it is close to the value (0.64) considering a

194

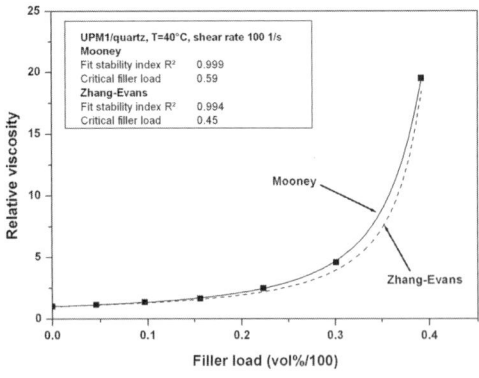

Fig. 5. Modeling the relative viscosity of the investigated UPM1-quartz-composites using the Mooney and Zhang-Evans approaches.

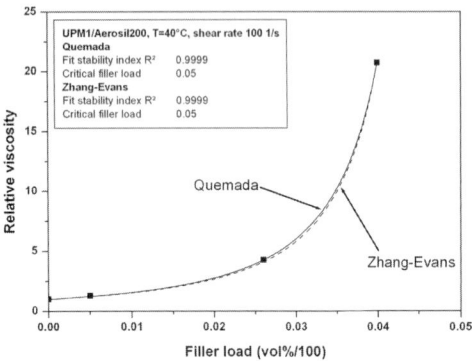

Fig. 6. Modeling the relative viscosity of the investigated UPM1-Aerosil200-composites using the Quemada and Zhang-Evans approaches.

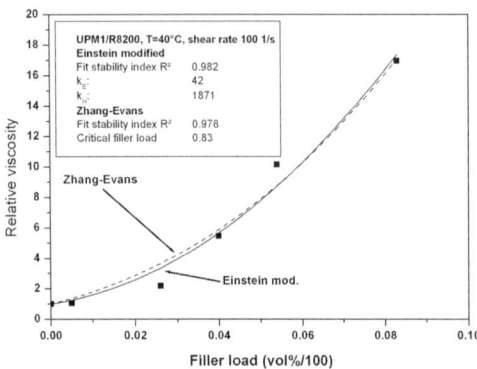

Fig. 7. Modeling the relative viscosity of the investigated UPM1-R8200-composites using the modified Einstein and Zhang-Evans approaches.

dens random package of the particles [1]. The change of the relative viscosity with nanosized filler load can be described with the established models as well (see Fig. 6 and Fig. 7). In case of composites containing the hydrophilic Aerosil 200 the Quemada and the Zhang-Evans approach can be used with reasonable quality following the stability index close to unity. An estimated critical filler load of 5 vol% appears to be realistic (see Fig. 6). The addition of the hydrophobic Aerosil R8200 to the polyester resin and the resulting impact on composite relative viscosity can be described using the modified Einstein approach and by the Zhang-Evans model. Despite a reasonable fit stability index around 0.98 the calculated critical filler load of 0.83 seems to be unrealistic and demonstrates the practical limit of the applied empirical models.

3. Results and Discussion

The strong influence of the particle size and the specific surface area on the flow behaviour of polymer-based composites has been demonstrated using different micro- and nanosized silica fillers. The change of the flow activation energy with filler load depends on the filler's surface polarity. The dependence of the relative viscosity upon the filler load can be described using different established empirical models.

Acknowledgements

The authors thank for the financial support by the European Commission within the 4M-Network of Excellence.

References

[1] German RG. Powder Injection Molding. Princeton: Metal Powder Industries Federation, 1990.
[2] Rath S, Merz L, Plewas K, Holzer P, Gietzelt T, and Hausselt J. Isolated metal and ceramic micro parts in the sub-millimeter range made by PIM. Adv. Eng. Mater. 7 (2005) 619-622.
[3] Merz, L, Rath S, Piotter V, Ruprecht, R. Hausselt J. Powder injection molding of metallic and ceramic microparts. Microsystem Technologies 10 (2004) 202-204.
[4] Einstein A. Eine neue Bestimmung der Molekül-dimension. Ann. Phys. 19 (1906) 289-306.
[5] Hanemann T. Influence of dispersants on the flow behaviour of unsaturated polyester-alumina-composites. Comp. A: Applied Science and Manufacturing 37(5) (2006) 735-741.
[6] Menges G. Werkstoffkunde Kunststoffe. Hanser Publisher, Munich 2002; p93.
[7] Hanemann T. Viscosity change of unsaturated polyester-alumina-composites using polyethylene glycol alkyl ether based dispersants. Comp. A: Applied Science and Manufacturing, 2006, in press.
[8] Hanemann T, Honnef K, Hausselt J. Rapid prototyping of microstructured ceramic and metal parts using reaction molding techniques. Proc. International Conference on Multi-Material-Micro-Manufacture (4M), 29.06.-01.07.2005, Karlsruhe, FRG.
[9] Leblanc J L. Rubber-filler interactions and rheological properties in filled compounds, Prog. Polym. Sci. 27 (2002) 627-687.
[10] Trunec M, Dobsak P, Cihlar J. Effect of powder treatment on injection moulded zirconia ceramics. J. Europ. Ceram. Soc. 20 (2000) 859-866.
[11] Dufaud O, Marchal P, Corbel S. Rheological properties of PZT suspensions for stereolitho-graphy. J. Europ. Ceram. Soc. 22 (2002) 2081-2092.
[12] Attarian M, Taheri-Nassaj E, Davami, P. Effect of ethylene-vinyl acetate copolymer on the rheological behaviour of alumino-silicate/polyethylene wax suspensions. Ceramics Intern. 28(5) 2002 507-514.

Micro EDM parameters optimisation

S. Bigot[a], J. Valentinčič[b,c], O. Blatnik [b], M. Junkar [b]

[a] Manufacturing Engineering Centre, Cardiff University, Cardiff, UK, e-mail: bigots@cf.ac.uk
[b] Faculty of Mechanical Engineering, University of Ljubljana, Slovenia, e-mail: lat@fs.uni-lj.si
[c] Corresponding author

Abstract

Electrical discharge machining (EDM) is an important process in the field of micro machining. However, a number of issues remain to be solved in order to successfully implement it in an industrial environment. One of these issues is the processing time. This paper investigates the optimisation of machining parameters for rough and fine machining in micro EDM. In one case, the parameters are selected to achieve the highest material removal rate (MRR). In the other case, the best surface roughness is targeted. Some of the main difficulties linked with micro EDM are caused by the high wear occurring on the electrode. The study focuses on a specific combination of electrode and workpiece material and proposes a typical method for micro EDM process optimisation.

Keywords: micro EDM, micromachining, roughing, finishing, micro holes and cavities

1. Introduction

Over the next four years, the Microsystems market, including Micro-Structure Technologies (MST) and Micro-Electro-Mechanical Systems (MEMS), is predicted to grow at a rate of 16% per year from $12 billion in 2004 to $25 billion in 2009 across a spectrum of 26 MEMS/MST products [1]. Conventional processes are increasingly being improved for use in micro-machining. The most common processes are micro milling, laser machining and more specifically micro EDM, which is being applied in many micro applications [2].

In EDM, the machining of conductive materials is performed by a sequence of electrical discharges occurring in an electrically insulated gap between a tool electrode and a workpiece. During the discharge pluses, a high temperature plasma channel is formed in the gap, causing evaporation and melting of the workpiece. Debris of material are removed by the resulting explosion pressure, enabling the machining of the workpiece [3]. The characteristics of the electrical discharge pulses are linked with a set of machining parameters, which control the energy and frequency of discharges and thus the power in the gap. Consequently, the chosen set of parameters affect the material removal rate (MRR), surface roughness and relative electrode wear rate. In the case of conventional sinking EDM, machining strategies using roughing and finishing paths are well established and a number of studies offer guidelines for machining parameters selection [4].

However, in micro EDM a number of issues remain to be solved. For instance, the processing time is significantly higher. The size of electrode used and the resulting high electrode wear makes conventional die-sinking methods inadequate. This has led to the development of a number of new EDM strategies for micro machining. One common approach is the micro EDM milling, where a cavity is produced using a single small electrode on a series of machining path [5].

This paper investigates the influence of various combination of machining parameters in micro EDM, in an attempt to optimise the MRR and surface roughness when using such approach, taking also into account the relative electrode wear.

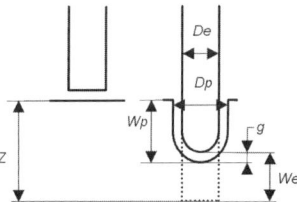

Fig. 1. Erosion process in drilling.

2. Experiment description

2.1. Machining set-up

The workpiece material selected for investigation was Tool steel P20. The electrode used was a tungsten carbide (WC) cylindrical rod with diameter $De=0.170$mm. The machining was performed using synthetic oil as dielectric.

Before each machining, the bottom of the electrode was EDM grounded flat. The rotating electrode was then used to drill a hole of diameter Dp, from the workpiece top surface down to a targeted depth $Z=0.5$mm. Figure 1 illustrates the drilling process affected by the wear, where Wp and We are the eroded lengths from the workpiece and electrode respectively and g is the bottom spark gap.

After all machining, the workpiece was cut along the middle of the holes using a wire EDM machine, allowing access to the holes profiles.

2.2. Design of experiments

In order to assess the effect of each machining parameter on the process, the Taguchi approach was used [6]. This method is a type of statistical technique called Design Of Experiments (DOE) that makes it possible to analyse the effect of more than one factor at the same time while reducing the number of experiment. Thus, using the Taguchi approach, the design of experiments and analysis of results can be done with less effort and expenses. However, since the method considerably reduces the number of experiments, quality loss of results could appear.

2.2.1. Output functions

In the proposed experiment, the main functions are material removal rate (MRR), electrode wear and surface roughness. MRR was calculated as a quotient of the volume Vp removed from the workpiece and the machining time. A rotating electrode was used to produce the holes, therefore their volumes can be estimated from their profiles. Digital images of the profiles were taken after machining and, based on each line of pixels composing the holes, the volumes were estimated using equation 1.

$$Vp = \sum_{i=1}^{N} \frac{\pi \cdot d_i^2}{4} \cdot K^3 \tag{1}$$

where d_i is the diameter of the hole at the i-th line of pixels, N is the number of lines of pixels composing the hole and K is the size of a pixel in mm.

The reduction of the electrode length due to the wear (We) was measured on the machine. This was achieved by assessing the difference, before and after machining, of contact point in the z axis between electrode tip and workpiece top surface. The wear of the tool electrode is usually given as the relative electrode wear (ϑ), which is a quotient of the volume removed from the electrode Ve and the volume removed from the workpiece Vp. In this experiment, in order to facilitate computation, the corner wear was considered negligible. Therefore the geometry of the worn electrode could be assumed cylindrical and Ve was easily assessed as a function of We.

The roughness was measured in Ra on the holes profiles using a white light interferometer microscope. And, because the surfaces are not flat, an algorithm supplied with the microscope was used to compensate automatically for curvatures and tilts.

2.2.2 Control factors

One crucial step in the Taguchi method is the identification of the control factors and of their values considered for investigation. These values (or factor levels) should be placed very carefully, since the Taguchi method defines the significant and optimal parameters only within the selected ranges.

In this study, the control factors selected for optimisation were the machining parameters:

- Electrode polarity, which can be either positive or negative;
- Peak current (i_e), which gives the highest electric current that can occur during the discharge (if no capacitors is used);
- Ignition voltage (u_i), which is the voltage generated in order to ignite the discharge;
- Pulse-on time (t_i), which is the duration of the impulse generated by the impulse generator;
- Pulse-off time (t_o), which is the time between two impulses;
- Capacitance C of the capacitors, which is included in the circuit in order to accumulate electricity and to increase the discharge current;
- Two parameters related with the servo system, namely the reference voltage related with the size of the front gap and the gain of servo system that defines the reaction time of the servo system to the conditions in the gap.

Table 1 Parameters Levels

	i_e [A]	u_i [V]	t_i [ms]	t_o [ms]	C [nF]	gap	gain
lower limit	0.8	60	1	2.4	/	50	2
middle value	1.4	80	2.4	13	2.7	65	5
upper limit	1.8	100	5.5	56	19.4	80	9

For each parameter, apart from the polarity, three values were selected for investigation. The range of values was identified empirically, taking into account on one hand the lowest power that can be supplied by the machine generator and on the other hand the highest power that the electrode can take before burning out.

The lowest electrical power the machine is able to supply in the gap is determined by the smallest values of peak current ($i_e = 0.8A$), pulse-on time ($t_i = 1ms$), and capacitance (no additional C), which were therefore used as lower limits. Also, a longer pulse-off time induces lower power in the gap and consequently results in a lower MRR. Thus, the upper limit for this parameter should be relatively high. However, because the micro EDM is already a slow process, this upper limit ($t_o = 56ms$) was selected in order to allow low power in the gap while keeping an acceptable MRR. The ignition voltage plays a role only in the discharge formation phase, i.e. formation of the plasma channel in the gap. It influences the frontal gap, the higher the ignition voltage the greater the front gap. Its lowest value ($u_i=60V$) was selected as lower limit.

The highest electrical power that can be used for machining is determined by the electrode material and diameter. Past a certain limit the electrode behaves as a fuse and burns. In this investigation, the upper limits for peak current, pulse-on time, capacitance and ignition voltage and lower limit for the ignition voltage, which does not cause the burning of the electrode, were found empirically by trial and error.

The ranges of the two parameters of the servo system and ignition voltage were selected empirically (comp $\in$ [50,80] and gain $\in$ [2,9]). These values are machine specific and are not given in physical units.

The selected levels for the machining parameters are given in table 1.

2.2.3. Set of experiments

Based on Taguchi approach, the experiments can be defined using the orthogonal array. For this study, the L_{18} orthogonal array [6] was chosen, where 18 combinations of parameters are proposed for investigation.

Table 2 L_{18} orthogonal array

No. of experiment	polarity	i_e [A]	u_i [V]	ti [µs]	t_o [µs]	C [nF]	gap [-]	gain [-]	MRR [mm³/min]	ϑ(-) [-]	Ra [µm]
1	+	0.8	60	1.0	2	/	50	2	0.0003	1.6162	0.34
2	+	0.8	80	2.4	13	2.7	65	5	0.0006	3.0760	1.14
3	+	0.8	100	5.5	56	19.4	80	9	0.0008	2.0715	0.8
4	+	1.4	60	1.0	13	2.7	80	9	0.0003	3.0782	1.51
5	+	1.4	80	2.4	56	19.4	50	2	0.0010	1.7383	0.72
6	+	1.4	100	5.5	2	/	65	5	0.0007	0.7540	0.62
7	+	1.8	60	2.4	2	19.4	65	9	0.0016	0.7588	0.73
8	+	1.8	80	5.5	13	/	80	2	0.0004	1.9202	0.41
9	+	1.8	100	1.0	56	2.7	50	5	0.0004	3.6376	1.62
10	-	0.8	60	5.5	56	2.7	65	2	0.0011	0.2151	0.34
11	-	0.8	80	1.0	2	19.4	80	5	0.0020	0.2847	0.56
12	-	0.8	100	2.4	13	/	50	9	0.0009	0.1247	0.46
13	-	1.4	60	2.4	56	/	80	5	0.0006	0.1988	0.22
14	-	1.4	80	5.5	2	2.7	50	9	0.0014	0.2124	0.97
15	-	1.4	100	1.0	13	19.4	65	2	0.0018	0.2954	1.36
16	-	1.8	60	5.5	13	19.4	50	5	0.0007	0.3916	1.07
17	-	1.8	80	1.0	56	/	65	9	0.0007	0.1941	0.58
18	-	1.8	100	2.4	2	2.7	80	2	0.0021	0.2584	0.44

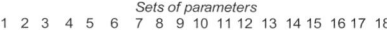
Sets of parameters
1 2 3 4 5 6 7 8 9 10 11 12 13 14 15 16 17 18

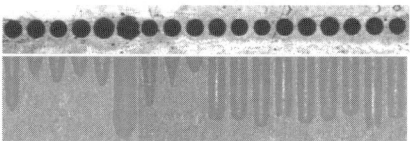

Fig. 2. Side and top view of the holes

The experiments were repeated three times, making a total of 54 machining. The values of the parameters for each machining are given in table 2 together with the average values of the obtained MRR, relative electrode wear and surface roughness.

3. Results and discussion

The first 18 holes are shown in Figure 2. From the picture, it can immediately be noticed that in contrast to conventional EDM machining, where positive polarity is preferable for machining hardened steel, negative polarity is more desirable in micro EDM.

3.1. Significant parameters

Based on the results shown in Table 2, the significance of each machining parameters to the MRR, relative electrode wear and surface roughness was calculated according to the ANOVA and F-test [7].

The principal of this method is to compare, for each output function, a calculated statistic F_0 for every machining parameter with a critical value derived from the Snedecor distribution [5], $F_{0.05}$ for parameters with three levels and $F_{0.005}$ for parameters with two levels. Machining parameters having F_0 greater than their critical value are considered significant to the main function and marked with asterisk (*) in the table 3.

In the given range of machining parameters, the polarity appears to be significant to the MRR and to the relative electrode wear, while the capacitance C shows significance to the MRR and the surface roughness.

3.2. Optimal parameters levels

According to the Taguchi method, in order to find the optimal machining parameters, the signal-to-noise ratio (S/N) [6] of each machining parameter level must be assessed for each output function. The highest S/N of the considered machining parameter levels indicates an optimal level.

The S/N values for MRR, relative electrode wear and surface roughness are given in Figure 3, 4 and 5 respectively.

Table 3 Machining parameters significance

	MRR		ϑ		Ra	
	F_0	$F_{0.005}$	F_0	$F_{0.05}$	F_0	$F_{0.05}$
polarity	9.76*	5.13	98.99*	5.13	1.38	5.13
i_e	0.02	4.46	0.11	4.46	0.99	4.46
u_i	2.03	4.46	0.17	4.46	0.80	4.46
t_i	1.83	4.46	1.36	4.46	1.47	4.46
t_o	2.79	4.46	1.93	4.46	1.81	4.46
C	5.59*	4.46	2.58	4.46	4.97*	4.46
gap	2.72	4.46	1.03	4.46	0.95	4.46
gain	0.43	4.46	0.91	4.46	1.37	4.46

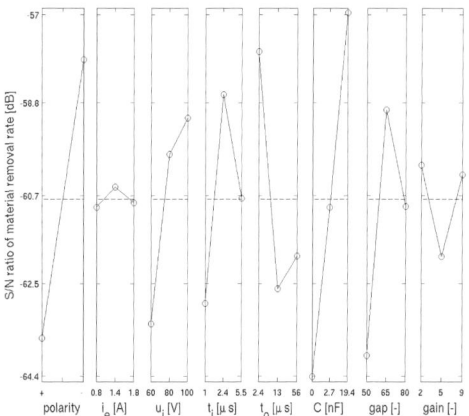

Fig. 3. MRR S/N ratios

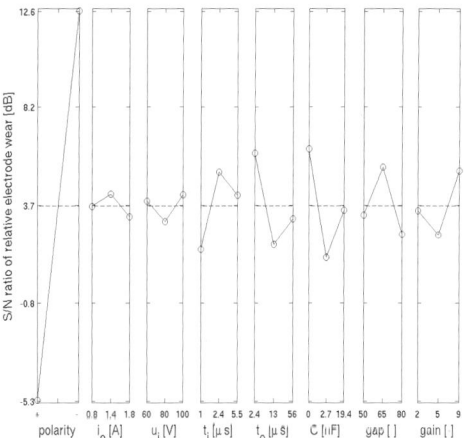

Fig. 4. Electrode wear S /N ratios

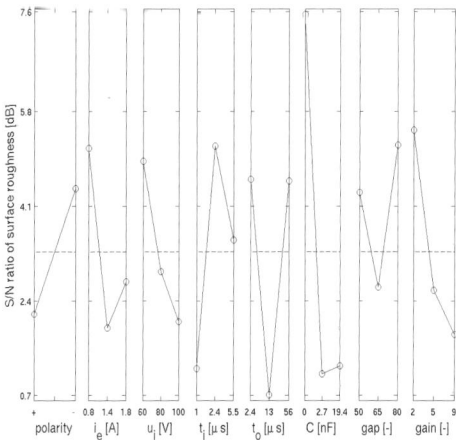

Fig. 5. Surface roughness S/N ratios

Table 4 Optimal machining parameters

optimum	polarity	i_e [A]	u_i [V]	ti [µs]	t_0 [µs]	C [nF]	gap [-]	gain [-]	MRR [mm³/min]	ϑ (-) [-]	Ra [µm]
		machining parameters							machining efficiency		
MRR	-	1.4	100	2.4	2.4	19.4	65	2	0.0024	0.1482	0.53
ϑ	-	1.4	100	2.4	2.4	/	65	9	0.0012	0.1409	0.22
Ra	-	0.8	60	2.4	2.4	/	80	2	0.0009	0.8579	0.20

Based on these figures, the theoretical optimal set of machining parameters can be assessed and are shown in Table 4.

Using these three optimal sets of parameters, another machining was performed on the workpiece and the resulting MRR, relative electrode wear and surface roughness can be found in Table 4.

The MRR achieved with the optimal set of machining parameters defined by the Taguchi method, appears to be better than the best results obtained previously. This is also the case for the surface roughness, where the optimal set of parameters gives the smoothest surface.

Thus, these two set of parameters can be used for roughing and finishing strategies in micro EDM, when using a similar electrode/workpiece combination.

The roughing technology is achieving a MRR three times greater than the finishing whereas the surface roughness it produces can be reduced by more than half using the finishing technology (table 4).

This should allow an improvement of the micro EDM processing time while producing smoother surfaces.

In cases where the electrode wear should be reduced to its minimum (to avoid frequent tool change for instance), the machining parameters optimised for the lowest electrode wear should be considered. However, the result obtained does not appear to be the best. The set of parameters number 12 (Table 2) seems to give a better result. This can be explained by the method used for electrode wear measurement, where only the length of the electrode wear was measured while neglecting the corner wear. Thus, the electrode volume wear assessment should be improved.

3. Conclusion

In this study, a method for machining parameters optimisation in micro EDM was proposed. Two sets of optimum parameters for roughing and finishing in micro EDM were successfully defined for a specific electrode/workpiece combination.

Further investigation will look at the improvement in processing time and surface roughness brought by using these sets of parameters when in a micro EDM milling strategy.

In addition, the results showed that in micro EDM, negative polarity is preferable for both roughing and finishing and that in the given range of machining parameters the capacitance is the most significant parameter for MRR and surface roughness.

Future work will focus on the use of various materials, electrode diameters and cavities shapes.

Acknowledgements

The authors would like to thank the European Commission, the Welsh Assembly Government and the UK Engineering and Physical Sciences Research Council for funding this research under the ERDF Programmes "Micro Tooling Centre" and "Supporting Innovative Product Engineering and Responsive Manufacture". And the EPSRC Programme "The Cardiff Innovation Manufacturing Research Centre". This work was also supported by Slovenian Research Agency and British Council in the frame "Partnership of Science". Also, this work was carried out within the framework of the EC Networks of Excellence "Innovative Production Machines and Systems (I*PROMS)" and "Multi-Material Micro Manufacture: Technologies and Applications (4M)".

References

[1]   NEXUS Market Analysis for MEMS and Microsystems III, 2005-2009, NEXUS Newsletter 01/2006.

[2]   E. Uhlmann, S. Piltz, U. Doll, Machining of micro/miniature dies and moulds by electrical discharge machining—Recent development, J. of Materials Processing Technology, 167 (2005) 488-493.

[3]   S.H. Huang, F.Y. Huang, B.H. Yan, Fracture strength analysis of micro WC-shaft manufactured by micro-electro-discharge machining, Int. J. of Advanced Manufacture Technology, 26 (2005) 68-77.

[4]   J. Valentinčič, M. Junkar, On-line selection of rough machining parameters. J. mater. process. technology, 149 (2004) 256-262.

[5]   Pham, D.T., Dimov, S.S., Bigot, S., Ivanov, A., Popov, K., 2004. Micro EDM – recent developments and research issues. Proceedings of the 14th CIRP International Symposium for Electromachining, ISEM 14, Edinburgh, UK.

[6]   Phadke MS. Quality engineering using robust design. Prentice-Hall International, Inc. 1989.

[7]   Montgomery DC. Design and Analysis of Experiments. New York: John Wiley & Sons, Inc., 2001. [1] Hurley E and Grant H. Probability maps. IEEE Trans. Reliab. R-124 (1995) 13-28.

Multi-Material Micro Manufacture
W. Menz, S. Dimov and B. Fillon (Eds.)

199

Hard tooling by µEDM-milling for injection molding

R. Jurischka[a], A. Schoth[a], C. Müller[a], D. Thiebaud[b], R. Gallera[b], H. Reinecke[a]

[a] *Department of Microsystems Engineering IMTEK, Laboratory for Process Technology*
University of Freiburg, 79110 Freiburg, Germany
[b] *Sarix SA, Micro EDM Technology, 6616 Losone, Switzerland*

Abstract

Today, microfluidic devices are becoming more and more important in life science, chemical analytic, and medical areas. For these applications disposable and low cost articles are highly convenient. Polymers are ideally suited for these applications due to their material properties and their applicability for high volume production with high accuracy. In this study, we optimized the µEDM-milling technology to fabricate mold inserts made from steel, including microfluidic structures, with an excellent surface quality. The microfluidic structures have channel dimensions down to 18 µm, aspect ratios of up to 4 and a surface roughness Ra < 70 nm. Additionally, demolding drafts of 2-6 degrees have been achieved, resulting in low demolding forces. These tools were characterized and polymer parts were reproduced by injection molding. The main advantage of the µEDM-milling technology is the rapid and direct structuring of tool steel for mold inserts with a 3D geometry and a long tool life time. These tools are used as mold inserts for injection molding, enabling flexible prototyping as well as the high volume replication of polymers.

Keywords: micro replication, injection molding, microfluidic, EDM, milling, steel, tool.

1. Introduction

The use of microfluidic devices in life science, chemical analytic and medical applications like point-of-care diagnostic has been strongly increasing through the past 4 years [1]. The market is demanding for disposable low cost products, which can be mass produced by industrial technologies. Polymers are preferably used for these applications due to their excellent solvent resistivity and biocompatibility as well as their possibility for high-volume production by replication technologies, mainly micro injection molding and hot embossing [2]. This paper presents an optimization of the micro electro discharge milling (µEDM-milling) technology to manufacture hard tools with microfluidic structures and excellent surface quality for micro injection molding. The µEDM-milling can structure quickly and directly tool steel for mold inserts with a long life time. In contrast to LIGA and bulk etching, µEDM-milling is a flexible machining technique to produce 3D structures in metals. This technology enables a rapid prototyping, a short development time, and facilitates the transition from the development phase to high-volume-production. We here show results obtained with the µEDM-milling technique in steel tool.

2 µEDM-milling technology

µEDM-milling is a evolution of EDM, and similar to numerically controlled milling, it uses a universal tool-electrode of simple shape (e.g. spherical or cylindrical), moving it along a complex path to machine hard materials like tool steel [3]. EDM is an electro discharge machining technique which removes continuously a small amount of workpiece material by controlling the discharge energy applied between tool electrode and workpiece [4]. Both electrodes are immersed trough a side flushing in a dielectric fluid (e.g. deionised water or oil) which electrically insulates and cools the electrodes as well as removes the particles out of the sparking gap. For this work we used a micro-EDM milling

machine SX-100 (Sarix SA, Switzerland) which integrates a 3D Micro EDM Milling CAM concept, enabling computer-aided manufacturing, and a Multi-Axis-Motion that enables contour-controlled machining according to a pre-programmed path in a simultaneous direction x, y, and z. It thereby achieves a high precision positioning accuracy of up to ± 2 µm and a resolution of 0.1 µm. By using standard tool electrodes with simple shape, the tool manufacturing costs are reduced to a minimum and additionally increase the degree of automation as well as a they allow for simplified process planning. The machining speed and surface finish can be adjusted by properly controlling of the micro milling and discharge strategies (shape pulse, energy). The µEDM-milling machine is illustrated in figure 1.

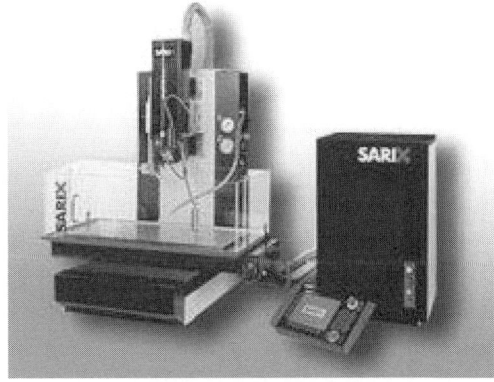

Fig. 1. High precision SARIX µEDM-milling machine with integrated CAM module.

Our approach is the achievement of an excellent surface quality with low roughness and burrs to manufacture polymer blood analytic chips for optical analysis.

3. Tool Fabrication

The overall fabrication sequence for creating a mold insert with microfluidic structures is illustrated schematically in figure 2.

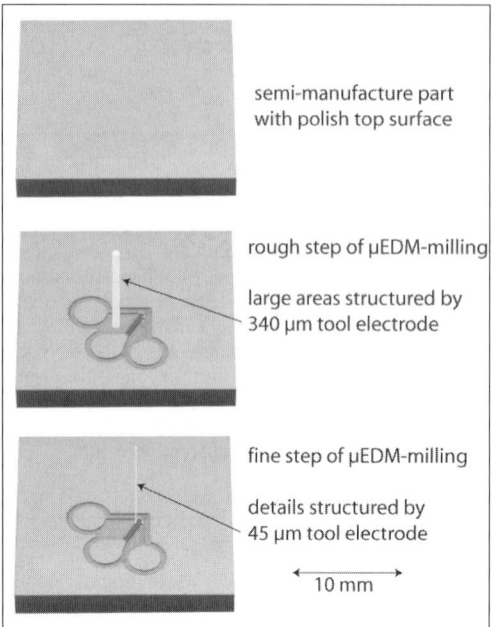

Fig. 2. Schematic process of tool fabrication.

Preliminarily, we fabricate semi-manufacture parts by CNC-milling and a following polish step to get a surface quality of about 15 nm. The µEDM-milling process is divided in two steps. First we use a carbide electrode with a diameter of 340 µm to machine large areas. Subsequently, the details are structured by fine machining with an electrode diameter of 45 µm. This electrode size was automatically prepared using a "built-in" micro wire EDM device (WEDG-wire electric discharge grinding unit). The fine machining operates by a new additional Micro-Fine-Pulse-Shape generator (SX-MFPS) to make small, deep and precise perfect round and shape holes as well as channels. The represented mold insert below was structured within 12 hours. Figure 3 is SEM micrographs of the micro structure on the mold insert.

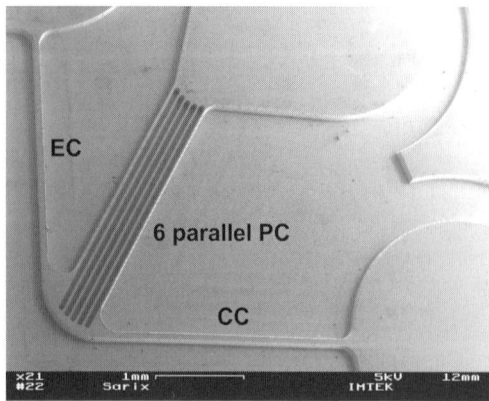

Fig. 3. Top view of the micro structure.

It depicts the structured geometry which results in reservoirs and micro channels after the replication process. The SEM micrograph in figure 4 shows a detail view of the bend region of the micro structure.

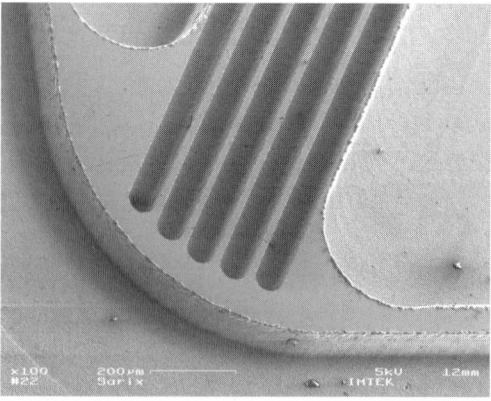

Fig. 4. Detail view of the bend region of the micro structure.

3.1 Characterization of the µEDM-milling structure

The height of the ribs and the surface roughness was measured by a tencor profilometer P-11. The ribs and reservoirs have a height of 96 µm ± 1 µm and roughness of the machined surface is about Ra 69 nm ± 2 nm. SEM pictures and roughness measurements show that there are no degradations of the polished surface quality of the semi-manufacture part due to the EDM milling process. The tencor profilometer P-11 can measure the height of the ribs and the surface roughness but it is impossible to measure the wall angle and the depth between the detail structures like the PC ribs (PC, EC and CC is defined in figure 2) due to the 60° angle of the measurement tip.

For a nondestructive characterization of the mold insert we cast the structured part with PDMS Elastosil RT 602 (Wacker-Chemie GmbH, Munich). This type of PDMS (polydimethylsiloxane) has no slimes and a low shrinkage of < 0.1%, so that we get an exact model of the mold insert in PDMS. This model allows making a cross section of the rib and channel structures easily. In figure 5, a cross section of the 6 parallel PC rib structures is represented.

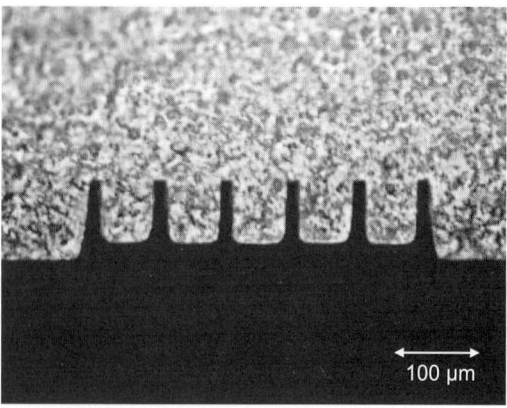

Fig. 5. Elastic impression of the mold insert by PDMS. Cross section of the 6 parallel PC rib structures.

The dimensions of the structures were measured by an Olympus BX61 microscope. They are summarized in Table 1.

Table 1. Structure dimensions measured by Olympus BX61

structure	width up [µm]	width down [µm]	edge radius [µm]	height [µm]	wall angle
EC	45.8	66.9	10	101.1	84°
CC	72.9	95.9	15	101.4	84°
PC1	15.4	23.3	10	99.5	86°
PC2	16.8	20.2	9	79.3	88°

The different µEDM-milling parameters (roughing and finishing) result in different heights of burr. In the first structuring step (roughing) we measured a burr of about 3 µm, but it could further be improved by a finishing step. On the middle PC ribs which are structured by the finishing step, the burrs have a height of only 1 µm (see figure 6).

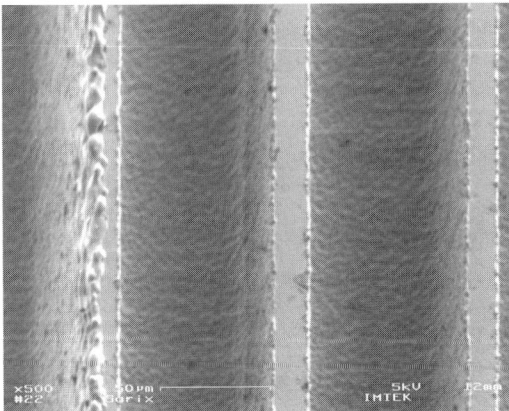

Fig. 6. Detail view of the PC rib structures.

4. Injection Molding

Injection molding is a key technology for the economic production of microstructures. This technology enables a low-cost mass production of micro structured parts [5, 6]. The characterized mold inserts can be replicated by injection molding. For this replication process, we constructed an injection tool in which we can easily and quickly change the mold insert. Figure 7 shows the injection molding tool with the exchangeable mold insert.

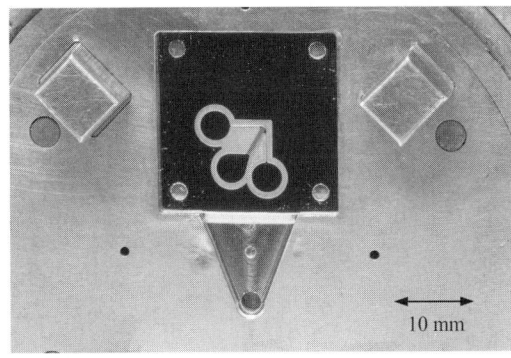

Fig. 7. Injection molding tool with exchangeable mold insert.

The injection molding machine (Microsystem 50, Battenfeld) used in this study is a fully electric machine for the precise production of micro structured moldings. For testing of mold inserts, an industrial grade COC (Topas 5013 and 6013, Ticona GmbH) is used to produce structured chips. The injection molding tool is designed with a flow runner. In the area of the runner, a pressure sensor is integrated for controlling process parameters. The molding tool is heated up to 120 °C. The process starts with the injection of the molten mass into the cavity using an injection plunger [7]. After injection, the post pressure avoids shrinkage during freezing of the polymer. To eject the molded chip, the molding tool is opened and the ejectors push out the chip automatically. After the ejection the next production cycle can start. The parameters of the injection molding process for producing the sample chips are listed in table 2.

Table 2. Parameters of the injection molding process.

Parameter	Quantity/value
max. injection pressure	700 bar
cycle time	40 sec
mass temperature	300 °C
tool temperature	120 °C
injection volume	440 mm³

In figure 8 a top view of an injection molded plastic chip consisting of COC polymer is represented.

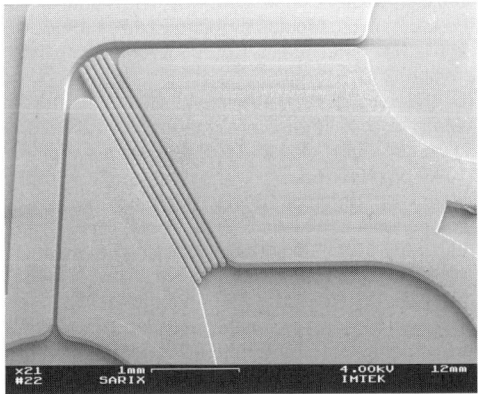

Fig. 8. SEM micrograph of the replicated plastic chip.

202

Figure 9 shows a cross section of some parallel PC channels. The mould was filling completely.

Fig. 9 Cross section of the PC channels.

4. Conclusions

For smooth production cycles and an economically efficient manufacturing process, success depends to a large extent on the life time of the tools employed. The µEDM-milling process enables the fabrication of mold inserts consisting of special hot work tool steel like, which is recommended for injection molding. The µEDM-milling process could create high aspect ratio 3D structures with an excellent surface quality, small burr, and rib widths down to 15 µm.

For the development of different microfluidic devices, the specific target material can be used right from the start of the evaluation process. The µEDM-milling technology combined with injection molding allows for cost effective and flexible rapid prototyping, and it can be used for mass replication of micro structured parts.

References

[1] NEXUS-Studie, 2002-2005, EU-commission.
[2] V. Piotter, T. Benzler, T. Haneman, R. Ruprecht, H. Woellmer, J. Hausselt, „Manufacturing of micro parts by micro moulding techniques", Proc. Euspen conf. 1999, pp. 494-497.
[3] W. Meeusen, D. Reynaerts, J. Peirs, H. Van Brussel, V. Dierickx, W. Driesen, "The Machining of Freeform micro Moulds by Micro EDM", Proc. MME 2001 (Micromechanics Europe Workshop), Cork, Ireland, 2001, 46-49.
[4] D. Reynaerts, W. Meeusen, H. van Brussel, „Machining of three-dimnsional microstructures in silicon by EDM", Sensors and Actuators A, Vol. 67, 1998, pp. 159-165.
[5] Hong L, Senturia S D, Molding of plastic components using Micro-EDM Tools, IEEE/CHMT Electronics Manufacturing Technology Symposim, 1992, pp. 145-149.
[6] Michaeli W, Rogalla A, Spennemann A, A new machine technology for the injection moulding of micro parts, Proc. of Euspen, 2000, pp. 141-148.
[7] Spennemann A, Michaeli W, Rogalla A, Process analysis and machine technology for the injection molding of microstructures, Proc. of ANTEC, 1999, 57, pp. 768-773.

Multi-Material Micro Manufacture
W. Menz, S. Dimov and B. Fillon (Eds.)
© 2006 Elsevier Ltd. All rights reserved

Characterization of Molecularly Imprinted Polymers as Novel Materials for Biochemical Sensors

Neda Haj Hosseini[a,b], Cristina Rusu[a], Anatol Krozer[a], Sjoerd Haasl[a], Kristina Reimhult[a], Jan-Olof Lindgren[a], Peter Enoksson[b], Lei Ye[c]

[a] *Imego AB, Arvid Hedvallsbacke 4, 411 33 Göteborg, Sweden*
[b] *MC2, Chalmers Univ. of Technology, 411 33 Göteborg, Sweden*
[c] *Applied Biotechnology, Lund Univ., PO Box 11, 221 00 Lund, Sweden*

Abstract

We study the behavior of Molecularly Imprinted Polymers (MIP) as novel synthetic receptor layers in biochemical sensors. Electropolymerization was used as an easy to implement method for polymer deposition and integrated impedance spectroscopy to monitor MIP behavior towards target molecules. Contrary to what is reported in the literature we consider that it is neither easy nor reliable to attribute the impedance changes to target binding/elution. However, alternative and/or more precise characterization methods are needed to track nano-scale changes in MIPs. Thus we are developing an optical set-up for detection of static deflections on the micro-cantilevers as a result of surface stress induced by changes in MIPs. Using an optical set-up, changes in the deflection of MIP-covered micro-cantilevers when exposed to the target molecules could be measured.

Keywords: Molecularly Imprinted Polymers, Impedance Spectroscopy, Micro-cantilevers, Biochemical sensors

1. Introduction

Biosensors integrate specific receptor molecules with signal transduction devices. They utilize various biological molecules or biomimetic functions (such as antibodies, enzymes, nucleic acids…) which are capable of recognizing a specific target molecule. However, many difficulties such as instability against high temperature, pH and organic solvents exist for their practical use. Furthermore, in many cases it is hard to find and purify a natural candidate which possesses the desired properties.

A biomimetic material called Molecularly Imprinted Polymer (MIP) is under development for sensing applications. MIPs are tailor-made materials containing molecule-specific cavities that mimic the behavior of natural receptor binding sites. They may be developed for a variety of target molecules. These polymers are prepared in the presence of a template molecule that interacts with the polymer network, usually via non-covalent interactions. After polymerization, the template is removed by washing and the polymer exhibits the ability to recognize the template with a high degree of selectivity (Fig. 1).

MIPs have many advantages over natural molecules including less preparation time, less cost and higher environmental stability. MIPs can be used in a wide range of applications and depending on their type and volume, MIPs can have thousands or millions of binding sites whereas biological receptors typically have a few binding sites.

However, the functional mechanism of MIPs is not fully known and there are controversial ideas of its ability to replace the natural materials. Thus, developing different methods for MIP characterization is required.

In this paper the progress status of two different characterization methods are presented. In one method the protocol for the electrochemical deposition of MIPs is developed. Cyclic voltammetry (CV) is used to monitor the electrodeposition procedure. Integrated electrochemical impedance spectroscopy was used as a method to study MIP performance after deposition. In the second method, static deflection of MIP-covered micro-cantilevers due to the MIP induced surface stress is measured optically.

2. Formation and characterization of Molecularly Imprinted Polymers using electrochemical methods

2.1. Deposition method

When preparing MIPs for sensing applications one often needs thin films. In this case several preparation methods are used. One method is to start polymerization at the surface. This is achieved by depositing initiator at the surface using suitable linker molecules, e.g., alkanethiols for Gold substrates or silanes for Si.

The process is governed by the electrode potential and by the reaction time, which allows us to control the thickness of the resulting film. The desired molecule is usually bulk-entrapped in a polymer matrix which grows onto the electrode surface from the solution containing the dissolved monomer and the desired molecule. The monomer is electrochemically oxidized at a polymerization potential giving rise to free radicals. These radicals are adsorbed onto the electrode surface and subsequently undergo a wide variety of reactions leading to the polymer network.

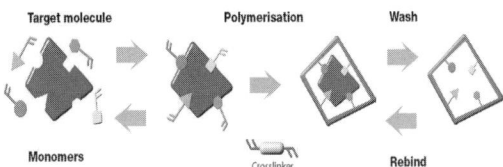

Fig. 1. The functional mechanism of Molecularly Imprinted Polymers.

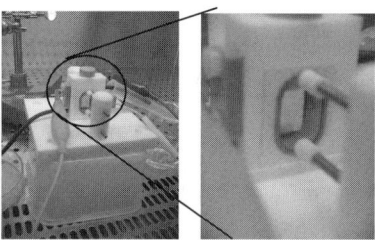

Fig. 2. Home-made electrochemical cell with glass window on one side and gold-coated electrode for MIP deposition with the option for surface Plasmon resonance detection on the other side.

A homemade electrochemical cell was used for the electropolymerization (Fig. 2). The gold electrodes were cleaned in TL1, UV-ozone and electrochemical cycling (CV) was performed in 1M sulphuric acid. Then the solution was changed to 10 mM acetate with monomer (54 ml 100 mg/ml) and 5 mM target molecule. We have chosen sorbitol as the target molecule. By sweeping DC voltage we monitored the current (CV). A low current implies the formation of a continuous MIP film. The MIP deposition was performed by electropolymerization following the procedure by Cheng [1].

Applying an AC-voltage (-0,044 to 0,756 V) we monitored the film impedance in the frequency range of 10 to 10^4 Hz. The most distinguishable difference was seen at the frequency of 10 Hz.

2.2 Electropolymerization Results

Cyclic voltammetry of the deposition of two polymers is shown in Fig. 3. The larger the hysteresis, the longer the polymer chains [2]. The dark curve shows a film of longer polymer chain and lower cross linking density than the lighter-coloured curve. The final current is of similar magnitude for both films.

Performance of the deposited MIP film was monitored by the impedance spectroscopy in acetate buffer solution. Impedance of the electrode was measured before and after electropolymerization. Following the film deposition, target molecules (sorbitol) were washed out of the polymer by water. The polymer was exposed to acetate acid buffer with different concentrations of sorbitol afterwards. The impedance was subsequently measured to monitor the MIP behaviour towards the target molecule (Fig. 4).

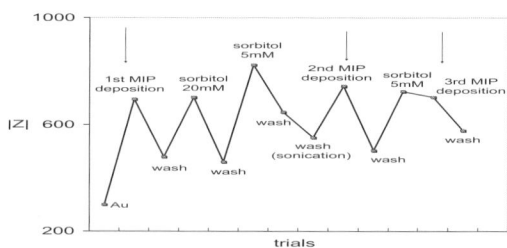

Fig. 4. Change of impedance at 10 Hz frequency due to MIP deposition, template molecule (sorbitol) wash out and its exposure to MIP.

In this figure the change of impedance with MIP deposition, template molecule elution and its exposure to MIP can be observed. In Fig. 4, arrows show an additional polymer deposition.

The electrical charge involved during electropolymerization is shown in Fig. 5. Electrical charges reaching the exposed surface of the electrode results in the formation of a new film; thus, formation of the additional films indicates the deterioration of the previously available film.

To investigate the causes of poor adhesion, the electropolymerization was delayed by 5 and 60 minutes to observe the effect of electrode surface cleanness on the deposited polymer (Fig. 6). By time, the monomers in the solution cover the surface of the electrode and prevent the charges from reaching the electrode surface; thus, the polymer formed would suffer from a poor adhesion.

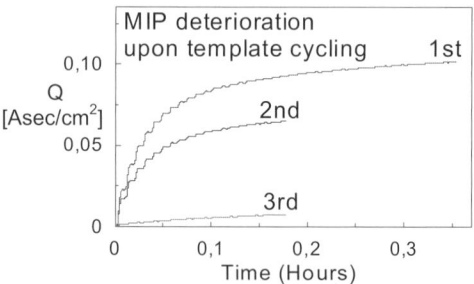

Fig. 5. The charge involved during electropolymerization. Each curve indicates a new polymer deposition.

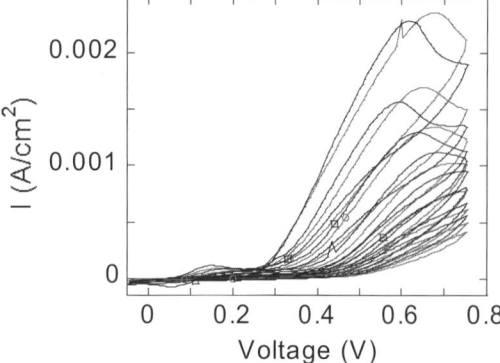

Fig. 3. The cyclic voltammetry showing MIP electropolymerization.

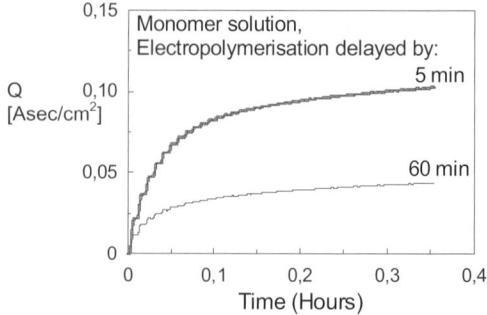

Fig. 6 The effect of electrode surface cleanness on MIP deposition

3. MIP Characterization using micro mechanical transduction

Micro-cantilevers as a type of micro-mechanical transducers offer the advantage of sensitivity and ability to operate in liquid solution making them suitable for biochemical sensors [3]. Molecularly Imprinted Polymers have not been yet characterized by micro-cantilever sensing, to our knowledge.

Deflection of a cantilever can generally be associated with the change in either the mass or surface stress which is measured in dynamic or static mode respectively. Surface stress can be produced by several causes: Thermal effects as a result of chemical reactions, energy due to adsorption processes or expansion of the polymer because of swelling or absorption of chemical molecules [3].

We monitored the deflection in the static mode due to the availability of the appropriate cantilevers and the fact that the mass change due to the binding of template molecules to the polymer would be smaller than the detection limit. Thus we aim to investigate the possibility of detection of surface stress induced deflections on the micro-cantilevers as a result of changes in MIP due to binding/elution of target molecules.

3.1. Measurement Set-up

The silicon micro-cantilevers with a spring constant of 0.03 N/m (500 µm x 100 µm x 1µm, Arrow™ TL8Au NanoWorld AG) are covered with 5nm/30nm Ti/Au. MIP was grafted on the gold using a UV-sensitive initiator [4]. The MIP was designed for theophylline as the target molecule in these experiments.

The optical set-up (fig. 7) uses in-plane movement of the laser and detector. The detector is kept stationary at a known angle and the laser moves. The angle of the maximum light intensity recorded on the detector is taken as the angle of reflection. The deflection angle can be calculated from the reflection angle using simple geometries.

The displacement in the z direction can be calculated from the angle of deflection using the equation:

$$z = R(\beta^2/2) = L.\beta.pi/360 \quad (1)$$

where L is the approximate length on the cantilever where the laser beam is focused (fig.8).

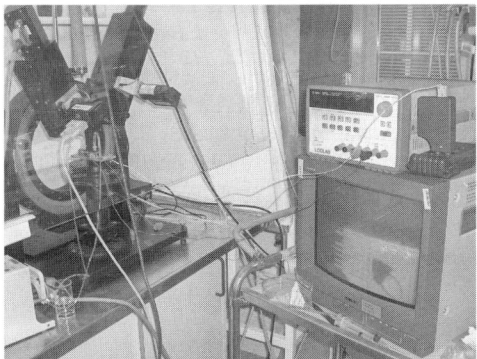

Fig. 7 The optical set-up, where the cantilevers are viewed by a CCD camera.

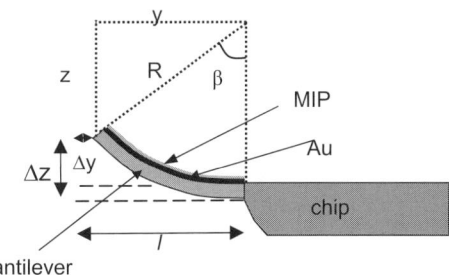

Fig. 8. Sketch of our cantilever and the geometrical relation of measured angles (β) to displacement (z)

One chip with bare gold and one with MIP-covered cantilevers are placed in a quartz flow cell (from Hellma GmbH) with a volume of 10^{-3} dl and are measured under the same conditions to provide a differential monitoring. The optical system was characterised for sensitivity and repeatability due to temperature, vibration, drift and other possible noises. After allowing the cantilevers to stabilize (2-3 hours), repeatability measurements were done on the bare and MIP-covered cantilevers to characterize the stability of the measurements in air and water. Measurements were done on the chip's base (as reference point), in the middle and tip of cantilever 20 times.

3.2 Results

By measuring on several different bare gold and MIP-covered cantilevers, both with optical deflection method and white light interferometry (Wyko NT1100), we found an upward bending in the bare gold cantilevers and a downward bending in the MIP-covered ones. The downward bending of MIP-covered micro-cantilevers is a result of compressive stress induced by the polymer. The bending of two typical chips is shown in Fig. 9.

Our system is able to measure changes down to 0,07 deg (~200nm) within 95% interval of confidence with a standard error of 0,008 deg (~30nm).

Stability measurements on both the bare gold chip and MIP-covered chip are shown in fig. 10. The standard deviation is 0,03 which counts for 2-10% of the total deflection.

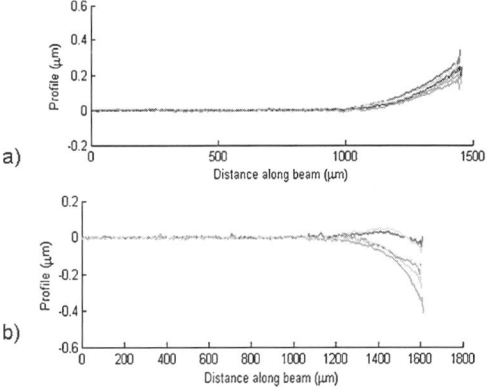

Fig. 9 Typical bending of the a) bare gold b) MIP-covered cantilevers measured by white light interferometry.

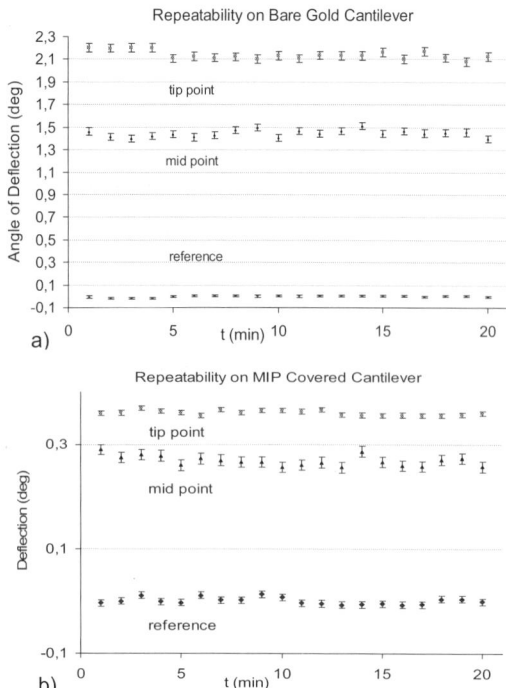

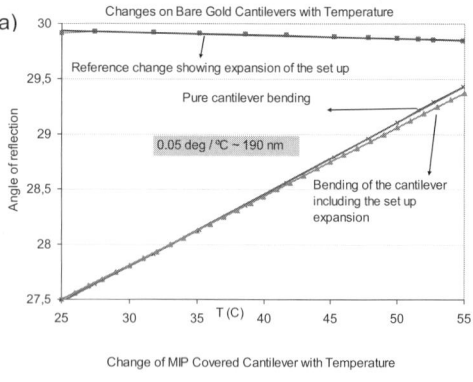

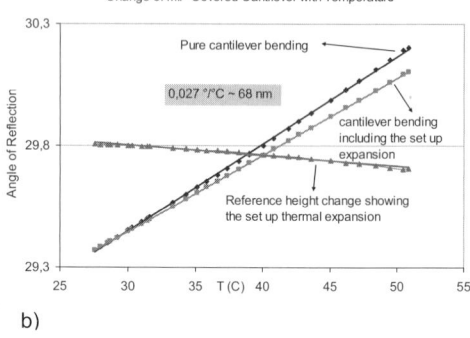

Fig. 10. Stability of cantilevers in water at room temperature for 20 measurements a) on bare gold cantilevers b) MIP-covered cantilevers.

Fig. 11. Change in the angle of reflection (bending of the cantilever) versus temperature for a) bare Gold cantilever and b) MIP-covered cantilever

To assess the temperature sensitivities of the bare gold and MIP-covered cantilevers, we heated up the whole cell. The cell undergoes a thermal expansion which should be deducted from the bending of the cantilever to obtain the net deflection angle of the cantilever. Since we calculate the relative changes of the angle and not the absolute value, the angle of reflection gives us the same values as the angle of deflection for sensitivity measurements. The gold cantilever deflects 0,05 deg/°C equal to 190nm and the MIP-covered cantilever deflects 0,017 deg/°C equal to 68 nm in water (Fig. 11).

When exposed to the target molecule, we see a downward bending on the MIP-covered cantilevers. After performing a target molecule wash out, we observed an upward bending on the MIP-covered cantilever. Leaving the polymer in liquid for a longer period (several days), a sharp upward bending in the cantilever was observed. This might be due to the polymer peel-off.

3. Conclusion

In this effort, MIP formation is investigated and two methods for studying MIP layers are presented. In both methods we have been able to measure the changes in MIP layer as a result of addition/elution of the target molecules.

We could observe clear changes in the polymer by impedance spectroscopy, though MIP performance was not perfectly reproducible from one experiment to another. Possible reasons include: insufficient gold electrode surface cleanness during electropolymerization and poor film stability.

Further work to improve the electrodeposition procedures is in progress.

An optical deflection detection set-up is developed as an alternative method to study the behavior of MIP by micro-mechanical means. The MIP-covered cantilevers show to have a downward bending as a result of a compressive stress induced by the polymer. They also show less variations compared to bare gold ones in the stability and temperature sensitivity measurements. Exposing the MIP-covered cantilevers to the target molecules, a change in bending is observed.

More experiments on studying MIP behavior as a result of binding to target molecules by micro-cantilevers are in progress.

References

[1] Z. Cheng, E. Wang & X. Yang, *Biosensors & Bioelectronics*, 16, 2001, p. 179.

[2] I. Losito, F. Palmisano & P. G. Zambonin, *Anal. Chem.*, 75, 2003, p. 4988.

[3] N.V. Lavrik, M.J. Sepaniak, P.G. Datskos, Cantilever Transducers as a platform for chemical and biological sensors, *Review of Scientific Instruments*, 75 (7), 2004.

[4] Wang, H.Y.;Kobayashi,T.;Fujii, N, *J. Chem. Technol. Biotechnol.* 1997, **70**, 355-362.

Structuring of phosphorescent pigments by two-step hot embossing for signaletic applications

M. Sahli[a,b,c], F. Legay[a], C. Roques-Carmes[a], C. Khan Malek[b] and J.C. Gelin[c]

[a] Laboratoire de Microanalyse des Surfaces (LMS), ENSMM, 25030 Besançon cedex, France.
[b] Laboratoire FEMTO-ST, CNRS UMR 6174, Département LPMO, 25044 Besançon cedex, France.
[c] Laboratoire FEMTO-ST, CNRS UMR 6174, Département LMARC, 25030 Besançon cedex, France.

Abstract

The hot embossing process is carried out in order to complement the panel of methodologies associated with the utilization of phosphorescent pigments for photoluminescent safety systems. The basic material, in compound form, is a strontium aluminate doped with Eu^{2+} and Dy^{3+} ions. Materials with these specific properties are protected from mechanical contact as well as from interactions with the environment, particularly atmospheric humidity to which these pigments are sensitive. Transparent structures with phosphorescent small-sized cavities were manufactured by a two-step hot embossing process developed for this application. The novelty of this work lies in applying the hot-embossing process to this new class of materials for safety applications. The modelling associated with this application allows the description of the velocity distribution in the various parts of the mould during the hot embossing process. Moreover it explains the interest of using sequential steps for testing signaletic applications.

Keywords: two-step hot embossing process, phosphorescent compound, glow in the dark material, strontium aluminate

1. Introduction

Fluorescence and phosphorescence are two phenomena based on luminescence emission [1]. They differ by the persistence time of visible afterglow when the photonic excitation source is switched off. The luminescence phenomenon which persists after the end of the excitation is called phosphorescence. It corresponds to the return of the electrons to the ground state from a metastable level. The pigments responsible for this property are either zinc sulfides doped with copper (Cu^{2+}) or strontium aluminates doped with Europium (Eu^{2+}) and Dysprosium (Dy^{3+}). They are commercialized as pigment particles of 10 - 20 µm size, and incorporated with different concentrations into polymers (70 % masterbatch, 5 - 25 % compound). The most often used polymers are polycarbonate (PC), poly(methylmethacrylate) (PMMA), polypropylene (PP), and ethyl vinyl acetate (EVA) or acrylonitrile butadiene styrene (ABS).

Recognized applications principally concern pictograms or safety systems after an accidental light switch off. Processes used in the realization of panels or signs are currently painting, serigraphy, as well as various forming techniques (injection, extrusion). Printing processes, which are more cost-effective, create an additional thickness which is often a limiting factor because the phosporescent product may be eliminated by abrasion. Moreover, dissolution can occur when in contact with environmental humidity as is the case of strontium aluminate. This is the reason why in many applications it is necessary to provide a protection of the phosphorescent product with regard to the outside environment. This protection can be achieved by simply using phosphorescent pigments inclusion. However a more efficient local marking on size-limited areas is obtained by locally incorporating the pigments in these specific areas. This process can be implemented by filling cavities with a syringe.

However, this technology, though easy to implement, does not provide a sufficient accuracy for microtechnical applications. This is the reason why hot embossing technologies [2] were favoured to ensure an optimised filling of small-sized cavities. It is this process, which we use for application purposes in our approach, that we will describe along with the study of the luminescent properties as a function of the size of phosphorescent areas and also the concentration of pigments selected for this application.

2. Modelling

The model which has been selected to describe, the flow of the polymer between two parallel plates on one hand, and on the other hand, the filling of the cavity located in an axisymmetric position It is based on the existence of an asymmetrical flow at imposed temperature and pressure between two level plates of circular shape separated by an initial gap h_0. The geometry of the system imposes that $u(r,z,t)$ and

$w(z,t)$ The upper plate moves at a velocity $v = \dot{h}(t)$ according to the Oz axis. The lower plate is fixed. The cavity which is present in the mould at an axisymmetric position is a cylindrical shape with a diameter $\varnothing = 2a$ and a height $= X_0$ (see Fig. 3a). In a system with cylindrical coordinates (r, Θ, z), we can consider, with certain assumptions (inertia and gravity forces are neglected) that the Navier-Stokes equations can be written as:

$$-\frac{\partial p}{\partial r}+\eta\left(\frac{\partial^2 u}{\partial r^2}+\frac{1}{r}\frac{\partial u}{\partial r}-\frac{u}{r^2}+\frac{\partial^2 u}{\partial z^2}\right)=0 \tag{1}$$

$$-\frac{\partial p}{\partial z}+\eta\left(\frac{\partial^2 w}{\partial r^2}+\frac{1}{r}\frac{\partial w}{\partial r}+\frac{\partial^2 w}{\partial z^2}\right)=0 \quad| \tag{2}$$

By considering the variables associated with the quantities u and w as well as the equation of continuity the velocity field becomes:

$$u=-r\left(\frac{\dot{\varepsilon}_{zz}}{2}\right) \tag{3}$$

Carrying over equation (3) in equations (1) and (2):

$$\frac{\partial p}{\partial r}=\eta\frac{\partial^2 u}{\partial z^2}=-\eta\frac{r}{2}\frac{\partial^2 \dot{\varepsilon}_{zz}}{\partial z^2} \tag{4}$$

$$\frac{\partial p}{\partial z}=\eta\frac{\partial^2 w}{\partial z^2}=\eta\frac{\partial \dot{\varepsilon}_{zz}}{\partial z} \tag{5}$$

Eliminating p between the two equations leads to:

$$\frac{\partial^3 \dot{\varepsilon}_{zz}}{\partial z^3}=0=\frac{\partial^4 w}{\partial z^4} \tag{6}$$

This expression can be integrated to obtain a velocity profile in polynomial form:

$$w(z)=Az^3+Bz^2+Cz+D \tag{7}$$

This expression can be used by taking into account the boundary conditions on u, $\partial u/\partial z$, and w.

The velocity vector in any point of the polymer distribution (see Fig. 1) can be written:

- In a first zone as :

$$w(z,t)=\dot{h}(t)\left[\frac{2z^3}{h^3}-\frac{3z}{2h}-\frac{1}{2}\right] \tag{8}$$

$$u(r,z,t)=3r\dot{h}(t)\left[\frac{z^2}{h^3}-\frac{1}{4h}\right] \tag{9}$$

- In a second zone as :

$$w(z,t)=\frac{2\left[\dot{h}(t)-\dot{X}(t)\right]}{h^3}z^3-\frac{3\left[\dot{h}(t)-\dot{X}(t)\right]}{2h}z-\frac{\left[\dot{h}(t)+\dot{X}(t)\right]}{2} \tag{10}$$

$$u(r,z,t)=3r\left[\frac{\dot{h}(t)-\dot{X}(t)}{h}\right]\left[\left(\frac{z}{h}\right)^2-\frac{1}{4}\right] \tag{11}$$

Where $\dot{X}(t)$ is the filling velocity of the cavity.

On these bases, one can demonstrate that the flow in the cavity (third zone) can be written as:

$$\bar{w}(r,z,t)=\dot{X}(t)\left[\frac{z}{X_0}-1\right]\left[\left(\frac{r}{a}\right)^2-1\right] \tag{12}$$

In addition the law characterizing this flow is supposed to depend little on the position of the microcavity [2] [3].

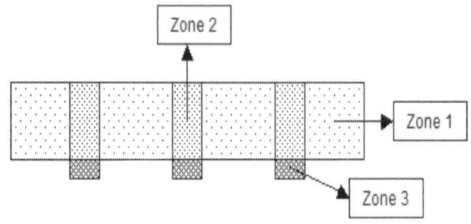

Fig.1. Schematic representation of the various zones corresponding to the differentiated flow of polymer.

These equations allow to determine the pattern of velocity profile represented in figure 2:

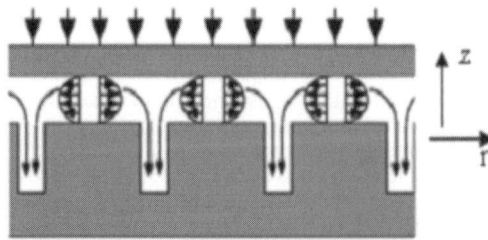

Fig.2. Schematic representation of the velocity profiles for the various zones indicated in Fig. 1.

An application utilizing bi-materials requires hot embossing utilizing various complementary elaboration steps.

3. Materials and methods

3.1 Materials

The samples exhibiting areas with luminescent properties were prepared from the selection of:

- a compound in granule form composed of a phosphorescent pigment (strontium aluminate doped with Europium and Dysprosium) included in a transparent acrylic polymer (PMMA) with a concentration of 25 %,

- a structural material of the same nature as the former one (amorphous PMMA) or of a different nature but recognized as compatible during preliminary tests using the bi-injection process. Cyclo-olefin copolymers (COC) and polyoxymethylene (POM) have been selected.

3.2 Mould

The upper part of the mould with an elliptical shape of dimensions a = 20 mm, b = 15 mm is elaborated in iron-chromium steel (40 Cr Mn Mo 8 + S). The lower part of the mould in the same material includes cavities of cylindrical shape of dimensions Ø = 4 mm, X_0 = 1 mm.

These cavities are laid out either in a central position or symmetrically on the large axes of the ellipse with respective distances of 10 mm on the Ox axis and 7.5 mm on the Oy axis. The cavities are located on the lower part of the mould, the upper part having the required geometrical shape. Ejectors allow the release of the samples from the mould.

3.3 Hot embossing

The selected process to produce the samples can be divided into different stages.

In the *first step* filling the cylindrical cavities is performed with an excess of the glow-in-the-dark PMMA compound (see Fig. 3b). The useful part of the mould is then heated up until reaching the desired temperature (Tg + 20°C) to allow the compression of the polymer at a pressure of P = 4 bars using a plate of cylindrical shape. The parameters for moulding phosphorescent PMMA compound (called material A) (see Fig. 3b') were similar to those for moulding PMMA. Cooling takes place while maintaining the pressure. The release from the mould facilitates the removal of excess material to obtain the cylindrical shapes of material A which are then reinserted later on into the cavities.

In a *second stage* the classical process of hot embossing is performed by replacing the plate ensuring the compression during the first stage by the final upper mould part with elliptical shape (see Fig. 3c). The associated temperature and pressure depend on the nature of the polymer selected for the support part. Three materials (called material B) were used for the support: PMMA, COC and POM and in the three cases the part made in PMMA compound adhered well to the polymeric support

In the *final stage*, the phosphorescent parts (A) were inserted in the substrate holes (see Fig. 3d) and the whole system was brought to a temperature lower than the lowest glass transition temperature of both materials.

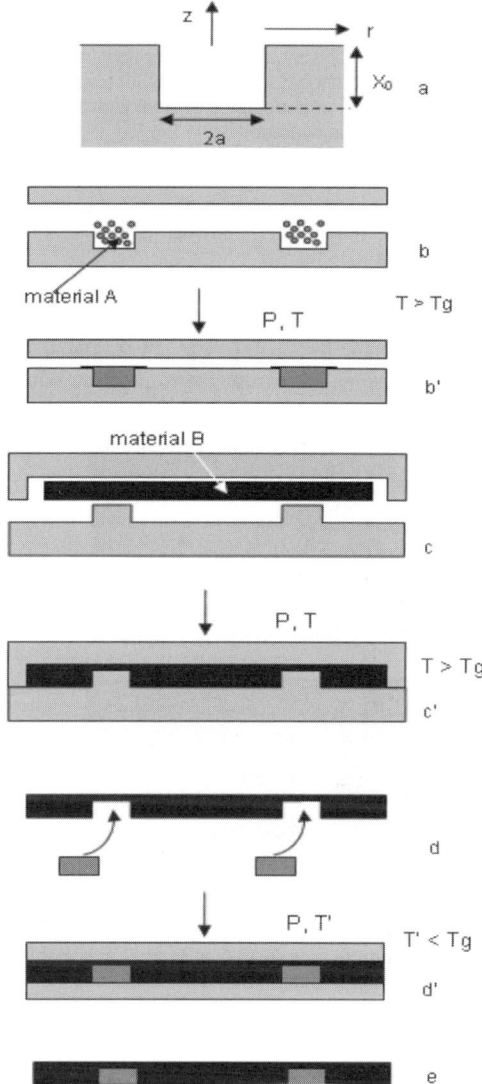

Fig.3. Process flow diagram illustrating the various stages of the process to manufacture a system with localized phosphorescent properties.

4. Experimental data

Structures made of two materials A and B with localised glow-in-the-dark properties could be produced via the selected experimental technologies. In a feasibility study material A is confined within the cavities whereas material B is used as a binder, as illustrated on Fig. 4.

In a finalised applicative approach the cavities can be structured to include various shapes such as logos, pictograms or sign posts.

The luminescent areas can be protected from the outside environment.

The excitation results from the ambient light source, preferentially a LED which is included in these components with luminescent properties. In this case, the luminescent phenomenon is similar to

210

the photoluminescence despite its longer effect. The properties of use of the components could thus be apprehended, which include the afterglow recorded with a luxmeter device (LMT model B250L).

The curve of luminescence intensity decay after different exposures is presented in Fig. 5.

These properties were almost unchanged after having maintained the component at the temperature of 100°C under an atmosphere of saturated vapour during one hour after having protected the phosphorescent parts.

If the properties of use deteriorate, the emissive part can be protected by a cover of material B.

Moreover, it was shown that the optimized size of the pigments is close to 20 μm and that as expected higher concentrations give better performances

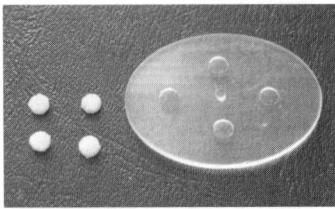

Fig.4. Various steps illustrating the two-step hot embossing process used to elaborate a component exhibiting phosphorescent areas to obtain a glow-in – the-dark pattern.

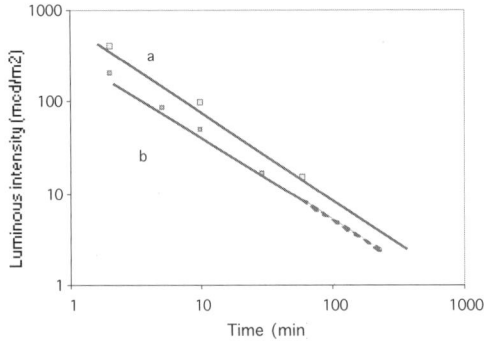

Fig.5. Variation of luminous intensity after excitation
(a) lab conditions (1000 lux, t=5 min)
(b) soft light environment (23 lux, t = 60 min)

5. Conclusions

The originality of this work lies in employing the well known hot embossing process as an alternative technique to pattern phosphorescent composite materials for a new applicative field in miniaturization of marking signs, such as safety systems.

A two-step hot embossing process was used to manufacture components with localized luminescent properties. These components have potential applications in the field of safety systems associated with a visual recognition of various micro-sized patterns. The protection of luminescent products with respect to the humidity and abrasion is ensured by the bimaterial structure of the component. The luminescence and photoluminescence properties are not modified when an amorphous polymer is selected for the substrate of the component. This type of process was applied to the realization of components presenting areas with thermochromic or photochromic pigments [5], or more generally any compound with inclusions of pigments with known visual properties such as pigments with an iridescent effect. In the same manner, the shape of the substrate can be curved if the component is utilized as a protection cover. The process used here with small-sized cavities can clearly be extended to cavities in the micro-range, should the application necessitate it. In complement one must retain that the phosphorescence phenomenon is dependent on the size of the part rich in phosphorescent pigments.

ACKNOWLEDGEMENTS

The Regional Council of the Franche-Comté is greatly acknowledged for providing the PhD grant for M. Sahli. This work was performed within the framework of the 4M Network of Excellence "Multi Material Micro Manufacture: Technology and applications (4M)" (EC funding FP6-500274-1; www.4m-net.org).

REFERENCES

[1]. Murayama Y., Takeuchi N., Aoki Y., Matsuzawa T., United States Patent n° 5,424,006(1995) «Phosphorescent phosphor».

[2]. Heckele M. and Schomburg W. K., *Review on micro-molding of thermoplastic polymers*, J. Micromech. Microeng, 14, 1-14 (2004).

[3]. Sahli M., Roques-Carmes C., Duffait R. and Khan Malek C., *Study of the rheological properties of poly(methylmethacrylate) (PMMA) and cyclo-olefin-copolymer (COC) to optimize the hot-embossing process*. Proc. Int. Conf. on Multi-Material Micro Manufacture (4M), W. Menz et S. Dimov (Eds), 29th june-1st July 2005, Karlsruhe,Germany. 83-86 (2005).

[4]. Sahli M. et al.. On the use of the hot embossing process for the reproduction of the surface topography of mould microreliefs (to be published).

[5]. M. Sahli. Thesis in progress

Multi-Material Micro Manufacture
W. Menz, S. Dimov and B. Fillon (Eds.)

An ultrathick SU-8 UV lithographic process and sidewall characterization

K. Jiang1, C.-H. Lee and P. Jin

Centre for MicroEnigineering and NanoTechnology, The University of Birmingham, Birmingham, U.K. B15 2TT

Abstract

This paper presents a UV lithographic process for fabrication of ultrathick SU-8 micro structures and sidewall surface characterization. The UV lithographic process has enabled the thickness of the SU-8 structures to be increased from 240 um given in the datasheet to 1000 um, and thus massively increases the application range. In developing the UV lithographic process, the best softbake and UV exposure times have been searched and tested in experiments. The UV light transmission spectrum has been analyzed. Then the best results have been produced on the research basis. The straight sidewall and 40:1 aspect ratio SU-8 structure images are presented in the paper. As SU-8 microstructures are suitable for microfluidics and bio-compatible, the sidewall surface roughness is characterized using AFM. The roughness contour of the sidewall shows that the surface topography is similar throughout the depth. The average roughness Ra is 46.46 nm. Other surface parameters, such as Rq, Rp-p, Rpk and Rsk, are also obtained and analysed. The implication of the smooth surface roughness of SU-8 structures to their applications is discussed in terms of transmission efficiency, the changes in friction to flowing liquid in a microchannel and the changes in the surface tension and capillary effect.

Keywords: Ultrathick SU-8, UV lithography, sidewall surface roughness, softbake

1. Introduction

Non-silicon microcomponents are developing rapidly in recent years and this trend has shown its strength in various areas. For example, polymer-based microfluidic devices are more promising because of the availability of different types of polymers, low cost and ease of fabrication using LIGA, embossing, casting, injection molding and imprinting process [1]. Among the popularly used polymers, SU-8 has been gaining a lot of attention as a material of choice for fabrication of microfluidic devices due to its superior chemical and mechanical properties in addition to the ease of fabrication [2], as well as its advantages of rapid prototyping and low costs over silicon DRIE process [3, 4]. Deep SU-8 structures have been used in RF MEMS for producing filters [5]. A highly notable application area of SU-8 is to make micromoulds, and then different material components can be produced from the moulds. It is a well known process that metallic micro components can be made through electroforming [6] and using metallic powder through injection moulding [7, 8]. Deep SU-8 micromoulds have also been used for making PDMS and ceramic structures [9]. Presently, SU-8 is gaining ground in finding more applications and gradually replacing those which used to be the applications of silicon components. Ultra thick SU-8 microfluidic structures are one of the areas attracting many interests in recent years and research work in producing microreactor [1], micronozzles, microcolumns [10] and multiple layer devices [11] can be found. Research has also found that SU-8 is bio and protein compatible, which makes it more suitable for lab-on-chips[11] and implant devices[12].

This paper presents an ultra thick SU-8 fabrication process for structures up to 1000 μm using UV lithographic process. The process was produced as a result of a rigorous study on the softbake and UV exposure steps. A mask of quartz coated with chromium was used in the investigation for high resolution. The improved UV lithographic process has been used to produce 40:1 aspect ratio structures in 1000 μm depth. As SU-8 has been widely used in microfluidics, such as in mass spectrometry analysis and drug mixing, the sidewall surface roughness of the SU-8 devices have been characterized to provide information for future simulation and experiments in using SU-8 devices. The sidewall roughness of Si devices fabricated using DRIE has been measured fro comparison. It is found that SU-8 sidewall surface is smoother and more even from top downwards than a Si device. The study results show that SU-8 is particularly suitable for making microfluidic devices.

2. Soft bake and dealing with cracking

SU-8 100 and 2100 from MicroChem have been used in the experiments. Both of the resists have high viscosity value and are transparent. The high viscosity, 51500 cSt in comparison to 375 cSt of AZ 4620, makes SU-8 physically possible for building up thick microdevices, while its high transparency nature makes little UV absorption. The low UV absorption property enables a uniformed exposure of the photoresist up to much bigger range compared with other thick photoresists. The ideal vertical sidewall profile is obtained based on the low UV absorption property. However, the transparency of SU-8 deteriorates as the layer gets thicker. It becomes more evident when the thickness is over 500 μm.

1 Corresponding author: k.c.Jiang@bham.ac.uk

Only a few references can be traced on ultra-thick SU-8 UV transmission spectrum [13]. Most of the references found so far provide only the transmission data of an SU-8 layer of a certain thickness, prepared with a constant prebake time and measured at a given wavelength [14, 15]. Proper soft bake time is regarded by some researchers as one of the most important control factors for the photoresist process [16]. Most investigations on the effect of the prebake time are focused on reducing the remaining solvent in SU-8 to improve the fabrication quality [17]. Such study has identified the minimum prebake time for vaporizing all solvent in SU-8. Cui, et al [17] recommended that a prebake time of about 30 h is necessary to vaporize the solvent for an SU-8 layer of 1000 μm. In this paper, the effects of longer the prebake times are discussed in terms of absorption property.

The UV light absorption property of SU-8 photoresist has been investigated by measuring the transmission using a HITACHI UV-3100 Spectrophotometer in the range of 360 nm to 460 nm with a 1 nm increment. The specimens are prepared using commercial SU-8-50 from MicroChem Corp on Corning glass substrates. The thickness of the SU-8 specimens is 1 mm, and a bare glass substrate is used as a reference in the measurement. The specimens are baked at 95°C for a time varying from 10 to 40 hours.

The transmission spectra shown in Fig. 1 indicates that when the soft bake time is long, the transparency property deteriorates and more UV light will be absorbed during the exposure time. Penetration depth is usually used to describe the absorption depth. The penetration depth is defined as the depth in which the incident light is decayed to a value that corresponds to a fraction of 1/e of the incident light. For an SU-8 film of a penetration depth thickness, the layer can be exposed uniformly at the given wavelength [14]. A highly transparent SU-8 layer effectively extends the penetration depth, and leads to producing a high aspect ratio feature. In reference to the transmission spectrum plot in Fig. 1, it can be observed that an SU-8 layer with a short soft bake time has higher transmittance than those with a long soft bake time. Therefore, a short soft bake time is identified as a method to reduce the UV light absorption. Keeping high transparency of the SU-8 after soft bake is important for the fabrication of the thick square coaxial transmission line.

Cracking is commonly observed in SU-8 structure fabrication, and has been also studied in this project. Cracking is considered to be caused by the internal tensile stress in an SU-8 structure. Factors for such problem include sudden temperature increase in a hot plate, high stress concentration as a result of improper mask and structure design, and big temperature difference during a complete SU-8 process. Cracking resulted due to the first two causes can be greatly improved by ramping temperatures up and down each time, placing component patterns evenly throughout the wafer in the mask design, and using arcs instead of sharp corners in the device design. The third cause to the cracking may be more significant than the first two and much attention should be paid to. In fact, the coefficient of thermal expansion of SU-8 is 50x10-6/K, about 12 times as much as that of Si, which is 4.2 x10-6/K. The higher the temperature is used during the soft bake time, the more internal stress will exist in the SU-

8 once the temperature drops down. In addition to the internal stress caused by the temperature change, the shrinkage of the SU-8 due to liquid evaporation usually adds more internal stress to the SU-8 layer. This is because with liquid evaporation, the volume of the SU-8 is reduced, but this reduction in volume is resisted by the rigid wafer, causing internal stress across the SU-8 layer. In most cases, the stress will not cause cracks immediately. However, when the exposed wafer is immersed in the developer, the mechanical property of the SU-8 begins to compromise considerably and the cracking problem begins to surface. Therefore, keeping a low temperature during the soft bake step will help reduce the internal stress and reduce cracking.

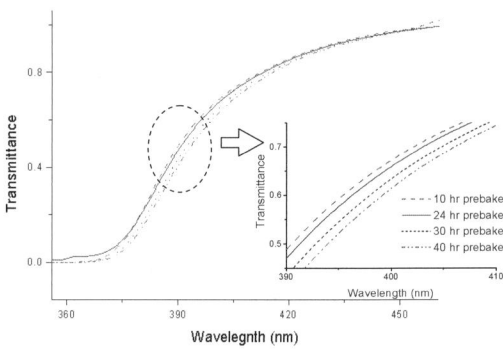

Fig. 1. UV Transmission spectra of unexposed ultra thick SU-8 layers coated on micro slide

3. Exposure and fabrication results

A series of experiments have been carried out and the optimized SU-8 structure fabrication processes have been identified for different thickness of SU-8 layers. The UV light source is a combination of g-line (436 nm), h-line (405) and i-line (365 nm). The optimized process for producing a 700 μm thick SU-8 layer consists of the following steps. (a) A Si wafer is cleaned with acetone and DI water, and then baked 200oC for 20 minutes. (b) SU-8-50 is deposited onto the wafer by direct casting and a scraper is used to spread it over. The deposited SU-8 on the wafer is then left for 20 minutes to get flat. (c) The wafer is baked first at 65oC for 2h, and then at 95oC for 15h. Afterwards, the temperature is ramped down to room temperature at a rate of 3oC per minute and the coated wafer is left there for another 15 minutes. (d) A 1.5 mm thick quartz mask coated with bright chromium is used for high resolution fabrication. (e) The coated wafer is exposed with the energy density 1712Mj/cm2. (f) The post exposure bake is carried out first at 65oC for 15 minutes and then at 90oC for 25 minutes. Then the temperature is ramped down to room temperature at a rate of 3oC per minute. (g) The exposed wafer is immersed in EC solvent, supplied by Chestech, for 1.5h at room temperature and agitated gently using a magnetic stirrer. (h) The wafer is rinsed with IPA to remove the residual developer. (g) The SU-8 structures are released in KOH solution at 40oC in 1 hour.

In comparison with the ultrathick SU-8 process proposed in [18], the optimized process differs in several aspects. First, the soft bake temperature is kept in maximum 95oC rather than 120-130oC. The

213

relatively low soft bake time helps reduce the internal stress of the SU-8 structures, reduce and eliminate cracks of the structures. Second, a metal mask is used to increase the edge quality of the structures. Third, a Si wafer, rather than a glass substrate in [18], is used as the substrate in the process. The optimized process has led to obviously improved fabrication results. Figure 2 shows an SU-8 layer with a smooth and vertical finish on the edge, which is important to the quality of filtering. The highest aspect ratio presented in [14] is 15:1. The aspect ratio achieved by using the optimized process, however, has been much increased. Figure 3 shows one of the experimental results produced by using the optimized process. The structure is a cross of 25 μm in thickness and 1000 μm in height, which is an aspect ratio of 40:1. The thickness of 1000 μm is sufficient for the filter applications. The sidewall angle in these experiments is controlled within 90±0.1 degrees, which is a significant improvement compared with those reported so far [18] and [19]. A two layer SU-8 master mould has been used to replicate a PDMS micro fluidic device, as shown in Fig. 4.

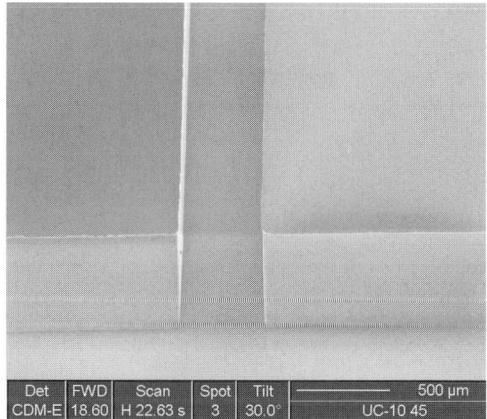

Fig. 2. An SU-8 layer of 700 mm thickness, fabricated for the filter

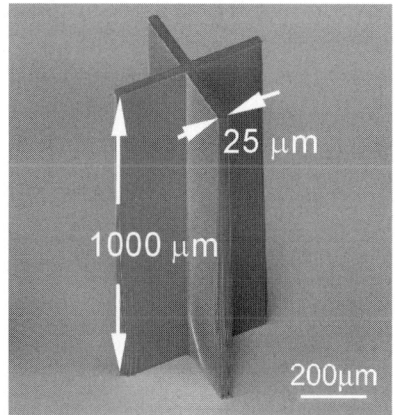

Fig. 3. An SEM image of a high aspect ratio cross.

Fig. 4. A two layer microfluidic device in PDMS fabricated from an SU-8 master mould.

4. Sidewall surface roughness measurement

Surface roughness has significant effects on the performance and reliability of a micro device where moving contact is necessary, such as gear transmission (Taz [20], Lee [21] and Bhushan [22]). Liquid distribution in a static microfluidic device is influenced by the capillary force and surface tension which, in turn, largely depends on the surface roughness (Myshkis [23]). In a flowing microchannel, surface roughness results in unsteady secondary flows (Winseler [24]). Fully understanding of the surface roughness of a microfluidic device will help in design and simulation of the device and ultimately the efficiency of the device. For these reasons, sidewall roughness of the deep SU-8 structures has been characterized and presented.

4.1 topographic parameters

Surface roughness measurements on the sidewalls of both silicon and SU-8 structures were carried out using an atomic force microscope (AFM, Burleigh Personal SPM). Measurements were taken at selected depth using a 20 nm radius silicon tip on each microstructure. Scan areas were 70x70 μm with a resolution of 256x256 points and 2 substeps.

Surfaces could be characterized using amplitude parameters, including root mean square roughness (Rq), peak to peak roughness (Rpp), skewness (Rsk) and kurtosis (Rku) (Whitehouse [25]):

Root mean square roughness is defined as the root mean square of the vertical departure of the profile from the mean line:

$$R_q = \frac{1}{A} \int_A^0 z^2 dx \qquad (1)$$

Peak to peak roughness is the vertical difference between the highest and the lowest points on the sample surface. Skewness measures the asymmetry of the profile. Surfaces with positive skewness are dominated with peaks. On the other hand, surfaces with negative skewness are dominated with valleys:

$$R_{sk} = \frac{1}{AR_q^3} \int_A^0 z^3 dx \qquad (2)$$

Kurtosis shows the degree of spikiness or bluntness of the profile:

$$R_{ku} = \frac{1}{AR_q^4} \int_A^0 z^4 dx \qquad (3)$$

where A is the area of the surface and z = f(x) is the profile from the mean line. A Gaussian surface, which results from a large number of recurring processes at discrete points, such as surface peening and electropolishing (Bhushan [26]), has a skewness of 0 and kurtosis of 3.

4.2 Measurement results

The SU-8 microcomponents fabricated using the above UV lithographic process has been repeatedly examined. Figure 5 shows the sidewall images of SU-8 structures scanned with the AFM at depths 50 mm and 600 mm. The sidewall surfaces of the SU-8 structure at various levels do not show much difference. The surface is relatively smooth with random peaks.

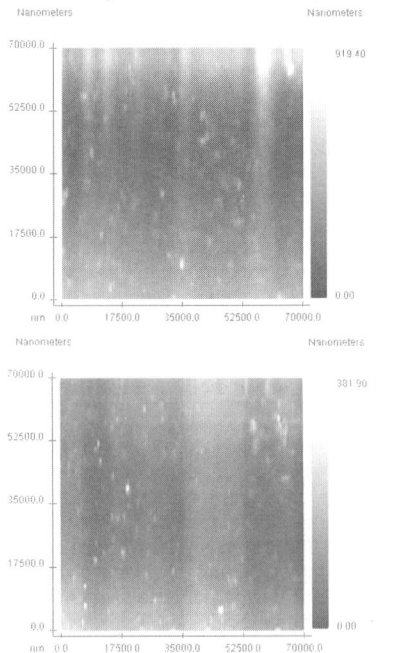

a. At 50 μm in depth.   b. At 600 μm in depth.
Figure 5. AFM sidewall images of SU-8 microstructure at various depth.

The sidewall roughness of SU-8 microcomponents has also been cross examined using Taylor Hobson Form Talysurf to verify the observations from AFM. Again the measurements were carried out at different depths of the sidewalls, similar to the measurements under AFM. Figure 6a shows an SU-8 sidewall roughness at 50 μm depth. The measured area is smooth with a few waves. A few small spikes exist, which are corresponding to the while spots in the AFM images of Fig. 5a. The peak values of the ridges can be found at 0.65 μm with reference to the scale on the right of the plot. Figure 6b shows the SU-8 sidewall roughness when measured at 450 μm depth. The small waves are still there, but the average level is lower than Fig. 6a, which indicates that

the sidewall has a little bit angle. Spikes become more obvious. In order to give a perspective view of the SU-8 sidewall roughness in comparison to microdevices of other material, silicon microstructures fabricated using DRIE have been measured. Fig. 6c shows the sidewall of a Si device measured at 5o μm. The area is full of ridges and the peak roughness can be found around 3.5 μm, some 5.5 times higher than SU-8 at a similar position. Fig. 6d shows the measurement at 600 μm on the same Si device. It is a transition area where ridges are gradually turning into tracks. The peak roughness is about 2.0 μm and the area looks smoother than the upper area shown in Fig.6c.

SU-8 surface

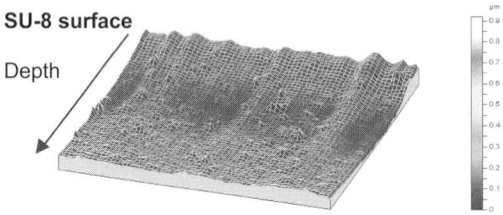

a. An SU-8 sidewall at 50 μm depth. The scale shows the peak value is about 0.7 μm.

SU-8 surface

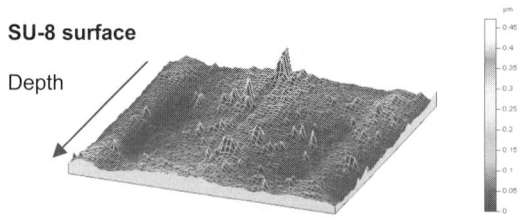

b. An SU-8 sidewall at 450 μm depth. Although the scale shows the peak value is about 0.35 μm, most areas stay under 1 mm. The spike surface nature is fully illustrated.

Si surface

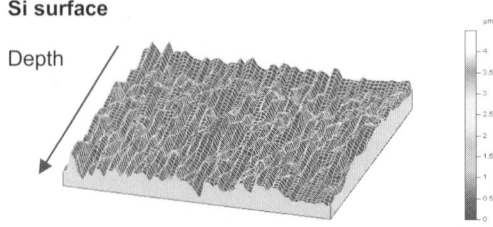

c. A Si sidewall at 50 μm depth. The scale shows the peak value is about 3.5 μm.

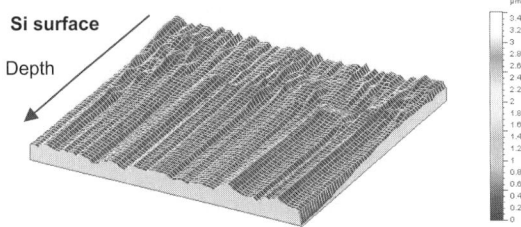

Si surface

Depth

d. A Si sidewall at 450 μm depth. The scale shows the peak value is about 2.0 μm and surface is smoother than at the top area.

Figure 6. Si and SU-8 sidewall surface roughness topographic plots at 50 μm and 450 μm depths.

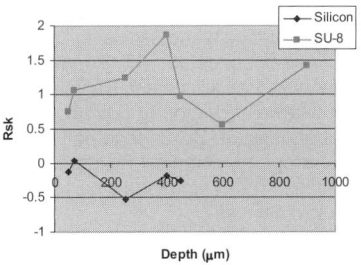

c. Skewness R_{sk}.

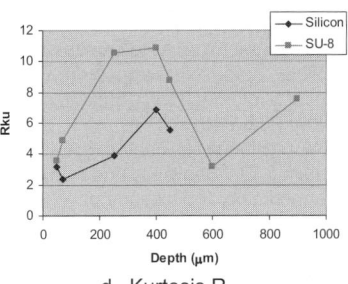

d. Kurtosis R_{ku}.

Figure 7. Surface roughness parameters as a function of depth.

The roughness parameters have been plotted. Figure 7 shows the values of Rq, Rp-p, Rsk and Rku against the depth on the sidewalls of SU-8 and Si structures. All of the surfaces have non-Gaussian characteristics. SU-8 surfaces are significantly smoother than silicon surfaces. In the middle section of the sidewall of the silicon structure, the Rq and Rp-p are the lowest and striate occur. On SU-8 sidewalls, the Rq and Rp-p are found to be similar at different depths with the lowest at the middle. Silicon surfaces are nearer to Gaussian surface than SU-8 surfaces, with near-zero Rsk as well as Rku ranging from 2.39 to 6.84 nm. SU-8 surfaces are dominated with peaks and positive Rsk and Rku. The variation of the sidewall roughness can be observed from Figure 7, where Rq and Rp-p of silicon sidewall varies considerably. In contrast, SU-8 exhibits more consistent Rq and Rp-p values. Both SU-8 and silicon show variance in Rsk and Rku values. The variation is also shown in the standard deviation in Table 1.

Table 1 shows sidewall measurement details on SU-8 components. The UV lithographic process on SU-8 results in an average roughness Ra at 46 nm, which is about 4 times smaller than the average roughness Ra=258 nm of the silicon sidewall fabricated using DRIE.

Table 1. Amplitude parameters taken from AFM scans on sidewalls of SU-8 structures, in correspondence to the graphs in Figure 6.

Depth	Ra (nm)	Rq (nm)	Rp-p (nm)	Rsk	Rku
50	109.93	135.57	919.40	0.75	3.53
70	75.35	97.47	750.90	1.07	4.88
250	25.18	33.61	512.10	1.24	10.58
400	18.89	28.00	341.90	1.86	10.91
450	25.00	32.79	470.60	0.97	8.76
600	37.66	46.80	381.90	0.56	3.17
900	33.23	46.08	567.60	1.42	7.59
Average	46.46	60.05	563.49	1.12	7.06
Standard Dev.	33.63783	40.70961	206.2191	0.433469	3.233399

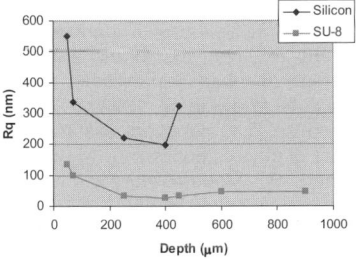

a. Root mean square roughness R_q.

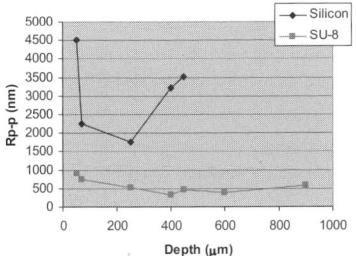

b. Peak to peak roughness R_{p-p}

5. Conclusions

The paper presents a UV photolithographic process for fabrication of ultrathick SU-8 structures. The fabrication process is a result of a rigorous study on the softbake and UV exposure steps. The UV lithographic process is capable of producing up to 1000 μm thick structures with very straight sidewall and 40:1 aspect ratio features. Then, the sidewall surface roughness characterization is presented. Repeated measurements have been made and surface roughness parameters have been produced, including Ra, Rq, Rp-p, Rsk, and Rku. For comparison, the sidewall surface roughness of

216

Si devices have been mearued. The analysis shows that the silicon surface fabricated using DRIE is rougher than that of SU-8 fabricated. The Rq value of SU-8 is 5.4 times smaller than silicon and Ra values are similar on a SU-8 sidewall from top to bottom. The excellent surface roughness of SU-8, together with its good mechanical property, chemical resistant property and bio compatible property, makes it a particularly suitable for microfluidic devices.

References

[1] Yujun Song, Challa S S R Kumar and Josef Hormes, "Fabrication of an SU-8 based microfluidic reactor on a PEEK substrate sealed by a 'flexible semi-solid transfer'(FST) process", Journal of Micromechanics and Microengineering, 14 (2004) 932–940

[2] Lee K Y, LaBianca N, Rishton S A, Zolgharnain S, Gelorme J D, Shaw J and Chang T H P, "Micromachining applications of a high resolution ultrathick", PR J. Vac. Sci. Technol. B 13 3012–6, 1995

[3] Ayon A A, Braff R A, Bayt R, Sawin H H and Schmidt M A 1999 Influence of coil power on the etching characteristics in a high density plasma etcher J. Electrochem. Soc. 146 2730–6

[4] Jackman R J, Floyd T M, Ghodssi R, Schmidt M A and Jensen K F 2001 Microfluidic systems with on-line UV detection fabricated in photodefinable epoxy J. Micromech. Microeng. 11 263–9

[5] K. Jiang, M.J. Lancaster, I. Llamas-Garro and P. Jin, "SU-8 Ka-Band Filter and Microfabrication", Journal of Micromechanics and Microengineering, vol 15, pp1522-1526, 2005

[6] Y E Chen, Y K Yoon, J Laskar and M Allen, "A 2.4 GHz integrated amplifier with micromachined inductors", IEEE MTT-S Digest, TUIF-32, pp 523-526, 2001

[7] J.-S. Kim, K. Jiang, and I. Chang, "A Net Shape Process for Metallic Microcomponents Fabrication Using Al and Cu Micro/Nano Powders", Journal of Micromechanics and Microengineering, vol 16, pp 48-52, 2006

[8] L-A Liew, W Zhang, V M Bright, L An, M L Dunn and R Raj, "Fabrication of SiCN ceramic MEMS using injectable polymer-precursor technique", Sensors and Actuators A 89 (2001) 64-70

[9] J S Kim, K Jiang and I T H Chang, "Making Alumina Microcomponents from Al Powder", in press, Journal of Key Engineering Materials, 2006

[10] C-H Lin, G-B Lee, B-W Chang and G-L Chang, "A new fabrication process for ultra-thick microfluidic microstructures utilizing SU-8 photoresist", Journal of Micromechanics and Microengineering, 12 590-597, 2002

[11] J Carlier, S Arscott, V Thomy, J C Fourrier, F Caron, J C Camart, C Druon and P Tabourier, "Integrated microfluidics based on multi-layered SU-8 for mass spectrometry analysis", Journal of Micromechanics and Microengineering, 14 619-624, 2004

[12] G Kotzar, M Freas, P Abel, A Fleischman, S Roy, C Zorman, J M Moran, J Melzak, "Evaluation of MEMS materials of construction for implantable medical devices", Biomaterials, 23. 2737–2750, 2002

[13] Z.G. Ling, K. Lian, L.K. Jian, "Improved patterning quality of SU-8 microstructures by optimizing the exposure parameters", Proc. SPIE, Vol. 3999, 1019-1027, 2000.

[14] N. LaBianca, and J. D. Gelorme, "High aspect ratio resist for thick film applications," Proc. SPIE, Vol. 2438, 846-852, 1995.

[15] K. Y. Lee, N. LaBianca, S. A. Rishton, S. Zolgharnain, J. D. Gelorme, J. Shaw, and T. H. P. Chang, "Micromachining applications of a high resolution ultrathick photoresist," J. Vac. Sci. Technol. B, 13, 3012-3016, 1995.

[16] J. Zhang, K.L. Tan, G.D. Hong, L.J. Yang, and H.Q. Gong, "Polymerization Optimization of SU-8 Photoresist and its Applications in Microfluidic Systems and MEMS", J. Micromech. Microeng., 11, 20-26, 2001.

[17] Z. Cui, D.W.K. Jenkins, A. Schneider, and G. Mcbride, "Profile Control of Su-8 Photoresist Using Different Radiation Sources", Proc. SPIE, Vol. 4407, 119-125, 2001.

[18] Lin, G.B. Lee, B.W. Chang, and G.L. Chang, "A new fabrication process for ultra-thick microfluidic microstructures utilizing SU-8 photoresist" Journal of Micromechanics and Microengineering, vol 12, pp590–597, 2002

[19] Z. Cui, D.W.K. Jenkins, A. Schneider, and G. Mcbride, "Profile Control of Su-8 Photoresist Using Different Radiation Sources", Proc. SPIE, Vol. 4407, 119-125, 2001.

[20] Tas, N., Sonnenberg, T., Jensen, H., Legtenberg, R., Elwenspoek, M., Stiction in surface micromachining, Journal of Micromech. Microeng, 6, 385-397, 1996.

[21] Lee, C.H., Jiang, K.C., Jin, P., Prewett, P.D., Design and fabrication of a micro Wankel engine using MEMS technology, Microelectronics Engineering, 73-74, 529-534, 2004.

[22] Bhushan B and Koinkar V N 1996 Sensors Actuators A 57 91–102.

[23] Myshkis, A.D., et al, Low-gravity fluid mechanics, Pringer-Verlag, 1986

[24] Winzeler, H.B., Belfort G., Enhanced performance for pressure-driven membrane processes – the argument for fluid instabilities, J. Membrane Science, 80 (1-3), 35-47, 1993.

[25] Whitehouse, D.J., Handbook of surface metrology, institute of physics publishing, 14-48, 1994.

[26] Bhushan, B., Modern tribology handbook, 47, 1997.

Multi-Material Micro Manufacture
W. Menz, S. Dimov and B. Fillon (Eds.)

Fabrication of nano-dimensional features in FOTURAN using focused ion beam technology

P.T. Docker, J. Teng, P. D. Prewett, K. C. Jiang1

MicroEngineering and Nano technology Group, Department of Mechanical Engineering and Manufacturing
University of Birmingham, Birmingham, B15 2TT, UK

Abstract

This paper details the findings of work carried out to determine the feasibility of manufacturing nano-dimensional features in photoetchable glass (Foturan [TM]) using focused ion beam technology. To date the standard processing techniques for producing features (UV lithography and UV lasers) in this material are limited by grain structure giving minimum feature size typically 10 microns. Focused ion beam technology (FIB) offers two potential advantages: there is no additional processing such as wet etching in hydrofluoric acid and there is the potential to manufacture nano-dimensional features, smaller than the micron scale which can be achieved using UV patterning with wet etch. This relies on the high resolution of the FIB system - <10nm at 30keV beam energy. A major obstacle is the local charging which occurs when an insulator such as Foturan [TM] glass is subjected to FIB irradiation. This was overcome by the use of electron charge neutralization. TRIM code Monte Carlo simulations of the ion implantation process clarify the ion sputter and implant mechanisms occurring during the FIB etch process.

Keywords: Photoetchable glass, Foturan, Nano-scale features, Focused ion beam patterning.

1. Introduction

The photoetchable glass (FOTURAN [TM]) is promising new material for the manufacture of micro and nano featured devices. It is a lithium aluminum silicate type glass with silver oxide, as a nucleating agent with the vital Inclusion of positively charged Cerium ions[1,2,3] for light sensitivity. It has a number of advantages regarding feature patterning and material properties[4,5.]. There is a requirement for components to be made in glass and ceramics with micron dimensions. Conventional machining techniques such as drilling, grinding, milling or moulding cannot achieve such micron features. Etching technology is also limited with glass as it has an amorphous structure and therefore etches isotropically preventing the fabrication of close packed dimensionally accurate microfeatures. This led to the development of the photoetchable glass FORTURAN [TM] by Schott Glaswerke, which can facilitate anisotrophic etching. Currently this glass is processed using three steps. A glass wafer is exposed to ultra-violet light, typically in the 250-350nm range, through a photomask. The exposure of the wafer in the prescribed area causes a release of electrons from the donor ion Ce++. These electrons then combine with Ag+ ions to form Ag atoms. The wafer is then placed in a furnace at 600 degrees centigrade where the exposed parts are crystalised causing the atoms of silver to agglomerate into small fragments. Further heat treatment leads to lithium metasilicate being formed. The final stage of processing is to etch the wafer using hydrofluoric acid. The exposed areas etch preferentially over the unexposed areas by a ratio of 20:1. Typical minimum hole feature sizes are 10 microns (see Ruf et al [6]). The material has excellent thermal and chemical resisitivity for many MEMS applications, such as chromatographic columns, sensors for aggressive chemicals, high temperature environments, filters, cooling systems, medical implants and inkjet printer heads[4].

Recent work has explored the potential of using UV excimer lasers[7,8,9] to expose FOTURAN [TM] wafers using a direct write technique, thus eliminating the requirement for masks. Resolution of 10 microns has been reported.

Gomez-Morilla et al have patterned FOTURAN [TM] using proton beams with energies in the MeV range[10]. The ultra high energy of their beams and the low atomic number of protons produces ion ranges of the order of 60-80 microns, exposing the glass to these depths. Feature resolution is reported to be on the micron scale. Due to the very high energies used in this system and the low mass of the H protons, there is no removal of surface atoms by sputtering.

The focused ion beam technology used in this work employs gallium ions which are approximately seventy times heavier than the protons used in the work of Gomrez-Morilla and with only 30keV beam energy the system will sputter etch the glass. Using the TRIM[11] software, the mean projected range of 30keV Ga+ ions in FOTURAN [TM] is predicted to be 50nm. The FIB direct write approach patterns the glass by a damage/displacement mechanism where as the proton implantation method patterns by creating Ag sites, as with UV lithography and UV laser technologies. The FIB system lends itself more to nanometer scale focusing because of its high resolution focused beam and the sophisticated scan algorithms available on commercial tools.

2. Equipment

The work described in this paper was carried out using a Strata DB235 FIB/SEM workstation from FEI Inc. This is a dualbeam system with coincident electron and ion beams integrated into a single piece of equipment. The sample holder is located on a

1 Corresponding author K.C.Jiang@bham.ac.uk

motorized stage, which allows for movement in the X, Y and Z-axes, with rotation and tilt.

Imaging can be done solely via the SEM, which reduces exposure to the ion beam, which mills the substrate continuously during live imaging. Immediate SEM imaging after FIB processing prevents exposure to atmospheric contaminants between etch and inspection cycles. The SEM incorporates a high brightness field emission electron source for low energy imaging of insulating samples with low electron damage. Energy Dispersive X-ray analysis (EDX) provides elemental analysis.

The pattern contron system is the Elphy Quantum from Raith GmbH. This lithography attachment, produces micro and nano structures by means of electron beam writing via the SEM or a focused ion beam system.

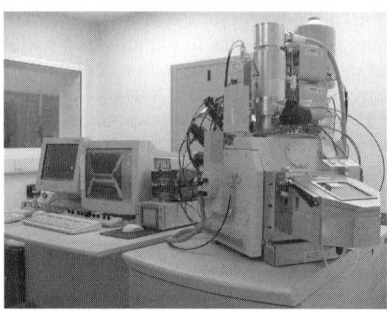

Figure 1. DB Strata 235 Dualbeam FIB/SEM system showing loadlock and SEM column. The FIB column is hidden from view.

A key component of the system is an electron flood gun charge neutralisation unit which allows controlled ion beam irradiation of the glass sample. When an insulator such as FOTURAN is irradiated with and ion beam, it charges positively at the ion impact location. This surface charge disrupts the incoming beam causing deflection, destroying the fidelity of the pattern.This can be overcome by use of the electron flood system to neutralise the charge build up.

3. Analysis

As stated earlier, the irradiation of the glass with a MeV proton produces damage at a depth of many tens of microns, leaving the surface untouched. This is confirmed by modelling the exposure of FOTURAN to a H proton MeV beam using TRIM software. The composition of FOTURAN glass was obtained from the manufacturers and is as follows (see table 1)

Table 1 Composition break down for FOTURAN glass

material	%	material	%
SiO_2	75-85	ZnO	0-2
Li_2O	7-11	Sb_2O_3	0.3
K2O	3-6	Ag_2O	0.1
Al_2O_3	3-6	CeO_2	0.015
Na_2O	1-2		

In performing the TRIM calculation, we neglected all elements present at less than 1% atomic concentration. The composition of FOTURAN was entered into the software as shown in Table 2.

Table 2 Approximate elemental composition of FOTURAN for TRIM code calculations

Element	%	Element	%
O	60	Al	4
Si	25	Na	1.3
Li	7	Zn	1
K	3		

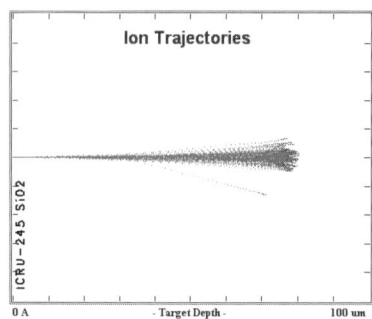

Figure 2 Output from TRIM for protons at 2.5MeV entering Foturan glass

As seen from Figure 2, the proton penetration is almost 100 microns. The proton beam energy is deposited with this range, far too deep to produce sputtering. This was confirmed by use of the sputter yield analysis module of the code. Figure 3 shows the ion penetration range for the 30keV Ga^+ ions using our FIB system for patterning FOTURAN.

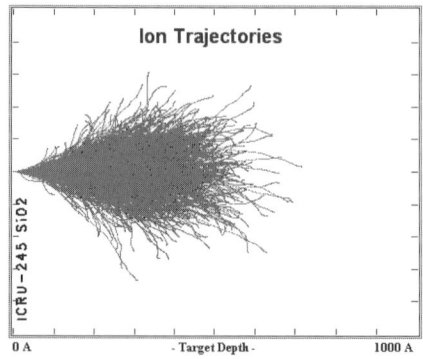

Figure 3 Predicted Ion penetration of gallium ions into FOTURAN at 30keV beam energy, calculated using TRIM code.

In contrast to the proton case, loose their energy within 30-40 nm. This results in a damaged region of implanted ions more than three orders of magnitude shallower than in the proton case. This is associated with significant sputtering of the surface of the glass with a predicted yield 5 atoms per ion.

4. Experiments

FIB sputter etching of a simple test pattern was carried out using a 30 KeV ion gun at 50 pA beam current, 0.1 second dwell time and 50% overlap. Milling for each line was carried out for 2 minutes each.

Top date 4 slots have been FIB milled in FOTURAN glass, all of which are sub micron (Figure 4) and a few microns deep. Using the electron flood charge neutralization system in conjunction with the Elphy-Quantum lithography unit, the DB235 Strata FIB/SEM system, excellent pattern and dose control were achieved, with submicron pattern placement accuracy.

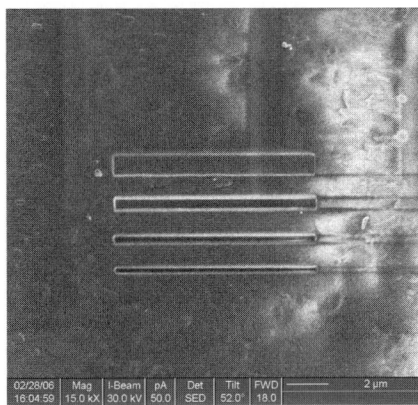

Figure 4 Features milled in FOTURAN glass using FIB. The slots are 10 microns long with widths of 217 nm, 342, 550 and 680 nanometres.

5. Conclusions

The work carried out for this paper shows it is possible to manufacture microfeatures in FOTURAN glass using FIB technology with a beam of 30keV gallium ions. Initial experiments confirmed the dominance of sputtering which provides a primary means of etching. However, it must be recognized that ion implantation damage is also occurring, albeit at a depth more than three orders of magnitude shallower than is the case for MeV proton beams. This means that further, wet etch processing may be possible following initial FIB etching. The very shallow range of the gallium implant effect will confine this chemical etch to a few tens of nanometers, providing excellent depth control for the fabrication of surface relief nanostructures. This, together with reducing the feature dimensions into the nanoscale regime will be the focus of future work.

6. Acknowledgments

We would like to thank Mikroglas for supplying us with FOTURAN™ wafers to make this work possible

References

1 Borelli NF, Morse DL, Bellman RH, Morgan WL, 'Photolytic technique for producing microlenses in photosensitive glass', Appl. Opt. 24,p 2520-2525 (1985)

2 Smith PG,'Some light on Glass Technol. 20, p149- (157 1979)

3 Hulsenberg D, Bruntsch R,'Glasses and glass-ceramics for application in micromechanics', J. Non-Cryst. Solids, 129 p199-205 (1991)

4 Dietrich TR, Braham A, Diebel J, Lacher M, Ruf A, 'Photoetchable glass for microsystems: tips for atomic force microscopy', J.Micromech.Microeng., 3, p187-189, (1993)

5 A, Diebel J, Abraham M, Dietrich TR, Lacher M, 'Ultra long glass tips for atomic force microscope', J. Micromech. Microeng. 6 p254-260 (1996)

6 Dietrich TR, Ehrfeld W, Lacher M, Speit B, 'Fabrication technologies for Microsystems utilizing photoetchable glass' Microelectric Engineering 30 p497-504 (1996)

7 Livingston FE, Helvajian H, 'The variable UV laser exposure processing of photosensitive glas-ceramics: maskless micro-to meso-scale structure fabrication', Applied Physics A-materials science and processing, 81(8), p1569-1581, (2005)

8 Cheng Y, Sugioka, Midorikawa k,'Microfabrication of 3D hollow structures embedded in glass by femtosecond laser for lab-on-chip applications', Applied surface science, 248(1-4) p172-176 (2005)

9 Kim J, Berberoglu H, Xu XF, Fabrication of microstructures in photoetchable glass ceramics using excimer and femtosecond lasers', Journal of microlithography and Microsystems, 3(3) p478-485 (2004)

10 Gomez-Morilla, Abraham MH, de Kerckhove DG, 'Micropatterning of Foturan photosensitive glass following exposure to MeV proton beams', J. Micromech. And microeng. 15 p706-709 (2005)

11 Biersack JP, Haggmark LG, 'A Monte Carlo computer program for the transport of energetic ions In amorphous targets', Nucl. Instrum Methods, 1980, 174, 257-69

Multi-Material Micro Manufacture
W. Menz, S. Dimov and B. Fillon (Eds.)

Influence of the process parameters on the microstructure of screen-printed $Ba_{0.6}Sr_{0.4}TiO_3$ (BST60) thick-films on alumina-substrates

F. Paul [*], J.R. Binder, A. Berto, G. Link, H.-J. Ritzhaupt-Kleissl

*Forschungszentrum Karlsruhe, Institute for Materials Science III, Hermann-von-Helmholtz-Platz 1,
76344 Eggenstein-Leopoldshafen, Germany*

Abstract

This article briefly discusses some aspects of the influence of process parameters on the microstructure of screen-printed BST60 thick-films on polycristalline alumina substrates. The focus of the experiments is directed toward the reduction of porosity by means of preparation without using sintering additives or composite techniques. The influence of paste composition, sintering temperature, cold isostatic compaction and microwave sintering on the morpholgy of the thick-films has been examined. The conclusion drawn from these results is, that the porosity of the films cannot be reduced by variation of paste composition or sintering temperature alone. Cold isostatic compaction of the green films leads to an improved homogeneity of the films, but not to a considerable reduction of porosity. Additional sintering by microwave might offer the possibility to reduce porosity of BST thick-films.

Key words: BST, thick-film, screen printing, ceramic powder, microstructure, microwave sintering
PACS:

1. Introduction

The system $Ba_{1-x}Sr_xTiO_3$ (BST) shows tunable behavior of its permittivity, when an electric DC-field is applied to it [1]. This opens up the field for a variety of potential applications in tunable radio frequency devices. Among these are phase shifters, tunable capacitors (varactors), RF-filters, modulators to mention only some. These applications require the implementation of the device's functional part as a planar film. The films are usually integrated into the device with thicknesses below 500 nm using standard thin film processing techniques like physical vapor deposition (PVD) or metal-organic-chemical-vapor-deposition (MOCVD). Both techniques require highly sophisticated and expensive equipment and furthermore show weakness in accurately controlling stoichiometry or dopants in the

resulting films. In contrast to the expensive thin film preparation methods the inexpensive thick-film deposition method screen-printing newly offers the possibility to create structured films with a thickness down to 0.1 μm [2]. However, so far screen-printing of thin films is limited to precious metals and furthermore, screen printed oxide thick-films of BST-, $BaTiO_3$ (BT-) or comparable systems show a considerable amount of porosity [3–13]. Remaining porosity can be beneficial for tunability on the one hand [14], but can probably induce a negative influence of environmental factors (e.g. humidity from processing) on dielectric loss, which results in a decreased figure of merit (FoM) on the other hand [15]. Increasing the sintering temperature to lower the amount of porosity can lead to a solid state reaction between the film and the substrate, especially in the case of BST-films on alumina substrates [4]. These points show, that the possibilities to influence the microstructure and through this

* corresponding author: florian.paul@imf.fzk.de

the dielectric properties of screen-printed BST-thick-films are very limited, if sintering aids or dopants are not to be used. Additionally, Astafiev et al. state that, under certain states of mixture, a dense composite of a lossy, tunable and a low-loss, non-tunable material does not improve the FoM [16], either. However, closing the open porosity of screen printed thick-films is of crucial interest for the performance of screen printed BST thick-films.

From the described point of view, this article will briefly discuss some aspects of the influence of process parameters on the microstructure of screen-printed BST thick-films on polycrystalline alumina substrates.

2. Experimental

The ceramic powders necessary for preparation of the screen-printing pastes were synthesized via an adapted sol-gel process based on the respective alkaline earth acetates and Ti-iso-propanolate. Details of the powder synthesis are published in a forthcoming publication. The resulting aqueous sols were freeze-dried or spray-dried immediately after preparation to form a white precursor powder. After drying the precursor powders were decomposed thermally (calcined) under ambient atmosphere at 700 °C for 1 h at a heating rate of 5 °C/min, using a Linn VMK 1400 chamber furnace and alumina crucibles. The cooling rate was the natural cooling rate of the furnace, but not faster than 10 °C/min.

The powders gained through freeze-drying were used for rheological experiments. They were ground 2 h in the mentioned attritor mill after calcination. The agglomerated powders $(2 \times 90$ g) were suspended in 2×250 ml acetone (p.A., Merck) and mixed with Mg-stabilized ZrO_2 grinding-spheres (950 g) of approximately 1.5 mm diameter. The resulting suspensions were ground at 600 rpm in an attritor (Netzsch) consisting of a stirrer made out of a stainless steel shaft and Mg-stabilized ZrO_2 paddles in a water-cooled alumina beaker. The suspensions were separated from the milling spheres with a nylon sieve and afterwards merged. The solid content of the resulting suspension was determined by gravimetric means. The suspension was divided up into 17 portions and mixed with the surfactant (Hypermer KD 1, Uniquema/ICI, dissolved in acetone, p.A.), terpineol (anhydrous, Fluka) and ethyl-cellulose (Ethocel, Fluka, Data given by the manufacturer: 5-15 mPas in a 80:20 toluol-ethanol

solution), dissolved in acetone (p.A., Merck) (in this order) according to the compositions given in table 1. They were mixed in a dissolver at 2000 rpm for approx. 2 h. Thereafter the acetone was evaporated and the resulting raw paste was treated several times on a roller mill (Exakt 50, Koenen) with rollers of alumina to form the final screen-printing paste. The viscosity of the pastes at 20 °C and at a shear rate of 100 1/s was measured with a Paar-Physika rheometer (MCR-300) fitted with a cone-plate system (opening angle: 2 °).

The powders gained through spray-drying were ground 7 h in an attritor mill (Netzsch) after calcination. Alternatively they were ground in the dry state for 1 h in a mechanical mortar mill with a rotating pestle and mortar made out of achat. For attritor grinding the agglomerated powders (100 g) were suspended in 200 ml ethanol (p.A., Roth) and mixed with the Mg-stabilized ZrO_2 grinding-spheres (950 g) of approximately 1.5 mm diameter. The resulting suspension was also ground at 600 rpm in a water-cooled alumina beaker. Afterwards the resulting suspension was dried in a rotary-film-evaporator and the milling spheres were separated from the powder in a sieve with stainless steel cloth with 71 μm mesh-size.

To form screen printing pastes from the ceramic powders gained through spray-drying, they were mixed in a dissolver with anhydrous terpineol (Fluka) containing surfactant (Hypermer KD 1, ICI). After 1 h of mixing a solution of ethyl cellulose (Ethocel, Fluka,Data given by the manufacturer: 5-15 mPas in a 80:20 toluol-ethanol solution) dissolved in terpineol was added as rheologic additive and was further mixed for 1 h. Finally the resulting raw paste was treated several times on the roller mill to form the final screen-printing paste.

The thick-films were printed with a semi-automatic screen-printing machine (Ekra, type M2) through a 22.5 ° sieve with a mesh size of 44 μm (325 mesh/inch). The substrates were polycrystalline alumina (Rubalit 710, CeramTec) with a thickness of 635 μm ± 50 μm and a density of 99.6 % of the theoretical density. After printing the films were dried for at least 8 h at 60 °C in covered petri-dishes and a drying oven with circulating air.

To reduce the porosity caused by cracks from drying and to increase homogeneity, some of the green thick-films were cold isostatically pressed together with the substrate at 400 MPa in evacuated and sealed bags of PE. All green thick-films were sintered afterwards under ambient atmosphere usually

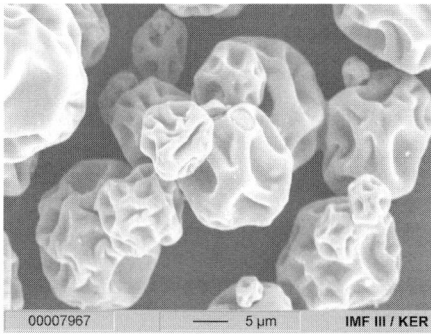

Fig. 1. SEM image of the ceramic powder gained through spray-drying of an acetate based sol after calcination for 1 h at 700 °C.

at 1200 °C for 1 h at a heating rate of 5 °C/min, in a Linn VMK 1400 chamber furnace. The cooling rate was 5 °C/min.

Alternatively the films were sintered for 20 minutes at 1200 °C in a 30 GHz Gyrotron powered furnace operated in continous wave mode. Details about the sintering method are given by Link et al. [17]. The heating rate was 30 °C/min and the cooling rate was the natural cooling rate of the furnace (up to 50 °C/min). The substrate was packed horizontally between two slabs of Zirconia, which act as a susceptor. The thermocouple to control the temperature of the sample was placed on top of the BST film and was in direct contact with the BST-film. The whole package was isolated against thermal loss by porous plates made out of Mullite.

Porosity was determined by optical evaluation of SEM images taken from the surface of the sintered thick-films. This was carried out using the software analySIS 5.0 (Soft Imaging System, Germany).

3. Results and Discussion

3.1. *Influence of Milling*

The initial morphology of the powders from which the screen-printing pastes are derived from has a crucial influence on the microstructure of the sintered thick-films. In figure 1 the morphology of the ceramic starting powder derived through spray-drying are shown. The agglomerates have a "dried berry"-like morphology, which is due to the spray-drying process of small droplets of sol. The primary particle size is in the range of 40 nm. The BET surface of the agglomerates is in the range of

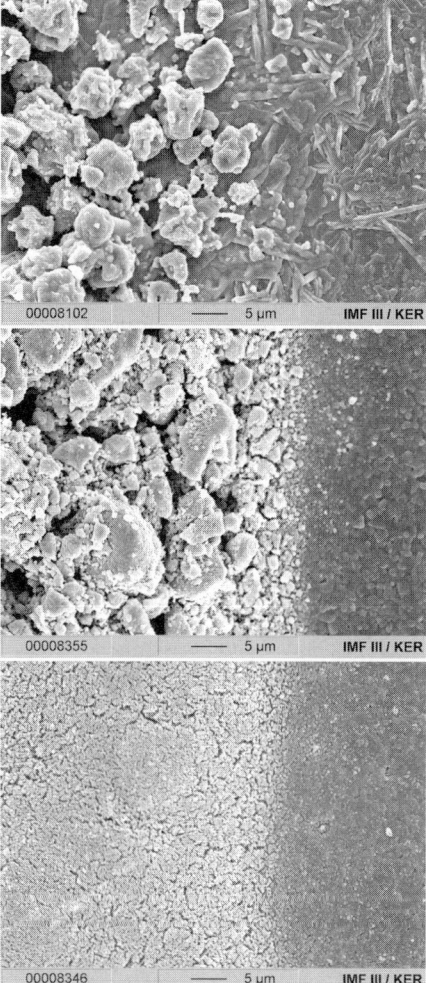

Fig. 2. SEM-images of the surface of sintered thick-films derived from spray-dried powder with the film on the left and the alumina substrate on the right. *top*: no milling at all, sintered 1h at 1250 °C; *middle*: 1 h milling with a mechanical mortar mill, sintered 1h at 1200 °C; *bottom*: 5 h milling in an attritor, sintered 1h at 1200 °C.

20 m^2/g before and around 32 m^2/g after grinding in the attritor-mill.

The thick-films resulting from spray-dried powders after sintering for 1 h at 1250°C or 1200 °C are shown in fig. 2. The films from pastes which contain unground, spray-dried powder (top) show only a "rock pile"-like morphology, consisting of a porous and loose collection of nearly unaltered agglomerates from the starting powder. The agglomerates themselves, however are very dense with virtually no remaining porosity, indicating a high sintering activity of the primary particles. Furthermore these results show that densely sintered structures can result, when the particles are already closely packed in the green state. The films from the slightly ground powder (middle) show a similar morphology and a minor increase in homogeneity. The films resulting from attritor-milled powder exhibit a more pronounced homogeneity and a smooth surface. However obvious cracks from drying, which are inhomogeneously distributed remain together with a considerable amount of porosity (around 35 %). The cracks go from the surface through the whole of the film down to the substrate, and therefore are not a surface-phenomenon of the films.

3.2. Influence of Compaction before Sintering

The cracks, which remain after drying of the films can be reduced by cold isostatic pressing (CIP) of the green films after drying. The resulting microstructures with and without CIP prior to sintering can be seen in fig. 3. The film, which had been pressed shows nearly no remaining cracks after sintering. However, the amount of porosity in the sintered films cannot be reduced to values below approx. 25 % by CIPing the green films, due to the low ceramic powder content in the green films.

3.3. Influence of Paste Composition

From the compositions of the screen-printing pastes listed in the table 1 the six graphs in fig 4 were modeled with the help of the commercial software MODDE 6.0 (Umetrics, Sweden). The used model was a CCC-model (central composite circumscribed) with 17 experiments. The two parameters viscosity and porosity were considered to be independent and thus were modeled separately. The powder, which was used to carry out these experiments was derived from the freeze-dried precursor. It showed BET surfaces of 6 m^2/g before and 15 m^2/g after attritor-grinding. The primary particle size was about 40 nm. Film thickness varied between 5 μm and about 20 μm depending on the solid content of the paste.

The following conclusions can be drawn from the modelled behaviour:
– *Viscosity* is influenced mainly by the content of ethyl-cellulose. It depends on the content of ethyl-cellulose by a square-law.

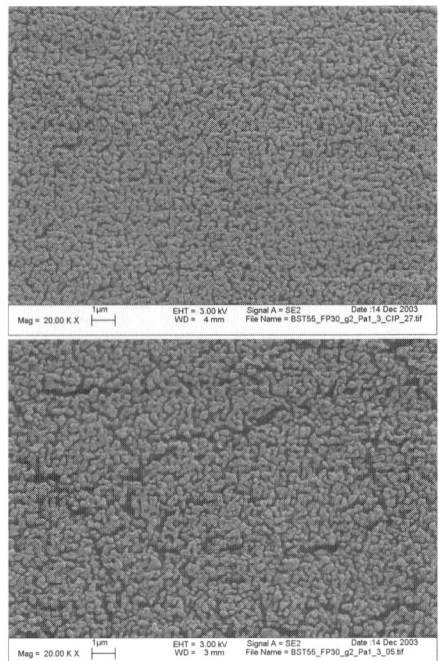

Fig. 3. SEM-images of the surface of sintered thick-films (1200 °C, 1 h) derived from a freeze-dried powder. *top*: CIP (400 MPa) in the green state; *bottom*: same print, no CIP in the green state.

– Viscosity depends on the content of ceramic powder by a linear-law. The influence of the ceramic powder content on the viscosity is much weaker than that of ethyl-cellulose.
– The influence of the surfactant is nearly negligible, but has a very small and linear influence on the viscosity.
– *Porosity* can be influenced by the composition of the screen-printing paste only within very narrow limits, which range around 30 % porosity. It is influenced by the ceramic powder content and the content of ethyl cellulose. The higher they are, the lower the amount of porosity, on the one hand and the higher the viscosity becomes, on the other hand. With an increasing content of ceramic powder and ethyl cellulose the viscosity increases in such a way that the paste cannot be printed properly any more. This means, that the viscosity is higher than the optimal viscosity range for printing with approx. 5-10 Pa·s.

These results indicate, that it is probably impossible to get a dense BST-thick-film only by varying the amounts of screen-printing paste components

Table 1

Nominal compositions for the pastes for experiments about viscosity of the pastes (at 100/s) and porosity of the sintered thick-films (1200 °C, 1 h) carried out with the help of the programme MODDE 6.0. Statistic outliers were not used in the model to create the graph in fig. 4.

exp.-No.	order	used in model	ceramic (Vol %)	ethylcellu-lose (Vol %)	surfactant (Vol %)	viscosity at 100/s (Pa·s)	porosity (%)
1	5.	yes	10	2,5	0,4	1,99	37,2
2	8.	yes	20	2,5	0,4	8,04	30,0
3	13.	yes	10	7,5	0,4	14,10	32,0
4	7.	yes	20	7,5	0,4	14,50	28,3
5	10.	yes	10	2,5	1,5	1,96	35,0
6	4.	yes	20	2,5	1,5	6,75	29,5
7	12.	yes	10	7,5	1,5	9,25	30,5
8	17.	yes	20	7,5	1,5	22,10	23,4
9	14.	yes	6,59	5	0,95	3,67	39,1
10	11.	yes	23,41	5	0,95	7,42	30,6
11	9.	yes	15	0,795	0,95	0,84	31,7
12	16.	yes	15	9,205	0,95	29,53	24,8
13	2.	no[1]/no[2]	15	5	0,025	7,26	24,0
14	15.	yes	15	5	1,875	8,60	28,2
15	6.	no[1]/yes[2]	15	5	0,95	5,35	32,5
16	1.	yes	15	5	0,95	2,49	30,6
17	3.	yes	15	5	0,95	3,98	29,4

[1] in model of viscosity;   [2] in model of porosity

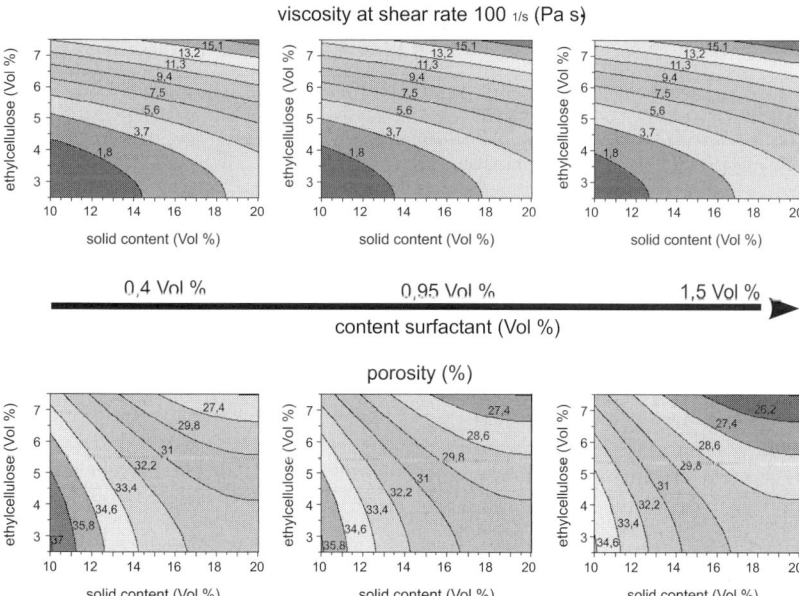

Fig. 4. Contour plot of the two separately modeled dependencies of viscosity and porosity from the BST-solid content ethyl cellulose content and surfactant content of the paste.

(solvent, surfactant and additive for rheology) without using sintering aids or composite techniques. Responsible for this behavior is the low ceramic content in the pastes, which could not be increased, due to the rising viscosity. Moreover the sintering temperature is limited as will be discussed in section 3.4.

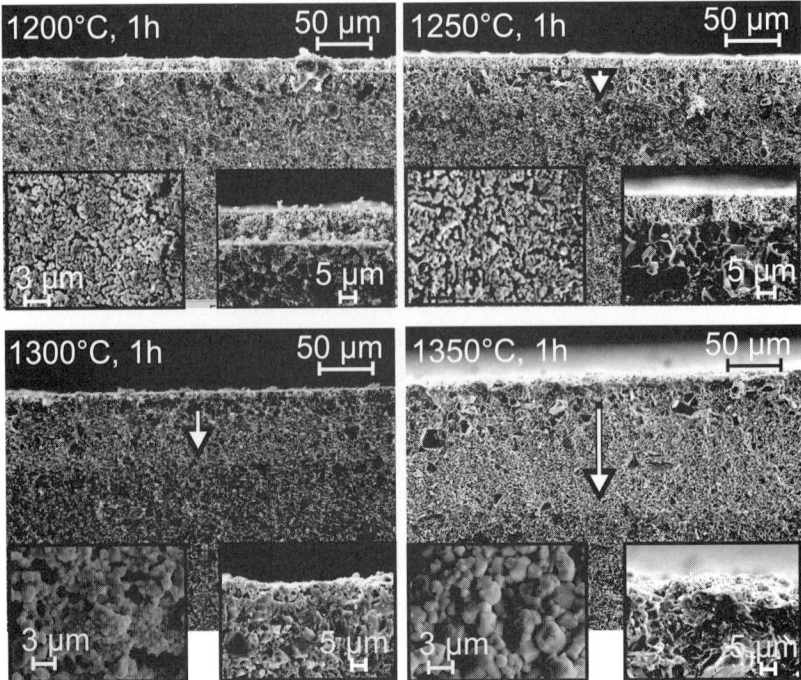

Fig. 5. SEM images of cross sections (fraction surface) of BST-thick-films on alumina substrate, sintered at different temperatures for 1 h at a heating and cooling rate of 5 K/min. All sintered films derive from the same green film (no CIP) from a freeze-dried powder. The arrows indicate the interface between the altered alumina microstructure and the part of the alumina substrate, which has not been influenced by the reaction with BST.

3.4. Influence of Sintering Temperature

One possibility to reduce porosity in sintered ceramic bodies is to increase sintering temperature. In the case of BST on alumina substrates this is not possible due to the severe interactions between BST and the substrate. The series of SEM images in fig 5 illustrate the proceeding degradation of the BST-film with increasing sintering temperature. The higher the sintering temperature, the deeper the reaction zone (light-gray area) travels into the substrate and the more the BST-film is degraded and becomes coarse grained. Though, the density of the thick-films cannot be improved.

The EDX line scan shown in fig. 6 suggests that mainly the Titanium and Barium cations diffuse into the Al_2O_3. Su and Button [4] find similar results. Sengupta et al. [18] find different second phases consisting of various Bariumalumotitanates and BST when reacting BST-powder with A_2O_3-powder. The EDX line scan of the sample sintered at 1300 °C suggests, that Ba and Ti diffuse about 25 μm into

the subtrate, whereas the fraction surface of the same sample altered through interdiffusion in fig 5 appears to be about 50 μm thick. This discrepancy cannot be explained, but might be due to a higher sensitivity of the grain boundaries of the microstructure to second phases in contrast to the sensitivity of the used EDX-equipment, which might not be sufficient to detect small variations in composition.

Again, porosity can be reduced to values roughly around 30 %, but not much below this value without sacrifying the functional film.

3.5. Influence of the Sintering Method

In fig. 7 SEM images of cross sections of conventionally and microwave sintered BST-thick-films on alumina are shown. The remaining porosity in the microwave sintered film is apparently lower compared to the conventionally sintered film. Additionally, the grain size is increased in the microwave sintered film. Moreover the mechanical stability of the microwave sintered thick-films was observably

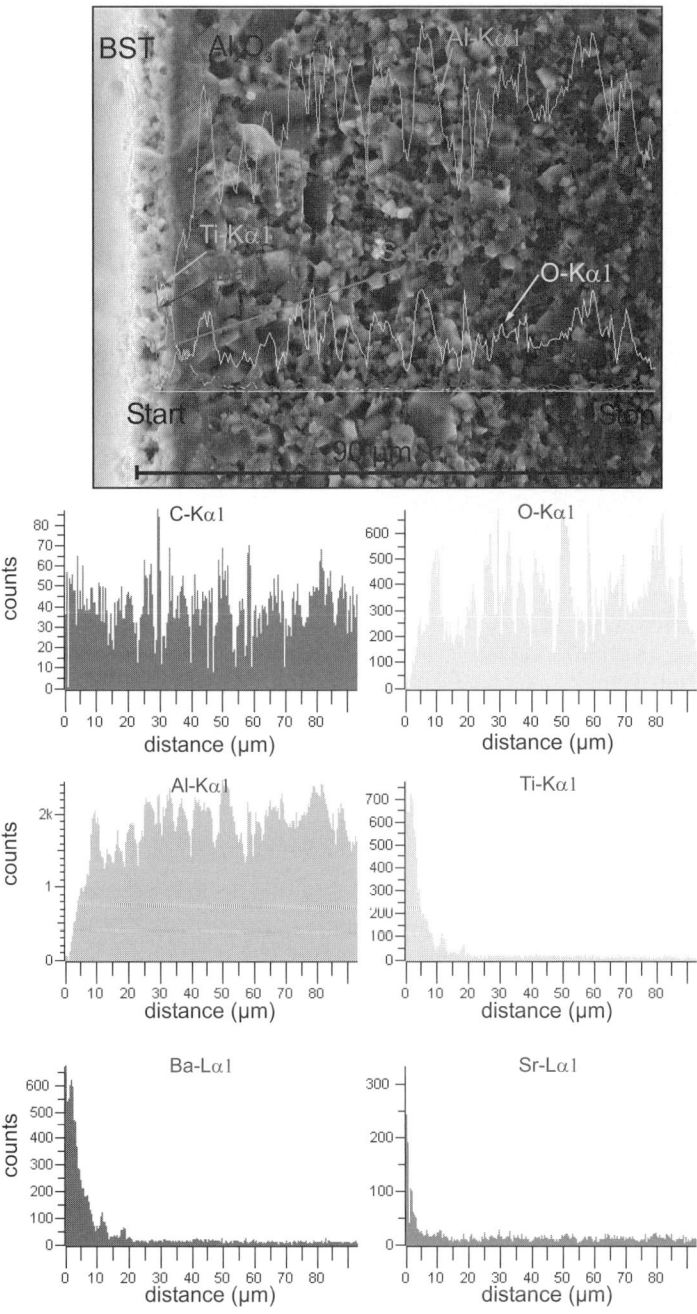

Fig. 6. EDX line scan through a cross section (fraction surface) of the BST-film on alumina substrate sintered for 1 h at 1300 °C.

higher, than that of the conventionally sintered films. However, porosity can not be reduced to zero and further examinations have to be carried out to proof these preliminary results.

Fig. 7. SEM images of cross sections (fraction surface) of sintered BST-thick-films on alumina. *top*: sintered conventionally, 1 h at 1200 °C in a chamber furnace; *bottom*: sintered 20 min. at 1200 °C using a 30 GHz microwave furnace.

Nevertheless this shows the potential of microwave sintering, to reduce porosity in screen printed thick-films. Unlike the other methods mentioned in this article, sintering by microwaves might offer the possibility to generate more densely sintered thick-films without the addition of sintering aids or dopants. RF-properties of microwave sintered thick-films are assessed momentarily and will be published in an oncoming article.

3.6. *Conclusion*

The conclusion drawn from these results is, that the porosity of the films cannot be reduced by variation of paste composition, sintering temperature or cold isostatic compaction of the green films alone. Additional sintering by microwave might offer the possibility to reduce porosity of BST thick-films. Further investigations have to be carried out to confirm this effect.

The thermal expansion mismatch between BST and Alumina has been neglected in this study. Nevertheless this is an important issue concerning the fabrication of crack-free films.

References

[1] A. K. Tagantsev, V. O. Sherman, K. F. Astaviev, Ferroelectric Materials for Microwave Tunable Applications, Journal of Electroceramics 10 (2004) 1–62.

[2] H. Fuchs, K. Abt, Metallo-Organic Preparations for Electronic Components, Ceramic Forum International/Berichte der deutschen keramischen Gesellschaft 82 (4) (2005) E17–E21.

[3] C. M. Ditum, T. W. Button, Screen printed barium strontium titanate films for microwave applications, Journal of the European Ceramic Society 23 (2003) 2693–2697.

[4] B. Su, T. Button, Interactions between barium strontium titanate (BST) thick films and alumina substrates, Journal of the European Ceramic Society 21 (2001) 2777–2781.

[5] C. M. Ditum, T. W. Button, Microstructural and electrical properties of $Ba_{0.5}Sr_{0.5}TiO_3$ in bulk and thick film form, Key Engineering Materials 206-213 (2002) 1263–1266.

[6] B. Su, T. Button, The processing and properties of barium strontium titanate thick films for use in frequency agile microwave circuit applications, Journal of the European Ceramic Society 21 (2001) 2641–2645.

[7] F. Zimmermann, M. Voigts, W. Menesklou, $Ba_{0.6}Sr_{0.4}TiO_3$ and $BaZr_{0.3}Ti_{0.7}O_3$ thick films as tunable microwave dielectrics, Journal of the European Ceramic Society 24 (2004) 1729–1733.

[8] B. Stojanovic, V. Foschini, C.R. Pavloviv, Barium titanate screen-printed thick films, Ceramics International 28 (2002) 292–298.

[9] B. Stojanovic, C. Foschini, V. Pejovic, Screen printed barium titanate thick films prepared from mechanically activated powders, Key Engineering Materials 206-213 (2002) 1425–1428.

[10] L. Simon, S. Le Dren, P. Gonnard, PZT and PT screen-printed thick films, Journal of the European Ceramic Society 21 (2001) 1441–1444.

[11] E. S. Thiele, D. Damjanovic, N. Setter, Processing and properties of screen - printed lead zirconate titanate piezoelectric thick films on electroded silicon, Journal of the American Ceramic Society 84 (12) (2001) 1863–1868.

[12] S.-G. Lee, C.-I. Kim, B.-C. Kim, Dielectric properties of screen-printed (Ba, Sr, Ca)TiO_3 thick films modified with Al_2O_3 for microwave device applications, Journal of the European Ceramic Society 24 (2004) 157–162.

[13] A. F. L. Almeida, P. B. A. Fechine, J. C. Góes, M. A. Valente, M. A. R. Miranda, A. S. B. Sombra, Dielectric properties of $BaTiO_3$ (BTO)-$CaCu_3Ti_4O_{12}$ (CCTO) composite screen-printed thick films for high dielectric constant devices in the medium frequency (MF) range, Materials Science and Engineering B 111 (2-3) (2004) 113–123.

[14] F. Zimmermann, M. Voigts, C. Weil, Investigation of barium strontium titanate thick films for tunable phase shifters, Journal of the European Ceramic Society 21 (2001) 2019–2023.

[15] F. Paul, Dotierte $Ba_{0.6}Sr_{0.4}TiO_3$-Dickschichten als steuerbare Dielektrika, Ph.D. thesis, Universität Freiburg (2006).

[16] K. Astafiev, V. Sherman, A. Tagantsev, Can the addition of a dielectric improve the figure of merit of a tunable material ?, Journal of the European Ceramic Society 23 (2003) 2381–2386.

[17] G. Link, L. Feher, M. Thumm, H.-J. Ritzhaupt-Kleissl, R. Böhme, A. Weisenburger, Sintering of Advanced Ceramics Using a 30-GHz, 10-kW, CW Industrial Gyrotron, IEEE Transactions on Plasma Science 27 (2) (1999) 547–554.

[18] L. Sengupta, S. Stowell, E. Ngo, Barium strontium titanate and non - ferroelectric oxide ceramic composites for use in phased array antennas, Integrated Ferroelectrics 8 (1995) 77–88.

Multi-Material Micro Manufacture
W. Menz, S. Dimov and B. Fillon (Eds.)

Self-assembled nanostructured polymer films from concentrated solutions of block copolymers

N. Didier[a], P. Panine[b]

[a] *Ecole Nationale Supérieure de Physique de Grenoble, 38 402 St-Martin d'Hères, France*
[b] *European Synchrotron Radiation Facility, 38 043 Grenoble, France*

Abstract

We are presenting in this paper a simple method to orient auto-assembling systems. We detail the production process of oriented nanostructured polymer films by shearing a concentrated solution of block copolymers. The most appropriate tool to simulate the applied strains during the film formation is a rheometer with a cone-plate geometry. We use the technique of Small Angle X-ray Scattering (SAXS) on the high brilliance beamline ID02 of the European Synchrotron Radiation Facility (ESRF) to analyse in situ the structure at large scale. By coupling on-line rheology with SAXS (Rheo-SAXS), we can determine simultaneously the structure and its stability associated with the shear rate. Hence, the solution is spread using a film applicator within the same mechanical strain deformation scheme. Once dried, films are observed by SAXS. Finally, a detailed analysis of orientation distribution function obtained from the SAXS diffusion pattern on each film provides a quantitative value of the level of alignment of the polymeric chains.

Keywords: block copolymer, shearing, orientation, nanostructure.

1. Introduction

Driven by the increasing interest in nanotechnology, materials that spontaneously form ordered nanostructures are becoming an important area of investigation. Block copolymers have been receiving considerable attention over the last few years due to their property to self-assemble into structures with periodicities on the nanometer scale [1,2]. These new materials are synthesized according to recent progresses in physical chemistry.

The remarkable rich spectrum of morphologies of block copolymers is directly linked with the wide variety of monomers and their relative lengths. One of the distinct assets of block copolymers solutions is the possibility to align their ordered structure to produce "single-crystal-like" materials [3].

However, we want to emphasize that the operational method described in this paper is very different from usual (block copolymers) techniques. Here, the polymer sample is already in the self-assembled state and we aim to directly and efficiently orient the nanostructure, skipping the usual time consuming steps such as drying and annealing.

Under the influence of applied flow fields, it is possible to control the orientational order of the macromolecular chains by changing the shear rate. Spreading block copolymers solutions with the optimal shear rate allows the production of solid polymeric films in which the nanostructure extends to macroscopic scales. This simple polymer processing provides the control of nanostructures orientation and may give a large flexibility on the final morphology of nanostructured materials.

2. Experimental

2.1. Block copolymer sample

In our investigation we used the Polystyrene-*block*-polybutadiene-*block*-poly(methyl methacrylate) block copolymer (SBM). Our specimen is characterized by the following weight proportions: PS 34%; PB 19%; PMMA 47% and the total molecular weight: 65 kg/mol. Much information about these monomers is avaible in [4]. This block copolymer was provided by Pr V. Abetz from Bayreuth University, Germany. The samples consist in solutions of SBM diluted with toluene at the weight concentration of 40%. All observations were made at room temperature, except for three film castings done at the temperature of 4°C.

2.2. Off-line rheological study

In the first place we made an off-line rheological study to observe the rheological response of the sample under shear stress on a large shear rate domain. The shear geometry of the rheometer used was of cone-plate type. A particular external geometry limits solvent evaporation. The sample is then sheared by the relative motion of the cone at 143 μm above the plate. The range of shear rate explored begins at 0.0005 s^{-1} and ends at 2000 s^{-1} where the sample starts spreading by Weissenberg effect. The shear stress versus shear rate behavior (see fig. 1) suggests us to focus our attention on shear rates between 100 s^{-1} and 300 s^{-1}.

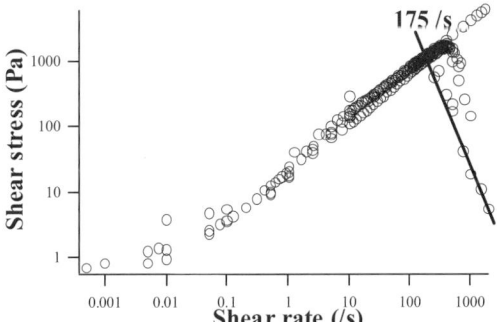

Fig. 1. Rheological response of the solution

2.3. On-line rheological study

To analyse the structure at large scale of the sample, we use the Small Angle X-ray Scattering (SAXS) technique on the high brilliance beamline ID02 of the ESRF. Coupling on-line rheology to SAXS (Rheo-SAXS, [5]), we can determine simultaneously the morphological structure and its stability associated with the shear rate. As we use a cone-plate geometry with a tiny gap (143 μm), it is necessary to reduce the size of the beam. This is not trivial and requires the high brilliance of synchrotron radiation. Diffusion patterns are then recorded at the following shear rates: $0~s^{-1}$, $1~s^{-1}$, $10~s^{-1}$, $50~s^{-1}$, $100~s^{-1}$, $150~s^{-1}$, $200~s^{-1}$, $300~s^{-1}$, and finally at $0~s^{-1}$ ten minutes after stopping.

2.4. Film formation

Polymer films are formed using the film coater Erischen 509 MC-I with the film applicator Erischen 288. This commercial instrument has to be adapted to spread 100 μm thick polymer films on mica windows. Mica windows (thickness: 15 μm, diameter: 16 mm) allow us to observe directly the films on the beamline without deterioration. Fifteen speeds are available from $2.5~mm.s^{-1}$ to $80~mm.s^{-1}$ corresponding to shear rates from $25~s^{-1}$ to $800~s^{-1}$. For each speed, we spread two films at room temperature in a laminar air flow extractor. Three samples are formed at 4°C. Films are then dried at room temperature and at controlled atmosphere for two weeks.

3. Results

3.1. Morphology analysis

In this work, the information on the nanostructure orientation comes from the bidimensional diffusion patterns. Indeed, it provides the morphology and a quantitative value of its alignment and orientation. The morphology of the sample is characterized by the curve $I(q)$ (fig. 2) where q is the diffusion vector norm (see fig. 3a). For each value of q, $I(q)$ is the mean value over Φ from 0 to 2π of the intensity $I(q,\Phi)$, excluding the beamstop. For Rheo-SAXS experiments with a shear rate above $100~s^{-1}$ and for all films, $I(q)$ characteristics are composed of a series of peaks (one to four peaks). These series include a fundamental peak at $q^* = 0,148$ nm^{-1} (corresponding to $d^* = 2\pi/q^* = 42$ nm) and, when present, peaks at $q = 2q^*$, $q = 3q^*$, $q = 4q^*$... which is the mark of a lamellar morphology. It was found that the shear stress induces a change in the nanostructure, further reported in [1, 6].

Scattering patterns stemming from the on-line rheological study Rheo-SAXS are presented in fig. 3. The quantitative analysis of alignment and orientation is described in the following section.

Concerning the films, lamellas are parallel to the mica window and perpendicular to the spreading direction (see fig. 3a). The macromolecular chains are hence oriented in the direction of the spreading direction.

3.2. Anisotropy analysis

Order parameter can be derived from the orientational distribution function [7]. First momentum is called P2 and is defined as :

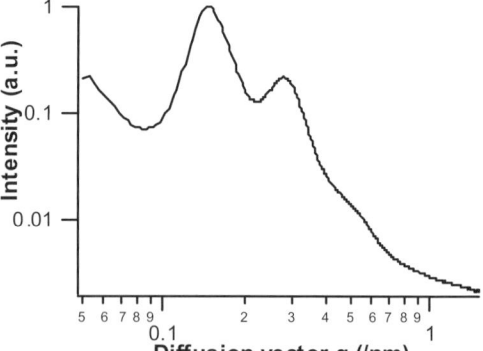

Fig. 2: $I(q)$ for a spreading speed of $30~mm.s^{-1}$

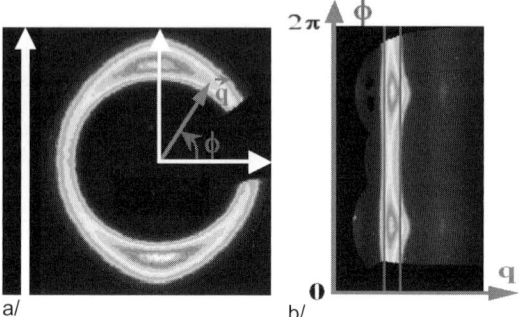

Fig. 3: Coordinates transformation from a/ cartesian coordinates to b/ polar coordinates. The white arrow on the left indicates the spreading direction.

$$P_2 = 1 - \frac{3}{2}\frac{\int_0^{\frac{\pi}{2}} I(\phi)\left(\sin^2(\phi) + \sin(\phi)\cos^2(\phi)\ln\frac{1+\sin(\phi)}{\cos(\phi)}\right)d\phi}{\int_0^{\frac{\pi}{2}} I(\phi)d\phi}$$

where Φ is the angle defined on fig. 3a. To calculate $I(\Phi)$, it is necessary to make a change of coordinates from cartesian coordinates to polar coordinates (see fig. 3b). It is then possible to determine the width of the peaks. For each angle Φ, $I(\Phi)$ is the mean value of the intensity $I(q,\Phi)$ over q on the thin band containing the peaks (illustrated on fig. 3b by the two parallel lines). An example of $I(\Phi)$ curve is represented on fig. 4. Due to the presence of the beamstop, the integration in the P2 parameter goes from π to 2π (actually only to ~356° because of the beamstop). For a perfect alignment oriented in the direction $\Phi = \pi/2$ we have: $P_2 = -1/2$.

The evolution of the order parameter as a function of the shear rate for Rheo-SAXS experiments is represented on fig. 5. A shear rate of $100~s^{-1}$ is sufficient to obtain an alignment. Eventually, the induced alignment is stable because the order parameter is conserved after shear stopping. This is a needed condition for manufacturing nanostructured films to conserve alignment and orientation at large scale during drying.

The order parameter of the films is plotted versus the spreading speed on fig. 6. This demonstrates that even at the lowest spreading speed ($2.5~mm.s^{-1}$) films are oriented. This alignment is observed on the whole surface of the films. The best long scale alignment is found for a speed of $30~mm.s^{-1}$ or $70~mm.s^{-1}$. These

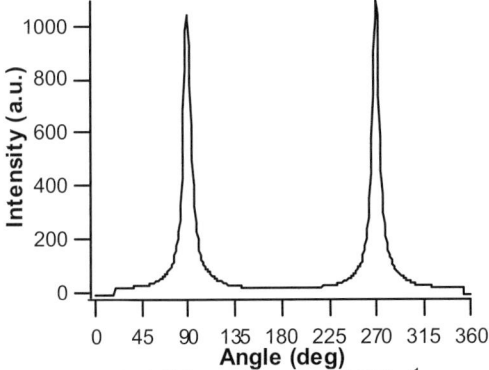

Fig. 4: I(Φ) for a shear rate of 300 s^{-1}

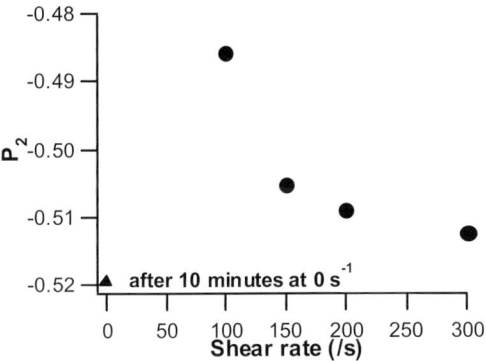

Fig. 5: Order parameter for Rheo-SAXS experiments

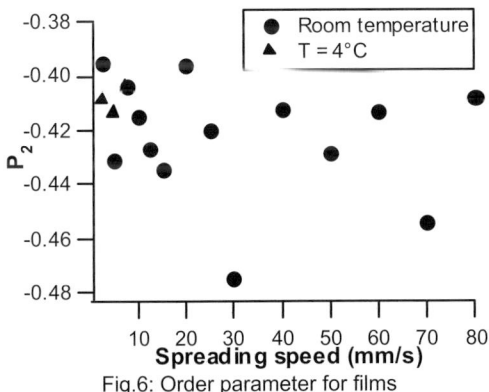

Fig.6: Order parameter for films

particular speeds may be linked to the relaxation time of the macromolecular chains.

4. Discussion

4.1. Hydrodynamically favored phase transition

During the in situ rheological analysis using Rheo-SAXS, the sample exhibits a shear induced phase transition. Below a shear rate of 100 s^{-1} the I(q) graph is composed of one peak at $q' = 0.193$ nm^{-1}. Above this shear rate, I(q) is made up of the series q^*, $2q^*$, $3q^*$, $4q^*$... characteristic of lamellar morphology. Furthermore, the relaxation of the q' peak is very fast: there is a total relaxation within one seconde after stopping shearing at an initial shear rate of 100 s^{-1}. We suggest that this may be the signature of a

hydrodynamically favored phase transition near to the order-disorder transition concentration at room temperature (see [8] for deeper explanations).

4.2. Improvements

The quantitative analysis proves the long scale alignment and orientation with a lamellar morphology. However, the comparison between the order parameter of the solution using Rheo-SAXS and the order parameter of the films highlights the loss of orientation during drying. Order parameter values plotted on fig.6 show that spreading and drying at a temperature of 4°C is not sufficient to limit thermal disorder. An interesting improvement would be to accelerate the evaporation of the solvent. For instance, evaporation could be done under low vapor pressure of solvent and at limited elevated temperature. Finally, the film applicator can be easily optimized to increase productivity.

5. Conclusion

In this paper, we present how to manufacture nanostructured films by shearing a solution of block copolymers with a film coater. With a spreading speed from 2.5 mm.s^{-1} to 80 mm.s^{-1}, we obtain oriented nanostructured polymer films with macroscopic mono-domains of lamellar morphology with a period of 42 nm. The quantitative analysis shows that alignment is reasonable from the slowest spreading speed. Improvements can be done to further enhance the orientation. The simplicity of this processing method allows to expect an easy adaptation to other morphologies such as cylinders or hexagonal compact spheres with interesting photonic crystal applications.

Acknowledgements

The authors thank Pr V. Abetz (GKSS, Institute for Polymer Research, Geesthacht, Germany) for providing the polymer sample and Pr J.-P. Cohen-Adad for lending the film coater. They also thank T. Narayanan for the support on the beamline, P. Boesecke for batch analysis of 2-D X-ray data, J. Gorini for his kind day-to-day help on ID02 and ESRF support groups.

References

[1] N. Hadjichristidis, S. Pispas, G. A. Floudas, *Block copolymers: synthetic strategies, physical properties and applications*, Wiley Interscience (2003).
[2] V. Abetz and T. Goldacker, Macromol. Rapid Comm. 21, 16-34 (2000).
[3] Mingshaw W. Wu, Princeton Univ. PhD thesis (2005).
[4] J. Brandrup, E. H. Immergut, E. A. Grulke, *Polymer Handbook*, Wiley Interscience, Fourth edition (1999).
[5] P. Panine, M. Gradzielski and T. Narayanan, Rev. Sci. Instrum. 74, 2451-2455 (2003).
[6] Z-R. Chen, J. A. Kornfield, Polymer 39, 4679 (1998).
[7] M. Deutsch, Phys. Rev. A 44, 8264-8270 (1991).
[8] P. Panine and N. Didier, Hydrodynamically favored phase transition in self-assembled block copolymers: application to film casting, *these proceedings*.

Process Modelling and Simulation

Mechanistic modelling of the micro- end milling operation

L. Uriarte[a], S. Azcarate[a], A. Herrero[a], L.N. Lopez de Lacalle[b], A. Lamikiz[b]

[a] Fundación Tekniker, Avda. Otaola 20, 20600, Eibar, Spain
[b] Dpto. Ing. Mecanica, ETSII, Alameda Urquijo s/n, 48013, Bilbao, Basque Country, Spain

Abstract

The paper describes a mechanistic model to predict the micro--milling cutting forces and to estimate the tool deflection and the real tool-path during the micro--milling process. Beginning from the "conventional" end milling cutting force model, several modifications are proposed to adapt it for the prediction of the micro--milling cutting forces, which is tha aim of this research. A variety of end mill shapes are considered in the geometric part of the model. The paper presents the experimental validation using two-flute carbide micro-- end-mills with diameters from 0.1 to 0.4 mm. Finally, the consistency between the simulated and measured cutting force is shown as main conclusion.

Keywords: Micro--milling, Modelling

1. Introduction

The current trend towards product miniaturisation is leading to a major increase in micro--technologies. The market is demanding industrial technologies for high yield manufacturing of products at a reasonable price. These two demands can be satisfied by the use of replication technologies which highly depend on the quality and performance of tooling technologies. For this purpose, micro--milling is still the most suitable technology when 3D complex micro-- structures are required as in micro---moulds and dies. Although this technique is similar to conventional scale milling from the operational point of view, the great reduction in dimensions means that cutting phenomena and mechanisms appear that are hardly ever encountered on a conventional scale. This scale reduction can be seen In some of the usual parameters of micro--milling: feed per tooth less than 1 μm, depth of cut 2 - 15 μm, spindle rotational speed more than 50000 rpm and tool diameter less than 0.3 mm. The milling machine itself must also be specific to this application and designed and built to Ultra--precision requirements.

Important efforts are being done to improve micro--milling accuracy [1], productivity, and for downscaling the current capabilities. Previous studies [2] have shown that main error sources in a final milled micro--part are the tool deflection due to cutting conditions, tool wear especially in hard metals, and burr formation [3].

Some of the investigations, including the one presented here, are focused in the precise force estimation, in a similar way than conventional end milling, which has been extensively researched [4-6]. However, there are several phenomena in micro-milling that prevent the results of conventional milling from being applied to it directly. Specific models have been developed that consider one or various of the basic differences that arise from the drastic reduction in size:

1. It cannot be assumed that the micro-structure of the workpiece material is homogeneous [7], as tool size is becoming smaller its effect becomes more important. In this work we consider tools bigger than ⊘0.1 mm, and for simplicity we have not taken it into account.
2. The effect of the cutting edge radius is not negligible [8]. It affects the chip forming mechanism, determining the transition between cutting and ploughing conditions [9].
3. As a result of high tool compliance [9] and the relative size of the cutting edge radius the associated dynamic effects, i.e. forced vibration and regenerative chatter, differ from conventional milling with regard to stability conditions.

On the basis of the high degree of tool compliance, some authors propose a new analytical model of cutting forces [10] that calculates chip thickness according to the actual tool trajectory, which in turn is obtained from the instantaneous balance between cutting force and the elastic restitution force of the system, which is determined mainly by the tool stiffness.

Many of these investigations have come up against problems at the experimental validation stage, given the difficulty of monitoring the extremely small magnitudes, particularly forces and displacements, with a large bandwidth. In most cases experimental validation is carried out using tools bigger than 0.5 mm in diameter. The study presented here was carried out using tools with diameters from 0.1 to 0.4 mm.

Previous studies of the working team [11] are focused in the improvement of the milling of precision moulds (conventional size) with slender tools, which is conceptually similar in many ways to the specific problems of micro-milling. They have provided the basis for the work presented here.

2. Processes: description and capabilities

2.1. Basic end milling model

In spite of the efforts dedicated to models based on FEM, the most used models for the conventional milling process are mechanistic. These kinds of models based their idea on dividing the cutting edge into small discrete elements and applying to each of these elements simple mathematical expressions for the cutting force estimation. Once the forces of each discrete element are calculated, they are added up along the cutting edge obtaining in this way the resulting cutting force. Most of the models have focused on the end milling, considering the case of a rigid tool. They characterize the shearing and ploughing effects separately by the respective specific cutting and edge force coefficients [1,11].

Figure 1 shows the geometric discretization of tool cutting edge during micro--end milling. The model includes the following geometries: cylindrical, ball and bull-nose end mills.

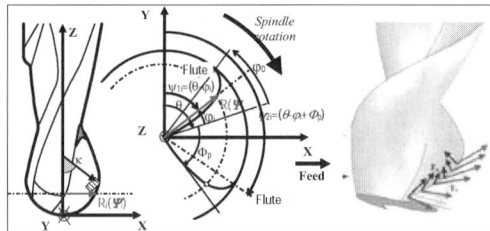

Figure 1 – Modelization of tool cutting edge.

In the force modelling, the cutting force in each direction is function of the uncut chip thickness, t_n, the length of each discrete element, dS, and the chip width, db, so they must be known at first to identify the cutting force. The resulting forces will be the addition of the shearing and friction force components. The contribution of each component is ruled by empirical coefficients, which depend on the workpiece material and the tool geometry. The three projections of the cutting force are shown in (1):

$$dF_t(\theta, z) = K_{te}dS + K_{tc}t_n(\psi, \theta, \kappa)db$$
$$dF_r(\theta, z) = K_{re}dS + K_{rc}t_n(\psi, \theta, \kappa)db \qquad (1)$$
$$dF_a(\theta, z) = K_{ae}dS + K_{ac}t_n(\psi, \theta, \kappa)db$$

Where the tangential cutting force $dF_t(\theta, z)$, radial cutting force $dF_r(\theta, z)$ and axial cutting force $dF_a(\theta, z)$, are represented by using the specific cutting force coefficients K_{tc}, K_{rc} and K_{ac}, and the specific edge force coefficients K_{te}, K_{re} and K_{ae} respectively.

2.2. Identification of specific force coefficients

Once t_n, db and dS are known, the tangential, radial and axial cutting forces of each cutting edge element used in the general model equation (1) are calculated. In order to obtain the resulting cutting force it is necessary to project the cutting forces components in a fixed reference system. Then, the components dF_x, dF_y and dF_z for a tooth of the end mill are determined by equations (2).

$$dF_{xj}(\theta(z)) = -dF_{rj} \cdot \sin \kappa_j \sin \Psi_j$$
$$- dF_{tj} \cdot \cos \Psi_j - dF_{aj} \cdot \cos \kappa_j \sin \Psi_j$$
$$dF_{yj}(\theta(z)) = -dF_{rj} \cdot \sin \kappa_j \cos \Psi_j \qquad (2)$$
$$+ dF_{tj} \cdot \sin \Psi_j - dF_{aj} \cdot \cos \kappa_j \cos \Psi_j$$
$$dF_{zj}(\theta(z)) = dF_{rj} \cdot \cos \kappa_j - dF_{aj} \cdot \sin \kappa_j$$

To obtain the resulting force, it is necessary to perform a numerical integration along the cutting edge engaged in the machining process. Therefore, the boundaries of the integration are estimated based on the engagement conditions of the mill on the part. The resulting cutting force for a cutting edge is calculated by integrating equations (2) within the z variable between 0 and z_{tsup}.

By using previous integration components Fx, Fy and Fz are determined, for each cutting edge of the tool. Since in most cases the tool has more than one cutting edge, the total cutting force is obtained by adding the force of each of them as shown in equation (3).

$$F_x(\theta(z)) = \sum_{j=1}^{N_t} F_{xj}(\theta(z))$$
$$F_y(\theta(z)) = \sum_{j=1}^{N_t} F_{yj}(\theta(z)) \qquad (3)$$
$$F_z(\theta(z)) = \sum_{j=1}^{N_t} F_{zj}(\theta(z))$$

The specific cutting force coefficients depend on the part material and the substrate, coating, rake angle and helix angle of the tool. The relationship among the coefficients and these parameters is very complex and there are no general rules that can be used to predict their values. Therefore, it is necessary to estimate both shear and edge specific cutting coefficients for each tool-material couple by means of characterization tests, using the measured cutting forces as input data. By applying an inverse method, the coefficients are obtained by least square error fitting. The characterization was carried out by horizontal slot milling test with different cutting conditions, avoiding or minimising the effect of tool deflection and minimum chip thickness.

In a first step the materials were characterized by constant values of shear and edge coefficients. A second round of tests considered using polynomial shear coefficients dependent on the position z of the cutting edge elements. The tests have confirmed that the approximation obtained using the linear shear specific coefficients is good enough for obtaining cutting forces estimations [11].

2.3. Cutting edge radius and wear effect

The previous model is applicable when cutting condition is acting. However, due to the cutting edge radius and its fast wearing when hard metals are machined, it is usual that uncut chip thickness was below the minimum chip thickness (h_{lim}), so no chip is formed [9], predominating the ploughing condition. In this study, the minimum chip thickness is assumed to be a fixed ratio (45%) of the cutting edge radius. Below this value, the cutting forces are calculated by (5), and no chip is formed, so the value of the uncut chip thickness, t_n, is accumulated until cutting condition was achieved: $t_n > h_{lim}$.

$$dF_t(\theta, z) = K_{ntc}t_n(\psi, \theta, \kappa)db$$
$$dF_r(\theta, z) = K_{nrc}t_n(\psi, \theta, \kappa)db \qquad (4)$$
$$dF_a(\theta, z) = K_{nac}t_n(\psi, \theta, \kappa)db$$

where the specific cutting force coefficient of the tangential force takes the same value than in cutting condition ($K_{ntc} = K_{tc}$), and the specific cutting force coefficients of the radial and axial forces take the following values: $K_{nrc} = 2.6 \times K_{rc}$ and $K_{nac} = 2.6 \times K_{ac}$ [9]. The measured values of cutting edge radius are 1-2 µm for a new tool, but this value quickly grows up to 10-15 µm when hard metals are machined (Figure 2).

Another effect of wearing is clearly notice in the rounding of the tool tip corner. In order to take it into account, for simulation purposes, even in case of cylindrical end mills they are modelled as bull nose mills.

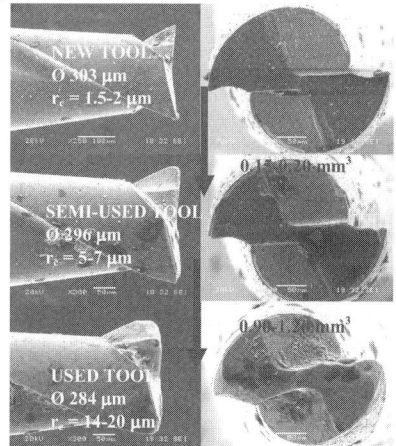

Figure 2–Cutting edge radius and tool tip corner wear

2.4. Tool deflection effect

The influence of the tool itself in the total compliance of the machine-tool-workpiece system varies from 86% for a tool diameter of 0.3 mm and 98% for a tool diameter of 0.1 mm [5]. For such reason, the static stiffness included in the model is the one corresponding to the tool, and only in the radial plane XY. At each instantaneous tool position, the actual chip thickness is achieved when the restoring force in the tool due to its deflection is balanced with the cutting force acting on the radial plane XY. The balance is obtained by means of expression (5)

$$K_x \delta_x = F_x(f_z - \delta_x, a_r - \delta_y)$$
$$K_y \delta_y = \Gamma_y(f_z - \delta_x, a_r - \delta_y) \quad (5)$$

where K_x and K_y are the tool static stiffness measured at the tool tip, F_x and F_y are the cutting force components in X and Y directions, δ_x and δ_y are the tool deflection in X and Y directions, f_z the feed per tooth and the a_r the radial depth of cut. It has been verified [5] that expected tool deflection is in the order of 2-5 μm for similar tool sizes and materials to those used in this research.

The tool deflection effect together with the cutting edge radius wearing increases the influence of the minimum chip thickness, therefore the cutting regime is continuously changing between shearing and ploughing regimes, producing the interrupted cutting phenomena. Also it is experimentally verified that the tool runout measured without engagement is substantially covered up by the tool deflection. Figure 3 shows an schematic view of the model after being adapted for micro-milling.

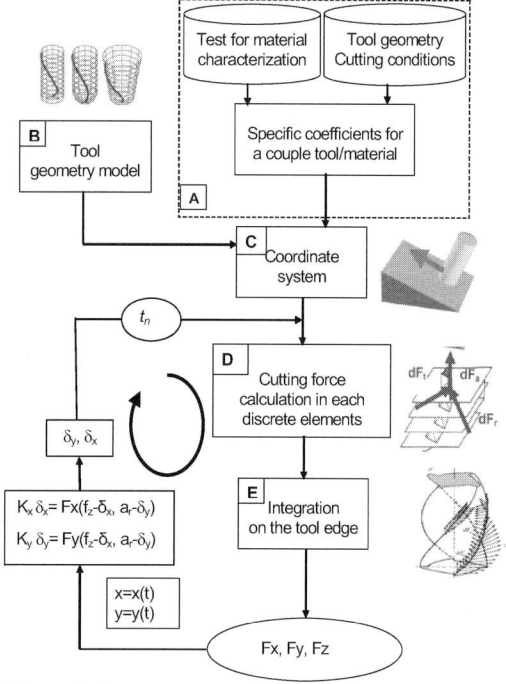

Figure 3– General scheme of the micro-milling model

3. Experimental procedure

3.1. Tools tested

The intention is that tests should be performed on the most widely used types of tool: carbide micro-grain grade end mills (type K10), in this case by M.A. Ford™, multi-layer with TiAlN coating (Figure 4).

D	d	Lc	L	Z	β
3	0.1	0.3	0.45	2	30º
3	0.2	0.6	1.00	2	30º
3	0.3	0.9	1.15	2	30º
3	0.4	1.2	1.40	2	30º

D = Shank diameter   d = tool tip diameter   Lc = Cutting length
L = length with Ød   Z = number of flutes   β = helix angle
All dimensions in mm

Figure 4 – Micro-milling tools analysed.

3.2. Material

The workpiece material used in the experimental tests was H13 steel hardened to 60 HRC. It is an specific steel for mould inserts, combining high hardness, wear and corrosion resistance. It is very fine grained (2-5 μm) in order to minimize the grain size to chip thickness size-effect [12]. In any case, the associated spring-back effect is considered within the minimum chip thickness condition.

4. Experimental and simulated results

First set of tests was done to identify cutting force coefficients by slot milling. The bigger tool, diameter 0.4 mm, was used in order to minimize tool deflection. In each set of tests rotational speed and axial depth of cut were maintained constant, and nine different values of feed per tooth, from 0.15 to 1.25 μm/tooth, were used. Similar set of test were done for different spindles speeds within a range from 30000 to 60000 rpm, and different depth of cut from 5 to 20 μm. Force was measured with a dynamometric plate.

4.1. Example 1
One of this experimental cases is shown in Figure 5. It corresponds to a 0.4 mm tool, 30000 rpm, 10 μm depth of cut, 0.75 μm feed per tooth, 12 μm estimated cutting edge radius and 0.15 μm of estimated runout.

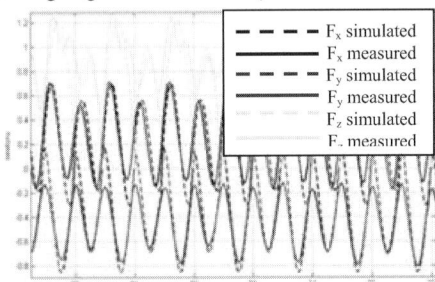

Figure 5 – Measured and simulated forces (example 1)

4.2. Example 2
Figure 6 shows another experimental case, corresponding to a 0.2 mm tool, 60000 rpm, 9 μm depth of cut, 0.2 μm feed per tooth, 12 μm estimated cutting edge radius and 1 μm of estimated runout.

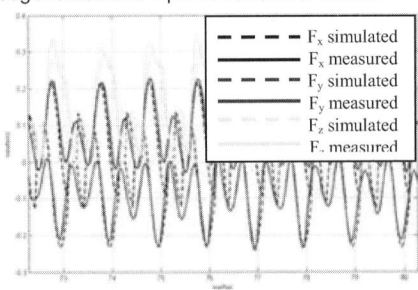

Figure 6 – Measured and simulated forces (example 2)

5. Conclusions and weak points

In the research, the "conventional" end milling model was adapted for micro-milling purposes. Main modifications were to model the cutting edge radius and the large tool deflection effects. The methodology to calculate the cutting force coefficients is proposed and verified with experiments. With these cutting coefficients in each direction, the predicted cutting force was estimated theoretically. The simulated cutting force shows a good consistency with the measured cutting force. However there is still a huge work to be done to characterize the tool wear in micro-milling, and all the dynamic effects like chatter.

References

[1] Weule, H., Huntrup, V. Tritschle, H., 2001, Micro--cutting of steel to meet new requirements in Miniaturization, Annals of the CIRP, 50/1:61-64.

[2] Uriarte, L., Zatarain, M., Santiso, G., Lopez de Lacalle, L.N., Albizuri, J., 2005, Evaluation of dynamic parameters of micro-milling tools smaller than 0.3 mm in diameter, 4M Conf. Proc., 325-328.

[3] Lee, K., Dornfeld, D.A., 2002, An experimental study on burr formation in micro--milling aluminium and copper, Trans. NAMRI/SME, 30:255-262.

[4] Altintas, Y., 2000, Manufacturing Automation, Cambridge University Press.

[5] Smith, S., Tlusty, J., 1991, An overview of modelling and simulation of the milling process, Trans. ASME, Journal of Engineering for Industry, 113:169-175.

[6] Van Luttervelt, K., Childs, T.H.C., Jawahir, I.S., Klocke, F., Venuvinod, P.K., 1998, Present situation and future trends in modelling of machining operations, Annals of the CIRP, 47/2:587-626.

[7] Vogler, M.P., 2003, on the modelling and analysis of machining performance in micro--end milling, PhD Thesis, University of Illinois.

[8] Lucca, D.A., Seo, Y.W., 1993, Effect of tool edge geometry on energy dissipation in ultra--precision machining, Annals of the CIRP, 42/1:83-86.

[9] Kim, C.J., 2004, Mechanisms of chip formation and cutting dynamics in the micro--scale milling process, PhD Thesis, University of Michigan.

[10] Bao, W.Y., 1999, Tool cutting force modelling and wear estimation of micro--end-milling operations, PhD Thesis, University of Florida, Miami.

[11] Lamikiz, A., Lopez de Lacalle, L.N., Sanchez, J.A., Salgado, M.A., 2005, Cutting force integration at the CAM stage in the high-speed milling of complex surfaces, Machine Science and Technology, vol.9:411-436.

[12] Bissaco, G., Hansen, H.N., De Chiffre, L., 2005, Micro-milling of hardened tool steel for mould making, Journal of Materials Processing Technology 167:201-207.

Multi-Material Micro Manufacture
W. Menz, S. Dimov and B. Fillon (Eds.)

On forces and interactions at small distances in micro and nano assembly process

D. Dantchev, K. Kostadinov

Institute of Mechanics, Bulgarian Academy of Sciences, Acad. G. Bonchev St. Bl. 4, 1113 Sofia, Bulgaria

Abstract

We present a short review on the quantum and the thermodynamic Casimir effect and show its practical relevance when considering objects placed away from each other at a distance below a micrometer range. Emphasis is made on the existing experimental verifications of the available theory as well as on the unresolved problems of the theory and the experiment. Everywhere where possible we try to point the potential practical application of the results described not only restricted to micro and nano assembly.

Keywords: Casimir effect, micro & nano assembly, gripping force, micro fluidic interaction

1. Introduction

If two perfectly conducting uncharged plates are arranged so that they are facing each other (see Fig. 1) in "empty space" (vacuum) they attract each other with a force per unit area (pressure) that at zero absolute temperature $T = 0$ is given by

$$F_{Casimir} = -\frac{\pi^2 hc}{240L^4},\qquad(1)$$

where h is the Planck's constant and c is the speed of light. Here L is the distance between the plates. *At distances below a micrometer the Casimir force becomes the strongest force between two neutral objects.* Indeed, at separations of 10 nm the Casimir effect produces the equivalent of 1 atmosphere of pressure.

The fact that such a force exists is called the *Casimir effect*, while the force (the pressure) itself is called the *Casimir force* (pressure) after the Dutch theoretical physicist Hendrik Casimir who first derived the above result in 1948 [1, 2].

The same type of phenomena exist also when the intermediate substance between the surfaces is not vacuum but critical or a correlated fluid (say helium, liquid crystals, etc.). In this form the effect was discussed by M. E. Fisher and de Gennes in 1978 [3].

According to our current understanding, the *Casimir effect is a phenomenon common to all systems characterized by fluctuating quantities on which external boundary conditions are imposed.* When the fluctuating medium is vacuum and the important fluctuations are the quantum ones of the electromagnetic filed the effect is called *the quantum mechanical (or "classical") Casimir effect.* When the fluctuating medium is a fluid or magnet, one speaks about *the statistical mechanical (or thermodynamic) Casimir effect.*

Within the statistical mechanics the system is characterized by a given thermodynamical potential – say the free energy $F(T,L)$, or the grand canonical potential $\Omega(T,L)$. Let the system is with film geometry with surface area A (see Fig. 2). Then, if $f^{ex}(T,L) = f(T,L) - Lf_{bulk}(T)$ is the excess free energy, *the thermodynamic Casimir force* is defined via

$$F_{Casimir}(T,L) = -\frac{\partial f^{ex}(T,L)}{\partial L},\qquad(2)$$

Here $f(T,L) = \lim_{A\to\infty}[F(T,L)/(k_B TA)]$ is the total free energy density (per unit area and per $k_B T$), while $f(T,L) = \lim_{L\to\infty}\lim_{A\to\infty}[F(T,L)/k_B TAL)]$ is the bulk free energy density. If the system is at its bulk critical

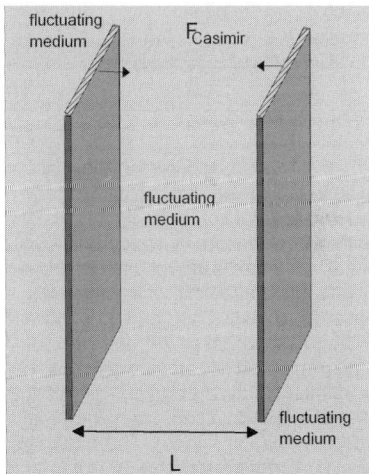

Fig. 1. The theoretical set-up envisaged in the calculation of Casimir. The fluctuating medium can be either vacuum (then one speaks about the quantum mechanical Casimir effect) or a critical fluid (then one speaks about the statistical-mechanical Casimir effect).

temperature T_C, the total free energy can be decomposed in the form

$$f(T_c,L) = Lf_{bulk}(T_c) + f^1_{surface}(T_c) +$$
$$+ f^2_{surface}(T_c) + L^{-(d-1)}\Delta^{(1,2)},\qquad(3)$$

where $f^1_{surface}(T_c)$ and $f^2_{surface}(T_c)$ are the surface free energies characterizing the both surfaces confining the system, $\Delta^{(1,2)}$ is the so-called *Casimir amplitude* and d is the dimensionality of the system. From the above two equations one immediately derives, in full analogy with the quantum Casimir effect, that [4]:

$$F_{Casimir}(T,L) = -L^{-d}\Delta^{(1,2)} . \qquad (4)$$

Near T_c one has that $F_{Casimir}(T,L) = -L^{-d}X^{(1,2)}(atL^{1/\nu})$, where $X^{(1,2)}$ is a *universal scaling function* that depends on the bulk and surface universality classes of the considered system [4], d is the dimensionality of the system and ν is the critical exponent that characterizes the behavior of the bulk correlation function near T_c.

The quantum-mechanical Casimir effect is the most famous mechanical effect due to vacuum fluctuations. It affects the work of micromashines. The understanding and the quantitative knowledge of the forces between two surfaces facing each other at nano or micro distances is of a crucial importance for the design of any micro- and nano-electromechanical

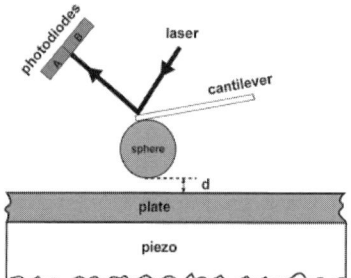

Fig. 3. This experiment measures the Casimir force between a metallized plate and a metallized sphere fixed to the tip of the cantilever of an atomic force microscope. When the sphere is brought near to the plate, an attractive Casimir force causes the cantilever to bend. This bending is monitored by bouncing a laser off the top of the cantilever and using photodiodes to record the reflected light. Application of voltage to the piezo results in the movement of the plate towards the sphere. The experiments were done at a pressure of 50 mTorr and at room temperature.

systems (NEMS and MEMS devices). For example, as MEMS devices are fabricated on the micron and submicron scale, the Casimir force can cause the tiny elements in a device to stick together.

The thermodynamic Casimir force reflects the behavior of a micro fluid and its characteristics – the order parameters profile, the thickness of wetting films, etc. It enters into the balance of the forces that determine the thickness of the wetting film. Its sign and value depends on the properties of both surfaces confining the micro fluid, on if they are imposing the same type boundary condition on the fluid, etc.

Until nowadays the experimentally observed quantum-mechanical Casimir force is always a force of attraction between the material bodies, while for the thermodynamic Casimr force both the cases of

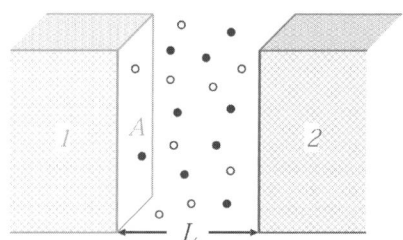

Fig. 2. Statistical-mechanical system that model either a) a simple liquid (the black points) in equilibrium with its vapor (the white points), or b) binary liquid mixture consisting of two type of particles represented by the full and the empty circles.

attraction as well as of repulsion have been observed.

The importance of the investigations on the Casimir effect is clearly viewed in the large number of the relatively recent review article devoted to the subject. For the quantum Casimir effect one can consult [5,6,7,8,9,10] while the developments in the field of the thermodynamic Casimir effect are elucidated in [4,10,11,12].

The first practical applications of the Casimir effect are already in discussion. In the current article we will try to outline some of them. This is done in Section **3**. Before that in Section **2** we summarize some basic experimental and theoretical results that are aimed to help understanding the effect and to grasp the scope of its importance. The article closes with Section **4** "Concluding remarks" where we summarize unresolved questions, which are currently an issue of research or a topic of controversial discussions.

2. *Basic experimental results and facts*

The first attempt to check experimentally the predictions of Casimir was performed as early as in

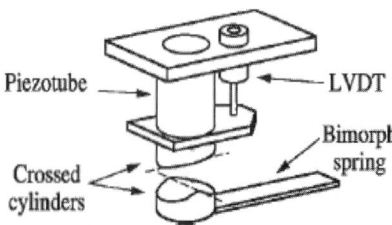

Fig. 4. Simplified view of the force measuring device. The position of the upper surface is controlled with a motorized stage (not in the figure) and a piezoelectric tube, while the response of the lower is detected with the bimorph transducer, acting as the measuring spring. The LVDT (linearly variable displacement transducer) is used to monitor the nonlinear expansion of the piezotube. The radius of curvature of the cylindrical surfaces is 10 mm.

1958 by Sparnaay [13]. Because of the large margin of error in this (and other) early measurements it is, however, now generally accepted that only the observations of Lamoreaux [14] in 1997 (half a century after the prediction) constitute the first definite experimental proof of Casimir's prediction of 1948. In the theoretical set-up envisaged by Casimir one has to ensure that the metallic plates are exactly parallel to each other, as the Casimir force is very sensitive to changes in distance. Since, as it turns out, it is very difficult to maintain parallelism at the requisite accuracy (better that 10^{-5} rad for 1 cm diameter plates) Lamoreaux measured the Casimir force between a spherical lens of radius R=2cm and an optical quartz plate. In such geometry the Casimir force is given by the expression

$$F_{Casimir}(d) = -2\pi R\left[\frac{1}{3}\frac{\pi^2}{240}\frac{\hbar c}{d^3}\right], \qquad (5)$$

where d is the separation at the point of closest approach between the sphere and the plate. Distances $d \in [0.6 - 6]\mu m$ were considered. The lens and plate were connected to a torsion pendulum - a twisting horizontal bar suspended by a tungsten wire - placed in a cylindrical vessel under vacuum. When Lamoreaux

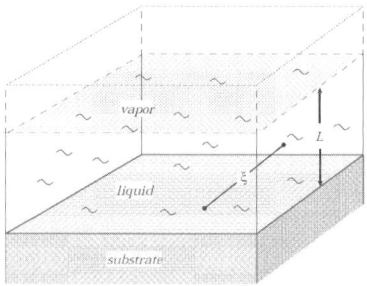

Fig. 5. The typical experimental set-up for measuring the Casimir force in a fluid. The surface of a substrate is placed at height H above the level of a reservoir of a fluid that is in coexistence with its vapor. A thin liquid film with thickness L forms on the substrate. The Casimir force enters into the force balance that determines the thickness L of the film. Measuring L as a function of the temperature (and eventually other parameters) one can study the behavior of the Casimir force.

brought the lens and plate together to within several microns of each other, the Casimir force pulled the two objects together and caused the pendulum to twist. He found that his experimental measurements agreed with theory to an accuracy of 5%. Inspired by Lamoreaux's breakthrough, Mohideen and coworkers [15, 16] attached a polystyrene sphere 200 μm in diameter to the tip of atomic force microscope (see Fig. 3). In a series of experiments they brought the sphere, which was coated with either aluminum or gold, to within 0.1 μm of a flat disk, which was also coated with these metals. The resulting attraction between the sphere and the disk was monitored by the deviation of a laser beam. The researchers were able to measure the Casimir force to within 1% of the expected theoretical value. The same precision has been achieved also by T. Ederth [17] (see Fig. 4) who has also used an atomic force microscope to study the Casimir effect. He measured the force between two gold-coated cylinders that were arranged at 90° to each other and that were

as little as 20 nm apart. For such geometry the Casimir force is given by the expression

$$F_{Casimir}(L) = -\frac{\pi^3 \sqrt{R_1 R_2}\hbar c}{360 L^3}, \qquad (6)$$

where R_1 and R_2 are the radii of the cylinders. The only recent experiment that replicates the Casimir's original set-up of two plane parallel plates was reported in [18]. There the force between a rigid chromium-coated plate and the flat surface of a cantilever made from the same material that were separated by distances ranging from 0.5-3 μm was measured. The researchers found that the force agreed to within 15% of the expected theoretical value. We will finish this small review by noticing that in order to achieve such a good agreement between the experiment and the theory one needs to take into account the corrections in the analytical expressions that are due to the finite temperature, the finite conductivity of the plates and the roughness of the surfaces. For details we refer the reader to the review articles cited above. We would like here only to stress that experimentally observed quantum-mechanical Casimir force was always a force of attraction. Theoretical conditions have been formulated when this force is to be expected to be repulsive [19, 20] (see also the discussion in [21, 22, 23]). Unfortunately till nowadays no experimental verification to the predictions presented [19, 20] is available.

Now we present a short review on the existing experimental results for the thermodynamic Casimir effect. Let us stress from the very beginning that here the situation both with respect to the theory as well as with the experiment is much less satisfactory.

In [24] one considers the critical Casimir effect in binary wetting films, whereas in [25] the effect is observed in 4He films near the superfuid transition.

In [24] the authors present the first experimental determination of the universal function $X(+,-)$ for critical binary liquid mixtures adsorbing from the vapor onto a noncritical substrate where opposite boundary conditions hold at the substrate-liquid and liquid-vapor surfaces of the adsorbed film. They studied the adsorption behavior of two different critical binary liquid mixtures, methanol+ hexane (MH) and 2-methoxy-ethanol + methylcyclohexane (MM), onto a Si wafer placed at a height H above the liquid-vapor surface of the critical liquid mixture (see Fig 5). In a MH mixture hexane adsorbs at the vapor surface while methanol adsorbs at the Si surface. In a MM mixture methylcyclohexane adsorbs at the vapor surface while 2-methoxy-ethanol adsorbs at the Si surface. In both cases the film can be modeled as a slab of thickness L with constant dielectric constant. For such a film the total free energy per unit area $f_L(T,H)$ can be written as

$$f_L(T,H) = \sigma_{ls} + \sigma_{lv} + \rho g L H \qquad (7)$$
$$+ W/L^2 + A\exp(-L/\delta) + L^{-2}k_B T_c X_{ex}^{(+,-)}(L/\xi_\pm),$$

where σ_{ls} and σ_{lv} are the surface tensions between the substrate and the liquid, and between the liquid and the vapor, respectively; $\rho g L H$ is the gravitational contribution to the free energy, ρ being the film density; W is the Hamaker constant characterizing the dispersion interaction, while $A\exp(-L/\delta)$ models the effect of a hard wall (substrate) that structures the fluid over a (molecular) distance δ. Considering L as a variational parameter one determines the equilibrium thickness L by minimizing $f_L(T,H)$ with respect to L. This leads to

$$k_B T_c X^{(+,-)}(x) = \rho g H L^3 - 2W + \frac{AL^3}{\delta}\exp(-L/\delta), \quad (8)$$

where the universal function $X^{(+,-)}(x)$ of the critical Casimir force is related to $X_{ex}^{(+,-)}(x)$ via

$$X^{(+,-)}(x) = 2X_{ex}^{(+,-)}(x) - x\frac{d}{dx}X_{ex}^{(+,-)}(x). \quad (9)$$

Having determined the parameters W, A and δ, one can evaluate also $X^{(+,-)}(x)$ from the t and H dependence of the thickness L. The data for the two mixtures are similar. In MH, for T sufficiently far from $T_c = 307\ K$, the thickness L has approximately the same values above and below T_c, being relatively temperature independent; L *increases sharply as* $T \to T_c^+$ *and exhibits a maximum below* T_c, at $T = 294\ K$ (in agreement with the expectations based on the $2d$ Ising model, see Fig. 12.4 in [4]).

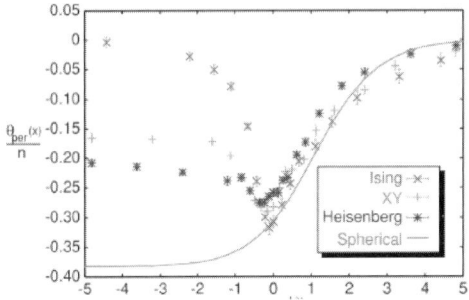

Fig. 6. The scaling function of the Casimir force $\theta_{per}(x)$ for the 3-dimensional $O(n)$ models under periodic boundary conditions. The case $n = 1$ corresponds to the Ising model (normal fluid or a binary critical mixture), $n = 2$ – to the XY model (superfluid helium) and $n = 3$ - to the Heisenberg model (ferromagnetic system). The case $n = \infty$ corresponds to the so-called spherical model that can be solved exactly [36, 37].

It should be noted that the pure components (hexane and methanol) do not exhibit unusual behavior in the vicinity of T_c; thus the behavior of the mixture around T_c is indeed primarily due to the Casimir force.

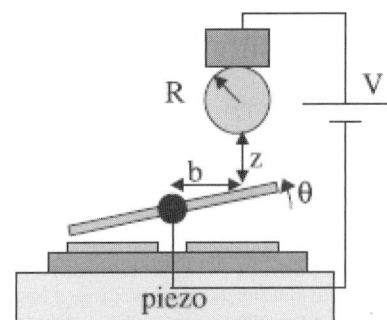

Fig. 7. The principal experimental set-up in [42, 43]

A similar experiment has been performed by Garcia and Chan [25] in which the thinning of a 4He film adsorbed on a Cu surface has been investigated near the superfluid transition. The effect is due to the critical fluctuations which develop in the system in this temperature regime. Note that in this case the both boundary conditions at the two interfaces of the wetting layer are believed to be well approximated by Dirichlet boundary conditions. Therefore, since $X^{(D,D)}(x) < 0$ one expects a critical *thinning* of the wetting layer thickness which is indeed observed in the experiment. In the experimental realization one measures the capacitance of 5 copper capacitors at different elevation H in a cell containing liquid 4He at the bottom. The films are formed on the surfaces of the capacitors. Measuring their capacitance at different temperatures one obtains the thickness L of the film on them as a function of T. However, the experimental estimate of $X^{(D,D)}(x)$ does not show the expected data collapse for $T < T_c$. The magnitude of $X^{(D,D)}(0)$ increases almost linearly with H but there is no clear explanation of this effect. An improvement is obtained if a correction factor accounting for the roughness of the Cu surface is introduced. Much better is the situation above T_λ where the data collapse seems to work, and there is a reasonable agreement between the experiment and the theoretical predictions [11,12]. Finally, it is worth mentioning that away from the transition region the superfuid film is noticeably thinner than the normal fluid film, as one would expect from the presence of Goldstone modes [34, 35, 36, 37, 38, 39, 40].

One might ask why one does not attack the problem of the Casimir force via computer simulations. The answer simply is that till recently there was now operator of the local variables available the average of which to be equal to the Casimir force. For the case of general $O(n)$ systems (where $n = 1$ corresponds to a simple fluid or to the binary liquid mixture, $n = 2$ represents superfluid helium and $n = 3$ corresponds to a ferromagnetic system) such an operator for the case of periodic boundary conditions has been proposed in [38]. The Monte Carlo results of this (stress-tensor) approach are presented in Fig. **6**.

3. Speculations on possible applications of the Casimir effect

Distances below a micrometer are important in nanoscale structures and micro-electromechanical systems (MEMS).

There the Casimir force can cause the tiny elements in a device to stick together - as reported recently in [41]. But the Casimir force can also be put to good use. For example, in [42,43] Capasso and his group showed how the force can be used to control the mechanical motion of a MEMS device (see Fig 7). They suspended a polysilicon plate from a torsional rod - a twisting horizontal bar just a few microns in diameter. When they brought a metallized sphere close up to the plate, the attractive Casimir force between the two objects made the plate rotate. They also studied the dynamical behavior of the MEMS device by making the plate oscillate. The Casimir force reduced the rate of oscillation and led to nonlinear phenomena, such as hysteresis and bistability in the frequency response of the oscillator. According to [42,43] the behavior of the system agreed well with the theoretical calculations. Finally, let us also mention the theoretical study of the Casimir piston that can eventually also find a proper type of application [44].

4. Discussion and concluding remarks

We have demonstrated the importance of the Casimir effect for the interactions between bodies separated at distances below a micrometer. Despite of the progress there is a variety of questions and problems that are still to be resolved. Let us just mention the question about the hypothetical existence of repulsive quantum mechanical Casimir force which is of obvious practical importance. In the thermodynamic Casimir effect one still has to take into account in the theoretical calculations the van der Waals tails of the fluid interactions [45] – both between the fluid particles as well as between the fluid and the substrate molecules. Extension of the stress-tensor approach to other boundary conditions that will allow modeling the force under realistic boundary conditions is also highly desirable.

Acknowledgements:

This work is partially supported by the Institute of Mechanics – BAS, Bulgarian Fund for Scientific Research F1402, Bulgarian Ministry of Education and Science under the Ro- MiNa project TH1308/03 to which the authors are expressing their acknowledgements.

We are also very grateful to the EC contribution of 4M NoE No. 500274 of NMP2-CT-2004 and HYDROMEL No.026622-2 of NMP3-IP-2006.

References

[1] H. B. G. Casimir, *Proc. K. Ned. Akad. Wet.* **B 51**, 793 (1948).

[2] H. B. G. Casimir, *Physica* **19**, 846 (1953).

[3] M. E. Fisher and P.-G. de Gennes, *C. R. Acad. Sci. Paris Serie* **B 287**, 207 (1978).

[4] J. G. Brankov, D. M. Dantchev, and N. S. Tonchev, *Theory of Critical Phenomena in Finite-Size Systems Scaling and Quantum Effects* (World Scientific, Singaporo, 2000).

[5] V. M. Mostepanenko and N. N. Trunov, *The Casimir effect and its applications* (Moscow, Energoatomizdat, 1990, in Russian); English version: (New York, Clarendon Press, 1997).

[6] M. Kardar M. and R. Golestanian, *Rev. Mod. Phys.* **71**, 1233 (1999).

[7] M. Bordag, U. Mohideen, V. M. Mostepanenko, Physics Reports **353**, 1 (2001).

[8] K. A. Milton, *The Casimir Effect: Physical Manifestations of Zero-Point Energy* (Singapore: World Scientific, 2001).

[9] A. Lambrecht, *Phys. World* **15** (9), 29 (2002).

[10] K. A. Milton, *J. Phys. A* **37**, R209 (2004).

[11] M. Krech and S. Dietrich, Phys. Rev. Lett. **66**, 345 (1991); Phys. Rev. A **46**, 1886 (1992); **46**, 1922 (1992).

[12] M. Krech, *Casimir Effect in Critical Systems* (Singapore: World Scientific, 1994); M. Krech, *J. Phys.* **C 11**, R391 (1999).

[13] M. J. Sparnaay, *Physica* (Utrecht) **24**, 751 (1958).

[14] S. K. Lamoreaux, *Phys. Rev. Lett.* **78**, 5 (1997), erratum **81**, 5475 (1998).

[15] U. Mohideen and A. Roy, Phys. Rev. Lett. **81**, 4549 (1998).

[16] B. W. Harris, F. Chen, and U. Mohideen, *Phys. Rev. A* **62**, 052109 (2000).

[17] T. Ederth, *Phys. Rev.* **A 62**, 062104 (2000).

[18] G. Bressi, G. Carugno, R. Onofrio, and G. Ruoso, *Phys. Rev. Lett.* **88**, 041804 (2002).

[19] T. Boyer, *Phys. Rev.* **A 9**, 2078 (1974).

[20] O. Kenneth, I. Klich, A. Mann, and M. Revzen, *Phys. Rev. Lett.* **89**, 033001 (2002).

[21] D. Iannuzzi and F. Capasso, *Phys. Rev. Lett.* **91**, 029101 (2003).

[22] O. Kenneth, I. Klich, A. Mann, and M. Revzen, *Phys. Rev. Lett.* **91**, 029102 (2003).

[23] M. P. Hertzberg, R. L. Jaffe, M. Kardar and A. Scardicchio, *Phys. Rev. Lett.* **95,** 250402 (2005).

[24] A. Mukhopadhyay and B. M. Law, *Phys. Rev. Lett.* **83**, 772 (1999).

[25] R. Garcia and M. H. W. Chan, *Phys. Rev. Lett.* **83**, 1187 (1999).

[26] A. Mukhopadhyay and B. M. Law, *Phys. Rev.* **E 62**, 5201 (2000).

[27] R. Garcia and M. H. W. Chan, *Physica* **B** 280, 55 (2000).

[28] R. Garcia and M. H. W. Chan, *J. Low Temp. Phys.* **121**, 495 (2000).

[29] R. Garcia and M. H. W. Chan, *Phys. Rev. Lett.* **88**, 086101 (2002).

[30] T. Ueno, S. Balibar, T. Mizusaki, F. Caupin, and E. Rolley, *Phys. Rev. Lett.* **90**, 116102 (2003).

[31] S. Balibar and R. Ishiguro, *Pramana J. Phys.* **64**, 743 (2005).

[32] R. Ishiguro and S. Balibar, *J. Low Temp. Phys.* **140**, 29 (2005).

[33] M. Fukuto, Y. F. Yano, and Peter S. Pershan, *Phys. Rev. Lett.* **94**, 135702 (2005).

[34] H. Li and M. Kardar, *Phys. Rev. Lett.* **67**, 3275 (1991).

[35] H. Li and M. Kardar, *Phys. Rev.* A 46, 6490 (1992).

[36] D. M. Danchev, *Phys. Rev.* **E 53**, 2104 (1996).

[37] D. M. Danchev, *Phys. Rev.* **E 58**, 1455 (1998).

[38] D. Dantchev and M. Krech, *Phys. Rev.E* **69**, 046119 (2004).

[39] R. Zandi, J. Rudnick, and M. Kardar, *Phys. Rev. Lett.* **93** 155302 (2004).

[40] D. Dantchev, M Krech, and S. Dietrich, *Phys. Rev. Lett.* **95,** 259701 (2005).

[41] E. Buks and M. L. Roukes, *Phys. Rev.* **B 63**, 033402 (2001).

[42] Chan H. B., V. A. Aksyuk, R. N. Kleiman, D. J. Bishop, F. Capasso, Science **291**, 1941, 2001.

[43] Chan H. B., V. A. Aksyuk, R. N. Kleiman, D. J. Bishop, and F. Capasso, *Phys. Rev. Lett.* **87,** 211801, 2001.

[44] Hertzberg M. P., R. L. Jaffe, M. Kardar, and A. Scardicchio, Phys. Rev. Lett. **95**, 250402 (2005).

[45] Dantchev D., H. W. Diehl, and Daniel Grüneberg, Phys. Rev. **E 73** 016131, 2006.

Multi-Material Micro Manufacture
W. Menz, S. Dimov and B. Fillon (Eds.)

Design procedure of planar compliant microgrippers with flexural joints

F. Székely, T. Szalay

*Budapest University of Technology and Economics, Department of Manufacturing Engineering,
Egry Street 1, Budapest, 1111, Hungary*

Abstract

The overall design scope of most microgrippers developed in recent years lacks a systematic mechanism design approach. Accordingly, the main objective of this investigation is to establish a new design concept in order to enhance the design scope of microgrippers. Using a systematic design procedure which particularly stresses planar compliant mechanisms, this study presents a two-fingered microgripper which has parallelogram mechanism. By the help of this construction the gripping surfaces stay in parallel position during its motion. The preliminary Finite Element (FEM) simulation results are in good agreement with the expected kinematic motion. Moreover, the stress analysis also points out that the relationship between the direction of driving force and orientation of deflected compliant joints is one of the crucial factors for designing the compliant microgripper mechanism. Hence, the mechanism design concept presented in this study can be integrated e.g. into the design of micro-scale actuating devices for Micro-Electro-Mechanical Systems (MEMS).

Keywords: microgripper, compliant mechanism, laser cutting

1. Introduction

The practical applications of MEMS are limited due to the difficulty of construction of three-dimensional (3-D) structures. An efficient microassembly technique is needed to integrate two-dimensional microparts into 3-D structures. Recently, researchers began investigating the assembly of microstructures. Lee et al. [1] presented the micropeg manipulation with a compliant microgripper, and Hui et al. [2] proposed a new method of microassembly of complex 3-D microstructures using a pop-up mechanism.

Topology optimization, which was originally proposed by Maxwell and Michell [3,4], is a special case of configuration design where we search not only the optimum member sizes and joint positions but also determine how many joints should be in the structure and how these joints should be connected by members [5]. Due to the difficulty of building actual structures, Dorn suggested a new approach [6], that defines a set of admissible joint positions for a planar truss structure. This concept was extended by Dobbs [7], which includes multiple loading conditions and the elastic design of planar trusses. Another approach was presented by Sheu and Schmit [8], in which they began with a reasonably large set of separate possible configurations, and then designed each structure subjected to stress constraints only. Taking the minimum weight structure from the entire set, they added displacement constraints and designed this structure subjected to the combined stress and displacement constraint set.

More recent work in topology optimization was conducted by Bendsoe and Kikuchi [9]. They used the homogenization method that designs the size and shape of holes in the structure. This method requires a special element formulation. A related method, known as the density-based approach, treats the density of finite elements as design variables and adjusts Young's modulus of the elements as a variable [10].

2. Theory

According to Her and Midha [11], compliant mechanisms represent a class of mechanical systems which make use of flexible beams in their designs, as opposed to the exclusive use of rigid-body members. Since the compliant mechanism is capable of providing a certain degree of controllable constraining movement, it can be physically regarded as a mechanism whose motion is not achieved via kinematic joints as in a conventional rigid-body mechanism, but by the deflection of part of the mechanism's microstructure. Fig 1. shows the geometry and significant dimensions (r, t, b and h) of a typical flexure hinge which is adopted as a compliant revolute joint in the design work.

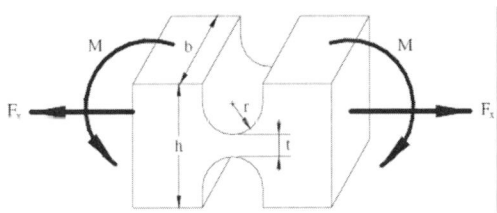

Fig. 1. The generalized flexure hinge model.

Classical analytical schemes for the displacement analysis of a flexure hinge have been comprehensively studied by Paros and Weisbord [12]. The spring constant or the bending stiffness, K_Θ, is given as follows:

$$K_\Theta = \frac{2 \cdot E \cdot b \cdot t^{2.5}}{9 \cdot \cdot r^{0.5}} \qquad (1)$$

where the subscript b denotes bending, and E is Young's modulus of the material.

3. Design

The planar two-fingered microgripper mechanism proposed in this study is a compliant microstructure made of a single material, and composed of flexible linkages and micro-joints. In other words, the microgripper is an integrated construction which can provide the desired gripping movement with no clearance or backlash at the micro-joints. The structural part of the mechanism merges the two functional regions of microgripper devices, namely the actuation source component and the gripping component.

This study basically derives the core design of the compliant microgripper from that of the conventional rigid body macrogripper mechanism. Using the typical rigid–compliant mechanism transformation process developed below, this paper successfully generates several compliant microgrippers suitable for further development into actual devices. The compliant microgripper employs a linear actuating element as a shape memory spring, while its gripping part involves individual revolute joints (hinges) or prismatic joints, or combinations of both (see Fig. 2)

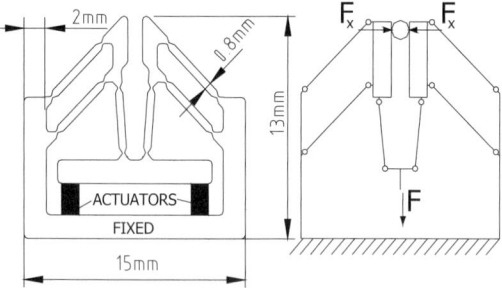

Fig. 2. Flexure hinge based mechanism of microgripper and its equivalent kinematic model.

4. Transformation of compliant mechanism

At present, the technical literature focuses primarily on the kinematic design of conventional macro-scale gripper mechanisms. In the field of MEMS microgripper development, very few studies have explored suitable methodologies for converting rigid kinematic designs into compliant kinematic mechanisms. It appears that there are two practical reasons why this should be the case, namely, the infeasibility of many designers successfully merging the common mechanism design methodologies which incorporate the inter-disciplines of micro-system technologies and conventional macromechanisms, and secondly, the difficulties inherent in implementing the technology required to realize the device with the desired performance characteristics. Therefore, the following paragraphs explore a microgripper design process which couples the technological problems involved in the complex micro–macro domain.

In rigid-body kinematics, mechanisms achieve motion through the use of kinematic pairs, such as revolute pairs (hinges), prismatic pairs (sliding joints), cam pairs and gear pairs, etc.

To construct the common kinematic relationship of a rigid–compliant mechanism, a transformation process is developed which converts the rigid body joints into corresponding compliant joints with similar actuation performance characteristics. This transformation process allows designers to quickly construct the basic configuration of compliant mechanisms. The process is presented in Fig.3.

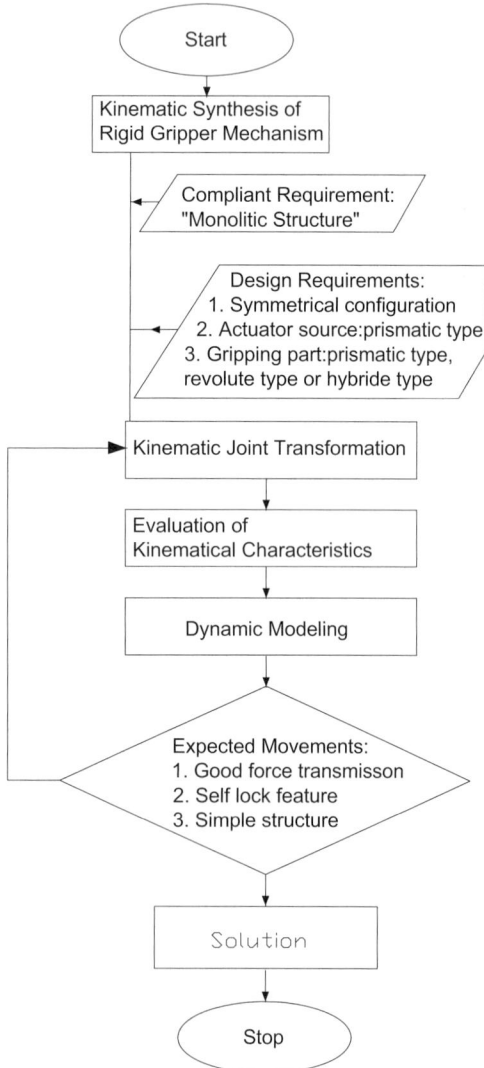

Fig. 3. Systematic transformation of rigid mechanisms into their compliant counterparts.

5. Pseudo-Rigid-Body-Model (PRBM)

The flexure-based parallelogram mechanism was analytically studied using a method called pseudo-rigid-body-model for analyzing compliant mechanisms. The model was derived from the flexure-based mechanism by replacing flexure hinges with rotational springs with equivalent spring stiffness constant calculated from the flexure hinge equations [12]. Since the mechanism is symmetric, only half portion is analyzed and the resulting PRBM is shown in Fig. 4. It is assumed that all the flexure hinges in the mechanism have same dimensions and hinge-stretching is neglected. Also it is assumed that no loss of motion due to bonding between actuator and the microgripper mechanism.

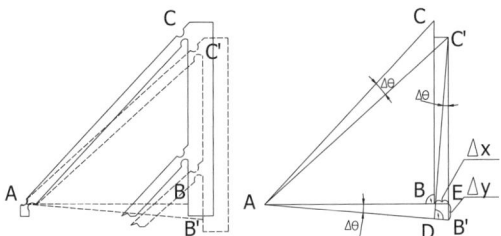

Fig. 4. Half-symmetric portion of the flexure-based mechanism and its PRBM.

Fig. 2. shows the initial (ABC) and final (AB'C') position of the gripping link when subjected to an input force F which results an angular displacement $\Delta\Theta$.
Geometrical conditions:

$$\Delta\Theta \leq 5° \Rightarrow \overline{DB'} \approx \overline{BE} = \overline{\Delta x},$$

$$\overline{AB} = \overline{BC} = \overline{B'C'}$$

From $\quad \Delta$ DB'C', $\sin\Delta\Theta = \dfrac{DB'}{B'C'},\quad$ for small

$\Delta\Theta$, $\sin\Delta\Theta \approx \Delta\Theta$;so, $\Delta\Theta = \dfrac{\overline{\Delta x}}{\overline{BC}}$ $\quad$ (2)

From $\quad \Delta$ AEB', $\quad tg\Delta\Theta = \dfrac{\overline{\Delta y}}{\overline{AB} + \overline{\Delta x}},$

for small $\quad \Delta\Theta$, $tg\Delta\Theta \approx \Delta\Theta$; $\quad$ so,

$$\Delta\Theta = \frac{\overline{\Delta y}}{\overline{AB} + \overline{\Delta x}} \quad (3)$$

From Eqs. (1) and (2), $\dfrac{\overline{\Delta x}}{\overline{BC}} = \dfrac{\overline{\Delta y}}{\overline{AB} + \overline{\Delta x}}$ or,

$$\frac{\overline{\Delta x}}{\overline{AB}} = \frac{\overline{\Delta y}}{\overline{AB} + \overline{\Delta x}} \Rightarrow \overline{\Delta y} = \frac{\overline{\Delta x} \cdot \left(\overline{AB} + \overline{\Delta x}\right)}{\overline{AB}} \quad (4)$$

Work done by input force F = Sum of potential energy in the rotational springs is presented in Eq. 5.

$$F \cdot \overline{\Delta y} = \frac{1}{2} \cdot K_{A,\Theta} \cdot \Delta\Theta_A^2 + \frac{1}{2} \cdot K_{C,\Theta} \cdot \Delta\Theta_C^2 \quad (5)$$

where, $K_{A,\Theta}$ and $K_{C,\Theta}$ are the rotational spring stiffness and $\Delta\Theta_A, \Delta\Theta_C$ are the rotational displacements. Since the flexure dimensions are assumed same for A and B,

$$K_{A,\Theta} = K_{C,\Theta} = K_\Theta\, and\, \Delta\Theta_A = \Delta\Theta_C = \Delta\Theta$$

$$F \cdot \overline{\Delta y} = K_\Theta \cdot \Delta\Theta^2 \quad (6)$$

Therefore, substituting Eqs. (3) and (4) in Eq. (6),

$$F \cdot \overline{\Delta y} = K_\Theta \cdot \left(\frac{\overline{\Delta y}}{\overline{AB} + \overline{\Delta x}}\right)^2 or,$$

$$F = K_\Theta \cdot \frac{\overline{\Delta y}}{\left(\overline{AB} + \overline{\Delta x}\right)^2} = K_\Theta \cdot \frac{\overline{\Delta x}}{\overline{AB} \cdot \left(\overline{AB} + \overline{\Delta x}\right)} \quad (7)$$

6. Estimation of gripping force from PRBM

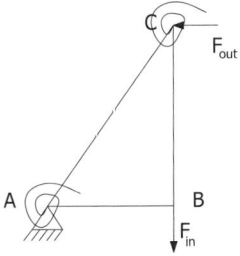

Fig. 5. Different forces acting on the gripper link.

From equilibrium condition, shown in Fig. 5

$$F_{in} \cdot \overline{AB} = F_{out} \cdot \overline{BC} + 2 \cdot K_\Theta \cdot \frac{\overline{\Delta x}}{\left(\overline{AB} + \overline{\Delta x}\right)} or,$$

$$F_{out} = \frac{F_{in} \cdot \overline{AB} - 2 \cdot K_\Theta \cdot \dfrac{\overline{\Delta x}}{\left(\overline{AB} + \overline{\Delta x}\right)}}{\overline{BC}} \quad (8)$$

7. Creating of the finite element model (FEM)

The given microgripper could be modelled in 2D. The thickness has been chosen for 0,1 mm. Polioximethylene (POM) has been chosen as the material of microgripper and the average strength characteristics of POM materials have been given in the fine element software, which are summarized in Table 1. Damage of objects must be avoided during grip, it explains the selection of POM.

250

Table 1

The average strength characteristics of POM materials

Property	Description	Value	Units
EX	Elasticity modulus	$2.4 \cdot 10^3$	Mpa
NUXY	Poisson's ratio	0.35	-
GXY	Shear modulus	$8.9 \cdot 10^2$	Mpa
DENS	Mass Density	1200	kg/m^3
SIGXT	Tensile strength	$5.17 \cdot 10^2$	Mpa
SIGYLD	Yield stress	$2.069 \cdot 10^2$	Mpa

Since contact surfaces are extraordinarily small therefore a relatively small tightening-up force can make a big surface pressure. The gripping force is located on the jaws because these forces keep the gripper in opened position when operating forces are ended, and these forces load the gripper in the aggregate.

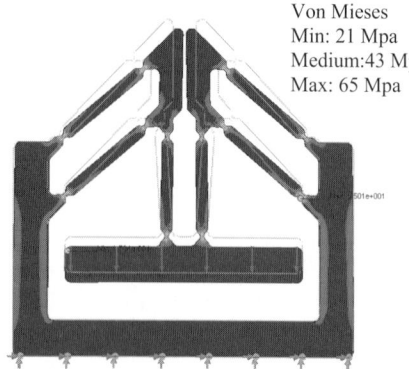

Von Mieses
Min: 21 Mpa
Medium:43 Mpa
Max: 65 Mpa

Fig.6. FEM simulation of the movement of the microgripper finger.

In the finite element model (see Fig. 6) it can be seen that the stress maximum (65 MPa) is on the flexible joints. The horizontal excursion of jaw is 800 µm. The operating force has been chosen for 0,15 N and the average strength characteristics are the same as in the previous case.

8. Fabrication of microgripper

Use of laser cutting is a perspective method for fabrication of this microgripper because forces do not load the gripper during the cutting, and so the utilization of available raw material is the most economical. The fabricated microgripper (see Fig. 7.) has waveless contour and small actual deviation.

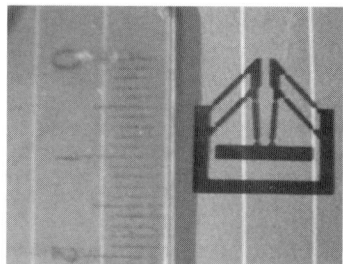

Fig.7. The fabricated microgripper.

9. Conclusion

By the help of design method presented in this paper microgrippers can be constructed systematically. The design scope is focused on a particular section of effectors used in microtechnics.

References

[1] W. H. Lee, B. H. Kang, Y. S. Oh, H. E. Stephanou, A. C. Sanderson, G. Skidmore, and M. Ellis, "Micropeg Manipulation with a Compliant Microgripper," In Proc. *IEEE Int'l Conf. on Robotics and Automation*, 2003.

[2] E. E. Hui, R. T. Howe, andM. S. Rodgers, "Single-Step Assembly of Complex 3-D Microstructures," *IEEE 13th Int'l Conf. on Micro Electro Mechanical Systems*, pp. 602-607, 2000.

[3] C. Maxwell, "Scientific Papers," Vol. 2, pp. 175-177, 1869.

[4] A. Michell, "The limits of Economy of Material in Frame Structures," *Philosophical Magazine*, Series 6, Vol. 8, pp. 589-597, 1904.

[5] G. N. Vanderplaats, *Numerical Optimization Techniques for Engineering Design*, McGraw-Hill, pp. 364-367, 1984.

[6] W. Dorn, R. Gomony, and H. Greenberg, "Automatic design of optimal structures," *Journal de mecanique*, Vol. 3, pp. 25-52, 1964.

[7] M. Dobbs and L. Felton, "Optimization of truss geometry," *ASCE Journal of Structural Division*, Vol. 95, pp. 2105-2118, 1969.

[8] C. Sheu and L. Schmit, "Minimum Weight Design of Elastic Redundant Trusses under Multiple Static Loading Conditions," *AIAA J.* Vol. 10, pp. 155-162, 1972.

[9] M. Bendsoe and N. Kikuchi, "Generating Optimal Topologies in Structure Design Using a HomogenizationMethod," *Comp. Meth. Appl. Mech. Eng.*, Vol. 71, pp. 197-224, 1988.

[10] O. Sigmund, "A 99 Line Topology Optimization Code Written in Matlab," *Struct. Multidisc. Optim.*, Vol. 21, pp. 120-127, 2001.

[11] Her I and Midha A 1987 A compliance number concept for compliant mechanisms, and type synthesis *ASME J. Mech., Transmissions Type Synthesis* **109** 348–55.

[12] Paros L M and Weisbord L 1965 How to design flexure hinges *Mach. Design.* **37** 151–6.

Finite element analysis of the micro glass moulding process

C. Brecher, M. Winterschladen, S. Lange, F. Klocke, G. Pongs, F. Wang

[a] *Fraunhofer Institute for Production Technology IPT,*
Steinbachstraße 17, 52074 Aachen, Germany

Abstract

The manufacturing of optics is an important field of technology and will serve key-markets in the future. The research activities of the Transregional Collaborative Research Centre "Process Chains for the Replication of Complex Optical Elements" SFB/TR4 of the Universities of Aachen, Bremen and Stillwater (USA) have the objective to lay the scientific foundations for a deterministic and economic mass production of optical and micro optical components with complex geometries, e.g. aspheric, non-rotational asymmetric or microstructured surfaces eventually superimposed on freeform geometries. The paper presents first principle simulation results of precision glass moulding for the analysis of micro form as well as nano surface quality manufacturing effects during the moulding process.

Keywords: Finite Element Analysis, Glass Moulding, Precision Optics

1. Introduction

Recently, aspherical lenses have been increasingly selected to replace spherical optics [1, 2]. If designed properly, optical devices using non-spherical elements or micro structured elements can help to reduce optical aberrations, internal reflections and subsequently minimize assembly requirements. Benefiting from continuing research and development, freeform optical surfaces and micro structured surfaces with dimensions of a few micrometers are now becoming a practical solution to many micro opto-mechanical designs.

As an alternative to the traditional glass lens manufacturing process, precision moulding of free form optics and micro structured optics can be an attractive approach [1-2]. The glass moulding process is a compression hot forming method in which a glass gob is pressed in a single operation into the shape of a finished lens. Annealing of the formed lens is then carried out to improve lens geometry. The glass lens manufacturing using the compression moulding method as compared to the traditional processes is an environmentally friendly process since it eliminates the use of grinding fluids and polishing slurries. Moreover glass lenses are preferred optical materials for compact micro design and better optical quality in harsh environments. To manufacture complex lenses and micro lenses with highest form accuracies the mould inserts must be fabricated with an adapted geometry to compensate the typical shrinking effects of the replication process. Today, to determine the suitable geometry of the mould inserts several iterations are necessary. That is due to the fact that the form deviations are too complex for a prediction in one step. For the machining of complex moulds in high quality several technologies are investigated in the SFB/TR4 [3, 4].

The superior goal of the presented work is to minimize the iterations of micro mould manufacturing in order to increase the efficiency of the fabrication process, to reduce the costs and the time to market. Therefore the precision glass moulding process is simulated by the use of Finite Element Analyses (FEA). This paper gives an overview of the process of glass moulding, the implementation of simulation model for micro optics and its boundary conditions. The micro relevant effects due to the demand of highest lens accuracies and micro structured surface designs are discussed.

2. Precision glass moulding

Precision glass moulding enables shaping of ready to use glass surfaces of large scale optics, of micro scale optics and optics with micro structured surfaces in one process step. This technology is qualified to mould inorganic glass melts in a temperature range between 450 to 800 °C (depending on optical glass material) to generate accurate optical surfaces. Machining of the optical surface afterwards is not necessary [5].

Both, patents and technical literature provide no consistent content about the manufacturing of optical components using precision glass moulding with form accuracies of few micrometers. Papers refer to different process methods although an optimum method for producing precise structured components can not be deduced. Most methods are suitable to mould rotationally symmetrical aspherical parts [6]. Fig. 1 shows the assembly of mould inserts with a diameter of 10 mm in the glass moulding machine as well as the machine set-up of the used glass moulding machine.

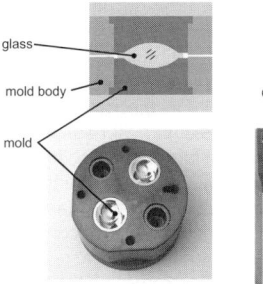
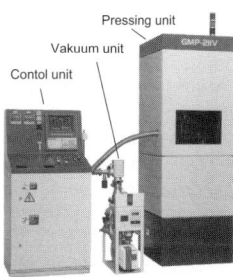

Fig. 1: Mould Design and Machine Set Up

The glass moulding process can roughly be grouped into six different stages as seen in Fig. 2. The process of moulding large scale optics as well as micro optics is in principle comparable. Differences can be found in heating time and pressuring time.

First, the glass gob is loaded into the lower mould precisely to realise constant starting conditions. In case of moulding micro optics or optics with structured surfaces, the precision of loading as well as the mass varieties of the glass gobs have a very high influence to the moulding results. After loading nitrogen flows through the mould assembly area to purge the chamber. Then the moulds and the glass gob are heated to a temperature slightly higher than the transition temperature of glass (Tg). In stage four, moulding of the glass lens takes place at a constant temperature. After pressing a slow cooling of the lens and the moulds is initialised. In stage five, the moulding process goes through a rapid cooling stage and the finished lens is released.

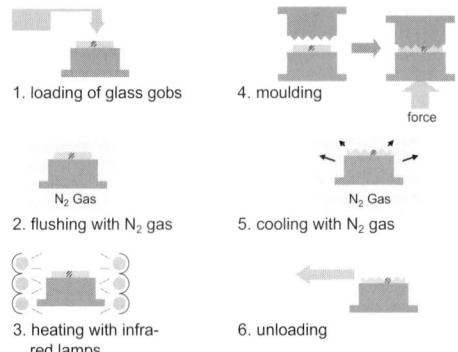

1. loading of glass gobs

2. flushing with N_2 gas

3. heating with infra-red lamps

4. moulding

force

5. cooling with N_2 gas

6. unloading

Fig. 2: Process cycle of precision glass moulding

The mould material has a great impact on precision glass moulding. Concerning the unsettled problem of adhering and sticking effects, past investigation were limited to certain soda lime glasses. Particularly both effects are very important for moulding of micro optics and large scale lenses with micro structured surfaces.

In contrast to conventional glass moulding the process time is substantially longer in precision glass moulding. Pure steel materials are not suitable for mould materials. The corrosion of the mould surfaces in conjunction with the extended contact time leads inevitably to sticking and adhering effects. Even specially developed glass moulding steel is inclined to sticking effects. This can cause destruction of the glass and mould surfaces by moulding micro optics or optics with structured surfaces.

The requirements in terms of heating and cooling cycles are particularly high. When the glass is heated in the press by the thermal conduction via the pressing tools, various heating rates are needed depending on the type of glass concerned. In case of internal and external heating, the pressed part is cooled within the machine after the pressing operations. The moulds undergo cooling functions in this time. The temperature control during cooling, especially in the range of the transition temperature of the glass, is essential.

The main goal of the precision glass moulding is the replication of components with complex optical geometries (micro optics, optics with free form surfaces or micro structured surfaces), in high precision at constant moulding conditions. With this technology form accuracies can be reached in the lower micrometer range (< 2 μm) in moulding large scale optics with outer dimensions greater 50 mm. The local form deviations of micro optics or optics with microstructures have to be below one micrometer.

To realise these accuracies the geometry of mould inserts has to be adapted to each the moulding conditions and to the moulding process. In this context the moulding process can be represented by a simulation model using finite element analysis, to increase the efficiency of the mould design and to decrease the iteration steps in predictively adapting the mould geometries to the process conditions.

3. Finite element analysis of precision micro glass moulding

In the following the primary effects in the simulation of precision glass moulding are described. This includes the general process design of moulding large scaled optics, micro optics and optics with micro structured surfaces.

The main components of a precision glass moulding machine are the upper and the lower mould, the glass gob as well as the heating system. These components can in principle be modelled by finite elements as shown in Fig. 3. The heating system is simplified by a cylinder and the power of the infrared lamps of the heating system is transferred to the cylinder. The pressing unit is designed by the upper and the lower mould. The geometry of the moulds as well as the geometry of the glass gob can be described with the NURBS data format. This data format is suitable for the design of all kind of surfaces.

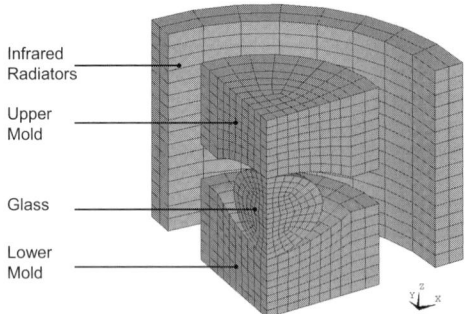

Infrared Radiators

Upper Mold

Glass

Lower Mold

Fig. 3: Finite element model of glass moulding

The main influences on form deviations between moulds and lens are shrinkage effects of glass. Optical glasses show a viscoelastic material behaviour. Under pressure the elastic deformation of glass takes place instantaneous while the viscous part of the deformation occurs over time. At high temperatures glass turns into a viscous fluid and at low temperatures it behaves as solid. Further, the material is restricted to be thermorheologically simple, which assumes that the material response to a load at a high temperature over a short duration is identical to that at a lower temperature but over a longer duration.

The used viscoelastic material model usually depicts the deformation behaviour of glass and is described by the generalized model of Maxwell. It can be build up with six parallel Maxwell elements being connected with a spring and a serial damper. The used material model in the simulation is shown in Fig. 4.

Thus the behaviour of Maxwell elements is described by the following equations:

$$\sigma(t) = \int_{-\infty}^{t} E\,e^{-\frac{t-t'}{\tau}}\,\dot\varepsilon(t')dt', \quad \tau = \frac{\eta}{E} \qquad \text{Eq. 1}$$

$$E(t) = \sum_i E_i\,e^{-\frac{t}{\tau_i}}\,(+E_\infty), \quad \tau_i = \frac{\eta}{E_i} \qquad \text{Eq. 2}$$

with:

t, t' : time; τ : shear stress; ε : extension;
η : dynamic viscosity; E : Young's-modulus

The used material model is representative for a first qualitative analysis of the deformation and the temperature distribution during the glass moudling.

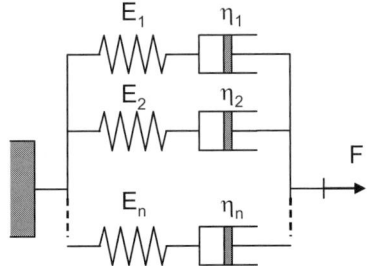

Fig. 4: Material model of glass using Maxwell elements

In addition to the complex material model of glass the calculation of the temperature distribution is one of the critical parts of the simulation since it has a strong influence on the deformation behaviour of the glass.

Using a principle data set the Young's modulus can be calculated by using the Kirchhoff's law. The behaviour of glass is time depending and is shown in Fig. 5 for different temperatures. The material behaviour of glass is the main influence on the shrinking effect of the glass lens, independent of the outer dimensions of the glass lens or of the micro sized design features of the optical surfaces.

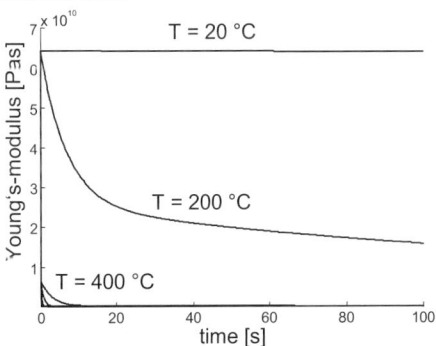

Fig. 5: Young's - modulus of glass

Fig. 6 shows the simplified process characteristics of the moulding cycle of a micro optic or a macro optic array. Subsequently the entire pressing system is heated with infrared heating elements until the forming temperature is reached (P1). During the simulation the glass gobs (pre-forms) are heated up without exerting pressure. The heating rate depends on the glass and mould material as well as the engaged heating principle. After reaching the forming temperature a soak time follows to balance temperature differences between the upper and lower glass surface. Due to the moulding force which is applied in the first moulding phase (P3) the glass pre-forms are deformed. In a second moulding phase on a lower force level temperature is slowly cooling down below Tg. In this resqueezing period local shrinkage in the glass is balanced. After cooling down below Tg the last cooling period starts up to unloading temperature.

In order to build up a finite element model that is able to analyze the replication process qualitatively, different principle models that describe the physical effects such as models with micro-structured surfaces have been solved. The simulation of moulds with micro

structured surfaces can be realised by the use of a sub-modelling technique. Therefore the physical effects in moulding of micro-structures are solved in a separate finite element model in an interaction with the simulation of the main geometry of the lens. With this technique micro structures with sizes below 1 μm are simulated with accurate results.

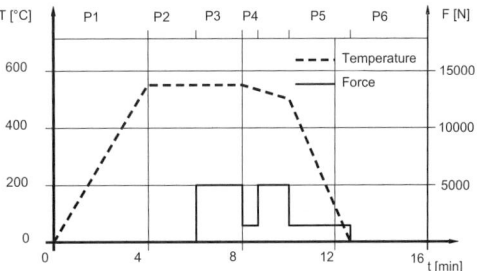

Fig. 6: Process of glass moulding

In addition to the viscoelastic material behaviour of glass the following physical effects are considered in the finite element simulation model:
- linear-elastic material behaviour of the mould inserts,
- contact mechanics,
- friction between mould inserts and glass gob,
- thermal expansion of glass and moulds,
- heat transfer (convection),
- temperature dependence on material values,
- heat capacity,
- heat conduction,
- thermal radiation.

The first results of the glass pressing simulation of a convex lens with two spheres are presented in Fig. 7. The temperature distribution after heating is shown on the left side of the figure. The lens temperature in the core of the lens is lower that on the lens surface. This temperature proportion depends on the gob geometry as well on the time period of the heating phases. During simulation the temperature distribution of the lens depends also on the geometry of the mould. For the realisation of quantitative results of the temperature distribution the results of the simulation have to be calibrated with measurements.

The results of the moulding process depend highly on the temperature distribution of the lens, due to its dependency on the Young's -Modulus of glass. The inner stress of the lens after moulding is shown on the right side of Fig. 7. The stress depends on the deformation degree and the flowability of the glass. The deformation degree is defined by the geometry of the gob and the moulds as well as by the moulding process.

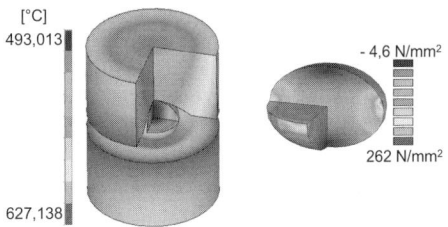

Fig. 7: Temperature distribution after heating (left), and stress distribution after moulding (right)

The simulation of smooth lens geometrie show the importance of the description of the complex material model of glass. Small changes of parameters of a Maxwell element cause a complex changing of the material behaviour of glass. The simulation of micro optics are even more ambitious. Therefore the use of two simulation separate models, one for the modelling of design features in millimetre range, one for the modelling of micro features is applicable.

The simulation results are very useful for the ramp up of the replication process. Simulations with varied process parameters assist the investigation of the influence of the process parameters to the form deviation and the reproducibility to realise a reduction of iteration steps. This enhancement seems to be possible by the use of compensation routines that analyse local form deviations.

4. Systematics of lens error compensation

The approach of the research project uses the simulation results for an analysis of the local form deviations of micro optics. To use the simulation results for the modification of the mould the lenses are segmented in short elements. For each segment a transfer function between the deformation of the replication and the simulation is be determined. Subsequently, finite element simulation are performed to find the ideal mould geometry. The dependence of the real mould geometry and the simulated mould geometry is defined by Eq. 3.

$$z_{i+1} = z_i + \frac{\Delta z_{iFEA}}{\Delta z_i}\left(z_{i+1_{FEA}} - z_i\right) \qquad \text{Eq. 3}$$

with

i : Iteration step;

z_i , z_{i+1} : Local mould geometry in z-direction;

$\dfrac{\Delta z_{iFEA}}{\Delta z_i}$: Transfer function;

$z_{i+1_{FEA}}$: Local mould geometry (modified with FEA).

This procedure considers that in addition to the qualitative analysis mentioned above the simulation results are useful for the geometry modification of the mould inserts by considering the following assumptions. The local deviations between simulation and reality only have an influence on a small area of the lens. Additionally the differences of the mould geometry between two iterations are very small compared to the geometrical sizes of the lens.

With the implemented finite element model the manufacturing process of glass pressing is analyzed to approximate the real pressing process. Next to the material behaviour of glass, the calculation of the temperature distribution during the pressing is very difficult to handle because of the detailed heat transfer coefficients of the different materials that are not known exactly. The significant reduction of the iteration steps while designing the mould can be achieved by using the simulation results for the optimization of the pressing process and for the geometry modification of the mould inserts. The results of this study clearly support this statement.

In the next step of the work the measured form deviations of manufactured lenses will be compared with the simulation results. Parallel the material properties will be investigated as well as the finite element model will be modified by enhancing the level of detail.

5. Summary and outlook

In this paper the precision glass moulding process is presented. This technology allows the manufacturing of complex glass lenses with freeform surfaces as well as micro optics. The form deviation between the geometry of the mould inserts and the finished lens is strongly depending on the moulding process and the moulding conditions. To calculate the form deviations the finite element analysis can be used. Therefore the moulding process is implemented in several simulation models in a high level of detail.

The complex material behaviour of glass is considered as a combination of different Maxwell elements. First simulation results show the high dependency of the moulding results and the solved temperature distribution. This is caused by the dependency of the Young's-modulus of glass and the moulding temperature. Both, solving the temperature distribution and the material behaviour of glass are the main effects that dominate the simulation results.

The simulation results are essential to calculate the form deviations of the manufactured micro optics compared with the given mould geometries. For this purpose the lens errors can be analysed systematically. Therefore form errors of moulded lenses are compared with simulation results. After each iteration both, real deformations as well as simulated deformations, are used to estimate the needed geometries of the mould inserts using local transfer functions. The goal of the work is the fundamental understanding of the moulding process as well as the simulation of the interaction of the process parameters to deliver a routine to predict form deviations and to reduce cost intensive iterations steps by 50 %.

In the future the material behaviour of glass as well as the calculation of the temperature distribution are investigated to realise a better approximation of the real form deviations of the lenses.

6. Acknowledgements

The work is funded by the German Research Foundation (DFG) within the Transregional Collaborative Research Centre SFB/TR4 "Process Chains for the Replication of Complex Optical Elements".

7. References

[1] A. Y. Yi, "Optical fabrication," Encyclopedia of Applied Physics, ed., Mallick, P. K., Marcel-Decker Publishers (2003)

[2] Ries, H.; Muschawek, J.: Tailored freeform optical surfaces, Journal Optical Society of America, 2002

[3] Weck, M.; Winterschladen, M.; Pfeifer, T.; Dörner, D.; Brinksmeier, E.; Autschbach, L.; Riemer, O.: Manufacturing of Optical Moulds Using an Integrated Simulation and Measurement Interface; Proceedings of SPIE, 5252 (2003).

[4] Weck, M.; Wenzel, C.: Adaptable module for five-axis polishing. Proceedings of the 4th International Topical Conference of euspen, 2003, pp. 153– 156

[5] Manns, P.: Blankpressen von optischen Komponenten für spezielle Anwendungen. Symposium: Neue Entwicklungen in der Glastechnologie, Berlin 1997

[6] Winter, S; Schaeffer, H.-A.: Effect of aggressive gases on the behaviour of glass surfaces in contact with mold materials. Glastechn. Berichte 61 (1988) 7, pp.184-190

Multi-Material Micro Manufacture
W. Menz, S. Dimov and B. Fillon (Eds.)

Unified approach for functional task formulation in domain of micro/nano handling operations

K. Kostadinov[a], R. Kasper[b], T. Tiankov[a], M. Al-Wahab[b], D. Chakarov[a], D. Gotseva[a,c]

[a] Institute of Mechanics, Bulgarian Academy of Sciences, Acad. G. Bonchev St. Bl. 4, 1113 Sofia, Bulgaria
[b] University "Otto-von Guericke" – Institute of Mobile systems, 2, Universitaetsplatz, Magdeburg, Germany
[c] Technical University, Faculty of Computer Systems and Control, 8, Kliment Ohridski Str., Sofia 1000

Abstract

A unified approach for functional task formulation is developed that allows communication between user and team of mechatronic engineers developing the mechatronic handling devices for requested task function. All-important parameters are set for realization of mechanics and control of desired process for automation. 3D variants are modeled in SDS 2004+ and simulated, making interactive closed loop with the user and better communication between different research and industrial domains for achievement of good results.

The possibilities of the developed iterative program for the task function formulation are illustrated with the results obtained for the case of micro/nano manipulations necessary for measurement of electrochemical impedance.

Keywords: task formulation approach, interactive program, micro/nano handling operations.

1. Introduction

Micro and nano manipulators are mostly used in biological and microelectronics research, cellular technology, chemistry and investigation of thin films, in atomic force microscopes (AFM) and scanning tunneling microscopes (STM). Most of the existing manipulators are man-operated or semi automated. Therefore, the accuracy and productivity depends on the operator.

The robots and mechatronic handling devices for micro/nano manipulations are applicable in many domains like electrochemistry, biology, microbiology, microelectronics, medicine, and microsurgery. They perform various operations like local surface scanning or cell manipulations [1-5]. The most of these tasks are hand manipulated by human. Sometimes these processes are very complex and very often they can't be overcome. Some operations can be performed one, two times per day and often the repeatability is unrealized. The most of these processes allow long time for manipulation that is impossible for human. Other tasks are accomplished in environment that is unhealthy. People possess hand skills, visual perceptions and other abilities. Operations that allow very precise motion control are difficult or impossible for human. He works very well in operations that require relative positioning, based on the visual or tactile closed loops. For precise positioning it is necessary to use robotized systems. Fine operations like microsurgery demand precise manipulation and human intervention.

Because of this reason it is necessary to automate similar manipulations. On this way it is possible to raise the efficiency and task quality. The specialists, performing these manipulations in most cases don't know the possibility of automation of the concrete process and this is the reason for the difficult relation between them and mechatronic engineers. It is necessary to create a unified language for communication between user and producer for accurate, qualified and fast formulation of all necessary parameters for development of the final robotized solution.

The increasing requirements of high precision equipment in the field of biology, electrochemistry, micro- and nanotechnology etc performing broad spectrum of handling and manipulation tasks is a prerequisite for searching of new concepts for fully automated or man-operated piezo-actuated mechatronic handling devices.

Unified approach for easy formulation of functional task is the aim to be developed in this paper. An interactive dialog between user and developer of any mechatronic handling device or robot able to perform the requested task function will be used to make easier the communication between those two different specialists for specification of all-important parameters.

2. Problem formulation.

Utilizing the technology capacity for rapid prototyping of mechatronic systems based on the piezo-structured ceramics and development of embedded control system is perquisite for development of technology for manufacturing of mechatronic handling device, able to perform a certain user requested micro- or nano-operation.

The past general use of piezo-actuators in a direction of motion (3, 3 or 3, 1) is to be extended for realization of piezo-structures for the operation tasks [6]. By structuring of ceramics, stacking and combination with further elements more piezo-actuators axis motions are subject for development.

To produce movement parameters by combination of the structuring possibilities with appropriate servo mechanism which is yet realizable with very complex and large systems is most promising design approach [7], for which the following four parameters are significant of determination of the reference task function of any mechatronic handling device:

First parameter is the stroke. The desired goal is an adjustment of the stroke of approximately 100 µm in all 3 directions X, Y and Z.

Second parameter is the force. It is well-known that stacks actuators produce forces within the kN range and bending transducers deliver forces up to 1 N. Neither stacks actuators nor bending transducers, are here optimally suitable. The possibilities of the structuring of the piezo ceramics with consideration of the generation of the stroke can help here. It will be possible to optimize stroke and force.

Third parameter is the size. The actuator for the movement generation must be integrated into the mechatronic handling device. The building area is limited. Thus both piezo plates and piezo disks can be used as raw material. The characteristics of piezo materials are to be selected in such a way to reach desired parameters for stroke and force.

Fourth parameter is the load speed. The structure of the piezo ceramic must be designed in such a way that the desired dynamics of the movement can be achieved. The intended speeds lie within the range of approximately 200μm/ms. Thereby also here an optimization is necessary. Emphasis is the join action of piezo ceramic and servomechanism system. The servomechanism system has to be adapted itself to the desired parameters for stroke and force.

The aim of this paper is combining the achievements and the knowledge expertise of teams experienced in development of piezo-structured mechatronic systems with 1 DOF and robot systems with more than 3 DOF for micro- and nano-manipulations to develop an unified approach for functional task formulation in domain of micro/nano handling manipulations in order to design a mechatronic handling devices with optimized structure meeting the requirements of users from different application fields as:

- Industrial:
-Positioning and mounting of optoelectronic devices (e.g. Laser diodes), micro- sensors and machines (i.e. autonomous systems for the maintenance of small pipes and arteries, micro manufacturing systems).
-Positioning and placement of sensors for diagnostics of microelectronic elements and devices.
- Medicine:
  - Micro operations, e.g. eyes, brain, etc.;
  - Sorting of cells for diagnosis
- Research:
  - Moving or treating microorganisms;
  - Operation of probe for electrochemical impedance measurement;
  - Sorting of cells and cell penetration;
  - Analysis of DNA
  - Manipulating samples in STM and AFM.

Requirements defined with the help of potential user will build the reference for synthesis, optimization and test operation of the following tasks as examples:

a. For cell manipulations. Typical manipulations as cell penetration, cell sorting and moving or treating microorganisms will be analyzed in order to specify requirements to mechatronic handling device for bio research applications. Critical issues here are speed of cell penetration, necessary stroke for penetration or operation and orientation of the end-effector, working space and space of the device itself, sensing and approach for teleoperation control able to offer a minute performance of the desired operation.

b. for electrochemical impedance application techniques

The desired operation of the probe for electrochemical impedance technique will be analyzed in order to achieve positioning and dynamic requirements for the right measurement task execution.

They are consisted of 2 scanning rotation motions and translational one for fine adjustment of the probe initial position. Here the regional scanning robot structure with 3 DOF (translation) will be necessary.

3. Unified Approach for functional task formulation in domain of micro/nano manipulations.

A suitable database for automate processes is created. First it is necessary to use the domain of application for final task localization.

Important parameters are determined for the optimized mechanical system.

The size of the system is the parameter that defines the working field of the mechanical engineer. The height, length and width are presented. On this way the user specifies for him the possible and suitable system mounting into his experimental stage.

The second parameter is the force. In most cases it is applicable on the motion system tool. The aim is to investigate the dynamic characteristics and to evaluate the stiffness and type of the regulated mechanism that have to be used for reduction of the system backlashes.

The third parameter is the velocity of the driving tool. The user has to specify the speed of motion for the investigated manipulation. This parameter is very important in cell injection where the velocity of the injection/sorting operation has to be under control. The speed influence also of the mechanical system and often it can cause system destruction.

The next parameter is the working volume of the system. The user sets the maximal range of motion on X, Y, Z axes and the producer could realize mechanically the system possibility to make motion in whole defined working space.

The definition of the range of motion on X, Y, Z-axes help to specify the system degrees of motion.

The orientation of the driving tool and system resolution is also important parameters that have to be done.

The control realization of the mechanical system depends of the parameters mentioned above like range of motion, resolution and speed. According to these characteristics the developer offers various suggestions for driving, using step motors or piezo-elements.

It is very important to specify also the type of system closed loop. The developer has to define the sensing part and its resolution. It is necessary to set the parameter for reading.

Like whole formulation of the given robotized solution it is important to define the type of communication between the user and the system. Many approaches for communication exist like autonomous, tele-operated [8-9] or hybrid approach [10-11]. The connection between the system and the computer is possible via LPT, Serial Port, USB Port, according to user request.

After the formulation of the electromechanical specifications it is necessary to present some variants of the developed system, modelled in SDS 2004+ software. On this way the user can see the design and he can make the best choice for his manipulation task.

In conjunction with presented mechanical variants it is presented the motion simulation of each system. On this way with 3D modelling and simulation it is possible to obtain a closed loop between the operator and the electromechanical team of engineers.

The developed software allows the developer to make some correction if the final solution does not fit the user requirements or is not so optimized.

257

The best choice of suggested automate system can be paste like file on the computer.

Like an example of this unified approach for formulation of functional task an automated process for quality surface investigation, measuring the electrochemical impedance is developed. The code is open and it is possible to add new robotized solution of different manipulation tasks in domain of chemistry, microbiology, cell manipulation, microelectronics.

4. Interactive program for specifying the requirements to the mechatronic handling device.

An interactive program is developed for formulation of functional task in domain of micro/nano manipulations. It is made using Microsoft Visual Studio.NET 2003 program package and Framework 1.1 platform. The block scheme (fig. 1) of the developed software presents the communication and transfer of all data. A tri-layer window database application is developed. The firs layer presents the graphical user interface (GUI) that creates the client/server window interface, controlling the information. The window dataset makes a database and the following components. The Questions, Answers, Movies realize access and data change. This information is transferred to the second layer, called Business logic. Here it is made a data evaluation. If the client or expert has an access to the server a reference to the used database is created.

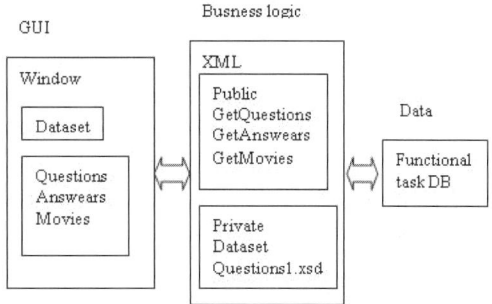

Fig. 1. Block scheme of three layer data application

This application is bilateral and therefore an answer from the functional database is returned to the busness logic. The visualisation is made into the graphical user interface with the coordinated solution.

This window application is developed for two different users: expert that create possible functional tasks and client that investigate these tasks and choose one of presented functional task formulations.

The expert can create a personal account. He can enter to a specific window editor with the possibility to write questions and answers. On this way he makes a database of different tasks formulations and finally he generates a suitable window form for client manipulation presented on the Fig. 2. This information is transferred to the client who also creates a personal account.

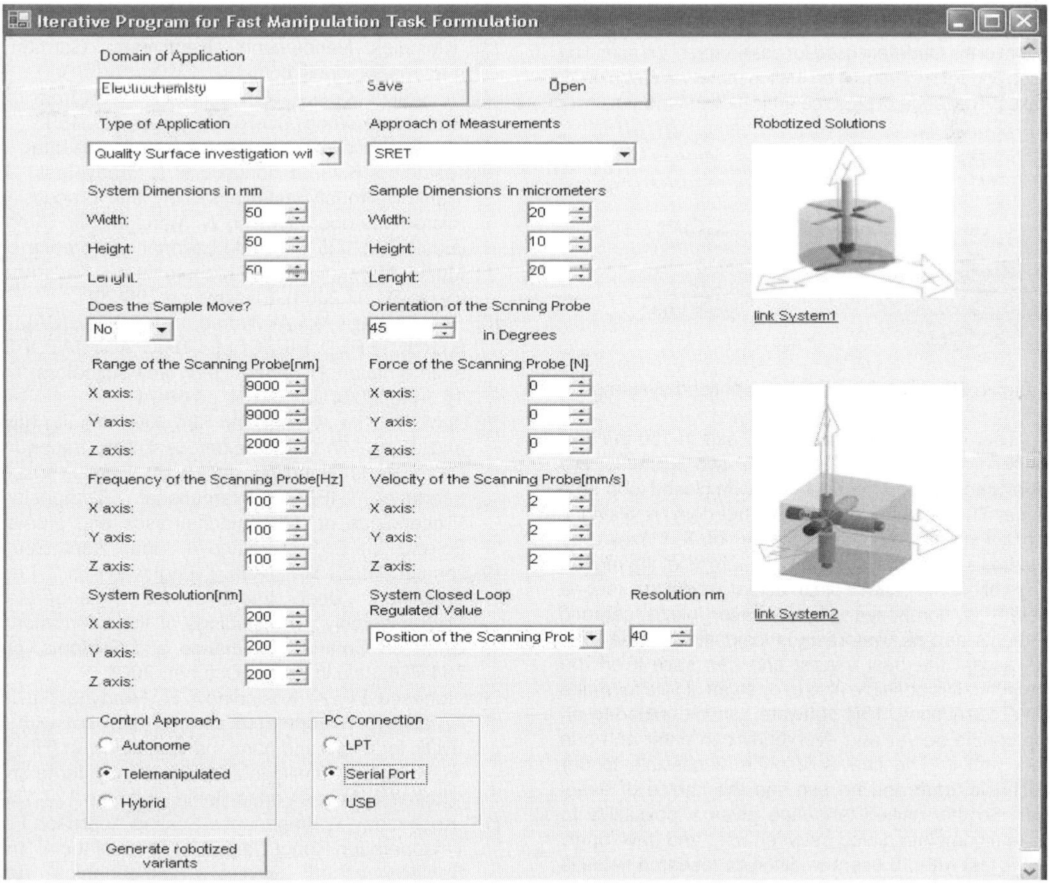

Fig. 2. User window form for fast functional task formulation.

258

At first the user has to define the domain of application. Then an interactive window is opened and it is possible to set all-important parameters for the desired manipulation. When he makes your personal choice he transfers it back to the expert who investigates it and if this solution is not existing before into the database he creates new animation 3D variants and sends them. If this solution is presented into the database he transfers stored 3D variants to the client.

The user can see all robotized variants, created according to parameters mentioned before. Each variant is modelled in SDS environment and interactive simulation presents the system motion (fig. 3).

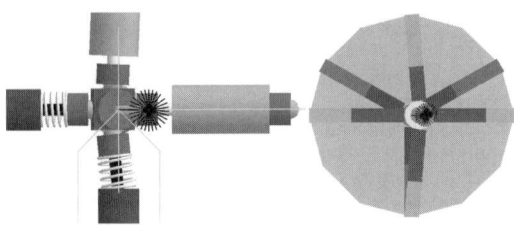

Fig. 3. SDS simulation of the system motion.

It is possible to see all 3D views (fig. 4) of the presented mechatronic solutions. For localized electrochemical impedance spectroscopy two 3D models are presented. The first model is composed of three stack piezo actuators with range of motion equal to 8 μm. The maximal range of the system motion is 8 μm on Z axis and 40 μm on X, Y axes. The second model is composed of 3 piezo ceramic elements on the base and 4 piezo ceramics on the top of the system body. The piezo ceramic elements on the bottom of the model are used for positioning of the scanning probe and piezo ceramics on the top move the probe on X, Y axes. The range of piezo ceramic motion is 10 μm.

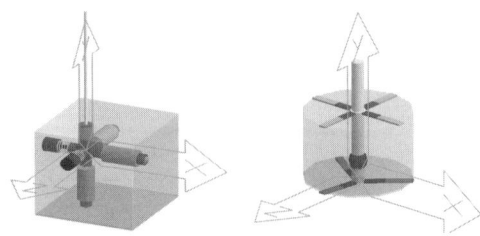

Fig. 4. 3D view of 2 modeled robotized systems.

Then the system motion on Z axis is 150 μm and on X, Y axes – 600 μm. The two systems are developed as micromanipulators with closed loop as a whole. They have 3 degrees of freedom, realising 1 translation on Z axis and 2 rotations on X, Y axes. The hysteresis of the first system is 10 – 14 % of the motion range of the used stack piezo actuators but the second system is composed of one layer piezo ceramic elements and its hysteresis is insignificant. The user can accept the best variant and can save it on the computer. It is possible also to open an old file for make some corrections. This software can be uploaded on appropriate server and everybody can enter and can work with it. The final solution is accessible to our mechanic team and he can see the choice of every user. So this unified language gives a possibility to make a complete dialog between user and developer.

A test with 10 users in different research fields is performed. 50% of them have a possibility to develop functional task formulation without this methodology and the others make this operation with developed software. But all of them find it very useful when they have to formulate the task in the robotic terms. All users declare getting knowledge for the task how it can be automated and what would be its specific features.

5. Conclusion:

A unified approach for functional task formulation is developed that allows communication between user and team of mechatronic engineers. All-important parameters are set for realization of mechanics and control of desired process for automation. 3D variants are modeled in SDS 2004+ and simulated, making interactive closed loop with the user and better communication between different research and industrial domains for achievement of good results.

The possibilities of the developed iterative program for the task function formulation are illustrated with the results for the case of measurement of electrochemical impedance.

Acknowledgements
This work is supported by DFG Project MeCHaPiCS, Bulgarian Ministry of Education and Science under the Ro- MiNa project TH1308/03 to which the authors are expressing their acknowledgements.
We are also very grateful to the contribution of 4M NoE No. 500274 of NMP2-CT-2004.

References
[1] Burleygh Instruments Inc., http://www.burleigh.com
[2] Physikinstrumente GmbH, Hexapod 6 Axis Parallel Kinematics Robot M850, www.physikinstrumente.com
[3] Kleindiek Nanotechnik, Reutlingen, Germany, http://nanotechnik.com
[4] Dr. Volker Klocke, Nanotechnik, Motion from the Nanoscale World, CD-ROM Version 1.5, 1998
[5] http://www.cellrobotics.com/workstation/cs.html
[6] Kostadinov K. Gr., F. Ionescu, R. L. Hradynarski , T. Tiankov, Robot based assembly and processing micro/nano operations, in W. Menz and St. Dimov (Eds.), 4M2005 First International conference on Multi- Material Micro Manufacture (Karlsruhe, 29.06.-01.07.2005), Elsevier, pp.295-298.
[7] R. Kasper, M. Al-Wahab, MECHANICALLY STRUCTERED PIEZOELECTRIC ACTUATORS, 9th International Conference on New Actuators, 14-16 June, 2004, BREMEN, pp.68-71
[8] Deok-Ho Kim, Kyunghwan Kim, Keun-Young Kim, and. Sang-Min Cha, "Dexterous Teleoperation for Micro Parts Handling based on Haptic/ Visual Interface", IEEE International Symposium Proceedings on Micromechatronics and Human Science, pp. 211-217, Nagoya, Japan, Sept., 2001.
[9] Song, E.-H., Deok-Ho Kim, Kyunghwan Kim, J. Lee, "Intelligent User Interface for Teleoperated Microassembly", Proceedings of the International Conf. on Control, Automation and Systems, pp. 784-788, Jeju in Korea, October, 2001.
[10] Ionescu Fl., K. Kostadinov, R. Hradynarski, Iv. Vuchkov, Teleoperated control for robot with 6 DOF for micro and nano manipulations, in [Hrsg.: R. Kasper...] Intelligente technische Systeme und Prozesse, Logos Verlag-Berlin, 2003, pp.127-132.
[11] Kostadinov K., Impedance Scaling Approach For Teleoperation Robot Control, in Proc. of the 4th Int. Conference on Bionics, Biomechanics and Mechatronics, 2004, Riga, pp.71-74.

Multi-Material Micro Manufacture
W. Menz, S. Dimov and B. Fillon (Eds.)

Thermally induced stresses in an adhesively bonded multilayer structure with 30-micron thick film piezoelectric ceramic and metal components

R.Jourdain and S.A.Wilson

Materials Department, School of Industrial and Manufacturing Science, Cranfield University, Cranfield Bedfordshire, MK43 0AL United Kingdom

Abstract:
In this study, the thermal stresses acting in a multilayer structure built up by successive bonding and machining processes are investigated. These arise mainly from the bonding process itself due to the wide mismatch in the coefficients of thermal expansion for the ceramic and metal components. Subsequent machining of the bonded ceramic down to 30 microns thickness leads to a redistribution of the internal stress which must be monitored closely as it influences both the structural integrity and the overall performance of the finished piezoelectric device. Previously reported analytical models are compared to new numerical simulations using either shell elements or two dimensional elements. High aspect-ratio, symmetric and non-symmetric structures are studied to show the changes in stress throughout the fabrication process. The finite element models allow detailed investigation of the stresses in all of the component parts and in the bonding layers. The influence of bond thickness is investigated in the range 0.5 to 10µm. This work shows that even at low bonding temperature the adhesive layer is under quite severe stress. This can be tuned by increasing the bond thickness to achieve high structural integrity and moreover the ceramic is engineered into compressive pre-stress suitable for high performance micro-actuators.

Keywords Bimorph; PZT; actuator; adhesive bonding; micro-fabrication; FEA; ultra-precision grinding, MEMS

Introduction

Adhesive bonding is assuming fundamental importance for fabrication of 3D micro electro-mechanical devices because of its ability to assemble component parts made from different materials. The subject of this study is a new technique which combines adhesive wafer bonding and ultra-precision machining of ceramics with standard micro-fabrication techniques to produce new devices on a scale that is difficult to achieve by other means. Such devices include piezoelectric micro-actuators, configured as unimorph or bimorph cantilevers and diaphragms. In common with other assembly techniques such as fusion bonding, anodic bonding, low-temperature glass bonding and eutectic bonding, this procedure induces residual thermal stresses in the multilayer structure that must be evaluated to achieve a device with good structural integrity. Applications for the devices include micro-actuators: micro-valves[1], head sliders[2], switches[3], micro-pumps[4] and micro-mirrors[5].

Thermal stresses arise due to thermal expansion coefficient mismatches between the different components. In the case of the adhesive bonding technique these parameters can be tuned within a limited range to suit the requirements of the composite structure. A piezoelectric ceramic for example bonded to a metal can be engineered into a compressive stress which has potential benefits in terms of increased robustness and product lifetime. Moreover residual compressive stress can increase the performance of multilayer cantilevers[6] (bending and blocking force).

The advantages of adhesive bonding include: bonding of very different types of material, relative low temperature bonding and high, consistent bond strength. Unlike some other bonding techniques, surfaces do not necessarily need to be scrupulously clean, but the component surfaces are expected to be wettable by the adhesive to ensure a void-free, strong and reliable bond using the thinnest adhesive layer needed.

The aim of this study is to predict the residual stress within all components throughout the fabrication process for a multilayer structure 80µm thick. Bond thickness is investigated in the range from 0.5 to 10µm and the evolution of highly localised stresses is taken to be a key issue for fabrication of high efficiency devices. Both numerical and analytical models are presented and these are evaluated in conjunction with experimental work. Hence the locations of the areas under critical stress are identified and the residual stress values are predicted. Finally the stress changes in each component due to material removal are evaluated. Numerical modelling was carried out using ANSYS version 8.

I.a) Numerical versus analytical modelling

Multi-layer structures made of smart materials are widely used in engineering applications. Analytical analysis is used to provide information on stresses, strains and deformations. This permits evaluation and optimization of the structural design but these models tend to be limited to basic shapes.

Bonded bi-material model

From an overall perspective it can be said that normal stresses in the components are responsible for the ultimate and the fatigue strength of the bonded materials. However, it is the interfacial shearing and peeling stresses which govern the structural integrity. Unlike the standard bi-material model first developed by Timoshenko[7] in 1925, the bonded bi-material components have different distribution of stresses. This is discussed in the lap joint theories attributed to Volkersen[8] and expanded by Goland and Reisner[9]. E. Suhir in 1986 presented two models, one for stresses in bi-metal thermostats[10] and one for stresses in adhesively bonded assemblies[11]. The first model describes the magnitude and the distribution of interfacial stress, while the second pays attention to the adhesively bonded bi-material assembly and to the attachment compliance.

In order to model our multilayer structure using finite element methods two geometries are considered (figure: 1). Two bonded bi-material analytical models are applied and compared to numerical analysis. The shear stress acting in the (5μm) bonding layer is displayed in figure: 2.

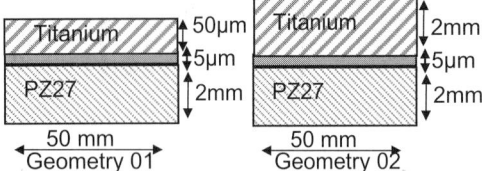

Figure 1: Geometry for FEA model assessment

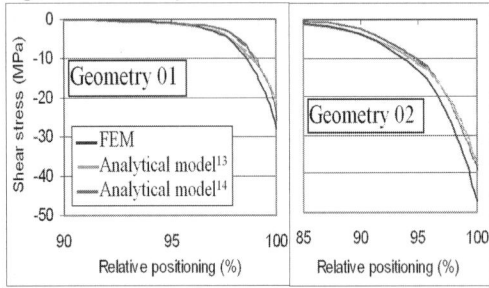

Figure 2: Shear stress acting in the bond layer

The numerical model shows good correlation with regard to shear stress. For the normal stress the numerical model underestimates the analytical results by about 10%.

I.b) Numerical model evaluation

The investigation of residual stresses in the real multilayer assembly is presented in this section. The model is first compared to experimental results, thereafter a detailed investigation is used to highlight areas under critical stress.

The bonding process is described in figure 3. The sandwich structure consists of five layers, two

piezoelectric ceramic discs (PZ27), a titanium shim and two bonding layers. The adhesive layer is spun onto coated ceramic discs at room temperature. Ceramic discs and titanium layers are heated at 190° Celsius and assembled by applying a consistent pressure during the bond curing. After a certain time, the pressure is reduced and the temperature is lowered to 25°Celsius.

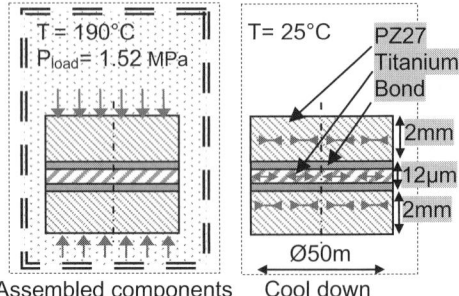

Figure 3: Primary bonding process

The numerical model initially sets a stress-free state in all of the materials at the bonding temperature and it then computes the thermal stress caused by the thermal expansion coefficient mismatch and difference of temperature. The material properties are displayed below.

Table 1: Material properties (*: non-poled material)

	C.T.E. *	C.T.E. *	v	E (Pa)
	0-150 °C	150-250 °C		
Ti	$9\,e^{-6}$	$9\,e^{-6}$	0.34	$100e^{+9}$
PZ27*	$2\,e^{-6}$	$1.5\,e^{-6}$	0.30	$76\,e^{+9}$
Bond	$52\,e^{-6}$	$52\,e^{-6}$	0.34	$2.6\,e^{+9}$

* CTE: coefficient thermal expansion

After bonding, one ceramic disc of the sandwich structure is machined down to 30μm. The geometry of the numerical model is then modified to represent the new configuration, giving a representation of the change in the stress regime due to machining. A cross section is displayed in the figure 4. This is established as the routine procedure for the fabrication process. One set of experimental and two sets of numerical results are compared in the table 2. Profile deformation U_{xx} and stress values are compared for two different bond thicknesses (4 and 1μm). Recorded data was carried out using a Dektak profilometer. Experimental and numerical results are presented in the table: 2.

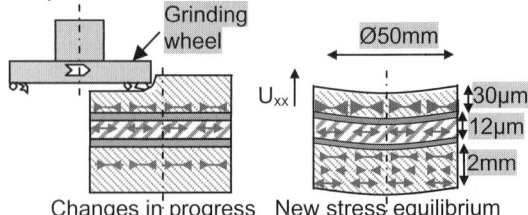

Figure 4: Stresses redistribution due to machining

Table 2: Numerical and experiment results for non-symmetric geometry described in the figure 4 (the materials considered are PZ27; Ti; BCB and the bonding temperature: 190°Celsius)

	Displacement U_{XX} (µm)		Von Mises stress (Pa)		
Bond thickness	4µm	1µm	Ti	PZ27 top	PZ27 bottom
Plane_42	11.7	10.7	131 E^6	-3.6 E^6	+1.7 to − 3.4 E^6
Shell_91	11.9	10.9			
Experiment: 12					

The tailored numerical model gives good correlation with experiment. Stresses in the components are still perfectly bearable.

II.a) Bond thickness investigation
This section highlights thermal stress change due to different bond thickness within the symmetric structure. The bond thickness is investigated in the range 0.5 to 10µm and the difference of temperature is 230°C. Results of the numerical modelling are focused on the area close to the edge and crossing all material components. Figure: 5 shows a path traced at 1µm from the edge, where results are collected from and presented in line graph 1.

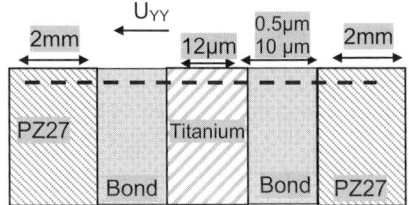

Figure 5 : Sandwich cross section and path investigated

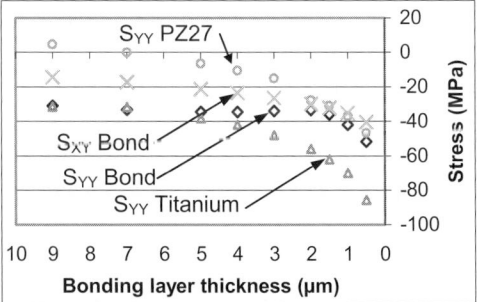

Line graph 1: Stress in the materials
The line graph 1 shows that maximum stress intensity S_{YY} in all the components increases when the bond thickness decreases. The bonding layer is under severe stress for the thinner bonding layers because both the shear stress and the peeling stress are relatively high. The numerical simulation shows the stress is about 40 and 50MPa for the shear and S_{YY} stresses respectively. We can say that the material used in this experiment is under sever stresses because it has a maximum tensile strength of 87MPa.

II.b.) Stress release investigation
As previously presented, the fabrication technique involves the thinning down to 30µm one piezoelectric ceramic disc. This material removal process is carried out by plunge grinding[12] and the machining process does not cause critical damage in the ceramic. Due to a loss of symmetry in the structure, the stress in the sandwich sample changes during machining and the multilayer disc bows. When it is released from the vacuum chuck, the induced thermal stresses are redistributed in the non-symmetric sandwich structure (see figure: 4).

An understanding of the stress release mechanism is essential for optimization of the fabrication process as it influences subsequent stages in the assembly of the complete device structure and also complete device performance. Results from the numerical model relating to the non-symmetric multilayer are compared and presented in Table 3 and line graphs 2-4 respectively. The stress intensity before and after thinning down of one layer of the piezoelectric ceramic is shown. As previously mentioned, special attention was given to edge of the multilayer assembly. Figure 6 shows the investigated paths.

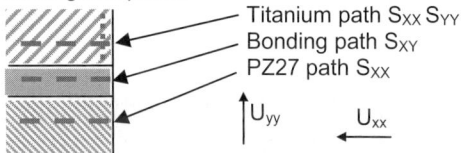

Figure 6 : Paths investigated

Table 3 shows the effect of the material removal process on the mechanical stresses of different components. With regard to S_{YY}, the stress in the titanium is decreased about 16 to 19 percent respectively for thin and thick bonds. The highest increase concerns the ceramic, where compressive stress is doubled when a large amount of material is removed.

Table 3: Stress change after material removal

Material	Stress direction	Thin bond Thick bond Change	
Titanium	S_{YY}	-16.4%	-19%
	S_{XX}	- 3%	
PZ27	S_{XX}	+ 50%	
Bond	S_{XY}	- 7%	

Thick ceramic layer (PZ27)
The top ceramic layer is now only three times thicker than the metal shim. The results highlight the stress increase in the thick ceramic disc by 50%. The line graph shows the stress for the bottom thick ceramic disc which is higher than that in the thin ceramic disc. Results show a consistent tensile stress value across the diameter at all bond layer thicknesses.

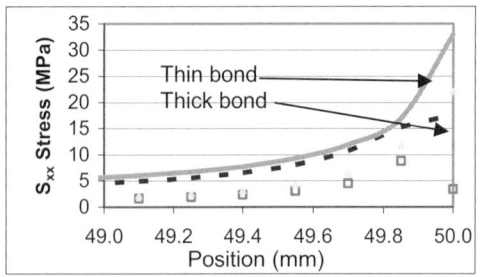

Line graph 2: Stress in the thick ceramic layer

Bonding layer

Average shear stress values across the diameter are considered in line graph 4. The thick bond causes an increase and the thin bond decreases the shear stress to around zero. With regard to the material removal the thick bond will not vary as much as the thin bond.

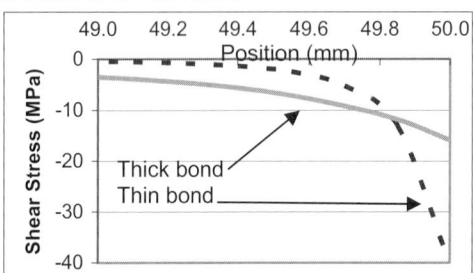

Line graph 3: Bond shear stress

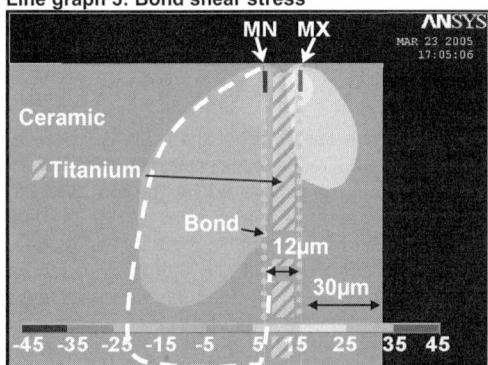

Figure 7: Shear stress MPa (Multilayer cross section)

The figure 7 shows the edge cross section of the non-symmetric multilayer structure. The maximum shear stress value is reached with the thinnest bonding layer (0.5µm). Highest stress is localised at the junction between ceramic and the bond (red and blue area). The dashed line shows the area of elevated stress when the bond layer is increased to 10µm.

Titanium shim

Line graph 4 shows that the tensile stress in the titanium shim is consistent across its diameter. The stress decreases within the last two hundred micro meters and the maximum stress intensity is consistent for any bond thicknesses. The material removal process causes the tensile stress to decrease, this is estimated at about 3%.

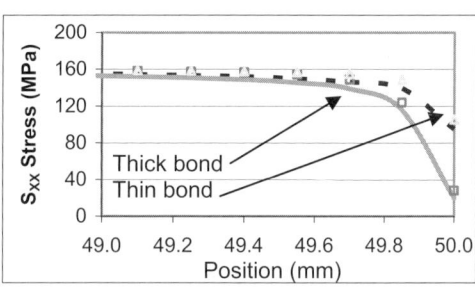

Line graph 4: Titanium tensile stress

Conclusion

A sophisticated numerical model has been developed. Stress changes due to fabrication process are predicted with regard to the bonding parameters or material removal process. This model can be tailored in order to accommodate different multilayer components. Understanding of the stress changes has led to optimisation of the fabrication process. High thermally induced stresses acting in the bonding layer are localised near the edge of the assembly at a distance comparable to the sandwich thickness. Critical stress intensity increases and the extent of the affected area decreases when the bonding layer becomes thinner.

Acknowledgement

This work was carried out in collaboration with BAE Systems Advanced Technology Centre as part of the European Growth Project – *'Aeromems II'* - Advanced Aerodynamic Flow Control Using MEMS.

References

[1] W.Van der Wijngaart, A.Thorsen; J. of Microelectro-mechanical Systems; Vol. 14; No 2, pp 200 -06, 2005
[2] Y. Jing, J. Luo, X. Yi, X. Gu; Source: Sensors and Actuators A (Physical), Vol. A116, No 2, pp 329-35, 2004
[3] H. Lee, J. Park; Microwaves and Wireless Components Letters, Vol. 15, No 4, pp , April 2005
[4] K. Junwu; Y. Zhigang; P. Taijiang; C. Guangming; Wu Boda Source: Sensors and Actuators A (Physical), Vol. 121, No 1, pp 156-6, 2005
[5] Y. Youngjoo, H.J. Nam, S.H. Lee; J. of Micro Electro Mechanical Systems, The 13th Annual Int. Conference pp 435-440, 2000.
[6] G. Li, E. Furman and G. Heartling; J. of the American Ceram. Society, Vol. 80, |6|, PP: 1382-88, 1977
[7] SP. Timoshenko; J. of Optical Society of America, pp 233-255, 1925
[8] O. Volkersen; "Luftfahrtforschung", Vol.15, pp 41, 1938
[9] M. Goland, Reissner; J. of Applied Mechanics, Trans ASME 11, pp 17-27, 1944
[10] E. Suhir; J. of Applied Mechanics, Vol. 53, No 3, pp 657-660, 1986
[11] E. Suhir; Elect. Pack. Mat. Science, Material Research Society, Symp. Proc., Vol. 72, pp133-38, 1986
[12] R. Jourdain, S.A.Wilson, R.W. Whatmore, Pro., euspen 4th International Conference, Proc. pp 137-38, Glasgow, 2004

Multi-Material Micro Manufacture
W. Menz, S. Dimov and B. Fillon (Eds.)

Modelling and simulation of piezoelectric actuation and reliability of micropumps

Meiling Zhu [a,*], Paul B. Kirby [a], Martin Richter [b], Yucel Congar [b], Alexander Diehl [c], Ralf Voelkl [c]

[a] *Nanotechnology Group, School of Industrial and Manufacturing Science, Cranfield University, UK*
[b] *Fraunhofer Institute for reliability and microintergration (IZM), Munich, Germany*
[c] *Chair of Manufacturing Technology, University Erlangen-Nuremberg, Germany*

* *Corresponding author address: Tel: +44 (01234) 750111 ext. 2580, email:m.zhu@cranfield.ac.uk*

Abstract

Modelling and simulation of a low cost micropump with promising mass production is described in this paper. The micropump has a very simple structure and therefore is very low cost effective. The pump consists of one plastic body, one metal diaphragm and three piezo PZT ceramics. The PZT ceramics glued on the metal diaphragm form three actuators and so the metal diaphragm bonded with the plastic body constitute three chambers, which are connected by interconnection channel in the plastic body. A 3D finite element analysis (FEA) model including an electromechanical coupled field simulation is set up and static analyses are preceded for different configuration under different applied voltage for understanding and optimization of the micropump performance. A significant dimensional design of piezo PZT actuators is obtained from simulation results. The ratio of the thicknesses of the piezo PZT ceramics to the metal diaphragm and the piezo in-plane dimension are suggested for obtaining a higher stroke for maximizing the pumping effect and lowering maximum stress in the metal diaphragm for concerning reliability issue.

Keywords: Micropump; FEA; Piezoelectric actuators; Metal diaphragm

1 Introduction

A 3D finite element analysis (FEA) model including an electromechanical coupled field simulation is set up and static analyses are preceded for different configuration under different applied voltage for understanding and optimization of the micropump performance. A significant dimensional design of the piezo PZT actuators is obtained from simulation results. The ratio of the thicknesses of the piezo PZT ceramics to the metal diaphragm and the piezo in-plane dimension are suggested for obtaining a higher stroke for maximizing the pumping effect and lowering maximum stress in the metal diaphragm for concerning reliability issue.

2 Design of micropump

Fig. 1 shows the disassembled and schematic views of the studied micropump. It consists of one plastic body, one metal diaphragm and three piezo ceramics. The metal diaphragm bonded with the plastic body form three chambers: two valve chambers, seated on the left and right, and one pump chamber, seated in the middle, and the three chambers are connected through the interconnection channel. Two fluid ports: inlet and outlet, connected with the valve chambers, are designed in the plastic body. Three piezo ceramics glued onto the metal diaphragm are used for three actuators of the micropump, later called piezo-metal actuators.

Upon a proper voltage applied on any one of the piezo ceramics, piezo-metal actuators can be deflected downwards. Fig. 2 shows the left piezo actuated by a positive voltage ($U_1 > 0$), leading to a deflection in such a way that the diaphragm closes the orifice in the plastic body. So the left valve is closed.

If the three piezo-metal actuators are actuated by three electrical signals with given phase shifts, for example, shown in Fig. 3, the piezo-metal actuators can generate a pumping effect in the micropump chambers.

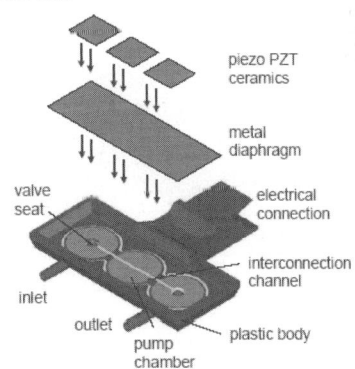

(a) Disassembled view

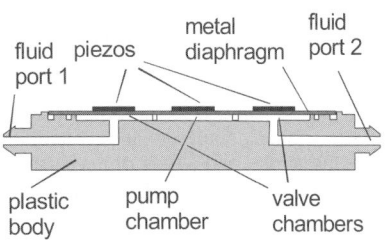

(b) Schematic view

Fig. 1 Micropump

264

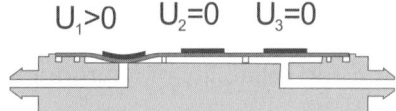

$U_1 > 0$ $U_2 = 0$ $U_3 = 0$

Fig. 2 Deflection of the actuated left piezo

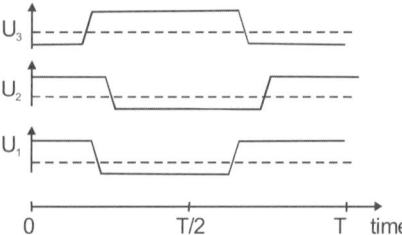

Fig. 3 Actuation signals for the pumping mode

3 Modelling approach based on FEA

Modelling and simulation of the piezo-metal actuator has been done here through the use of a commercial finite element software package ANSYS (ANSYS Inc. Canonsburg, PA) that includes an electro-elastic coupled analysis capability. In the actuation study, simulated results on the actuator stroke and maximum stress in the metal diaphragm have been obtained for understanding of pumping effect and concerning the reliability issue. In the design study, modelling of variations in geometrical structural parameters, such as the ratio of PZT thickness to steel diaphragm thickness (t_p/t), piezo ceramic in-plane dimension, and voltage applied on the PZT, have been performed to be used to optimise the actuator.

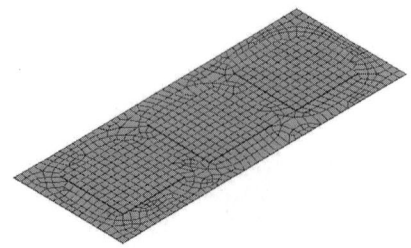

Fig. 4 FEA model

A finite element model (FEM), shown in Fig. 4, was built to perform 3D piezoelectric finite element simulation of the micropump. This 3D model uses the 8-node, hexahedral, coupled-field element SOLID5 for the piezoelectric materials and the 8-node, linear, structural element SOLID45 for the non-piezoelectric materials. Voltages of 0 volt and 100 volts were applied to the top and bottom actuator electrodes, respectively. Static analyses were performed to determine the stroke and stress of the micropump at different applied voltage and different geometric dimension.

4 Design analyses

The model very briefly described in the previous section was used to investigate effects of change in the structural parameters of actuators and the applied voltage on pumping effect. The simulated

results are shown in Fig. 5 to 13. The material parameters and geometrical dimensions used in the simulations are listed in Table 1 except where specifically mentioned in the text.

Fig. 5 shows the displacement distributions of the piezo-metal diaphragm when a voltage of 100 volts is applied to the left, middle and right piezo ceramics, respectively. Fig. 6 shows the deformation shape of the piezo-metal diaphragm when a voltage of 100 volts is applied to the middle ceramic. A very large stroke of about 62μm was achieved with these actuation units so it is possible for the micropump to pump a high volume of fluids through the chambers because the large stroke connects to high flow rate. Fig. 7 shows the stroke of the middle actuator at the centre of the diaphragm and the maximum stress in the metal diaphragm at the corners of piezo ceramic as a function of voltage applied to the middle piezo-ceramic. The higher the voltage applied, the higher the stroke of piezoelectric actuator and the higher the maximum stress the metal diaphragm is subjected to. It should be noticed that this conclusion is only applicable to the linear range of the piezoelectric material properties. Fig. 8 shows the stroke of actuator at the centre of the diaphragm and the maximum stress in the metal diaphragm at the corners of piezo ceramic as a function of piezo in-plane dimension. From Fig. 8, it can be observed that when the piezo in-plane dimension increases steadily, the stroke and the maximum stress increases steadily as well but when the in-plane dimension increase nearly (a little less than) to the chamber size, the stroke decreases a little but the maximum stress increases greatly. So the piezo in-plane dimension must be less than the chamber size. To have a high stroke and a low maximum stress, for example, for the geometrical dimensions of the micropump listed in Table 1, the piezo in-plane dimension is best to be less than 10mm×10mm, where the chamber size is 14mm×14mm with radius fillet of 4.8mm. Fig. 9 shows the stroke of the actuator at the center of the diaphragm and the maximum stress in the metal diaphragm at the corners of piezo ceramic as a function of the piezo ceramic thickness (t_p), where the metal diaphragm thickness (t) is fixed and taken to be 0.05mm. For normalization, the ratio of PZT thickness to metal diaphragm thickness (t_p/t) is taken to be variable in the x-axis of Fig. 9. It should be known that, in order to compare the effects of ratio of t_p/t on the stroke and the maximum stress, the strength of the electric field applied to the piezo-ceramics should be always kept the same value and so here taken to be 1000 volts/mm for all ratio of t_p/t in Fig. 9 and this leads to the applied voltage, which is linear to the ratio of t_p/t. From Fig. 9, it can be observed that, when t_p/t=1.4 or $t_p = 70\,\mu m$, the stroke reaches a maximum value, about 67 μm, and then starts to decrease while the maximum stress continues rising until t_p/t=2.4 or $t_p = 120\,\mu m$ and then reaches a levelled maximum value, about 166MPa. So the ratio of t_p/t should be design at t_p/t =1.4 or $t_p = 70\,\mu m$ to obtain maximum stroke and less stress for geometrical dimensions of the micropump listed in Table 1. But due to extremely difficulties in the fabrication of bulk piezoelectric material thickness less than 100 μm, the piezo ceramic thickness of 100 μm is designed in the studied micropump. It is worthwhile to mention that, for $0.6 < t_p/t, < 1.4$ or $30\,\mu m < t_p < 70\,\mu m$, if the piezo thick film is available with high performance forces proportionate to those in the bulk world, the piezo thick film is highly recommended for energy efficiency and maximization of the stroke of the micropump. For $t_p/t, < 0.2$ or $t_p < 10\,\mu m$, the stroke of the actuator is very small, about 17.5μm, not enough for pumping effect so it is not recommended to use thin film to make this micropump. Fig. 10 shows the maximum stress position in the metal diaphragm and piezo-ceramics.

Additionally, not only is the maximum stress in the metal diaphragm important but also the maximum strain and stress in the piezo ceramic are important reliability parameters. Therefore, the maximum strain and stress in the piezo-ceramics against applied voltage, against the piezo in-plane dimension, and against the ratio of thicknesses of the piezo ceramic to the diaphragm are respectively shown in Fig. 11, 12 and 13. From Fig. 11, 12 and 13, very similar conclusions can be drawn as from Fig. 7, 8 and 9. As a rough criterion, the compressive strain in the PZT should be not more than 1% and the tensile strain should be not more than 0.1%, otherwise, the PZT will break.

5 Conclusions and future work

A 3D finite element analysis (FEA) model is set up and static analyses are performed for different configurations under an applied voltage for understanding and optimization of the micropump performance and concerning reliability issue. Significant dimensional design strategy of the actuators has been obtained from the simulation results. One suggestion is, from the design point of view, to design the ratio of t_p/t properly. Another suggestion is to exploit the piezo thick film with high performance forces proportionate to those in the bulk world. This highlights the need for the piezo thick film with high performance.

Future work will be addressed on the effect of the bonding process on the performance of the actuator and also on modelling of the liquid flow with pressure distribution within chambers for effective actuation of the micropump. The second future work is very challenging[1].

For an estimation of the micropump lifetime, reliability investigations will be performed based on a sensitivity analysis via FE simulation. In a first step, critical parameters influencing life time of the micropump will be identified from simulation results. For the sensitivity analysis, FE-simulations will be performed varying these parameters. In the next step, the influence of scattering of the identified parameters and pump operation parameters are investigated by means of stochastical simulation. As a result, boundary values for critical parameters (e.g. glue layer thickness, geometrical tolerances, components mechanical properties) and pump operation parameters will be obtained.

Table 1 Material parameters and geometrical dimensions

Metal diaphragm dimensions	
Length l (mm)	44
width w (mm)	16.95
thickness t (mm)	0.05
Piezoelectric plate dimensions	
Length l_p (mm)	9.6
Width w_p (mm)	9.6
Thickness t_p (mm)	0.1
Metal diaphragm (Argeste)	
Young's modulus E_s (GPa)	200
Poisson ratio μ_s	0.3
Density (kg/m^3)	7800
Chamber	
Area $lc \times lc$ (mm×mm)	14×14
Fillet radius (mm)	4.8
Piezoelectric material from Block Technique PPK11	
Compliance constant	
10^{-12} (1/Pa)	15.9
	19.0
s_{11}^E	
s_{33}^E	
Density ρ_p (kg/m^3)	8100
Piezoelectric constant (10^{-12} C/N)	
d_{33}	680
d_{31}	350
Dielectric constant	
$\varepsilon_{11}/\varepsilon_0$	5000
$\varepsilon_{33}/\varepsilon_0$	5000
Applied voltage (volts)	100

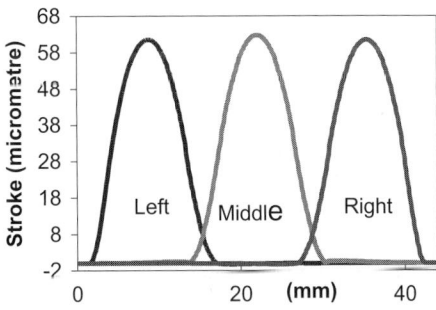

Fig. 5 Displacement distribution of the piezo-metal diaphragm when a voltage of 100 volts is applied to the left, middle and right piezo ceramics, respectively

Fig. 6 Deformation shape of the piezo-metal actuator when a voltage of 100 volts is applied to the middle ceramic (very large scaled)

266

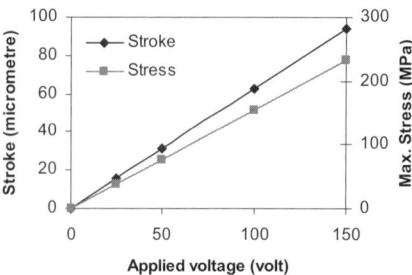

Fig.7 Stroke at the centre of the diaphragm and the maximum stress in the metal diaphragm at corners of the piezo-ceramic, shown in Fig.10, against applied voltage

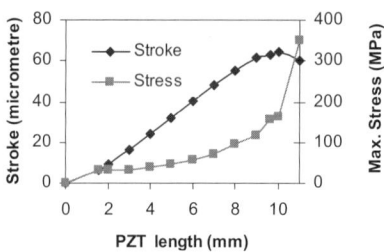

Fig. 8 Stroke at the centre of the diaphragm and the maximum stress in the metal diaphragm at the corners of the piezo-ceramic, shown in Fig.10, against the piezo in-plane dimension

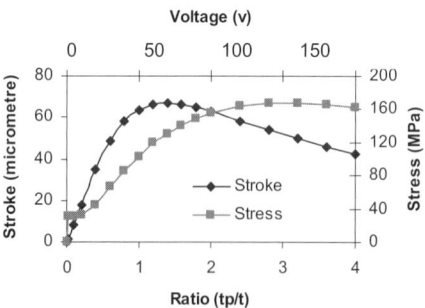

Fig. 9 Stroke at the centre of the diaphragm and the maximum stress in the metal diaphragm at the corners of piezo ceramic against the ratio of thicknesses of piezo ceramic to the diaphragm

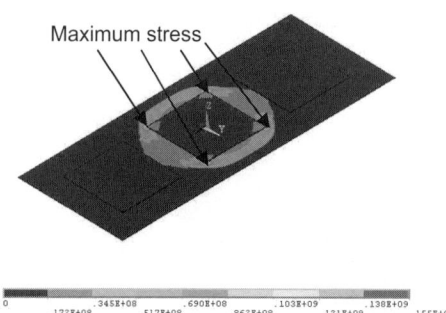

Fig. 10 Max. stress position in the metal diaphragm and PZT

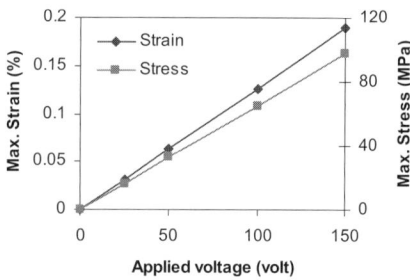

Fig. 11 The maximum strain and stress in the piezo ceramics against applied voltage

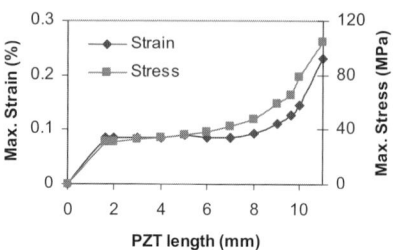

Fig. 12 The maximum strain and stress in the piezo-ceramics against the piezo in-plane dimension

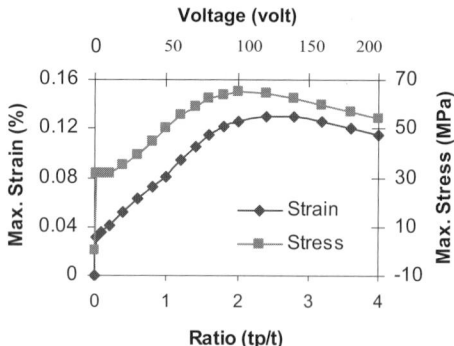

Fig. 13 The maximum strain and stress in the piezo ceramic against the ratio of thicknesses of the piezo ceramic to the diaphragm

Acknowledgements

We gratefully acknowledge the support from the Multi Material Micro Manufacture (4M) network.

References

[1] R. Linnemann, M. Richter, P. Woias, , A self priming and bubble-tolerant silicon micropump for space research, Proc. 2nd Round Table on Micro/Nano Technologies for Space (ESTEC), 15.-17.10.97, Noordwijk, The Netherlands (1997), pp. 83-90

Process Characterisation including Process Chains

Multi-Material Micro Manufacture
W. Menz, S. Dimov and B. Fillon (Eds.)

Micro-injection moulding: factors affecting the replication quality of micro features

B. Sha, S. Dimov, C. Griffiths and M. S. Packianather

The Manufacturing Engineering Centre, Cardiff University, Cardiff, CF24 3AA, United Kingdom.

Abstract

Micro-injection moulding is one of the key technologies for micro-manufacture because of its mass-production capability and relatively low component cost. The surface quality in replicating micro features is one of the most important process characteristics and constitutes a manufacturing constraint in applying injection moulding in a range of micro-engineering applications. This research investigates the effects of three processing and one geometric factor on the surface quality of micro features in three different polymer materials. In particular, the following factors are considered: barrel temperature, mould temperature, injection speed and placement density of micro features. In this investigation, the mould temperature was set in the conventional range. The study revealed that in general, increasing the barrel temperature, mould temperature and the injection speed improves the polymer melt fill in micro cavities. However, the effects of these factors on the process replication capabilities are not consistent for different polymer materials, and could be adverse in specific conditions. Varying the placement density of micro features does not affect the melt fills.

Keywords: micro-injection moulding; micro features; surface quality.

1. Introduction

There is an increasing trend towards product miniaturisation with the rapid development of micro-engineering technologies. The development of new micro devices is highly dependent on manufacturing systems that can reliably and economically produce micro parts in large quantities. In this context, micro-injection moulding of polymer materials is one of the key technologies for serial micro-manufacture.

Parts manufactured by micro-injection moulding fall in one of the following two categories. Type A are parts with overall sizes of less than 1 mm, while Type B have larger overall dimensions but incorporate micro features with sizes typically less than 200 µm. Currently, the following main groups of parts are manufactured successfully through micro injection moulding: optical grating elements, micro pumps, micro fluidic devices and micro gears. In all these applications the replication quality of parts' micro features is a factor determining the reliability of the selected manufacturing route. It depends greatly on their size, aspect ratio, and overall geometry [1].

Thus the quality of the replicated micro features is an important characteristic of any micro fabrication process that also determines the manufacturing constraints of a given process/material combination. Therefore, the factors affecting the quality of the parts produced by micro-injection moulding, for example their surface quality and edge definition, should be studied systematically. The flow of polymer melts in micro cavities is influenced by many factors that ultimately determine the replication capability of the process.

To improve the surface quality and the edge definition of injection-moulded micro-parts many research groups worldwide investigated different factors affecting the melt fill in micro-cavities. In particular, this includes research in process optimisation, material rheology, and tool design and manufacture. With regard to process optimisation, the melt and mould temperatures [2], and injection speed are considered as the main factors affecting the part quality in micro-injection moulding. However, with the increase of the mould temperature the cycle time also increases, which introduces stringent requirements in the technology for heating and cooling the tool. Consequently, this reduces the process output and hence increases the cost of the moulded parts.

Although it is considered that high settings of the process parameters are a good general strategy for producing quality parts, there are also some negative effects on the surface quality and the edge definition of micro features. In addition, the geometric configuration of micro features is an important factor affecting the replication quality. In this paper, the positive and negative effects of the barrel and mould temperatures, injection speed, and feature density on the quality of micro features are investigated.

2. Experimental Set-up

A test part with micro features in the form of circular "pins", "gear", and "fingers", as shown in Figure 1, was designed to conduct this experimental study. There are two types of pins having diameters of 100 and 150 µm, heights of 350 and 400 µm, and placement densities of 330 and 660 µm. The fingers also have two placement densities of 50 and 150 µm. The overall size of the part is bigger than 1 mm and therefore it belongs to type B micro-parts.

Two tooling inserts were manufactured in tooling steel (AISI 01), using a KERN HSPC micro-machining centre. The injection-moulding machine used in this study was a Battenfeld Microsystem 50. To measure and assess the quality of the produced micro-features, an optical measuring system, Mitutoyo Quick Vision, with an accuracy of ± 2 µm in the XY plane was employed.

3. Experiment Design

Three feature types, pins, gear, and fingers were selected to investigate the effects of process parameters on the quality of their replication. The effects of the different placement densities of the pins and fingers on the melt filling of the micro-features were also studied.

Semi-crystalline polymers, polypropylene (PP Sabic-56M10) and polyoxymethylene (POM C-9021), and an amorphous polymer, acrylonitrile-butadiene-styrene (ABS Cycolac-X17), were used to conduct the experiments. To study the filling behaviour of these three materials, three process parameters were used, barrel temperature (T_b), mould temperature (T_m), and injection speed (V_i). All control parameters together with their interactions were factors affecting the replication capabilities of the process. To assess their effects on the quality of micro-features, the Design of Experiment (DoE) method was applied. A two-level two-factor randomised full factorial design with four (2^2) experiments was used for PP and ABS, because T_b and V_i are considered as significant factors for these two materials [3]. For POM, T_m is also an important factor [3] and therefore a two-level three-factor randomised full factorial design with eight (2^3) experiments was employed for this material.

The experiments were conducted in a randomised sequence. These full factorial designs provided sufficient information about single-factor and two-factor interaction effects. Table 1 presents the process settings used for PP, POM and ABS during the experiments. Table 2 shows the matrices of the two DoE.

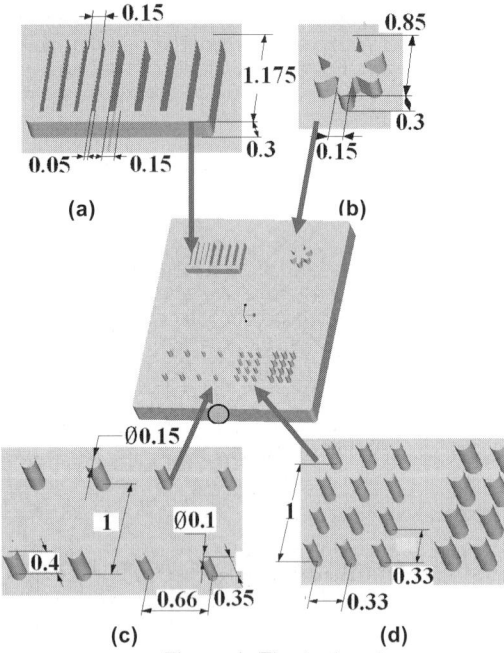

(a)   **(b)**

(c)   **(d)**

Figure 1. The test part

Note: The dimensions of the micro features are shown in mm. The position of the gate is marked with a circle.

Table 1 Processing parameters

Polymers	Level	T_b (°C)	T_m (°C)	V_i (mm/s)
PP	-	225	50	100
	+	255		200
POM	-	200	60	100
	+	220	120	200
ABS	-	258	75	100
	+	268		150

4. Experimental Results

4.1 PP

In Figure 2, it can be seen that the holes corresponding to pins with a diameter of 150 µm were completely filled by the melt for all parameter combinations. However, the holes with a diameter of 100 µm were not

fully filled when the barrel temperature and injection speed were set low. In this experiment, deep ring grooves were found on the surfaces of most pins. When the barrel temperature was set to 255°C, the filling of these arrays of micro holes improved. However, the ring grooves on the surface of the pins remained. At an injection speed of 200 mm/s the ring grooves still remained but they moved up in the pin. Finally, at the high settings (Exp. 4), the melt completely filled all micro holes and the ring grooves were shallower in depth and fewer in numbers. The feature density did not have an effect on the process replication capability.

Table 2. The DoE matrix

Exp.	Levels	
	T_b	V_i
1	-	-
2	+	-
3	-	+
4	+	+

(a) PP & ABS

Exp.	Levels		
	T_b	T_m	V_i
1	-	-	-
2	+	-	-
3	-	+	-
4	+	+	-
5	-	-	+
6	+	-	+
7	-	+	+
8	+	+	+

(b) POM

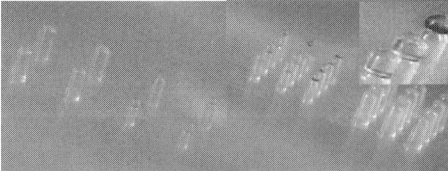

Experiment 1

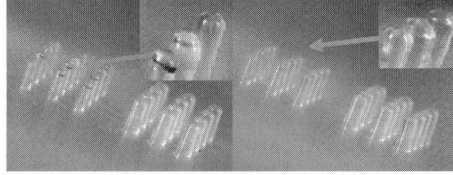

Experiment 3   Experiment 4

Figure 2. The micro pins in PP

The central cavities of the gears and some teeth were sufficiently filled with melt. Changing process parameters did not improve the replication quality.

Figure 3 shows that the grooves representing the fingers were filled with melt under all process settings except for their top edges. At an injection speed of 200 mm/s, the top surfaces of the fingers appear rougher and display more voids. A change of the gap between micro grooves did not have an effect on the replication quality.

4.2 POM

The pins with a diameter of 150 µm were completely filled by the melt in all experiments. In Exp. 1, the pins with a diameter of 100µm were shorter than designed. When the barrel temperature was set to 220°C, the depth of the melt fill increased. At a mould temperature of 120°C, the holes representing the 100 µm pins were completely filled.

However, when both the barrel temperature and mould temperature were set to their high settings (Exp. 4), the smaller holes were not completely filled but melt fill improved in comparison with the results in Exp. 2. When the injection speed was set to 200 mm/s (Exp. 5), the melt filled the smaller holes completely. By keeping the injection speed high and by increasing the barrel temperature to 220°C (Exp. 6), the smaller pins that are close to the gate could not be filled completely. Finally, at a high mould temperature and injection speed (Exp. 7 and 8), the melt filled all pins completely. The density of the holes in the cavity did not affect the replication quality.

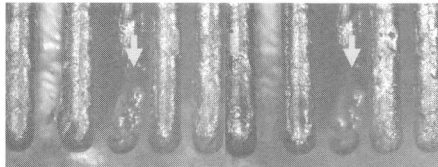

Experiment 1   Experiment 3

Figure 3. The micro fingers in PP

Note: The arrows indicate the finger that was damaged during the experiments.

Regarding the replication of the gear, as shown in Figure 4, at a barrel temperature of 220°C, the melt filled the cavity well except when all three process parameters were at their high settings (Exp. 8). When the mould temperature and/or injection speed were set high, the edge definition worsened. Even at the high settings of all process parameters, the edge definition did not improve.

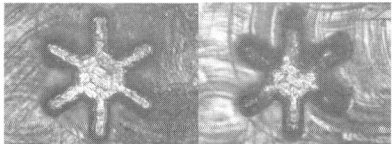

Experiment 2   Experiment 8

Figure 4. The micro gear in POM

Under all process settings the micro grooves representing the fingers in the cavity were filled completely. Changing the pitch between the micro grooves did not influence the melt fill.

4.3 ABS

Figure 5 shows that ABS melt filled completely the large pins under all process settings. The smaller pins with a diameter of 100 μm were not completely filled when the barrel temperature and injection speed were set to 258°C and 100 mm/s respectively (Exp. 1). By increasing the barrel temperature to 268°C (Exp. 2), the depth of the fill in the small holes increased. The melt filled completely most of the small pins at an injection speed of 150 mm/s even at a low barrel temperature (Exp. 3). At the high settings of all process parameters (Exp. 4), the melt did not fill completely all the small holes, and even the height of some pins was lower than those produced in Exp. 3. However, small holes were not filled completely when compared with the results in Exp. 1 and 2. Changing the pitch between the pins did not produce different results.

The central area of the micro gear cavity was filled relatively well under all parameter combinations. However, the edge definition was not satisfactory in all experiments, regardless of small improvements under settings 2, 3 and 4.

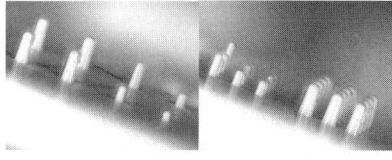

Experiment 1

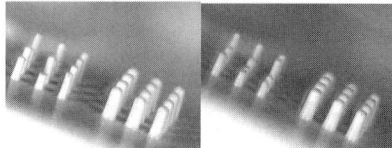

Experiment 3   Experiment 4

Figure 5. The micro pins in ABS

Figure 10 shows that like the gear, the edge definition of the micro fingers was not satisfactory. It should be noted that the definition was slightly better when T_b was set to 268°C (Exp. 2). However, when V_i was set to 150 mm/s (Exp. 3) the filling of the edges did not improve but worsened. Finally the fill quality in Exp. 4 was slightly better than that in Exp. 3. Changing the pitch between the micro grooves did not affect the melt fill.

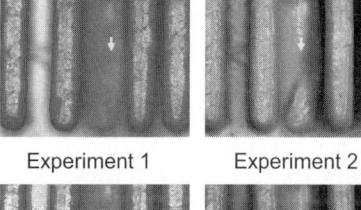

Experiment 1   Experiment 2

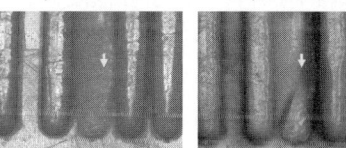

Experiment 3   Experiment 4

Figure 6. Micro fingers in ABS

Note: The arrow marks the damaged fingers during the experiments.

5. Analysis of the Experimental Results

5.1 Micro pins

An increase of T_b and V_i leads to a decrease of melt viscosity that improves the melt flow in micro cavities. This explains the better melt fill at the high setting of T_b and V_i. For PP, the formation of ring grooves on the pins may be explained with a temporary stop of the melt flow into the micro holes before the pressure in the cavity reaches its maximum value. This leads to a freeze of the outer layer of the melt, however the flow front continues to advance. The continuous build up of the pressure drives the melt flow further in the hole at the moment when the main cavity is already completely filled. As a result, the newly added melt and the frozen layer form the ring grooves. At V_i of 200 mm/s, there were less ring grooves or their position along the pins moved upwards close to the top. This indicates that the higher injection speed can transport more melt into the micro holes before the maximum cavity pressure is reached. Changes in T_b did not show any obvious effects on the formation of these ring grooves. However, by increasing T_b and V_i simultaneously the number of ring grooves decreased and also made the remaining ones shallower. Thus, the interaction of these two process parameters had a positive effect on the replication quality and improved the melt flow in the micro holes.

For POM, a high T_m decreased the thickness of the frozen layer and delayed the cooling down of the cavity so that more melt went into the micro holes. By increasing T_b, the melt fill close to the gate worsened at the high settings of T_m or V_i. This could be explained with the complex flow of material when melt is injected into an empty cavity. Normally, a low viscosity melt flows forward faster than a high viscosity one. Thus, a plausible reason for the above phenomenon could be that due to a higher T_b, the relatively low viscosity melt flowed at a higher speed and passed by the micro holes close to the gate when the melt was just injected into the empty cavity. Hence, this created a stagnant area around the gate that was filled only at the end of the injection stage. This could be the cause for the incomplete filling of those micro holes. A higher V_i resulted in more melt reaching the micro holes before freezing, and therefore more micro holes were filled at the high setting of T_b and V_i than when V_i was set low.

For ABS, raising T_b reduced the melt filling when V_i was set to150 mm/s. As the case with POM, the stagnant area around the gate could be the cause of the incomplete filling of the micro holes. In addition, the varying height of the pins indicates that the filling speeds also fluctuate in the cavity during the micro-injection moulding process.

5.2 Micro gear

The experiments show that different polymer materials had different responses to changes of the process parameters. For PP, an increase of the process parameters did not lead to obvious improvements of replication quality. This could be explained with the relatively large size of the gear, its lower aspect ratio and the good flow properties of PP that facilitates the melt fill. Therefore, the effects of changing the process parameter settings were less obvious.

For POM, the edge definition worsened when T_m, V_i or both were set high and T_b was set to 200ºC. One explanation for this could be that the expanding residual air, resulting from the high T_m in the corner area, further hindered the melt flow. A high V_i could cause more air to be trapped in the cavity and thus lead to incomplete fills in the corner areas. At T_b of 220˚C, the melt could fill the micro cavity, including the teeth and their edges, except under the high process settings (Exp. 8). This indicates that a melt with a low viscosity, resulting from a high T_b, could facilitate the air evacuation from the cavity corners and hence the filling of this area. However, setting T_b at 220 ˚C (Exp. 8) did not improve the edge definition, because of the negative interaction between the high T_m and V_i.

For ABS, the melt could not fill the corners of the gear cavity and even an increase of T_b and V_i only slightly improved the melt fill. This could be explained with the relatively high viscosity of the ABS melt [3]. In particular, it is difficult for a highly viscous melt to flow into the corners of micro cavities.

5.3 Micro fingers

Experimental results for PP show that as with the micro gears, due to the relatively large size and low aspect ratio of the fingers, their grooves were easily filled. Larger voids and more bubbles appeared at the edges when increasing V_i. It is known that as the injection speed increases the venting becomes more difficult, and bubbles

or incomplete fills may occur [4]. Hence, the foregoing phenomenon may be explained by the fact that the cavity could not be vented completely when V_i is high so that the residual air results in more bubbles and larger voids.

For POM, again due to the large dimensions and low aspect ratios of the features, the melt easily filled the micro grooves including the edges of the fingers under all process parameter combinations.

In the case of ABS, the edge definition improved only slightly at T_b and Vi of 268ºC and 100 mm/s respectively. This could be explained with a lower viscosity of the melt at a high T_b. A high V_i can hinder the air evacuation during the filling, and result in poor edge definition (Exp. 3). At high parameter settings (Exp. 4), the melt with a lower viscosity facilitates the air evacuation from the micro grooves, so that the replication quality, especially the edge definition, improved slightly.

7. Conclusions

This work investigates the effects of T_b, T_m, V_i, and also feature density on the replication capability of the micro-injection moulding process within the conventional mould temperature scope. The following conclusions could be drawn from this study.

• A higher V_i improves the filling of micro holes, however, it may also result in a poor surface quality and edge definition.

• A high T_b improves the melt flow, however, it could lead to stagnant flows in areas close to mould gates, especially in case of POM and ABS.

• In addition, by increasing T_b the edge definition improves when micro features are replicated in these two materials.

• A high T_m improves the replication quality of the pins in the case of POM, however, the results are not consistent when other micro features are considered.

• The melt flow is not sensitive to changes of the feature density.

Acknowledgements

The authors would like to thank the European Commission, the Department of Trade and Industry, the Welsh Assembly Government and the UK Engineering and Physical Sciences Research Council for funding this research under the ERDF Programme 53767 "Micro Tooling Centre" and the EPSRC Programme "The Cardiff Innovative Manufacturing Research Centre". Also, this work was carried out within the framework of the EC Networks of Excellence "Innovative Production Machines and Systems (I*PROMS)" and "Multi-Material Micro Manufacture: Technologies and Applications (4M)".

References

[1] Weber, L. and Ehrfeld, W. Micromoulding-market position and development. *Kunststoffe*, 1999, 89(10), 192-202.

[2] Yao, D. and Kim, B. Scaling issues in miniaturization of injection molded parts. *Manufacturing Science and Engineering*, 2004, Vol. 126, pp. 733-739.

[3] Sha, B., Dimov, S. S., Griffiths, C. A. and Packianather, M. S. Micro-Injection Moulding: Factors Affecting the Achievable Aspect Ratios. Int. Journal of Advanced Manufacturing Technology, 2006, Accepted for publication.

[4] Belofsky, H. Plastic: Product Design and Process Engineering, Hanser, New York, 1995.

Multi-Material Micro Manufacture
W. Menz, S. Dimov and B. Fillon (Eds.)
© 2006 Elsevier Ltd. All rights reserved

Laser polishing

T. Dobrev, D. T. Pham and S. S. Dimov

Manufacturing Engineering Centre, Cardiff University, Cardiff, CF24 3AA

Abstract

Laser milling with long pulses is generally associated with a high surface roughness, which prevents the broader use of this technology, especially in the manufacture of micro tools. In micro scale, achieving low surface roughness is very important, due to feature sizes. Laser polishing, also known as cleaning, is a novel technique for improving the surface quality of laser milled surfaces. It employs the same working principles as laser milling, but instead of having the target material in the focal point of the laser beam, an offset is introduced to the extent that there is not enough fluence for material removal. This paper reports on an investigation of the effects of laser polishing on two materials, industrial copper and stainless steel 316, used widely for the manufacture of micro cavities. The study provides an insight into the process behaviour during de-focused laser machining.

Keywords: micro, laser, polishing, cleaning, surface roughness

1. Introduction

Laser milling is a new and rapidly developing manufacturing technology. It is based on the laser ablation phenomenon where a target material is exposed to highly focused pulsed laser light [1]. Due to its high energy density, the incident laser heats up the target to high temperature and a small volume of material is instantly evaporated. The removal of material by scanning along the top surface of the target with a laser beam in this way is very similar to conventional milling.

A major application area for laser milling is micro-toolmaking, where the need for accurate and relatively small features is fuelled by the increasing demand for product miniaturisation. Micro-replication technologies, such as imprinting, micro-injection moulding, and hot embossing, introduce many new challenges that are not easy to address.

One such challenge is the roughness of micro-features. Laser milling of metal substrates can produce an average to high surface roughness, that usually requires further processing to make features suitable for micro-replication. The techniques for improving laser-milled surfaces could be grouped under two main categories. The first group includes techniques for post-process cleaning, for example by ultrasonic cleaning, de-oxidisation through pickling or electro-chemical polishing. The techniques for on-the-machine surface improvements form the second group. In this paper, a laser cleaning technique belonging to the second group is investigated. Laser cleaning, also known as laser polishing, is a process very similar to laser milling and usually can be performed on the same machine setup.

Laser cleaning has attracted a lot of attention from the semi-conductor industry, where it is used for the removal of micro-sized particles from the wafer surfaces [2-5]. The mechanisms of particle removal from a metal substrate are discussed by Hsu and Lin [2]. They focus on the creation of surface waves as the main particle removal mechanism in laser cleaning, and propose a model where the thermal stress waves are dominant in the cleaning process. Curran et al. [3] apply the laser beam at a glancing angle, and report an improvement of the cleaning efficiency of the process.

Laser cleaning of ablation debris was investigated by Coupland et al. [4]. The paper reported on the use of a CO_2 laser to clean the debris around laser-drilled vias in polyimide-based flex circuits. However, the cleaning process was taking too long compared to the drilling process.

Laser polishing of metal substrates was investigated by Shao et al. [6]. The materials studied were Fe, Al, Ti, and stainless steel 304 with dimension of the asperities in the sub-micron range. The thermo-evaporative polishing mechanism of different size asperities was investigated. The melting and/or evaporation of surface asperities led to improvements of the surface roughness through direct removal or truncation of their heights

The structure of the paper is as follows. The next section provides an overview of the laser polishing process. Then, the experimental set-up is outlined. Finally, the results of the experimental study are analysed and conclusions made about the process capability.

2. Process overview

Laser cleaning is a form of laser ablation performed on the laser milling system where instead of having the target material in the focal point of the laser beam, an offset is introduced to reduce the laser fluence [2][6]. As a result of this the substrate is only heated to reach a temperature close to the melting point of the particular material and thus to "flatten" any debris and other recast contaminants on the surface [2].

Figure 1 presents the set-up for a normal laser ablation operation, where the laser beam is focused on the top surface of the substrate. In this way a maximum power density could be achieved on the surface that is sufficient to raise the material temperature to its boiling point. Any deviation from the accurate focal distance normally results in a decrease of the volume of material that is removed from the target and hence this has a detrimental effect on the process performance. Thus, in normal conditions this is an undesirable effect that should be avoided. Therefore, an optical sensor system is integrated in the machine to measure the distance to the target surface and if required to modify the laser power output so that a constant material removal rate is maintained.

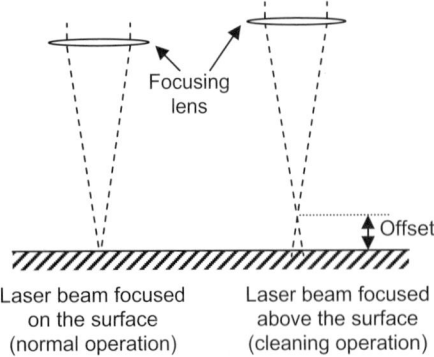

Laser beam focused on the surface (normal operation) Laser beam focused above the surface (cleaning operation)

Figure 1 Normal and cleaning operation of the laser milling system

However, to carry out laser cleaning the system should operate in this defocused state (see Figure 1). As a result of the offset, the laser beam does not reach a power density sufficient to perform material removal, while it is enough to soften the top-most surface layer and any debris that was deposited during the machining.

3. Experimental set-up

To study the processing capabilities of the laser cleaning technique a series of experiments was carried out to investigate its effects on the resulting surface quality. In particular, the resulting roughness on the flat bottom surfaces of the machined features is studied. The surface finish on the sidewalls and inclined surfaces is affected by other factors outside the scope of this research.

The materials selected for this investigation were stainless steel 316 (SS 316) and industrial copper and their physical and thermal properties are given in Table 1. They were chosen due to their different laser milling characteristics. While machining of SS 316 usually produces a better surface finish than copper, more material is re-deposited on the surface in the form of recast layers and debris during the cleaning process. To minimise these negative effects on the final surface roughness during the experiments, both materials were cleaned in an ultrasonic bath before any measurements were performed. In addition, to avoid any influence from the original surface topography, the samples were machined to R_a under 0.5 µm.

The process variable that affects the laser cleaning performance is the offset of the focal spot from the top-most surface of the substrates. In this experimental study, the focal offset was varied from -3 mm to 3 mm. As a reference for assessing the surface finish improvement, on each test workpiece a patch was machined with an offset equal to zero to represent the normal machining conditions during laser milling. To find the "optimum" processing windows for both materials a series of setting-up machining sessions were carried out on the

samples. The optimisation started with some already available parameter values for machining a sequence of layers. After every 4 layers, the machined depth was measured and the average material thickness removed per layer calculated. Through this iterative procedure the process was tuned for laser milling layers with a thickness equal to 2 µm. The process settings that were identified in this way for both materials are given in Table 2.

Table 1 Material properties for copper and SS 316 [7]

Material properties	Copper	Stainless steel 316	Units
Density @ 21°C	8800	8238	kg/m³
Thermal conductivity @ 21°C	52	13.4	W/m K
Specific heat capacity @ 21°C	420	468	J/kgK
Melting point	1300	1670	K
Boiling point	2843	3173	K

Table 2 Laser milling settings for the selected materials

Parameter	Industrial Copper	Stainless steel 316
Laser flashlamp current, I [%]	59.6	71
Avg. laser power P_{av} [W]	2.36	7
Frequency f [kHz]	10	40
Scanning speed V [mm/s]	100	400
Pulse duration τ [µs]	10	10
Hatching distance [µm]	10	10
Sample size, mm	2×2	2×2

The difference between the processing parameters for the two materials is due mostly to their thermal properties. In particular the thermal conductivity of copper is 4 times higher than that of SS 316. This can explain the considerable difference between their laser milling parameters. Specifically, the average laser power applied to machine copper is significantly less than that used for SS 316. The pulse frequency and scanning speeds for both materials also differ with a factor of 4. To laser mill SS 316 successfully a higher scanning speed setting of 400 mm/s was required in comparison to 100 mm/s applied to machine copper. Thus, to process SS 316 reliably the laser milling process needs more powerful pulses, with a shorter pulse period, and higher speed scans on the substrate surface. If only a single crater is considered, and neglecting the laser frequency and the scanning speed, to remove the same volume of material the average power should be four times higher in case of SS 316.

The experiments were carried out on a commercial laser milling machine with a microsecond Nd:YAG laser. The resulting surface roughness obtained by applying the processing parameters in Table 2 is discussed in the next section.

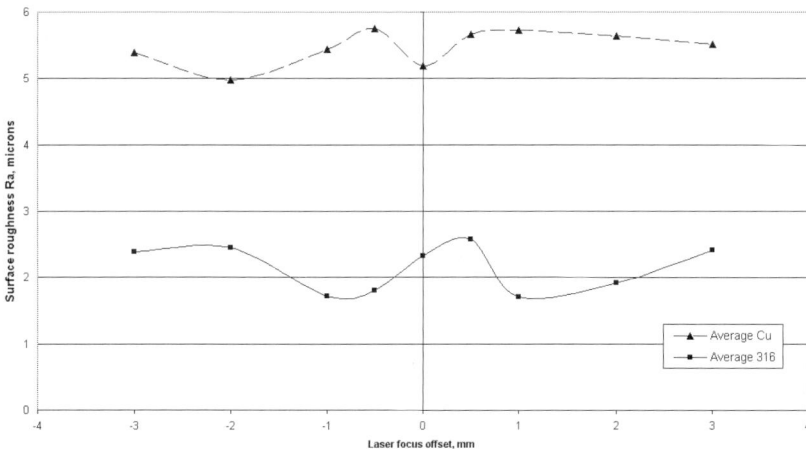

Figure 2 Surface roughness vs. the offset of the laser focal spot

4. Results

A total of 9 test patches were machined on each substrate. The roughness of the flat bottom surface of each patch was measured at two different locations using a MicroXAM surface profiler.

Figure 2 presents the average surface roughness corresponding to each patch measured on both substrates. As expected the average surface roughness of the patches machined on the copper substrates was higher than that achieved on the SS 316 substrate. The lowest roughness achieved on the copper and the SS 316 samples were R_a 4.98 μm and R_a 1.7 μm respectively. Also, the graphs in Figure 2 indicate different process behaviour when laser milling these two materials. The graph for the surface roughness measured on the SS 316 substrate is almost symmetrical around the zero offset point, while for copper it is asymmetrical.

The best roughness measured on the copper substrate, R_a 4.98 μm, was achieved with an offset equal to – 2 mm. However, this is only marginally better than the roughness resulting from the machining with a zero offset. In addition, it is important to stress that the results that were obtained on the copper substrate with an offset of + 2 mm were different from those achieved with an offset of – 2 mm. As it can be seen in Figure 2, the SS 316 graph could be considered symmetrical about the zero offset point, which should be expected from a symmetrical optical de-focussing process. The roughness of the machined patches on the copper substrate was so high that the effect of the de-focussing was actually opposite to what was expected. In particular, instead of improving the surface quality there was a marginal increase in the measured roughness. The other significant difference in the process behaviour observed during the experiments was that when machining copper the roughness was relatively low at zero offset, and then deteriorated with the increase of the offset. On the contrary, for SS 316 the roughness of the patch machined with a zero offset was relatively high and then the surface quality improved with the increase of the offset up to +/- 1 mm. In particular, the surface roughness improvements were around 27 % for the patches laser cleaned with offsets of 1 or –1

mm. However, this positive effect of de-focusing the laser beam did not continue with a further increase of the offset. The roughness of the patches processed with an offset of + 2 and - 2 mm was almost the same as that resulting from a machining without any de-focusing.

The effect of the laser cleaning on the surface roughness can be seen in Figure 3 which shows line profiles taken from the sample patches on the SS 316 substrate. The surface profile on the reference patch, machined without any offset, exhibits erratic and irregular characteristics compared to the much smoother curves resulting from laser cleaning with a 1 mm offset.

The results obtained on the copper substrate (see Figure 4) did not show such differences in line profiles as those observed on the SS 316 sample, however this could be attributed to an R_a figure almost 3 times higher.

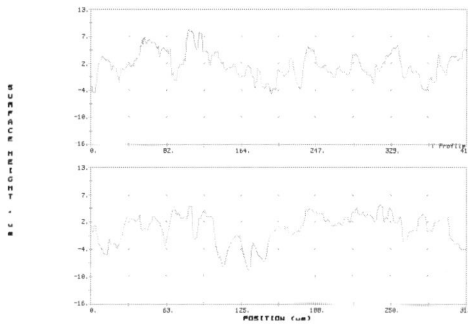

a) Before laser cleaning

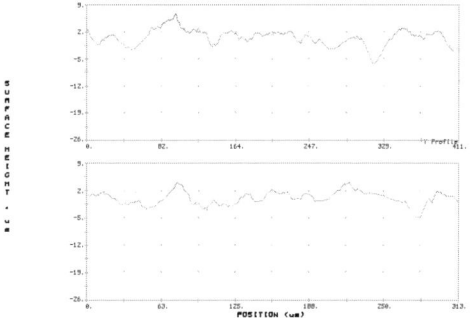

b) After –1 mm laser cleaning

Figure 3 Line profiles of the roughness measured on the SS 316 substrate

276

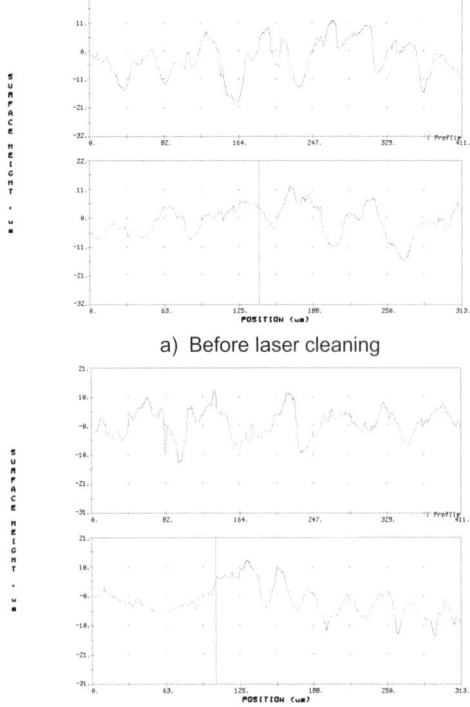

a) Before laser cleaning

b) After –2 mm laser cleaning

Figure 4 Line profiles of the roughness measured on the copper substrate

5. Conclusions

A technique for cleaning laser milled surfaces is investigated in this research. The technique utilises a de-focused laser beam to traverse/scan the surface in the same way as during normal laser milling. In this way the power density on the machined surfaces is reduced drastically. As a result the beam does not remove material from the substrate and the laser power is sufficient only to melt and smooth out any surface irregularities. However, as is the case with all laser machining processes, the results depend on the specific laser-material interactions and thus the effect of the laser cleaning will vary for different materials. In this study, the process behaviour when machining industrial copper and stainless steel 316 is investigated.

The same experimental set-up was used for both materials. The laser beam offset was varied from -3 to +3 mm. A patch machined with zero offset was utilised as a reference for assessing surface roughness improvements. The laser settings were selected for each material separately applying an iterative procedure commonly used for identifying a processing window for a stable and uniform laser milling process. However, it should be stressed that the process parameters were not optimised for achieving the best possible surface roughness, in order to resemble normal machining conditions.

The results obtained provide an insight into the process behaviour during de-focus laser processing. For copper, the process does not reduce the surface roughness, and thus laser cleaning has a small or nil effect. On the contrary, the processing of SS 316 leads to surface finish improvements in comparison with the roughness achievable without performing laser cleaning. In particular, during the experiments carried out a surface roughness improvement of up to 30% was observed.

Another difference in the process behaviour when machining these two materials relates to the results obtained with the same absolute offset value. Surface roughness graphs were expected to be symmetrical due to the symmetrical optical de-focussing process. However, this was the case only when processing SS 316. For copper, the surface roughness graph is asymmetrical, hence the process behaves differently.

Acknowledgements

The authors would like to thank the European Commission, the Department of Trade and Industry, the Welsh Assembly Government and the UK Engineering and Physical Sciences Research Council for funding this research under the ERDF Programme 53767 "Micro Tooling Centre" and the EPSRC Programme "The Cardiff Innovative Manufacturing Research Centre". Also, this work was carried out within the framework of the EC Networks of Excellence "Innovative Production Machines and Systems (I*PROMS)" and "Multi-Material Micro Manufacture: Technologies and Applications (4M)".

References

[1] Dobrev, T., 2005. *Investigation of laser milling process characteristics for micro tooling applications*. PhD thesis, University of Wales, Cardiff, Dec 2005

[2] Hsu H.-T. and Lin J., 2005. *Thermal–mechanical analysis of the surface waves in laser cleaning*. Int. J. of Machine Tools & Manufacture, Volume 45, 2005, pp. 979 – 985

[3] Curran C., Watkins K.G. and Lee J.M., 2001. *Effect of wavelength and incident angle in the laser removal of particles from silicon wafers*. 20th International congress on applications of laser and electro-optics ICALEO, Jacksonville, Oct 2-5, 2001

[4] Coupland K.,. Herman P.R. and Bo Gu, 1998. *Laser cleaning of ablation debris from CO_2 - laser-etched vias in polyimide*. App. Surface Science, Vol. 127–129, 1998, pp. 731–737

[5] Lee J.M. and Watkins K.G., 2000. *Laser removal of oxides and particles from copper surfaces for microelectronic fabrication*. Optics Express, Volume 7, No. 2, July 2000, pp. 68 - 76

[6] Shao T.M., Hua M., Tam H.Y., Cheung E.H.M., 2005. *An approach to modelling of laser polishing of metals*. Surface and Coatings Tech., Vol. 197, Issue 1, 2005, pp. 77 – 84

[7] Incropera, F. P., 2002. *Fundamentals of heat and mass transfer* / Frank P. Incropera, David P. DeWitt., ed. 5. New York : J. Wiley, 2002

Multi-Material Micro Manufacture
W. Menz, S. Dimov and B. Fillon (Eds.)

Size effects in the production of micro strip by the flat rolling of wire

K. van Putten, R. Kopp, G. Hirt

Metal Forming Institute, RWTH Aachen University, Intzestrasse 10, D-52056, Germany

Abstract

Comparison between down scaled flat rolling experiments of thin round wire and numerical simulation of those experiments have shown that the production process of manufacturing micro strip out of thin round wire is influenced by size effects. Decreasing the wire diameter into the microscopic domain is accompanied by decreasing yield stresses of the wire and turned out to be the dominating size effect for low rolling reductions. For high rolling reductions an increase of friction becomes of major influence. The plane strain compression test is applied as an experimental simulation of the rolling process. It is also used for further investigation of the determined size effects. An experimental set-up has been built, that allows scaled experiments on small and miniaturised specimens maintaining geometric similarity. Micro structural effects, such as the Hall-Petch effect and effects on the evolution of the surface roughness have been determined with this plane strain compression set-up.

Keywords: rolling, plane strain compression test, size effects, surface roughness, FEM simulation

1. Introduction

Micro strip is defined as small and thin strip with its width in the lower millimetre domain and its thickness in the sub-millimetre domain. Strips with such a small cross sectional area are located in, or close to the microscopic domain.

In 1936, R.V. Jones reported about the rolling of micro strip out of thin round wire [1]. At that time, the development of some specific scientific instruments like galvanometers, bolometers and thermoelements, required a very thin strip only a few millimetres wide which were produced sporadically and in low volume.

Nowadays, the rise of medical diagnosis systems and the development of medical micro devices lead to the application of micro strips in diverse medical equipment. Applications of micro strips can be found in the field of minimal invasive surgery, artificial organs and implant developments as well as the application of biochemical diagnostics. Besides that micro strip is also applied in modern electronic products and precision engineering applications.

There are several methods to produce small and thin strip. This article is about the production of micro strip out of extruded, hot rolled or cold drawn and soft annealed, precursor wire which is rolled flat by parallel cylindrical rolls. During the rolling pass the round cross sectional geometry of the precursor wire is formed into a flat strip with an approximately square cross section with rounded, so called mill, edges.

The production process depends on several complex parameters, partly due to the high three-dimensionality of the internal deformation imposed during the rolling process, in combination with the initial conditions of the precursor wire. Until now, the parameters to control the process are mainly determined empirically. Unfortunately, it turned out that below certain wire and strip dimensions, the common techniques to do this are no longer effective.

When the product dimensions become small enough, the process switches from the macroscopic to the microscopic domain. Here the material does not behave like a continuum anymore and process parameters may be influenced by size (or scale) effects.

To determine the size effects for the rolling of micro strip out of wire, the rolling process itself as well as the plane strain compression test, used for the determination of material properties, are scaled with (partial) similarity [2]. Differences between the scaled processes maintaining geometric similarity are indicative of size effects.

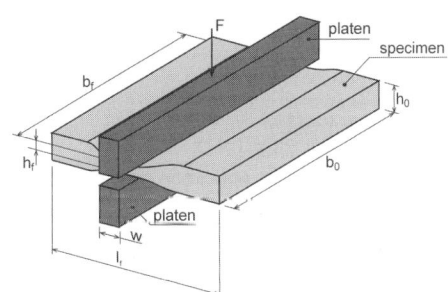

Fig. 1. Principle of the plain strain compression (PSC) test.

In the following, the experimental results performed by the plane strain compression of small specimen and the determined size effects in the flat rolling of thin round wire are presented.

2. Plane strain compression test

Plane strain compression (PSC) testing is used to determine the yield curve of a material, by the compression of a rectangular specimen between two accurately aligned platens. The specimen is compressed across its width by the narrow platens that are wider than the strip (see Figure 1). Assuming ideal plane strain, i.e. no lateral spread, the deformation of the specimen occurs in the direction of the platen motion and in the length direction of the specimen. The deformation is (desirable) symmetric and after deformation the specimen is characterised by a deformed region between the tools with (almost) rigid regions outside this deformed region.

2.1. Experimental set-up

In order to perform PSC tests scaled with similarity a precision PSC device with exchangeable compression platens is built. This PSC device is integrated in an alternating tensile and compression testing machine.

To ensure a good alignment, so that even the compression of the smallest specimen is accurate, the compression platens are integrated in a four pillar die set. The friction of the die set is reduced to a minimum by a shaft guiding with linear ball bearings.

The compression forces are measured by either a 100, 50 or 1 kN load cell (accuracy class 0.02) which can be exchanged according to the expected forces. Inductive displacement transducers with 1 μm accuracy measure the specimen thickness during the experiment.

In order to enable the scaling of the PSC test, sets of compression platens with various width w are available (see Table 1). These grinded platens are hardened and have a hardness of 54 HRC.

2.2. Specimen preparation

All specimen are made out of pure copper (OF-Cu) rolled sheet material. To obtain 90° specimen corners, the specimen with 4.0, 2.0 and 1.0 mm thickness are milled and not cut. Conventional machining of the specimen with 0.5 mm thickness resulted in specimen of poor quality. To overcome the negative effects of burrs and a poor quality of conventional cut faces and to obtain specimen with 90° corners, the specimen out of 0.5 mm sheet material are cut by wire EDM.

The length (l_f) of the specimen is parallel to the rolling direction (RD), the specimen breadth (b_0) is equal to the transversal rolling direction (TD). The specimen thickness (h_0) agrees to the normal direction (ND) of the rolled sheet. The rolled sheet did not go through any heat treatment.

Soft annealing of the specimen as well as varying the mean grain size is done by annealing the specimen at temperatures varying from 300 to 900 °C in a high vacuum oven for different times.

Table 1
Compression platens and specimen dimensions scaled with geometric similarity.

m_l	w (mm)	h_0 (mm)	b_0 (mm)	l_0 (mm)
1	6.00	4.00	30.00	60.00
1/2	3.00	2.00	15.00	30.00
1/4	1.50	1.00	7.50	15.00
1/8	0.75	0.50	3.75	7.50

2.3. Determination of Hall-Petch effects by PSC tests

In order to study the effect of the ratio of grain size diameter to specimen thickness, PSC test on 2 mm thick specimen with different grain sizes were performed. It is expected that the effect will be according to the Hall-Petch effect [3]. However, the Hall-Petch effect has never been obtained by PSC testing of small specimens. Additionally such experiments will prove if the built experimental set-up is suitable for measuring micro structural influences.

For ductile polycrystalline materials of conventional grain size ($d_k > 1$ μm) it has been found that the yield

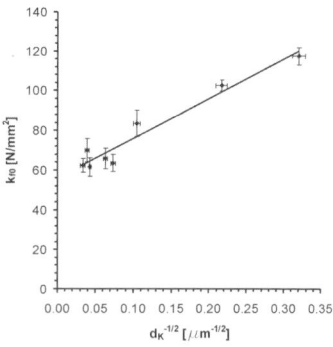

Fig. 2. Hall-Petch effect determined with aid of PSC testing of 2 mm thick OF-Cu specimen.

stress is a function of grain size. Based upon the dislocation theory it might be concluded that, especially at the transitional stage from elastic to plastic deformation, where the grain boundaries form the major obstacle for the movement of dislocations, the stresses (lower yield stress or 0.2% proof strength) will depend on the mean grain size. The empirically determined Hall-Petch equation expresses the grain size dependence of the yield stress:

$$\sigma = \sigma_0 + k_y \, d_k^{-0.5} \qquad (1)$$

The constant factor k_y is the so called Hall-Petch constant and σ_0 represents the shear stress opposing dislocation motion.

The grain sizes are measured by applying the mean lineal intercept method in the RD-ND cross sectional surface. The lower yield stress is obtained from the yield curves determined by the PSC of the specimen.

Figure 2 shows the relation between lower yield stresses with their accompanying standard deviations and the measured mean grain size diameters with their accompanying standard deviations. Nearly all related pairs of standard deviations in Figure 2 are crossed by the regression line calculated by the method of least-squares best fit. From the regression analysis it follows that, the Hall-Petch constant k_y equals 0.199 MPa√m and σ_0 measures 56 MPa.

2.4. Size effects in surface evolution

After the PSC of specimens with higher grain size diameter to specimen thickness ratio a change in surface roughness can be observed. From macroscopic metal forming processes it is known that the surface roughness of a specimen compressed by any tool reduces if the tool itself has a smooth surface [4]. The grains in the deformed zone of the specimen that are near the surface are unable to deform freely (as in e.g. tensile testing), they have to fulfil the boundary constraint imposed by the tool.

In spite of that, the surface roughness of the small PSC specimen with higher grain size diameter to specimen thickness ratio has increased. The surface roughness of the compressed zone even exceeds the surface roughness of the tools by several orders of magnitude. The surface roughness R_a, R_t and R_z of the compression platen measures 1.26, 11.7 and 8.8 μm respectively for all tools. For a 4 mm thick specimen with d_k/h_0 of 0.17 the R_t and R_z value are equal to 190

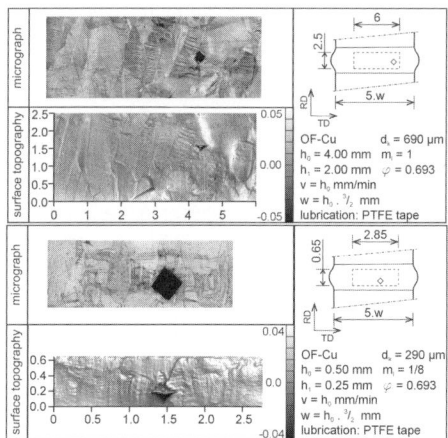

Fig. 3. Correlation between surface topography and microstructure.

and 78 μm and for a 0.5 mm thick specimen with d_k/h_0 of 0.58 R_t and R_z measure 123 and 47 μm.

A comparison between the surface topography, measured by a opto-electronic 3D measurement system and a micrograph of the compressed specimen surface, shows a clear correlation between the surface topography (or roughness) and the grains in the specimen (see Figure 3). The single grains from the micrograph can clearly be recognised in the surface topography.

The influence of the Teflon lubricant is not completely clear yet. It might be, that the lubricant, as known from macroscopic forming with abundant lubrication, forms a viscous layer that enables to deform like a free surface. But preliminary PSC experiments with MoS_2 and without lubrication have also shown an increasing surface roughness for specimen with higher ratio of d_k/h_0.

3. Size effects in the rolling of micro strip

In order to investigate if the production process of the rolling of micro strip out of thin round wire is subjected to size effects a set of experiments was performed. Within the experiments the rolling process was scaled out of the macroscopic into the microscopic domain.

Because it is not possible to change the rolls of the high precision rolling mill, the conditions for exact geometrical similarity could not be fulfilled. Therefore numerical simulations with aid of the finite element method are used as reference. The FEM simulations do not include any size effects, so that different trends between experiment and simulation are indicative of size effects.

3.1. Resistance to forming and contact area

The resistance to forming of the rolling process was chosen as the variable for comparison between experiments as well as for comparison with finite element simulations. It is defined by the quotient of the rolling force F and the contact area A_d between the cylindrical roll and the flat rolled wire:

$$k_w = F/A_d \qquad (2)$$

The contact area represents a surface in three

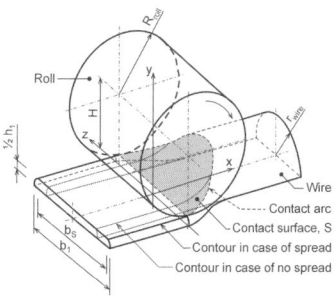

Fig. 4. Model of the flat rolling of a wire.

dimensions (see Figure 4). Due to symmetry only half of the model is presented in Figure 4. The cylindrical rolls with radius R_{roll} deform the wire with round cross sectional area of radius r_{wire} into a strip with a rectangular cross section b_1 to ½ h_1. If there is no spread at all during the flat rolling of a wire, the contact surface between roll and wire is, in a mathematical sense, equal to half of the collective casing surface between two crossing cylinders. In case of spread, an additional term can be added. The contact area during flat wire rolling can be calculated by adding the spread term to the part formulated for flat wire rolling without spread. In this way the surface area of the contact is described by an integral equation. This integral equation describes a complete elliptical integral and cannot be solved analytically. The contact area is calculated by solving the integral with aid of numerical analysis in MatLab. Comparison of calculated and measured surfaces proved the correctness of the mathematical calculation [5].

3.2. Experimental flat rolling of thin round wire

Soft annealed OF-Cu bars cut out from wire with 4.0, 2.0, 1.0 and 0.5 mm in diameter were rolled flat with a reduction of 25, 50 and 75% respectively. For each combination of wire diameter and reduction five experiments were performed. The mean value of the resistance to forming out of these five experiments is used for comparison.

3.2.1. Experimental set-up

A precision rolling mill, built in 2003, with 140 mm roll diameter of the Prymetall company was used for the experiments. During the rolling of the OF-Cu bars the rolling force was measured. There were no forward and backward tensions applied on the copper bars and rolling speed was reduced to a minimum of about 0.75 m/min. The strip thickness h_1 and breadth b_1 were measured afterwards so that the contact area could be calculated.

3.2.2. Specimen preparation

Bars of 1.5 m length were cut from 4.0, 2.0, 1.0 and 0.5 mm diameter OF-Cu wire. The bars were annealed for 30 min. at 400 °C and straightened afterwards. The stresses induced during the straightening were reduced by a second heat treatment for 15 min. at 150 °C.

3.2.3. Experimental results

The resistance to forming in dependency of the wire diameter for each combination of wire diameter and reduction is plotted in Figure 5. The combinations with the same reduction 25, 50 and 75% are marked

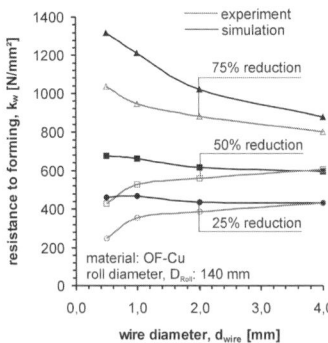

Fig. 5. Resistance to forming k_w for the flat rolling of pure copper (OF-Cu) wires with various wire diameters d_{wire}.

with open circles, squares and triangles respectively and connected by grey coloured curves.

3.3. Numerical simulation of the experiments

The physical experiments were numerically simulated with aid of the finite element program PEP/Larstran-Shape. There is no inclusion of any size effects, simulations are done by commonly used continuum methods with classical yield curves and a constant friction coefficient of 0.225 for all simulations. The simulation assumed a purely mechanical process without any thermo mechanical coupling.

Due to symmetry only half of a wire was modelled with a length of 3 times the contact length between roll and flat rolled wire. The volume was meshed by using 8 node hexahedron elements. The roll is modelled as a non-deformable rigid body. The yield curve of the pure copper was implemented by means of a table, the Young's modulus, density and Poisson's ratio used were 100000 N/mm², 8890 kg/m³ and 0.30 respectively.

As in the physical experiments, the resistance to forming is calculated. Therefore the calculated roll force as well as the breadth of the strip were extracted from the simulation results. The resistance to forming in dependency of the wire diameter are plotted in Figure 5. The combinations with the same reduction 25, 50 and 75% are marked with filled circles, squares and triangles respectively and connected by black coloured curves.

3.4. Comparison between simulation and experiment

Figure 5 shows that in the macroscopic domain, namely for 4 mm wire diameter, there exists a quite good agreement between experiment and simulation. However, with decreasing wire diameter the deviation between simulation and experiment increases. For 25 and 50% reduction there exists a completely different trend between simulation and experiment! Here the simulation calculates an increasing resistance to forming where the experiment shows a decreasing trend. The differences between simulation and experiments indicate that the experiments are influenced by size effects.

3.5. Analysis of the determined size effects

The different trends between simulation and experiment can be partially explained by two size effects that are described in literature: First, the yield stress reduces with decreasing specimen size [6, 7]. Second, the friction coefficient increases with decreasing specimen dimensions [7].

The effect of decreasing yield stress will dominate for the rolling experiments with 25 and 50% reduction. Eventual frictional effects will be minimal because of the relative small contact area between roll and wire. The decreasing yield stress will cause the decreasing resistance to forming with decreasing wire diameters.

However for the rolling experiments with 75% reduction frictional effects will be of major importance. The ratio of contact area and deformed volume underneath the roll has increases with decreasing wire diameter. This will cause the increasing resistance to forming with decreasing wire diameters.

4. Conclusions

The plane strain compression test can be scaled with full similarity. Micro structural effects, such as the Hall-Petch effect can be measured with aid of plane strain compression testing of small specimen. The surface roughness after the compression of the small PSC specimen is influenced by the grain size diameter to specimen thickness ratio and the topography of the roughness correlates with the micro structure.

Comparison of physical experiments with numerical simulations have shown that the production of micro strips by the flat rolling of thin round wire is influenced by size effects. It is most likely that the decreasing yield stress and the increasing friction with decreasing specimen size are the main contributors to those influences.

Acknowledgements

The authors gratefully acknowledge the financial support of the Deutsche Forschungsgemeinschaft (DFG) within the priority program (SPP) 1138: "Modelling of scaling effects on manufacturing processes". The authors also would like to express their thanks to Wieland-Werke AG for providing the copper wire and sheet material as well as Prymetall for the given possibility to use their precision rolling mill to perform the experiments.

References

[1] Jones R.V. The production of metallic films. Proc. Phys. Soc. 39, 1927, 282 - 288.
[2] Hergemöller R. Anwendung der Ähnlichkeitstheorie auf Probleme der Umformtechnik. PhD. Thesis, RWTH Aachen University, Germany, 1982
[3] Petch N.J. The cleavage strength of polycrystals. J. Iron Steel Inst. 13, 1953, 25 - 28.
[4] Lange, K. Umformtechnik – Grundlagen (2nd edn). Springer Verlag, Berlin, 2002.
[5] Van Putten K., Kopp, R. and Hirt G. Scale effects in the rolling of micro strip. Proc. 2nd Coll. of DFG Priority Program Process Scaling, Bremen, Germany, 2005, 41 – 50.
[6] Kals R.T.A. Fundamentals on the miniaturisation of sheet metal working processes. PhD. Thesis, Friedrich-Alexander University Erlangen-Nürnberg, Germany, 1999.
[7] Messner A. Kaltmassivumformung metallischer Kleinstteile. PhD. Thesis, Friedrich-Alexander University Erlangen-Nürnberg, Germany, 1997.

Multi-Material Micro Manufacture
W. Menz, S. Dimov and B. Fillon (Eds.)

Manufacturing and characterization of water repellent surfaces

A. De Grave, P. Botija, H.N. Hansen, P.T. Tang

Department of Manufacturing Engineering and Management, Technical University of Denmark
Produktionstorvet, Building 427 S, DK 2800, Lyngby, Denmark

Abstract

The leaves of some natural plants show a micro structure giving them the capacity of being cleaned from any undesired particles by rainfall. Thorough studies of the physical laws that lay behind this phenomenon, known as the lotus effect, has been conducted throughout the years, in order to obtain a set of useful characteristics for such surfaces. The problem of adapting this behaviour to artificially roughened surfaces could be addressed by providing design criteria for water-repellent and self-cleaning surfaces. They could then be engineered according to the actual performances desired for them. Using an enhanced design criteria, the production of patterned micro structured surfaces following two different process chains is reported in the present paper. The first is a combination of laser manufacturing and hot embossing on polystyrene. To compare geometry and functionality, a non-silicon based lithography technique using copper laminated epoxy is described as a second approach. Hydrophobization of some surfaces was attempted. Results of characterisation are shown. Drop deposition tests are performed on the obtained surfaces following two methods.

Keywords: water repellent surfaces, manufacturing processes chain, characterisation

1. Introduction

It is widely known that the surface of plant leaves sometimes causes high water repellency. Water which is applied to such surfaces forms small droplets rolling free over the leaves. This so-called Lotus-Effect [1] or superhydrophobicity depends on a combination of two scale level of microstructures on the surfaces and of the chemical properties of the surface (for example, a hydrophobic wax layer). Since this phenomenon relies on the physicochemical properties of the leaves it can be transferred to engineered surfaces, in a biomimetic approach. In this article mainly the manufacturing of a geometrical micro structure will be investigated. The study leading to the design criteria is not the scope of the paper. Two techniques will be described to achieve a surface with topology in accordance to these design criteria. The geometrical characterization of the surfaces and the tests for water repellency by measuring contact angles and surfaces tilt angles are then proposed.

2. Proposed design criteria

Numerous articles have addressed the theory behind the Lotus-Effect. The first work on the field are from the mid 30s with Wenzel [2,3] and Cassie [4]. Later the theory expanded to energetic approaches as well as purely mechanical approaches [5]. The phenomenon is very multidisciplinary and it is out of the scope of this article to summarize the complexity of the different views. A wide review of theory and experimental work has been carried out and used in order to create a suitable and systematic design criterion to achieve controlled wettability in [6]. Indeed, to provide the theoretical basis for the development of design criteria, different hypothesis regarding specific requirements that such surfaces were investigated. In different occasions more than one solution was presented for the same problem, each deriving from a different theoretical approach. In

order to clarify which of those hypothesis matches reality better in each case, it was necessary to compare each of them with data from real experiments.

The aim of the whole project in [6] was to create a controlled and systematic geometry. Water repellent manufactured surfaces can be divided in two different types, according to the methods to obtain them: fractal surfaces or patterned surfaces [7,8]. In the present article, only patterned surface are considered. The design criteria refer to the final topographical and physicochemical properties of the surfaces. A more sophisticated geometric criteria is introduced in [6] and lead to geometrical parameters such as: the side angle (Ψ), feature tops angle (Φ), the feature shape and disposition pattern (created in accordance to the effect of contact line in sliding of droplets described in [9]), the top to bottom surface ratio (r) and the feature height (h). The feature tops angle Φ represents the angle between the top shape and the pillar side, its value is 90° if the top is flat. In order to achieve this geometry two manufacturing process chains were investigated. One is composed of laser drilling of stainless steel foils, later on used as a hot embossing tool to obtain polystyrene surfaces. The other one is lithography and etching of a copper laminated epoxy wafer, then chemically hydrophobized.

3. Manufacturing of the surfaces

3.1. Laser and hot embossing

A femtosecond laser was chosen to be used for machining holes. 50µm and 100µm thick stainless steel AISI 316L (Fe/Cr18/Ni10/Mo3) foils were used as starting material. The laser used was a Ti:Sapphire CPA femtosecond laser with λ=775nm, τpulse<150fs, P=0.5W at 6kHz, Epulse~80mJ, M2<1.3 (almost gaussian). The scanner head was a Raylase Turboscan galvanomotor, mounted x & y mirrors. Lenses are telecentric, f=80mm, giving a spot size of ~50μm.

Square arrays of circular holes were machined on four different areas of 10mm x 10mm each, combining two different spacing between the holes and two different foil thicknesses. Distances between centres of two consecutive holes were the same in both directions of each square array. After each array was produced, the holes on both sides of the foil were observed and measured with an optical microscope, finding that their dimensions were consistant on all the four foils. It was also observed that the holes were not circular, but slightly elliptical. The average diameter of the holes on the top surface (entry of the laser beam) was 55μm, and between 20 and 25μm for the exit holes. Even though, due to the gaussian shape of the laser beam, the holes are not perfectly conical, from this data (hole diameter on both sides and foil thickness), an approximate value for Ψ (angle between the bottom of the foil and the side of the hole) can easily be calculated on each foil. The foil cross section is presented schematically in Fig. 1 and Table 1.

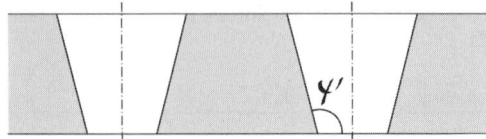

Fig. 1. Cross section of drilled holes in stainless steel foils.

Table 1. Angle Ψ of the machined holes in stainless steel foils.

Foil	Calculated Ψ (°)
SF 50/60 and SF 50/80	106,5 to 109
SF 100/60 and SF 100/80	98,5 to 100

Polystyrene (PS) was chosen for the hot embossing process because, even though it does not have as small surface energy as other polymers like fluorocarbons, it is expected to replicate the mould with high accuracy. The polystyrene was a granulated high-molecular-weight amorphous polystyrene (Novacor 777D-300) with a glass transition temperature of 97ºC. A pre-shape is used before embossing the final micro structure. Process parameters are 160ºC under no pressure for 10min and then under approximately 4MPa for 5min. This pre-shape consisted of four discs attached to a larger and much thinner disc. To stamp the micro structure, the four steel foils with the holes machined with the femtosecond laser were placed on top of a protective aluminium foil. The previously obtained pre-shape was inserted again in a steel plate with four holes and positioned on top of the foils. The same process parameters were used. In order to study the repeatability of the hot embossing process, the whole process was carried out five times with each foil. Thus, five micro structured surfaces were manufactured in polystyrene with each steel foil, the surfaces resulting in discs of 25mm of diameter with a micro structured area of 10mm x 10mm.

3.2. UV lithography and etching

The wafer consists on a thin (2 to 7μm in thickness) layer of negative photoresist coated on top of a 17μm thick layer of copper. The underlying substrate is fibre reinforced epoxy. In order to pattern the photoresist a lithography mask has been designed. Different patterns are drawn on 40 different regions of the wafer. The patterns consist of arrays of lines (stripes) and arrays of circles (square and hexagonal arrays, different diameters of circles). The patterns can be seen on Fig. 2. Important parameters follow:

- for the stripes: stripe width, distance between beginning of one stripe and beginning of the next one.
- for the circles arrays: distance between centres of circles, diameter of the circles, geometric patterns.

The wafer with the patterned photoresist layer was then subject of a wet etching process, thus producing a microstructure according to the designed mask. The wafer was etched in a solution of 250g of ferrichloride $FeCl_3$-$6H_2O$, in 1000ml of water, for 80min. On copper, such acids act isotropically. This allows to obtain features characterized by an acute angle in the edge of the feature top (Φ), which was one of the objectives of the design. The wafer was then cleaned with acetone, water and air.

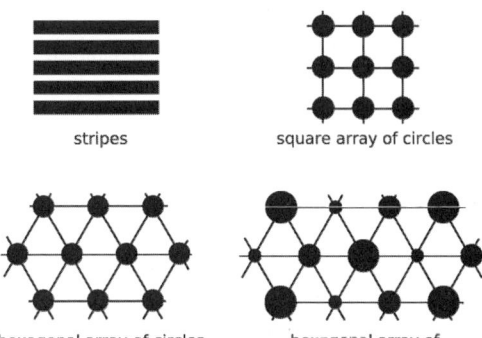

Fig. 2. Patterns on the mask

stripes

square array of circles

hexagonal array of circles

hexagonal array of mixed diameter circles

4. Geometrical characterization of the surfaces

In order to measure the geometry of the surfaces produced, both an optical and a mechanical measurement device were used. The optical device used was a 3D laser profilometer, that scans the surface in parallel lines obtaining height information about the surface Z(X,Y). It works on the basis of an autofocusing principle similar to the one known in CD/DVD players. The mechanical profilometer works on the basis of mechanical contact between a stylus and the surface. Due to the tip geometry and dimension there is a lower limit as to the features that can be measured using this instrument.

4.1. Polystyrene surfaces

Due to the low reflectivity of polystyrene, many problems were found to perform scans with the laser profilometer. Therefore, scans of a resolution of 200points/mm in each direction over areas of 200μm x 200μm were performed as standard on all the surfaces on which it was possible. Three extra scans of the same kind were performed on different points of these four chosen surfaces in order to obtain standard deviation of the instruments. In all the data files obtained, a plane correction was performed to eliminate the inclination of the sample. Three scans were performed on those surfaces when possible.

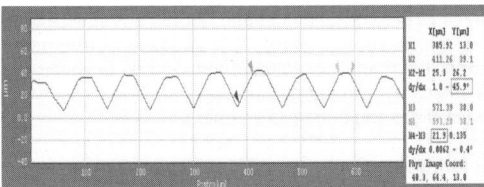

Fig. 3. SPIP [10] processed profile of PS 50/80 (1).

Furthermore, profiles were obtained (vertical sections, as seen in Fig. 3). From them, step height (top-to-bottom distance), top diameter and side wall inclination angle of the pillars can be obtained. This was done with profiles taken in different directions in order to gather minimum and maximum values of these features. This parameter corresponds to Φ as seen on Fig. 1. It can also be related to Ψ. The step height was obtained by calculation on a profile according to the ISO 5436 norm. Furthermore the parameter r, ratio between the total area and the projected area, was systematically calculated.

4.2. Etched copper wafer

Regions of the wafer were measured with the 3D optical profilometer since the copper has an appropriate reflectivity. The area scanned varied from one surface to another, depending on the dimensions of the measured geometry. It was in every case an area wide enough to include a whole number of features. The data was extracted in the same way described for the polystyrene surfaces, except for the fact that only one value for the top diameter and side angle were necessary. This is due to the fact that the etched copper surfaces presented circular (non elliptic) features.

4.3. Results

The two types of surfaces obtained can be seen in Fig. 4. A short sample of the resulting data from the characterization process is gathered in Table 2. Among other parameters: feature height (h), r (ratio between the total and the projected areas), maximum and minimum top diameter (Md and md), maximum and minimum side angles (Ma and ma in deg.) were considered.

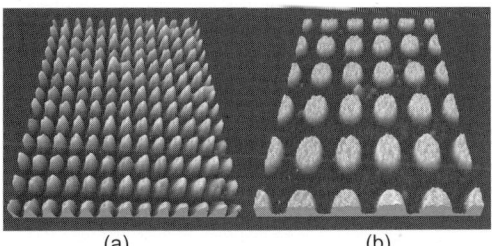

(a)   (b)

Fig. 4. rendered 3D surfaces from 200µm x 200µm, (a) PS 50/80 and (b) copper etched.

Table 2. Examples of measurements for PS 50/60 (2) and PS 50/80 (1), and two different regions on the copper wafer (3) and (4)

#	h µm	r -	Md µm	md µm	Ma °	ma °
1	31.0	1.760	35	25	75	60
2	24.7	1.934	30	30	65	60
3	15.9	2.580	-	40	-	70
4	13.2	1.750	-	23	-	60

The results of the characterisation can be summarized as follow:

- All surfaces show a pattern of micro pillars that can be considered regular, in the way that the repeatability of the pillars in terms of disposition and shape correspond to what was expected in the design criteria.
- The pillars are not considered conical shaped but a maximum and a minimum top and base diameter can be defined, hence leading to Φ.
- Inclination of the sides of the pillars can be defined and has a maximum and a minimum value along their perimeter, it is related to Ψ.

These results allowed a quantitative assessment of the correlation beetween the aimed geometry and the obtained geometry.

5. Contact angle and tilt angle measurements

All known available effective methods to apply a surface treatment to the surfaces made in polystyrene (to decrease their surface free energy) would affect the patterned micro structure. The geometry of the micro pillars would be strongly altered or completely erased due to its small dimensions. Therefore, the surfaces manufactured in polystyrene were not further hydrophobized. On the other hand, the copper wafer was immersed in a solution of 25ml of Tetrabutyltitanate diluted in a total volume of 250ml with 1-methyl-3-butanol, for 30s in order to obtain a coating by a very thin (<1µm) hydrophobic layer.

Contact angles were measured with an optical contact angle measurement system [11]. It has a powerful zoom-lens, an integrated dosing system and an integrated video system (see Fig.5 for examples). The dispensing procedure and the measurement procedure are computer controlled. It also measures the contact angle on both sides of the drop. The measurements were performed according to two controlled deposition procedures:

- Advancing deposition by approaching the needle to the surface and injecting liquid until a certain volume. The procedure is expected to render an angle close to the maximum possible on the rough surface, the drop forming on the surface.
- Gentle deposition of a drop.

Furthermore, after depositing each drop, the surface was tilted in order to measure the necessary tilt angle for inducing drop movement. Before each measurement, a reference measurement was taken using a smooth surface of the same material. The characterization process and the testing process were as systematic as possible. The same methodology was followed for all the polystyrene surfaces and for all the regions of the copper wafer. To describe a superhydrophobic surface the contact angle must be greater than 150°.

284

The smooth polystyrene surface used for measuring the characteristic contact angles of the material was one of the pre-shapes obtained in the embossing process. Six measurements where taken by gently depositing 10μl drops from above, as a controlled procedure. The average contact angle is 130.4° which results in a wetted contact, this was done three times, each in a different location of the surface. After each drop deposition the surface was tilted in order to measure the necessary tilt angle to induce drop movement. In all cases the drop did not move even if the surface was completely vertical.

Regarding the etched copper wafer after being hydrophobized, measurements of the characteristic angles were taken in a un-structured surface of the same material and the same intrinsic roughness as in the top of the features. This was done in the corners of the wafer, where the not-patterned photoresist protected the copper from etching. The measurements on the rough hydrophobized copper surfaces were performed according to the two controlled deposition procedures. Examples of average measured contact angles (average of 6 measurements) are from 76.9°±0.5° to 143.8°±1.0° for the stripped area of the copper wafer, and from 91.3°±1.0° to 117.9°±4.2° for the square array of circle. For all the drops deposited on the wafer, no movement was observed even if the surface was completely vertical, even in the case of high contact angles (> 90°).

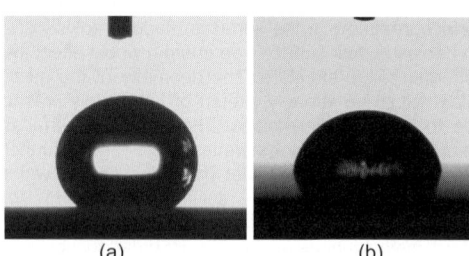

(a) (b)

Fig. 5. Example of images for contact angle measurement. (a) drop has a volume of 10μl and lays on the surface PS 100/60 (2); (b) drop on a patterned copper wafer.

6. Conclusion

A design criteria to obtain patterned manufactured surfaces geometry showing water repellency was tested. Two manufacturing process chains were used. In one case polystyrene was formed by hot embossing using a laser drilled foil of stainless steel. In the other case a copper wafer was selectively etched and hydrophobized following a patterned mask. The surfaces were characterized geometrically and showed a good correspondence with the nominal structures.

The surfaces were tested by depositing water onto them. Two types of deposition techniques were used: advancing angle and gentle deposition. Contact angles were measured on each surface and showed that no superhydrophobicity was achieved, although a clear influence of the deposition method was proven.

The reason for not obtaining water repellency is believed to be the lack of proper chemical properties of the surfaces as the geometry showed expected characteristics. Indeed the hydrophobization of the copper wafer turned out to be insufficient.

Acknowledgements

The authors thank Ph.D. Martin F. Jensen, Danish Technological Institute, for his assistance with the femto second laser. The authors furthermore thank the Dept. of Micro and Nano Technology, Technical University of Denmark, for assistance with the contact angle measurements.

References

[1] Von Baeyer HC. The Lotus Effect. Sciences (2000), 40(1), pp 12-15
[2] Wenzel RN. Resistance of solid surfaces to wetting by water. Industrial and Engineering Chemistry, 28, 8, August 1936, pp 988-994.
[3] Wenzel RN. Surface roughness and contact angle. Journal of Physical Colloid Chemistry, 53, 9, December 1949, pp 1466-1467
[4] Cassie ABD, Baxter S. Wettability of porous surfaces. S. Transac. Faraday Soc. 40, 1944, pp 546-551
[5] Extrand CW. Criteria for ultrahydrophobic surfaces. Langmuir (2004), 20, pp 5013-5018
[6] Botija PA. Establishement of microstructured surface for achieving Lotus Effect. Msc Thesis, DTU 2005
[7] Schrauth AJ, Saka N, Suh NP. Development of nano-structured hemocompatible surfaces. Proceedings of the 2nd International Symposium on Nanomanufacturing, Korea, November 2004, pp 460-465
[8] Lee J, He B, Patankar NA. A roughness-based wettability switching membrane device for hydrophobic surfaces. J. Micromech. Microeng. 15, 2005, pp 591-600
[9] Yoshimitsu Z, Nakajima A, Watanabe T, Hashimoto K. Effects of Surface Structure on the Hydrophobicity and Sliding Behavior of Water Droplets, Langmuir, 18 (15), 2002, pp5818-5822
[10] The Scanning Probe Image Processor, Image Metrology A/S - http://www.imagemet.com/
[11] DSA10 (Krüss GmbH, Germany), Software Release DSAI V1.80 - http://kruss.fr/instruments/

285

Challenging the sustainability of micro products development

A. De Grave, S.I. Olsen

Department of Manufacturing Engineering and Management, Technical University of Denmark
Produktionstorvet, Building 427 S, DK 2800, Lyngby, Denmark

Abstract

Environmental aspects are one of the biggest concerns regarding the future of manufacturing and product development sustainability. Furthermore, micro products and micro technologies are often seen as the next big thing in terms of possible mass market trend and boom. Many questions are raised regarding the impact of size for recycling or environment. Indeed micro production is often seen as environmental friendly thanks to the small amount of material used. Such a statement can be misleading. In this article EcoDesign or Design for Environment (DFE) and Life Cycle Assessment (LCA) will be presented, together with some tools used in order to implement the concepts into design activities. Micro injection moulded components and MEMS are used as examples. Specificities of micro products will be investigated through a categorization in three levels: the end products, the whole production chain and the intermediate parts which can be in-process created. Possible future trends for micro products development scheme involving environmental concerns are given.

Keywords: Sustainability, Design for Environment, Life Cycle Assessment

1. Introduction

The common stereotypes in the field of environmental impact of micro product is that the small size induce less material, less production energy and less material to waste, hence being more environmentally friendly. Is this stereotype valid, and are there any benefits in using a sustainability method for micro products development? It is the aim of this paper to challenge this perception and to emphasize the need to investigate these issues early in the product development using the tools discussed.

It is not the purpose of this article to focus on societal (for example, considering to use out of date computers where the application does not require calculation power) and economical issues although they are undoubtedly a big part of the equation in sustainability. Such issues have been discussed in [1]. Micro injection moulded components (such as parts used in hearing-aid devices) and MEMS devices are used as recurrent examples throughout the article.

When looking at any technology three different levels are interesting to consider in the complete product development scheme:

- The final product, defined as the product close to the end user. It is fairly easy to definite it in the case of hearing-aid equipment, but less evident in the case of a MEMS RF-switch used in a mobile phone antenna.
- Intermediate parts that are not included in the final product: they represent part of the waste from manufacturing. Examples of such parts can be injection moulded feeding system or parts specially used for ease of handling and assembly or the part of a silicon wafer which is grinded before packaging, the thickness being required for mechanical stability during production.
- The production system, it is mainly the manufacturing chain but also the recycling chain.

This categorization is especially pertinent in the case of micro products where the intermediate parts can represent up to 98% of the one of the product components, as for example for the case of micro injection moulded components. It is not easy to give such a figure for macro scale injection moulded components because of the wide variety of applications but it does seldom reach such extremities.

2. EcoDesign for micro products

EcoDesign and Design For Environment (DFE) are synonymous. They take into account recycling, remanufacturing, reuse and LCA during the design phases of the product. It is widely acknowledge that the more you know your product the less you can change it, this classical remark is also valid in the case of DFE and has been pointed out in [2]. One of the key ideas in sustainability and EcoDesign is to reduce materials use. It can be done either through process optimisation, re-use of the product or part of it, or when this it is not possible, through considering recycling material. Re-used parts can be in the same field (broken parts being replaced by re-manufactured parts, upgrade of systems...) or in some other unrelated fields (a common example is the protection of boats in harbour by use of old car/truck tires). DFE of course involves other objectives such as for example that the manufacturing chain should aim at causing the least possible environmental impact, an issue which is more specifically addressed by the concept of Environment Benign Manufacturing (EBM) [3].

From [4], DFE relies on eight axioms:
1. manufacture without producing hazardous waste
2. use clean technologies (not in the "clean room" meaning)
3. reduce product chemical emissions
4. reduce product energy consumption
5. use non-hazardous recyclable materials
6. use recyclable material and reuse components

7. design for ease of disassembly
8. product reuse or recycle at end of life

Although originating from DFE of electronic devices, it is quite clear that these axioms are of a level of abstraction high enough not to be technology dependent. Certainly they can be applied in the case of such micro systems as MEMS, manufactured using VLSI technologies. Although MEMS are just a portion of the diversity of micro products they attract a lot of media attention and are often seen as examples.

Many tools within the DFE family have been developed through the years, from either academia or industry. Examples are Design for Disassembly (DFD) (and therefore Design for Assembly (DFA)), the Eco-function matrix or Design for Recycling. A thorough overview of them is given in [1]. These tools are not specific to particular technologies although electronic and mechanical industries have been known for developing different set of tools. In the case of micro products, often being at the crossing between miniaturisation of mechanical manufacturing and unstandardised use of microelectronic manufacturing processes, both fields should be investigated. Indeed for example, studies between automotive industry and computer industry show that the complete life cycle of the computer has not been documented to the same extend as it is the case for cars [5].

2.1. Micro scale specificities

When looking at the specificities of micro products, several issues arise.

It has already been pointed out in [6] that one of the difficulties of micro products development is a lack of knowledge, both on manufacturing processes, solution principles and development methods. This lack of knowledge applies even more in the field of reliability and end of life behaviour.

In the case of product re-use, disassembly and/or reassembly, handling and assembly is a big issue that still to a high extent remains to be solved at micro scale. Examples of research in these areas can be found in [7]. The maturity of the knowledge in the field of handling and assembly for micro components is still far from being suitable for industrial applications. In the case of a MEMS device, the concept of disassembly doesn't really apply. Indeed these devices are often monobloc or embedded thus making it difficult to separate the parts and/or materials. MEMS are manufactured layer by layer and thus it is very difficult if not impossible to separate one layer of material from another. In the case of two components injection moulded parts the same concept apply as there is a strong bond between the two polymers, or between the metal insert and the polymer. Indeed it is the goal of such a technology to create this strong bond, hence going against DFD principles.

Remanufacturing poses even more of a problem than reassembly due to the relative repositioning and the skimming/trimming inherent to the process. The amount of material may even not be sufficient to allow a remanufacturing operation. On one hand it is perhaps impossible to apply remanufacturing concepts to micro components but these concepts can possibly be applied on the intermediate parts. For a typical micro injection moulded part, the runner system and other wasted parts can amount to 95 to 98% of the total mass as picture on Fig.1, due to the minimal volume of a shot possible on the available hardware. The effect of recycling of polymer, both with [8] and without fibre reinforcement have been studied. Even if the mechanical characteristics of a part change with the number of injection cycle, it can be possible to design a piece with a mechanical limit that allows a number of recycling cycles. Stress-strain curves can be seen on Fig.2.

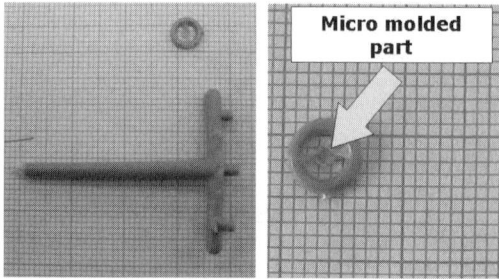

Fig. 1. Relative size of micro injection moulding components

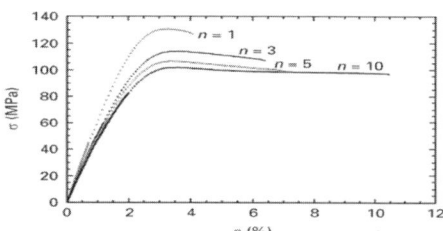

Fig. 2. Stress-strain curves of PEEK10 at different numbers of injection cycles from [8]

After the 10th injection it is shown that there is a loss in mechanical properties (mainly due to fibre degradation) but these characteristics are still in the range of use for some other design. Moreover, by starting with mechanical properties higher than necessary it is possible to stay within expected limits after a number of injections. Using carbon nanotubes reinforced polymer could possibly solve this problem, since nanotubes are less likely to be damaged in such a process.

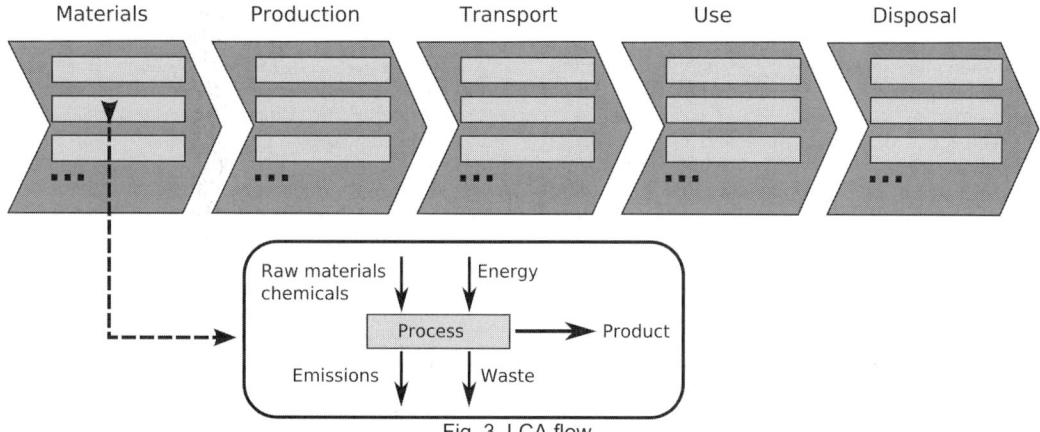

Fig. 3. LCA flow

3. LCA for micro products

A strong tool for the environmental assessment of products in product development is Life Cycle Assessment (LCA) the principles of which also forms the backbone of DFE, EcoDesign, etc. [1]. LCA has been developed to analyse and assess the environmental impacts attributable to a product through the whole life cycle of that product, i.e. extraction of resources, conversion into materials, production, use and disposal as well as transport and infrastructure. LCA therefore covers not only the production but all supportive functions during the product life.

Thus it is necessary to gather information concerning energy and materials use as well as emissions and waste for all processes in the product life cycle, as illustrated in Fig. 3.

The inclusion of supportive functions demanded by a product in the assessment extends to include all impacts that a company (or a product) is responsible for – not only within the manufacturing chain but in all products life stages. In a quest for sustainability these are important aspects to include in product development, since this is the stage at which decisions can be taken concerning the environmental impacts of the product.

In LCA the environmental impacts of different options for e.g. materials choice, disassembly options etc. can be compared or environmental improvement options can be identified or it can be identified whether a choice to reduce environmental impact in one part of the life cycle creates a larger environmental impact in other parts of the life cycle. A more detailed explanation of the LCA-methodologies can be found in [1].

In micromanufacturing LCA has predominantly been used in the MEMS sector. The rapid development of technologies and limited availability of data in the MEMS industry makes full blown LCAs difficult and rather quickly outdated. An example is the manufacture of a PC for which the energy requirement in the late 1980s were app. 2150kWh whereas in the late 90s efficiency were improved and only 535kWh were necessary [9]. Using old data could result in erroneous results. Looking at the overall environmental impact this fourfold increase in efficiency has been overcompensated by an increase in number of sold computers from app. 21mio to more than 150mio [9]. The latter provides an example of a rebound effect showing that economic and social aspects may have a huge implication for the overall environmental impact.

A major trend is that shrinking product dimensions raise production environment requirements to prevent polluting the product. It involves energy intensive heating, ventilation and air conditioning systems. Clean room of class 10.000 for example requires app. 2280kWh/m^2 per year whereas a class 100 requires 8440kWh/m^2 per year. The same increase in requirements is relevant for supply materials like chemicals and gases. The demand for higher purity levels implies more technical effort for chemical purification, e.g. additional energy consumption and possibly more waste. Most purification technologies are highly energy intensive, e.g. all distillation processes which are often used in wet chemical purification, account in total for about 7% of energy consumption of the U.S. chemical industry [10]. Chemicals used in large volumes in semiconductors industry are hydrofluoric acid (HF), hydrogen peroxide (H_2O_2) and ammonium hydroxide (NH_4OH). These materials are used in final cleaning processes and require XLSI grades (0.1ppb). Sulphuric acid is also used in large amounts, but it is a less critical chemical and mainly requires an SLSI level purity [10].

Micromanufacturing of other types of products also puts higher requirements on the quality and purity of the materials. Additionally, a considerable amount of waste is produced. As shown in the example on micro injection moulding up to 98% of the material used may be intermediate parts. If not properly accounted for in the planning and design recycling of this waste may not be possible due to requirements to and reduction of the material strength. Another implication of both the higher materials requirements and miniaturisation in general is that more expensive and rare materials may be employed. Very often the extraction and purification of such materials have higher environmental impacts than those of more abundant materials. For example the energy use for extraction of palladium and accompanying processes is app. 1000 times higher than the energy use for the corresponding processes for steel. And again gold requires app. twice the energy of palladium [11].

As mentioned miniaturisation also cause new problems in electronics recycling due to difficulties of disassembly. Take back will hardly be possible. If electronics are integrated into other product they need to be compatible with the recycling of these products (established recycling paths) [9].

3.1. Micro scale specificities

When introducing a life cycle perspective on the environmental aspects of micro products, some trends emerge.

The focus of material recycling in the case of micro products could be shifting from the end product to the recycling of the intermediate parts, as seen in the example of the micro injection moulded parts. Although, often materials used in micro products are new, hence another lack of knowledge in the recycling chain. In many case purer material might be needed, hence involving a more complicated and energy consuming chain.

The production of raw material is also likely to be more constraining. For example the grain size of crystalline materials is a very important parameter in many manufacturing processes. Often the smallest grain size is sought, as for example in micro forming. Producing such raw material can be more energy consuming.

In the production line, environmental parameters such as temperature, hydrometry and air purity needs to be better controlled. For example, as stated previously, maintaining the clean level of a class x clean room is a very energy consuming process.

4. Conclusions

Using examples from the semiconductor industries and micro injection moulding it has been indicated that micro production not per definition can claim to be more environmentally friendly. To ensure a sustainable development of micro products environmental issues needs to be addressed using tools like EcoDesign, DFE and LCA. A few topics have been identified as the possible objects for a focus on sustainability in micro products development.

In terms of production chain it has been stated than even though less material is used, more energy is likely to be required. It is important to consider a complete LCA of the chain as the ratio between the product and the production systems is changed.

From the products side, the size of the end product often makes it difficult if not impossible to re-use or remanufacture, moreover materials and manufacturing techniques induce difficulties in the separation for recycling. While there is a tendency to begin using materials with a higher environmental burden (in their upstream processes), at the same time these materials cannot be reused.

Considering the intermediate parts, it can be a possible way to work toward increased recycling opportunities for these. Indeed more leverage can be found here as the scale is conventional. DFE for intermediate parts (DFEI ?) could be a path for further investigation, using the broad knowledge corpus in macro scale to serve the micro scale products development.

References

[1] Hauschild M, Jeswiet J, Alting L, From Life Cycle Assessment to Sustainable Production: Status and Perspectives. Annals of CIRP 54/2/2005, pp 535-555

[2] Hauschild M, Wenzel H, Alting L, Life Cycle Design – a route to the sustainable industrial culture?, Annals of CIRP 48/1, pp393-396

[3] Gutowski, T., C. Murphy, D. Allen, D. Bauer, B. Bras, T. Piwonka, P. Sheng, J. Sutherland, D. Thurston, E. Wolff. "Environmentally Benign Manufacturing: Observations from Japan, Europe and the United States," Journal of Cleaner Production, Vol. 13, 1-17, 2005

[4] Hill B, Industry's integration of environmental product design, 1993 IEEE International Symposium on Electronics and the Environment.

[5] "The 1.7 Kilogram Microchip: Energy and Materials Use in the Production of Semiconductor Devices," by Eric Williams, Robert Ayres, and Miriam Heller [Williams 2002]

[6] Hansen HN, De Grave A, Bissacco G. Micro product development methods – how do we focus on the right issues? Proceedings of the 4M conference 2005, pp 439-442

[7] Van Brussel H, Peirs J, Reynaerts D, Delchambre A, Reinhart G, Roth N, Weck M, Zussman E. Assembly of Microsystems. Annals of CIRP 49/2/2000, pp 451-472

[8] Sarusua JR, Pouyet J. Recycling effects on microstructure and mechanical behaviour of PEEK short carbon-fibre composites. Journal of Material Science vol32 (2), pp 533-536

[9] Schischke,K.; Griese,H, 2004: Is small green? Life Cycle Aspects of Technology Trends in Microelectronicss and Microsystems. http://www.lcacenter.org/InLCA2004/papers/Schischke_K_paper.pdf

[10] Plepys, A., 2004: The environmental impacts of electronics. Going beyond the walls of semiconductor fabs. IEEE International Symposium on Electronics and the Environment, 2004.Conference Record.

[11] EDIP data base, 2005: Data base for the LCA-methodology EDIP (Environmental Design of Industrial Products) implemented in to the tool GaBi.

Multi-Material Micro Manufacture
W. Menz, S. Dimov and B. Fillon (Eds.)

Process parameter analysis on surface roughness and process forces in micro cutting

J. Fleischer[a], G. Lanza[a], M. Schlipf[a], J. Kotschenreuther[a], J. Peters[a]

[a] *Institute of Production Science (wbk),Universität Karlsruhe (TH), Germany*

Abstract

High precision engineering has a great technological potential regarding the manufacturing of microtechnical products. Due to its flexibility and the possibility of producing complex three-dimensional geometries in a broad variety of different materials, micro cutting is of special importance in the already mentioned field. However, milling and turning in micro dimensions follow special rules what is caused by size-effects. Successful micro cutting depends on reliable processes and therefore on the knowledge about parameter adjustments and process characterization. By means of micro cutting test series and statistical analyses, effects and interactions of process parameter variations for work piece material, cutting edge radii, cutting speed, and depth of cut were identified and mathematically quantified. The results show a significant influence of the mentioned factors on the response variables. Thus, a linear model for specific cutting force and surface roughness is proposed. Furthermore, the findings are compared to the empirical cutting model of Victor-Kienzle in macro dimensions.

Keywords: Micro Cutting, Process Characterization, Size Effects

1. Motivation

Micro production with annual forecasted growth rates of 20 % [1] has to be indicated as one of the key technologies of the 21st century. High precision engineering uses alternative manufacturing techniques like milling, turning, drilling, etc. and is characterized by its flexibility (short processing time for small and medium sized batches) and the possibility of producing complex three-dimensional geometries in a broad variety of different materials.

The machining of complex features in wear resistant material like metals or ceramics renders high precision engineering and especially micro milling so interesting for micro production. When the depth of cut decreases to the order of micro meters some unexpected phenomena concerning the needed cutting force or the quality of the machined surface appear. These phenomena, which are not foreseen by conventional scaling for cutting technology in macro dimensions, are summed up as so-called size-effects.

Therefore, this research aims at identifying and classifying the effects and interactions of process parameters considering the example of micro turning. The experimental and theoretical approach of the research is presented in chapter 2 of the paper. In chapter 3 the results of different parameter adjustments regarding surface roughness and specific cutting force are described and afterwards discussed in chapter 4. This research concludes with an overview of possible future analyses in the field of micro cutting in chapter 5.

2. Approach

The approach of the research is twofold. At first, test series with different parameter adjustments were executed. Hence, the experimental setup is described in paragraph 2.1. In a second step the measurement results were statistically analyzed and evaluated. The theoretical proceeding is presented in paragraph 2.2.

2.1. Experimental setup

An experimental setup for micro cutting has been developed using a high precision micro milling center Kugler MicroMaster 2 which has a controllable C-axis able to spin at up to 400 rpm and an absolute position accuracy < 1 µm (see Fig. 1).

The chisel is a standard item to accept indexable

Fig. 1. View of experimental setup

inserts CCMW 12 04 08 with a neutral angle of inclination. The tested tool materials were uncoated cemented carbide inserts, which show cutting edge radii of 5 and 50 µm. Geometries of the cutting tools were in both cases rake angle $\gamma = 0°$ as well as clearance angle $\alpha = 7°$.

The cutting speed was varied in a turning process in which the ring shaped work piece was rotating at cuttings speeds of 50 m/min, 100 m/min and 200 m/min. The depth of cut has been varied between 50 µm and 1 µm.

Ring-shaped work pieces were attached magnetically on the C-Table. Work piece materials which have been taken into consideration are Armco-Iron in three different heat treatment states, showing grain sizes of 30 μm, 100 μm and 200 μm, as well as AISI O2 in three treatment states: normalized, quenched and tempered and soft annealed. In addition, AISI 1045 has been investigated as reference material for the modeling of the cutting forces. Process forces have been measured with a measuring platform type Kistler MiniDyn 9256A1. Surface roughness has been determined by a confocal white-light microscope type Nanofocus μ-surf.

2.2. Statistical analysis

The objective of the analysis is to determine whether there is an influence of work piece material, cutting edge radius, cutting speed, and depth of cut on the specific cutting force and the surface roughness. Furthermore, the quantification of an interaction of the given variables in a mathematical model is targeted. See Table 1 for an overview over the model variables.

Table 1
Overview of the model variables

Variable	Symbol	Kind
Specific cutting force	k_c	Response
Surface roughness	R_z	Response
Work piece material		Explanatory
Depth of cut	a_p	Explanatory
Cutting speed	v_c	Explanatory
Cutting edge radius	r_β	Explanatory

All response variables are quantitative, while the explanatory variables (qualitative and qualitative) are mixed. This setup can be handled within a multiple regression analysis, as described by Draper and Smith [2]. Calculations and plots were made with the statistical software "R" [3].

The two response variables had to be treated separately, because different work piece materials were used in the experiments.

In a first step the distributions of the responses were examined in order to eliminate possible outliers. Specific cutting speed and surface finish are both non-negative values, so the assumption of a log-normal distribution seems reasonable. This is of special importance as test values for classical linear models depend on the assumption of normality. After a log transformation of the values a normal distribution fitted well in both cases. A performed Shapiro-Wilk and an Anderson-Darling normality test [4,5,6] did not reject normality. In a further step, the depth of cut was log-transformed as well in order to achieve a linear relationship to the response variables and constant variance of the model's residuals [7].

On this basis a linear least squares model was fitted to the log-transformed responses. The explanatory variables for the model were chosen starting with a single variable and adding step by step further ones. Furthermore, the model was checked for interdependencies by adding respective terms to the mathematical function. After the introduction of a new

term and a verification of the model's residuals for normality, homoscedasticity and autocorrelation, its influence on the response was tested for significance. Non-significant terms were removed again.

3. Results

3.1. Surface roughness

On the surface roughness significant influence of the depth of cut, cutting speed, and material as well as significant interdependency between cutting speed and material could be found in the data. The influence of the cutting edge radius was not significant, so it was not included in the regression model.

The model equation and a plot of the regression for Armco Iron as work piece material are shown in Eq. 1 and Fig. 2, and for AISI O2 in Eq. 2 and Fig. 3, respectively. Because of the higher values for the cutting speed compared to depth of cut in the equations its coefficients appear very small.

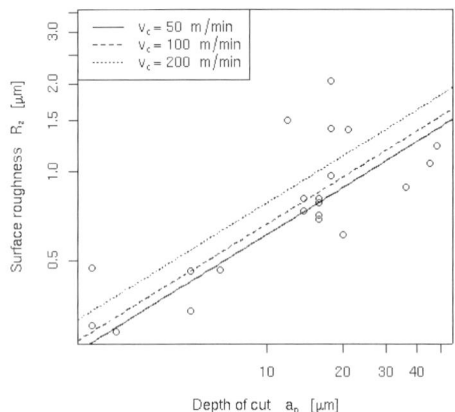

Fig. 2. Surface roughness for Armco Iron on a double logarithmic scale

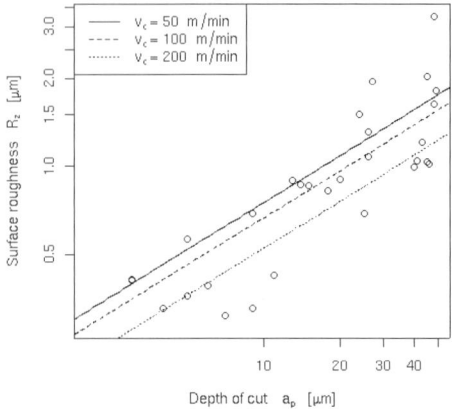

Fig. 3. Surface roughness for AISI O2 on a double logarithmic scale

The multiple correlation coefficient R^2 of the model is 0.74, which means 74% of the data's total variation about the mean is explained by regression.

$$\log(R_{z,ARMCO}) = -0.78 + 0.53 \log(a_p) + 0.0008\ v_c \quad (1)$$

$$\log(R_{z,AISIO2}) = -0.60 + 0.53 \log(a_p) - 0.001\ v_c \quad (2)$$

The data shows an increase of surface roughness with increasing depth of cut. Surface roughness is higher for AISI O2 than for Armco Iron. Due to interdependencies, the effect of the cutting speed depends on the material.

3.2. Specific cutting force

The data showed significant influence of the depth of cut, the cutting edge radius, and the cutting speed on the specific cutting force. Only data with AISI 1045 as work piece material was used. Model equation and plot can be found in Eq. 3 and Fig. 4. The model's multiple correlation coefficient R^2 is 0.89.

$$\log(k_c) = 2.99 - 0.23 \log(a_p) + 0.006\ r_\beta + 0.0007\ v_c \quad (3)$$

Plot and equation show a decrease of the specific cutting force with depth of cut, but an increase with the cutting edge radius and the cutting speed.

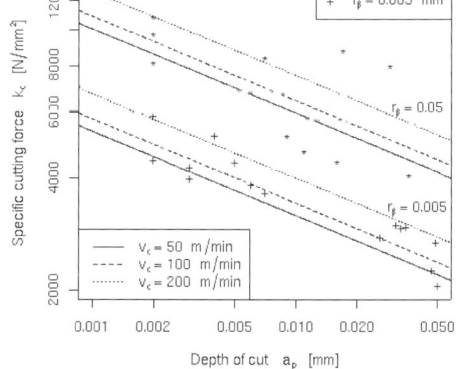

Fig. 4. Specific cutting force for AISI 1045 on a double logarithmic scale

4. Discussion

4.1. Summary

The results concerning surface roughness showed a strong and uniform dependency of depth of cut on both investigated work piece materials. In both cases, the ascent of these plots is indicating improper parameters for this process in an exponential manner. Additionally, at high depths of cut, a relatively high scattering of the R_z-values can be observed. This can be attributed to the fact of build-up edges, which assume the task of the real cutting edge. These build-up edges collapse periodically at about 1 kHz [8]

leaving residues on the machined surface. A further increase of the cutting speed could solve this problem, but can not be investigated due to limitations of the machine equipment.

However, the influence of cutting speed is rather small and non-uniform for the scrutinized materials. Indeed, the difference between various cutting speeds is rather small so that a definitive statement on the influence of the cutting speed in combination with the work piece material can not be made until there is more data available in the further course of the research.

Concerning the specific cutting force, depth of cut has a significant influence on the specific cutting force as already known from macro cutting [9, 10, 11]. However, there are two additional minor influences. One is cutting speed, resulting in thermal loss of cohesion at higher speeds and thus reduced cutting forces. The second is the cutting edge radius which at small depths of cut creates an effective rake angle which is strongly negative resulting in higher cutting forces. The larger cutting edge radius was about as large as the highest depth of cut, thus the chip was eventually exposed to the actual rake angle of the insert or to an effective rake angle in the vicinity of zero, which explains the scattering of the larger cutting edge radius at high depths of cut.

4.2. Comparison Micro to Macro Results

Fig. 5 shows a plot of the Victor-Kienzle model for specific cutting force with the regression lines from Fig. 4. As can be seen, the line for the Victor-Kienzle model is slightly steeper than the ones of the used regression model. However, the value of its gradient (-0.25) lies well within the 95% confidence interval for the gradient (in respect to the depth of cut) of the regression model, which is [-0.28, -0.16].

The regression model is able to explain more of the variation of the given data since it includes two additional terms for the cutting edge radius and the cutting speed. It can easily be seen that these corrections are of major importance since the R^2 of Victor Kienzle being 0.38 is lower than the R^2 value of the regression model of 0.89.

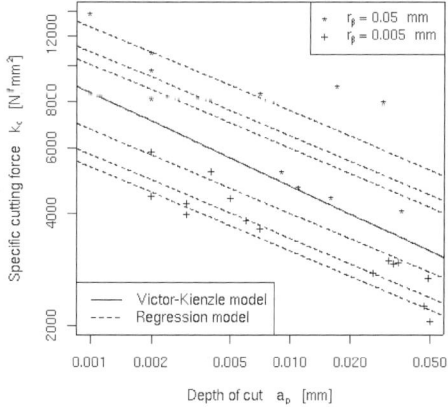

Fig. 5. Comparison with the Victor-Kienzle model for specific cutting force on a double logarithmic scale

4.3. Limitations

The analyzed data was slightly unbalanced in the explanatory variables. This mainly affects the choice of the model variables: the precise p-values of the statistical tests for influence of a model term have to be regarded with caution. This is why model variables were not only chosen by their p-values but also by graphical examination of their influence on the model's residuals.

5. Conclusion and Outlook

In the field of high precision engineering knowledge of process parameter adjustment is of particular importance in order to guarantee reliable manufacturing processes and thus high quality products. Due to size-effects process characterization and modeling from macro dimensions cannot be transferred to micro dimensions. By means of test series and statistical analyses the aim of this research was to identify and classify the effects and interactions of process parameters considering the example of micro turning.

Process parameter variations for work piece material, cutting edge radius, cutting speed, and depth of cut showed a significant influence that is not taken into consideration by the original model of Victor-Kienzle. Linear models for the calculation and prediction of specific cutting force and surface roughness in micro dimensions were presented and compared to conventional cutting scaling.

For the future, further parameters such as grain size, coolant, and initial temperature have to be examined. Regarding process characterization in particular, the effects and interactions on surface roughness and process forces will be analyzed.

It is intended to extend the Victor-Kienzle approximation in order to overcome the actual problem of validity only within defined boundaries of depth of cut. So far, for different depths of cut different main values of the specific cutting force (e.g. $k_{c1.1}$) need to be implemented into the formula. The new approach is to provide a model that is valid throughout the entire range of depths of cut. Within the scope of the research the effects of cutting in the vicinity of the minimum uncut chip thickness need to be analyzed in order to qualify the new formula to deal with effects like ploughing or material displacement [8].

Moreover, the actual results from micro turning have to be assigned and adapted to the micro milling process in order to set up complete specific cutting force and surface roughness models for micro milling processes.

Acknowledgements

The authors wish to thank the German Research Foundation (DFG) for their support within the Collaborate Research Center 499 (SFB 499) and the Priority Program 1138 (SPP 1138).

References

[1] Hesselbach, J.: Academy Journal, 2004
[2] Draper, N.R., Smith, H.: Applied Regression Analysis, Wiley & Sons, 1981
[3] R Development Core Team: R: A language and environment for statistical computing, R Foundation for statistical computing, Vienna, Austria, ISBN 3-900051-07-0, URL http://www.R-project.org, 2005
[4] Royston, P.: An Extension of Shapiro and Wilk's W Test for Normality to Large Samples, Applied Statistics, 31, 1982
[5] Royston, P.: Algorithm AS 181: The W Test for Normality, Applied Statistics, 31, 1982
[6] Thode Jr., H.C.: Testing for Normality, Marcel Dekker, N.Y., 2002
[7] Carroll, J.R., Ruppert, D.: Transformation and Weighting in Regression, Chapman and Hall, 1988 (pp. 121-124)
[8] Albrecht: New Developments in the Theory of the Metal-Cutting Process, Part I, Transactions of the ASME, pp. 348-357
[9] Kienzle O., Victor H.: Die Bestimmung von Kräften und Leistungen an spanenden Werkzeug-maschinen, VDI-Z 94, 1952, pp. 299-305
[10] Richter, A.: Die Zerspankräfte beim Drehen im Bereich des Fließspans, Wiss. Z. der TH Dresden, 1953, p. 651
[11] Taylor, F.W.: On the art of cutting metals, Trans. ASME 28, 1907, pp. 31-350

Multi-Material Micro Manufacture
W. Menz, S. Dimov and B. Fillon (Eds.)

On the use of hot embossing for the reproduction of the surface topography of mould microreliefs

M. Sahli[a,b,c], C. Millot[a], C. Roques-Carmes[a] , C. Khan Malek[b] and J.C.Gelin[c]

[a] *Laboratoire de Microanalyse des Surfaces (LMS), ENSMM, 25030 Besançon Cedex, France*
[b] *Laboratoire FEMTO-ST, CNRS UMR 6174, Département LPMO, 25044 Besançon Cedex, France*
[c] *Laboratoire FEMTO-ST, CNRS UMR 6174, Département LMARC, 25030 Besançon Cedex, France*

Abstract

This article describes the quality of reproduction of microcavities structured into mould substrates which were filled by the imposed flow of amorphous polymeric materials. The rheology of the selected materials (cyclic olefin copolymer, COC) depends on the experimental parameters (temperature, pressure) used for the hot embossing process. To support the experimental data, the polymers were qualified by their melt flow index, flow index and consistency. The efficiency of the filling procedure into microcavities with smaller and smaller size is described using the potentialities of a customized Scanning Mechanical Microscope (SMM).

Keywords: cyclic olefin copolymer (COC), rheology, viscosity, squeeze flow, hot embossing.

1. Introduction

The hot embossing process [1] is a moulding technology which is more and more frequently used in the fabrication of polymeric micro-components and systems, in particular for Micro-Opto-Electro-Mechanical Systems (MOEMS) and microfluidic systems [2].

One advantage which is exploited in these applications is the quality of the forming technique which allows an accurate reproduction of the surface states at the micro-and nano-scales. The technique consists in imposing a compression to the substrate, in general an amorphous polymer, above the glass transition temperature (T_g) in order to decrease the viscosity of the polymer. The rheological parameters which control the behaviour of the flow of the polymer into the microcavities of the mould not only determine the filling ability but also the conformity to the cavity facets. The conformity relates to the final shape of the replicated part and also to the quality of the surface state.

In this work, the quality of the filling process as a function of the location of the microvavities on the mould is investigated by a three-dimensional (3D) characterization method at the microscale. A customized tool allows displaying and characterizing the 3D roughness, waviness and surface shape [3] [4].

2. Materials and methods

2.1 Materials

Cyclic olefin copolymers (COC – Topas™ from Ticona) of amorphous structure, density equal to 1.02. Two grades, 6013 and 5013, were selected. Table 1 summarizes the main data on the polymers, relevant to the replication process by hot embossing. The values of the melt flow index gathered in this table were determined at 260°C under a static load of 2.16 kg in accordance with the norms [5].

Table 1
Data corresponding to the COC materials under study.

Type	COC6013	COC5013
Melt flow index [g/10 min]	14	48
Glass transition temperature T_g [°C]	130	130
Young's modulus [MPa]	2900	3200
Mean coefficient of linear expansion [°C^{-1}]	$0.6.10^{-5}$	$0.6.10^{-5}$

The rheological characteristics of these materials for temperatures equal or higher than 220°C were reported in the reference [5]. These amorphous polymers as characterized by their differentiated fluidity index present equivalent flow indexes but a specific consistency [5]. This consistency is linked to the viscosity of the polymeric materials. The radial flow imposed by squeezing the polymer displays a pseudo-isotropic character.

Experiments to determine the data corresponding to lower temperatures using a viscosimeter dedicated to this metrology are presently under way [6].

2.2 Hot embossing

To reproduce the hot embossing process, we used an instrumented tensile-compressive testing machine (Instron 6025, see Fig. 1). This machine is equipped with two parallel and coaxial circular plates of 120 mm diameter. On the upper plate, micro-imprints were realized using a Vickers hardness indentor. A cross-shaped pattern consisting of a series of Vickers hardness pyramid indentations of increasing size separated by a distance of 5/10 mm was reproduced into the mould along two perpendicular directions (See Fig. 2). These indentations were realized using loads of 1, 2, 3, 5, 10, 15.75 and 20 kg respectively. The pyramids have a square base with a dihedral angle between two opposite faces whose value is equal to 136°. This angle corresponds to a ratio between the base area and the lateral area equal to 0.9272. The cross shaped pattern was replicated all over the mould with a spacing equal to 15 mm, with the "origin" pattern at the centre of the mould.

The mould and the 20 mm x 1mm x 1 mm wide polymeric plate which served as imprinting medium were positioned between the two plates of the machine. The set-up was temperature controlled and both the imposed force (F) and the temperature (T) during the experiment could be recorded.

Fig. 1. Instrumented set-up used to perform the hot embossing process.
1 – Upper plate
2 – Thermocouple
3 – Lower plate

The machine was programmed to deliver a constant compression force, the value of which was controlled by a force sensor.

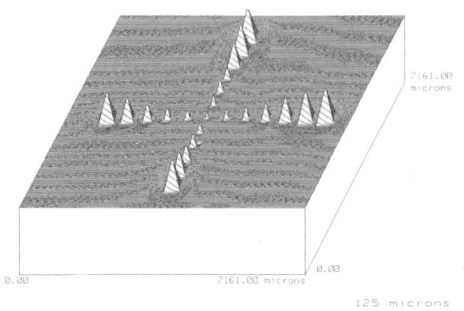

Fig. 2. 3D-map of the mould surface showing the pattern made of a series of Vickers imprints of decreasing size positioned along two perpendicular directions (the data were numerically inverted).

2.3 Scanning Mechanical Microscope (SMM)

The SMM was selected to compare the metrological measurements carried out on the mould and its replica.

The acquisition system of the SMM is made up of two components:
- A measurement unit based on a diamond tip (lateral resolution 1 μm) connected to an inductive sensor. The signal output of the sensor is amplified and converted (A/D).
- A displacement unit, based on two stepping motors, disposed perpendicularly and controlled by a dedicated logic board.

Acquisition data are performed after each incremental displacement, once the stepping motors have come to a full stop.

The vertical resolution of each point z (xi, yi) of the surface is about 1/100 μm.

3D maps of N^2p^2 points are represented with a perspective angle. In addition, hidden line removal is performed to better render the surface geometry. The acquired data can also be represented as a contour map.

A customized software allows several treatments to be realized on the data: pre-processing treatments (least mean square fitting, high or loss-pass filtering, polynomial filtering…), statistical treatments, special treatments (developed length or area, developed volume), spectral treatments, fractal analysis. Moreover extended capability for interfacing with an image analyser is possible.

As an example, Table 2 gathers geometrical, surface and volume descriptive data of a series of micro-hardness imprints on the mould, where z_{max} represents the maximum range, S the projected area and V the volume of the imprint.

Table 2
Metrological data associated with a series of Vickers hardness imprints on the mould.

Ref	d (μm)	z_{max} (μm)	S (10^3 μm^2)	V (10^6μm^3)
1	119	12.8	5.5	0.025
2	170	22	15.4	0.114
3	228	27.7	23.3	0.22
4	283	37.1	39.8	0.5
5	431	56.2	90.2	1.65
6	535	70	134.7	3.1
7	611	76.2	157.8	4

3. Experimental data

Only the data corresponding to the experimental parameters at T = 150°C (T_g + 20°C) and F = - 80 N are presented in this paper. In practice the mould was heated to reach the temperature of 150°C associated with an initial compression of -10 N. As soon as the temperature was reached, the value of the compression pressure was increased to the nominal value of the experiment. The holding time in these conditions was of 5 minutes, before cooling down (see Fig.3).

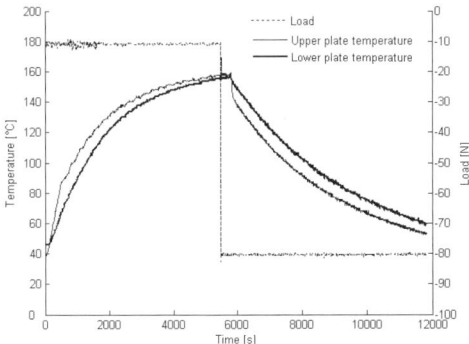

Fig.3. Evolution of the experimental parameters
chosen for the hot embossing process.

The quality of the replication process is first
determined by several comparative observations as
is illustrated in figure 4. Fig. 4 represents a replica of
a branch of the series of Vickers hardness imprints
(see Fig. 2). It was observed that the material with
the highest fluidity index (COC 5013) almost
identically reproduces the geometrical shapes of the
mould when filling it (see Table 3).

The values recorded in Table 3 confirm these
findings and point out that the filling ratio is close to
100% for the three smallest imprints realized
whichever polymer was selected. We need to point
out that as observed for the same pattern on the
same sample, the filling can present an anisotropy
which is a function of the orientation of the branches
to be filled with respect to the radial deformation as
has been confirmed on data collected in Table 4.

Table 3
Metrology of the volume (10^6 µm^3) of the imprints
realized on the mould and comparative data on the
replica using two different grades of COC polymers.

Imprint	Mould	COC 5013	COC 6013
1	0.025	0.025	0.025
2	0.114	0.114	0.114
3	0.22	0.22	0.22
4	0.5	0.5	0.42
5	1.65	1.65	1.20
6	3.1	3.1	2.17
7	4	4	3.43

Table 4
Metrology of the volume (10^6 µm^3) of imprints
realized along two orientations on the replica made
of the 5013 COC polymer.

Imprint	Mould	COC 5013 (// radius)	COC 5013 (⊥ radius)
1	0.025	0.025	0.025
2	0.114	0.114	0.111
3	0.22	0.22	0.20
4	0.5	0.5	0.37
5	1.65	1.65	1.32
6	3.1	3.1	1.98
7	4	4	2.76

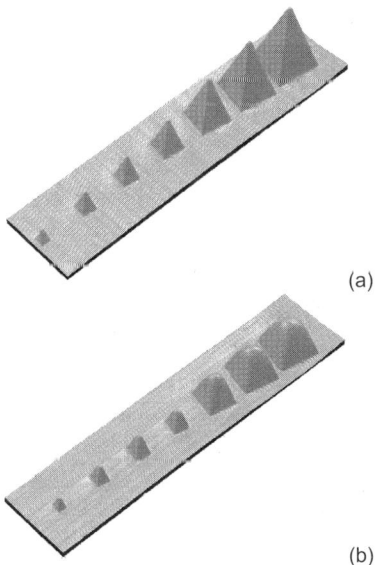

(a)

(b)

Fig. 4. Comparison of the series of Vickers hardness
imprints replicated into two different polymers, COC
5013 (a) and COC 6013 (b).

To improve these data, the specific analysis of
each imprint compared to its mould was performed
and is recorded in figure 5.

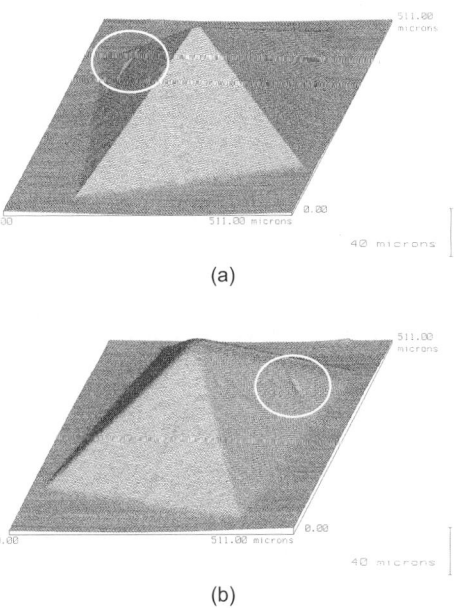

(a)

(b)

Fig. 5. Comparison of a single imprint of the mould
obtained after numerical inversion of data (a) and of
the replica made in COC 5013 polymer (b). The
circle indicates a defect in the mould and the
corresponding defect in the replica.

We can notice that the reproduction of surface defects such as a scratch is conformal even if the filling of the cavity was not optimized. Moreover, metric, surface or volume parameters resulting from the selected metrology can be comparatively presented on similar bases.

4. Conclusion

The quality of the filling of micro-structured patterns of a mould by amorphous polymers was investigated using an apparatus reproducing the hot-embossing process.

A good quality of filling was observed for the patterns with the smallest size while on the contrary a degradation of this quality occurred as soon as the size of the patterns increased. This effect seems to be linked with the competitive kinematics of the radial and elongational flow. These results were confirmed by other experimentations which are not presented here but were conducted both on the same materials at higher temperatures (200, 225 et 250°C) and on acrylic polymers qualified by their fluidity index. Finally we need to consider that this hot-embossing replication process operating at relatively low temperatures is perfectly adapted to the reproduction of superficial surface states.

Surface or volume data of the imprints were recorded. They facilitated the description of both the filling quality and the reproducibility of the differences in height. The lateral resolution was within the micrometer range while the vertical resolution was within a few hundredth of micrometers.

Acknowledgements

The Regional Council of the Franche-Comté region is greatly acknowledged for providing the PhD grant for M. Sahli. We would also like to thank Pascal Robinet for his help in the compression tests and the Ticona Company for gracefully providing us with the COC pellets.

References

[1] Heckele M. and Schomburg W. K. Review on micro-molding of thermoplastic polymers, J. Micromech. Microeng, 14 (2004) 1-14.
[2] Becker H. and Heim U. Silicon as tool material for polymer hot embossing. Twelfth IEEE International Conference on MEMS. '99 (1999) 228 – 231.
[3] Wehbi D., Quiniuo J.F., Roques-Carmes C., Rôle de la rugosité des surfaces en science des matériaux. Traitements thermiques (1984) 222 – 88.
[4] Wehbi D., Clerc M.A., Roques-Carmes C., Three dimensional quantification of wear tracks. Wear 107 (1986) pp. 263 – 378.
[5] Sahli M., Roques-Carmes C., Duffait R., Khan-Malek C., Study of the rheological properties of poly(methylmethacrylate) (PMMA) and cyclo-olefin-copolymer (COC) to optimize the hot-embossing process. Proc. Int. Conf. on Multi-Material Micro Manufacture (4M), W. Menz et S. Dimov (Eds), 29th june-1st July 2005, Karlsruhe,Germany (2005) 83-86.
[6] Sahli M. PhD thesis in progress.

Multi-Material Micro Manufacture
W. Menz, S. Dimov and B. Fillon (Eds.)

Mechanical properties and bending behaviour of metal foils

A. Diehl, U. Engel, M. Geiger

Chair of Manufacturing Technology, University of Erlangen-Nuremberg
Egerlandstr. 11, D-91058 Erlangen / Germany

Abstract

Ongoing miniaturisation in various technical fields like electronics industry or micro systems technology is requesting precise forming processes for the production of small and thin components. Regarding these fields, thin metal foils are being used in micro-electro-mechanical systems, electronic components (e.g. leadframes) and in medical devices. As well as micro bulk metal forming processes, metal foils with thicknesses in the range of microns are subjected to the so called size effects. Previous investigations on various metal forming processes have shown the share of surface grains to be a decisive factor for the forming behaviour as it can be observed at micro foil forming. Besides this general valid size effect, every forming process inhibits specific size effects. In the case of bending processes large strain gradients are present, influencing the bending parameters and process accuracy (e.g. spring back).

In the present paper, the mechanical properties and the bending behaviour of metal foils with thicknesses ranging from 25 µm to 500 µm are discussed in dependence of material properties, microstructure and foil thickness. Fundamental experiments are being performed, providing an experimental basis for future development of theoretical models describing strain gradient dependent forming behaviour of foil bending processes.

Keywords: metal foils, mechanical properties, spring back

1. Introduction

Forming technology is predestined for an economic production of large required with high accuracy. Therefore more and more effort has been put into the investigation of the forming behaviour of thin metal foils in the last decade due to ongoing miniaturisation especially in electronic industry. There, metal foils are being used mainly for micro-electro-mechanical systems and electronic components (e.g. lead frames)

It was found by several authors [1,2] that scaling down process and part dimensions to the range of micrometers is affected by size effects in various fields, e.g. the mechanical properties, friction behaviour and shape evolution. These effects have only minor influence on conventional macro processes but must be taken into account for the design and production of micro parts.

A well known size effect is caused by the increasing share of surface grains on the overall volume when specimen dimension is scaled down. These grains exhibit lower yield strength due to fewer constraints in contrast to grains positioned within the material, leading to a decreasing material strength. This effect is generally valid for all microforming processes and can be sufficiently described by the surface layer model [3].

Besides this effect there are other size effects, having significant influence only on specific processes. In the case of micro sheet bending processes these strain gradients are present. It has been shown, that large strain gradients lead to an increase in material strength [4], which can be explained by the formation of geometrically necessary dislocations based on Ashby's dislocation theory [5]. An approach for the analytic description of this effect is the strain gradient plasticity theory [4], which describes the forming behaviour of materials in dependence of the present strain gradient.

An important issue of the current research is the prediction of the forming behaviour of micro products by means of finite element simulation, enabling accurate process design with only few iterative loops. To reach this goal it is necessary to provide a broad range of experimental results and a reliable determination of size dependent mechanical properties which finally can be used to improve FE simulation of microforming processes.

In the present paper the mechanical properties and the spring-back behaviour of copper and aluminium foils with thicknesses ranging from 25 µm to 500 µm are discussed in relationship to microstructure and process parameters.

2. Experimental Procedure

2.1. Materials Selection

Two materials were used for the experiments, SE-Cu 58 and Al 99.5. Both are single phase materials, enabling fundamental research without additional effects resulting from interactions between different phases having different mechanical properties. However, copper and aluminium exhibit a significant difference regarding the stacking fault energy (Cu. 60 mJ/m^2; Al: 200 mJ/m^2), leading to a higher primary density of imperfections (e.g. dislocations and twins) in the copper crystal structure, thus affecting the mechanical properties.

2.2. Microstructure

In micro sheet metal forming the ratio of foil thickness t_0 to grain size L_G is a decisive factor for the occurrence of size effects. The mean grain size of the investigated foils was varied applying different heat treatments. For both materials a fine grained and a coarse grained microstructure was adjusted leading to different t_0/L_G-ratios (Table 1). For the calculation of the

t_0/L_G-ratios, the real foil thickness t_0 was measured. Since the grain size in direction of the foil thickness is restricted, the grain size values were determined along the foil plane.

Table 1: Microstructural parameters of the invest-tigated foils with nominal thickness $t_{0,\,nom}$

$t_{0,nom}$	t_0/L_G (fine grained)		t_0/L_G (coarse grained)	
	Al 99.5	SE-Cu 58	Al 99.5	SE-Cu 58
25 μm	2.0	1.1	0.8	0.5
50 μm	2.5	1.5	1.5	0.6
100 μm	5.6	2.7	2.0	1.0
200 μm	9.9	7.1	5.4	2.3
500 μm	12.3	9.1	8.9	6.9

The t_0/L_G-ratio represents the average number of grains existing in the thickness direction. With a decreasing t_0/L_G-ratio the influence of individual mechanical properties of the grains in the plasticised area on the forming behaviour is increasing. In Fig. 1 the forming area of a bent coarse grained copper foil ($t_0 = 100$ μm) is displayed, featuring only one grain.

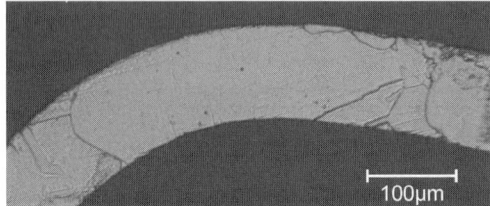

Fig. 1: Bending area of a coarse grained copper foil ($t_0 = 100$μm) with only one grain through the thickness.

2.3. Mechanical Properties

The mechanical characterisation of the investigated foils was done by tensile tests. Since the 500 μm Al 99.5 foils were delivered by a different supplier, thus revealing different mechanical behaviour compared to the other Al 99.5 foils, they are not discussed in the following sections.
While there was no significant influence of foil thickness on flow stress of both the fine grained copper and aluminium material, the coarse grained materials showed a decreasing flow stress with decreasing foil thickness. In Fig. 2 the flow curves of the copper foils are displayed.

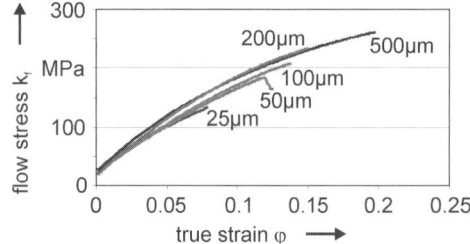

Fig. 2: Flow curves of SE-Cu 58 foils with coarse grain structure for different foil thicknesses

This behaviour is due to the increasing share of surface grains on the overall volume as explained in section 1. From Fig. 2 it can be seen, that the strain to

fracture is decreasing as well with decreasing foil thickness. This was observed for both materials and microstructures. A possible explanation for these findings is the increasing inhomogeneity of the microstructure with decreasing t_0/L_G-ratio yielding lower elongation to fracture. Hence for thin foils, tensile tests provide reliable data only for small plastic deformation. In the case of coarse grained aluminium foils the smallest elongation to fracture (φ_{fr} app. 0.01) was observed.

The significant influence of the t_0/L_G-ratio on the plastic deformation behaviour is shown in Fig. 3, where the flow stress k_f of the copper foils is plotted for different the t_0/L_G-ratios at a true strain φ of 0.08.

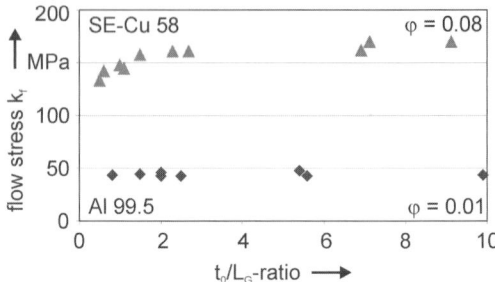

Fig. 3: Dependence of flow stress of SE-Cu 58 foils on the t_0/L_G-ratio (true strain $\varphi = 0.08$) and of Al 99.5 foils (true strain $\varphi = 0.01$)

A critical t_0/L_G-ratio ≈ 2 can be identified from Fig. 3. For smaller t_0/L_G-ratios, the flow stress is decreasing significantly.
Due to the low elongation to fracture of the coarse grained aluminium foils the flow stress could only be compared at a true strain φ of 0.01 (Fig. 3). In this range no significant influence of the t_0/L_G-ratio was observed.
Furthermore, a significant difference between fine and coarse grained foils can be observed for both materials regarding work hardening behaviour. As can be seen by example for SE-Cu 58 foils in Fig. 4 the strain hardening exponent n of coarse grained material exceeds the n-values of fine grained material for all foil

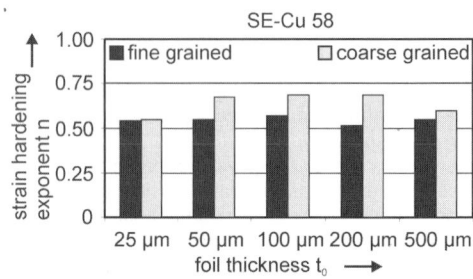

Fig. 4: SE-Cu 58 strain hardening exponent n in dependence of the foil thickness of SE-Cu 58 foils

thicknesses. It must be noted, that the absolute flow stress is larger for coarse grained material as shown before.
An explanation for the observed difference in the strain hardening behaviour could be found in the larger initial area of grain boundaries of fine grained material compared to coarse grained material. During uniaxial plastic deformation of a material the grains rearrange

according to the direction of the applied load, so that the slip systems of the grains are orientated in the direction of maximum shear stress. A higher potential energy within the material due to a larger area of grain boundaries is reducing the work needed for rearrangement of grains and thus leading to a lower increase in strain hardening.

2.4. Bending Experiments

An established method for the investigation of size effects occurring in micro forming processes are scaled experiments according to the theory of similarity. Applying this concept, a tool was developed (shown schematically in Fig. 5) allowing scaled free-bending experiments where all parameters (foil thickness t_0, bending radii r_d, punch radii r_p, punch velocity v_p, bending width W) are adapted by a scaling factor λ.

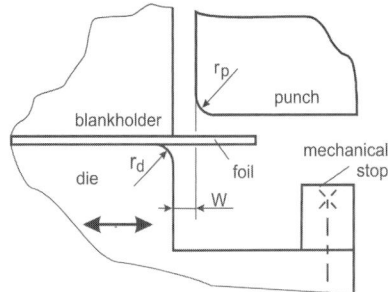

Fig. 5: Schematic set-up of the free bending tool

The foil thicknesses investigated in the bending experiments are in the range between 25 µm ($\lambda = 0.05$) and 500 µm ($\lambda = 1$). The parameters used for the scaled experiments are shown in Table 2.

Table 2: Parameters used for the scaled experiments

mater.	λ	r_p; r_d	W	v_p
Cu	1	1250 µm	600µm	10 mm/min
Cu,Al	0.4	500 µm	240 µm	4 mm/min
Cu,Al	0.2	250 µm	120 µm	2 mm/min
Cu,Al	0.1	125 µm	60 µm	1 mm/min
Cu,Al	0.05	62.5 µm	30 µm	0.5 mm/min

The maximum strain ε_{max} (eq. 1) on the upper surface of the bending area depends on the foil thickness t_0 and the bending radius of the neutral layer r_m.

$$\varepsilon_{max} = (\lambda \cdot t_0)/(2 \cdot \lambda \cdot r_m) \qquad (1)$$

Due to the small bending width (W = 1.2·t_0) the bending radius r_s of the foils surface in contact with the die is assumed to be equal to the die radius. Thus the bending radius of the neutral layer can be calculated by eq. 2.

$$r_m = r_s + t_0/2 \qquad (2)$$

Since both, the foil thickness and the bending radius are scaled by λ, ε_{max} remains constant ($\varepsilon_{max} \approx 0.167$) for all foil thicknesses regardless the scaling factor of eq. 1.

For the investigation of the influence of maximum strain on the bending behaviour, additional experiments with copper foils ($t_0 = 25$ µm; 100 µm; 500 µm) were performed using different die radii and thus different maximum strain values (Table 3).

Table 3: Parameters for bending experiments with varying ε_{max}

t_0	$r_{p,1}$; $r_{d,1}$	$\varepsilon_{max,1}$	$r_{p,2}$; $r_{d,2}$	$\varepsilon_{max,2}$
500µm	500 µm	0.333	2500 µm	0.091
100µm	125 µm	0.286	500 µm	0.091
25µm	125 µm	0.091	250 µm	0.048

The bending process was recorded via CCD-camera placed in front of the bending tool. From the recorded pictures (see Fig. 6) both bending and spring-back angle were determined automatically by means of image processing.

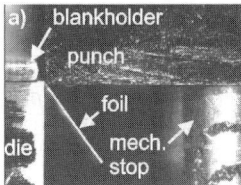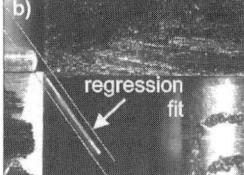

Fig. 6: Recorded (a) and processed (b) CCD image with regression fit of the foil slope

The parameter of interest is the spring-back angle α_s as a function of the foil thickness. According to common bending theory the spring-back angle in scaled experiments remains constant in absence of size effects and differences in material properties. Regarding mechanical properties, for coarse grained material decreasing flow stress with decreasing foil thickness was observed. Hence a decreasing spring-back with decreasing foil thickness is predicted for the coarse grained foils by macro bending theory.

The results of the scaled bending experiments for both materials are shown in Fig. 7, where the spring back angle α_s is plotted against the scaling factor λ in case of a bending angle $\alpha_b = 45°$.

While a constant increase of the spring back angle with decreasing λ can be observed for the aluminium foils, the spring back angle of the copper foils is decreasing with decreasing λ ranging from $0.2 \leq \lambda \leq 1$. If $\lambda < 0.2$ the spring back angle is increasing again. For both materials the spring back angle of fine grained material exceeds the spring back angle of coarse grained material.

These results show the contrary influence of the share of surface grains and the present strain gradient on the forming behaviour. Both are increasing with decreasing λ, leading to a decrease in material strength resulting from the surface layer effect and a parallel increase in material strength due to the strain gradient.

For copper foils and $\lambda \geq 0.2$ the influence of the surface grains is dominating, hence a decrease of spring back can be observed. If $\lambda < 0.2$ the influence of the increasing strain gradient can not be neglected any more and the spring back is increasing. This influence is larger for coarse grained material compared to fine grained material, leading to stronger decrease of spring back for $\lambda \geq 0.2$ and a smaller increase of spring back for $\lambda < 0.2$ if decreasing scaling factors are considered.

For aluminium foils the influence of the surface grains is not as distinct as for the copper foils. This has also been observed for the mechanical properties. Regarding these foils the influence of the strain gradient is dominating the spring back behaviour. Thus the

300

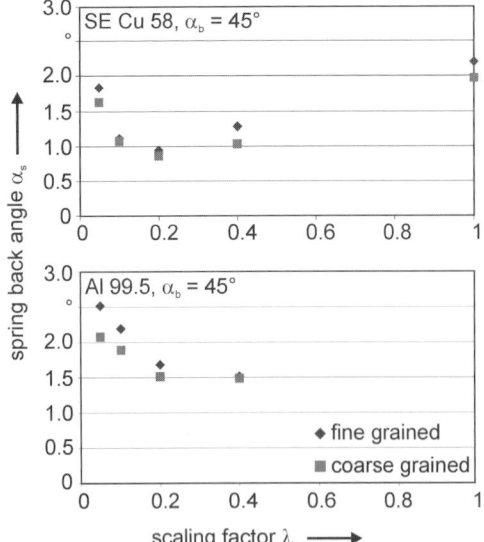

Fig. 7: Spring back angle in dependence of the scaling factor for a bending angle of 45°

spring back is increasing continuously with decreasing λ for λ < 0.4. As explained above, the 500 μm (λ = 1) Al 99.5 foils, representing the macro reference material were delivered by a second supplier. Since they show a different behaviour regarding the spring back and mechanical properties compared to the thinner foils they are excluded from further discussion. However, for λ > 0.4 larger spring back is expected, due to the decreasing influence of strain gradient, but further experiments are needed to investigate this behaviour in more detail.

Comparing the spring behaviour of copper foils in dependence of the maximum strain ε_{max} a significant influence of the foil thickness can be observed (Fig. 8). While the spring back angle is increasing with increasing maximum strain for foil thicknesses of 500 μm, the opposite behaviour can be observed for thin foils with t_0 = 25 μm. For foils with 100 μm thickness no significant influence of maximum strain on spring back angle is detectable. These observations indicate again two contrary effects.

An increasing maximum strain leads to an increasing plastified area, thus the elastic share within the forming area of the bent arc is decreasing. Since spring back is caused by elastic unloading, α_s is decreasing with decreasing elastic area. This effect is dominating the bending behaviour of 25 μm foils. The

increase of α_s for the 500 μm foils is due to an increasing bending arc with increasing die radius. The total spring back is the sum of the spring back of each incremental arc, leading to an increase of α_s of the 500 μm foils.

3. Conclusions and outlook

The t_0/L_G-ratio has a strong influence on the mechanical behaviour of metallic foils. It has been shown, that for copper foils the flow stress at a strain of 0.08 is decreasing with decreasing t_0/L_G-ratio. This behaviour was not observed for aluminium, where the determination of the flow curve of thin foils was only possible for small strains. The strain hardening exponent of coarse grained foils is larger compared to fine grained material due to differences in the microstructural forming behaviour.

The spring back behaviour of the investigated foils is indicating different effects of an increasing share of surface grains and an increasing strain gradient with decreasing scaling factor. For copper foils the increasing share of surface grains is dominant for scaling factors larger than 0.2. Going to smaller scaling factors an increasing influence of the strain gradient can be observed. Regarding aluminium foils the influence of the strain gradient is decisive, leading to a constant increase of the spring back angle with decreasing scaling factor.

More experiments are needed for the determination of mechanical properties at even higher strains and under biaxial load to provide fundamental data for FE-modelling. The strain gradient dependent forming behaviour has to be included in simulation tools in order to improve predictability of micro sheet metal forming processes.

Acknowledgements

The authors gratefully acknowledge the support from the German Research Foundation (DFG). Also, this work was carried out within the framework of the EC Network of Excellence "Multi-Material Micro Manufacture: Technologies and Applications (4M)".

References

[1]   Geiger M., Kleiner M., Eckstein R., Tiesler N. and Engel U.: *Microforming*. Annals of the CIRP 50 (2001) 445-462.

[2]   Raulea L.V., Goijaerts A.M., Govaert L.E. and Baaijens F.P.T.: *Size effects in the processing of thin metal sheets*. J. Mater. Proc. Technol. 115 (2001). 44-48.

[3]   U. Engel, A. Meßner, M. Geiger: Advanced Concept für the FE-Simulation of Metal Forming Processes for the Production of Microparts. In: Altan, T. (Edtr.): Advanced Technology of Plasticity 1996, Proc. Of the 5th ICTP, Oct. 7- 10, 1996, Columbus, Ohio USA, Vol. II, p. 903 – 907.

[4]   Fleck N.A. and Hutchinson J.W.: Strain gradient plasticity. Adv. in Appl. Mech. 33 (1997) 295-361.

[5]   Ashby M.F.: The deformation of Plastically Non-homogeneous Materials. Philosophical Magazine 21 (1970) 399-424.

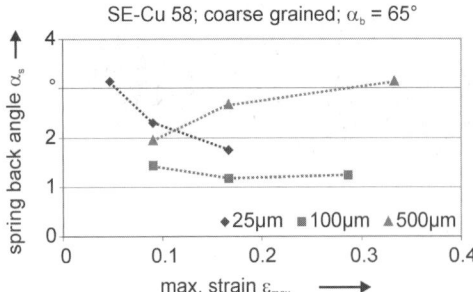

Fig. 8: Influence of ε_{max} on the spring back behaviour

Experimental study on the free-sintering process of a micro-porous HDPE membrane

D.B. Trifonov[a], Y.E. Toshev[b]

[a] *Institute of Information Technologies, Bulgarian Academy of Sciences, 1113 Sofia, Bulgaria*
[b] *Institute of Mechanics and Biomechanics, Bulgarian Academy of Sciences, 1113 Sofia, Bulgaria*

Abstract

The paper presents an investigation of the possibilities for optimization of the free-sintering process of a micro-porous HDPE membrane to be used mainly for fine bubble aeration. The HDPE membranes have priority significance in fine bubble operation in wastewater treatment applications. Precision molded porous plastic media have many applications in the industry, medicine and consumer products market. Of the other hand fine pore HDPE membranes, developed of special HD-PE material, provide excellent chemical and thermal properties for the most challenging environments. Various types of laboratory experimental tools for forming the micro-porous HDPE membranes at various process conditions were developed. The two-stage compression molding methods were used for forming the porous membranes. Approach for changing the process parameters for the sintering of the membranes after the compression molding was developed. Some experiments were accomplished using different variants of the technological equipment at different combinations of the sintering process parameters. The possibilities for obtaining of micro-porous membranes with different pore size and minimum pressure loss of the compressed air flow, passing through the membrane were investigated.

Keywords: micro-porous HDPE membrane, free-sintering, fine bubble aeration

1. Introduction

Sintered porous plastic media are an economical solution with many applications: filtration, fluidization, aeration and diffusion. Precision molded porous plastic media have many applications in the industry, medicine and consumer products market. Typical application examples are capillary filming/diffusion, coarse bubble aeration, water treatment, disposable elements for filtering compressed air, fine bubble aeration, waste water treatment, fluidization and aeration of bulk solids, fluidizing, air slide conveying, flux aerators - electronic applications, medical products, pen nibs, petrol filtration - feed line, process water filtration, silencing duties-compressed air pumping of gases, vacuum valve - bulk solids discharge control, vapor diffusion, vent filters - air cushions and vacuum cleaning - vent exhaust filtration [1, 2, 3].

One of the main features of HDPE fine pore membranes is the achievement of low level of the pressure loss at different flows of the dispersed air from the membrane. Of the other hand fine pore HDPE membranes, developed from special HDPE material, provide excellent chemical and thermal properties for the most challenging environments [4, 5, 6, 7].

Many materials are produced by powder processes. Nearly all production processes in powder technology include a sintering step at high temperature. The reason may be that the material is difficult to melt and cast. Another reason for using powder technology may be that it is economically attractive to make complex shaped structural parts by powder compaction and sintering. Further, some polymers, such as HDPE, UHMWPE and PTFE, are processed by pressing and sintering. Finally, powder metallurgy is applied to obtain high-grade materials with well-controlled, uniform microstructure.

Throughout all HDPE porous plastic runs an intricate network of open-celled, omni-directional pores. These pores, which can be made in average pore sizes down to one micron, give porous plastics their unique combination of filtering capability and structural strength. Uniaxial powder compaction in a die usually does not yield a homogeneous density distribution in the pressed part (the so-called green body). Rather, the green density is more or less inhomogeneous depending on the part geometry, the tool design and the friction between powders and dies wall. As a result, the part undergoes shape distortions during sintering, as regions with different green density shrink at different rates and by different amounts. It is not unusual that cracks develop during pressing and sintering. Sinter distortions may also be a consequence of gravitational forces or friction on the support during firing. Since, however, the requirements for close geometrical tolerances increase continuously, an extra processing step, such as hard machining or sizing may become necessary, if the distortions are too large [8, 9, 10, 11, 12].

The purpose of the present paper is to investigate the possibilities for developing of experimental laboratory equipment for micro-porous HDPE membranes forming and optimization of the compression molding and sintering processes.

2. Experimental study and results

2.1. Porous HDPE membrane forming

HDPE is an economical, high impact, lightweight thermoplastic which characteristics include excellent chemical resistance, high tensile strength, resilience, stress crack and abrasion resistance. It is easily machined and requires welded or mechanical style of construction. Despite the numerous attractive features of HDPE, its industrial production and application is limited mainly by the specialized machines and equipments necessary for its processing. It is one of the few materials that are not adapted for processing on conventional equipment. The present article is grouped around two of the most important processing steps in powder technology, namely powder die compaction and sintering.

The investigations were provided with HDPE "BUPLEN 6", that has a molecular mass of 1, 5 million and bulk weight up to 0, 18 gr/sm^3.

The technology proposed for the design of micropore membranes of HDPE includes two variants:
- sintering of pure HDPE with low direct compression molding;
- sintering of mixtures of HDPE and dissoluble filler with direct compression molding and with two-stage compression molding;

2.2. Sintering of pure HDPE

Porous HDPE media is achieved by a special sintering process, in which the granules of plastic are heated until their surfaces soften, and fuse at the contact faces, while retaining more or less their original shape. The shaped body created thus has open continuous pores, whose size and number depends on the sintering conditions and the size of the polymer particles selected. The pore size is mainly determined by the size and shape of the plastic granules employed.

The preliminary work showed, that the material agglomerates at negligible mechanic action of pressure. The temperature of melting is about 180^{0}C. It sinters well at different low pressures and at free heating. The final density of the product depends on the pressure applied during cold compaction of the HDPE material and its thermal processing and cooling. When cold pressure above 300 kg/sm^2 is applied, it is not suitable for sintering without pressure due to its clearly expressed tendency to cracking. At decrease of the molding pressure its density can be diminished to approximately 0,3 gr/sm^3, when it becomes gas-permeable. A disadvantage of this method is the impossibility to control the pores size, their repeating and also to define the porosity of the product with respect to obtaining products with defined hydrodynamic resistance.

2.3. Sintering of mixtures of HDPE and dissoluble salt

In order to obtain a micro-pore membrane with a diameter of 200 mm and thickness of 10 mm, a method of a direct compression molding and a method of two-stage compression molding were applied. The direct compression molding process was used to form a finished component from raw resin. The concept of the process is similar to block or sheet molding except on a smaller scale. Instead of using flat platens, the plungers have the contour of the components being formed. The plunger is contained in a sleeve that has the same inner diameter geometry as the profile of the component being formed. To mould a cold tablet by use of direct compression molding, the mixture of powder was placed into cavity ring and than the male moving punch is placed into the ring and the material is compressed while a cold tablet with thickness of 10mm is achieved.

The mold for direct compression molding is presented on Fig. 1 and Fig. 2.

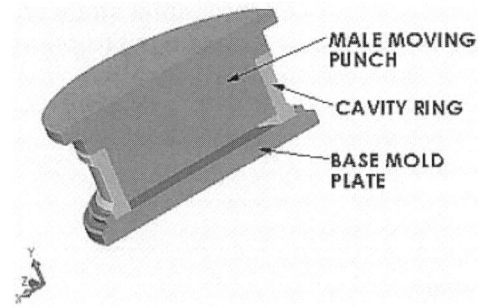

Fig. 1. 3D CAD model of direct compression mold

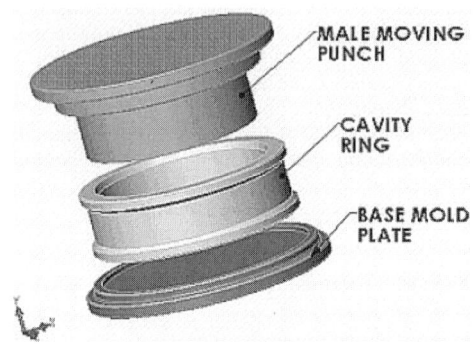

Fig. 2. View of the 3D CAD model of direct compression mold

The pressure stroke of 70 mm is used. Stabile tablets are obtained at pressure from 10 up to 400 kg/sm^2. The minimal hydrodynamic resistance, achieved in this method is 700 mm water column. That is why the control and the repeatable type of the main parameters are not guaranteed. The disadvantage of the direct compression molding is the impossibility to receive a uniform pore distribution on the surface of the membrane.

When designing determined pores with the help of dissoluble filler and the use of two-stage compression molding, it was possible to create a porous media with different hydrodynamic resistance at equal degree of dispersing of the passing gas and uniform distribution of the pores through the surface of the membrane. The process consists of cold compaction into the cavity ring at low pressure and after that the final compression stroke on the material to achieve a cold tablet with 10 mm thickness. Filling the powder into the cavity ring is a very complex process, in which inhomogeneous filling densities as well as totally unfilled areas can occur, especially if the cycle times are shortened more and more in order to increase the trough put. In practice, the initial packing density depends on the kinematics of the die filling process.

The flow diagram of the process is presented on Fig. 3. The mold for two-stage compression molding is presented on Fig. 4 and Fig. 5. Direct relationship has been established between the size and type of the filler, its concentration, temperature, pressure and sintering time, cooling, thermal processing and chemical processing of dissolving, which enables alteration in the range of the pores from 35 up to 300 microns.

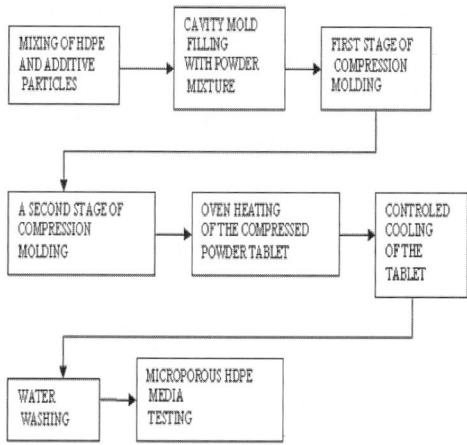

Fig. 3. Process diagram of micro-porous HDPE membrane forming

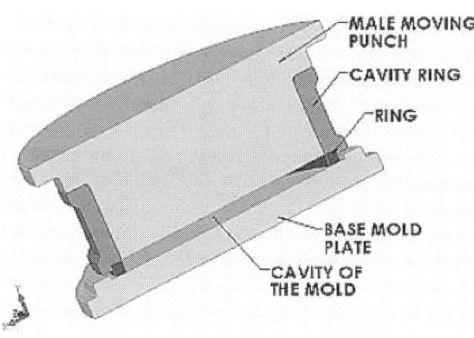

Fig. 4. 3D CAD model of a two-stage compression mold

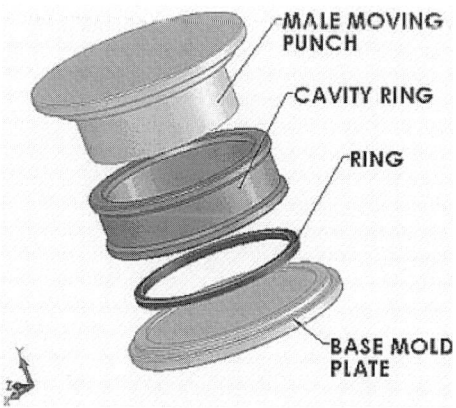

Fig. 5. View of 3D CAD model of a two-stage compression mold

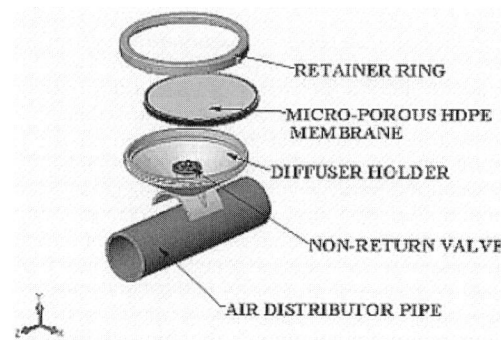

Fig. 6. View of 3D CAD model of a diffuser with porous HDPE membrane.

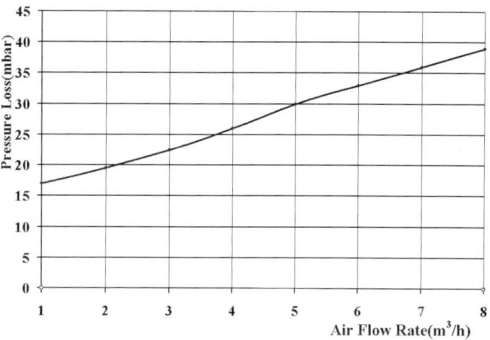

Fig. 7. Pressure loss microporous membrane testing

The defined in size dissoluble hard fillers are included in the content of the system and subjected to machine mixing for a time of 39 min. This time depends on the type of the mixer and it could be altered.

The mixtures obtained are cold pressed under pressure within the range of 50-100 kg/sm^2. The tablets obtained are thermally processed at temperature of 220^{0}C for approximately 120 min. The separation of the dissoluble components is done by extracting under vacuum, or by water washing under ordinary conditions.

The best results are obtained when applying hard dissoluble fillers, enabling the alteration of the bulk weight of the compound and the reaching of 2,9 gr/sm^3 at pores size of 100 microns. When altering the pores size, the optimum is displaced.

To verify the level of the porous membrane pressure loss, an experimental work was provided on a laboratory installation. Fig. 6 shows an example of HDPE membrane application in the fine bubble diffuser construction. Fig. 7 presents the final result of the membrane pressure loss level that is suitable for use in fine bubble aeration applications.

3. Conclusions

In this paper, investigations on various types of laboratory experimental tools for forming the micro-porous HDPE membranes at various processes are presented. The direct compression molding, the two-stage compression molding and an oven with possibilities for changing of process parameters for the sintering of the membranes after the compression molding were developed. The best results were obtained when the two-stage compression molding technique for cold pressing of the mixture of a HDPE and additive particles was applied and then the received tablet is sintered in a specialized oven at controlled processing parameters. A manufacturing process allowed us to successfully design a strong HDPE porous membrane that is superior to conventional products with a low degree of distortion and an even pore size, low pressure drop and high collection efficiency was achieved. Controlled particle size distribution and a precise temperature cycle were used to produce a wide range of interconnected cell structures and specific pore size ranges. These pore structures were designed and controlled to be used in a variety of applications such as filtering, fine bubble aeration, diffusing or venting gases and water treatment.

Acknowledgements

We are grateful to FP6 NoE 4M.

References

[1] Babcock R W and Stenstrom M. Proccedings of the 66th Annual conference, Anaheim, California U.S.A., October 3-7 (1993) 217-224.

[2] Groves KP et al. Water Environ. Res., 64, (1992) 691.

[3] Sprull R. Proceedings of the South Carolina Environmental Conference, March 17-20, (2002), 45-49

[4] Trifonov D. et al. IIT/WP-168B (2003), 77-83

[5] Trifonov D. Ecological Engineering and Environment Protection, 3, (2004), 50-57.

[6] Trifonov D. et al.Cybernetics and Information Technologies, 5, (2005), 126-133.

[7] Trifonov D. Et al. Ecological Engineering and Environment Protection, 2, (2003), 14-19.

[8] A. A. Keller, Britta G. Bierwagen. Sci. Technol. 35 (2001) 1875-1879.

[9] K. Damaka*, Abdelmoneim Ayadia, Philippe Schmitzb, Belkacem Zeghmati. Desalination 168 (2004) 231–239.

[10] Continuum Scale Simulation of Engineering Materials: Fundamentals – Microstructures – Process Applications, edited by Dierk Raabe, Franz Roters, Frederic Barlat, Long-Qing Chen (Wiley-VCH, Verlag GmbH & Co, (2004)

[11] Masao N. Nitto Denko, Technical Report 39 (2001).

[12] Splinter A. et al. Proceedings of the 11th International Conference on Solid-State Sensors and Actuators, Munich, Germany, June 10 – 14 (2001), 156-159

Multi-Material Micro Manufacture
W. Menz, S. Dimov and B. Fillon (Eds.)
© 2006 Elsevier Ltd. All rights reserved

Hybrid tooling: a review of process chains for tooling microfabrication within 4M

S. Azcarate[1], L. Uriarte[1], S. Bigot[2], P. Bolt[3], L. Staemmler[4], G. Tosello[5], S. Roth[6] and A. Schoth[7]

[1] Fundación Tekniker, Avda. Otaola 20, 20600, Eibar, Spain
[2] M.E.C., School of Engineering, Cardiff University, Cardiff CF24 0YF, Wales, UK
[3] TNO Industrial Technology, De Rondom 1, 5600 HE Eindhoven, The Netherlands
[4] HSG-IMAT, Breitscheidstr. 2 b, 70174, Stuttgart, Germany
[5] Department of Manufacturing Engineering and Management, DTU, 2800 Kgs. Lyngby, Denmark
[6] Bayerisches Laserzentrum GmbH, Konrad-Zuse-Str. 4-6, D-91052 Erlangen, Germany
[7] IMTEK, University of Freiburg, Georges-Koehler-Allee 103, EG-79110, Freiburg, Germany

Abstract

The current paper is based on the information gathered by the *"Processing of Polymers"* Division (Task 4.2 *"Hybrid Tooling"*) and *"Processing of Metals"* Division (Task 7.2 *"Tooling"*) within 4M Network activities. The aim of the task involves a systematic analysis of the partners' expertise in different technologies for processing tooling inserts for further replication in polymers. Firstly, the 4M partners current capabilities in individual tooling processes is briefly presented, and also the expected capabilities for year 2010 are analysed for each of the following processes: micromilling, micro-wire electrodischarge machining (μWEDM), micro sinking electrodischarge machining (μSEDM), laser micromachining, electrochemical micromilling (μECM), and electrochemical milling with ultrashort pulses (ECF).

Later the concept of 'hybrid tooling' as different process chains for tooling fabrication is introduced. Several examples of 'hybrid tooling' within 4M partners are presented. Considered materials are nickel for electroforming, stainless steel for ECF, and tool steel for the other processes. The paper results provide a global comparison between the previously mentioned processes, the current limitations of these technologies concerning feature sizes, surface finish, aspect ratios, etc. have been identified. The main conclusion drawn is the imperative requirement to combine individual processes ('hybrid tooling') to produce 3D free-form microshapes for tooling purposes.

Keywords: Hybrid tooling, Process chain, Micromanufacturing

1. Introduction

The use of micro products has been strongly increasing through the past 5 years and market surveys predict an even larger increase [1]. The market is demanding industrial technologies for manufacturing large numbers of products at a reasonable price. These two demands are satisfied by the use of replication technologies, mainly micro-injection moulding and hot-embossing. On the other hand, the mould and die industry in Europe has more than 5.500 companies, with an average size of 23 employees and total sales of 10.000 M€ [2]. Therefore, mould making is an important industry in Europe, dominated by SMEs, and having a strong competition from USA, Japan and other Asian countries.

All reviews of microtechnologies agree in the two following statements:

- Mass production of micro-parts must be based on replication technologies.
- Quality and performance of a replicated micro-part depend mainly on the quality and performance of the corresponding micro-mould.

Previous works [3] within 4M Network have reviewed micro-manufacturing processes for tooling from an individual point of view. Section 2 of this paper shows a summary of it. However, they are quite limited in order to machine 3D free-form micro structures in a wide range of materials, mainly hard metals for tooling purposes, as is one of the goals of 4M microtechnologies.

In most of the application cases, a combination of some of them are required to produce micro-tooling, and this combination leads to the apparition of specific problems and methodologies to overcome them. These particularities are analyzed and some of them

emphasized due to its singularity. Data and figures presented below reflect experience within 4M partners, and seek merely to estimate the state of the art in micro-tooling fabrication.

2. Present capabilities and future trends of individual processes

The data contained in Table 1 summarise the information collected from 4M partners in order to know the limits of each process for producing mould inserts for plastic replication. Although in specific cases these limits are surpassed, the figures correspond to most common materials used for mould inserts. LIGA and LIGA like techniques are not included.

Table 2 summarizes the expected new limits in the same processes for year 2010 according to 4M partners comments. Although no disruption is expected the trends are: the improvement of accuracy, aspect ratio and miniaturization, and nearly no productivity improvements are expected.

Specific remarks are stated for each process. In micromilling the tool is the critical item, its material, size, accuracy, stiffness and coatings might enhance the process capabilities. However in μWEDM the wire size is not the limit but the sensitivity of the whole wire tension and guiding system. For μSEDM restrictions are based on electrode rigidity, spark gap and size of the debris. More research into mechanism of material removal and modelling of the erosion process is needed to improve the predictability of the results. Also, the control systems driving the movements of the electrode could be easily improved using technologies available for other processes; this could dramatically improve currently achievable accuracies and dimensions.

	µmilling	µWEDM	µSEDM	µLASER	µECM	ECF
Maximum mould insert external dimensions (mm)	200x200x200	200x200x100	300x300x150	200x200x200	200x200x30	20x20x10
Maximum micro structured area (mm)	100 x 100	200 x 200	200 x 150	200 x 150	40 x 30	20 x 20
Minimum tool size (mm)	0,1	0,02-0,05	0,05	-	-	-
Minimum feature dimension (µm)	50 - 100	20-50	10 - 50	5	50	10
Accuracy (µm)	3-10	2	3	1	8	2
Aspect ratio	2-5	10-50	3-10	2-50	-	10
Roughness (µm Ra)	0,2	0,1	0,1	1	0,1	
Material removing rate (mm³/min)	20	0,1	0,01	0,04	-	6×10^{-6}

Table 1 – Summary of 4M partners present capabilities of individual processes

	µmilling	µWEDM	µSEDM	µLASER	µECM	ECF
Maximum mould insert external dimensions (mm)	300x300x200	200x200 x100	300x300x160	200x200x200	300x300x30	100x100x50
Maximum micro structured area (mm)	150 x 150	200 x 200	200 x 150	200 x 150	100 x 100	100 x 100
Minimum tool size (mm)	0,04	0,01	0,005	-	-	0,0005
Minimum feature dimension (µm)	10 – 25	5-20	4 – 10	1	10	0,5
Accuracy (µm)	1	0,5	0,5	0,5	2	0,5
Aspect ratio	5-10	100	50	> 50	20	100
Roughness (µm Ra)	0,1	0,05	0,05	0,4	0,05	
Material removing rate (mm³/min)	20	0,1	0,01	0,04	1	4×10^{-3}

Table 2 - Summary of 4M partners expected limits for year 2010

In laser micromachining restrictions are coming from laser spot and affected laser spot as well as from focusing angle of the beam. In µECM restrictions are coming from electrode size mainly and lack of basic knowledge of the process.

3. Hybrid Tooling: process chains for tooling

Hybrid tooling was defined by *"Processing of Polymers"* Division as *"the capabilities of producing a mould insert combining two or more processes in sequence"*. Most of cases are based on the combination of conventional and energy assisted processes like micromilling, µEDM, µECM or laser. Typical clean room technologies are inherently sequences of processes. What is not usual is the combination between technologies of both groups [4]. Here below are pointed out some examples of 'hybrid tooling' within 4M partners.

3.1. Case 1: µEDM + USM

First case comes from MEC and consists of the use of ultra sonic machining (USM) together with the micro EDM process for the machining of any conductive material. Initially, the electrode could be ground down to a specific shape and size applying ultrasonic vibrations to the grinding device, while controlling vibration directions. After that, machining of the workpiece could

be done in a similar way by using ultra sonic vibrations on either the electrode or the workpiece. Consequently, the material removal rate is increased and surface roughness and defective layer are decreased.

3.2. Case 2: Micromilling + electroforming + etching

This case comes from DTU and combines material-removal by mechanical processes (micromilling), with electroforming (Alternative A in the figure shown below) and etching, and should be considered as a viable alternative to direct mould insert machining (alternative B).

Manufacturing alternatives & shape relationships

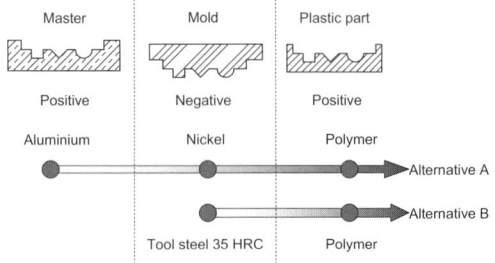

Figure 1 – DTU alternative versus direct machining

The process chain starts with the micromilling in a 'soft' material like aluminium or zinc, following by the electroforming on milled surfaces to form a nickel layer with the negative geometry. It requires a conducting surface for plating purpose. And finally the dissolution of aluminium by selective etching to set free the nickel insert and thereby negative geometry. It requires a true selective etching process that only attacks the base material: aluminium or zinc, but not nickel.

Its main advantage is based on the easier machining of a positive structure in 'soft' material, with minimum tool wear and good surface quality. It allows to combine a true 3D process as micromilling with clean room technologies. It is critical a good cleaning and activation between steps 1 and 2, and also to achieve a stable and low stress electroforming process. Other cases with a similar configuration chain are presented by DTU as shown in Tables 3, 4 and 5.

In this case to overcome the limits imposed by the complexity of some features it is necessary to use an indirect way for producing mould inserts.

Case 2.2 : Si etching + electroforming + etching	
Process 1:	Advanced silicon etching of a silicon wafer
Process 2:	Electroforming to form the negative tool insert
Process 3:	Selective etching of silicon without damaging the master structure
Remarks	It is much easier to dissolve silicon than aluminium since silicon is very pure and dissolved readily in potassium hydroxide. A metallization step is required between process 1 and 2, in order to make the silicon master sufficiently conducting for the electroforming process.

Table 3–Si etching+electroforming+selective etching

Case 2.3 : Laser + electroforming + etching	
Process 1:	Laser micromachining in polymer
Process 2:	Electroforming to form the tool insert
Process 3:	Selective etching
Remarks	Polymers like AB3 are easily dissolved in acetone which will not attack electroforming metals as Ni or Cu. Also a metallization step is required to make the polymer master sufficiently conducting for electroforming.

Table 4 – Laser +electroforming+selective etching

Case 2.4 : Photolithography + electroforming + photoresist removal	
Process 1:	Photolithography in SU8
Process 2:	Electroforming to form the tool insert
Process 3:	Photoresist removal without attacking the electroformed metal
Remarks	Potential problems with side wall angles and the removal of the SU8. Solving the photoresist removal by introducing intermediate layers will often result in problems elsewhere.

Table 5 – Photolithography+electroforming+removal

3.3. Case 3: Micromilling + ECF

This case comes from HSG-IMAT and combines material-removal by mechanical processes (micromilling), with ECF. In principle the roughing and bigger shapes (up to 100 µm) of the tool insert are machined by micromilling, usually in steel; and then the ECF process is used where higher resolution and sharper edges are required. Its main advantages are the small tools available (up to 5-10 µm), no tool wear and no generation of burrs. However the inclusions in the steel should be small.

Within 4M activities a test part was done consisting in five ribs each 100 µm depth and 1 mm long, and width from 20 to 100 µm (Figure 2). The outer contour was micromilled and the grooves between ribs were ECF-processed into stainless steel 1.4301.

In cases like this one, where no burrs are mandatory, first is used a process with high material removing rate and later the tool is finished by a slow but without wear process.

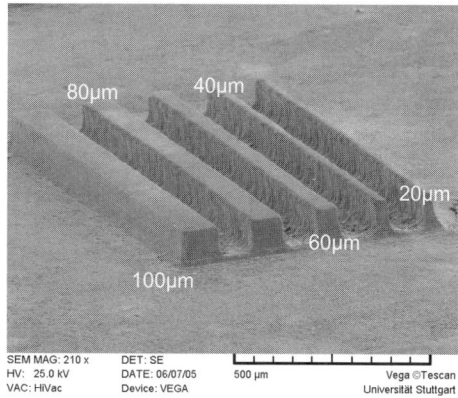

Figure 2 – 4M test part by micromilling + ECF

3.4. Case 4. Micromilling + UV-LIGA + Electroplating

Next cases come from IMTEK and were applied for the fabrication of tooling for mass production of LabonChip. In both cases the starting process is the micromilling. In the first case the base material is brass, and second step corresponds to UV-LIGA in order to produce 'hybrid tooling' of small and big features. This second process is used to overcome the limitations of micromilling in terms of low accuracy, high surface roughness and size restriction because of the tool size. The requirements are an ultraprecision CNC milling machine, with an interchangeable clamping system, and a modified UV-LIGA process. Final step is the electroplating to produce a Ni mould. A modified version of the process is required. This combination of processes is a good way to produce moulds for prototyping with very good quality.

The second case coming from IMTEK also starts with the micromilling in this case in stainless steel, following by electrochemical milling. The objective of this process is to produce small 3D ribs less than 20 µm, with a distance of 70 µm, and aspect ratio of 5. It is impossible to produce it directly by micromilling due to the high roughness and high machining forces typical in micromilling, and the restricted tool size. For such combination are required an ultraprecision milling and ECF machines, and the corresponding clamping and measurement systems. It has been used to produce moulds for mass production, but still need for improvement.

3.5. Case 5: Metal Foil LOM + Diffusion welding

Last case is a singular example of 'hybrid tooling', combining a manufacturing technology with a welding process. A detailed process description can be seen in [5] by BLZ. It is useful not only for tooling purpose but also for final parts production, specially suitable for microfluidic devices. Laminated Object Manufacturing (LOM) uses a laser beam (pulsed Nd:YAG) to cut the desired shape in a metal foil, in this example in steel. Later a diffusion welding process is used join the foils. The objective is to enhance the mechanical stability towards the properties of the used material. So, if the joining process is good, the strength of the metal foil LOM part can nearly reach the strength of the base material. The welding process requires a high temperature press, inert atmosphere, and a powder bed in a pot for realizing complex geometries. However, the diffusion welding process of complex geometries is currently under development.

Next Table summarizes present and expected capabilities of the LOM for year 2010. Regarding the maximum external dimensions of the mould insert, the size is actually limited by the furnace system for post-processing (diffusion welding).

In this case, when large mould inserts are needed with no high accuracy LOM + Diffusion welding could be the unique option.

	Present	Year 2010
Maximum mould insert external dimensions (mm)	120 x 150 x100	150 x 150 x200
Maximum micro structured area (mm)	120 x 150	150 x 150
Minimum tool size (mm)	10x10x10	5x5x5
Minimum feature dimension (mm)	5	2
Accuracy (µm)	10	5
Aspect ratio	0,1	0,5
Roughness (µm Ra)	10	5
Material removing rate (mm³/min)	Metal Foil LOM is an additive process	

Table 6 - Summary of LOM capabilities

4. Conclusions

We can state that 'hybrid tooling' should be considered not like an alternative but like a need for the production of tool inserts with real 3D free-form microfeatures.

A review of 'hybrid tooling' capabilities within 4M partners has been presented, which can be considered in a certain way like a state of the art.

Some the shown technologies are very well established with presently industrial application, even continuous improvements are developed. Some other are more in a research phase looking for industrial partners to make the step towards industrialization.

In any case, the most attractive advantage of such combination of processes is the chance to produce 3D free-form micro structures in hard metals with an accuracy, resolution and quality only available for clean room technologies.

Acknowledgements

We are grateful to all the contributing members of the Polymers and Metals Clusters of the 4M NoE.

References

[1] Masuzawa T. State of the art in micromachining. CIRP Annals 49/2/2000.

[2] The International Special Tooling & Machining Association (ISTMA) Annual Report 2004.

[3] Uriarte, L., Ivanov, A., Oosterling, H., Staemmler, L., Tang, P.T., Allen, D. "A Comparison between Microfabrication Technologies for Metal Tooling", 4M 2005 Proc., pp.351-354, ISBN 0.080-44879-8.

[4] Bissaco, G., Hansen, H.N., Tang, P.T., Fugl, J. "Precision manufacturing methods of inserts for injection molding of microfluidic systems". ASPE Spring Topical Meeting on Precision Mecro/Nano Scale Polymer Based Component & Device Fabrication, Int. Proceed. ASPE. Vol.35, 2005.

[5] Prechtl, M., Seidel, S., Otto, A., Roth, S. "Microfluidic Devices Made by Automated Metal Foil LOM", 4M 2005 Proc., pp.389-392, ISBN 0.080-44879-8.

Multi-Material Micro Manufacture
W. Menz, S. Dimov and B. Fillon (Eds.)

Adapting ECF to steels used for micro mould inserts

L. Staemmler[a], K. Hofmann[b], M.-H. Kim[b], D. Warkentin[a], H. Kück[a,b]

[a]Hahn-Schickard-Institute for Micro Assembly Technology (HSG-IMAT), Stuttgart, Germany
[b]University of Stuttgart, Institute of Micro- and Precision Engeneering (IZFM), Stuttgart, Germany

Abstract

Electrochemical machining with ultra short voltage pulses (ECF) is an innovative technique to machine electrochemically active materials particularly very hard materials at micrometer feature size. Since the ECF technique is an electrochemical process neither mechanical forces nor thermal load are applied to workpiece or tool. For that reason ECF it is an ideal technique for the production of microstructures. Especially the use of steel as the workpiece material makes the ECF technique a promising technique for the production of micro mould inserts, since steel is resistant against the wear that occurs due to the injection process. Therefore the abilities and the limits of the ECF process for different types of tool steels are shown and the process parameters have been optimised. Acetic acid has been proven to be a suitable electrolyte for the ECF-process of tool steels. In this electrolyte tool steels like 1.2767 or 1.2312 can be processed with the ECF technique at a high quality. But in low-alloyed steels like 1.1730 only a poor quality can be obtained.

Keywords: ECF, micro mould, steel

1. Introduction

The production of microstructures demand highly precise techniques due to the small size of the structures. Good results have been obtained by the LIGA technique. But also with high speed cutting (HSC) milling machines which can have an accuracy in the micrometer range microstructures can be achieved by using milling cutters with diameters of 80 μm and below. Also EDM is capable of producing structures in the 10 μm-range. Nevertheless the production of microstructures is extremely time-consuming.

It therefore makes sense to reproduce the microstructure, for example by injection moulding. With a single microstructure several thousand parts can be made. This demands a mould that withstands the abrasive properties of the injected polymer. The lifetime of nickel moulds which are made with the LIGA technique is often too short for an effective use of theses moulds. Hard materials like steel, which are therefore normally used for making these moulds, cannot be processed by the LIGA technique. Because EDM brings a huge thermal load to the workpiece and HSC is not capable of obtaining patterns with dimensions much below 100 μm, electrochemical milling with ultrashort voltage pulses (ECF, where F is the German acronym for milling) is a powerful alternative for the production of the microstructures in the mould [1]. Because it is an electrochemical process neither thermal load nor mechanical forces act on the tool, hence no tool wear arises.

In the ECF process short voltage pulses in the range of 10 ns to 1000 ns are applied between tool and workpiece to dissolve the workpiece locally [2]. Therefor the workpiece as well as the tool are submerged in an appropriate electrolyte. In the idle state both the workpiece and the tool are hold at a constant cathodic potential to prevent corrosion. For the dissolution process the pulses are applied additionally to the existing potentials. An electrode that is submerged in an electrolyte forms a double layer capacitance at its surface. During the pulse these capacitances are charged over the resistance of the electrolyte. The value of this resistance depends on the length of the path of the electric current in the electrolyte. Thus a large separation between tool electrode and workpiece leads to a slow charging of the capacitance whereas a small separation leads to a fast charging (fig.1). Since charging takes place only during the time of the pulse the double layer capacitance is charged to a potential high enough for an anodic dissolution of the work piece only in a close proximity around the tool, the so called working distance. Surface areas of the workpiece that are further away from the tool are not affected by the ECF process. This leads to a confined milling with a high spatial resolution. Nevertheless the size of the working distance depends in a first approximation linearly on the pulse width. Hence by changing the pulse width the working distance can be adjusted.

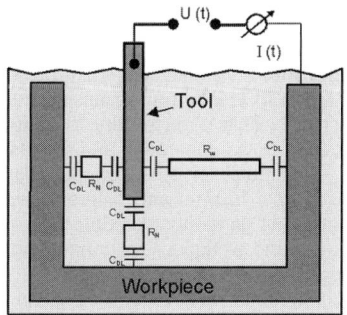

Fig. 1. Sketch of the electrochemical cell in the ECF process.

Since the pulse polarizes only the workpiece anodically the tool itself is not dissolved by the process.

If the feed rate is higher than the dissolution rate the tool comes into contact with the workpiece forming a short circuit, the so called contact. This is detected by the ECF machine and the motion of the tool is stopped. Because this event appears mainly at inhomogeneities of the steel, a strategy of tool movements is started to remove the contact and continue the ECF process.

IZFM together with HSG-IMAT set up an ECF machine. Free form surfaces can be obtained with tools with diameters down to 20 µm and below. Good results have been achieved in stainless steel using an electrolyte containing hydrofluoric acid (s. fig. 2).

The aim of the experiments shown here is to ECF-process tool steels that are typically used for producing mould inserts. It is also shown that acetic acid instead of HF is suitable for the ECF process. For these steel-electrolyte systems optimised sets of ECF parameters had to be found. The set of parameters that leads to a small number of contacts and at the same time to a small working distance is optimal for a fast and accurate ECF process.

2. Experimental

All ECF experiments shown here were carried out on the named ECF machine. It consists of an xyz-stage (Walter Uhl, Aßlar, Germany) with a travel range of 100 mm in each direction. A PC software interprets CAD/CAM-data files and drives the stage accordingly. Due to the fact that the electrochemical parameters have to be adjustable during the ECF-process special CAD/CAM-codes are used to set the values via this software. The feed rate for all experiments shown here was set to 0.3 µm/s.

The electrochemical cell is made from PTFE and it is provided with a silver/silver-chloride (Ag/AgCl) reference electrode and a platinum counter electrode. The workpiece is mounted to the bottom of the cell and connected from the outside. A feedback loop controls the level of the electrolyte using an attached photo sensor.

The potentiostat and the pulse generator were especially developed for the ECF process (IZFM and ECMTEC, Holzgerlingen, Germany). The potentiostat is equipped with an input that stops the potential control keeping the potential constant while the tool is in contact with the workpiece. The pulse generator is designed to fit into the processing head of the ECF machine. This is necessary to keep the distance between pulse generation and tool as short as possible. The pulse on-time can be adjusted between 10 ns and 1600 ns and different periods can be set. The pulse generator also detects the contacts between tool and workpiece by measuring the current through the tool.

All experiments were processed in an aqueous solution of 1 M CH_3COOH. For the workpieces samples of tool steels 1.1730 1.2312 and 1.2767 with a diameter of 19 mm and a height of 5 mm were used. Each workpiece was prior to the experiment finished by wet-grinding to 2500 grit silicon carbide paper and rinsed with deionised water.

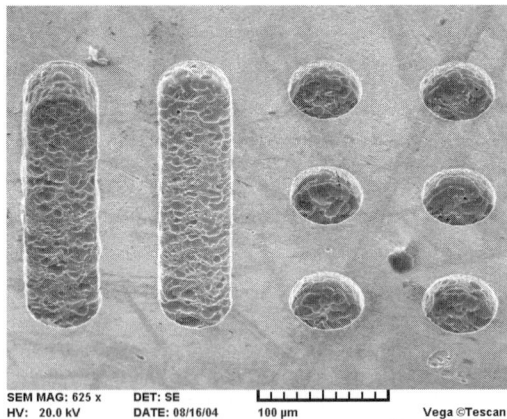

SEM MAG: 625 x	DET: SE	
HV: 20.0 kV	DATE: 08/16/04	100 µm
VAC: HiVac	Device: VEGA	
		Vega ©Tescan Universität Stuttgart

Fig. 2. Grooves and holes in stainless steel 1.4301 produced with the ECF technique in HF-based electrolyte.

The tools are made from tungsten wire. They are etched by ECM grinding to achieve a diameter of typically 40 µm [4]. Afterwards the tip shape is formed using the ECF process. This is done by applying an inverted pulse. This forming is done already in situ in the ECF machine prior to the processing of the workpiece.

To find out whether a steel-electrolyte system is suited for the ECF technique a first test array was processed. The array is set up of 24 holes 30 µm deep milled by ECF technique with different sets of parameters. The parameters varied are the tool potential U_{Tip}, the pulse amplitude U_m, the pulse width B and the current through the counter electrode I_{CE} that keeps the workpiece at its cathodic potential. During the milling the number of contacts KT was counted for each hole.

3. Results

Figure 3 shows an optical micrograph of an array processed into the tool steel 1.2767 using the CH_3COOH-based electrolyte. In the first two rows a current I_{CE} of 2 mA was used whereas in the second two rows I_{CE} was 0.1 mA. In row one and three the pulse width B was set to 500 ns in row two and four B

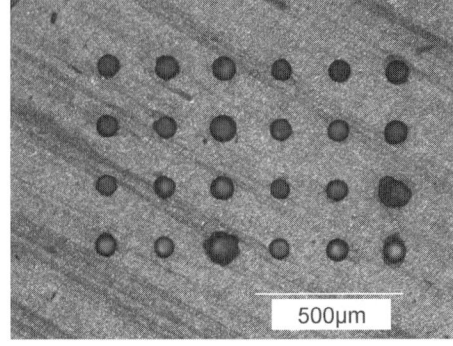

500µm

Fig. 3. Array of 24 holes ECF processed into steel 1.2767 in CH_3COOH-based electrolyte.

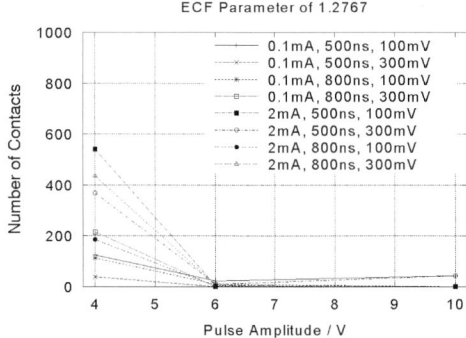

Fig. 4. Number of contacts as a function of the pulse amplitude for 1.2767.

was 800 ns. In the left three columns U_{Tip} was 100 mV, in the three right columns U_{Tip} was 300 mV. The pulse amplitude for both blocks of the three columns was set to 4 V, 6 V and 10 V, respectively.

Even from this optical image (fig. 3) it can be seen that the ECF process is strongly dependent on the used set of parameters. This can also be seen in fig. 4. Here the number of contacts is plotted as a function of the pulse amplitude for all sets of parameters.

It is remarkable that the number of contacts for a pulse amplitude of 6 V and above is negligible

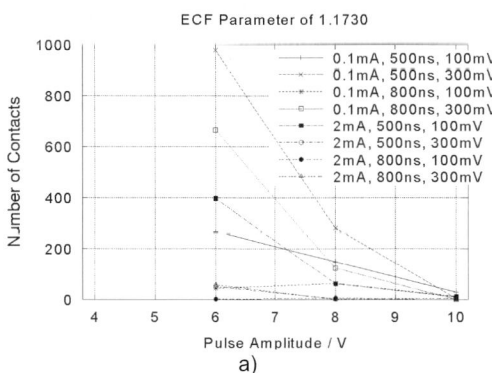

a)

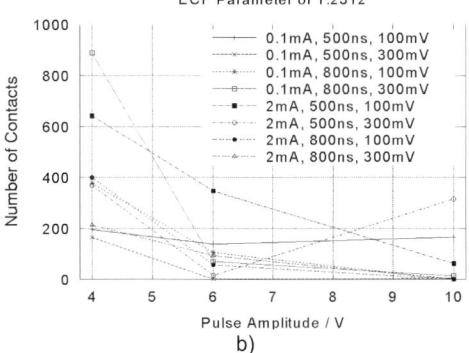

b)

Fig. 5. Number of contacts as a function of the pulse amplitude for a) 1.1730 and b) 1.2312.

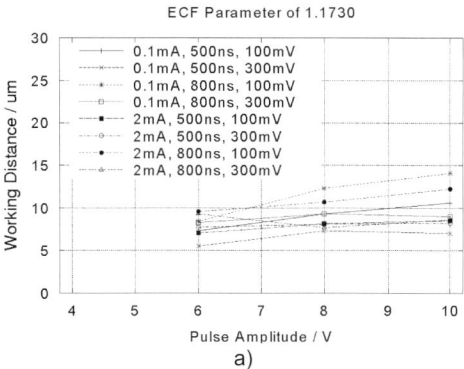

a)

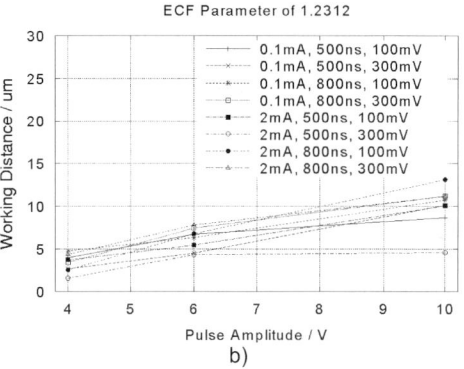

b)

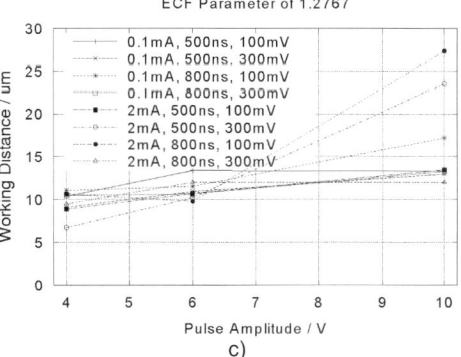

c)

Fig. 6. Working distance as a function of the pulse amplitude for a) 1.1730, b) 1.2312 and c) 1.2767.

whereas for an amplitude of 4 V it increases dramatically. This is true for all systems tested, even though not as obvious in the case of 1.2312 (fig. 5).

No ECF processing was possible with a pulse amplitude of 4 V in the case of 1.1730. For that reason a set of 6 V, 8 V and 10 V was chosen.

Only for 1.2312 for an amplitude of 10 V and a current I_{CE} of 2 mA, a pulse width B of 500 ns and a tool potential U_{Tip} of 300 mV the number of contacts increases again. This may be due to a large unsolvable inclusion that hinders the ECF process.

As can be seen in fig. 6 the working distance in all of the steel-electrolyte systems increases with respect to the pulse amplitude. While for the 1.1730 and 1.2312 these distances are in the same range, it is remarkable that for the 1.2767 the distance is up to a factor of two larger and much more dependent on the set of parameters. This may be explained with the smaller amount of manganese in the 1.2767 (0.15 % compared to 0.8 % for the 1.1730 and 1.4 % for the 1.2312) leading to less MnS inclusions. This comparably higher working distance and the fact that the number of contacts at an amplitude of 4 V is still only two thirds of the number of contacts for the 1.2312 indicates that a smaller pulse amplitude would be adequate for the 1.2767.

If one compares in fig. 6 the working distances for a single set of parameters it can be seen that the working distance for the 800 ns pulse width leads to an higher working distance as for the 500 ns pulse width at the same pulse amplitude. This is as expected and confirms the theory.

In tab. 1 the sets of parameter that lead to the best results are listed. With these set the following experiments have been carries out.

If fig. 7 nine holes are shown which are ECF processed with the optimal set of parameters for the 1.2312 (fig. 7 a) and the 1.1730 (fig. 7 b) respectively. It is obvious that in the case of the 1.2312 all holes have the same shape, i. e., the ECF process worked out reproducibly. This is not the case for the 1.1730 (fig 7 b). Here the shapes vary. Especially the hole at the left bottom is much smaller. This is due to the influence of inclusions, precipitations and grain boundaries in this low-alloyed steel.

Although further experiments are needed to fully explain the role of inclusions in the case of 1.1730, good results are obtained in the case of 1.2312 and 1.2767.

4. Conclusion

The ECF process was applied to different steel-electrolyte systems. It can be observed, that quite a number of systems can be treated with the ECF process using acetic acid. Nevertheless does the set of the applied parameters have a large influence of the achieved results. For all shown steels in this electrolyte a pulse amplitude above 6 V leads to a negligible (for the 1.1730 at least reasonable) number of contacts whereas a pulse of 4 V and below makes ECF processing difficult if not impossible.

It is shown that for all investigated systems the working distance increases with both the pulse width and the pulse amplitude.

Even if the ECF processing of a steel-electrolyte system with an appropriate set of parameters is possible the resulting holes may have a not circular shape. This is probably due to the grain size of the steel and the composition of the grain boundary. Further experiments will give an answer to this phenomenon.

Acknowledgements

This research project (AiF FV-Nr.: 180 ZN) was aided by budget funds of the Bundesministeriums für Wirtschaft und Technologie through the Arbeitsgemeinschaft industrieller Forschungsvereinigungen „Otto von Guericke" e. V. (AiF). We also like to thank R. Schuster (TU Darmstadt) and T. Gmelin (ECMTEC) for the helpful discussions and comments.

Tab. 1. Sets of parameter leading to best results.

	1.1730	1.2312	1.2767
U_{Tip}	300 mV	100 mV	300 mV
U_m	8 V	6 V	6 V
I_{CE}	2 mA	2 mA	2 mA
B	800 ns	800 ns	800 ns

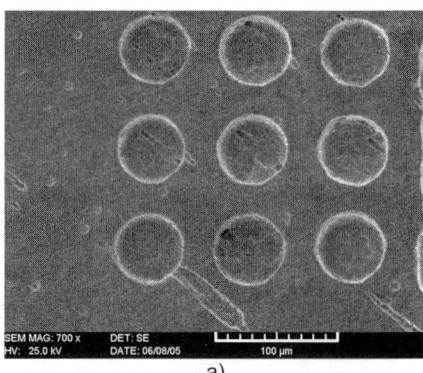

a)

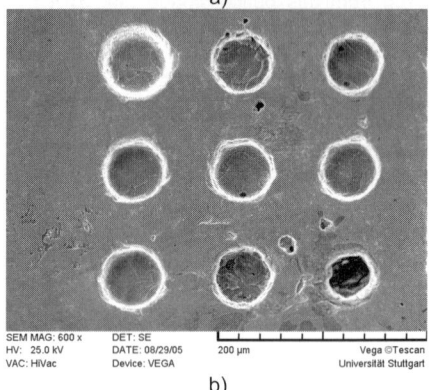

b)

Fig. 7. Holes ECF processed in a) 1.2312 and b) 1.1730.

References

[1] Schuster R, Kirchner V, Allongue P, Ertl G. Electrochemical Micromachining, Science 289 (2000) 98-101.

[2] Kock M, Kirchner V, Schuster R. Electrochemical micromachining with ultrashort voltage pulses—a versatile method with lithographical precision, Electrochimica Acta 48 (2003) 3213–3219

[3] Cagnon L, Kirchner V, Kock M, Schuster R, Ertl G, Gmelin WT, Kück H. Electrochemical Micromachining of Stainless Steel by Ultrashort Voltage Pulses. Z. Phys. Chem. 217 (2003) 299–313.

[4] Hacker B, Hillebrand A, Hartmann T, Guckenberer R. Preparation and characterization of Tips for scanning tunneling microscopy of biological specimens, Ultramicroscopy 42-44 (1992) 1515–1518

Microfluidics on foil

T. Velten[a], H. Schuck[a], M. Richter[b], G. Klink[b], K. Bock[b], C. Khan Malek[c], S. Polster[d], P. Bolt[e]

[a] *Fraunhofer Institute for Biomedical Engineering (IBMT), 66386 St. Ingbert, Germany*
[b] *Fraunhofer Institute for Reliability and Microintegration (IZM), 80686 Munich, Germany*
[c] *Laboratoire FEMTO-ST/Dpt. LPMO, UMR CNRS, 25044 Besançon, France*
[d] *Bayrisches Laserzentrum, 91052 Erlangen, Germany*
[e] *TNO Science & Industry, Eindhoven, The Netherlands*

Abstract

The concept of microfluidics on foil opens up new opportunities for combining the advantages of having a flexible substrate with reel to reel processing which have the potential to be the basis for extremely cheap micro products. To reach this goal foil substrates must be combined with micro manufacturing technologies well adapted to these substrates. Some technologies are already available, some are subject of current research and some still have to be conceived. Here, technologies like reel to reel embossing, reel to reel laminating and laser ablation/cutting as well as laser welding will be discussed. A polymer electronics based alcohol sensor is presented and aspects of combining polymer (opto-)electronics and microfluidics on one foil based substrate will be discussed.

Keywords: microfluidics, foil, reel to reel

1. Introduction

Nowadays, most micro structures are realised on rigid substrates like silicon, glass, ceramics, metals or polymers. Most available tools for micro machining are adapted to these substrates which often come in the shape of wafers. The disadvantage of the traditional technologies and equipment is their restriction to relatively small areas and planar substrates as well as the high price per micro machined surface area. An economic use of these technologies is only possible if the surface area per device is in the range of a few square millimetres. But there are many applications like e.g. microfluidic applications where lateral dimensions in the range of 100 µm and areas per device of about 2000 mm^2 are typical values. Using conventional substrates and technologies would lead to very expensive devices. Even the use of polymer-based substrates and replication techniques like hot embossing, casting of poly(dimethylsiloxane) (PDMS) or micro injection moulding as an alternative fabrication technology would be too expensive for many applications. An example is medical applications like point of care testing (detection of proteins or nucleic acids in body fluids) where cheap disposables are needed. Another example is polymer electronics like polymer circuits [1], displays based on organic light emitting diodes (OLEDs) or opto-(polymer)electronic sensors. Reel-to-reel fabrication is presently being used in a number of these applications and can be transferred to the manufacture of multi-functional flexible plastic microsystems.

2. Technologies for micro machining of foils

2.1. Roller hot embossing

Roller embossing of polymer foils in a reel-to-reel process is in principle a well known process. But up to now, only a few attempts are known where this technique has been exploited for applications in the field of micro system technology [2]. An industrial application of reel-to-reel hot embossing is the fabrication of polymer foils for car interior lining. For this application, the feature size is in the range of several 100 µm with rather low requirements for contour accuracy. For micro system applications the process of stamp hot embossing is well known. This process leads to very accurate replication even of structures in the sub-micron range. The most severe disadvantage of stamp hot embossing is the long cycle time associated with the heating and cooling phases which is in the range of 10 minutes.

Fig. 1. Experimental set-up for reel-to-reel hot embossing of foils.

For many microfluidic applications the feature size is in the range of 50-200 µm. An accurate replication in the sub-micron range is not needed. Here we report on attempts to use reel-to-reel hot embossing as a technique to fabricate microfluidic structures with cycle times known from conventional reel-to-reel embossing where a feed rate of several 10 meters per minute is a typical value.

First experiments have been performed with a modified commercial gravure printing machine. The print roller has been replaced by an embossing roller

314

and an infrared heater for heating up the polymer foils has been added. This experimental set-up is shown in Fig. 1.

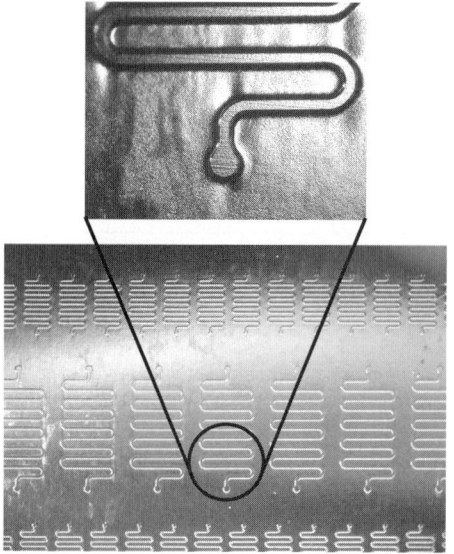

Fig. 2. Microfluidic structures on an embossing roller.

A commercial embossing roller with microfluidic channel structures as depicted in Fig. 2 has been used. The height of the structures is 80 µm. Materials like polycarbonate (PC), cyclic olefin copolymer (COC), poly(ethylene terephthalate) (PET) and polyvinyl chloride (PVC) have been used for the embossing of microfluidic channels. Among the tested materials, PC seemed to be the most tolerant to deviations from the optimal process parameters.

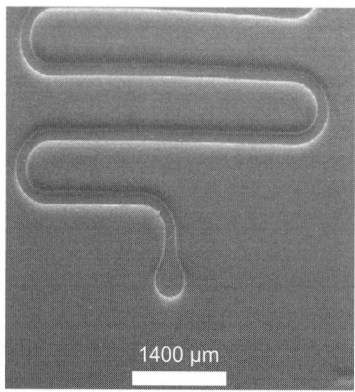

Fig. 3. Hot embossed microfluidic structure on a 500 µm thick foil of PC.

Fig. 3 shows the result of a roller embossing process on a 500 µm thick foil of PC. The result of the embossing process does not only depend on parameters like feed rate, foil temperature, embossing roller temperature and contact pressure but also on the hardness of the impression roller. First experiments showed a tendency that soft impression rollers lead to better results and especially to a better flatness of the embossed foils than hard ones.

It has been noticed that all fabricated micro channels show rounded edges. This seems to be a typical phenomenon for reel-to-reel hot embossing processes. We assume that this is due to the short time of intimate contact between the embossing roller and the foil substrate which is only a fraction of a second. Possibly, this time is not long enough to assure that the heated polymer can flow into the corners of structures on the embossing roller.

Another technique, ultraviolet (UV) embossing or "cold embossing", can also be applied to foils. Here, a substrate with a coating of UV-curable materials is pressed against a micro structured embossing mould, hardened under irradiation with UV and demolded. UV embossing offers several advantages. Firstly, UV curing is very rapid. Secondly, unlike thermal embossing, UV embossing is usually done at room temperature and low pressure, with little shrinkage of the pattern. It also allows the patterning of delicate substrates such as temperature sensitive materials particularly relevant for bio-MEMS. It may also be done on one or both sides of the substrate. Also, in contrast to photolithography, which is a batch process involving exposure and wet development in a clean room environment, UV embossing is used as a roll-to-roll process for the production of binary and continuous-relief microstructures for diffractive optical and micro-optical elements, for example for optical security features with structure size down to a few hundred nanometres [3].

2.2. Lamination

Generally, lamination of substrates is carried out either as direct sealing of polymer foils or by the use of an adhesive coating. Direct sealing can be used for some thermoplastic polymers like polyethylene (PE), PVC or PC. The process is normally done under high pressure and temperature e. g. for fabrication of smart cards. It uses closed hydraulic presses and it is a quite time consuming process due to the necessary heating and cooling times. A more rapid way to join polymer films is hot- roll lamination.

Fig. 4. Machine for laminating foils is a reel-to-reel process.

This method has a quite long tradition in electronics in applying dry photoresist films for fabrication of printed circuit board technology. Currently, this process is also used in an equivalent way to pattern metallised foils with a reel-to-reel photolithography process. Another application is the passivation of flexible systems. Widespread use of such a lamination process is made in the production of Smart Labels, where a cover foil is laminated to the substrate to protect the devices in a cost-effective way.

Typical processing speeds for these roll-based techniques are in the range of several m/min and are therefore rather limited in the appliance of temperature and pressure. Therefore reel-to-reel lamination uses foils which are coated with an adhesive. Normally, a hot-melting material is used to facilitate fast processing. Fig. 4 shows a machine for laminating two foils in a reel-to-reel process. One of the foils is coated with an adhesive.

2.3. Laser ablation

Compared to conventional methods, laser micromachining has many advantages as it is a precise, fast, flexible and contactless technique that can be used to structure a variety of materials. This technique is particularly well suited for rapid prototyping or for rapid fabrication of small and medium lot sizes. Direct ablation using UV lasers is particularly attractive for micromachining polymers, and is increasingly used in manufacturing microfluidic bulk substrates. Direct laser machining can produce microfluidic chips in their entirety or correspond to a step in a chain of hybrid fabrication. In particular, it can be used to generate the microfluidic structures such as channel network and reservoirs by direct laser ablation into the polymer foil or/and delineate microelectrodes in thin metal films within the microchannels, as well as etch through multilayer stacks of different materials for fluidic and electrical interconnection to define access holes and vias. In the case of polymeric foils, laser micromachining is well adapted to reel to reel production. A variety of resistive sensors as micro-heaters and thermo-elements on polyimide foils are presently being produced by this method [4].

2.4. Laser welding

Laser plastics welding has proven its reliability in numerous applications where rigid parts are welded together ([5], [6]). For the purpose of welding of foils the energy deposition can follow the transmission welding process, where the laser energy is irradiated through one of the welding partners and absorbed in the second layer. As most polymer materials are transmissive for wavelengths around 1 μm the lasers commonly used for this process are CO_2, diode and Nd:YAG lasers. The absorption in the second part is realised by means of additives. The principle of this technique is depicted in Fig. 5.

An example where this technique has been applied successfully is shown in Fig. 6. The upper part consists of unmodified PC. The lower part consists of PC with 0,1% carbon black in order to enhance light absorption. The resulting weld seam has a width of 34 μm. Alternatively, the precise energy deposition into the welding zone can be achieved by focussing the laser with great aperture angle onto the parting plane between the foils. Using a CO_2 laser will result in superficial heating of the first layer. The weld between the foils can be realized via heat conduction into the second part.

All of these methods are capable of producing miniature welding seams with widths in the range of micrometers to tenths of millimetres, depending on the quality of the beam source and the focussing optics.

Nevertheless, the welding speed can reach more than 100 mm/s depending on the system technology in use which opens up the possibility of high throughput production for cheap products, for instance reel-to-reel production.

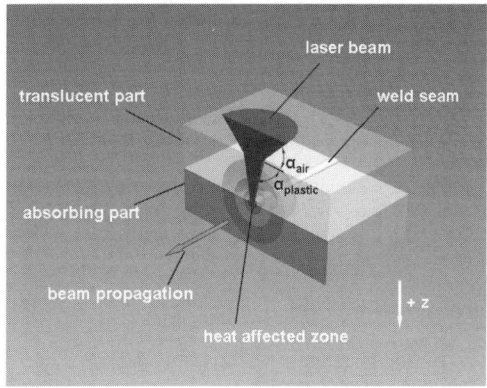

Fig. 5. Principle of laser penetration welding.

Establishing a weld between flexible materials issues new requirements for the clamping technology. The necessary force has to be exerted exactly at the welding zone because of the flexibility that prevents transmission of torsion moments. Therefore clamping elements have to be transmissive. Preliminary research activities have proven laser welding in combination with a non tactile clamping technique via air pressure to be apt even for joining high temperature materials like polyether-ether-ketone (PEEK).

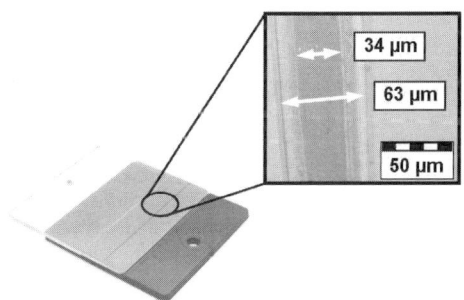

Fig. 6. Example for transmission welding of unmodified PC to PC with increased absorption. The zoomed area shows the weld seam which has a width of 34 μm.

To ensure that the plastic parts are not damaged thermally during the laser welding process heat generation and fluxes must be controlled. A thermal imaging device is used to watch the welding process and the resulting heat affected zone (see Fig. 7). In the case of miniaturized parts and welding geometries this insight into critical regions can be transferred into special process parameterisation and cooling devices [7]. By evaluating temperature gradients an overheating of critical geometries can be avoided. Furthermore, this knowledge is the base for increasing productivity by means of optimized energy deposition.

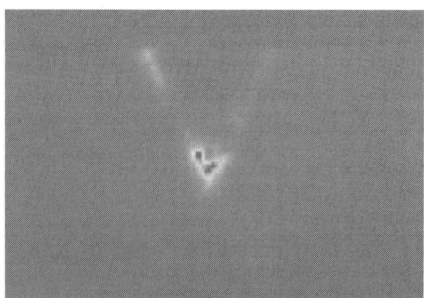

Fig. 7. Side view of the temperature distribution during welding. Welding is performed at the point of intersection of two laser beams.

2.5. Micro-thermoforming

Truckenmüller et al (2002) [8] developed a micro-thermoforming process for low-cost manufacturing of single-use, plastic hollow microfluidic structures, which combined in one step process thermoforming microchannels into a thin plastic and simultaneously welding to a sheet with fluidic ports. Short cycles in the order of seconds could be achieved as only a small temperature cycle was needed to heat up the two films. Demoulding of microstructures was also found to be much easier for micro-thermoforming than for injection moulding and hot embossing, because the welded films are very flexible. The long-term objective of their work was to produce microfluidic devices from two rolls of polymer materials so the final devices could be just cut off from a long band produced in a quasi-continuous process. Hermetically sealed capillary electrophoresis (CE) channels made in polystyrene (PS) with integrated electrodes were produced.

2.6. Screen-printing

Screen-printing is a commonly used technique in electronics assembly and printed circuit board technology which has also been applied to polymer electronics [1] for the gate electrode printing of PEDOT-PSS, for printing of passivation and protection layers based on dielectric polymer materials as well as for printing of the second interconnection layer based on conductive polymers filled with conductive particles, like silver flakes or nano silver particles. Further more, it can be used to print contact pads as well as electrodes for low cost sensors as can be seen in Fig. 8. Screen printing can be performed with resolutions typically at about 100 µm line space and with a registration of about 50 µm for adjusted printing to a given pattern. The resolution can be extended down to 20 µm line patterns with special screens with high-dense mesh structures and a special surface treatment. Faster screen-printing can be facilitated with rotary-screen printers. Such equipment is used for screen printing of radio frequency identification (RFID) antennas based on similar conductive filled polymers as described above. Screen printing spacer layers can also be used to define micro-fluid channels between two foil substrates. Usually the spacer layer is printed onto the first foil substrate. Depending on the viscosity of the materials used to form the spacer and the required depth of the micro-fluidic channels it is possible to print several times. A second foil is then laminated or glued on top of the screen-printed spacer. Even though

resolution and registration are limiting the dimensions of micro-fluidic channels, it is possible to fabricate channels down to 0.1 mm width. The channel height strongly depends on the viscosity and the number of printing steps. Typically the height is in the range of approx. 20 µm per step for high viscosity materials.

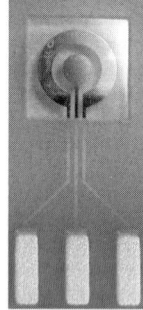

Fig. 8. a) Screen printed contacts on an electro-chemical foil sensor, b) screen printed inter-digital capacitor on foil, with screen printed contact pads. Both disposable sensors are fabricated on low-cost PET substrates.

3. Polymer opto-electronic sensors on foil

Opto-electronic polymers transfer light into electricity and vice versa. Applications, mostly in the development stage, are e.g. flexible displays based on organic light-emitting diodes (OLEDs) and photo-voltaic cells based on photo-diodes. These are large area applications, which are expected to be manufactured on foil by reel-to-reel technologies. This is of interest for micro-fluidic applications in foil, because OLEDs and photodiodes can also be used as optical sensors when a selective media is positioned in between [9].

Advantages of opto-electronic polymer sensors are the potential (i) to fine tune the spectrum of the LEDs or photodiodes, with regard to the sensor material, (ii) to manufacture these in high volumes with reel-to-reel (embossing, lithographic, jetting, printing and laminating) techniques and (iii) to integrate these in micro-fluidic applications in foil.

Fig. 9 depicts as example an alcohol sensor. The optical transmission between the LED and photodiode is affected by a (reversible) alcohol sensitive coating on the waveguide surface. A second uncoated waveguide is used for reference measurements.

This sensor was manufactured by two shot moulding of the polymer waveguides and substrate, selective metal plating of the substrate and thin film applied opto-elelectronic polymers. Waveguide material was COC, substrate material was LCP (Liquid Crystal Polymer) LDS (from Ticona) which was plated by laser direct structuring. The alcohol sensitive coating was organic solvent based and was applied by drop-casting.

The photodiode and LED devices were a stack of (i) a glass layer with a sputtered transparent conductive indium doped tin-oxide (ITO) film, structured by a photo-lithographic process, as anode, (ii) spin coated layers of a transparent conductive polymer (PEDOT) and the actual opto-electronic polymer (derivative of poly(phenylene vinylene)) and (iii) a thermally evaporated metal cathode of barium-aluminium alloy and (iv) an aluminium encapsulation.

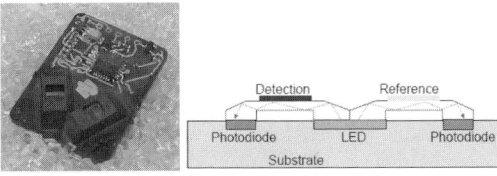

Fig. 9. Dual channel polymer waveguide alcohol sensor with organic opto-electronics. Left: photograph; right: schematic.

Similarly to displays, it is possible to make sensors with arrays of OLEDs or photodiodes. By varying the spectral bandwidth and intensity or sensitivity of the OLEDS or photodiodes and/or using additional layers with different filters or selective coatings, an optical micro-array sensor can be made for quantitative analysis of media with single or multiple targets, for example gasses, or (labelled) nucleic acids or proteins in a fluid.

The combination or integration of these sensors with or into micro-fluidic structures on foil could result in cheap disposable cartridges for drug, pathogen or chemical tests.

Many challenges remain though, such as the lateral shielding of light beams which may be scattered in fluidic or polymeric layers. For example by means of optical components (lenses) which need to be integrated in the pursued layer architecture. The micro fluidic channels are usually much deeper (tens of microns) than the structures needed for OLEDs and photo diodes (a few micron deep). Other challenges are to provide sufficient protection (for example by barrier layers) against oxidation of the opto-electronic polymers, to electrically connect the OLEDs and photo-diodes, to integrate (polymeric) electronics, to print or jet a large variety of sensor materials

4. Conclusions

Several technologies for micro patterning of foil based substrates are already available. The same applies to applications of foil-based micro systems. There are also first attempts to establish technologies which allow the micro patterning of polymer foils in a cost-effective reel-to-reel process. What is still missing is a complete process chain for the fabrication (e.g. via roller embossing, UV embossing or laser ablation) and covering (e.g. via lamination or laser welding) of microfluidic channels as well as the singularisation of microfluidic systems. For many applications, there is furthermore a need to realise electrical or optical components like electrodes or waveguides on the same substrate. Thus, it would be desirable to include such technologies into the complete reel-to-reel process chain.

All the above described technologies need to be harmonised with each other with regard to common feed rate, temperatures and substrate thickness, to name just a few. New reel-to-reel technologies need to be developed in order to be able to fabricate micro-electro-opto-fluidic systems on foils in a cost-effective way.

As a conclusion, it can be stated that there is a strong need for further research in the field of microfluidics on foil.

Acknowledgements

Thomas Velten would like to thank the Fraunhofer-Gesellschaft for financially supporting the reported research in the field of roller embossing.

Also, the writing of this paper was carried out within the framework of the EC Network of Excellence "Multi-Material Micro Manufacture: Technologies and Applications (4M)".

References

[1] Burghardt, M. ; Liemann, G. ; Klink, G.; and Bock, K. Evaluation of reel-to-reel processes for polymer electronics. Proc. 3rd Int. IEEE Conf. on Polymers and Adhesives in Microelectronics and Photonics, "Polytronic 2003", 21-23 Oct. Montreux, Switzerland, 2003, pp. 287-293.

[2] Michael Schuenemann et. al. Packaging of disposable chips for bioanalytical application. Electronic Components and Technology Conference, 2004, pp. 853-861.

[3] M. T. Gale. Replicated diffractive optics and micro-optics. Optics and Photonics News, 2003, pp. 24-29.

[4] D. Meier and J. Kickelhain. Sensor and sensor elements manufacturing: laser direct patterning (LDP) for reel to reel processing to generate high throughput. http://www.lpkf.com/applications/laser-micromachining/thin-layer-structuring/index.htm.

[5] Vetter, J. Neue Technologien zum Strahlungsschweißen von Kunststoffen. Tagungsband des 7.Erlanger Seminars LEF 2004, Meisenbach Verlag, 2004.

[6] Polster, S., Wiengarten, M. Neue Möglichkeiten beim Laserstrahlkunststoffschweißen. In Geiger, M.; Otto, A. Laser in der Elektronikproduktion & Feinwerktechnik Tagungsband des 6. Erlanger Seminars LEF 2003 Bamberg, 2003, pp. 99-115.

[7] J. Zettner, B. Spellenberg, Th. Hierl. Wärmefluss-Prüfung zur Qualitätssicherung von Schweißverbindungen In der Automobilindustrie, DGZfP-Thermografie-Kolloquium, Stuttgart, 2003.

[8] Truckenmüller, R.; Rummler, Z; Schaller, Th.and Schomburg, W. K. Low-cost thermoforming of microfluidic analysis chips. J. Micromech. Microeng. 12, 2002, pp. 375-379.

[9] Patent WO2005015173, H. Schoo et al., Sensor comprising polymeric components, 2005.

Application of SOI based sensors for MEMS wafer-level packaging

Farzan Alavian Ghavanini[1], Cristina Rusu[2], Katrin Persson[2], Peter Enoksson[1]

[1]Chalmers University of Technology, Solid State Electronics MC2, SE-41296 Göteborg, Sweden
[2]The Imego Institute. Arvid Hedvalls backe 4, SE-41133 Göteborg, Sweden

Abstract

A method to evaluate MEMS wafer-level packaging by using sensors based on Silicon-on-Insulator (SOI) is developed. The SOI-MEMS technology has been used to create arrays of micromechanical stress and pressure sensors in the device layer of an SOI wafer. These sensors arrays will be used to evaluate the encapsulation characteristics of wafer-level packaging, by revealing the bonding-induced stress distribution as well as bonding quality. Mechanical resonance frequency and Q-factor of the sensors have been derived theoretically and were compared to the results of the Finite Element Modelling (FEM). The pressure sensor is designed to give Q-factor above 1000 at the pressures below 1 mbar and static capacitance of few pF. The resonance frequency of the stress sensor is designed for ca. 45 MPa as estimated for anodic bonding. The design, simulation and fabrication of the sensors are presented in this paper.

Keywords: wafer level packaging, anodic bonding, resonators, stress sensor, pressure sensor, SOI

1. Introduction

Bonding is in general used to form a large number of Micro Electro Mechanical Systems (MEMS) structures like sensors / actuators [1]. Bonding can be also utilized for packaging and encapsulation on wafer scale (or zero-level); and at die-level, for instance by flip-chip bonding. Wafer level bonding allows a higher flexibility of the process than die-level because the environment can be controlled (pressure, gasses). Therefore, wafer-level packaging is an important step to obtain functional and reliable devices.

Temperature, electrical field or pressure are typically used in most of the bonding technologies such as silicon direct bonding (800-1100°C), anodic bonding (or electric field-assisted bonding 200-1000 V, 180-450°C), thermo compression bonding (300-400°C, 10-100 MPa) etc. Thus, the differences in Coefficient of Thermal Expansion (CTE) between the bonded wafers introduce stress in the final structure.

In order to investigate the characteristics of packaged MEMS and its influence on the device functionality, two important parameters of the bonding process need to be characterised: the distribution of the stress induced on the silicon wafer caused by the bonding process (for example, to a cap wafer) and the bonding quality (hermeticity).

We have chosen resonating cantilevers (clamped-free beam) and bridges (clamped-clamped beam) to detect pressure and stress, respectively, because of their relatively simple fabrication process, well-established theoretical models, and their high-resolution output. Mechanical resonance frequency and Q-factor of the sensors have been derived theoretically and were compared to the results of the Finite Element Modelling (FEM) using FEMLAB software.

Silicon bulk micromachining methods have been used to form an array of micromechanical pressure and stress sensors in the SOI wafer's device layer. This array of sensors will be used to evaluate the encapsulation characteristics of the whole wafer by revealing the bonding-induced stress distribution as well as the hermeticity. The sensors' characterization before bonding is performed by electrostatic excitation and capacitive detection using the Third-Harmonic Method [2].

The design, simulation, and fabrication of the sensors are presented in this paper.

2. Design

The pressure sensor (by measuring the pressure dependent Q-factor) designed in this work is a clamped-free cantilever connected to a perforated plate with large surface area (fig. 1a and fig. 2) for achieving high detecting capacitance. Having the seismic plate perforated provide us decreased squeeze film gas damping as well. It also makes the sacrificial etching easier during the fabrication. The sensor is

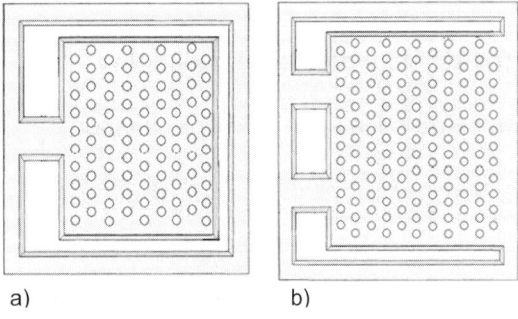

a)   b)

Fig.1: Sensors have perforated plates to reduce squeeze damping and making under-etch easier (orifices are not to scale). a) Clamped-free beam. b) Clamped-clamped bridge.

designed to give Q-factor above 1000 at the pressures below 1 mbar and static capacitance of few pF. Quality factor (amplitude) of the vibration which encodes the pressure was found theoretically by solving the structural mechanic equations [3]. The effect of orifices in the perforated plate [4] combined with the modified Reynolds equations describing the squeeze film damping effect [5] of the trapped gas between the vibrating plate and the substrate was also taken into account.

The resonance frequency of the clamped-clamped bridge (fig. 1b) depends on the stress induced to the sensor. The stress estimation, for example, for an anodic bonding to LTCC at 420C is ca. 45 MPa. The same structural mechanic equations can be used for the stress sensor but with different boundary conditions that satisfy the clamped-clamped situation. In addition to the equations used for the pressure sensor, one must take in consideration the buckling phenomena that might happen in the case of compressive stress.

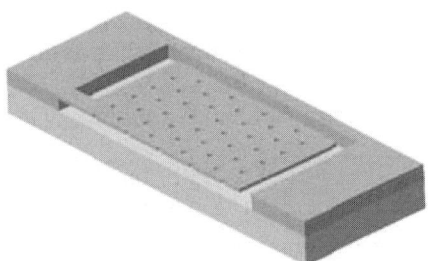

Fig.2: The pressure sensor cross section.

4. Simulation

A finite element model using FEMLAB multiphysics environment, based on its MEMS solid stress-strain module has been built.

Squeeze-film damping effect was added to the model by applying modified Reynolds equation as the boundary condition to the surfaces that are subject to gas damping [6] (fig. 3). Next to that, the model was revised to take into account the effect of orifices in the perforated plate [5]. Time dependent analysis was chosen to solve the model for quality factor and resonance frequency of the sensors. Q-factor of the pressure sensor was calculated from the damping ratio which itself, was found indirectly from time-varying output graph at different pressures.

A set of different axial forces were applied to the model to find corresponding shifted resonance frequency in the stress sensor. Additionally, as the compressive stress was the case in our study, numerical buckling analysis was performed to avoid entering the buckling region (fig. 4).

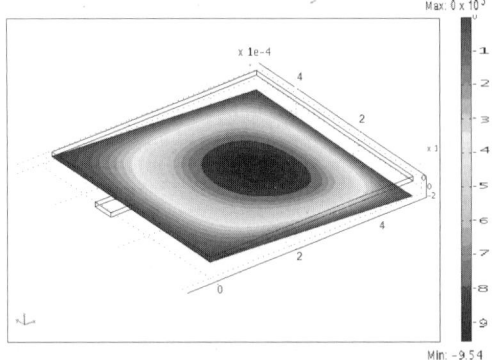

Fig.3: Pressure gradient under the resonating plate of the pressure sensor cause by squeezed film damping (FEMLAB simulation).

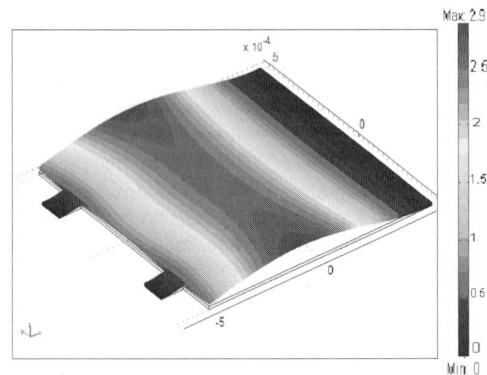

Fig.4: Stress distribution in the stress sensor clamped at both ends under an axial force (FEMLAB simulation).

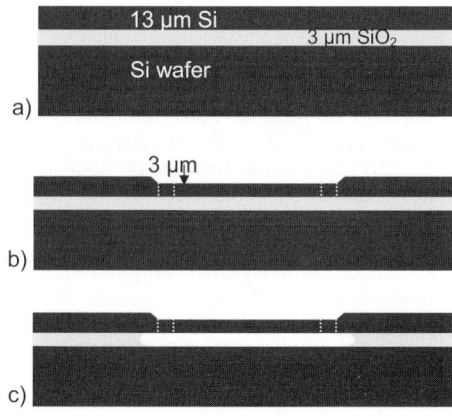

Fig. 5. Sketch of sensor processing:
a) start SOI with 13µm device layer and 3µm thermal SiO$_2$;
b) lithography and 3µm Si wet etch by KOH for electrode gap; lithography and Si dry etch for perforation in the plate;
c) release of resonating structures by HF etch of thermal SiO$_2$.

5. Fabrication and measurements

The designed sensors were fabricated using bulk micromachining processes by two lithography masks on 4-inch SOI wafers having 13µm of device layer and 3µm of buried thermal silicon dioxide layer (fig. 5a). Each sensor was created inside a 3um shallow trench to allow further bonding to encapsulating wafers (fig. 5b). Then, the resonating plates were released by removing the underlying oxide layer in buffered HF solution (fig. 5c).

The last step is the most critical since capillary forces of rising liquid causes "pull-down" of the microstructures during drying step (fig. 6a). The microstructures may remain stuck to substrate even after dry. The cause can be solid bridging, van der Waals force, electrostatic force, hydrogen bonds, etc. By using isopropanol alcohol direct after HF solution instead of water rinsing, we could increase the yield dramatically (fig. 6b). SEM photos of final pressure- and stress sensors are shown in fig. 7.

Having the sensors built on SOI wafers enables us to characterize them even, before bonding to any cap wafer containing excitation / detection electrodes. In this case, the substrate layer can act as excitation / detection electrode if a 'common electrode' measurement method such as 'Burst Mode' [7] or 'Third-Harmonic Method' [2] is used. In this experiment, we have adopted the second method, mostly because of its simple and fast implementation.

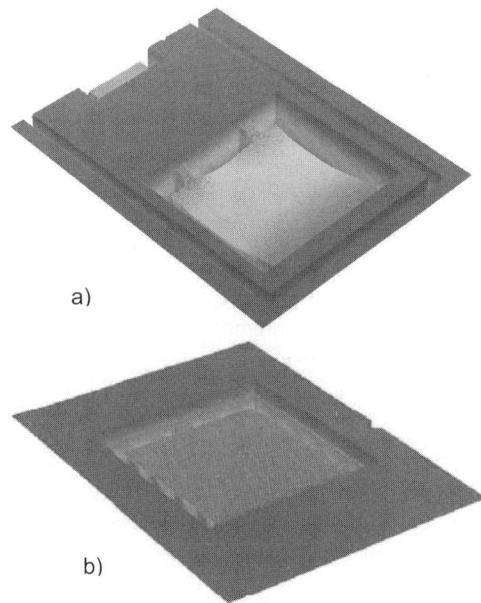

Fig.6: Sensors' structure after release a) stuck to the substrate; b) freestanding (WYKO optical profiler).

Figure 7. SEM photo of a) pressure sensors with 12 µm perforation, b) stress sensor with 6 µm perforation.

6. Conclusion and outlook

SOI based stress and pressure sensors for investigating the quality of MEMS wafer-level packaging have been designed, simulated and fabricated. The preliminary measurements on the pressure sensor characterization were performed using substrate layer as excitation / detection electrode. Further work will continue with the bonding of these SOI wafers to different encapsulating materials such as Si, glass, LTCC etc. A wafer map is then obtained from the stress and pressure measurements over the whole wafer.

References

[1] P. Enoksson, C. Rusu, A. Sanz-Velasco, M. Bring, A. Nafari, S. Bengtsson, Wafer bonding for MEMS, *Proc. 8th International Symposium On Semiconductor Wafer Bonding Science & Technology Applications*, 207th Electrochemical Society Meeting, 15-20 May 2005, Quebec City, Canada.

[2] Farzan Alavian, Henrik Rödjegård, Peter Enoksson, An-easy-to-implement method for evaluation of capacitive resonant sensors, *Proc. 16th MicroMechanics Europe Workshop 2005*, p 276-279, 2005, Sweden.

[3] M.-H Bao, Handbook of Sensors and Actuators-8; Micromechanical Transducers, *ELSEVIER Press*, 2000.

[4] D. Homentcovschi, R.N. Miles, Viscous damping of perforated planar micromechanical structures, *Sensors and Actuators A*, **119** (2005), p. 544-552.

[5] Minhang Bao et al., Modified Reynolds' equation and analytical analysis of squeeze-film air damping of perforated structures, *J. Micromech. Microeng.* **13** (2003), p.795-800.

[6] COMSOL FEMLAB documentation, "Squeezed film gas damping in an accelerometer".

[7] T. Corman, P. Enoksson et al., Burst technology with feedback-loop control for capacitive detection and electrostatic excitation of resonant silicon sensors, *IEEE trans. on electron devices*, **47**(11), 2000.

Multi-Material Micro Manufacture
W. Menz, S. Dimov and B. Fillon (Eds.)
© 2006 Elsevier Ltd. All rights reserved

Protective coating of zinc and zinc alloys for industrial applications

Y. Toshev[a], V. Mandova[b], N. Boshkov[a], D. Stoychev[a], P. Petrov[a], N. Tsvetkova[a], G. Raichevski[a], Ch. Tsvetanov[a], A. Gabev[c], R. Velev[b], K. Kostadinov[a]

[a] *Bulgarian Academy of Sciences(BAS) - Institutes of Mechanics, Physical Chemistry and Polymers;*
[b] *TDP Group Ltd, Plovdiv,*   [c] *Gabex Ltd, Sofia.*

Abstract

Zinc and zinc-based electrogalvanized coatings have attracted remarkable interest because of the increasing demand for layers with better corrosion resistance. A new approach for improvement of the protective ability is the incorporation in the metal matrix of polymeric nanoparticles. The latter affect positively the protective properties of the nanocomposite coatings and result in increasing of their corrosion resistance in 5 % NaCl solution compared to the pure zinc and Zn-Co.

The zinc coating process using the developed saturating mixtures excludes the phases of evaporation and condensation and the process is defined as hard-phase zinc coating. The hard phase zinc coating decreases the coating time 1, 5-2 times and powder dissipation with 20-25% compared to vapor-phase zinc coating. It does not require cooling the product together with the powder at the end of the process. The hard phase zinc coating has a high degree of automation and provides uniform coating of long-sized products and allows creating high-performing equipment. The prevention of the superficial roughness, during the hard phase zinc coating, leads to higher adhesion of varnishing coverings on the zinc-coated surface in comparison with vapor-phase zinc-coating.

Keywords: RAS activities, zinc coating, corrosion protection, nano composite layers

1. Introduction

TDP Group Ltd. and Gabex Ltd. are technology oriented companies. Their mission is to produce zinc coatings on ferruginous products for corrosion preventio utilizing thermo diffusion or electrodepositing methods.

The most recent TDP Group enquiry was for methodological and examination support to create a more active zinc powder, capable of working at higher temperatures when the zinc is steamed away.

BAS team has consulted TDP Group Ltd. about:
- State of the art for methods and technologies for activation of zinc powder with a size less than 100 microns;
- Appropriate technologies/methods for increasing the active phase of zinc powders;
- Methods for increasing the temperature of the zinc powder when it is steamed away and related appropriate technological means;

Experimental investigations are performed upon the elucidation of the technological physic-chemical characteristics of the zinc layer deposition. Analysis of some probes (Fig.1.) not covering the quality requirements leads to following conclusions:

a. Furnace hermeticness is very important improving the quality of zinc coating. At low hermeticness the zinc in the surface coating is less (49% - in the case of attached data) than necessary (min 75%) (Table 1);

b. The structure of deposited zinc is in the phase, which makes the coated iron surface accessible for the environment influence.

c. Carbon is discovered in some probes which change the technology regime - the structure of the deposited zinc and its activity. The reasons for that could be fall into the furnace lubricants from the details production technology and wood shavings used for their cleaning out.

The "Gabex Co" has worked in the last 10 years almost exclusively in the field of creating of special chemical products for the plating industry, including galvanizing techniques as well as the prevention of metals and metal alloys against corrosion. BAS has consulted "Gabex Co" about:

- chemical methods to obtain and characterize (SEM and light scattering) nano particles of co-polymers with sizes of 50-200 nm.

- adoption of the preparation (process for producing) stable electrolytes containing some various concentrations of the nano particles;

- optimizing the electrolysis technological regimes (cathode current density, duration of electrolysis, mixing etc.) for the electro deposition of qualitative bright nano composite zinc coating with improved protection capability;

- methods and electrolytes with low concentration of Cr for chemical passivity (chromating) of the coating with the aim of additional increase of their protection.

Table 1. XPS investigations of the element distribution (in at %) in the surface layers of deposited Zn coating at low furnace hermeticness

File 1	Y22669		Al Kalpha-1486.6 eV		
	C1s	O1s	Fe2p	Zn2p	Cl2p
A (cps*eV)	5835,28	21499,59	87413,83	76125,48	2872,82
at%	3	11,1	45,11	39,29	1,5
E(ev)	285	~529	710,7	1021,8	198,8
			D~13,4	D~23,1	D=1.60

2. Hard phase zinc coating

Three methods for diffusion zinc coating are distinguished: from vapor, gas and liquid phase [1].

When zinc coating is in vapor phase the saturation is done in powdered mixtures, consisting of zinc powder or zinc powder with addition of zinc chloride or ammonium chloride, or inert materials (oxides of Al, fire

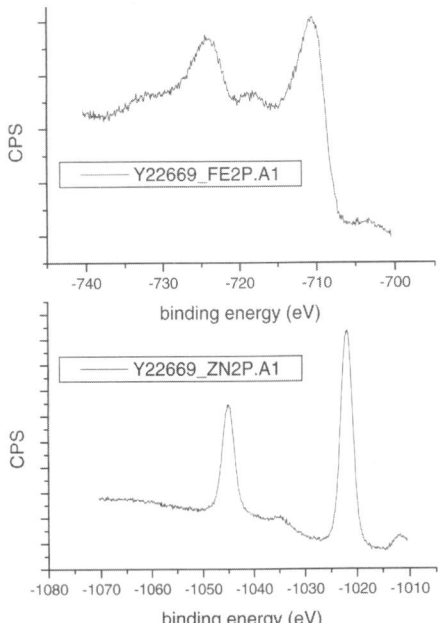

Fig.1. XPS Spectra of 2 Fe2p and Zn2p from the specimen of coated with Zn steel substrates at low furnace hermeticness

resistant clay, quartz sand). This method can be realized using contact and contactless technology. Following the contact technology zinc evaporation is done in the space of the reaction near the place where the zinc powder touches the surface to be coated. The contactless technology presumes zinc evaporation to be done at a certain distance away from the surface to be coated. Presently in manufacture is spread mainly the contact technology for diffusion zinc coating using supplementary additions to the main mixture, consisting of zinc powder and inert materials. There are mixtures for one and multiple time use with different methods of repeated application of the mixture. The basic disadvantage of the vapor-phase zinc coating is the irregular coating of the product (especially large-sized products and little articles in a large coating space), as well as the process instability (difficulty in controlling the coating thickness with high precision) seriously harden the selling of products worked in this technology.

According to us, the enumerated disadvantages of the vapor-phase zinc coating are due to the technology of forming the coverings using the given method and the complexity of regulating the technological parameters during the process.

It is known that the coating forming, using the phase method, consists of the following steps vapor[1]:

a. Zinc evaporation forms the surface of the particles in the powder.

b. Condensation of zinc vapors on the product surface.

c. Mutual diffusion between the zinc and product material.

In general the speeds of evaporation (I_v) and condensation (I_c) are determined by the convergent equations of Herz-Knudsen:

$$I_v = \alpha_v (P_e - P)/(2\pi mkT)^{1/2}$$

where: P_e – pressure of the saturated zinc vapors, P – pressure of the zinc vapors in the system, α_v – coefficient of evaporation, m is the mass of the gaseous atom or molecule, T the temperature and k the Boltzmann constant.

$$I_c = \alpha_c (P - P_e)/(2\pi mkT)^{1/2}$$

where α_c – coefficient of condensation.

The pressure of the saturated zinc vapors grows logarithmically with the rise of temperature and, as a result, we observe growth in the speed of evaporation with the rise of temperature. It must be noticed that evaporation takes place when the zinc vapors pressure is lower at the process temperature that the saturated zinc vapors pressure at the given temperature, and just the opposite – condensation takes place when the zinc vapors pressure in the system is higher than the saturated zinc vapors pressure for the temperature reached at the product surface. The rise of temperature on the product surface troubles the process of condensation, for it requires rise of the zinc vapors pressure over the saturated zinc vapors pressure at that temperature in order to have condensation. Hence, the process may be stable if it is possible to maintain the necessary temperature gradient between the zinc powder and the product. In this case the powders have a higher temperature and there goes intensive evaporation, and the temperature on the product surface is lower, and the actual pressure of the zinc vapors on the surface turns out to be higher than the pressure of the saturated vapors for the temperature of the products surface. However following the contact technology it is impossible to keep even the sign of the temperature gradient during the whole process.

Relatively stable results on thermo-diffusion zinc coating are obtained when coating products fully sunk in powder at static conditions. Here, when heating the product, the saturating powder heats earlier and creates enough stable temperature gradient. This way we obtain stable thermo-diffusion zinc coating. This technology was used for pipe zinc coating in the Nikopolsk and Rustav (Russia) metallurgical factories.

When rotating products while coating and powder and product temperature are almost the same the only thermodynamic agent stimulating zinc transfer in the form of vapor from the particles of the powder to the product surface is the difference in the pressure of the saturated zinc vapors over the convex surface of the particles and the flat product surface. In practice this seems to be insufficient and we observe the dark surface of the product saying that the process flows with insufficient transfer of zinc to the surface for coating. This is proved by the presence of considerable amounts of iron on the surface of such a covering and covering consisting mainly of G-phase.

It must be noticed as well, that the pressure of saturated vapors on the concave surface is lower than the pressure of saturated vapors on the convex surface at the given temperature. As a result hollows on the surface of the product are coated more intensively, what also makes it difficult to coat mounting products with small thread.

The evaporation coefficient α_v is of the great importance for the stability of the saturating capability of mixtures. The coefficient of evaporation of real zinc surfaces in dependence on the state of the surface oxide layer may vary from 10^{-6} to 10[1-2]. In this sense, the single-time-use saturating mixtures, used from certain manufacturers, are interesting. It seems, that in those mixtures, during the process of heating the surface of the particles in the powder is completely covered with a layer of passive chemical compounds what results in the future unusability of the powder.

The company "TDP GROUP" Ltd. produces saturating mixtures, consisting of zinc powder whose particles are covered with a thin layer of zinc oxide. This results in evaporating coefficient of the particles of the mixture equal to 0,001 of the theoretical evaporation speed at temperature 380 - 450°C. But the porous structure of the oxide layer and the considerable excess in the overall surface of the powder particles over the surface to be coated determines the sufficient transfer of zinc via superficial zinc diffusion from the top of the particles during the time of contact. As a result a high-quality δ-phase coating is obtained. This zinc coating can be done at high speed at enough low temperatures of 380-400°C.

In such a way, using the saturating mixtures of the "TDP GROUP" Ltd. company, the only method for forming the thermo-diffusion zinc coating is the diffusion of zinc from the surface of the particles to the surface of the product and the process control is realized with a few parameters (temperature and duration of the contact between the saturating mixture and the surface of the product). The exclusion of the hard controlled process evaporation and condensation from the manufacturing technology allows augmenting the level of the process automation of thermo-diffusion zinc coating in powdered contact mixtures and defining the process as hard phase zinc coating.

Thermo-diffusion zinc coating–basic differences in the technology when using different saturating mixtures:

- Saturating mixture from "TDP Group" Ltd TU 1721-001-5143849-2001 is a homogenous mixture without supplementary activators and additives. All the other mixtures are multi-component and they cannot guarantee constant quality of the reacting materials. Thanks to special processing the zinc powder becomes hard flammable and pollutes the environment less.

- The consumption of saturating mixture for coating 1 ton of product does not exceed 25 kg for coating thickness of 40 μm, and for all other mixtures the quantity is over 30 kg.

- "TDP Group"'s mixture can be used many times, while all others – only once.

- The coating thickness can be controlled with the process duration time and with changing the mixture quantity.

- Coating time – 35-60 minutes. Furnace cooling is not required after the end of the process. All other coatings are obtained for a time more than 2,5 – 3 hours and they require furnace cooling to room temperature.

- The main difference is in the product appearance – "TDP Group"s mixture creates bright-colored parts with excellent decorative quality, contrary to the gray-color obtained using the other mixtures.

- The coating obtained with the mixture TU 1721-001-5143849-2001 does not require post-processing (passivation) and the expensive devices thereby.

- Metallographic studies have proved that the coating obtained using the mixture of "TDP Group" LTD, are more uniform and without cracks, typical for coatings obtained with other mixtures.

Hence, the usage of the saturating mixture TU 1721-001-5143849-2001, produced by "TDP Group" LTD leads to decrease of the expenses for zinc coating and increase of the quality of the coating.

Although the process of obtaining coatings by precipitation from vapor-phase using contact technology seems to be simple, its regulation and control in order to obtain uniform and stable coating is hard enough. The main steps when forming coatings via vapor-phase technology are [2]:

1. Zinc evaporation
2. Condensation of zinc vapors
3. Zinc diffusion in the product material.

The used contemporary technologies and equipment do not offer the possibility of reliable control of the processes evaporation and condensation of zinc on the surface of the product.

The zinc coating process using the saturating mixtures of "TDP GROUP" LTD excludes the phases of evaporation and condensation and the process is defined as hard-phase zinc coating.

3. Electrodeposition, corrosion behavior and surface morphology of zinc and zinc-cobalt nanocomposite coatings

Four main types of coatings are investigated:

- Zn galvanic coatings electrodeposited from a starting electrolyte (SE) containing $ZnSO_4.7H_2O$ (175,0 g/l); $(NH_4)_2SO_4$ (25,0 g/l); H_3BO_3 (30,0 g/l) and additives AZ1 and AZ2 [3].

- Zn-Co(~ 3 wt%) galvanic alloy coatings from a slightly acidic electrolyte (SE-1) containing $ZnSO_4.7H_2O$ (100,0 g/l); $CoSO_4.7H_2O$ (120,.0 g/l); NH_4Cl (30,0 g/l); H_3BO_3 (25,0 g/l) and additives ZC1 and ZC2 [4].

- Zn composite galvanic coatings, electrodeposited from a SE, with an addition of 1 g/l of stabilized polymeric micelles from commercially available [poly (ethylene oxide)-b-poly (propylene oxide)-b-poly (ethylene oxide)] - PEO-PPO-PEO - three block copolymers;

- Zn-Co composite galvanic alloy coatings, electrodeposited from a SE-1, with an addition of 1 g/l of stabilized polymeric micelles from commercially available [poly (ethylene oxide)-b-poly (propylene oxide)-b-poly (ethylene oxide)] - PEO-PPO-PEO – three block copolymers.

Electrodeposition conditions are cathode current density 2 A/dm^2, temperature of the electrolyte 22° C, zinc anodes. The samples of each type of coatings could be additionally treated by immersing in a solution for iridescent yellow chromating film [5] commonly used for pure zinc coatings - chromating time 20 s.

5. Corrosion behavior of electro-deposited Zn and Zn-Co coatings

Polarization resistance (Rp) measurements are used to determine the protective ability of electro-deposited coatings since the registered Rp values are contrariwise proportional to the corrosion current (higher polarization resistance means lower corrosion current). Polarization resistance is defined as the resistance of the specimen to oxidation during the application of an external potential. The corrosion rate is directly related to the Rp and can be calculated from it. The results obtained are presented in Fig. 2 [6].

The electrodeposited Zn and Zn-Co composite coatings (samples 2) demonstrate close, although lower Rp values compared to the pure galvanic coatings – samples 1. The chromate Zn composite represents the highest Rp value and protective ability. As expected, the chromate metal coatings provide better protective ability against corrosion (higher Rp values) compared to the non-chromate ones. During the chromating process the co-polymer particles are

326

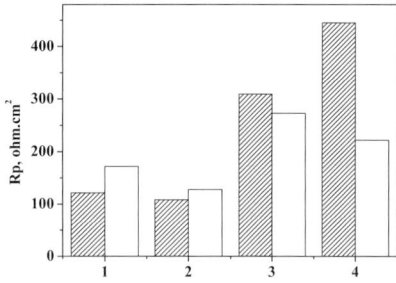

Fig.2. Rp values of Zn-based electrodeposited coatings
(hatched areas–Zn; white–Zn-Co):
1 – non-chromate coatings; 2 – nanocomposite coatings; 3 –
chromate coatings; 4 – chromate nanocomposite coatings.

included in the structure of the passive film causing formation of additional micro-cracks (Fig.3a & 3b). They form local places where the metal destroys more rapidly and on these areas more corrosion products with lower product of solubility appear that ensures higher protective ability. In the nanoparticles form a barrier layer that impedes the case of non-chromated samples the polymeric penetration of the corrosion medium in the depth of the layer (Fig 4a). In some cases they tear off the surface forming holes with different sizes and causing local decrease of the current density following by filling of the holes with corrosion products thus ensuring protecting barrier (Fig.4b). Principally, this process needs more time and could be observed after prolonged period.

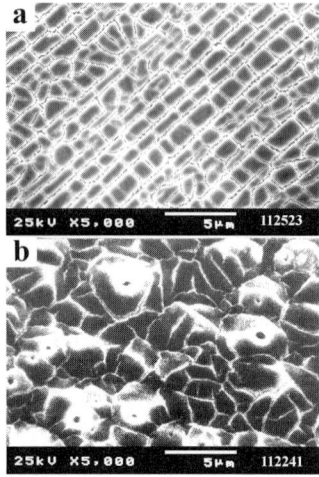

Fig.3. SEM microphotographs of:
a. chromated Zn; b. – chromated composite Zn.

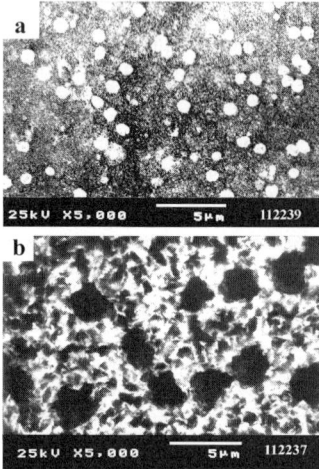

Fig.4. SEM microphotographs of: a. chromated Zn;
b. composite Zn in 24 h immersion in 5% NaCl solution

6. Conclusions

The hard phase zinc coating decreases the coating time 1,5-2 times and powder dissipation with 20-25% compared to vapor-phase zinc coating. It does not require cooling the product together with the powder at the end of the process. The hard phase zinc coating has a high level of automation and provides uniform coating of long-sized products and allows creating high-performing equipment. The prevention of the superficial roughness, during the hard phase zinc coating, leads to higher adhesion of varnishing coverings on the zinc-coated surface in comparison with vapor-phase zinc-coating.

Electrodeposited zinc layers are widely used in practice to protect steel details. The traditional method to improve the corrosion resistance of these coatings is their treatment in chromate or phosphate solutions. A new tendency is the usage of electrochemical nanotechnologies to obtain composite layers with incorporated nano particles of copolymers. Their presence in the metal matrix spread the corrosion processes on a greater surface and this holds up the destruction in the depth of the coating.

The results reported in this investigation confirm the fact that the electrodeposited Zn-Co coatings present higher corrosion resistance (higher Rp values) compared to the pure Zn coatings. A difference exists when the samples are chromated because the chromate Zn-Co shows worse characteristics. Including of organic polymer nano particles in the metal matrix

decreases the Rp values of Zn and Zn-Co coatings, although this parameter increases strongly when the samples are chromated, especially for Zn composite. Another interesting peculiarity is the assumption about the existing of local stresses in the surrounding zones of the included particles in the metal.

Generally, it was established that the incorporation of polymeric nano particles into the metal matrices increases their protective ability, especially after additional passivation treatment.

Acknowledgements

We are very grateful to the contribution of Bulgarian Innovation Fund under the TD-Zinc Project Nr.111 by and 4M NoE No. 500274 of NMP2-CT-2004 for the RAS activities.

References

1. Renshaw W., R.Chen (1997), Thermodynamics of zinc evaporation/condensation/oxidation in the snout, BHP Coated Steel Research Laboratories (internal report)
2. Williams J.J., Surface reactions of zinc vapour with steel relevant to the Zn-55%Al-1,5%Si hot dip metal coating process, PhD thesis, 2005, University of Wollongong
3. Bulgarian Patent Nr. 54483(1981)
4. Boshkov N., K. Petrov, S. Vitkova, S. Nemska, G. Raichevski, Surface and Coating Technology, **157** (2002), 171-178.
5. Bulgarian Pattent Nr. 44463 (1979)
6. Koleva D. N. Boshkov, G. Raichevski, L. Veleva, Transactions of IMF, **83**, Nr.4 (2005) 188-193.

Multi-Material Micro Manufacture
W. Menz, S. Dimov and B. Fillon (Eds.)

A comparative study of three technologies for producing castings with micro/meso-scale features

J-F. Charmeux[a], R. Minev[a], S. Dimov[a], E. Minev[a] and U. Harrysson[b]

[a]Manufacturing Engineering Center, Cardiff University, Cardiff, CF24 3AA, UK
[b]Fcubic, Kallarlyckevagen 6, 42935 Kullavik, Sweden

Abstract

The paper investigates the capabilities of three different process chains for vacuum investment casting of parts with micro/meso-scale features. In particular, the capabilities of two layer-based manufacturing technologies, ThermoJet and PatternMaster, and a new direct shell printing technology developed by Fcubic are studied. The first two technologies create patterns out of a thermoplastic material that are suitable for casting parts utilising the classical two-stage lost-wax process while the 'Fcubic' process produces directly a casting tree in zirconia ceramics. The tests were carried out on a gravity casting machine on which overpressure/vacuum could be applied to facilitate the replication of components with micro/meso-scale features.

Keywords: investment casting, 3D printing, metal micro-components.

1. Introduction

Vacuum Investment Casting (VIC) is a promising technology for producing high quality metallic components within close dimensional tolerances. The constant evolution and improvement of the Rapid Prototyping (RP) pattern building techniques broadens their application area and makes possible their use for precision investment casting of components incorporating micro features. So far, micro-gears, micro-turbine parts, and cast metal fibres with a diameter of 100μm and maximum aspect ratio of 90 have been successfully produced using Gold and Palladium based alloys [1-2]. Lost patterns used for such castings were produced by applying micro-manufacturing technologies, in particular LIGA, laser ablation, and injection moulding. Because VIC is a multi stage process, the production of high quality castings requires the sources of errors associated with each individual step to be studied and their effects reduced. The lost pattern accuracy is dictated by the technological capabilities of the RP process. Studies comparing the accuracy of different RP pattern-making processes were reported, however they did not consider components with micro/meso-scale features [3].

The aim of this research was to perform a benchmarking between three different VIC process chains for producing castings with micro/meso-scale features. In particular, the capabilities of two layer-based manufacturing technologies, ThermoJet and PatternMaster, and a new direct shell printing technology developed by Fcubic [4] were studied.

2. Experimental Methodology

2.1 Materials

The lost patterns were built in thermoplastic materials using two RP machines, ThermoJet (ThJ) and SPI PatterMaster (PM). Castings were produced using Aluminium LM25 and Zinc ZA12 alloys in a VIC unit (MCP equipment) applying vacuum and overpressure. The investment material used to fabricate the "casting trees" was a gypsum-bonded silica. The ceramic direct shell material used in the Fcubic process was a mixture of zirconia standard powder with particles sizes from 20 to 70 μm and 8% fine grained ZrO_2 infiltrated with $ZrYS_4$ after a heat treatment at 900°C.

2.2 Design of patterns

The test part "Pyramid" (Fig.1), was a structure of 10 steps with dimensions from 11.5x11.5 mm downscaled to 300x300 μm by a factor of 1.5. This unconstrained geometry allowed the metallic part to shrink freely. The pyramids were used to perform an assessment of:

- Geometrical accuracy;
- Positional accuracy;
- Surface roughness.

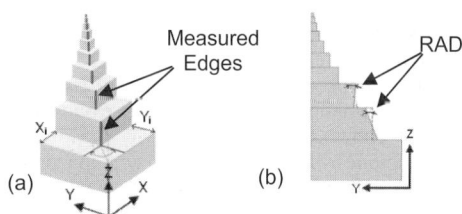

Fig. 1 Test part; a) X_i and Y_i of measured edges;
b) RAD between two successive edges in YZ.

Ten castings, five in LM25 and five in ZA12, were produced by applying each of the three studied process chains, 30 castings in total.

2.3 Measurement procedure

The analysis of dimensional accuracy was performed on an optical measurement system, Mitutoyo Quick Vision Excel. Each pyramid was analysed at the pattern stage and after casting so that any variation due to the VIC multi-step process could be investigated. The measurements of the pyramids were carried out in two planes, ZX and ZY. In particular, the distance between the edges of two consecutive steps, X_i and Y_i, in XZ and YZ was measured as shown in Figure 1. The nominal dimensions were: 3.83, 2.56, 1.7, 1.14, 0.76, 0.5, 0.34, 0.22, and 0.15 mm. Edges were approximated with a line constructed by measuring 10 points on each of them. The data collected from the samples in their pre- and post- casting state were first compared and then statistically analysed by applying a paired t-test procedure. In this way the process chains' consistency in terms of accuracy and variability with level of significance $\alpha=0.05$ was assessed. The capabilities of three processes were finally compared using a performance index (PI) over all dimensional ranges. The index represents each specified process indicator (e.g. roughness) as a percentage of the maximum values achieved for each of them during the experiment.

X_i and Y_i for every two consecutive edges were used to evaluate the linear dimensional accuracy of the pyramid steps. The measurement data for each step, nine data sets in total, was processed separately and then their arithmetic means and standard deviations (SD) compared. Similarly, the relative angular deviation (RAD) between every two consecutive edges represented by their respective approximation lines was analysed as shown in Fig. 1.

In addition, the spreading of the available data points around the theoretical line representing each edge of the pyramid was determined. Then, the diameter of the smallest cylinder that bounded these data points was estimated. This diameter was used as an indication for the edge straightness.

Finally, roughness tests were performed using the Mitutoyo SJ-400 tester. Measurements were taken along X and Y, X and Z, and Y and Z in planes XY (Top surface), XZ ,and YZ, respectively.

3 Results and discussion

3.1 Visual inspection of castings

All castings produced from ThJ and PM were successful. However, in the case of the Fcubic process chain, only four pyramids were successfully cast while the others exhibited missing steps. Studying the failed samples, it seemed that the molten alloy agglomerated with the mould material and thus created a coarse metallic structure surrounding the pyramid. Different factors could have contributed to the incompleteness of some Fcubic castings:

- The poor accessibility to the pyramid tip cavity from the bottom open shell during the cleaning and inspection stages.
- A difference in the air permeability of the zirconia shells compared with the gypsum-bonded silica investment that could affect the effectiveness of the flask air evacuation under casting conditions. The trapped air can create a back-pressure effect and thus prevent the molten alloy from filling the smallest features of the moulds.

3.2 Analysis of Linear Dimensions (LD)

Because of the similar behaviour of both alloys, only the results for Zn were displayed on the charts. The deviations from the nominal dimensions in percentage were plotted as a function of the step size. Fig. 2 displays the results from the carried out measurements in the YZ plane of the ThJ and PM patterns and also for the castings produced by all three process chains.

The following observations could be made by analysing the charts:

- There is a close similarity of the charts' profiles for the pre and post casting states.

- The deviations of the dimensions that are smaller than 340 µm are more than 10% while for the bigger steps of the pyramid they are within 3-5%.

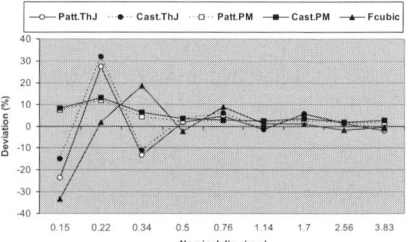

Fig. 2. Deviation in percent.

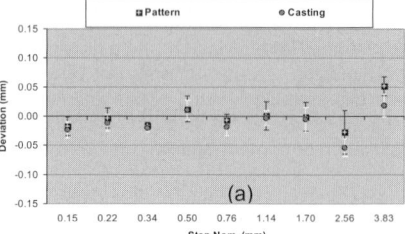

(a)

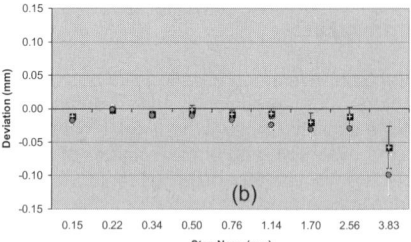

(b)

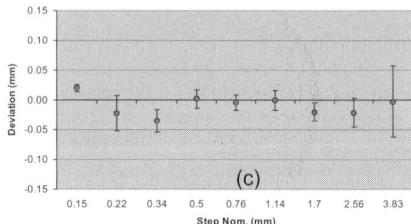

(c)

Fig. 3. Absolute deviations: a) ThJ; b) PM; c) Fcubic.

- The PM patterns and castings exhibited the most consistent results among the three processes.
- The consistency of the ThJ patterns and casting was good for the dimensions in the range of 3.83-0.5 mm and it worsened bellow 0.5 mm.

In micro-manufacturing, the deviations of micro-features from their nominal dimensions in absolute values could be more important in studying the accuracy of a given process. Thus, to investigate the influence of the three building processes on the final accuracy of the castings, the length-scale dependence of the absolute deviations was analysed. The comparison of the absolute deviations of the micro/meso-scale features from their nominal dimensions showed that the accuracy of the ThJ patterns and castings was within ±50µm (Fig.3a). The accuracy of the larger steps of the test part showed deviations in some cases exceeding 100 µm. The main reason for this could be sought in the volume shrinkage of the thermoplastic material. This shrinkage lead to a significant shape distortion and an accuracy deterioration that affected both linear and angular dimensions.

The influence of the metal shrinkage during the solidification on the casting accuracy is illustrated in Fig.4. The chart shows clearly that similar to the shrinkage of the thermoplastic material, the metal shrinkage is more pronounced for the bigger step of the pyramid where a relatively large volume of material shrinks. The difference between the behaviour of the Al and Zn alloys was not significant although Al exhibited a slightly bigger shrinkage due to its higher coefficient of thermal expansion and solidification volume reduction.

The p-values obtained for the paired t-test and their analysis at the α-level showed that only the features with dimensions bigger than 1.7 mm ($p<0.05$), a volume difference bigger than 55 mm^3, underwent statistically significant changes for both alloys. The steps/features were less affected due to their relatively small volumetric changes at the pattern-making and casting stages.

Based on the results provided in Fig.4, the shrinkage allowance for LM25 and ZA12 could be estimated for the investigated test part. In particular, this allowance would be between 1 and 2% for features larger than 1.7 mm and would not be dependant linearly on the sizes. These values were about three times higher than those recommended for macro castings and had a significant effect on the part accuracy due to the relatively small features scale.

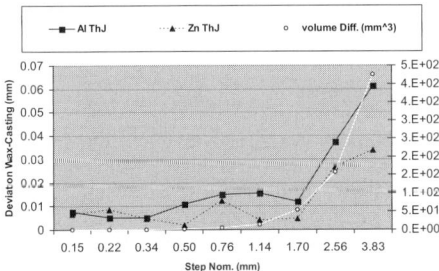

Fig. 4. Linear deviation due to metal shrinkage during solidification.

The accuracy of the Fcubic process was similar to that of the ThJ process chain. In particular, the accuracy of the micro/meso-scale features was within ±50μm (Fig.3). However, the accuracy of the steps/features of the Zn castings with dimensions bigger than 340μm was much better than that achieved with Al. These results could be explained by the better castability of Zn alloys and also the lower heat dependence of the Fcubic shells during the casting process.

The PM process chain produced samples with accuracy in the range of 0 to -50 μm as shown in Fig.3. The chart depicts a linear dependence between the deviations and the nominal dimensions that indicates a better repeatability than the other two studied processes.

Based on PI, Fcubic was ranked the first among the three investigated processes, followed by PM and ThJ (LD in Fig.7).

3.3 Analysis of Angular Dimensions

The mean and SD of the (RAD) for each sample group was plotted as a function of step's nominal dimensions (Fig. 5). Again, a close similarity of the RAD curves

before and after casting for the ThJ and PM process chains could be seen. The measurement results in both planes, YZ and XZ, were similar. The steps at the top and on the bottom of the pyramids exhibited a higher deviation than those in the middle.

The analysis of p-values from the paired t-test demonstrated a resemblance to the results from the linear dimensional analysis.

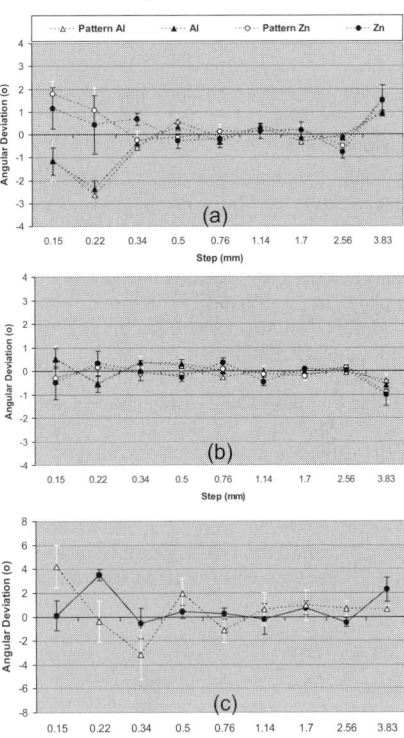

Fig. 5. Process RAD. a) ThJ; b) PM; c) Fcubic

In particular, as was the case with the linear dimensions, the metal shrinkage affected only the accuracy of the steps with dimensions bigger than 1.7 mm, $p<0.05$, while the others did not exhibit a statistically significant difference.

The angular deviations of the Fcubic samples were more scattered in comparison to those produced using the ThJ and the PM process chains as shown in Fig.5. Based on PI, PM was ranked the first, followed by the ThJ and the Fcubic processes (RAD in Fig.7).

3.4. Edge straightness

The edge straightness of the pyramid samples were within 0.03 mm ±0.1 mm for the ThJ and the PM process chains and 0.04 mm ±0.2 mm for Fcubic as shown in Fig.6.

The edge straightness of the ThJ and the PM patterns deteriorated with the increase of the step sizes which could be explained with the shrinkage of the thermoplastic material during the building of the pyramids. There were no such noticeable effects on the Fcubic samples and the edge straightness could be considered almost uniform throughout the steps of the pyramid. No statistically significant differences were found in the pre and post-casting states of the samples.

The comparison of PI for the three studied process chains showed that the accuracy of Fcubic was close to that of ThJ and PM (ES in Fig.7).

330

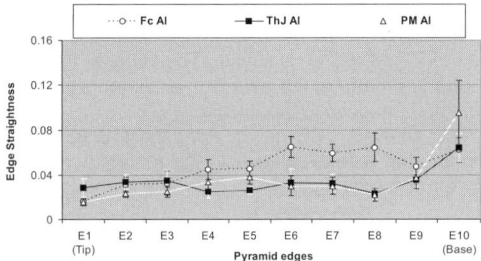

Fig. 6. Processes edge straightness.

3.5. Roughness

Table 1 displays the results from the roughness measurements on the castings produced using the three studied process chains. The results showed that there was no significant variability between measurements performed along different directions in one plane, e.g. along X and Z in the XZ plane, and also in different reference planes, XZ, and YZ. However, the measurements performed in the XY plane on the castings produced by the ThJ and PM process chains exhibited a smaller roughness. The PM samples had a smaller deviation in XY due to the milling of the top surface after the building of each layer. The Fcubic samples exhibited the smallest variability of the roughness based on the measurements taken in all three planes, XY, XZ, and YZ. At the same time the PM castings showed the highest variability and the results in the Z direction were close to that obtained on the Fcubic samples.

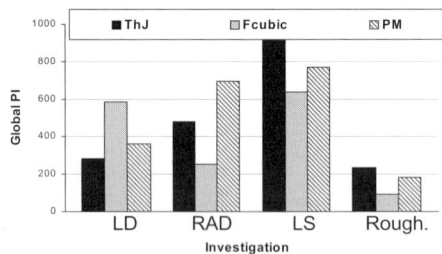

Fig.7. Processes global PI using Zn.

However, it should be noted that a layer thickness of 38 µm was selected for building the PM patterns and thus to achieve a building speed similar to that of ThJ which was at the expense of the samples' roughness. The reason for the different roughness in the Z direction in the XZ and YZ planes and that achieved in XY was the step effect in ThJ and PM resulting from the use of layer-based additive processes. The ink jet technology employed in Fcubic did not suffer from this effect.

Another interesting observation that could be made was that the surface finish of the Al samples in the XY plane was much better than that achieved on the Zn samples. This could be explained with the better fluidity of Zn than Al reproducing more closely the internal mould surface.

The analysis of PI for all studied process chains showed that ThJ performed better than PM and Fcubic (Rough. In Fig.7).

4. Conclusions

The results from this experimental study showed that when PM was set to match the building speed of ThJ, both processes produced castings with a similar

quality. The Zn castings manufactured using the Fcubic process chain showed a good dimensional accuracy and a consistent roughness.

The accuracy of ThJ and PM patterns was the key factor for producing high quality castings because the metal shrinkage did not affect significantly the accuracy of the features that were smaller than 1.7 mm.

Table 1. Roughness Results

Planes/ Directions		Average			SD		
		YZ & XZ / Z	YZ & XZ / X&Y	XY/ X&Y	YZ & XZ / Z	YZ & XZ / X&Y	XY/ X&Y
Al ThJ	Ra	3	2	1	0.8	0.4	0.2
	Rz	13	10	6	3.3	3.2	1.9
Zn ThJ	Ra	2	2	2	0.5	0.7	0.6
	Rz	12	10	12	2.0	3.1	3.1
Al Fc	Ra	5.4	6.2	4.7	1.3	0.6	1.0
	Rz	24.4	28.0	23.4	5.2	3.4	3.7
Zn Fc	Ra	6.2	5.3	5.5	0.9	0.8	2.0
	Rz	29.4	25.9	26.5	2.4	3.2	7.3
Al PM	Ra	4.5	2.5	0.8	0.9	0.9	0.1
	Rz	23.3	14.4	5.1	3.8	3.8	1.0
Zn PM	Ra	5.3	2.7	1.2	1.7	1.0	0.2
	Rz	26.6	14.5	7.9	6.4	4.2	1.1

Further research is required to assess the effects of the metal shrinkage on the accuracy of the features bigger than 1.7 mm.

Acknowledgments

The authors would like to thank the European Commission, the Welsh Assembly Government and the UK Engineering and Physical Sciences Research Council for funding this research under the FP5 Project "A Novel Manufacturing Process Fastfab, Effective Innovation Transfer of New Production Technology", the ERDF Programmes "Micro Tooling Centre" and "Supporting Innovative Product Engineering and Responsive Manufacture", and the EPSRC Programme "The Cardiff Innovative Manufacturing Research Centre". Also, this work was carried out within the framework of the EC Networks of Excellence "Innovative Production Machines and Systems (I*PROMS)" and "Multi-Material Micro Manufacture: Technologies and Applications (4M)".

References

[1] G. Baumeister, K. Mueller, R. Ruprecht, J. Hausselt. Production of metallic high aspect ratio microstructures by microcasting. Microsystem technologies 8 (2002): 105-108.

[2] G. Baumeister, R. Ruprecht, J. Hausselt. Replication of LIGA structures using microcasting. Microsystem technologies 10 (2004): 484-488

[3] Dickens P.M et al, Conversion of RP Models into investment castings, RP Journal (1995),1,4-11.

[4] Fcubic, High Precision Ink Jet Manufacturing, www.fcubic.com, © 2005 fcubic ab • Källarlyckevägen 6 • 429 35 kullavik • Sweden

Multi-Material Micro Manufacture
W. Menz, S. Dimov and B. Fillon (Eds.)
© 2006 Elsevier Ltd. All rights reserved

Influence of EDM machining on surface integrity of WC-Co

Mohammad Reza Shabgard, Atanas Ivanov, Andrew Rees

Manufacturing Engineering Centre, Cardiff University

Abstract:

This paper studies the effects of ultrasonic vibration of the tool electrode on the surface integrity of tungsten carbide (WC-10%Co) in the electro-discharge machining (EDM) process. Scanning electron microscopy (SEM) with energy dispersive X-ray (EDX), optical microscopy, micro hardness testing and white light interferometery were employed in the investigation. The paper studies the composition, number and size of cracks on the surface layer, the topography of the machined surface and thickness of the defective layer.

Keywords: Electro-discharge machining (EDM), Ultrasonic assisted EDM, Ultrasonic vibration of tool, Tungsten carbide (WC-Co), Surface integrity, Surface topography.

1. Introduction

The surface integrity of a machined surface is becoming more and more important to satisfy the increasing demands of sophisticated component performance, longevity, miniaturisation and reliability. Surface integrity is defined as the inherent or enhanced condition of a surface produced in a machining or by other surface generating operation. The nature of the surface layer has been found in many cases to have a strong influence on the mechanical properties of the part.

Tungsten carbide (WC-Co) is one of the materials widely used in die and punch manufacturing with expected application growth in the rapidly developing micro world. It is also considered as one of the most suitable electrode materials for micro EDM process. With its exceptional hardness, wear resistance and high mechanical strength it is becoming very desirable for a number of applications. To overcome technical difficulties in machining this material, non-traditional machining methods are employed, among which, electro–discharge machining (EDM) has been the most suitable. There is no direct physical contact between the electrodes in EDM process and therefore no direct mechanical forces are exchanged between the workpiece and the tool. Therefore the process has the capability of machining tungsten carbide (WC-Co) regardless of its relatively high hardness [1].

Much work has been reported in ultrasonic assisted EDM (US/EDM) of steel and some ceramics and composites. Kremer [2] reported that ultrasonic vibrations significantly improve the discharge efficiency, in the EDM machining. Murti [3] showed that ultrasonic assisted EDM of steel significantly reduced inactive pulses in the process. Zhixin et al. [4] reported increase of MRR when machining advanced ceramics, by ultrasonic assisted EDM. Lin and Yan [5] showed higher MRR and the elimination of the recast layer using the same method. The aim of this paper is to investigate the affect of the ultrasonic vibration of the electrode on the surface integrity of tungsten carbide (WC-10%Co) in the electro-discharge machining (EDM) process.

2. Experimental setup

The experimental set-up is shown schematically in Fig.1. The average current was measured by an ampere meter connected in series with the spark gap and the average voltage was measured by a voltmeter connected in parallel to the spark gap. An oscilloscope (Hitachi VC-6524) was employed to capture and store random frames of gap voltage variations, which were transferred and stored on a PC.

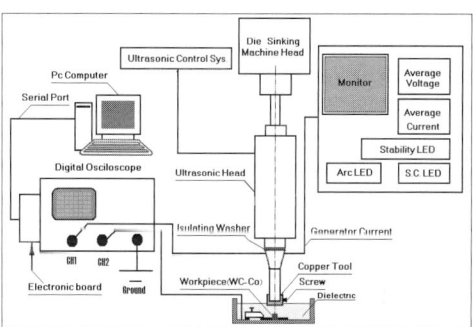

Fig.1. Experimental set-up

Fine grain Tungsten Carbide (WC-Co) composite with 10%wt Co content (ISO K15-30) of 10mm diameter and 6.5 mm height was used as a workpiece for the experiments. Table 1 gives the mechanical properties of the workpiece material and process variables and the experimental conditions are given in Table 2. The experiments were conducted in a submerged dielectric bath with normal submergible flushing. The electrode material was forged commercial copper machined to a cup shape (fig. 1). After the machining of the samples they were cut with wire EDM along the cylinders axes in two halves, hot-mounted with bakelite resin, polished, etched and prepared for investigation of the structural changes by SEM with energy dispersive X-ray (EDX). The Vickers hardness was measured on each machined surface using a diamond indenter (Mitutoyo HM-122 Digital Micro-Hardness Testing Machine, load of 0.5 kg). The surface roughness parameters R_a, R_z, R_k and R_{max} were measured by using surface roughness measuring equipment (Mahr-Perthometer M2). 3D surface topography was scanned using white light interferometer Surface Mapping Microscope (MicroXAM, TMC, USA).

Table 1. Mechanical properties of the workpiece material

Nominal compos.(by weight)	90%WC-10%Co
Grain size	Fine
ISO range	K15-K30
Hardness (HV30)	90.7-91.3
Density gr/cm3	14.6
Transve.Strength(MPa)	3100
Compre.Strength (MPa)	5170
Modulus of Elasticity(GPa)	620

Table 2. Experimental conditions

Variables	Values
Max. Amplitude of ultrasonic vibration (µm)	5 (Approx.)
Frequency of ultrasonic Vibration (KHZ)	≈25
Open circuit voltage (V)	120
Gap average voltage (V)	40
Workpiece	Ø10mm , 6.5mm height
Tool	Ø24mm O.D.& Ø14mm I.D. Cup with bottom thickness of 2.5mm
Tool polarity	Positive
Dielectric	Kerosene 85% & transformer oil 15%
Pulse durations Ton (µs)	1,5,10,20,30
Pulse off time (µs)	10 constant
Pulse currents (Amp)	11, 18, 25,32 and 40
Flashing type	Normal submerged
Machining mode	Butt mode
Output power of transducer	200 (W)

3. Effect of ultrasonic vibration on surface topography

There is a significant difference between EDM machined surfaces with and without ultrasonic assistance in regards to arcing and crater size.

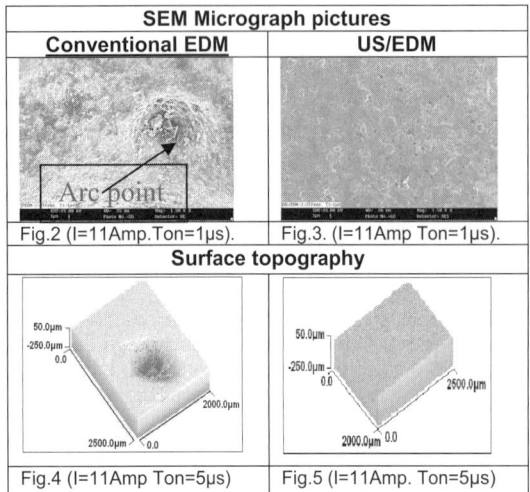

SEM Micrograph pictures	
Conventional EDM	**US/EDM**
Fig.2 (I=11Amp.Ton=1µs).	Fig.3. (I=11Amp Ton=1µs).
Surface topography	
Fig.4 (I=11Amp Ton=5µs)	Fig.5 (I=11Amp. Ton=5µs)

Figures 2 to 5 show the topography of machined surfaces with and without ultrasonic assistance for certain current and pulse duration settings. The depth of crater that has been created by arc pulses on the surface during conventional EDM process is deeper

than 180µm, whereas arcing was not observed during ultrasonic assisted EDM. It is hypothesised that the ultrasonic vibration of the tool, creates compressive and refractive wave fronts, micro jet bubbles and intensive ejecting micro streams [6], which can act as a pump, causing better debris evacuation from the spark gap.

One of the most important parameters of surface topography is surface roughness. Fig.6 and 7, show the surface roughness (R_{max}) of EDM machined surfaces, with and without ultrasonic vibration of the tool for different settings of current and pulse duration. The surface roughness achieved by US/EDM is marginally bigger than the surface roughness achieved by conventional EDM. A possible explanation is cavitation and higher energy discharge pulses produced from the ultrasonic vibration of the tool in the dielectric liquid. These factors contribute to more molten material being removed from the spark gap, consequently surface roughness is bigger.

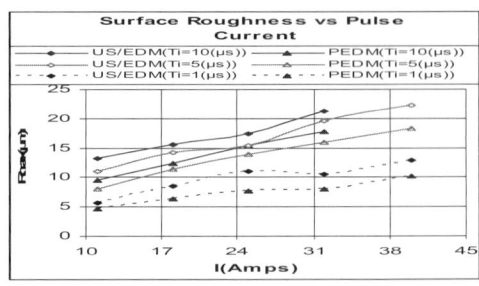

Fig. 6 Surface roughness (Ton=1.5 and 10µs).

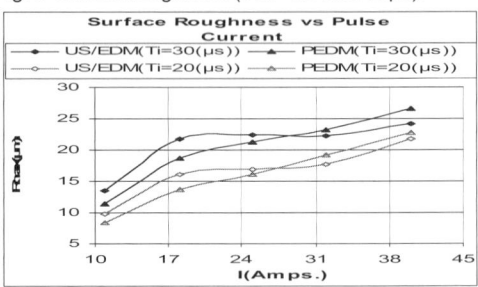

Fig. 7 Surface roughness (Ton=20 and 30 µs)

Cavitation produces positive and negative pressure into the craters. Negative pressure creates an evacuation or suction force and helps the bulk boiling mechanism and causes more super heated molten material to explode into the dielectric. The positive pressure creates an impact force on the molten puddle and causes further molten material to be dispersed out of the puddle. When comparing the US/EDM process and the conventional EDM process, if all other input parameters and setting conditions remain constant, more molten material is ejected by every discharge pulse, the craters become deeper and the surface roughness of the machined surface is increased during the US/EDM process. In this case the material removal rate (MRR) also increases up to 4 times [7].

4. Effect of ultrasonic vibration on Micro-cracks

In the EDM process the shallow thermal impact of a spark is accompanied by the very rapid quenching rate of the bulk of the material in the dielectric. These thermal waves and expansions and contractions of resolidifed and heat affected material are the main reason for the appearance of the micro cracks. Some of the cracks may extend into the base material, causing metallurgical alterations such as rehardened and tempered layers, heat-affected zones, etc. Figs.8 to 13 show the cross-sections of the machined surfaces by conventional EDM and US/EDM under different settings of current and pulse duration. On low discharge energy (I=11Amp, Ton=1µs) there were no cracks on the cross-sections of the machined surfaces machined by the two processes. A distinct difference can be seen with medium discharge energy of the pulse (I=18, 25, 40 Amps, Ton=1µs.). EITHER NO DEFECTS or only very small cracks were observed in the ultrasonic assisted EDM samples, whilst on the conventionally EDM machined surface the cracks were severe and on some settings the length of radial and transverses cracks can exceed 420µm and 800µm respectively (fig.10). When the current and especially the pulse duration were increased, the number and the size of cracks increased and there was no substantial difference between the conventional and ultrasonic assisted EDM process results (fig.12 and 13).

A possible explanation for this difference is that in the EDM machining of WC-Co composite the melting and evaporation temperatures of cobalt are low, 1320°C and 2700°C respectively whereas the melting and evaporation temperatures of WC are high, 2800°C and 6000°C respectively. This means that before the melting of the WC the binder material Co is melted and evaporated. However the thermal expansion and contraction coefficient of Co, WC and WC-10wt.%Co is, 1.4e-5[/k], 5.0e-6 [/k] and 5.2e-6 [/k] respectively. With the fairly high thermal expansion and contraction coefficient of Co and the difference between the other component (WC), it can create high thermal tension stresses during resolidifycation and quenching, exceeding the fracture strength of the material in the crater, causing the cracks on the surface layer.

SEM micrograph (cross-sections of machined surfaces)	
Conventional EDM	US/EDM
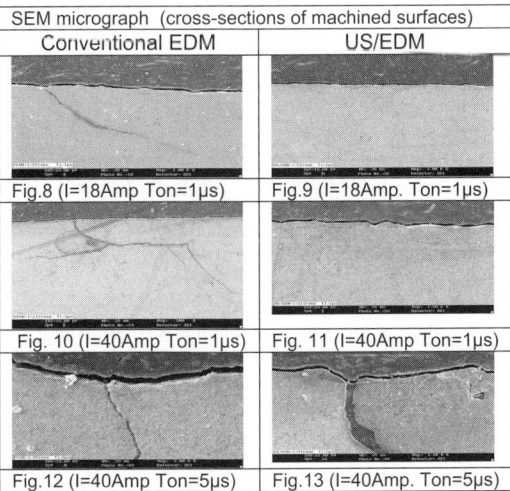	
Fig.8 (I=18Amp Ton=1µs)	Fig.9 (I=18Amp. Ton=1µs)
Fig. 10 (I=40Amp Ton=1µs)	Fig. 11 (I=40Amp Ton=1µs)
Fig.12 (I=40Amp Ton=5µs)	Fig.13 (I=40Amp. Ton=5µs)

With low and medium energy settings during US/EDM process, the material removal rate is higher, as explained above, because more material is splashed out of the crater due to the cavitation, which

means that the volume of resolidified material in each crater is less; therefore the thermal stress created is less which will lead to less cracks on the surface layer. The increased current and pulse duration time increases the energy of the discharge spark which causes the spark gap to also increase [9]. The increase in distance between the electrode and the workpiece is an obstacle for the ultrasonic waves to remove the increased amount of molten material in the craters. Therefore larger amount of molten material re-solidifies in the craters. The high thermal expansion coefficient of Co creates high thermal tension forces, exceeding the fracture strength of the new brittle layer.

Another explanation could be when high current and pulse duration settings are used, the energy of the sparks is high enough to melt the WC grains. In this case it is possible that there is a formation of brittle compounds like cobalt carbide (Co_2C and Co_3C) and W_2C, Co_3W_3. The result is the deterioration of the mechanical properties and increase in brittleness of the binder material (Co) [8].

5. Effect of ultrasonic vibration on micro hardness

The micro hardness was measured from the machined surface into the depth of the material on the cross section of the samples to investigate the change in micro hardness. Figs. 14 and 15 show the microhardness results of the two EDM processes for two current and pulse duration settings.

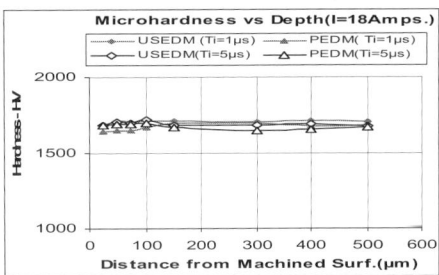

Fig.14 Cross-section micro hardness (I=18Amp)

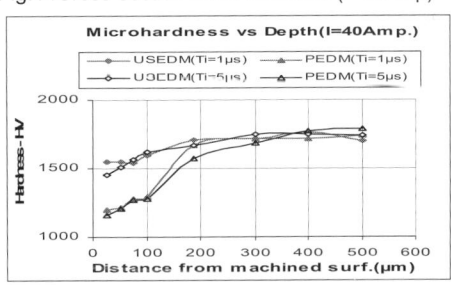

Fig.15 Cross-section micro hardness (I=40Amp)

It can be seen from Fig.14 that with low to medium energy settings, there is no significant difference in micro hardness between the two processes in depth. This means that the two processes either do not leave a re-cast layer or the size of such layer is very small. In Figure 15 (higher energy settings) there is a noticeable variation in the micro hardness results in the depth of the sample for the two processes and for the two pulse duration settings. In this case the hardness of the machined surface layer is lower than the hardness in depth of the material. However the hardness of the samples produced by conventional EDM are lower than the hardness of the samples produced by ultrasonic assisted EDM. This shows that ultrasonic assisted EDM creates smaller recast layer than the conventional EDM process. Hardness of Co, WC and (WC-10wt%Co) are 1024VH, 2700HV, 1500HV respectively. Therefore with the increase of the cobalt percentage in the recast layer the hardness of surface layer decreases.

6. Effect of ultrasonic vibration on the defective layer

In the EDM machining of WC-Co, during the time-off of the pulse, bulk boiling and shock wave of plasma bubble, cause the relatively loose WC grains to dislodge from the molten crater as shown in Fig.16 and 17 at the marked locations. According to the results of EDX analysis, Cobalt percentage at the marked locations is 0.8%, and 4.1% respectively, whereas the average Cobalt percentage of base material is 10% (Table.1.). When the EDX analyses of the whole area shown on fig.16 was carried out, the results show that the average Cobalt content of the surface layer is 18%. This means that the dislodging of WC grains from the machined surface causes the Cobalt content in the recast layer to increase Therefore after the EDM machining of WC-Co, the composition and consequently some of the mechanical and physical properties of machined surface layer may change.

| SEM micrograph of machined... | |
Surface	Cross section
Fig.16 (I=11Amp. Ton=1µs)	Fig.17 (I=11Amp Ton=1µs).

When ultrasonic assisted EDM is applied for the machining of WC-Co, the cobalt percentage of the machined surface layer is less than conventional EDM. The results of microhardness tests show that the hardness of the surface layer in conventional EDM machining is lower than US/ED-machining. A possible explanation is that the higher MRR of ultrasonic assisted EDM leaves a reduced amount of re-solidified material on the surface layer. Resolidified material has higher thermal stress, higher percentage of Cobalt, different composition, physical and mechanical properties from the rest of the material. The higher Cobalt percentage and the thickness of the defective surface layer explain the main difference in the microhardness measurement results.

7. Conclusions

1. Significant differences can be observed between the surface integrity characteristics of conventional EDM machined surface and ultrasonic assistant EDM machined surface of tungsten carbide (WC-10%Co).
2. Ultrasonic vibration of tool creates cavitation which in turn produces hot spot, core plasma, electric current and intensive micro-jets, causing better flushing conditions in the spark gap and reduces undesirable unstable pulses and as a consequence better surface topography is achieved.
3. Surface roughness achieved through ultrasonic assisted EDM is marginally higher than the roughness achieved by the conventional EDM process due to better evacuation of the molten material from the crater.
4. Ultrasonic vibration of the tool, when applied to the finishing mode machining, decreases the thickness of the heat affected zone (HAZ) and recast layer. It also reduces the numbers, size and depth of radial and transverse cracks appearing on the machined surface.
5. Ultrasonic assisted EDM on a smaller scale alters the micro hardness of the machined surface.
6. The decreased thickness of the recast layer, with US/EDM, prevents the alteration of physical and mechanical properties on the surface layer.

8. Acknowledgment:

The authors would like to thank Metallurgy laboratory of Almaseh Saz Factory   Tabriz University Iran for support in preparing the samples. The authors would like to thank the European Commission, the Welsh Assembly Government and the UK Engineering and Physical Sciences Research Council for funding this research under the ERDF Programmes "Micro Tooling Centre" and "Supporting Innovative Product Engineering and Responsive Manufacture" and the EPSRC Programme "The Cardiff Innovative Manufacturing Research Centre". Also, this work was carried out within the framework of the EC Networks of Excellence "Innovative Production Machines and Systems (I*PROMS)" and "Multi-Material Micro Manufacture: Technologies and Applications (4M)".

References:

[1] A.Abdullah, "Voltage injection and performance evaluation in EDM", PhD Thesis, The Victoria University of Manchester, Vol.1, March (1989).
[2] D.Kremer, et al "Effects of ultrasonic vibrations on the performances in EDM", Annals of the CIRP Vol.38/1(1989), pp. 199-202.
[3] V.S.R.Murthy, et al "Pulse train analysis in ultrasonic assisted EDM",int. J. Mach. Tools Manufact. Vol. 27, No. 4, (1987), pp. 469-477.
[4] J.Zhixin, et al "Study on new kind of combined Machining and electrical discharge machining", Int. j. Machin Tools, Manufact. Vol.37, No.2 (1997), pp.193-199.
[5] Y.C.Lin, et al "Machining characteristics of titanium alloy(Ti-6Al-4V) using a combination process of EDM with USM", Journal of Material Processing Technology 104 (2000), pp.171-177.
[6] F.R.Young, "Cavitation", Published by Imperial College Press, (1999).
[7] A.Abdullah, M.R.Shabgard, "Effect of ultrasonic vibration of tool on electrical discharge machining of tungsten carbide (WC-Co)", International Manufacture Engineering Conf., Iran, (2005).
[8] Y.Fukumiya, et al "Thermal stability and hardness of metastable Co-C composite alloy films", Material Science and Engineering, A312 (2001) 248-252.
[9] Philip T. Eubank, et al ,Theoretical models of the electrical discharge machining process, Journal of applied Physics, 73 (11), 1993, 7900-7909

Multi-Material Micro Manufacture
W. Menz, S. Dimov and B. Fillon (Eds.)

Large area plastic replication with modular molding tools

M. Heckele[1], C. Mehne[2], R. Steger[3], P. Koltay[3], D. Warkentin[4]

[1] Institute for Microstructure Technology (IMT), Karlsruhe Research Centre (FZK), Germany
[2] Institute for Microstructure Technology, University of Karlsruhe, Germany
[3] Department of Microsystems Engineering (IMTEK), Lab for MEMS Applications, University of Freiburg, Germany
[4] Hahn-Schickhard Institute for Microassembly Technology (HSG-IMAT), Stuttgart, Germany

Abstract

Hot embossing of polymer microstructures has evolved over recent years from a simple lab technique to an established Research and Development (R&D) technology.

A multitude of technical applications demonstrate the high performance of this replication technique, which can provide more than first prototypes. Admittedly this technology is not recognized as a manufacturing process by industry. In this publication molding tool is presented which by its modular character can be easily integrated into existing hot embossing machines. Especially into a newly developed machine, which has been specially designed for the integration into a process chain. This modular molding includes a customized tool as well as a microstructured mold insert. The mold insert comprises discrete modules which can be changed for different applications or in case of partial damage. Finally the operator panels and equipment controls are based on standard PLC controls, which are known in manufacturing, and in this manner hot embossing also optically becomes more trustworthy.

The function and link-up of these components has been demonstrated using a "Dispensing Well Plate" with dimensions of 127.8 mm x 85.5 mm, to produce ambitious through hole nozzle components. With this continuous process chain it has been shown that hot embossing is ready for industrial implementation.

Keywords: Replication, Polymer, Tool, Mold Insert, Hot Embossing, Dispensing Well Plate, Through Holes

1. Introduction

Hot embossing of polymer microstructures is a polymer replication technique which is not only suitable for the fabrication of fast first prototypes from a multitude of micro structured mold inserts [1]. The low forming speed allows the replication of high micro structures with large aspect ratio. It is also possible to structure multi layers from polymers with different properties, e.g. for optical wave guides. Furthermore the integration of metallic layers into polymer micro-structures by hot embossing is feasible. These examples illustrate the high standard in development which is achieved in hot embossing, compared to the beginnings of polymer micro system technology, when hot embossing of simple plastic components often was made by simple presses for a prove of principle.

With the Jenoptik HEX series, developed with IMT / FZK, a high performance hot embossing machine has been on the market for 10 years. While many important successes have been demonstrated in R&D labs (over 50 installations worldwide), however to date there has been no break-through as a manufacturing technique. The reason for this is that manual sheet material feed and embossed part removal restrict or prevent inte-gration with a fully automated process chain. To over-come this challenge Wickert Maschinenbau [2] and IMT developed a new hot embossing machine designed for automated industrial applications. A hot embossing tool was developed suitable for this machine, especially for double sided, large area replication. Different demolding features allow for automatic operation with a handling system for part removal. Even though this tool has become quite large and heavy by integrating many sophisticated features special insulation techniques enable efficient heat management. In this way it is possible to shorten heating considerably and cooling times, increasing efficiency necessary for manufacturing and assembling complex microsystems.

2. Hot Embossing Machine

Characteristic for the new hot embossing machine WMP 1000 is the enlargement of the embossing force to 1000kN (Fig. 1). Up to now machines with a maximum force of 250kN were available. This high force is necessary due to the enlargement of the replication area up to 8" (200mm in diameter) and is reason for a new construction.

Fig. 1: Wickert WMP 1000 for 8" replication.
A hydraulic driven hot embossing machine designed for use in industrial environment

Major modification to previous concepts is the

336

change from a servo-electrical drive to a hydraulic drive. Together with the high force of 1000 kN a high accuracy is required. Closing of the embossing stampers requires a touch force of only some 100N. These two contradictory specifications are realized by the use of a second hydraulic cylinder for the lower range. This combination of two cylinders allows a large range of movement with rough and fine adjustment. For opening and closing speeds until 1000mm / min can be set. During the embossing process the stampers can be closed with a minimum speed of only 0,06 mm/min. The mechanical construction is dominated by a four column guidance to achieve a high mechanical stability and low torsion. For a flexible use of the machine and the integration of multiple tools the machine has a generous free working. A clamping area of 650*550mm² with a height of 1250mm gives free access to the tool. This is an important fact for changing tools and working with handling systems.

A PLC (Programmable Logic Controller) system provides a comfortable communication environment for the operator. This operator system looks similar to systems currently used for industrial press and punching machines- thus hot embossing equipment becomes more user friendly.

3. Hot Embossing Tool

Important part of a modular concept for our hot embossing equipment is the tool. In a very complex construction features for double-sided hot embossing are integrated to allow short cycle times. Development of a cost-efficient process with short cycle times was deemed necessary to meet the demand for disposable components e.g. for microfluidic analysis chips [4].

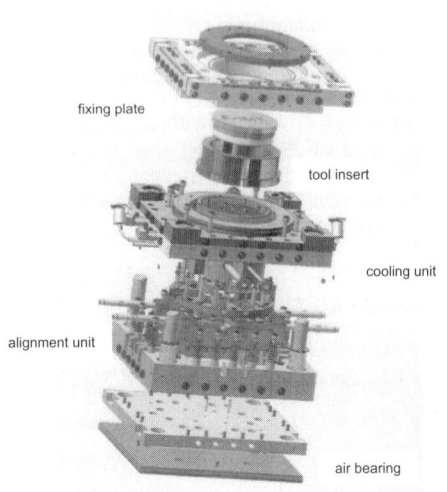

fixing plate

tool insert

cooling unit

alignment unit

air bearing

Fig. 2: Sophisticated hot embossing tool for precise double-sided replication, lower part

The tool and process concept covered the development of appropriate mechanisms for accurate orientation of the tool halves, defect-free demolding of the large-area and usually flexible molded parts, and efficient temperature control of the tools to minimize cycle times [2].

Based on the process technology requirements determined, a tool setup of two symmetrical tool halves

and a positioning table was developed and implemented. The tool halves each consist of a tool frame of constant temperature, a tool insert of minimum mass that is subjected to thermal cycling, and a cooling block kept at constant temperature. During heating, the tool insert is thermally insulated from the cooling block. In this way, short heating times are reached. During the embossing and cooling phase, the tool insert is coupled thermally to the cooling block such that heat can be removed rapidly (Fig 2). As a result of the short times needed for heating and cooling the tool during embossing, cycle time is reduced considerably.

The tool halves are additionally provided with plates, by means of which the semi-finished product is fixed during heating and cooling and shrinkage is avoided. Using a demolding drive, the fixing plates can be moved accurately and independently of the tool opening movement. This allows to apply the semi-finished product to the mold inserts or to separate it from them for demolding. Via separate vacuum and pressurized air connections in the upper and lower tool, independent evacuation of the cavity or controlled separate pressurized gas supply to the upper or lower side of the semi-finished product is ensured. In this way, the large-area molded parts that have been microstructured on both sides can be demolded easily.

The alignment unit is equipped with an air bearing and flexible joints. Consequently, alignment is not influenced by friction or play of the joints. Accurate orientation of the tool halves is achieved by piezo actuators driving the table and micrometer-resolution induction coil sensors integrated into the tool frame.

4. Modular Mold Inserts

For hot embossing a new approach for molding through holes was used. Because the height of the nozzle pins exceeds the depth of the cavity during molding, the pins immerse into the counter surface of the substrate plate. To prevent damage to the pins the hardness of the mold insert has to be higher than the hardness of the substrate plate. Therefore steel was used for the mold insert and brass for the substrate plate [3].

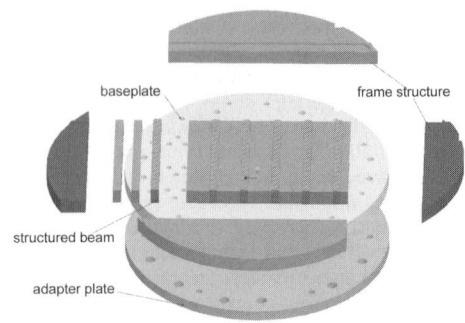

baseplate

frame structure

structured beam

adapter plate

Fig. 3: Concept of the modular mold insert

The outer dimensions of the mold insert were adapted to the new hot embossing tool developed at IMT. In this tool mold inserts with a diameter of 154 mm are used, so it was possible to emboss structures on an area with dimensions of standard microplates (127.8 mm x 85.5 mm). Due to the dimensions of the mold insert a modular concept was developed (Fig. 3).

The mold insert was built up from 24 beam-shaped sub units that were mounted in a frame structure and fixed onto a baseplate. The frame structure is used to clamp the single beams together and to minimize the gap clearances. The rigid baseplate ensures the close tolerance in height and the stability of the mold insert that is needed during the hot embossing process of the through holes. The mold insert was finally produced by HSG-IMAT. For High Speed Cutting (HSC) this modular concept is a great advantage, because the long machining time is split into several shorter runs and thereby the risk of damage or failure can be reduced considerably. In addition each sub unit can easily be replaced in case of damage during the molding process. In a first step a mold insert was used with 6 structured and 18 unstructured beams that can be easily replaced by structured ones.

5. Applications

Microfluidic "Dispensing Well Plates" (DWP) are an appropriate method for precise and parallel handling of nanoliter volumes [5]. The DWP technology is characterized by arraying microfluidic dispensing units consisting of a reservoir, a capillary connection channel and a nozzle. For this nozzles through holes must be made. To create such through holes some modifications of the hot embossing process are necessary. In principle the hardness of the mold insert and the substrate plate must be adapted, thereby it is possible to displace the polymer melt completely between the tool and the counterplate, while the protruding nozzle pins are immersed some microns into the counter plate without being damaged.

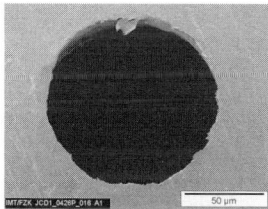

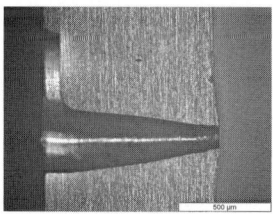

Fig. 4: (a) Embossed through hole with a diameter of 100 μm; (b) Cross section of an embossed through hole.

For this molding process different material combinations were tested [6]. It was determined, that the combination of steel for the mold insert and brass as counterplate is well suited. Another solution is the usage of steel for the mold and a stack of a steel foil and a Polytetrafluoroethylene foil (PTFE) as substrate. By choosing the optimized thickness of the steel and PTFE foils the resilience of the substrate can be adjusted and through holes with good edge quality can be achieved (Fig. 4).

The variables of the molding process must be controlled very accurately, especially molding temperature and molding time are important. For hot embossing of the DWP structures Polymethylmethacrylat (PMMA) and Cyclic Polyolefines (COC) was used. For PMMA (G77Q11) the molding temperature was 165°C and the embossing time was approximately 8 minutes. COC (Topas 5013) was embossed at a molding temperature of 165°C and a molding time of 12 minutes to get the best results.

6. Conclusion

The integration of the hot embossing technique into a a manufacturing environment has been demonstrated. Hot embossing machine and tool have been designed especially for industrial applications. As well tool and mold inserts are part of a modular concept and can be varied depending on the application. The tool presented is the high end version and can be replaced easily by a cheaper tool for single sided replication. Also the mold insert can be modified or partially replaced in case of damage. With the integrated handling system the replicated parts can be automatically transported to the next fabrication step for coating or assembly.

Acknowledgements

The work was partially funded by Landesstiftung Baden Württemberg

References

[1] Heckele, M.;Schomburg, W.K; (2004), Review on Micro Molding of Thermoplastic Polymers, Journal of Micromechanics and Microengineering", 14, R1-R14

[2] Herzinger, S, Heißprägeanlage für Mikrostrukturen, in „μFEMOS MikroFertigungstechniken für hybride Sensoren", Hrsg.Bär, M., Schriftenreihe des Institutes für Angewandte Informatik, Bd. 10, Universitätsverlag Karlsruhe, 2005, p. 27-31

[3] Dittrich, H., Heckele, M., Schomburg, W.K; Werkzeugentwicklung für das Heißprägen beidseitig mikrostrukturierter Formteile, Dissertation, Universität Karlsruhe, Forschungszentrum Karlsruhe, Wissenschaftliche Berichte FZKA 7058, 2004

[4] Warkentin, D., Hiltmann, K.; Untersuchung der Formgebungsverfahren von mikrofluidischen Strukturen aus thermoplastischen Kunststoffen für Anwendungen bei Drop-on-Demand und Jet-on-Demand Systemen, Abschlussbericht AiF13437N, 2005

[5] Koltay, P., Steger, R., Bohl, B., Zengerle R.; The Dispensing Well Plate: A Novel Nanodispenser for the Multi Parallel Delivery of Liquids (DWP Part I) Sensors & Actuators A, 2004, 116; pp 483-491

[6] Mehne, C.; Heckele, M.; Steger, R.; Koltay, P.; Zengerle, R.; Warkentin, D; Hot Embossing of Large Area Microfluidic Devices with Through Holes; Proceedings on the 6th International Conference european society for precision engineering and nanotechnology, may 2006, Baden (Austria), Vol. II, pp. 28-31

Multi-Material Micro Manufacture
W. Menz, S. Dimov and B. Fillon (Eds.)

Microforming of titanium – forming behaviour at elevated temperature

B. Eichenhüller, U. Engel

Chair of Manufacturing Technology, University of Erlangen-Nuremberg, Germany

Abstract

Titanium is applied in a wide range of technical applications, especially when high strength in combination with light weight or corrosion resistance is demanded. Titanium is also a popular material for implants in the field of medical technology due to its specific properties like the superb biocompatibility, the elastic behaviour matching that of the human bone, the radiological transparency and the galvanic neutrality. For the production of these partly small-sized implants manufacturing methods with high accuracy and close tolerances are necessary. When forming processes are used to manufacture the miniaturised medical components the size effects occurring with miniaturisation have to be considered. The size effects are amongst others responsible for an increased scatter of process parameters and a reduced accuracy, thereby reducing the process stability. Also the limited formability of titanium and titanium alloys at room temperature is a drawback of forming microparts made of titanium. The main objective of the present study is to investigate the forming behaviour of titanium at microscale and enlarge the formability by means of forming at elevated temperature.

Keywords: microforming, elevated temperature, titanium

1. Introduction

Titanium belongs to the group of lightweight materials and is often used for technical products where the combination of both reduced weight and high mechanical strength is required. This is probably the most important reason why titanium is a common material in the areas of aviation and aerospace. Other industries are using titanium for nautical and offshore applications, reactors for chemical system engineering and food industries, because of the excellent corrosion resistance. Especially in medical technology titanium has shown its potential being an appropriate material for implants [1]. In particular, miniaturised implants (some examples are shown in Fig. 1) are used as plates, screws and nuts, artificial joints, prosthesis or middle ear implants. Other components like pacemakers' housings or artificial heart valves are also made from titanium due to its superb biocompatibility. Especially for implants inside the oral cavity titanium can be used as an alternative material due to the

galvanic neutrality and the radiological transparency. Furthermore, it is well suited as implantation material, compared to the usually used metals or ceramics, because of the low specific weight, the high strength and the better matching elastic modulus compared to the human bone [2].

In general, titanium alloys exist in three different types α-, β- and (α+β)-alloys. Commercially pure titanium (Ti c.p.) and α-alloys feature a hexagonal closed packed (hcp) lattice structure, characterised by a limited formability at room temperature. In contrast, β-alloys feature a body centred cubic (bcc) lattice and thus a better formability at room temperature compared to α-titanium, but still lower than the one obtained by using other metals. To achieve a better formability and to enlarge the forming limits, forming processes using titanium or its alloys are usually performed at temperatures above 500°C [1].

In this work, investigations on the forming behaviour of titanium at microscale as well as on the occurring size effects have been done using upsetting tests of small specimens performed at elevated temperature. In this case, the temperature chosen is above room temperature, but well below recrystallisation temperature. Thus, hardening effects still takes place [3], combining the advantages of cold forging - high accuracy and surface quality - and hot forging where the main advantages are the reduction of the required forming force, caused by thermally activated additional slip systems and the improved formability. The materials used in this research are Ti c.p. (grade 2) and the (α+β)-alloy TiAl6V4, which represent more than 80 % of the annual titanium production and are mainly used for the production of smallest medical parts.

When parts with main dimensions in the range of few millimetres are being produced by forming methods, so-called size effects appear, preventing the transfer of process knowledge from conventional length scale to microscale. As results of these size effects an increased scatter of typical process parameters and an unpredictable shape evolution is being observed. These

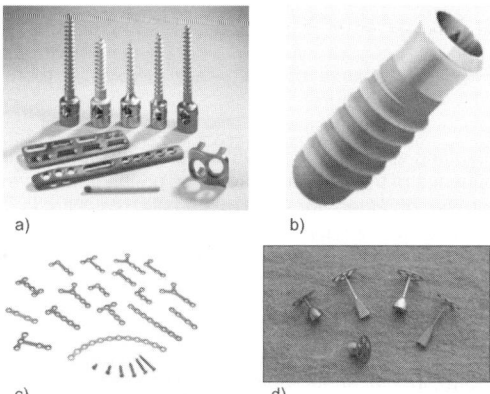

Fig. 1. Examples of implants made of titanium (source: a) AHC Oberflächentechnik Holding GmbH, c) Asanus Medizintechnik GmbH, d) Spiggle & Theis Medizintechnik GmbH)

are reducing the process stability and preventing forming technology from being used for the production of microparts. Some parameters, like the grain size or the surface roughness, are mostly scale invariant thus having different influence on the forming process at different length scales.

The main idea of forming titanium at elevated temperature is therefore to overcome existing application limits and reduce the impact of the size-effects on the forming process. Previous investigations using the materials CuZn15 and X4CrNi18-10 [4] have successfully shown the potential and applicability of this approach, using parts with main geometrical dimensions in the sub-micrometer scale [5].

2. Experimental set-up

The investigations shown in this paper are performed at different temperature levels using micro upsetting test with cylindrical specimen made from titanium and titanium alloys. The geometrical dimensions of the specimen have been chosen as 0.5 mm in diameter and 0.75 mm in length to be clearly at microscale. As it is one objective of this research to investigate the scatter of process parameters the influence of the experimental setup (shown in Fig. 2) on the forming process has to be minimized. This has been reached by using a high-precision testing system (MTS Synergy 100; maximal testing load of 500 N) with an additional high-resolution distance measuring devise. Impacts caused by the operator are eliminated by using an automated test procedure which executes all interactions between the handling system, the heating device and testing machine itself. The handling of these smallest specimens, with a total weight of less than 1 mg, is done using a vacuum gripper, mounted on a translation stage to transport the parts from the magazine to the upsetting dies. The small dimensions, combined with the low weight and the high temperatures of the heated dies make this handling system indispensable for the positioning of the specimens in the testing machine. Furthermore, close tolerances applied to the manufacturing process of the specimens ensure a minimised impact of the specimens' geometrical dimensions on the results of the forming process.

In contrast to conventional forming processes the microparts cannot be heated to the forming temperature outside of the tool. The high ratio of surface to volume favours the heat dissipation with the result that the parts will be cooled down before starting the forming process. In order to attain a well defined temperature within the part, the tools are heated up by special heating sleeves. Thus the small parts are heated by conduction. As it has been shown by previously performed finite element

simulation of the heating process, the temperature distribution inside the specimen is homogeneous and stationary within a few seconds. [4]

Putting in mind that the size effect on material flow is mainly controlled by the ratio between grain size and specimen dimension, there are different possibilities to investigate this effect. The straightforward approach is to scale down the process. However, in this case the size effect on friction has to be considered. In order to exclude this effect i.e. to ensure constant friction conditions, another way is to keep the specimen dimension unchanged and to vary the grain size. In this kind of physical modelling the fine grain and coarse grain material represent the macro and micro case, respectively [6]. If the geometry is unchanged, similar friction conditions in the interface between tool and workpeace can be assumed. To avoid process scatter caused by non-uniform lubricant distribution, the experiments are carried out without lubricant.

Additionally to the analyses of the process characteristics, like the force-displacement, metallographic analyses of polished cross-sections have been performed at the titanium specimens after the forming process, in order to gather additional information regarding the work hardening and thus the local forming behaviour of single grains.

3. Results

The first investigations have been performed in order to get more detailed knowledge on the flow curve in dependency on the mean grain size, as this is an important criterion for the design of metal forming processes and for further investigation on the forming behaviour. In Fig. 3 the flow curves of Ti c.p. and TiAl6V4 gathered at room temperature with upsetting tests for different average grain sizes L_K with an

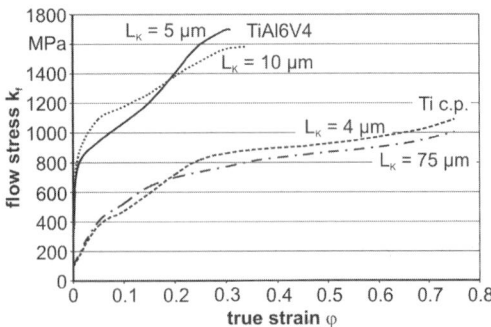

Fig. 3. Flow curve of Ti c.p. and TiAl6V4 at room temperature

upsetting velocity of v = 0.05 mm/s are displayed. The influence of the different grain sizes on the flow curves can be obtained for both materials, resulting in a decreasing flow curve with increasing mean grain size. A more important effect is obvious when comparing the process scatter of the upsetting test results for different mean grain sizes and materials. The maximum observed coefficient of variation of the flow stress determined by using Ti c.p., TiAl6V4 and CuZn15 is shown in Fig. 4 for both, fine and coarse grained material. While the process scatter in case of Ti c.p. and TiAl6V4 does not significantly increase with increasing grain size, previous research using CuZn15 has shown different forming behaviour [7]. By

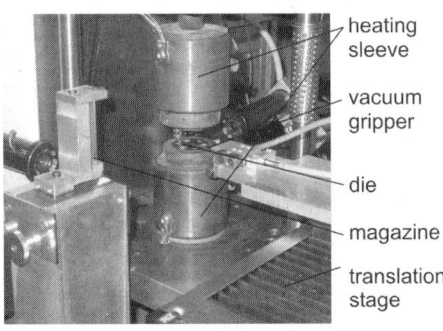

Fig. 2. Set-up of the testing machine

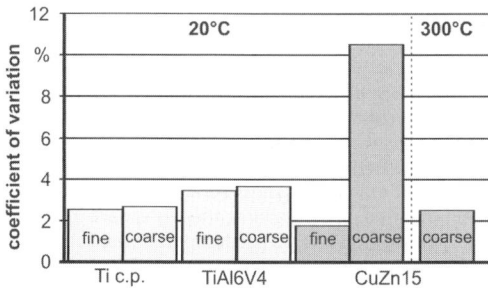

Fig. 4. Maximal coefficient of variation of the flow stress depicted in Fig. 3

increasing the average grain size of CuZn15 from 4 µm (fine) to 200 µm (coarse) the maximum coefficient of variation increase from approximately 2% to more than 10%. Compared to that findings, when increasing the average grain size of Ti c.p. from L_K = 5 µm to 75 µm no significant increase of the coefficient of variation is observable. One possible explanation therefore can be given by having a closer look on the lattice structure, the texture and the number of grains being located in the forming area. In case of titanium the texture caused by previous forming processes (e.g. wire drawing) of the raw material can be intensified by subsequent recrystallisation annealing and grain growth. The direction of the main crystal orientation within the titanium specimens is therefore more directional compared to the one from specimens made of CuZn15. As no significant increase of the process scatter is observable at room temperature, the coarse grain structure and thus the reduction of the number of grains within the forming area has negligible influence on the scatter. Additional investigations at elevated temperature have shown only small influence on the process scatter - compared to CuZn15 formed at 300°C – and thus have not lead to an improvement of the process stability.

Another limiting aspect is the formability when performing the tests at room temperature. While the specimens made of Ti c.p. can be formed at room temperature without fracture until the maximum testing force of 500 N is reached, TiAl6V4 meets its forming limit long before. The upsetting test aborts either when fracture is detected, or when the maximal testing force is reached. Fig. 5 shows the flow curves of fine grained TiAl6V4 at different temperatures. By increasing the temperature from 20°C up to 600°C two effects can be observed, a decreasing flow stress and an increase in formability, both already known from conventional scale. At a true strain of φ = 0.3 the flow stress drops

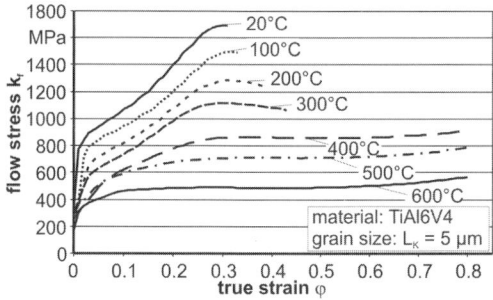

Fig. 5. Flow curve of fine grained TiAl6V4 at different temperatures

from approximately 1700 MPa at 20°C to 490 MPa at 600°C, this is a reduction of 71 % compared to the flow stress at room temperature. After passing a peak at φ = 0.3 the flow curves at a temperature between 100°C and 300°C show a descending course until they reach the end of their formability and fail by shear fracture. One possible explanation of this behaviour can be beginning recovery processes. At a forming temperature above 400°C and φ > 0.3, the course of the flow curves converges to a nearly horizontal line. The flow stress is not increasing significantly during the upsetting test, indicating, that equilibrium between work hardening on the one hand and recovery processes on the other hand is reached.

The forming limits of fine grained (average grain size L_K = 5 µm) TiAl6V4 are shown in Fig. 6 by plotting the average reachable true strain against the different forming temperatures. As far as the temperature is

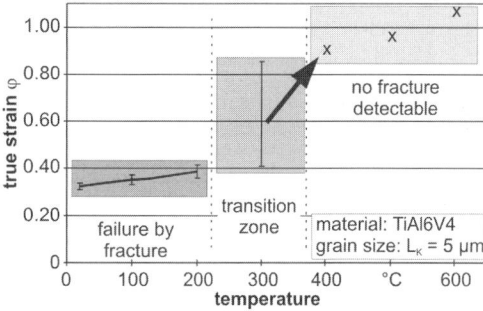

Fig. 6. Fracture behaviour of fine grained TiAl6V4 in dependence on temperature

below 200°C, all specimen fail within a small range by shear fracture enclosing the typical angle of 45° to the load direction (see Fig. 7 a). The formability and thereby the forming limit of the $(\alpha+\beta)$-alloy TiAl6V4 is increasing slightly from a true strain of φ = 0.33 where the first breakage of a specimen is detected at 20°C up to a true strain of φ = 0.41 where the first specimen at 300°C fails. At this temperature two-thirds of the specimens fail at a true strain of φ = 0.43 similar to specimens formed at lower temperature, but one third can be formed without fracture until the maximum testing force of 500 N is reached. Above a temperature of 400°C, all specimens can be formed until the maximum force is reached without detectable failure. Thus, the crosses in Fig. 6 do not represent the forming limits, but mark the reachable true strain at the corresponding temperature and when the maximum testing force is reached. Fig. 7 shows the polished micrograph cross-sections of a specimen formed at 300°C and a true strain of φ = 0.42 (Fig. 7 a) and a specimen at 600°C and a true strain of φ = 1.08 (Fig. 7 b). The crack running from lower left corner to the upper

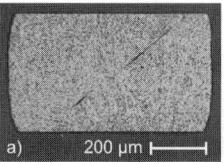 

Fig. 7. Upsetted specimens (TiAl6V4, L_K = 5 µm)
a) fracture at φ = 0.43 and 300°C
b) no fracture at φ = 1.08 and 600°C

342

right corner of the specimen formed at 300°C is clearly visible, while the specimen formed at 600°C shows no crack at a true strain being 2.5 times larger than one of the broken specimen. With increasing temperature a rise of the true strain and thus the formability is noticeable. At a temperature below 200°C the formability is limited to a small true strain. A transition zone, making it nearly impossible to predict the forming behaviour and fracture, exists between 200°C and 400°C. By increasing the forming temperature above 400°C, a significant increase of the formability and thus the forming limit can be reached.

A similar behaviour can be observed using TiAl6V4 with an average grain size of L_K = 10 µm, as shown in Fig. 8. Similar to the fine grained material, fracture does occur at low temperatures, but the temperature range

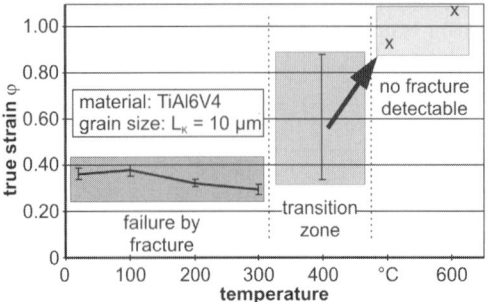

Fig. 8. Fracture behaviour of coarse grained TiAl6V4 in dependence on temperature

where all specimens fail by shear fracture increases to 300°C compared to 200° C for fine grained TiAl6V4. Furthermore the formability, indicated by the true strain, is slightly decreasing with increasing temperature between 20°C and 300°C. The transition zone, where some specimens fail, can be detected at a temperature between 300°C and 500°C. A further increase in the temperature a forming behaviour comparable to that of fine grained TiAlV6 is observed with forming limits caused by the maximum force of the testing machine. The metallographically prepared cross-sections show the formed specimens with the fracture (Fig. 9 a), running from the lower left to the upper right corner, at 400°C and the specimen formed until the maximum force is reached (Fig. 9 b) with no sign of failure. This shows clearly that elevating the forming temperature in case of TiAl6V4 leads to an enlargement of the forming limits. The temperature range for a successful forming is strongly depending on the average grain size of the specimens, as it can be observed for the fine and coarse grained TiAl6V4. A grain size difference of only 5 µm is responsible for a shift of the transition zone of 100°C.

4. Conclusions

Upsetting tests show that the scatter of the flow curves is not significantly increasing when scaling down the process using Ti c.p. and TiAl6V4 in contrast to CuZn15 or X4CrNi18-10. An increase of the forming temperature to reduce the influence of the size effects is therefore not necessary. As the most important aspect of these investigations, the significant increase in the forming limit at elevated temperatures, compared to room temperature, has been shown. This behaviour is already known from conventional scale, but the experiments have shown that even a small change of the grain size features a significant impact on the forming behaviour of the specimens. With increasing grain size a higher forming temperature is necessary to reach reproducible forming results with high true strains. This behaviour should be investigated in more detail, using additional microforming processes like can backward, full forward or lateral extrusion on the other hand.

Acknowledgements

The authors gratefully acknowledge the financial support of the German Research Foundation (DFG), which supported the research project "Warm forming to improve the formability in the field of microforming". Also, this work was carried out within the framework of the EC Network of Excellence "Multi-Material Micro Manufacture: Technologies and Applications (4M)".

References

[1] Breme J.: Titanlegierungen in der Medizintechnik. In: Peters M., Leyens C., Kumpfert J. (eds.): Titan und Titanlegierungen, DGM (1996), pp. 205-227

[2] Frommeyer G.: Herstellung, Eigenschaften und Auswahl von Titan und Titanlegierungen. Quintessenz, Jubiläumsausgabe (1998)

[3] ICFG Document No. 12/01: Warm Forging of Steel. In: International Cold Forging Group. Meisenbach, Bamberg, 2001, pp. 5-7

[4] Engel U., Egerer E., Geiger M.: Production of Microparts by Cold and Warm Forming. In: Proceedings of the CIRP Seminar on Micro and Nano Technology 2003, pp. 69-72

[5] Geiger M., Kleiner M., Eckstein R., Tiesler N., Engel U.: Microforming. Annals of the CIRP, Vol. 50/2 (2001), pp. 445-462

[6] Geißdörfer S., Engel U.: Mesoscopic Model - Simulation of Size Effects in Microforming, In: Proceedings of the 10th International Conference Metal Forming 2004, pp. 699-703

[7] Egerer E., Engel U.: Process Characterization and Material Flow in Microforming at Elevated Temperatures. Journal of Manufacturing Processes 5(2004)1, pp. 11-16

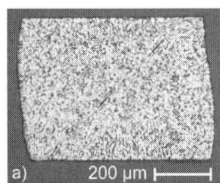

Fig. 9. Upsetted specimens (TiAl6V4, LK = 10 µm)
a) fracture at φ = 0.38 and 400°C
b) no fracture at φ = 1.06 and 600°C

Systems : Novel Product and System Designs

Ultrasonic plasticising for micro injection moulding

W. Michaeli, D. Opfermann

Institute of Plastics Processing at RWTH Aachen University, 52056 Aachen, Germany

Abstract

Ultrasonic energy is already used in the field of polymer welding. Research conducted at IKV shows a potential to use ultrasonic plasticising to generate melt for micro injection moulding. A test unit has been built to prove its potential as a plasticising unit. Based on these results the ultrasonic plasticising has been integrated into the micro injection moulding process. This leads to a significant reduction of the cycle time. Further research will focus on determining processing parameters for different materials and to further optimise the equipment.

Keywords: micro injection moulding, plasticising, ultrasonics, melting

1. Motivation

Injection moulding of single micro parts with shot weights of less that 10 mg poses a great problem. Screw based plasticising is limited by the minimum screw diameter of nowadays 14 mm. Newest developments show plasticising units with a screw diameter of only 12 mm. This diameter can be seen as the minimum for the use with standard pellets.

Every screw contains a certain amount of melt within its flight. In micro injection moulding this amount is very large in comparison to the actual shot weight. That leads to a long residence time of the polymer. Therefore the minimum screw diameter limits the achievable accuracy and the minimum shot weight. In order to make single micro parts producers sometimes design moulds with extremely big sprues to meet the minimum shot weights of their injection moulding machines [1]. Figure 1 shows parts for a toy train which possess a sprue that weights about 85 % of the shot weight.

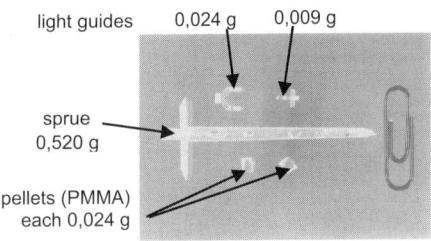

Fig.1 Example for a micro parts with a big sprue [Märklin]

A possible alternative for plasticising is the use of conductive heating in combination with a plunger [1, 2, 3]. A prototype of a micro injection moulding machine using this concept was build and investigated at IKV [1]. The set-up can be seen in Figure 2. The plunger diameter can be chosen as small as required to assure the exact dosing of the melt. In this case the injection plunger has a diameter of 2 mm. The maximum shot volume of the machine is about 0,1 cm³. Other institutions also do research in the area of plunger plasticising with plunger injection [4]. Heat conduction in polymers is poor compared to other

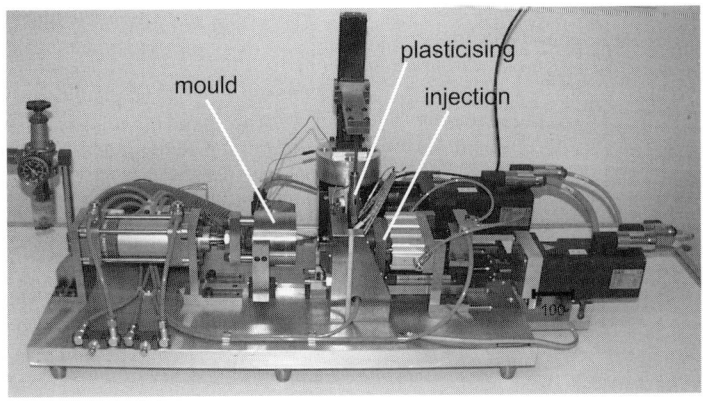

Fig. 2 Set-up of the prototype of the micro injection moulding machine

materials. Therefore plasticising by conductive heating is time consuming and can extend the cycle time [5]. To improve the efficiency of plasticising the use of ultrasonics is seen as a promising alternative.

2. Ultrasonic plasticising

Ultrasonics are used successfully for welding and riveting in plastics processing, since it enables to produce fast and neat bondings. The process times are in the range of seconds. The energy transfer from the ultrasound emitting sonotrode to the polymer is very fast. The plasticising in the joining zone is based on dissipation of ultrasonic energy within the polymer [6, 7]. During the ultrasonic welding process only a small area between the parts, which have to be joined, is plasticised.

In order to investigate the capability of ultrasonics to operate as a plasticising means a test unit was developed. The set-up of the test unit is shown in Figure 3.

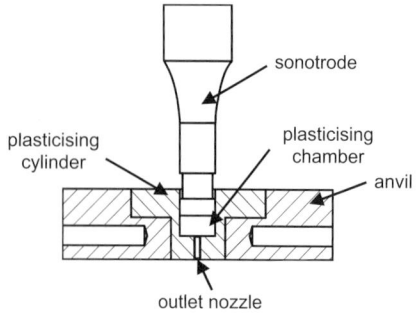

Fig. 3 Set-up of the test unit

The polymer in the form of pellets or plastic sheets is filled into the plasticising chamber. Then the sonotrode is pushed downwards and starts transmitting ultrasonic waves into the polymer. The resulting melt is pushed through the outlet nozzle by the downward movement of the sonotrode. The temperature of the exiting melt is measured by a pyrometer. The actual set-up is shown in Figure 4. The test unit is set underneath a conventional ultrasonic welding press.

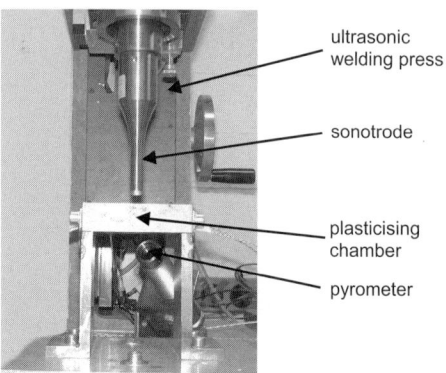

Fig. 4 Ultrasonic plasticising unit

During the tests different machine parameters were measured, including the ultrasonic generator output, the amplitude and the movement of the sonotrode. Also recorded is the trigger power, at which the emission of ultrasound starts, the holding force during the ultrasound emission and temperature of the melt when exiting the test unit. With these results a processing window as seen in Figure 5 is defined.

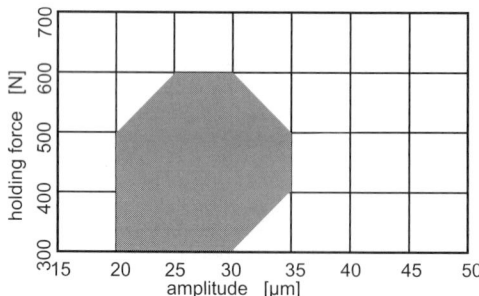

Fig. 5 Process window for processing POM

The material used in these preliminary tests is POM (polyoximethylene). The plasticising takes less than three seconds for an amount of melt of about 500 mg. To collect enough measuring data the amount of material plasticised is bigger than needed for a typical micro part. The first test results clearly show the potential of plasticising by ultrasonics. The plasticising time can be reduced to about one second (Figure 6).

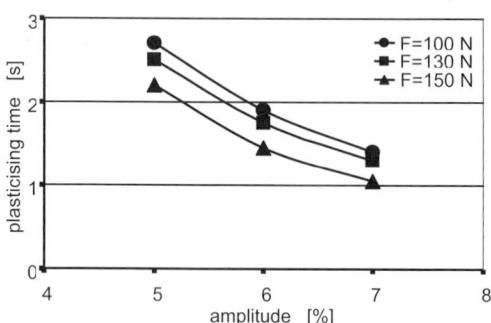

Fig. 6 Plasticising time depending on amplitude and holding force

The amount of energy transmitted into the material sometimes leads to thermal degradation of the polymer. Optimal plasticising results are achieved using medium settings. This leads to plasticising times of around two seconds to prevent thermal degradation. The plasticising results regarding homogenising and morphology of the molten mass have been evaluated by microscopy. As shown in Figure 7 the material has crystallized with a regular and homogeneous structure.

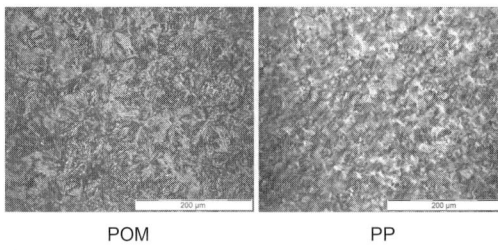

POM   PP

Fig. 7 Morphology of plasticized samples

3. Ultrasonic plasticising with direct injection

Ultrasonic plasticising can be used to directly produce plastics micro parts on the ultrasonic welding press. It is an interesting possibility to use the sonotrode as an injection plunger itself. The material is plasticized and injected. This general set-up of this idea is shown in Figure 8. Instead of pellets a plastics sheet is used. This method has the advantage that dosing the necessary amount of polymer is easier. The amount of melt generated during plasticising is determined by the diameter of the sonotrode and the thickness of the plastics sheet. The cavity is attached under the outlet nozzle. The melt generated during plasticising is pushed into the cavity by the downward movement of the sonotrode and forms a micro part.

The sonotrode pushes downwards and starts to plasticise the sheet (step 1). During the plasticising the sonotrode injects the melt into the cavity which is located underneath the plasticising chamber (step 2). After the plasticising the sonotrode is moved upward (step 3) and the part can be demoulded (step 4). Meanwhile the plastic film is moved in position to provide new material for the next plasticising (step 5) [8].

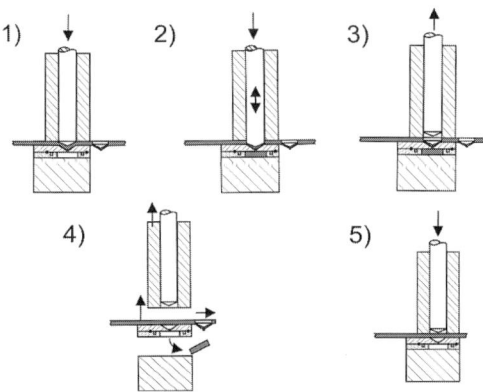

Fig. 8 Ultrasonic plasticising with direct injection

The actual test unit is shown in Figure 9. It is fitted under the ultrasonic welding press. The demoulding is accomplished by retracting the cavity with pneumatic cylinders. The mouldings produced using this method show sink marks on their surface. These are the result of shrinkage due to a missing holding pressure. This is because of the technical limitations of the welding press. The achievable maximum welding force is

limited. Therefore the maximum injection pressure is approx. 15 bar in the plasticising chamber. Due to the loss of pressure from the plasticising chamber to the cavity the cavity pressure is even lower (approx. 5 bar).

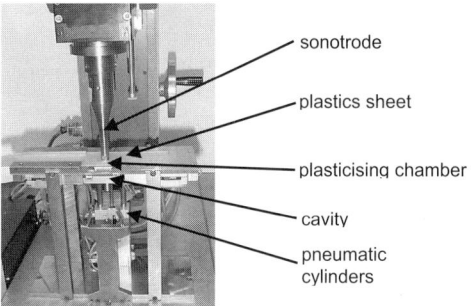

Fig. 9 Ultrasonic plasticising with direct injection

The low injection pressure and the missing holding pressure lead to a poor moulding quality. The resulting shrinkage causes a low geometric precision and low surface quality. The part quality cannot be compared to parts moulded using an injection moulding machine. The ultrasonic plasticising with direct injection offers the possibility to produce simple parts with small shot weights using straight forward technology (Figure 10). With some limitations the quality is acceptable. The possibility of a redesign of the welding press to increase the injection pressure has to be evaluated in further investigation.

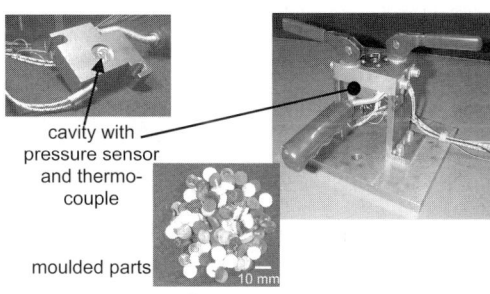

Fig. 10 Direct injection and moulded parts

4. Integration into the injection moulding process

The plasticising by ultrasonics offers the possibility to integrate this system into micro injection moulding machines. Instead of the material hopper or the plasticising cylinder the plasticising chamber is attached to the injection moulding machine (Figure 11). The polymer is plasticized by the ultrasonic welding press and transferred into the injection chamber by the downward movement of the sonotrode. After the plasticising process the melt is located in front of the injection plunger and is injected by the machine. With the injection moulding machine the necessary injection velocity and injection pressure to mould micro parts with high precision is ensured. The problems described above with the direct injection can be avoided.

348

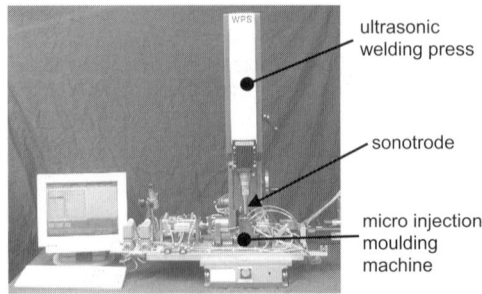

ultrasonic welding press

sonotrode

micro injection moulding machine

Fig. 11 Micro injection moulding machine with ultrasonic plasticising

5. Conclusion

Ultrasonic plasticising can be used to generate the small amounts of melt necessary in micro injection moulding. Due to the fast energy dissipation of ultrasonics in comparison to heat conduction or screw based plasticising, a reduction of cycle time is possible. The method can be integrated into a conventional micro injection moulding process. Based on these results the ultrasonic plasticising with direct injection has been investigated. As described above there are limitations to the parts quality. This method cannot be seen as an alternative to micro injection moulding, but there may be certain areas where the achieved quality is acceptable.

6. Future research

Since the plasticising by ultrasonics proved to have potential in micro injection moulding, further research is planned. Up to now the research was very basic and mainly a proof of concept. In the future detailed investigations will focus on the theoretical background of the effects taking place during the melting phase. Also the fusibility of different polymers will be analyzed. The influence of fillers on the plasticising behaviour will be analysed. Aim is to gain knowledge of this new plasticising process and to introduce this innovation in micro injection moulding.

Acknowledgements

The investigations presented in this paper received financial support by the Deutsche Forschungs-gemeinschaft (DFG), to whom we extend our thanks.

References

[1] Spennemann A. A New Machine and Processing Technology for the Injection Moulding of Micro Parts, Dissertation, RWTH Aachen, 2000

[2] Michaeli W. and Spennemann A. New plastification concepts for micro injection moulding. Proceedings Micro System Technologies, pp. 303-308, 2001

[3] Michaeli W. Haberstroh E. Gärtner R. Lützeler R. Schulz J. Wehr H. Ziegmann C. Mikrotechnik – Neue Anwendungsfelder, Maschinen- und Verfahrenstechnik. 21. Internationales Kunststofftechnisches Kolloquium, Aachen, 2002

[4] Jüttner, G.: Plastifiziereinheiten für kleinste Schussgewichte. Kunststoffe 94 (2004) 1, pp. 53-55

[5] Michaeli W. Gärtner R. Opfermann D. New plastification concepts for micro injection moulding. Proceedings of euspen International Topical Conference, 2002

[6] Potente, H.: Zur Frage der Energieumwandlung beim Ultraschallschweißen von Thermoplasten, Plastverarbeiter 22 (1971) 8, S. 556-562; 9, S. 653-658

[7] Ritter, J. Untersuchungen zur Energieumwandlung und zum Schwingungsverhalten des Systems Sonotrode, Fügeteile und Amboß beim Ultraschallschweißen ausgewählter Thermoplaste, Technische Universität München, Dissertation, 1986

[8] Michaeli W. Opfermann D. Gärtner R. Ultrasonic Plastification Concepts in Micro Injection Moulding. Proceedings of the euspen International Topical Conference. Aachen, May 19th-20th, 2003

Multi-Material Micro Manufacture
W. Menz, S. Dimov and B. Fillon (Eds.)
349

The micro-machining evaluation of non-metallic materials – by a fluid guided laser

Paul C. Snowdon[a], David Wood[a], Paul G. Maropoulos[b]

[a] School of Engineering, Durham University, South Road, Durham, DH1 3LE, UK
[b] Department of Mechanical Engineering, University of Bath, Bath, BA2 7AY, UK

Abstract

The waterjet-guided laser technique was originally developed to reduce the heat-affected zone near the cut, but many other advantages were observed due to the use of a waterjet rather than an assist-gas stream applied in classical laser cutting. This paper reports on the application of a fluid guided laser system, as an agile tool for the micro-machining of MEMS substrates, and materials associated with the expanding MEMS industry. An increasing number of applications require a more superior machining quality than can be achieved using standard/classical laser cutting. The materials machined during these evaluations include, gallium arsenide (GaAs) & silicon wafers, ceramic packaging (alumina) and kapton. An Nd:YAG laser operating at 1064 nm (infra red) and frequency doubled 532 nm (green) was employed for the machining of the various non-metallic materials.

Keywords: Fluid guided laser, Laser Microjet®, MEMS, micro-machining

1. Introduction

The field of laser machining is extremely diverse, and virtually all commercially available high power lasers have applications for which they are particularly suited [1]. Termed the Laser-Microjet® or LMJ by its developers Synova SA, this hybrid technology combines a high power laser and a low pressure water jet. The high power pulsed laser has pulse durations varying between 200 nanoseconds and 1 milliseconds, this laser is coupled within a guiding beam of water to simultaneously cut and cool the work-piece, as shown in Fig. 1. During the laser cutting procedure the low-pressure water jet performs three primary functions [2];

1. as a light guide for the laser beam
2. allows cooling of the work piece
3. removes the process debris

The laser beam is focused in a nozzle while passing through a pressurised water chamber [3]. The diameter of the waterjet nozzle, which ranges between 25μm and 75μm, directly controls the cutting diameter of the laser. Control of the geometry of both chamber and nozzle are crucial for efficient coupling of the laser beam to the water jet. Finally, the water exits the nozzle as a low-pressure jet of between 100 bar and 400 bar.

2. Defining the experiments

In any systemic process, repeatability is a high priority requirement. The ability to precisely control a machining parameter is crucial to the accurate reproduction of a desired structure. This series of experimental evaluations was designed to give an insight into the potential capabilities of the Laser-Microjet® (LMJ) as an agile machine tool for the following application areas within the MEMS industry:

- Patterning of MEMS structures on silicon.
- Cutting of various materials useful in the IC and MEMS industries, e.g. polymers, ceramics.
- Dicing of integrated circuits on a variety of substrates.

Particular interest was directed towards speed, cleanliness of cut, and repeatability. The minimum feature size is much greater than that from deep reactive ion etching (DRIE-the major silicon patterning tool), however the LMJ could have advantages in other respects.

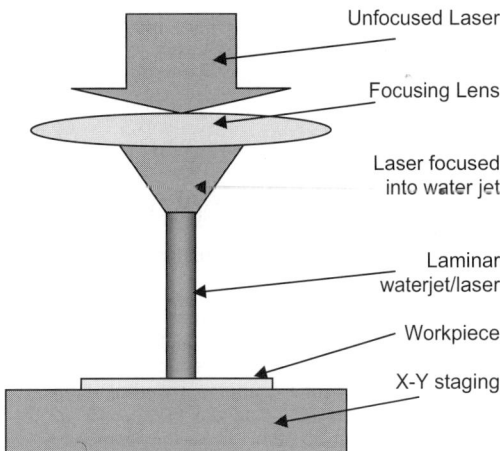

Fig. 1. The basic operating principle of the LMJ.

3. Holding the workpiece

The evaluations commenced with the machining of the silicon wafers that were mounted to standard dicing frames and this procedure was used throughout the evaluations. The frames contained a UV activated LaserTape® [4] that was developed specifically for this process. The dicing frames were then mounted on a vacuum chuck within the LMJ. Table 1 lists the materials used during this series of experimental evaluations.

Table 1.
Materials machined during evaluations.

Evaluation number	Deposited material descriptions
1	2 mm Via hole (Silicon)
2	Multiple 1 mm Via holes (Silicon)
3	Spirals (Silicon)
4	Pillars (Silicon)
5	Polyimide (Kapton)
6	Polyethylene Tetraphthalate (PET)
7	Ceramic chip packaging
8	Gallium Arsenide wafer

4. The Evaluations

4.1. Evaluations 1 & 2

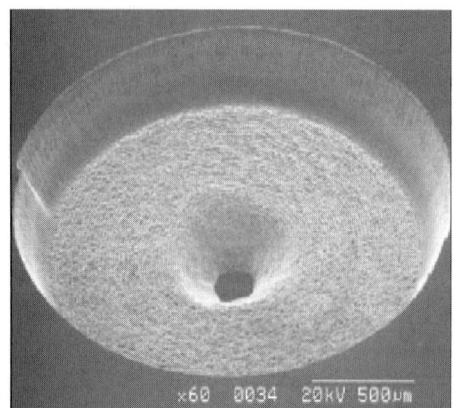

Fig. 2. Image of 2mm diameter via hole.

The machining procedure to create via holes was based on a spiral cutting raster pattern to gradually reduce the wafer thickness. By reducing the diameter slowly to a zero centre diameter a via hole could be created. The machining time was 90 seconds for the 2mm via hole and 33 seconds for the 1mm via hole, shown in Fig. 2. and Fig. 3. respectively. The results

show an excellent repeatability in that the via holes are virtually identical to each other. They show a consistent round shape, a clean cut and a relatively debris free surface, considering there has been no dedicated cleaning procedure.

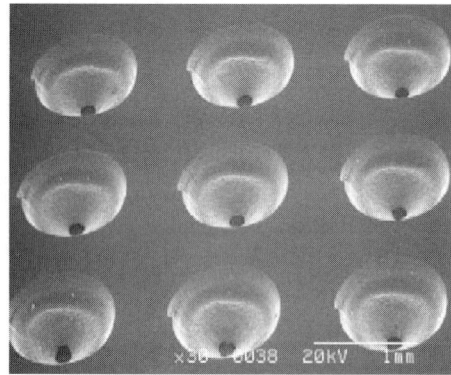

Fig. 3. Image of 1mm diameter via hole array.

Because of the way that the LMJ functions, i.e. it uses the water jet as a fluid waveguide, once the water jet is disturbed the laser becomes ineffective as a cutting tool. This is a limiting factor to the achievable depth to diameter ratio and explains the sloping nature of the walls. The machine manufacturers, Synova SA, claim a surface roughness of 3μm peak to trough and this appears to be the case and is shown in Fig. 4. Fig. 5 shows the back face (unpolished surface) of the silicon wafer showing through hole of the via hole.

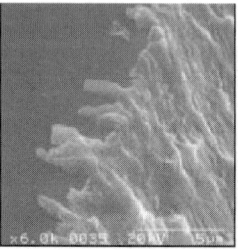 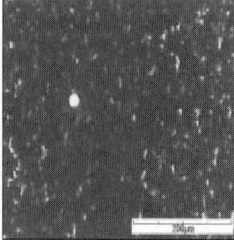

| Fig. 4. Shows the rim of the 1mm diameter via hole and debris contamination. | Fig. 5. Image of a via hole taken from back face of the wafer. |

4.2. Evaluation 3

The machining process "Spirals" is based on the same cutting program as the "Via holes" but with a significantly larger pitch. Of special interest was the width of the remaining part of the silicon wafer forming the wall of the spiral to make it as thin as possible with a high aspect ratio. This experiment would examine the LMJ's ability to machine areas while leaving structures untouched, and its ability at the machining of arbitrary contours, facilitating the production of multi-project wafers and materials.

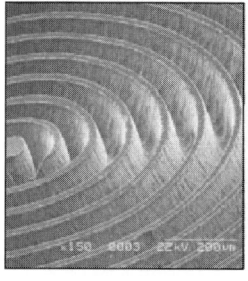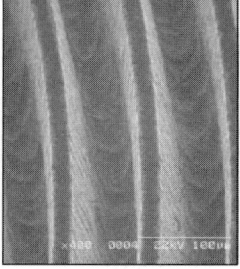

Fig. 6. SEM image of the sidewalls of the spiral.

Fig. 7. SEM image the spiral showing the bottom of the cut.

After the cutting process was complete, the laser was switched off and the program rerun using only the water jet. This was done to clean out the channels of any remaining loose debris. The examination showed that the water-jet had effectively cleaned the channel and that the walls of the spiral were smooth and consistent in dimension.

4.3. Evaluation 4

The machining process "Pillars" removed a large area of surface material, leaving pyramid shaped pillars with a top surface area of approximately 10µm x 10µm and 120µm in depth. The consistency and repeatability of the machining process is evident in Fig. 8, and Fig. 9. shows the cleaning process was again successful as there was no debris apparent during the Scanning Electron Microscope (SEM) examination. This cleaning procedure described in the previous experiment was also conducted on this evaluation.

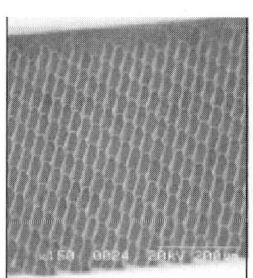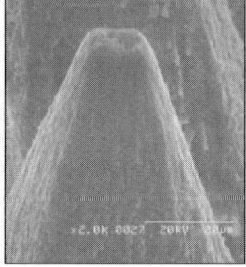

Fig. 8. Image of the silicon pillar array.

Fig. 9. Single silicon pillar.

4.4. Evaluation 5

It was already understood that Polyimide (kapton) could be machined by means of a sub 400nm laser [5, 6], and several papers and reports are available on

this subject. However this series of evaluations was completed with the 532nm Q-switched green laser. Fig. 10. shows the Polyimide (kapton) was cut cleanly with good repeatability, with heat-affected zone restricted to approximately 250µm from the cut due to the immediate cooling effect of the water jet. The weave pattern shown in the photograph is from the supporting LaserTape®.

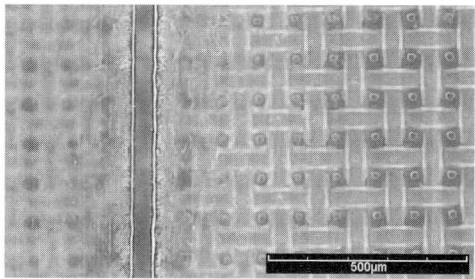

Fig. 10. 40µm cut in sample 5, kapton film.

4.5. Evaluation 6

The clear Polyethylene Tetraphthalate (PET) was left uncut and unmarked by the LMJ due to the laser energy at 532nm not being absorbed by the material. This was the only unsuccessful experimental evaluation from this series of trials.

4.6. Evaluation 7

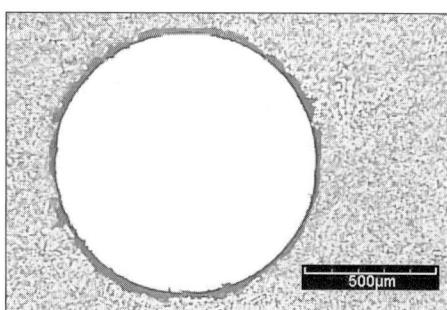

Fig. 11. Ceramic packaging with 1mm Ø hole.

Fig. 11. shows a 1mm diameter hole machined through 1mm thick ceramic (Alumina) chip packaging with gold plating. Several sections of the plating have peeled at the edge of the hole; this is due to some inconsistency of the plating process.

4.7. Evaluation 8

Gallium arsenide (GaAs) as a MEMS substrate is a brittle, difficult- to-process material [7] and the machining of GaAs, by diamond saw or by conventional laser, releases arsenic into the atmosphere. This is in the form of either dust or as arsine gas. The LMJ, however, contains the Arsenic within the water, which can be filtered out and disposed of in an environmentally controlled manner [7, 8]. The photographs of the cut featured Fig. 12. and Fig. 13. is of excellent quality, being debris free, clean and burr free.

Fig.12. shows a cross street pattern.

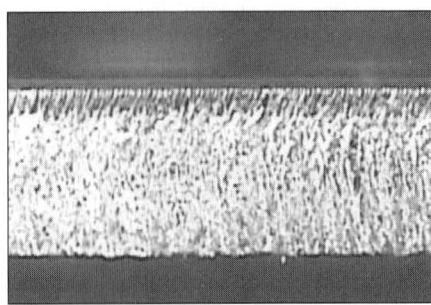

Fig. 13. Image of GaAs cut edge.

5. Conclusions

As stated at the beginning of this report, this series of evaluations was designed to give an insight into the potential capabilities of the Laser-Microjet® as an agile machine tool for different applications within the MEMS industry [9]. The results shown in the images in section 4 of this report all show a successful conclusion apart from the Polyethylene Tetraphthalate experiment. The cuts are clean, reliable, accurate, and show negligible heat damage. In particular, the cutting of complicated shapes such as spirals and pillars give excellent results, achieving faithful consistent pattern transfer. The water-jet guided laser is especially adapted for thin wafers, due to the absence of thermal damage, mechanical stress, and excellent cutting quality i.e. parallelism, smoothness [10].

The minimum feature size is limited when compared to Deep Reactive Ion Etching (DRIE), but if this is not an issue then the LMJ looks an attractive tool. It is direct write, fast, flexible and clean in both operation and in waste disposal [11]. The range of materials, which are suitable for the LMJ, is also great. The LMJ has proved to be an agile tool for the MEMS industry,

and future evaluations will lead to additional new applications. Synova SA has recently announced the potential introduction of a frequency tripled 355nm laser, which will further enhance the capabilities and flexibility of the LMJ.

Acknowledgements

The authors acknowledge the assistance and expertise of the Synova SA engineering staff. The authors are grateful to One North-East, via the Centre of Excellence CENAMPS, for the financial support of this project.

References

1. Henry, M., et al. *Laser Milling – A Practical Industrial Solution for Machining a Wide Variety of Materials*. in *The 5th International Symposium on Laser Precision Microfabrication*. 2004. Nara, Japan.
2. Sibailly, O., F. Wagner, and B. Richerzhagen, *Laser micro-machining in microelectronic industry by water jet guided laser*. The International Society for Optical Engineering, 2004. **5339**: p. 258-264.
3. Truskey, G.A., et al., *Total internal reflection fluorescence microscopy (TIRFM)*. Journal of Cell Science, 1992. **103**: p. 491-499.
4. Wagner, F.R., et al., *Water jet guided laser versus saw dicing*. The International Society for Optical Engineering, 2003. **4977**: p. 75-77.
5. Chang, W.S., et al., *Photothermal three-dimensional fabrication of polymers using diode-pumped solid state lasers*. Journal of Microlithography, Microfabrication, and Microsystems, 2004. **3**(3): p. 472-477.
6. Arnold, N. and N. Bityurin, *Model for laser-induced thermal degradation and ablation of polymers*. Applied Physics A: Materials Science & Processing, 1999. **68**(6): p. 615 - 625.
7. Perrottet, D., R. Housh, and B. Richerzhagen, *Laser-Microjet Ready for Dicing of GaAs Wafers*. Future Fab International, 2006. **20**.
8. Mayor, L., N.M. Dushkina, and R. Romanowicz, *Dicing of GaAs Wafers -Security and Yield Issues: What Dicing Technologies Have to Offer?*, in *Future Fab Intl*. 2003, www.future-fab.com.
9. Snowdon, P.C., D. Wood, and P.G. Maropoulos. *The Laser Microjet as an Agile Micro-Machining Tool*. in *MME Workshop*. 2005. Göteborg, Sweden.
10. Sibailly, O., F. Wagner, and B. Richerzhagen, *Laser-Edge Grinding of Thin Wafers with the Water-Jet-Guided Laser*. Future Fab Intl, 2004. **16**.
11. Kröninger, W., et al., *Stress Release Increases Advantages of Laser-Microjet Dicing*. Semiconductor International, 2005. **28**(4): p. sp4-sp8.

353

Low cost capacitive inclination sensors based on selectively metallized polymer

Benz Daniel [a]; Botzelmann Tim [b]; Kück Heinz [a,b];Warkentin Daniel [b]

[a] Institut für Zeitmesstechnik, Fein- und Mikrotechnik, Universität Stuttgart, Breitscheidstr. 2b, 70174 Stuttgart, Germany
[b] Hahn-Schickard-Gesellschaft, Institut für Mikroaufbautechnik, Breitscheidstr. 2b, 70174 Stuttgart, Germany
email: benz@izfm.uni-stuttgart.de, Phone:++49 711 121 3706, Fax: ++49 711 121 3705

Abstract

We report on an innovative low cost concept for micromechanical capacitive inclination sensors based on MID-technology (Moulded Interconnect Device) where three dimensional polymer devices are fabricated by injection moulding and covered by a structured metal layer using electroless plating and laser ablation. The sensor concept based upon an extended electrode design which combines two differential capacitors arrangements for a measurement range of ±180°. First MID-demonstrators were fabricated and characterized. The demonstrators show very promising properties. The sensors have a nice linear characteristic within ±45° and 45°-135° with a resolution of about 0,01° and a nonlinearity better than 0.5% FSO. At different angular rates a dynamical hysteresis occurs. The dynamical hysteresis is proportional to the rotational speed of the sensor. The temperature dependence is very low and almost linear. No failure occurred within first reliability tests of 100 sensors with thermal shock test and constant damp heat storage.

Keywords: inclination sensor, moulded interconnect device, non-silicon material

1 Introduction

Inclination sensors are used for many applications, like automotive-, automation- and consumer-engineering. Examples are the measurement of the spacial position of automobiles with intelligent hand brake systems or position control of robots, electric irons, mobile phones, etc..

The concept introduced in this paper based on MID-technology (Moulded Interconnect Device) where three dimensional polymer parts are fabricated by injection moulding and covered by a structured metal layer using electroless plating and laser ablation [1]. MID-technologies are already in use for the packaging of micro devices like infrared motion sensors and camera modules [2,3]. Recent investigations show that MID-technology can also be used to build sensors and actuators [4,5,6].

2 Concept for inclination sensor and principle of operation

Figure 1 shows an exploded schematic view of the sensor concept. The sensor consists of two MID-parts, a dielectric fluid and an ASIC for electrical readout.

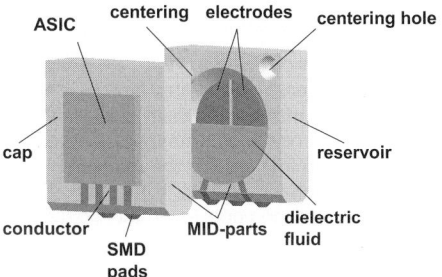

Fig. 1. Schematic view of MID-inclination sensor concept.

Both MID-parts form an enclosed cylindric cavity. Half of the cavity is filled with a dielectric fluid. For a ±90° inclination sensor two semi circular shaped electrodes 1a and 1b are manufactured on one front end of the cylindric cavity. One common circular shaped electrode 2 is manufactured on the opposite as indicated in the schematic view of figure 2.

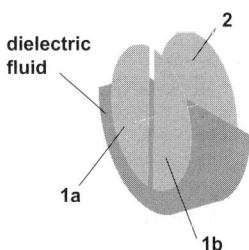

Fig. 2. Principle of operation of the ±90° inclinometer.

As can be seen in figure 3 the electrodes and the fluid form two variable capacitors C_1 and C_2. The capacitances of the two capacitors depends on the volume between the electrodes filled with the fluid. In horizontal position both capacitances are equal. If inclination of the housing occurs the fluid keeps its horizontal position. Therefore the capacitances change inversely. The change of the two capacitors can be measured with a differential capacitance to voltage converter. This electronic readout device (e. g. ASIC) can be mounted on the MID-parts of the inclinometer using flipchip-technology or wire-bonding [7,8]. A first demonstrator with electrodes on two Printed Circuit Boards showed the high potential of the concept [9].

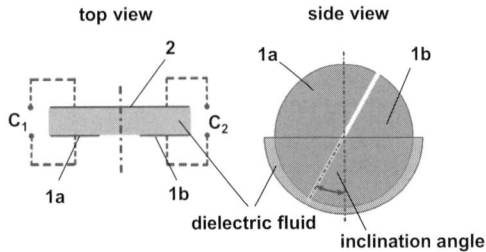

Fig. 3. Differential capacitance arrangement of the ±90° inclination sensor.

To increase the measurement range up to ±180° an extended design of electrodes was developed. The layout of the electrodes of an ±180° sensor is schematically shown in Fig. 4.

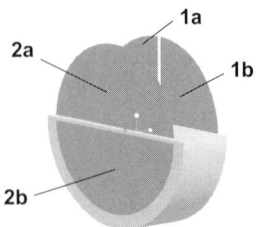

Fig. 4. Design of electrodes for ±180° inclination sensor.

In contrast to the ±90° electrode design the common electrode is also divided into two semi circular shaped electrodes 2a and 2b. The second pair of electrodes is rotated by 90°. By alternately combining the electrodes 1a-1b and 2a–2b the electronic readout creates two differential capacitance arrangements. Arrangement 1 with the variable capacitors 1a-2 and 1b-2 (analogous to the schematic in figure 2) and arrangement 2 with the capacitors 2a-1 and 2b-1. Therefore the sensor generates two output signals with 90° phase shift.

3 Fabrication of a Demonstrator

To demonstrate the principle of operation, a demonstration device was constructed. The housing for the cylindric cavity is based on MID-technology and consist of two parts; a reservoir and a cap. The reservoir and the cap of the demonstrator as shown in figure 5 are made from liquid crystal polymer using injection moulding. The cylindric cavity is manufactured in the reservoir element. Both parts have SMD-Pads for the connection to a PCB. The polymer parts are covered by copper using electroless plating. Afterwards the electrodes of the capacitors and conductors are manufactured using laser ablation. In a next step the remaining copper is reinforced with a nickel and a gold layer using electroless plating. Before sealing reservoir and cap the cavity is half filled with the dielectric fluid. Due to its dielectric constant of about 3 and its adjustable

viscosity a silicon oil is used. To seal the sensor parts and the cavity an adhesive is applied in an angular groove. So far the yield of the semi-automated fabrication process is about 95%.

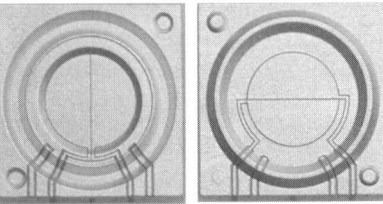

Fig. 5. Reservoir and cap of the inclination sensor.

For characterisation the sensor is mounted on a PCB and connected to the electronic readout using an isotropic conductive adhesive. Additionally an underfiller is applied for mechanical strengthening. A photo of the mounted device is shown in figure 6. Table 1 gives the details of the sensor specifications.

Fig. 6. Sensor mounted on PCB with electronic readout.

Table 1
Specifications of the inclination sensor

Technical specification	value
Overall dimensions [mm³]	12,3 x 12,5 x 2,6
Measurement span [°]	± 180
Diameter of electrodes [mm]	6
Gapwidth of electrodes [µm]	500
Silicon oil volume [µl]	10,5
Capacitance C [pF]	0,5
Change of capacitance ΔC/C of configuration 1 at 90°	90 %

4 Measurement Results

For characterization the demonstrator is mounted on an inclination test system. The demonstrator is inclined between ± 180°. The output signals of the electronic readout as well as the inclination of the test system are measured. Figure 7 shows the sensor output of both differential capacitance arrangements over the position of the test system.

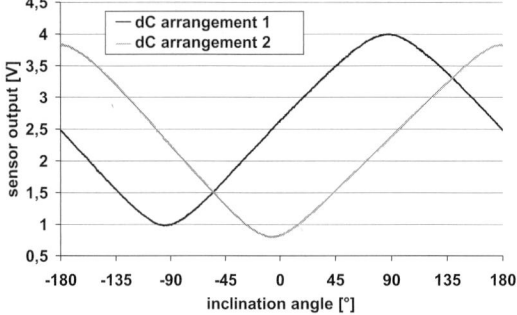

Fig. 7. Sensor output over inclination angle.

Figure 8 shows the characteristic of arrangement 1 between ±90° inclination and the characteristic of arrangement 2 between 0 to 180°inclination. As can be seen the sensor has a nice linear characteristic. Within ±45° for arrangement 1 and within 45° to 135° for arrangement 2 the nonlinearity is better than 0.5% of the Full Scale Output. The nonlinearity beyond this range indicate an incorrect filling level of the demonstrator and capillarity effects of the fluid. By combing both signals the nonlinearity is better than 0.5% FS0 over the whole measurement span of ±180°.

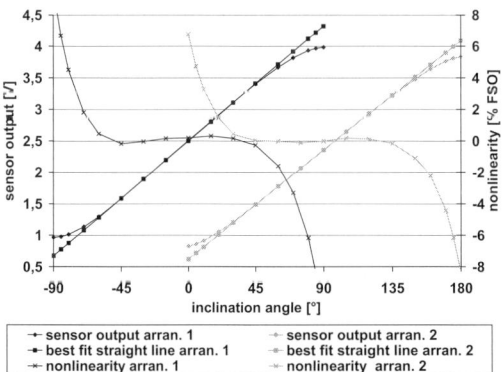

Fig. 8. Sensor output and nonlinearity over inclination angle.

The resolution of the sensor is about 0,01° and shows a high repeatability. Also the selectivity at cross inclination is better than 0.1% FSO/ °.

To investigate the dynamical behaviour the demonstrator is inclined between ±90° at different angular rates. Figure 9 shows the sensor output over inclination at different angular rates for electrode arrangement 1. Electrode arrangement 2 shows a similar characteristic. With increasing rotational speed one can observe a speed dependent hysteresis. This hysteresis is proportional to the rotational speed as indicated in figure 10. Basically the dynamical hysteresis depends on the viscosity of the dielectric fluid and the width of the cavity.

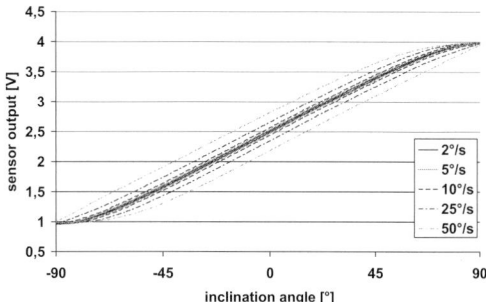

Fig. 9. Dynamical hysteresis over inclination at different angular rates.

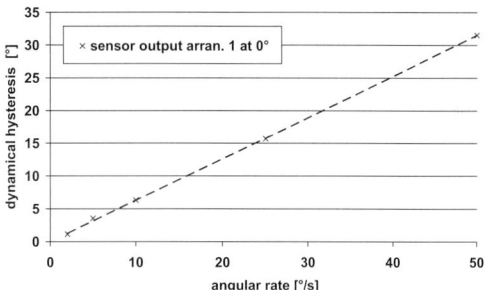

Fig. 10. Dynamical hysteresis over angular velocities.

Furthermore the response of the sensor on temperature was tested in a climatic chamber. Therefore the sensor output was measured at an inclination of 0° while the temperature changed between -40°C and 125°C. As indicated in figure 11 the sensor output changes about 40 mV over the temperature difference of 165°C.

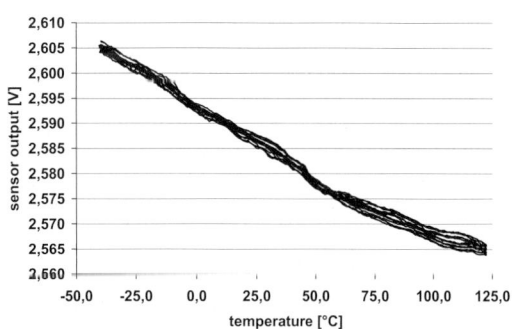

Fig. 11. Sensor response on temperature.

Table 2 summarizes the measured sensor data.

356

Table 2
Perfomance data of the demonstrator

Data	Value
Full Scale Output (FSO) [V]	approx. 3,00
Sensitivity [ΔV/°]	0,02
Resolution [°]	0.01
Nonlinearity max. [% FSO]	± 10
Nonlinearity between ± 45° of arrangement 1 [% FSO]	± 0,5

4. Environmental test

To show the reliability of these demonstrators thermal shock test and constant damp heat storage test were performed. After 1000 thermal shock cycles of 100 MID-demonstrators at temperatures from -40°C up to 85°C no sensor failed. Also demonstrators withstand 750 hours constant damp heat at 85°C / 85% relative humidity.

5. Conclusion and Outlook

We demonstrated a new principle of operation for capacitive inclination sensors to be fabricated in MID-technology. First demonstrators made of selectively metallized polymers were built, connected to the electronic readout and characterized. The sensors show a very promising characteristic.

An extended design of electrodes increases the measurement range of the demonstrator up to ±180° by combing two differential capacitance arrangements. Therefore the nonlinearity caused by an incorrect filling of the sensor is reduced.

To test the reliability of the sensors, thermal shock test and heat damp storage were performed with no failure of the test sample.

The next stage of development is the improvement of the current electrode design referring to the demonstrators characteristic. Also the sensor requires optimisation for a fully automated fabrication and assembly processes.

References

[1] Mitglieder der Forschungsvereinigung 3-D MID e.V.; *3D-MID Technologie, Räumliche elektronische Baugruppen, Herstellungsverfahren, Gebrauchsanforderungen, Materialkennwerte*, Carl Hanser Verlag, München, 2004.

[2] Yamanaka, H., et al., *MID Technology to Miniaturize Electro-Optical Devices*, MID 2000, 4. International Congress Molded Interconnect Devices, Erlangen, Germany, Sep. 27 28, 2000, 129 138.

[3] Yagi, Y., et al., *Micro Camera Module with MID Using Flip-Chip Mounting Technology*, MID 2002, 5. Internationaler Congress Molded Interconnect Devices, Erlangen, Germany, Sep. 25 26, 2002, 99 107.

[4] W. Eberhardt, Th. Gerhäußer, M. Giousouf, H. Kück, R. Mohr, D. Warkentin, *Innovative concept for the fabrication of micromechanical sensor and actuator devices using selectively metallized polymers*, Sensors and Actuators, A 97-98 (2002) 473 477.

[5] M. Arnold, W. Eberhardt, M. Giousouf, H. Kück, G. Munz, M. Münch, M. Oprea, D. Warkentin, *Fabrication of an Electrostatic Miniature Valve from Metallized Microinjectionmoulded Polymers*, MICRO.tec 2003, München, Germany, Oct. 13 15, 2003, 319 322.

[6] W. Eberhardt, H. Kück, R. Mohr, D. Warkentin, *Low Cost Accelerometers made from Selectively Metallized Polymer*, Eurosensors XVII, Guimaraes, Portugal, Sept. 21 24, 2003.

[7] U. Keßler, W. Eberhardt, H.Kück, U.Scholz, *Flipchip Klebetechnik auf MID*, Workshop "Innovative Anwendungen der MID-Technik", HSG-IMAT, Stuttgart, Germany, Oct. 10, 2003, 7 18.

[8] U. Keßler, W. Eberhardt, H. Kück, *Adhesive Technology for Flipchip Assembly on Moulded Interconnect Devices (MID)*, High Density Microsystem Design and Packaging and Failure Analysis (HDP'04), The Sixth IEEE CPMT Conference, June 30, July 03, 2004, Shanghai, China, 242 247.

[9] D. Benz, T. Botzelmann, H. Kück, D. Warkentin, *On low cost inclination sensors made from selectively metallized polymer*, Sensors and Actuators, A 123 – 124, 2005, p. 18 – 22.

Multi-Material Micro Manufacture
W. Menz, S. Dimov and B. Fillon (Eds.)

High resolution low cost optical angular resolver

V. Mayer, D. Warkentin, H. Kück

Hahn-Schickard-Gesellschaft, Institute for Micro Assembly Technology, Breitscheidstr. 2b, 70174 Stuttgart, Germany

Abstract

At HSG-IMAT a new concept for a high resolution and low cost optical angular resolver is developed. The key element of the sensor concept is a disc with a high precision solid measure which is fabricated using the well known manufacturing process for compact discs (CD-Technology). Using this process, it is possible to fabricate a high precision solid measure in high quantities and at low manufacturing costs. To detect the angular position, a laser beam is focussed onto the solid measure. The beam is reflected from the backside of the disc onto a photo diode. The light intensity is modulated by diffractive microstructures of the solid measure. With a first experimental setup, the functional principle of the sensor was verified and the signal modulation with high accuracy was demonstrated. With a 10μm laser spot on the disc, diffractive microstructures with a width of 10μm (equivalent to 20μm pitch) are easy to detect. The experiments show that even fields smaller than 6μm can be detected with the setup. Incremental and absolute encoded systems can be realised with the sensor concept. For serial production of the sensor a 3D packaging in MID-Technology is proposed.

Keywords: angular resolver, rotary encoder, optical sensor, low cost, compact disc, laser

1. Introduction

Angular resolvers are used in many industrial applications for the detection of the angular position of rotary motions. Examples are the detection of steering angle in cars for driving assistant systems [1] or the position sensing of rotating units in machines [2].

For a contact free determination of the angular position normally capacitive [3, 4, 5], magnetic-inductive [6, 7, 8] or optical [9, 10, 11, 12] sensors are used [13]. Optical and magnetic sensors have the highest economical weight [14]. While magnetic sensors with limited resolution are mostly found in the low cost sector and for absolute encoded systems, optical sensors are provided as absolute or incremental encoded systems and can reach highest resolutions. High resolution optical angular resolvers mostly work with very cost intensive solid measures made of glass with lithographic structured coatings. Thus traditional optical angular resolvers usually show the disadvantage that high resolution is hard to combine with low manufacturing costs. Based on this discrepancy a new concept for optical low cost high resolution angular resolvers is developed.

2. Functional principle

The fundamental idea of the sensor is to use a micro structured plastic disc with a metallic coating as solid measure, known from the compact disc (CD) or digital versatile disc (DVD) technology. With the well known manufacturing processes for CDs or DVDs, it is possible to fabricate a high precise solid measure in high quantities at very low manufacturing costs.

Fig. 1 shows the microstructures of a data or audio compact disc with bits and bytes. Fig. 2 shows the microstructures of a test disc of the innovative optical angular resolver with geometrically positioned structures.

The solid measure consists of micro patterned and unpatterned fields which are assembled in a circle around the rotary axis. For an incremental encoded

system, the patterned and the unpatterned fields alternate. The micro patterned fields are diffractive phase gratings consisting of longish cavities with a wavelength dependent depth.

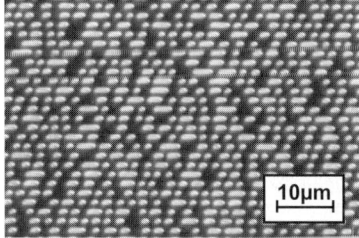

Fig. 1: Data/Bit structures of a compact disc

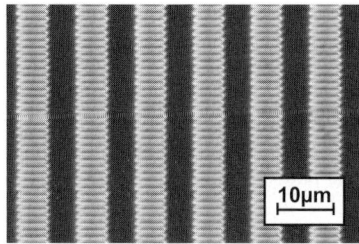

Fig. 2: Geometrical pattern of test disc for angular resolver

The optical sampling of the solid measure is done by a focused beam of a laser diode which is reflected by the back side of the solid measure onto a photo diode as indicated in Fig 3. The arrangement of laser diode, lens and photo diode is called the reading unit of the angular resolver.

The laser spot on the disc has about the same size as the patterned and unpatterned fields. The intensity of the beam in the zeroth order of diffraction is modulated by the patterned and unpatterned fields as indicated in Fig 4. When focused on an unpatterned field the laser beam is simply mirrored by the reflective coating on the backside of the disc onto the photo diode. The photo diode is detecting a high light intensity. When focused on patterned fields, the beam is inflected into higher orders of diffraction. The photodiode detects a low light intensity in the zeroth order of diffraction. A modulated intensity signal on the photo diode is the result when rotating the disc with the solid measure relative to the reading unit.

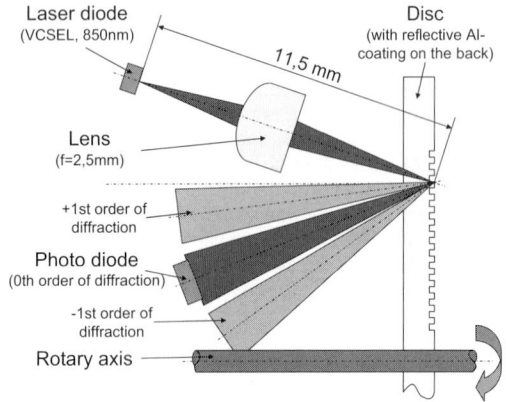

Fig. 3: Optical path of the reading unit

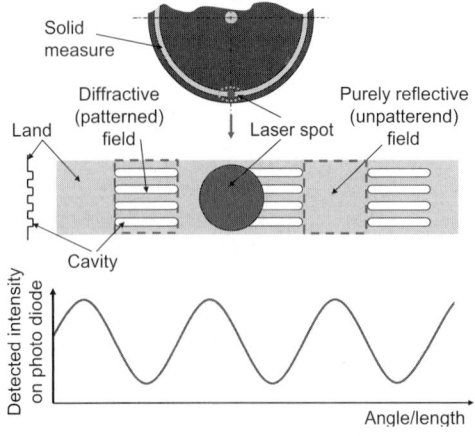

Fig. 4: Schematic view of the signal modulation

In case of an incremental encoded sensor, the detection of the rotation direction can be realized with two phase shifted reading units. One possibility to realize the phase shift is to use a double laser diode chip, see Fig. 5.

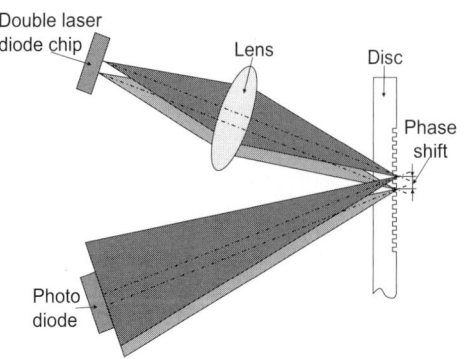

Fig. 5: Double laser diode chip for phase shift

The chip is assembled in a way that both beams are focused by a common lens onto a common photo diode. This results in two phase shifted laser spots on the disc. To obtain the two phase shifted signals from the photo diode, the laser diodes are driven alternately and the corresponding signals of the photo diode are used to form the two phase shifted signals for the detection of rotation direction. Since laser diodes and photo diodes are well known for their high switching frequencies, this can be done at high frequencies and high rotational speeds of the sensor.

With a double laser diode arrangement, a focus diameter of 10µm on the disc, a field width of 10µm and a disc with 50mm diameter, an incremental angular resolver with a resolution of about 0.01 degree can be realized by simply processing the harmonic output signal of the diodes as a digital signal. With an additional interpolation of the harmonic signals, the resolution could be easily improved by a factor of 100.

The proposed functional principle can also be used to realize an absolute encoded angular resolver.

One possibility for an absolute encoded system is to use different micro patterns for each field on the disc. Using those different micro patterns, different light intensities in the zeroth order of diffraction can be generated. An A/D-converter determines the intensity and therefore a corresponding absolute angle value.

Another possibility is to use a third kind of field that generates a third signal level of interference in the zeroth order. With this third, half digital signal level, the high levels of an incremental code can be superposed by a sequentially absolute binary code. Fig. 6 shows the concept.

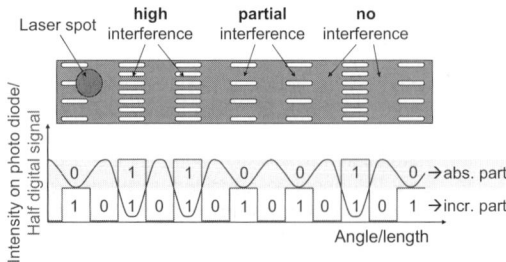

Fig. 6: Absolute sequentially encoding by a third signal level

3. Assembly concepts

Fig. 3 shows the three dimensional arrangement of the components. The three dimensional assembly and packaging of the components of the reading unit as well as the assembly of further components, e.g. electrical readout, can be realized using a plastic housing in 3D-MID-Technology (Moulded-Interconnect-Device-Technology) [15]. Using this technology, the optical system can be solidly arranged in the package.

Calculations of the fabrication and thermal tolerances show that displacements of the different components e.g. disc displacement or displacement of the optical parts can be neglected under certain circumstances [16, 17].

Fig. 7 shows the 3D concept of an angular resolver with a disc diameter of about 50mm in MID-Technology.

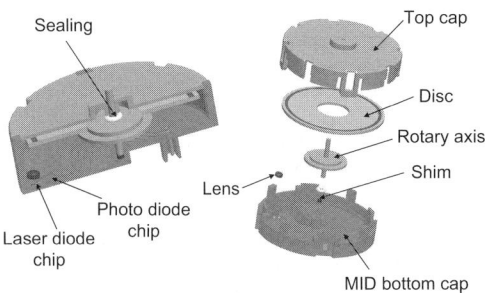

Fig. 7: 3D concept of an angular resolver in MID-Technology

4. Experimental setup

With a first simple experimental setup, the basic signal modulation has been verified [18]. The experimental setup has an optical path as shown in Fig. 3. A VCSEL laser diode with 850nm wavelength is used as a light source. The laser diode is focused by a glass lens, with 3mm diameter and 2.5mm focal length, to a laser spot diameter of about 10µm (1/e²) on the disc's backside. The total length of the optical path from laser diode to the backside of the disc is about 11.5mm. A simple IR selective photo diode is integrated in the setup, to detect the modulated zeroth order of diffraction.

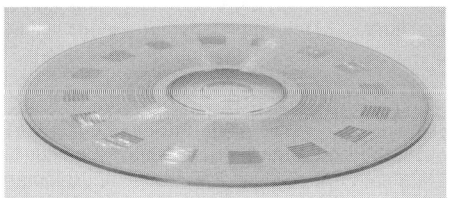

Fig. 8: Test disc for experimental setup

The test disc shown in Fig. 8 was produced with standard CD manufacturing equipment. Different incremental test structures with fields from 6 to 100µm width are on the test disc.

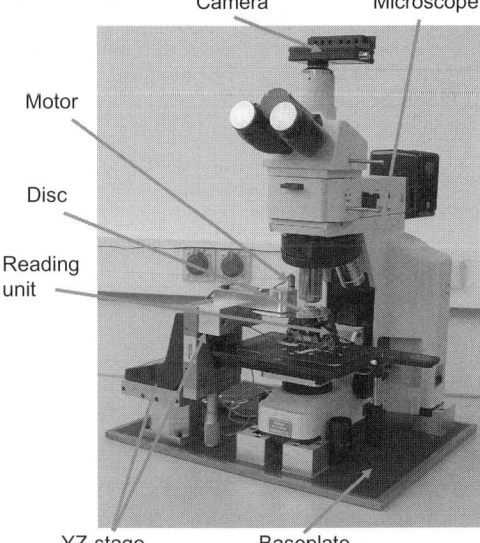

Fig. 9: Experimental setup

The whole experimental setup is shown in Fig. 9.

A light microscope is integrated in the experimental setup for the alignment of the components and the characterization of the laser beam or the micro pattern. The disc can be rotated by a motor with integrated gearbox.

The reading unit is placed on a xyz-stage of the microscope. It consists of an installation bracket and adjustable holders for the laser diode, the glass lens and the photo diode. The unit has also the possibility to add an aperture. Fig. 10 shows the reading unit in detail. Thereby the position of the optical elements is partially adjustable.

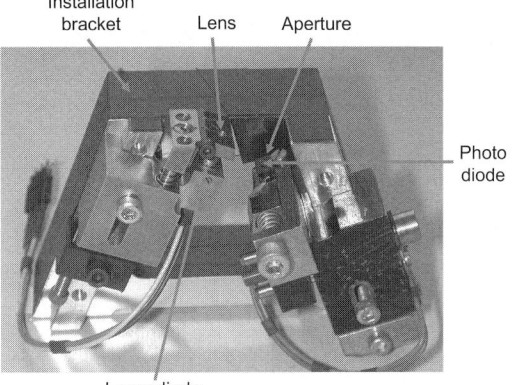

Fig. 10: Reading unit of experimental setup

5. Experimental results

With the described experimental setup, the basic signal modulation was verified [19]. Fig. 11 shows the signal modulation detected by the photo diode with an incremental pattern of 200µm pitch and a laser spot of about 10µm (1/e²) on the disc. With this very wide pitch, a very large signal modulation of more than 60 percent is achieved. There is an offset in the output signal of the photo diode because the cavity depth on the disc and ratio between land and cavity on the diffractive fields are suboptimal.

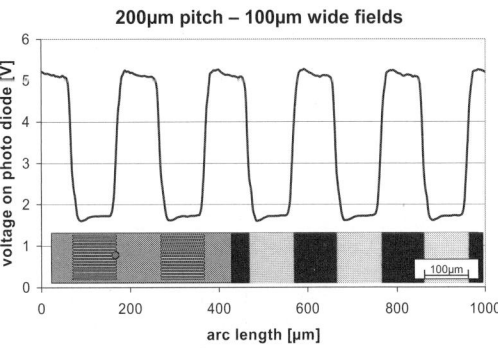

Fig. 11: Experimental sensor output with 200µm pitch

The signal modulation decreases for smaller pitches. Because of the Gaussian intensity profile of the laser beam even pitches smaller than 12µm (fields of 6µm) can be detected easily with the 10µm laser spot on the disc.

360

Fig. 12 shows the signal output for a 12μm pitch (equivalent to 6μm wide fields). The signal modulation on the photodiode is smaller and the offset bigger than in Fig. 10. But signal modulation is still more than 40 percent with very low noise and the sinusoidal signal shape provides best capabilities for further interpolation.

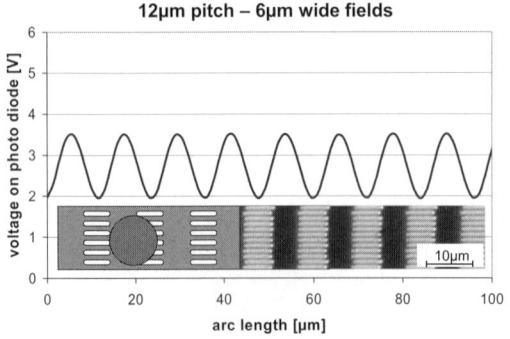

Fig. 12: Experimental sensor output with 12μm pitch

6. Conclusion

The functional principle of a low cost optical angular resolver with a compact disc as solid measure was demonstrated. The first experimental results with a simple experimental setup show the enormous capability and the large potential of this approach for an angular resolver. The new concept can be adapted and optimized to various applications. An example for an automotive application is the detection of steering angle. Also applications in automation industries e.g. position sensing of turning machines, or in consumer techniques e.g. rotary sensing of a joystick, are possible areas of application for the new sensor concept.

7. References

[1]   W. J. Fleming; Overview of Automotive Sensors; IEE Sensors Journal; Vol. 1, No. 4, December 2001
[2]   Hoffmann, J.:Handbuch der Messtechnik; Hanser-Verlag; München; 1999
[3]   X. Li, G. C. M. Meijer, G. W. de Jong, J. W. Spronck; An Accurate Low-Cost Capacitive Absolute Angular-Position Sensor with a Full-Circle Range; IEEE Transactions On Instrumentation And Measurement, VOL. 45, No. 2, April 1996
[4]   S. P. Cermak, G. Brasseur, H. Zangl, P. L. Fulmek; Capacitive Sensor for Incremental Angular Measurement; SIcon/02 Sensors for Industry Conference, Houston, Texas, USA, 19-21 November, 2002
[5]   G. W. de Jong, G. C. M. Meijer, K. van der Lingen, J. W. Spronck, A. M. M. Aalsma, D. A. J. M. Bertels; A Smart Capacitive Absolute Angular-Position Sensor; Sensors and Actuators A, 41-42 (1994), Page 212-216
[6]   Y. Kikuchi, F. Nakamura, H. Wakiwaka, H. Yamada; Consideration for a High Resolution of Magnatic Rotary Encoder; IEEE Transactions on Magnetics, Vol. 32, No. 5, September 1996
[7]   S. Günther; Magnetische Absolut-Multiturn-Drehgeber – Genau genug; IEE, 50. Jahrgang, 03/2005

[8]   Y. Grabowski; Anwendungsbereiche und Vorteile magnetischer Drehgeber; Special Antriebstechnik S2/2005
[9]   Epstein, H. C.; Optical and mechanical design trade-offs in incremental encoders; Proceedings of the tenth Annual Symposium on Incremental Motion Control Systems and Devices, Chicago, IL, June 1987, pp. 57-64
[10] K. Engelhardt, P. Seitz; Absolute, High-Resolution Optical Position Encoder; Applied Optics, Vol. 35, No.1, January 1996
[11] R. Sawada; Integrated Optical Encoder; Transducers `95 Eurosensors IX, The 8th International Conference on Solid State Sensors and Actuators, and Eurosensors IX Stockholm, Sweden June 15-29, 1995
[12] A. Kourilovitch, P. Bloechle; An Interference-Based Incremental Optical Encoder; Sensors Magazine Online; November 2000; www.sensorsmag.com
[13] M. Dimmler, C. Dayer; Optical Encoders for Small Drives; IEEE/ASME Transactions on Mechatronics, Vol. 1, No. 3, December 1996
[14] Marktübersicht – Drehgeber; GO-IEE, 2. Jahrgang, 01/2004
[15] Forschungsvereinigung räumlich elektronische Baugruppen 3-D-MID e.V. (Hrsg.): „Herstellungsverfahren Gebrauchsanforderungen und Materialkennwerte Räumlich Elektronischer Baugruppen" 3D MID, Handbuch für Anwender und Hersteller, 2. Erweiterte Ausg. Erlangen 1999
[16] Weber, J.: Untersuchungen zu einem optischen Drehwinkelsensor, Institut für Zeitmess-, Fein- und Mikrotechnik, Diplomarbeit Universität Stuttgart, 2004
[17] Müller, A.: Auslegung und 3D-Konzept eines optischen Drehwinkelsensors; Institut für Zeitmess-, Fein- und Mikrotechnik, Diplomarbeit Universität Stuttgart, 2005
[18] Seybold, J.: Konstruktion, Auslegung und Aufbau eines Demonstrators zur Verifikation des Funktionsprinzips eines optischen Drehwinkelsensor; Institut für Zeitmess-, Fein- und Mikrotechnik, Studienarbeit Universität Stuttgart, 2005
[19] Schneider, M.: Charakterisierung und Optimierung der Signalmodulation im optischen System eines Drehwinkelsensors; Institut für Zeitmess-, Fein- und Mikrotechnik, Studienarbeit Universität Stuttgart, 2005

Multi-Material Micro Manufacture
W. Menz, S. Dimov and B. Fillon (Eds.)

Micro EDM: accuracy of on-the-machine dressed electrodes

A. Rees[a], S.S. Dimov[a], A. Ivanov[a], A. Herrero[b] and L.G. Uriarte[b]

[a] *Manufacturing Engineering Centre, Cardiff University, CF24 3AA, U.K.*
[b] *Tekniker, Avda, Otaola 20, 20600 Eibar, Spain*

Abstract

The introduction of a technique for on-the-machine electrode generation, utilising Wire Electrode Discharge Grinding (WEDG) has broadened the application area of the µEDM process. This paper studies the capabilities of the WEDG process for manufacturing micro electrodes. The accuracy of the electrodes produced on-the-machine is investigated and a solution is proposed to improve the process. The experimental results revealed the inherent limitations of the WEDG dressing process and the effects of machine accuracy on electrode quality. The study shows that by employing an optical verification system the accuracy of the dressing process could be improved significantly.

Keywords: Micro EDM, micro-machining, EDM accuracy, WEDG, electrode dressing

1. Introduction

Micro Electrode Discharge Machining (µEDM) is one of technologies widely used for manufacture of microstructures and tooling inserts for micro-injection moulding and hot embossing. Originally µEDM was applied predominantly for producing small holes in metal foils. Due to the flexibility of the EDM process and its capability to produce complex 3D structures, currently the technology is employed in a number of applications. For example, µEDM is employed for manufacturing micro parts for watches, keyhole surgery, housings for micro-engines and also tooling inserts for fabrication of micro-filters, housings and packaging solutions for micro-optical, and micro fluidics devices. However, to broaden the application area of this technology a number of constraints remain.

During the µEDM process the volumetric wear, the ratio between electrode and workpiece wear, is relatively high and cannot be considered negligible [1]. Thus, to manufacture microstructures there is often a need to compensate the wear by replicating electrodes. To address this requirement, a technique for on-the-machine electrode generation was developed that utilised a technology called Wire Electrode Discharge Grinding (WEDG) [2]. The accuracy and repeatability of the µEDM process is still highly dependent on the WEDG process The electrode generation and re-generation is considered a key enabling technology for improving the performance of the µEDM process [5].

This paper studies the factors affecting the capabilities of the WEDG process for manufacturing micro electrodes. Especially, the influence of machine accuracy on the quality of electrodes produced on-the-machine is investigated and a solution is proposed to improve this process.

2. Factors affecting electrode quality

The electrodes that are commonly used for machining during the µEDM process are manufactured from tungsten (W) or cemented tungsten carbide (WC) [6]. The effect that the material micro structure has on the achievable minimum feature sizes in EDM has been studied by Kawakami and Kunieda [6]. The relationship between surface roughness and fracture strength of electrodes has also been investigated by Huang et al [7].

When machining strategies are studied in µEDM, the main focus of such investigations is on the technological parameters and their optimisation. For example, Kawakami and Kunieda studied the influence of open voltage, capacitance and polarity on the process performance [6]. Lim et al [8] researched the characteristics of the dressing process.

Pham et al [9] investigated the accumulation of errors in the µEDM drilling process and also factors affecting the accuracy of the machined holes. In addition, the sources of errors in setting up the dressing process on the machine are studied. In particular, a unidirectional approach for reaching the dressing position was proposed in order to improve the machine positional accuracy and repeatability. Also, it was suggested by optimising the approaching speed to minimise the time delay in registering the contact between the electrode and the running wire.

In this paper the same experimental set-up as in [9] is used to assess the capabilities of the on-the-machine electrode dressing process. In particular, the effects of machine accuracy on electrode quality are studied and a method is proposed for improving the accuracy of the WEDG dressing process.

3. Experimental Set-up

To study the factors affecting the quality of on-the-machine dressed electrodes, a WEDG system similar to that proposed by Masuzawa [2] is employed. Figure 1 shows the system design and its working principle. The continuously running wire is fed from a retaining spool into the system at a speed of 3 – 5 mm/s. The retention of the wire spool ensures that a constant tension is applied to the running copper and zinc alloy wire throughout the dressing process. Then, the Ø0.3mm running wire goes through a vibration damper. Next it travels around a fixed guide that ensures wire stability during the process. This is also the position where the electrode is brought into contact with the running wire. Finally the wire progresses further through one more guide before it is deposited.

362

In this research the same grinding unit is used on two µEDM machines, Agietron Compact 1 Micro die sink and Sarix Micro Hole Drilling machine. For all experiments the machine programmes were generated manually.

The electrode employed in this study was a Ø150µm WC rod, commonly used in µEDM machining. On both machines the rods are clamped in spindles. The technological parameters applied in the dressing processes are given in Table 1 and Table 2

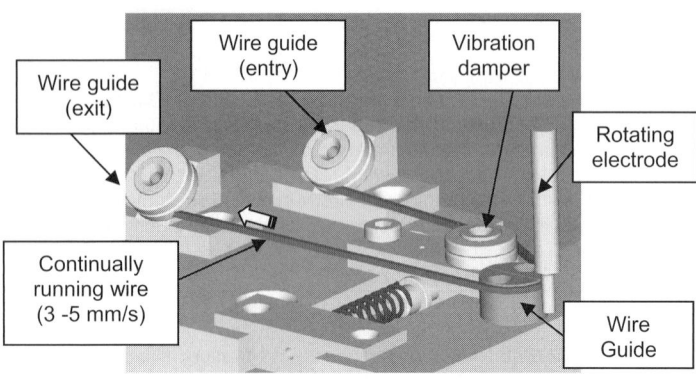

Figure 1 Experimental set-up

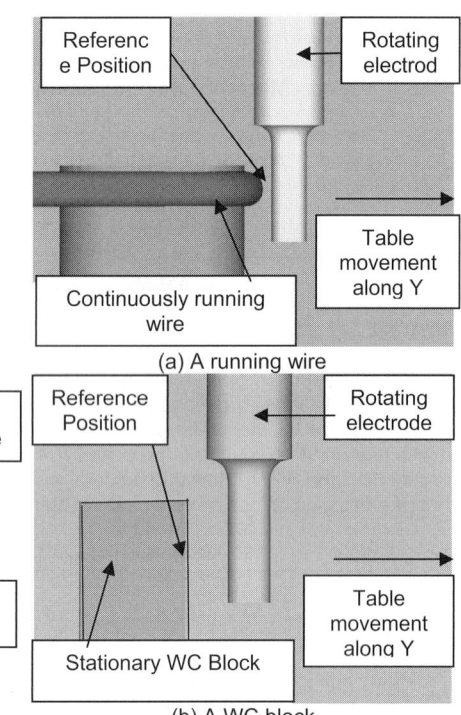

(a) A running wire

(b) A WC block

Figure 2 Experimental set-ups

Table 1. Technology parameters used on the AGIE machine

Technological parameter	Value
Polarity	Positive
Idle voltage	60 volts
Charging current	0.8 amps
Time ON	1.3µs
Time OFF	1.8µs

Table 2. Technology parameters applied on the Sarix machine (coded Values)

Technological parameter	Value
Polarity	Positive
Idle voltage	90 volts
Charging current	2 amps
Time ON	5µs
Time OFF	3.3µs

4. Machine accuracy effects

Before electrode dressing can start, the reference position of the running wire on the dressing device needs to be determined. This position is called a 'dressing position' in the machine set-up. In particular, to set up the process the electrical contact between the electrode and the running wire is registered by the machine CNC system. However, when the electrode approaches the dressing position, the checking up for an existence of a contact is not carried out continuously which leads to a time delay in registering it. In addition, the specific design of the WEDG unit affects the accuracy of the dressing operation.

To investigate the errors introduced by the adopted method for setting up the 'dressing position', experiments were carried out on a continuously running wire and on a WC block as shown in Figure 2.

20 tests were carried out using the same experimental set-up as in [9] on the Agietron Compact 1 Micro die sink. The lowest contact detection speed was selected, 1 mm/minute, and the dressing position was approached with movement along the Y axis. Also, during the experiments the machine was in its best state of thermal stability taking into account the results reported in [9].

Errors in setting up the 'dressing position' occurred as illustrated in Figure 3, due to the machine positioning accuracy and repeatability (ΔY_{pos}). The continuously running wire used by the WEDG unit had also introduced some errors (ΔD_w) that could be a result of the wire vibrations [6]. In addition, the diameter of the dressing wire is prone to a degree of variation.

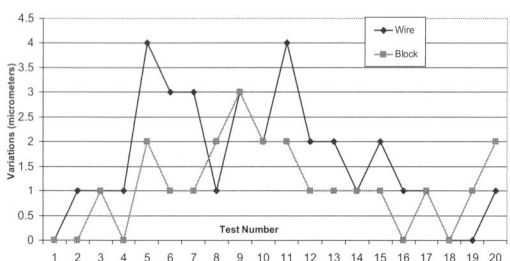

Figure 3 Variations in setting up the 'dressing position'

Figure 4 shows the dressing procedure used in this research. The dressing movement can be expressed using the following equation:

$$D_2 = D_1 - 2(Y_{pos} + S_g + D_w) \qquad (1)$$

where: D_1 is the initial diameter of the electrode, D_2 - the diameter of the dressed electrode, S_g - spark gap, Y_{pos} - machine movements and D_w – running wire diameter.

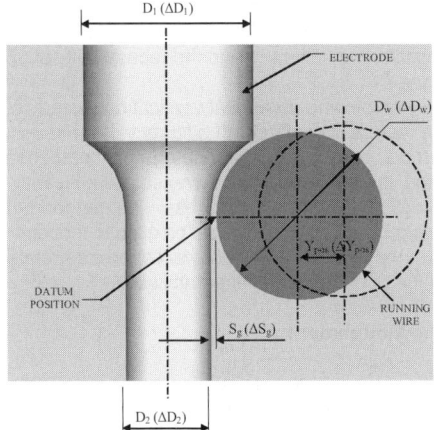

Figure 4 Dressing Procedure

These errors within the 'Machine-Electrode-Wire' (MEW) system can be expressed by the following equation:

$$\Delta D_2 = \Delta D_1 + 2 \left(\Delta Y_{pos} + \Delta S_g + \Delta D_w \right)$$

where: ΔD_2 is the deviation of the electrode diameter from its target value after the dressing, ΔS_g - variations of the spark gap, ΔD_w – variation of the wire diameter from their nominal values, and ΔY_{pos} - the accuracy of machine movements.

An experiment was carried out to assess the effects of ΔD_1, ΔS_g, ΔY_{pos} and ΔD_w on the dressing process. The experiment included dressing of 10 electrodes to the profile shown in Figure 5. The 'dressing position' on the running wire was set-up before each electrode machining. The technology parameters applied in the tests were the same as those used in the other experiments (see Table 1). All electrodes were produced using a three cuts' machining strategy.

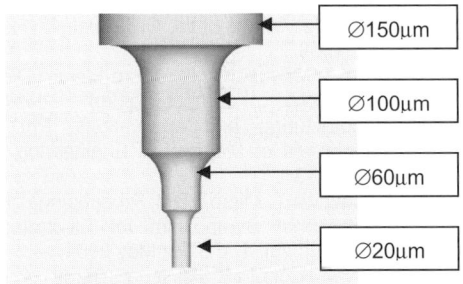

Figure 5, The profile of the dressed electrode

The results of the tests are shown in Figure 6.

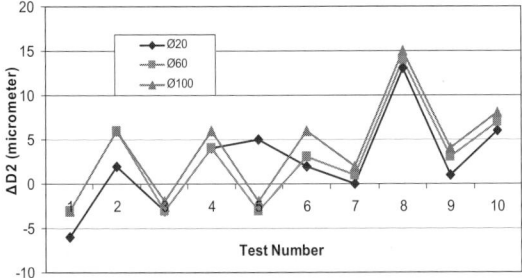

Figure 6 Variations of dressed electrodes

ΔD_2 is caused by several factors. In particular, the factors affecting the accuracy of the dressing process are the machine positional accuracy and repeatability, spark gap variations, machine temperature stability, and errors in detecting the dressing position. According to the performed experiment on Agietron Compact 1 Micro, ΔD_2 could be up to 21µm that is not acceptable for many applications. Therefore, an approach to improve the accuracy of the dressing process is discussed in the next section.

5. Process Improvements

To improve the accuracy of the dressing process it is required to compensate some of the errors within the MEW system. In this research the process improvements that could be achieved by utilising an optical verification method are discussed. The algorithm describing the compensation process utilising such an optical measuring device is presented in Figure 7. In particular, the device applied in this study is an Erowa 80x microscope as shown in Figure 8.

STEP1: Assemble the electrode
STEP2: Set the datum
STEP3: Machine the electrode to $D_2 + \Delta D_2$
STEP4: Measure the electrode diameter, D^M_2
STEP5: **WHILE** $D_2 - D^M_2$ is not within the D_2 tolerance
     DO
          Introduce a compensation, ΔC
          Machine the electrode to $D^M_2 - \Delta C$
          Measure the electrode, D^M_2
     END WHILE
STEP6: Stop the dressing process

Figure 7 Compensation procedure

To assess the capabilities of the proposed verification method, the optical device was fitted to the Agietron Compact 1 Micro die sink EDM machine and the compensation procedure in Figure 7 applied.

Figure 8 Optical measurement set-up

The experiment included dressing sequentially 10 electrodes to the profile shown in Figure 5 applying the technology parameters in Table 1. The electrodes were initially dressed to a diameter, $D_2 + \Delta D_2$, bigger than the final target diameters of D_2 for each of the steps, Ø100 µm, Ø60 µm or Ø20 µm, respectively. This was done to ensure that there was sufficient material left to compensate for any errors within the MEW system.

After the initial dressing, the electrodes were measured and if required, corrections, ΔC, were introduced. This procedure is interactive and continues until $D_2 - D^M_2$ is within acceptable limits. Due to the machine accuracy and repeatability, and errors in

364

optical measurements there will always be some deviation ΔD_2. The results of the experiment are given in Figure 9.

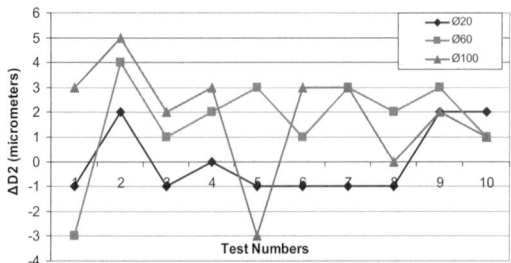

Figure 9 Variations of dressed electrodes

The variability previously witnessed in the dressing process was considerably reduced. In particular, ΔD_2 in the carried out ten tests was reduced from 21µm (see Figure 6) to 8 µm (see Figure 9). As already stated, this deviation is due mostly to the positional accuracy and repeatability of the machine.

To verify this, the experiments were repeated on a different µEDM machine. The second machine tool used in this study was a Sarix desktop EDM machine.

The experiment included dressing again 10 electrodes from diameters of 150 µm to 10 µm applying the same WEDG set-up as that on the first machine and the technology parameters in Table 2. Initially, 10 electrodes were produced without using the optical verification process. Then further 10 electrodes were dressed following the procedure in Figure 7.

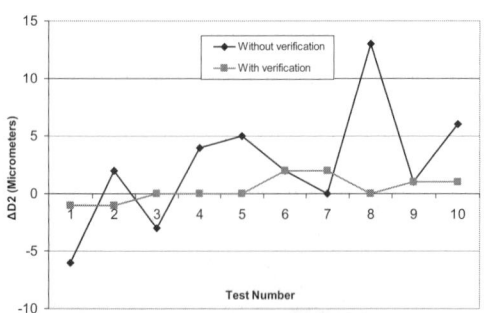

Figure10. Dressing with and without optical verification on the Sarix machine

The results of the experiments are provided in Figure 10. As expected the accuracy of the dressing process is very much dependent on the machine accuracy. The deviation of ΔD_2 decreased significantly when the proposed verification procedure was applied. In particular, ΔD_2 decreases from 9 to 3µm. The high accuracy of the dressing process is directly linked to the machine accuracy. This demonstrates that the proposed compensation method is highly effective and could be used to compensate many of the inherent process errors except those related to machine accuracy and repeatability.

6. Conclusions

This paper investigates the effects of machine accuracy on the quality of on-the-machine dressed electrodes. The results of the conducted experiments demonstrate the inherent limitations of the dressing process. The main factor affecting the quality of the electrodes produced on-the-machine using WEDG are the positioning accuracy and the repeatability of µEDM machines.

By employing an optical verification system, many of the factors that affect the accuracy of the machined electrode can be compensated. A compensation method is proposed that increases significantly the accuracy of the dressing process. The method proved to be highly effective and could be used to reduce many of the inherent process errors except those related to machine accuracy and repeatability.

Acknowledgement

The authors would like to thank the European Commission, the Welsh Assembly Government and the UK Engineering and Physical Sciences Research Council for funding this research under the ERDF Programmes "Micro Tooling Centre" and "Supporting Innovative Product Engineering and Responsive Manufacture". And the EPSRC Programme "The Cardiff Innovation Manufacturing Research Centre". Also, this work was carried out within the framework of the EC Networks of Excellence "Innovative Production Machines and Systems (I*PROMS)" and "Multi-Material Micro Manufacture Technologies and Applications (4M)"

References

[1] S. Bigot, A. Ivanov, K. Popov. A study of the micro EDM electrode wear. Proceedings of the First International Conference on Multi-Material Micro Manufacture, 4M, p 355 – 358, 2005

[2] T. Masuzawa, M. Fujino and K. Kobayashi. Wire electro-discharge grinding for micro-machining. Annals of the CIRP, vol. 34(1), p 431-434, 1985

[3] Zuyuan Yu, Takahisa Masuzawa, Masatoshi Fujino. 3-D Micro-EDM with Simple Shaped Electrode. Annals of the CIRP, vol. 47(1), p 169 – 172, 1998

[4] W. Meeusen. Micro-Electro-Discharge: Technology, computer-aided design & manufacturing and applications. PhD thesis, Department of Mechanical Engineering Leuven, June 2003

[5] Masuzawa.T; Micro-EDM, Proceedings of the 13th International Symposium for Electromachining ISEM XIII, vol. 1, p 3-19, May 2001

[6] T. Kawakami, M. Kunieda. Study on Factors Determining Limits of Minimum Machinable Size in Micro EDM. Annals of the CIRP, vol. 54 (1), p 167 – 170, 2005

[7] Sheng Ho Huang, Fuang Yuan Huang, Biing Hwa Yan; Fracture strength analysis of micro WC-shaft manufactured by micro-electro-discharge machining, Journal of Advanced Manufacturing Technology, vol. 26, n 1-2, p 68 – 77, 2004

[8] H.S. Lim, Y.S. Wond, M. Rahman, M.K. Edwin Lee. A study on the machining of high-aspect ratio micro-structures using micro-EDM. Journal of Materials Processing Technology 140, p 318 – 325, 2003

[9] Pham, D.T; Dimov, S.S; Bigot, S; Ivanov, A; Popov, K; A study of the accuracy of the Micro Electrical Discharge drilling process, Journal of Materials Processing Technology, vol. 149, n 1-3, p 579-584, Jun 10, 2004

Multi-Material Micro Manufacture
W. Menz, S. Dimov and B. Fillon (Eds.)

All polymer electrochemical sensor

M.Bengtsson[1], G.Thrastardottir[1], T.S.Hansen[2], O.Geschke[1]

[1] Department for Micro- and Nanotechnology, Technical University of Denmark, 2800 Kongens Lyngby, DK
[2], Department of Chemical Engineering, Technical University of Denmark, 2800 Kongens Lyngby, DK
Danish Polymer Centre

Abstract

The paper describes the development of an all polymer electrochemical sensor in poly(3,4-ethylendioxythiophene) (PEDT) for detection of hydrogen peroxide (H_2O_2). To increase the sensitivity of PEDT to H_2O_2 Prussian blue (PB) has been added to the polymer to act as an catalyst for reduction of H_2O_2. The sensitivity of the sensor with this addition was shown to increase up to 4 times. In order to use the sensor in a miniaturized flow format a flow cell was designed and tested with the same setup of polymer electrodes.

Keywords: conductive polymers, electrochemical sensors, PEDT

1. Introduction

Hydrogen peroxide is one of the most extensively investigated molecules with interests from different fields such as in clinical, food, pharmaceutical and environmental analyses. The interest also adheres from the fact that a highly sensitive hydrogen peroxide sensor is useful to fabricate biosensors for various substances by combining it with hydrogen peroxide-producing oxidases [1]. As a consequence there are several techniques developed for the monitoring of H_2O_2 such as chemiluminescence [2], fluorimetric [3] and electrochemical [4].

Most of these sensors use metal or carbon as electrode material. Specifically for hydrogen peroxide sensors the use of platinum is common due to its catalytic effect on the reduction reaction. The use of platinum however, has a lot of disadvantages especially in material cost, but also in the fabrication methods required and in the number of processing steps in the fabrication compared to an all polymer electrode setup. Metals in a sensor are also a drawback when it comes to disposability.

The goal of this project has been to replace the platinum working electrode with an electrode made of the conductive polymer poly(3,4-ethylendioxythiophene) (PEDT). PEDT is a polymer initially developed for antistatic coatings but is now found in several applications as all-organic field effect transistors, polymeric organic light-emitting diodes and polymer photovoltaic cells [5].

However, PEDT has not the catalytic properties for hydrogen peroxide as platinum and therefore Prussian blue was added Into the polymerisation mixture to act as catalyst since it has been shown that Prussian blue (PB) possesses excellent catalytic properties for H_2O_2 reduction [6,7,8,9].

Since it is desirable to make the analysis in a miniaturized flow format, a flow cell was designed with working and auxiliary electrodes inside the channel and a separate reference electrode. The flow cell was designed so that the flow channel ended in a large reservoir with sufficient space for an external reference electrode still sufficiently close to get a stable potential for the working electrode.

2. Materials and Methods

2.1. Fabrication of PEDT electrodes

PEDT was fabricated by mixing 6.5ml Baytron C, 220μl Baytron M, 150μl Pyridine and 2ml H_2O, which yields a PEDT layer with conductivity around 500-1000 S/cm. [10]. The solution was then spin-coated onto the PMMA (Nordisk Plast A/S) substrate with 500 rpm for 30s. After spin-coating the substrates was baked on a hotplate at 60°C for 5min after witch the salt was removed by washing and the substrate dried on the hotplate. The process resulted in a PEDT layer with a thickness of 500 – 800 nm.

Electrodes containing PB were fabricated by dispersing 2 g/L of small $Fe_4(Fe(CN)_6)_3$ particles in the polymerisation solution, which gave approximately 0.01-0.1 mg PB per area of PEDT.

2.2. Laser ablation of electrodes

Working and auxiliary electrodes were laser welded from a PMMA sheet with a PB doped PEDT film. The laser used was a CO_2 laser (Synrad, WA, USA) with wavelength of 10.6μm and the power was modulated to remove the PEDT and as little of the PMMA as possible to preserve an even surface for the thermal bonding. The working electrode was made 1.0mm by 0.4mm and the auxiliary electrode was made 20 times larger divided into two equal areas on each side of the working electrode. The distance between the electrodes was set to 2mm.

2.3. Fabrication of flow cell

A channel 400μm wide, 400μm deep and 30mm long was fabricated in PMMA by micro milling (MicroMill 2000, MicroProto Systems). At the end of a channel an 800μm hole was milled all through the wafer.
The channel wafer was subsequently bonded thermally to the electrode wafer at 110°C for 1h. On top of the channel wafer, over the through hole, a polycarbonate reservoir of 3ml was bonded with epoxy.

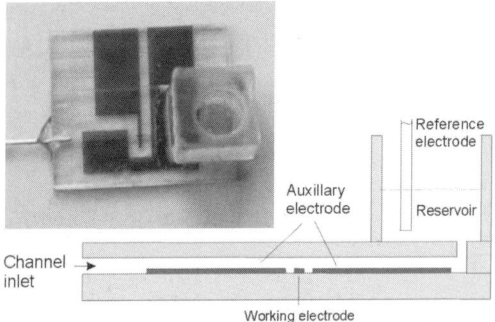

Fig 1. Photograph and schematic drawing of the flow cell sensor. A narrow channel passes the electrode system and ends in a large reservoir where the reference electrode is placed.

2.3. Experimental

2.3.1 Electrochemical Methods.

The electrochemical measurements were carried out in a three electrode system with a potentiostat (VMP, BioLogic Science Instruments) and an Ag/AgCl electrode as reference. Initially Cyclic Voltammograms (CV) were run in a 0,1 M KCl solution, from -200 mV to 600 mV with scan velocity of 50 mV/s For the common PEDT measurements the surface area of PEDT was 45 mm^2 and the auxiliary electrodes were made of gold.

The chronoamperometric measurements were made by immersing the electrodes in 10 ml of 0,1 M KCl, and applying 150 mV to the system. When the signal reached a stable baseline 10 µl of 30% H_2O_2 was added every 30 seconds, giving a concentration rise of 9.8mM for every addition. The system was thoroughly stirred all through the experiment and the concentration can be assumed to be the same in all the volume.

2.3.2 Flow cell measurements.

Phosphate buffer, 0.1M, pH 7, was pumped through the system with a flow rate of 10µl / min. 400mV vs. Ag/AgCl was applied on the working electrode. When the baseline was considered stable the flow was switched to 19.5mM H_2O_2 for 5 minutes and then the buffer flow continued.

3. Results

The results from the cyclic voltammograms are shown in figure 2. With the addition of Prussian blue in the matrix two significant peaks show up at -0,06 mV and 0,5 mV. The peaks correspond to the oxidation and reduction of Prussian blue [7].

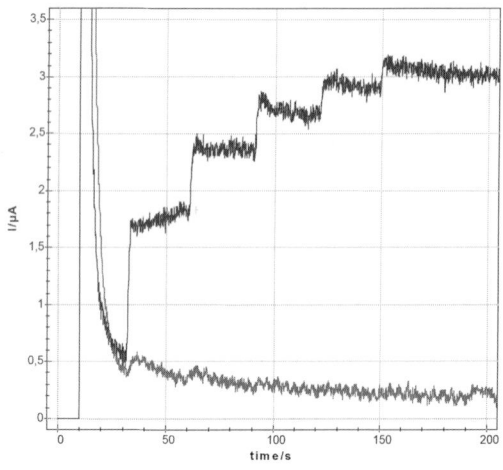

Fig. 3. Amperometric response of PEDT (red) and PEDT/PB (blue) electrodes for an increasing concentration of H_2O_2 .

Figure 3 shows that the results from the chronoamperometric measurements of PEDT, with and without the PB addition. The first injection of H_2O_2 is at t=30s, then injections are subsequently made at 30s intervals. It is shown that the response for PB doped PEDT is approximately 67 µA / M of H_2O_2, but only 17µA / M of H_2O_2 for pure PEDT. The corresponding sensibility of the Pt electrode towards H_2O_2, figure 4, is approximately 0,63 mA / M H_2O_2, as 15 µl of 2.4M H_2O_2 was added every minute giving a concentration rise of 3.7 µM for every injection.

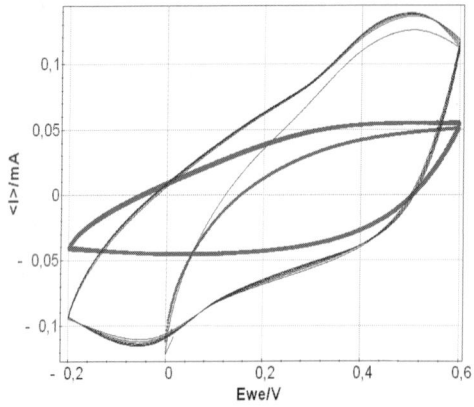

Fig. 2. Cyclic Voltammetry of PEDT (red) and PEDT/PB (blue) electrodes in KCl.

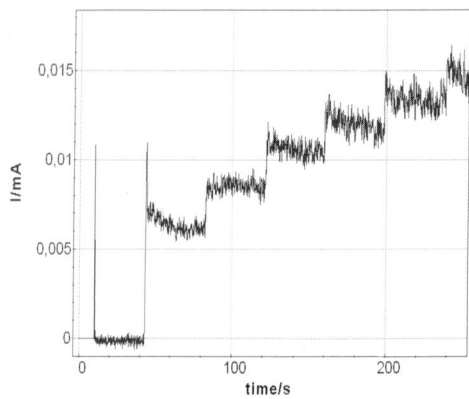

Fig.4 Amperometric response of Pt electrode to the injections of H_2O_2.

The results from the measurements on the flow cell are shown in figure 5. The sample plug, 30 µl should reach the cell at t=420s and 1140s. The flow rate is 10µl / s

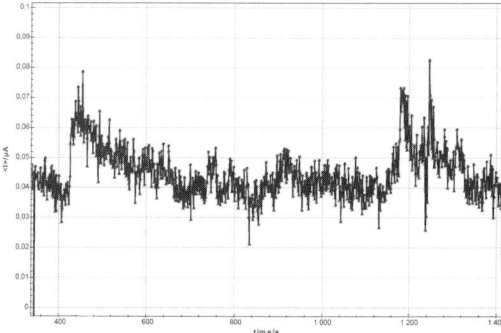

Fig. 5. Amperometric response from the sensor to two consecutive injections of H_2O_2 .

4. Discussion

The results clearly show that the Prussian blue is incorporated in the matrix while maintaining its catalytic effect on the hydrogen peroxide reduction. The PEDT electrode without PB shows a clear but very faint response to the H_2O_2 injections, the signal to noise ratio is 1.2. With the addition of Prussian blue in the polymer solution the response is approximately 4 times higher while maintaining the same noise level. Compared to the platinum electrode however, the signal is still significantly lower.

PEDT/PB catalytic activity towards H_2O_2 does diminish with concentration. This is because the "insoluble" form is used, which has limited stability at neutral pH [7]. The "insolubility" was also a problem in the spincoating process since there sometimes were particles in the film which caused an uneven thickness of the PEDT. This also indicates a risk of an uneven distribution of PB in the PEDT film. For a small electrode, like in the flow cell, the absence or presence of such large particles may totally alter the behaviour of the electrode.

The platinum electrode shows a much higher noise level, probably due to the electrode setup with platinum wires as both working and counter electrodes. The PEDT setup is more stable since the electrodes are fixed at a much smaller distance.

In the miniaturized system, the flow cell, the immediate response to the injected sample is clear. For higher concentrations of H_2O_2 however, the signal is rapidly decreasing in time. This phenomena is not significant for lower concentrations indicating that it adheres to all available Prussian blue being oxidized and thus the catalytic performance is lost. After the sample plug has passed the PB is slowly reduced and the catalytic performance restored for the next sample plug.

One drawback with the fabrication protocol is that the auxiliary electrodes contain PB why they also are catalyzing the reduction of H_2O_2 and thus increases the noise in the signal. Due to this, there is also a clear difference in the noise level in the signal from the injected sample plug compared to the buffer flow.

In our future work we will focus on optimizing the flow cell setup and further develop the system by immobilizing enzymes in a polypyrrole matrix onto the PEDT electrode, using the present setup to monitor the catalyzed reaction.

5. Conclusions

PEDT has been investigated as an electrode material for H_2O_2 sensing. To increase the catalytic effect of the electrode and thus the signal, Prussian blue was added into the polymer matrix with successful results.

References:

[1] J.Davis, D. Huw Vaughan, and Marco F. Cardosi, Enzyme and Microbial Technology **17**, pp1030-1035, 1995
[2] D. Price, R. Fauzi, C. Mantoura and P. J. Worsfold, *Anal. Chim. Acta*, 1998, **371**, 205.
[3] L.-S. Zhang and G. T. F. Wong, *Talanta*, 1999, **48**, 1031.
[4] H.Huang and P.K.Dasgupta, *Talanta,* 1997, **44,** 605-615].
[5] Groenendaal, L.B.; Jonas, F.; Freitag, D.; Pielartzik, H.; Reynolds, J.R., Adv. Mater. **12** (7), 481, 2000
[6] Arkady A. Karyakin, Olga V. Gitelmacher, and Elena E. Karyakina Analytical. Chemistry. 1995, 67, pp 2419-2423
[7] F. Ricci, G. Palleschi; Biosensors and Bioelectronics 21 (2005) pp. 389-407
[8] D. Pan, J. Chen, L. Nie, W. Tao, and S. Yao *Electrochimica Acta,* 49:795-801, 2004
[9] A. Karyakin , E. Karyakina, Sensors and Actuators B **57,** 1999, 268–273
[10] B. Winther-Jensen, D. W. Breiby and K. West. *Synthetic Metals,* 152:1-4, 2005

Multi-Material Micro Manufacture
W. Menz, S. Dimov and B. Fillon (Eds.)

369

A novel protective cover for microcomponents

Aleksandra Cvetanovic, Andreja Cvetanovic, Daniela Andrijasevic, Ioanna Gioroudi, Werner Brenner

Vienna University of Technology, Institute of Sensor and Actuator Systems, Floragasse 7/ 2, 1040 Vienna, Austria

Abstract

This paper presents a novel tool for assembly of micro particles in a SEM chamber. A new mechanismus, called Protective Cover for Microcomponents, which allows the positioning of the micro parts on the platform inside the chamber without gluing them on the platform is proposed.

It is well known the vacuum pump, especially turbo pump, that makes a vacuum in the SEM chamber, has higher sensitivity to foreign object damage. That is the reason why the micro parts are glued on the disks in the platform of the SEM chamber. Otherwise, the grippers that have to pick, to lift and place the particle in a desired system (position, orientation) are very fragile and can not overcome the adhesive force of glue and separating the particle from the disk. Besides, for the automated assembly process it is very important that the particles stay on the exactly defined location and position.

The showed system is standard and can be mounted without additional time, modification or expenses into the SEM chamber. It enables, on the one hand, easier manipulation of the micro parts, that need not to glue on the base and, on the other hand, opens the door for introduction more automation in the manipulation process in the SEM chamber by making a basis for modular assembly system.

Therefore the future constructing of a modular holder which can correspond to different requirements of the assembly of numerous microsystems will also be reported. Need for standardisation in micro world becomes more and more articulate. With a higher level of standardisation, the degree of automation grows up, too. The uniformity of parts, operations and tools is a strong support to an automation assembly system.

Keywords: micro assembly, SEM, micromanipulation

1. Introduction

1.1 Assembly in SEM chamber

The assembly in the SEM chamber has its application in the cases when it needs a great resolution, depth of field as well as large working distance, because there are advantages SEM to the light microscope.

Nevertheless, the restrictions by manipulation of micro parts, its positioning and orientation in the given system are big, too. We mention some: an evacuation process decelerates the work and complicates it, samples prepare (drying, sputtering) takes the time and many processes are not allowed: gluing (except special low outgasing adhesive), cell manipulation, magnetically processes or processes with evaporation.

Further difficulties by assembly of the microparts, as in SEM so under the light microscope, make an absence of the adequate standard tools for manipulation and the phenomena occuring in the range μm and nm in general. [1]

1.2 Problem of sucking off of not fixed micro particles

During the evacuation of the SEM chamber and, particulary in the beginning of process, it is possible that the microparts under manipulation are sucked into the pump, due to the produced air stream. This possibility increases as the size of microparts decreases. Yet if the aspect ratio of the particle is bigger and the particle lies on its wider side the mentioned possibility decreases. Furthermore, the occuring van der Waal's force is stronger in the case of more polished particles due

to adhesion. This means that manipulations with the gripper are more complicated. Therefore this phenomenon has to be prevented since it is highly undesirable.

2. System Overwiev

The novel system is presented in Figure 1 and consists of three units:

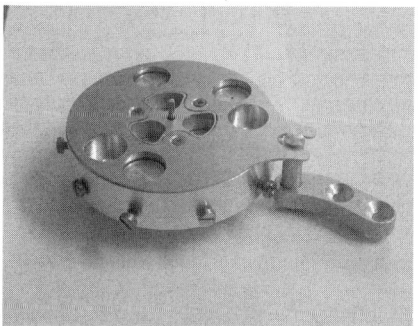

Figure 1

The platform: the sample disks made in the TU Vienna are compatible with SEM. Depending on the samples, the made disks have different shape (grooves, holes or smooth). The material of the disk is Aluminium, diameter of 65, 5 mm, thickness of 14 mm.

The novel cover plate with dimensional adjusted perforations: the design of the plate enables the user to open some disks while others remain covered. With an allocation, an optimal cover using without repeating "vent-pump" procedure of SEM is provided.

The holder system: with the help of this holder the perforated cover plate is fixed in relation to the rotable platform. Therefore the same disk can appear covered or open. When the platform moves in x, y or z direction, the whole system moves as well since everything is fastened on the same podium in the SEM chamber.

3. Working principle

3.1 Mounting and prepare

The system operation has as follows:

First, the particles are placed on the disks that will be covered with the cover plate. The cover plate is screwed in the middle together with the platform. After the vacuum in the SEM chamber is formed, the cover is opened by rotation of the platform below it. Then, the particles which were covered, are exposed to the electron beam, and can be picked up. All assembly tasks on any other disk are then performed accordingly to the procedure described above.

The system "cover plate-platform" has predefined tolerances thus preventing rapid pressure changes which could suck off the particles or disturb the stability of the electron beam. (Figure 4 and 5)

3.2 An efficiency analysis

Since it is very complicated to analytically determine when or even whether the particles will be sucked, a comparison case, with and without the protective cover, was examined. Therefore will here be compared case with and without protective cover.

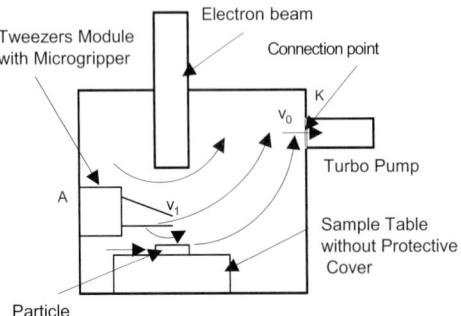

Figure 2: Chamber without protective cover

In general, the danger for the microparts to be sucked is greater if the air stream velocity v_1 is very high. The velocity v_1 is exact determined by velocity v_0. v_0 is the velocity of the air stream caused by pumping. Pumping is determined by the flow rate through the vacuum pump, or even better by the dynamics of the pressure drop at the connection point between the chamber and the vacuum pump. (See Fig.2) We shall try to define a preassure behaviour on that point. Equivalent electric circuit is shown on Figure 3.

B_1 and B_2 are the atmosphere and vacuum pump, respectively. The capacitor C_1 represents the chamber, the resistor R_1 is the resistance to air streaming through the pipe that leads from chamber

to pump (proportional to the pipe length and inversely proportional to its cross-section).

The battery B_1 is the atmosphere so that the tension in point A (Figure 3) corresponds to the atmospheric preassure. The battery B_2 is the vacuum pump and the tension in point K is the pressure at the exit of the chamber. The switch p represents the door of the SEM chamber. When the p is closed, the door of the chamber is open and the pressure in the chamber is equal to the atmospheric pressure ($u_k = u_a$). When the p is opened, the chamber door is closed (no contact between the chamber and the external environment) and evacuation starts. Precisely this moment is interesting to beginn the analysis of the system.

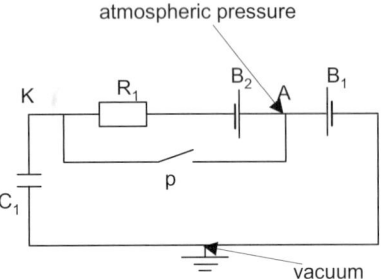

Figure 3: Chamber without protective cover

Analysis:

$u_k(t) + R\ i_{c1} = 0$, u_k - tension at point K

$i_{c1} = C_1 \dfrac{du_k}{dt}$, i_{c1} – current at point K

$u_k(t) + R_1\ C_1 \dfrac{du_k}{dt} = 0$

$u_k(t) = - R_1 C_1 \dfrac{du_k}{dt}$

$$-\frac{1}{R_1 C_1}\int_0^t dt = \int_{u_k(t=0)}^{u_k(t)} \frac{du_k}{u_k}$$

$$e^{-\frac{t}{R_1 C_1}} = \frac{u_k(t)}{u_k(t=0)}$$

$$u_k(t) = u_k(t=0)\, e^{-\frac{t}{R_1 C_1}}$$

$$\frac{du_k(t)}{dt} = -\frac{1}{R_1 C_1} u_k(t=0)\, e^{-\frac{t}{R_1 C_1}}$$

$$v_1 \leftrightarrow v_0 \sim -\frac{\partial p}{\partial t} = \frac{1}{R_1 C_1} U_a\, e^{-\frac{t}{R_1 C_1}} \qquad (1)$$

Judging from Eq.1, where for t=0 → exp=1, it is concluded that the possibility for the micro parts to be sucked off is maximum at the beginning of the evacuation process when the air flow rate is maximum.

While analyzing the following approximations are made: The air stream through the door of the SEMchamber is neglected. Air cooling due to expansion is not considered. The vacuum pump is ideal, there is no air drain and for t= ∞ the vacuum in the chamber could be absolut.

Comment: Since C_1 is constant, it is obvious that the risk of particles suction would decrease with increasing resistance R_1 (by decreasing the pipe cross-section or by its lenghten), but the evacuation time would be increased, a fact that is unacceptable.

Case 2 – SEM Chamber with protective cover

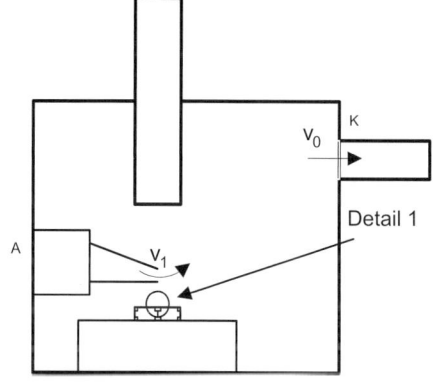

Figure 4: SEM chamber with protective cover

It is important to mention that the covering plate must not be sealed on the platform, because in that case it would cause rapid pressure changes (In the chamber there is vacuum, in magazine atmospheric pressure) which could suck off the particles or disturb the stability of the electron beam, when the cover opens. See Figure 4, Figure 5

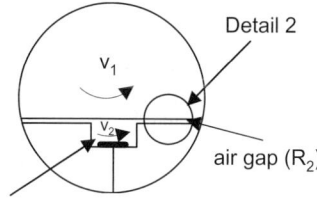

protected section (C_2)

Figure 5: Detail 1

The capacitor C_2 represents the protected (covered) section, the resistor R_2 represents the resistance to air stream through the channel that leads from the protected magazine to the chamber (proportional to the channel length and inversely proportional to its cross-section). (See Figure 6) Because the chanel that connects protective magazine is very narrow, the resistance to the air stream is very high, i. e. the current through the resistor R_2 will be low, so that the capacitor C_1 will

discharge faster then the capacitor C_2. That means that the derivation of the air preassure in response to time will be quicker in the SEM chamber then in the protected magazine and automatically it will be a rate of air stream through the magazine slower then in the SEM chamber. It is observed that in this case the danger of the particles to be sucked is less and therefore the air stream is faster then in the case where the protective cover is not present. When the chamber is evacuated (vacuum is established), no air stream is present anymore and the cover can be opened without exposing the micro parts in the danger of being sucked.

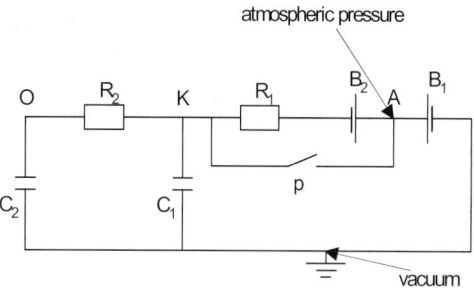

Figure 6: SEM chamber with protective cover

Analysis:

$$i_{c2} = -C \frac{du_{c2}}{dt} , \quad i_{c2} - \text{current at point K}$$

u_{c2} - tension at point K

$$u_{c2}(t) - R_2 i_2 = U_a \, e^{-\frac{t}{T_1}} , \quad T_1 = R_1 C_1$$

$$u_{c2}(t) + R_2 C_2 \frac{du_{c2}}{dt} = U_a \, e^{-\frac{t}{T_1}}$$

U_a - tension at point A (chamber door)

$$T_2 \frac{du_{c2}}{dt} = - u_{c2}(t) + U_a \, e^{-\frac{t}{T_1}}$$

$$u_{c2}(t) = f(t) \, e^{-\frac{t}{T_2}} , \quad T_2 = R_2 C_2 \qquad (2)$$

$$\frac{du_{c2}}{dt} = \frac{df(t)}{dt} e^{-\frac{t}{T_2}} - \frac{1}{T_2} f(t) e^{-\frac{t}{T_2}} \qquad (3)$$

(2), (3) → (1)
$$\frac{df(t)}{dt} e^{-\frac{t}{T2}} = \frac{1}{T_2} U_a \, e^{-\frac{t}{T_1}}$$

$$df(t) = \frac{1}{T_2} U_a \, e^{t(\frac{1}{T_2} - \frac{1}{T_1})dt}$$

$$\int_0^t df(t) = \frac{1}{T_2} U_a \int_0^t e^{t(\frac{1}{T_2} - \frac{1}{T_1})} dt$$

$$f(t) = U_a \frac{T_1}{T_1 - T_2} e^{t(\frac{1}{T_2} - \frac{1}{T_1})} + C \quad (4) \rightarrow (2)$$

$$u_{c2}(t) = (U_a \frac{T_1}{T_1 - T_2} e^{t(\frac{1}{T_2} - \frac{1}{T_1})} + C) e^{-\frac{t}{T_2}}$$

condition initial: $u_{c2}(t=0) = U_a \rightarrow$

$$U_a = (U_a \frac{T_1}{T_1 - T_2} + C) \qquad C = -U_a \frac{T_2}{T_1 - T_2}$$

$$u_{c2}(t) = U_A (\frac{T_1}{T_1 - T_{2)}} e^{-\frac{t}{T_1}} - \frac{T_2}{T_1 - T_2} e^{-\frac{t}{T_2}}) \qquad [2]$$

It should be noted that the pressure which could suck the particle is considerably lower when the particle is in the chamber, without the protective cover, particularly in the beginning, which causes more chances for suction of the particles. The pressure in the chamber decreases faster then in the protective magazine, as well. (Figure 7)

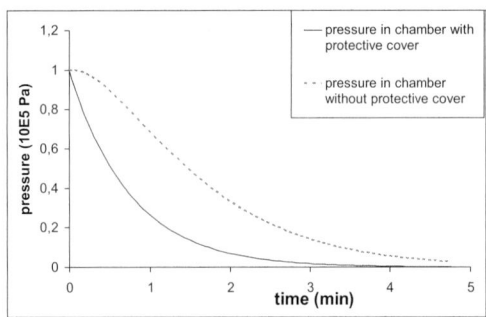

Figure 7: Comparison between case without protective cover and case with protectivecover

Under assumption that the gap width b = 5 μm (cover made by milling) and the another measered dimensions length l = 3 mm, disk diametar d_1 = 12,8 mm, magazine deepness h = 2 mm. There are always five disks exposed (opened) and five covered, in a appropriate way. The chamber dimension is a = 30 cm, the cover diametar D = 65,5 mm and the pipe diametar d = 100 mm and length L = 2 m.

It is possible the correlation between the analogical parametar in electric circuit:

$R_1 \leftrightarrow V_{pipe}$
$R_2 \leftrightarrow V_{gap}$

$$V_{pipe} = \frac{d^2 \pi}{4} l = 15700 \text{ mm}^3$$

$$V_{gap} = bD\pi d = 3,085 \text{ mm}^3$$

$$V_{pipe} / V_{gap} = 5,089 \ 10^3$$

$C_1 \leftrightarrow V_{chamber}$
$C_2 \leftrightarrow V_{magazine}$

$$V_{chamber} = a^3 = 30^3 \text{ cm}^3 = 27 \ 10^6 \text{ mm}^3$$

$$V_{magazine} = \frac{d^2_1 \pi}{4} h = 253,23 \text{ mm}^3$$

$$V_{chamber} / V_{chamber} = 0,105 \ 10^6$$

The surface finish of the covering plate and the platform defines the gap through the air streams and the velocity of the chamber evacuation. The smaller the particle, the better surface finising. The surface roughness coresponds to the particle dimensions. It is possible to make different covering plates with different surface quality for different dimensions of particles. Figure 8

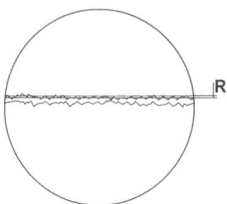

Figure 8: Detail 2

5. Experimental results

Figure 6 shows probe, a ferric powder 40-60 μm and a wire Φ = 15 μ on the disk before and after the two "pump-vent" cycles. There is no difference in the position of the objects.

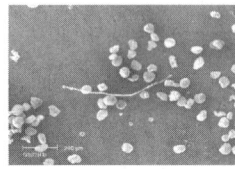 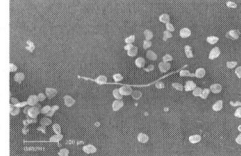

Figure 9: Before and after exposing

6. Summary

In this work we dealt with solving of the problem of adhesion of micro particles on the disk when a glue film is used in order to avoid sucking of the micro particles by the pump. A protective covering plate has been developed to enable the usage of a finest gripper. It has been proven that there is no risk of breaking during separation of the particles from the disk.

7. Reference

[1] S. Fatikow, J. Seyfried, St. Fahlbusch, A. Bürkle und F. Schmoeckel, Entwicklung flexibler Mikroroboter zur Handhabung von Mikroobjekten, 4. Chemnitzer Fachtagung Mikromechanik & Mikroelektronik, 1999
[2] Dr Branko Popovic, Osnovi Elektrotehnike 2, ISBN 86-395-0259-5, „Novi dani", Vojvode Brane 13, Beograd

Multi-Material Micro Manufacture
W. Menz, S. Dimov and B. Fillon (Eds.)

373

Micro injection moulding: the effects of tool surface finish on melt flow behaviour

C.A. Griffiths, S.S. Dimov and D.T. Pham

Manufacturing Engineering Centre, Cardiff University, Cardiff CF25 3AA, UK

Abstract

Micro-injection moulding is one of the key technologies for micro-manufacture and is widely used as a cost effective replication method for mass production. The capabilities of this replication technology have to be studied systematically in order to determine the process constraints. In this research, the factors affecting the flow behavior are discussed and special attention is paid to the interaction between the melt flow and the tool surface. Three different polymer materials, PP, ABS and PC, were utilised to perform moulding tests using cavities with the same geometry but different surface finish. Through a series of experiments the effects of the tool surface finish on the melt flow length and part quality were investigated by varying three process parameters, melt and mould temperature, and injection speed.

Keywords: micro-injection moulding, part quality, surface finish.

1. Introduction

The development of new micro devices is highly dependent on manufacturing systems that can reliably and economically produce micro components in large quantities. In this context micro-injection moulding of polymer materials is one of the key technologies for micro manufacturing.

Components manufactured by microinjection moulding fall into one of the following two categories. Type A are components with overall sizes of less than 1 mm while Type B have larger overall dimensions but incorporate micro features with sizes typically smaller than 200 μm. Currently, the following main groups of components are moulded successfully: optical grating elements, micro pumps, micro fluidic devices and micro gears. In all these applications the replication of component micro-features is a key requirements in selecting a given manufacturing route. It depends greatly on their size, aspect ratio and surface area [1].

The aspect ratio achievable while replicating micro features is one of the important characteristic of any micro fabrication process and determines the manufacturing constraints for a given process/material combination. At the same time, critical dimensions of microstructures are often related to required aspect ratios [2]. Thus, it is very important to study the factors that affect the replication capabilities of micro-injection moulding.

There are several alternative methods of manufacturing cavities for micro-injection moulding. By applying each of these methods a different surface finish could be achieved. Thus, the surface finish specified at the design stage in respect to the parts and the tool cavities should take into account the manufacturing constraints introduced by these tool-making processes. In micro tooling applications, the quality and topography of the machined surface could have a significant impact on their replication capabilities [3].

The surface finish of the cavities should reflect the part design requirements and may differ from that specified for the runner system. The high part to runner volume ratio means that a high percentage of the total shot volume that travels through the tool melt flow path does not require a specific surface finish. Therefore, the runner system is usually manufactured using the most cost-effective method. This is in spite of the fact that its surface finish could have a significant impact on the tool filling behaviour and part quality.

Research in micro-injection moulding has found that high melt and mould temperatures, and high injection speeds facilitate the filling of micro cavities [4,5]. In addition, high injection rates lead to high shear rates that could increase the friction between the mould surface and the melt flow in the cavities even further. This could result in a localised rapid heating to the point of material degradation. However, it could also be possible to benefit from this effect by reducing the initial melt temperature and thus compensating for the temperature increases produced by shear heating. Another positive effect of the shear heating could be an improvement of the melt filling of micro cavities.

This paper investigates the flow behaviour of the polymer melt in micro cavities with varying surface roughness levels

2. Experimental set-up

The design of the part used to analyse the interfacial interactions in micro-moulding is basically a series of runner sections through to a rectangular cavity as shown in Figure 1. The runner includes 4 unequal length sections totaling 40.8 mm. At the end of the runner system there is a rectangular section of 10 x 2.5 x 0.5 mm. The runner system has a square cross section with dimensions 0.5 x 0.5 mm. Due to its square cross-section, the runner system could be easily manufactured using wire electro discharge machining (EDM).

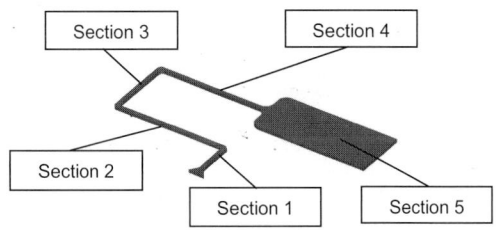

Figure 1. Test part

P20 Steel was used to produce the fixed and moving halves of the tool. The fixed half was fitted to the primary mould and the other half was fixed first to a secondary shim and then to the primary mould. Both halves were manufactured conventionally except for the cavity faces that were machined using wire EDM. This was required in order to achieve an identical surface finish on all four sides of the runner system and the part cavity. First, the fixed and moving halves of the tool inserts were machined to produce the front and back faces of the cavity as shown in Figure 2. Then, the shim was wire cut, see Figure 2b, to manufacture the side walls with the same surface finish.

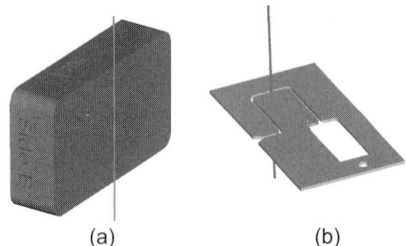

(a)　　　　　　　　(b)

Figure. 2. The wire EDM machining of (a) the fixed and moving halves of the tool inserts and (b) the side walls of the shim.

The 0.50 mm thickness shim was then fitted to the moving plate with screws and an epoxy adhesive that could maintain its mechanical properties up to 190 °C, as shown in figure 3.

Following these manufacturing steps three tools were produced that differed only in their surface finish. This was achieved by machining the parts of each of them by applying the same EDM technology.

A profiling microscope, Micro-XAM, was employed to measure the surface roughness of the wire cut parts. The surface roughness of the produced tools was Ra 0.07 μm, Ra 0.8 μm and Ra 1.5 μm respectively.

Finally, the moving and fixed halves were assembled to a primary mould tool and then inspected for parallelism and shut off of the mating faces.

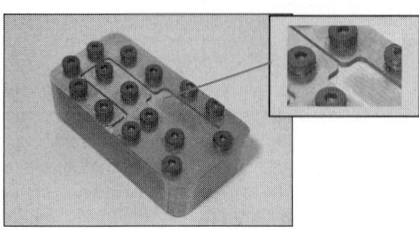

Figure 3. Insert assembly

Three commonly used materials in injection moulding, Polypropylene (PP), Acrylonitrile Butadiene Styrene (ABS) and Polycarbonate (PC), were selected to conduct the planned experiments. To characterise them, the Moldflow Viscosity Index (VI) was used, In addition, the molecular weight (M_w) and Molecular Number (M_n) of the three selected materials were tested using Gel Permeation Chromatography. In particular, the following three materials were chosen because their VI and molecular weights differed significantly.

- PP (C_3H_6) Sabic 56M10. It is a crystalline thermoplastic with M_w 261,000 and Moldflow VI (210) 0103 where the number in the brackets

refers to the material melt temperature [°C] and the other four digit number is its viscosity [Pa/sec] measured at a shear rate of 1000 [1/sec].
- ABS ($C_{15}H_{17}N$) DOW Magnum 8434. The selected material is an amorphous thermoplastic blend of 15-35% Acrylonitrile, 5-30% Butadiene and 40-60% Styrene with a total M_w of 169,000 and Moldflow VI (240) 0087.
- PC ($C_{16}H_{14}O_3$) DOW Calibre 303. It is a colourless amorphous polymer with M_w 26,000 and Moldflow VI (300) 0269

The injection-moulding trials were carried out on Battenfeld Microsystem 50.

3 Experimental design

The interfacial interactions affect the filling performance of micro cavities and therefore this experimental investigation was focused on the filling stage of the moulding process. Thus, in the carried out tests the process parameters affecting holding, cooling and ejection stages were set to their default values for the selected three materials.

The filling performance of micro moulds relies heavily on Injection speed (Vi) and the temperature control during the injection and is much less dependent on the holding pressure [6]. Therefore, the effects of Melt temperature (Tb) Mould temperature (Tm) and Injection speed (Vi) on the filling behaviour of cavities with different surface roughness were investigated in this study.

The Taguchi L9 orthogonal array was used to plan the experiments for each of the three materials. The selected three levels of Vi and tool surface finish were the same for all materials, while Tb and Tm levels were different for each material array. For each trial 10 mouldings were measured. The material array is given in Tables 1

Table 1. L9 orthogonal array for PP, ABS, PC

	Ra μm	T_b °C			T_m °C			V_i mm/s
		PP	ABS	PC	PP	ABS	PC	
1	.07	220	220	280	20	40	80	200
2		250	250	300	40	60	100	500
3		270	280	320	60	80	120	800
4	0.8	220	220	280	40	60	100	800
5		250	250	300	60	80	120	200
6		270	280	320	20	40	80	500
7	1.5	220	220	280	60	80	120	500
8		250	250	300	20	40	80	800
9		270	280	320	40	60	100	200

The melt temperature was controlled indirectly through Tb. Each polymer has a recommended processing window. In this research the selected three levels of Tb are maximum, minimum and medium temperatures in the range for each of the polymers.

In micro moulding polymer solidification time is much shorter than for that in conventional moulding and therefore the processing requires heated tools. Tm used in this research are the minimum, medium and maximum temperatures in the range for each material. By increasing Tm the bulk temperature of the polymer is kept sufficiently high to facilitate the melt flow during the filling stage. However, the effect of the Tm increase on cooling and cycle times is not analysed in this research.

The Vi has two main effects. It can help polymers to fill the cavities before the melt flow solidifies and also it can increase the shear rate of the polymer which results in shear heating. The three levels of Vi used in this research are given in Table 1. They were chosen taking into account the capabilities of Battenfeld Microsystem 50, especially its maximum injection speed of 946.4 mm/s over a stroke distance of 84 mm. However, it is worth noting that the high settings of Vi could lead to material degradation.

4 Experimental results

The effects of interfacial interactions on the flow length and the resulting part quality are analysed separately.

4.1 Flow length

Table 2 presents the flow length results obtained from all 27 trials. 10 parts were measured for each combination and the mean value was calculated. The combination of parameters in trial 3 of the L9 orthogonal array provided the best conditions for achieving a maximum flow length and thus the best filling of the cavities for all three materials. This result was expected from the process point of view because all controlled parameter were set at their highest values.

The results of other trials were not conclusive in regards to the minimum flow length achieved during the experiments. Unfortunately, from the carried out tests for all three materials it was not possible to identify a single process parameter as the main cause for the resulting low flow length.

Table 2. Flow length results

Trial	Levels				Mean value of the flow length [mm]		
	Ra	T_b	T_m	V_i	PP	ABS	PC
1	1	1	1	1	47.2	32.6	22.8
2	1	2	2	2	52.0	42.4	39.0
3	1	3	3	3	54.7	45.0	39.3
4	2	1	2	3	48.9	39.8	17.2
5	2	2	3	1	52.8	32.5	22.3
6	2	3	1	2	54.6	29.3	26.2
7	3	1	3	2	49.7	32.8	21.7
8	3	2	1	3	52.7	36.4	25.6
9	3	3	2	1	54.1	33.1	22.7

PP had the highest average flow length of 51.90 mm for all 9 tests. Test 1 resulted in the lowest flow length and the variation in length for all PP tests was 7.53 mm. This material filled the first four sections of the tool and only partially filled the fifth at all settings. In test 3 as it was already mentioned the highest flow length was achieved. When Tb was at its high settings the best filling results were observed, in particular tests 3, 6, and 9. At the same time the lowest filling results, tests 1, 4 and 7 were attained when Tb was at its lowest setting. Thus it could be concluded that Tb had the greatest effect on part filling.

ABS had the second highest average flow length of 36.04 mm for all three materials. The process conditions in test 6 resulted in the lowest flow length. The difference between maximum and minimum flow lengths for all ABS tests was 15.68 mm. In all experiments the first three sections of the test tool were filled and just partially the fourth except test 3. Only in

this test the melt flow reached section 5. It was not possible to identify a single factor affecting the flow filling length.

PC had the lowest average flow length of 26.34 mm for all three materials. Test 4 resulted in the lowest flow length. The difference between maximum and minimum flow lengths for all PC tests was 22.14 mm. Section 1 of the test tool was filled completely in all experiments. Only in tests 2 and 3, section 4 was filled partially. Again as it was the case with PP, Tb was the factor with the highest influence on the achieved flow length.

The experiments showed that PP was less susceptible to variations of the flow length as a result of changing process conditions. In the case of PC and ABS, the flow behavior was much more sensitive to changes of the process parameters and tool surface quality. However, any explicit relationship between the flow length and tool surface roughness was not observed.

A dependence between the molecular weight of PP, ABS and PC, and the resulting flow length was observed. In particular, the highest flow length was achieved with PP, the materials with the highest molecular weight and vice versa the lowest for the polymer with the lowest molecular weight, PC.

4.2 Part Quality

Figures 4, 5 and 6 present pictures depicting the quality of the moulded test parts. The images were taken using an optical measuring system. Significant variations of the melt flow behavior under the different trial settings were observed for the three investigated materials. The pictures provide a visual representation of flow directions, the existence of turbulence in the polymer flow, the presence of material phases within the part and melt fracture.

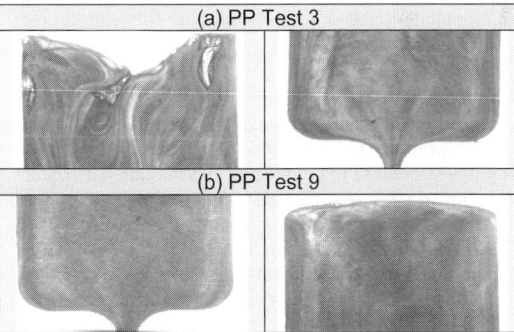

Figure 4. PP experiments

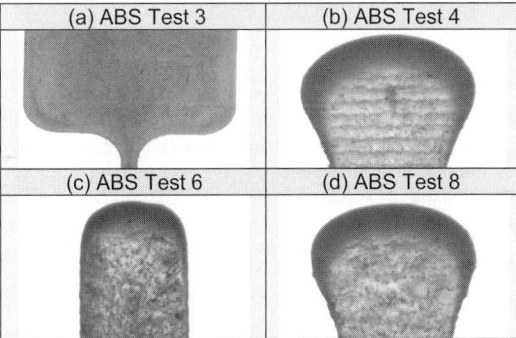

Figure 5. ABS experiments

376

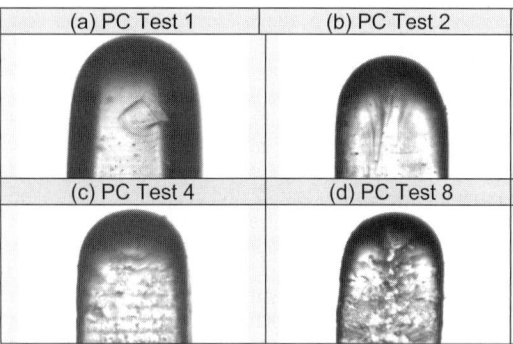

| (a) PC Test 1 | (b) PC Test 2 |
| (c) PC Test 4 | (d) PC Test 8 |

Figure 6. PC experiments

For PP, all experiments with low parameter settings illustrate a rounded melt front. In addition, all tests show evidence of jetting and fountain effects when the melt flow enters section 5 of the test tool. It can be seen that the flow turbulence increases with the increase of the distance from the exit point, while the flow path show signs of randomness at the high settings. Another effect that can be detected on the pictures is that the mixing of the polymer and the master batch colour die is better for the samples produced in the two cavities with a lower surface finish. This could be considered as evidence of higher turbulence compared to polymer flow in cavities with a higher surface finish. The importance of interfacial interactions is also evident from the trials with the highest settings of the process parameters in the cavity with the best surface finish. In particular, test 3 in Figure 4a shows an uneven melt front with evidence of melt fracture and gas traps. At the same time, this effect is less noticeable in the trials carried out in the other two cavities that have a lower surface finish.

The trials with ABS and PC show variations of the rounded melt front that are typical for a non Newtonian flow. The pictures of the PC samples, for example test 1 in Figure 6a, also depict the existence of rounded edges along the cavity walls. This is an indication of solidification of the melt flow without proper packing.

Some of the ABS and PC experiments show visual lines on the parts that are perpendicular and equally spaced to the melt flow, in particular the results from test 4 for both materials in Figures 5b and 6c. This is an illustration of the slip stick effect during the filling stage. For both materials this occurs at low Tb and high Vi that lead to high shear stress. However, no explicit relationship between the occurrence of the slip stick effect and the tool surface finish was observed.

All materials at high Tb showed signs of degradation with some evidence of a melt fracture.

5 Conclusions

Microinjection moulding is one of the key technologies for scale up manufacture of micro components. This research investigates the importance of interfacial interactions on the process replication capabilities. The following conclusions can be drawn from the study.

- The experiments performed with the highest settings of the controlled factors resulted in the highest flow length for all three materials.
- From the conducted experiments for all three materials it was not possible to identify a single process parameter as the cause of a low flow length.
- The flow length of PP was less susceptible to changes of the process parameters and tool surface quality in comparison to PC and ABS.
- There is a relationship between the tool surface finish and the level of turbulence in the melt flow. The trails for all three materials in the cavity with the highest surface finish indicate the existence of two distinctive phases in the polymer flow, while the patterns are mixed and not so clear for the other two.
- No explicit relationship between the occurrence of the slip stick effect and the tool surface finish was identified. On some of the ABS and PC samples, there are visual lines on the parts that are the result of the slip stick effect during the filling stage. For both materials this occurs at low Tb and high Vi that lead to high shear stress.

Acknowledgement

The authors would like to thank the European Commission, the Welsh Assembly Government and the UK Engineering and Physical Sciences Research Council for funding this research under the ERDF Programmes "Micro Tooling Centre" and "Supporting Innovative Product Engineering and Responsive Manufacture". And the EPSRC Programme "The Cardiff Innovation Manufacturing Research Centre". Also, this work was carried out within the framework of the EC Networks of Excellence "Innovative Production Machines and Systems (I*PROMS)" and "Multi-Material Manufacture Technologies and Applications (4M)"

References

[1] Webber L and Ehrfeld W. Micromoulding Market Position and Development, Kunststoffe, 89(10) (1999) pp 192-102.
[2] Dr Volker Piotter, Dr Thomas Hanemann, R Ruprecht and J Hausselt Micro Injection of Medical Device Components Report Institute of Materials Research III, Karlsruhe Research Center.
[3] T Dobrev D T Pham and S S Dimov A simulation model for crater formation in Laser milling. Multi Materials Micro Manufacture 2005 Elsevier Ltd.
[4] Yu Chuan Su, Jatan Shah and Liwei Lin. Implementation and analysis of polymetric microstructure replication by micro injection molding Institute of Physics publishing J.Micromech. Microeng.14 (2004 415-422).
[5] Masaki Yoshii and Hiroki Kuramoto Experimental study of transcription of minute width grooves in injection moulding Polymer Engineering and science, 1994,Vol.34, No.15.
[6] B.Sha, S.S. Dimov, D.T.Pham, and C.A. Griffiths. Study of Factors Affecting Aspect Ratios Achievable in Micro-Injection Moulding. Multi Materials Micro Manufacture 2005 Elsevier Ltd.

Multi-Material Micro Manufacture
W. Menz, S. Dimov and B. Fillon (Eds.)

Polymer optical micro-array sensor

P.J. Bolt[a], R. de Zwart[a], R. Houben[a], G. van Heck[a], F. Dennard[a], P. Rensing[a], J. van Veen[b], B. Langeveld[c], M. Koetse[c], H. Schoo[c] *

[a] TNO Science & Industry, P.O. Box 6235, 5600 HE Eindhoven, The Netherlands
[b] TNO Quality of Life, P.O. Box 360, 3700 AJ Zeist, The Netherlands
[c] Holst Centre, High Tech Campus 48, 5656 AE Eindhoven, The Netherlands

* corresponding author herman.schoo@tno.nl

Abstract

All-polymer sensors made by mass-production techniques are feasible due to the development of opto-electronic polymers. Using these materials, it is possible to put sensor and electronic functions on a foil by printing techniques, allowing sensors to be mass produced at low cost. However, before this will be reality, many steps have to be made. This paper describes one of these, namely the development of a demonstrator to show the feasibility of an optical micro-array sensor. Such a sensor comprises matrices of both organic light emitting and light responsive devices, which can be fine tuned with respect to their spectral response. This allows a large freedom for making sensors for different applications. The lay-out of the matrices with organic devices, the interlayer with active sensor materials, jetting of these materials on an interlayer and the interlayer itself will be presented. Future steps will be discussed.

Keywords: optical sensor, opto-electronic polymers, reel-to-reel, printing

1. Introduction

Sensor systems will be used more and more in our environment, houses, offices, and on our body to monitor wellbeing, to set a diagnosis and guard medical treatment or recovery. The overall size of the sensor market is estimated to be about 35 billion$ in 2005 with an average annual growth of 12% [1]. In order to serve these markets and to make these applications possible, new and especially affordable sensors are needed. This will rely on new ways of manufacturing, such as reel-to-reel processes. Of great interest are opto-electronic polymers which are suitable for making thin film devices by printing techniques.

Opto-electronic polymers transfer light into electricity and vice versa. Applications, mostly in the development stage, are e.g. displays based on organic light-emitting diodes (OLEDS) and photo-voltaic cells [2,3,4]. Typically these are large area applications, which are expected to be manufactured on foil by mass production technologies, such as printing and lamination.

Organic LEDS and photo-diodes can be used in optical sensors [5] according to the principle depicted in Fig. 1.

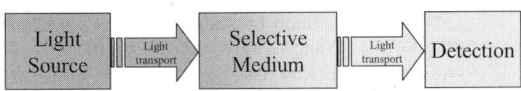

Fig. 1 Optical sensor principle.

The selective medium can be a coating that changes colour when in contact with a chemical. In the plastic waveguide sensor in Fig. 2 the amount of light that reaches the photodiode through the waveguide depends on whether the (detection) coating is in contact with alcohol or not. A specific advantage of polymer LEDS and photodiodes is the possibility to finetune their spectral response for the active sensor

coatings. OLEDS can be made that emit two distinct wavelengths and hence can be used as a single source for signal and reference [6]. This will increase the stability of the measurements.

This example uses a rigid polymer (moulded) substrate. If the LED, photodiode and detection layer would be applied on foil, geometries as depicted in Fig. 3 would be feasible.

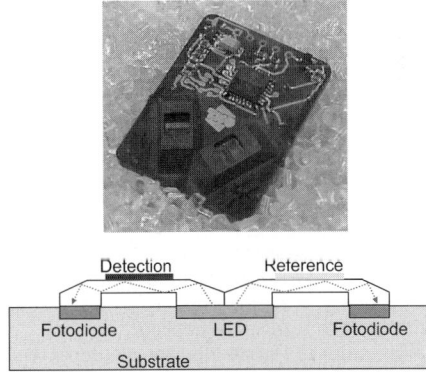

Fig. 2. Dual channel polymer waveguide sensor with organic opto-electronics.

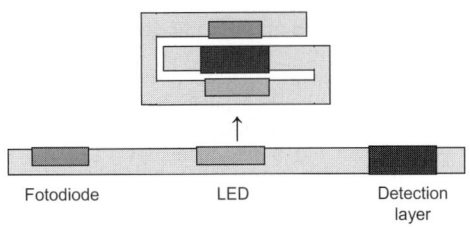

Fig. 3. Concept for a transmission sensor made on foil.

By using matrices of OLEDS and photo-diodes with different emission spectra and/or respectively intensity and sensitivity, a sensor could be made for quantitative analysis of media (gas, liquid) with single or multiple targets, for example labeled DNA or antibodies.

Such matrices resemble mini-displays and in principle such optical micro-array sensors could be produced at relatively low cost on mass scale in a similar fashion as OLED displays. These allow the use of a stack of matrices of devices (light emission, responsive materials, photodiodes), see Figure 4. Many sensor devices can be combined in one application. This can be used to increase the dynamic range (optimization of signal strength) and sensitivity range (multiple analyte).

The combination or integration of these sensors with or into microfluidic structures on foil, could result in cheap disposable cartridges for drug, pathogen or chemical tests.

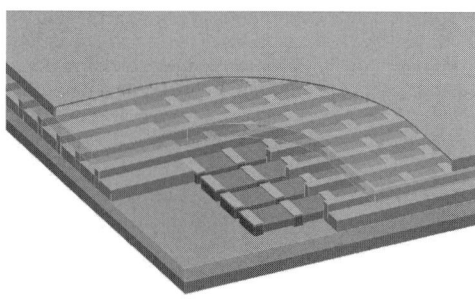

Fig. 4   Possible lay-out of an optical micro-array sensor with (layers with) matrices of (e.g. printed) polymer LEDS, photodiodes and filters and, optionally, optical elements with active sensor materials, over or between which a gas or liquid is led.

TNO demonstrated the feasibility of an electro-optical polymer based optical micro-array sensor by manufacturing matrices of LEDS and photodiodes and applying these in a complete NH_3 and ethanol vapour sensor device with active sensor materials and electronics.

The total sensor consisted of a layer with a matrix of polymer LED devices, a similar layer with photodiodes and an interlayer with transparent windows, with a sensitive coating.

In order to realize this, knowledge is required of polymer LED en photodiode devices, jetting of sensor materials, design and manufacturing of an interlayer with sensor material coatings and lateral optical separation, the sensor materials and electronics.

2. Polymer LEDs en photodiode arrays

A matrix of 10 x 10 pixels (OLEDS and/ or photodiodes) was the basis of the instrument. The glass substrate measured 30 x 30 mm and pixels were 1.6 x 1.6 mm. Fig. 5 depicts the design.

The ten (shared) vertical anodes consisted of indium doped tinoxide tracks, which were structured by photolithography. The cathodes were made by vapour deposition of 5 nm barium (for the OLEDS) or 1 nm lithiumfluoride (for the photodiodes) tracks, covered with a 100 nm aluminium layer. In Fig. 3, the cathode

tracks are horizontal, alternately connected from left and right side. The electro-optical polymer layer in the LED devices was a stack of 150 nm pedot layer with a 70 nm thick PPV (poly(phenylene vinylene)) -derivative coating, both applied by spin coating.

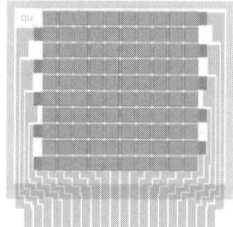

Fig. 5. Lay-out of the array-sensor

3. Interlayer with selective coatings for optical array

In between the OLEDS and photodiode matrices the specific vapour sensitive materials are placed. Principle is that the transmission spectrum of the coating changes due to reaction with vapour molecules. In the demonstrator, coatings were applied for detection of ethanol and ammonia (see Fig. 6). Both coatings are solvent based.

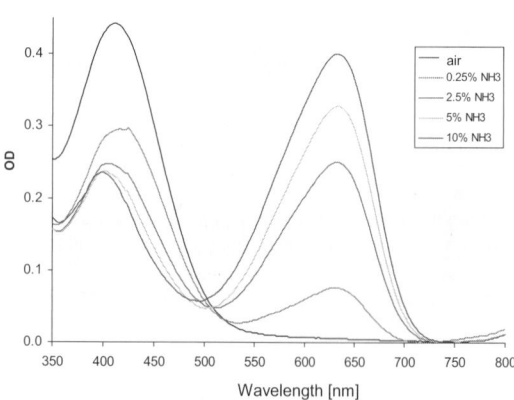

Fig. 6. Spectrum of active coating on exposure to NH_3.

These materials were applied on transparent windows in a frame of non-transparent and chemically inert LCP (Liquid Crystal Polymer), in order to prevent scattering of light in lateral direction (see Fig. 7). The windows were made out of an optical grade polymer.

The sensor materials were applied by a drop-on-demand jetting system. Because of a demand to process a large range of materials, with no restriction with regard to the solvent, thermal printing systems could not be used. More suitable are piezo-electric drop-on-demand systems. For these systems, the solvent type is of less importance. The material is ejected by a pressure wave which is generated by the movement of the piezo-element. Best option was to use a glass printhead because of its resistance for most solvents. A recurring problem is however cavitation, see Fig. 8.

Fig. 7. LCP frame with ethanol sensitive coating on transparent windows.

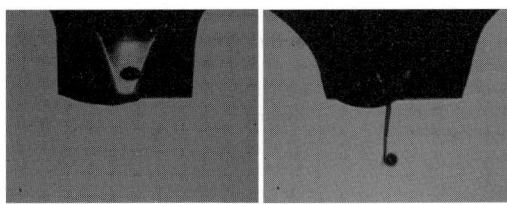

Fig.8. Print head with cavitation (left) or tail formation (right).

4. Demonstrator

For the integrated demonstrator a simplified lay-out was chosen: 9 x 4 LEDs or photodiodes, with a common cathode. Measurements with alcohol- and NH_3 sensor material showed a lower limit for optical detection of 0.1% (measured with reference channel). This is sufficient for the pursued application. The onboard sensor electronics transfers its data to a laptop by a USB connection.

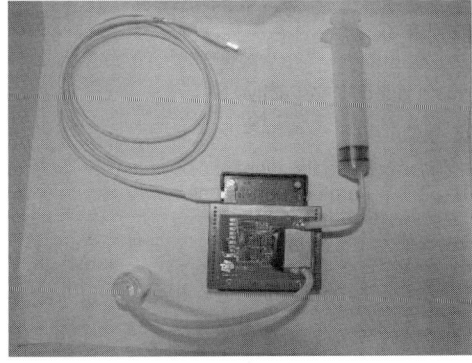

Fig. 9. Completed demonstrator.

5. Future

The demonstrator will be developed further. This includes a higher number of sensor elements and integration of more functions (without enlarging the total size). It will contain the sensor, a processor, wireless communication (ZigBee) and power supply.

In longer term however the goal is to manufacture the sensors on foil, with integrated:

- sensor arrays
- polymer electronics
- microfluidics
- wireless communication
- autonomous functioning.

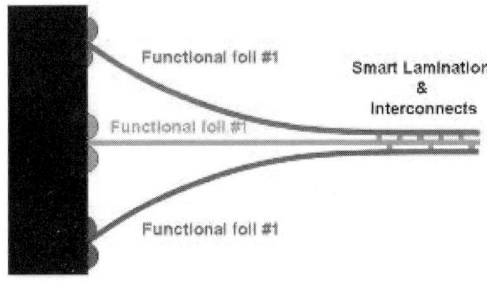

Fig. 10. Envisaged production of sensors-in-foil.

The sensor will be made by large area thin film techniques with generic process steps to integrate functionalities in foil, see Fig. 10. Different groups of functionalities (sensor array, power, communication, fluidics) will be put on separate foils which will be stacked and bonded ready-made.

Issues are:

Substrate manufacturing:
  - polymer substrate instead of glass. This requires barrier layers to protect the organic devices from water and oxygen,
  - Integration of fluidic and optical functions in one foil.
Printing of device structure:
  - opto-electronic polymers,
  - sensor materials,
  - optical & electronic structures.
Bonding of layers:
  - alignment of foils,
  - (electrical) interconnect of devices on separate layers.

Hence not only alignment of devices on foil, but also of foils (layers) relative to each other for lamination is an issue. And not only the precision of the printing or jetting process (positioning, dosing) and properties of the printed materials (high-viscosity, filled polymers, metals) has to be considered, but also the mechanical properties (stretching) and speed of the substrate.

These challenges will have to be outweighted by advantages such as high productivity (e.g. due to reel-to-reel set up) and ease to modify the products (related to printing techniques).

The application of optical sensor arrays on foil will not be based on low cost only. For example the flexibility of substrates is important for skin sensors [7], e.g. plasters to monitor body functions or bedsores.

The advantage of optical sensors is for the absence of direct contact between the vulnerable opto-electronic polymers and e.g. air or water, a disadvantage is the need for optical structures. Organic devices are being developed however which may be brought in direct contact with air or water, thanks to development of improved barrier coatings. This enables direct electronic sensor principles (without light-electricity transfer) in which the gate of a polymer transistor reacts to the environment. Although less sensitive than optical sensors, electronic polymer sensors offer more direct signal reading with simpler equipment, easy integration with solid-state electronics, miniaturization towards the nanoscale and low power consumption.

6. Summary and conclusions

Opto-electronic polymers have a high potential to be used in mass-produced optical sensors. TNO has made a step in the development of their application by showing the feasibility of an optical micro-array gas sensor. The sensor consisted basically of three layers with arrays of OLEDs, active sensor materials and photodiodes. This is a working reference for future device development, in which substrates will be truly flexible and large area manufacturing techniques will be used.

References

[1] Nanosensors: A Market Opportunity Analysis, Nanomarkets LC (2005).
[2] Krebs, FC (ed.), The development of organic and polymer photovoltaics, Solar Energy Materials and Solar Cells, 83 (2004) pp. 125-321.
[3] Burroughes JH, Bradley DDC, Brown AR, Marks RN, Mackay K, Friend RH, Burn PL, Holmes AB, Light-emitting diodes based on conjugated polymers, Nature, 347 (1990) pp 539-541.
[4] Sirringhaus H, Tessler N, Friend RH, Integrated Optoelectronic Devices Based on Conjugated Polymers, Science, 280 (1998) pp 1741–1744.
[5] Patent WO2005015173, Schoo H et al., Sensor comprising polymeric components (2005).
[6] Patent WO2005001945, Schoo H et al., Light emitting diode (2005).
[7] Someya T et al., A large-area, flexible pressure sensor matrix with organic field-effect transistors for artificial skin applications, Proc. Nat. Acad. Sciences, 101 (2004) 9966-9970.

Multi-Material Micro Manufacture
W. Menz, S. Dimov and B. Fillon (Eds.)

Design and testing of a d_{31} mode piezoelectric unimorph for direct power generation by coupling with a microcombustor

G. Liu[a], R.W. Whatmore[a,b]

[a] Nanotechnology Centre in the Department of Materials, Cranfield University, Beds, MK43 0AL, UK
[b] Present address: Tyndall National Institute, 'Lee Maltings', Cork, Ireland

Abstract

The design and testing of a piezoelectric unimorph disk for electric power generation is discussed. The motivation for the work is the increasing demand for extended life power supplies for use in portable electronic devices such as laptops and remote surveillance systems.

The unimorph comprises a piezoelectric ceramic bonded onto a steel disk, operating in d_{31} radial mode so that lateral stresses cause a charge difference through the thickness of the unimorph. The charge can then be extracted and conditioned to do useful work. This article highlights the design and the testing of the unimorph for two loading conditions. Firstly to apply a uniformly distributed pressure pulse to the underside of the disk using compressed air. Secondly to couple the unimorph to a 5mm^3 combustion chamber, this burns a stoichiometric mixture of hydrogen and oxygen. The overall pressure change due to the compressed air or combustion causes the unimorph to deflect up to 50μm and become stressed, in doing so generating electrical charge through the piezoelectric effect. Testing has shown that milli-Watts of power can be generated with this configuration of piezoelectric unimorph, found through the power dissipation across a known load.

Keywords: piezoelectric, d_{31}, unimorph, flexure, power generator, microcombustor

1. Introduction

Miniaturization and increased functionality of defence and consumer electronics has been one of the major factors for developing high density electrical power supplies. These require prolonged sources of electrical power from small and lightweight packages.

This work investigates how hydrocarbon and other gaseous or liquid fuels which possess high energy densities can be harnessed to produce energy. They have long been exploited as power sources through combustion for e.g. automotive applications. In this article, we report the design and testing of a piezoelectric unimorph disk, coupled with a microcombustor device fuelled on hydrogen and oxygen for direct conversion of the energy released by combustion into electricity.

Whalen et al [1] have tested an external micro heat engine. Heat is supplied externally from a resistive heater which causes a working fluid to expand. In this case, a membrane consisting of a thin film piezoelectric on silicon becomes stressed and converts the thermal energy input into an electrical output. The heat engine had a volume of ~0.6mm^3 with an output of 80mV, observed at 240Hz.

Lesieutre et al [2] have built a piezoelectric energy harvesting device which generates electricity from a piezo-bimorph attached to a vibrating structure. Up to 30mW could be harvested using a two-layer bimorph.

Several others have looked at the use of piezoelectric for power generation [3-5].

2. Unimorph Design

The piezoelectric effect is where a mechanical stress applied to a material with a specific crystalline structure develops an electrical charge. Piezoelectric materials can be polarised with a high voltage electric field so that directionally applied stresses can causes a

charge difference of a desired polarity.

The d_{31} mode of operation, discussed in this article is where the radial stress (X_1) applied to a piezoelectric disk causes a change in polarisation through its thickness (P_3), as illustrated in Fig. 1, also represented by Eq. 1.

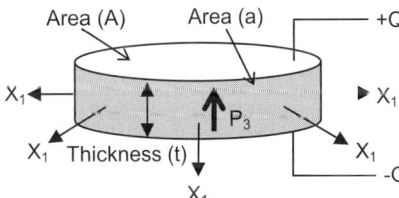

Fig. 1. A d_{31} mode piezoelectric schematic.

$$P_3 = d_{31} . X_1 \tag{1}$$

Where d_{31} is the piezoelectric charge coefficient expressed in Coulombs per Newton. The piezoelectric can be used to generate charge in d_{31} mode by fabricating a unimorph where the piezoelectric is bonded onto an elastic material. The perimeter of the elastic can be held rigid to form a flexible diaphragm. A load applied to the face of the unimorph causes the diaphragm to deform and stretch the piezoelectric in the radial, X_1 direction leading to charge developing on the major face. (See Fig. 2)

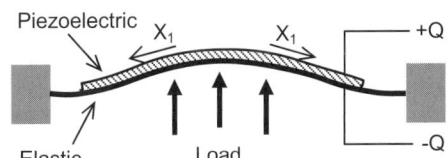

Fig. 2. A radially stressed (X_1) unimorph develops charge (Q) across electrodes on the major faces.

The unimorph consists of a thin steel disk onto which was bonded a piezoelectric ceramic (Ferroperm PZ27). The epoxy bond has negligible thickness enabling electrical contact to the bottom electrode via the steel disk. When the unimorph becomes deformed it undergoes both tensile and compressive stress. To optimise the piezoelectric output the neutral stress plane should occur at the interface between the piezoelectric (PZT) and steel. (See Fig. 3)

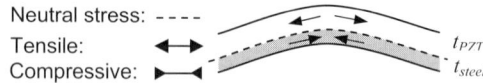

Fig. 3. Neutral stress plane in a deformed unimorph

The optimum thickness ratio t_r, between PZT and Steel was determined by Eq. 2.

$$t_r = \sqrt{E_{PZT} / E_{Steel}} \qquad (2)$$

Where E is the Young's modulus for each material, t_r is the ratio between t_{PZT} and t_{Steel} [6, 7]. In this case, t_r is ~0.5. Using these thicknesses ensures that the neutral stress plane resides along the interface of the piezoelectric and steel.

A 0.25mm thick stainless steel foil was used as an elastic material and a piezoelectric ceramic disk (38mm OD, 0.5mm thick) bonded to it with epoxy resin (EPO-TEK 301-302, Epoxy Technology). The steel disk also provides an electrical connection to the bottom electrode of the unimorph. The unimorph disk is shown schematically in Fig. 4.

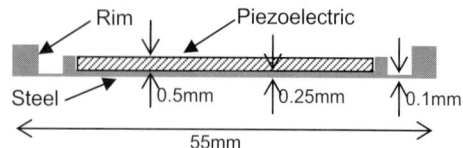

Fig. 4. A schematic of the unimorph disk with a thick support rim around the perimeter.

The formed diaphragm is thinned at its rim [8], to prevent the disk from being 'built-in' to the thick supporting ring. If it is, the piezoelectric disk will experience tensile and compressive stresses at different radii from the centre, reducing overall piezoelectric output.

3. Experimental

A prototype micro combustion device has been fabricated and contains a 4mm diameter chamber, 5mm³ in volume, where the top side is enclosed by the unimorph disk. The device setup (Fig. 5) includes two piezoelectrically actuated poppet valves, valves springs to close valves, a thermocouple, a spark ignition module for direct access into the base of the chamber, and a fibre-optic displacement probe to measure the deflection of the unimorph. The electronics are controlled by a data acquisition card (PCI-DAS6040, Adept Scientific) and LabVIEW (Version 7.0, National Instruments).

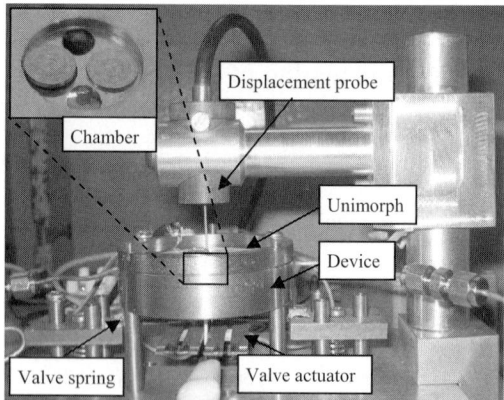

Fig. 5. Test setup showing combustion device with unimorph. Inset of 4mm diameter combustion chamber usually covered by the unimorph, with valve heads, spark from element and thermocouple tip.

Two setups were used to measure the electrical output from the unimorph disk for experimentation by applying a load using compressed air and combustion (see Fig 6 and 7). In both cases the gas valves allow gas into the cylindrical chamber.

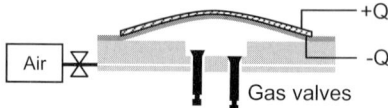

Fig. 6. Setup to apply a uniform distributed load to the unimorph with compressed air.

During pressure tests, compressed air was regulated from 40-100kPa, an impulse of pressurised air was vented into the chamber by the opening on the inlet valve. The pressurised chamber was then discharged by the opening of the exhaust valve.

The combustion tests require the metering of hydrogen and oxygen gases, achieved with mass flow controllers to obtain the correct stoichiometric mixture. Similarly the mixed gas was vented into the chamber at atmospheric pressure. Combustion was initiated using a custom spark plug which comprises a copper electrode housed in a ceramic insulator.

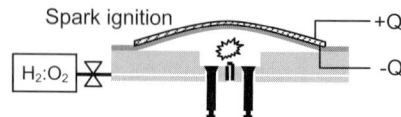

Fig. 7. Basic setup for loading the unimorph through combustion of hydrogen and oxygen

The bottom electrode of the piezoelectric unimorph was grounded to the device body. The top electrode was connected via a shielded cable to a 10kΩ load. The electrical output can then be extrapolated from the voltage across the load against time graph. Data was recorded via a data acquisition system and LabVIEW.

4. Results & Discussion

4.1 Pressure tests

The results for pressurizing the chamber using a compressed air feed have shown an electric output as a direct response to the stress caused by the pressure increase (Fig.8). The chamber was pressurized by the opening and closing of the inlet valves at the set pressure.

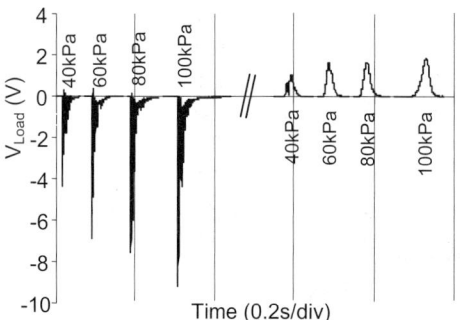

Fig. 8. Plot of unimorph voltage output for pressure testing, larger negative peaks correspond to pressurization, smaller positive to de-pressurization.

The corresponding peak output voltages and energy dissipated in the load by the voltage generated by the unimorph are shown in table 1. It can be seen that there is a relation between the inlet pressure and the output generated upon pressurization. This was not so evident during depressurization and was partly due to pressure leakage through the valves during the delay between inlet and exhaust of the compressed air.

Table 1
Unimorph electrical output from pressure testing

P (kPa)	Pressurization		Depressurization	
	V_{peak} (V)	Energy (μJ)	V_{peak} (V)	Energy (μJ)
40	-4.40	4.82	1.04	1.07
60	-6.90	15.05	1.64	3.01
80	-7.57	19.01	1.65	3.53
100	-9.10	24.46	1.86	5.34

The polarity of the signal was determined by the polarization of the piezoelectric unimorph disk. The respective charge and energy developed by the unimorph can be calculated from the area under the current vs. time graph. The peak power generated by the unimorph was found for pressurization, see Fig. 9.

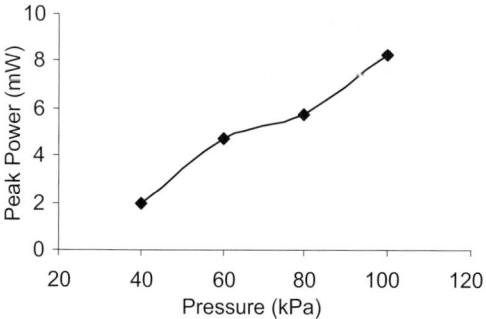

Fig. 9. Peak power dissipation on the 10kΩ load generated by unimorph due to pressurization tests.

It shows a linear relation between the pressure in the chamber and the peak power dissipated across the load, which is a direct representation of the power generated by the unimorph.

4.2 Combustion tests

The unimorphs electrical output was characterized through the combustion of a hydrogen-oxygen mixture at atmospheric pressure. A typical piezoelectric unimorph output is shown in Fig. 10, along with the temperature measurements. This shows a stoichiometric mixture of hydrogen and oxygen at an average flow of 5ml/min. A purge cycle of several seconds with both valves open was used to vent the chamber with a fresh charge of $H_2:O_2$. The fuel was then contained in the chamber by closing the exhaust valve first, immediately followed by closing the inlet valve.

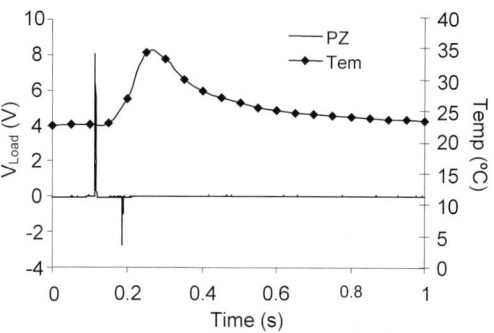

Fig. 10 Unimorph voltage output and thermocouple measurement, when burning a 2:1 mixture of $H_2:O_2$

The duration of the voltage peak due to combustion was approx. 5ms. The thermal mass of the thermocouple, closely coupled to the large thermal mass of the device body accounts for the small 12°C temperature change. The voltage and energy output from fig. 10 are listed in table 2, together with data for a higher flow rate.

Table 2
Unimorph electrical output for combustion testing

$H_2:O_2$ (ml/min)	Combustion		Exhaust	
	V_{peak} (V)	Energy (μJ)	V_{peak} (V)	Energy (μJ)
5	7.93	7.9	-2.78	1.27
10	9.06	11.6	-3.95	2.4

A comparison between the measured temperature increase against the charge produced by the unimorph during combustion and exhaust (i.e. when the pressure was released during the exhaust stroke) is shown in Fig. 11, obtained for different mixture ratios.

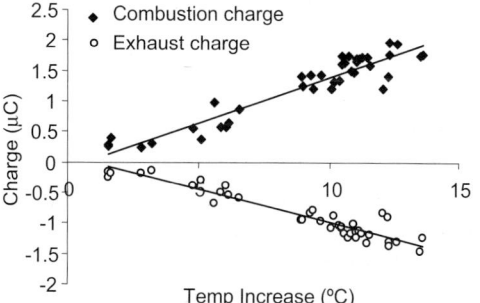

Fig. 11. The unimorph charge output against the respective temperature increase from combustion

384

A good linear correlation is shown between the amount of heat energy released from combustion (as measured by the temperature change) and the amount of charge produced. The output of the unimorph due to combustion was greater than the pressure release during the exhaust stroke. This is most likely a slight pressure drop immediately after combustion, due to a combination of heat drawn away by the device body and imperfect valve sealing. A small amount of energy loss may also be attributed to mechanical losses within the piezoelectric. The peak power dissipation in the 10kΩ load (see Fig. 12) shows a non-linear curve increasing with temperature.

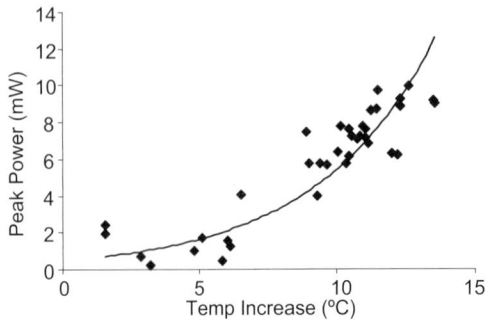

Fig. 12 Peak power dissipation on load as an indication of unimorph generative power

A point to note is that polarity of the voltage for combustion was positive (Fig. 10); opposite to the pressure test results (Fig. 8). This is due to a reduction in pressure during $H_2:O_2$ combustion.

During pressure testing the unimorph rises up ~50µm at 100kPa, whereas during combustion the unimorph drops by ~30µm. This suggests that the reaction of hydrogen-oxygen was quenched, and that the H_2 and O_2 reacted to produce H_2O, with a corresponding (33%) reduction in the molar volume.

Therefore the linear plot (Fig. 11) suggests that more complete combustion is occurring at certain ratios leading to higher volume reduction and higher temperatures.

The rapid loss of heat from the gases to the chamber of the device meant that the combustion flame could not sustain a net volume increase, but instead there was a decrease in volume. It is believed that the main reason for this was that the unimorph was not perfectly flat after fabrication. The resulting upward bowing (~30µm) in the disk produces a trapped volume between the unimorph and the combustor body which was comparable to that of the combustion chamber, but with a very large surface area to volume ratio.

This promotes the rapid loss of heat from the flame; therefore the efficiency of the device in its current form is poor, with the estimated chemical input energy in tens of milli-joules range, and only tens of micro joules are produced by the unimorph.

Nevertheless, the principle of coupling a piezoelectric unimorph disk to a microcombustion device in order to obtain a useful output of electrical energy directly from the chemical energy released by combustion has been demonstrated. It is anticipated that further improvements in device design and construction will considerably increase the efficiency and magnitude of the output that can be obtained.

5. Conclusion

This work has demonstrated the viability of this power generator using piezoelectric materials in a d_{31} mode unimorph configuration, coupled with a microcombustor. Initial testing has revealed encouraging results with peak power output of 10mW from an un-optimized device.

The large surface area to volume ratio of the present combustion chamber design has meant that hydrogen-oxygen combustion was quenched. Improved combustion efficiency would lead to a greater amount of heat being produced, and eventually to a volume increase due to gas expansion.

The potential for the microcombustor is very significant, current experiments have been at low frequencies ~5Hz, limited by the quenched flame.

The next step is to run the microcombustor with a better designed chamber to reduce heat loss, addition of a compression stroke, and run at the resonance frequency of the unimorph disk (~200Hz) which would significantly increase the unimorph output.

Acknowledgements

The author would like to thank HMGCC their support gained throughout this work.

The author would also like to acknowledge the support obtained from the EPSRC under a PhD studentship from the Doctoral Training Account at Cranfield University

References

[1] S. Whalen, Design, fabrication and testing of the P^3 micro heat engine, Sensors and Actuators A, 104, 2003, pp. 290-298.

[2] G. A. Lesieutre, G. K. Ottman, H. F. Hoffman, Damping as a result of piezoelectric energy harvesting, Journal of Sound and Vibration, 269, 2004, pp. 991-1001.

[3] U. Bonne, B.R. Johnson, Free piston piezoelectric generator, United States Patent No.0178702 A1, 2004.

[4] F. Lu, H. P. Lee, S. P. Lim, Modeling and analysis of micro piezoelectric power generators for micro-electromechanical-systems applications, Smart Materials and Structures, 13, 2005, pp 57-63.

[5] C. Keawboonchuay, T. G. Engel, Electric power generation characteristics of piezoelectric generator under quasi-static and dynamic stress conditions, IEEE Trans. on Ultrasonics, Ferroelectrics, and Frequency Control, Vol. 50, No. 10, 2003, pp 1377-1382.

[6] Q. M. Wang, Electromechanical coupling and output efficiency of piezoelectric bending actuators, IEEE. Trans. on Ultrasonics, Ferroelectrics and frequency control, Vol. 6, no. 3, 1999, pp. 638-646.

[7] M. R. Steel, F. Harrison, P. G. Harper, The piezoelectric bimorph: an experimental and theoretical study of its quasistatic response, Journal Physics D: Applied Physics, Vol. 11, 1978, pp. 979-989.

[8] C. Tanuma, Y. Suda, S. Yoshia, A highly sensitive displacing device using piezoelectric bimorph with double support structure, Proc. of the 4th meeting on Ferroelectric Materials and their Applications, Japanese Journal of Applied Physics, Vol. 22, 1983, pp. 154-156.

Multi-Material Micro Manufacture
W. Menz, S. Dimov and B. Fillon (Eds.)

Water jet machining of MEDM tools

O. Blatnik[a], H. Orbanic[a], C. Masclet[b], H. Paris[b], M. Museau[b], J. Valentincic[a],
B. Jurisevic[a] and M. Junkar[a]

[a] University of Ljubljana, Slovenia
[b] University of Grenoble, France

Abstract

This contribution presents an investigation about the possibilities of using Water Jet (WJ) technology in combination with Micro Electrical Discharge Machining (MEDM) for tooling production in micro manufacturing. In the first phase the tool copper used in MEDM is produced by WJ machining. Afterwards, the final tool in steel is produced by MEDM. Such kinds of tools intend to be used in processes like hot embossing, molding, and other replication technologies in the field of micro manufacturing. The first results are very promising and the proposed tooling strategy, which involves besides MEDM also WJ technology, shows a lot of potential especially in the design and developing phase of micro-fluidic devices.

Keywords: tooling, non-conventional technologies, MEDM, WJ, micro-fluidics

1. Introduction

Machining of tools used to produce micro components is usually expensive and time consuming. Thus, the tool used for replication technologies such as hot embossing or molding has to be well defined in advance. Once the tool is produced the corrections or modifications are difficult to implement.

The main objective of this contribution is to present a new tooling strategy based on the application of WJ machining, which would allow relatively quick and cost effective production of prototype micro components. In the first step the tool for MEDM is produced in copper by WJ technology. Then the copper tool is used by MEDM technology to produce the final tool in tool steel, which may be further used for replication processes such as hot embossing, pressure molding and others. The complete process chain is shown in Figure 1.

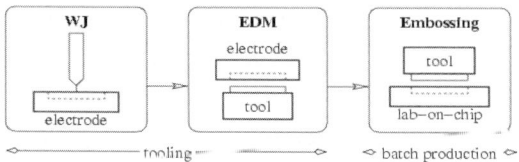

Figure 1. The production sequence of micro-fluidic channel.

The proposed tooling strategy offers high flexibility and cost effectiveness. Additionally, it provides more freedom and testing opportunities during the development of new micro devices. The most common used tooling strategy is direct manufacturing of the tool by micro milling. When the features of the tool are rather ribs then grooves, the tooling strategy proposed in Fig. 1 has a great advantage over micro milling tool manufacturing, which is the most common used tooling strategy. In the latter case, an end-mill with a relatively small diameter has to remove relatively big volume of the tool which is time consuming and not cost effective.

The main application field of the proposed tooling strategy is the design and development of micro-fluidic devices. Typically, these devices require a well-controlled geometry and surface roughness [1]. With this technology these devices can be manufactured relatively fast and in a cost effective way. Therefore many new concepts and designs can be experimentally validated during the development phase in order to improve the performance of the final product. In the actual context of R&D, flexibility in the manufacturing process enables variety and innovation in the design. The proposed tooling strategy consumes most of its machining time in MEDM machining while WJ machining accounts just for a small portion of the total machining time. However, facing this sequence of different processes, WJ machining of the MEDM tool has an important influence on the final result.

The paper is organized as follows: after an introduction to the proposed tooling strategy the second section describes the WJ machining process applied to the machining of the electrode. In the third section, MEDM is described, as well as the manufacturing of the tool by the MEDM process using the electrode made by WJ. The results are presented in the fourth section. They constitute the basis for the analysis and discussion that is held in the fifth section. Finally, conclusions are given on the proposed tooling strategy in the last, sixth section.

2. WJ machining of EDM tools

WJ machining is a non-conventional machining process in which a high-speed jet of water is used as the tool to remove the workpiece material by erosion. In order to generate such a tool, the water is first pressurized. The water pressure is usually set up to 400 MPa and is generated with a hydraulic intensifiers specially designed for this technology. To generate an effective WJ, a specially designed nozzle with an orifice made in a sapphire insert is used. Typically, the diameters of orifices used for this technology are between 0.3 and 0.08 mm, depending on the application. This technology is mostly used to cut softer materials like wood, plastics, rubber, etc. but can be also applied for cleaning and engraving in harder materials. For metals and other hard to machine materials Abrasive Water Jet (AWJ) machining is more appropriate due to a higher Material Removal Rate (MRR) comparing to WJ machining [2].

In spite of all the advantages that WJ technology brings when producing electrodes for MEDM, the main problem is how to control the penetration depth of the WJ into the workpiece material. Investigations in the field of AWJ milling [3, 4] show that it is difficult to predict and control the depth of penetration of the AWJ into the workpiece material.

In general, the depth of penetration depends on several process parameters. One of the most important and easiest to control is the traverse velocity of the cutting head, which defines the exposure time of the workpiece to the WJ. This means that only 2.5 D tools for MEDM can be produced with this technology. The difficulty of controlling the WJ penetration consequently causes that all features on the EDM electrode are deeper than required for the EDM process as shown in Figure 2. Thus the bottom of the slots of the EDM tool machined with WJ will not be functional surfaces.

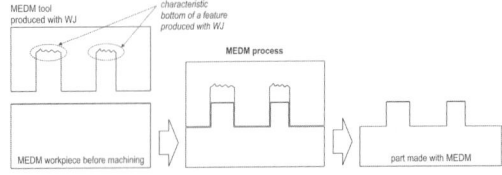

Figure 2. Specifics of MEDM tool production with WJ.

As a case study, the manufacturing of a tool for a lab-on-chip application was chosen as presented in Figure 3.

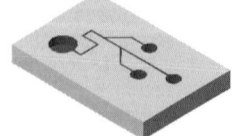

Figure 3. Lab-on-chip feature.

Machining of the MEDM tool took place on the OMAX type 2652A/20HP abrasive water jet cutting system powered by a Böhler Ecotron 403 hydraulic intensifier capable of reaching water pressures up to 410 MPa. In this case a WJ cutting head was used with the orifice diameter of 0.1 mm. According to previous experience the water pressure was set at 300 MPa, the traverse velocity of the cutting head was 10 mm/min and the stand-off distance between the cutting head and the workpiece was kept constant at 2 mm.

3. MEDM production of the embossing tool

Electrical Discharge Machining (EDM) is a machining technique through which the surface of a metal workpiece is formed by discharges occurring in the gap between the tool, which serves as an electrode, and the workpiece. The gap is flushed by the third interface element, the dielectric fluid. The process consists of numerous randomly ignited mono-discharges. During a discharge, a plasma channel is formed which serves as the current conductor and the heat generator. As a

consequence, a crater appears on the spot of the discharge. The size of the crater depends on the discharge energy, which can be set on the machine by adjusting the discharge current and its duration. The discharge voltage, which also determines the discharge energy, cannot be adjusted on the machine explicitly since it depends on the gap width between the workpiece and the electrode [5]. The MRR is determined by the crater size and the frequency of craters generation, i.e. discharge energy and the frequency of discharges. The latter is influenced by the discharge duration and the pulse interval between two discharges. The gap width between the workpiece and the electrode is in the range of 0.01 and 0.1 mm. The MRR is around 100 times higher on the workpiece than on the electrode.

The main difference between EDM process used for "macro" workpieces and MEDM is in the accuracy for the electrode feeding system, also called servo system, and machining parameters. The comparison is given in Table 1.

Table 1. EDM and MEDM machining parameters

mach. parameter	EDM	MEDM
discharge current	up to 200 A	up to 3 A
discharge duration	up to 1000 μm	up to 50 μm
pulse interval	up to 1000 μm	up to 200 μm

The MEDM machining is usually performed with rod electrode, whose path is controlled by CNC controller. The commercially available rod electrodes have diameters down to 0.05 mm, but usually bigger electrode diameters, for instance 0.15 mm, are often used due to limitations in electrode handling systems. Smaller electrode diameters are obtained by applying wire EDM grinding or etching after clamping of the electrode on the MEDM machine [6]. In this research, the tool for embossing of the micro-fluidic channels is made by employing a sinking MEDM, where the electrode has a negative shape of the required shape on the workpiece. The accuracy of the electrode shape is directly transferred into the workpiece if the orbital or planetary motion of the electrode (employing CNC controller) is not used. In this case of manufacturing the pin-like shape of the tool (Figure 1), the sinking EDM achieves much higher MRR than MEDM milling since the size of the machining surface of the electrode (Figure 4) is much bigger than the machining surface size of the rod electrode with a diameter less than 0.2 mm [7]. The following machining parameters were used on the machine IT Elektronika 200M-E: discharge current i_e=1 A, ignition voltage u_i=180 V, discharge duration t_e=8 μs and pulse interval t_o=36 μs.

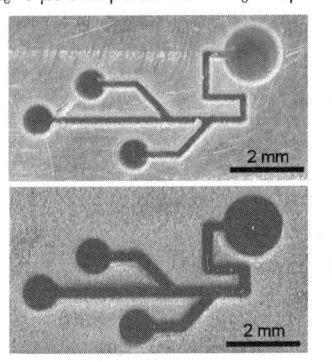

Figure 4: Electrode (a) before and (b) after the machining of the tool.

The wear phenomenon of MEDM process can be observed in Figure 4(b), while the tool produced by the given electrode is presented in Figure 5.

Comparing the machining times of the WJ, MEDM and embossing processes, the following conclusions arisen. In order to produce the electrode given in Figure 3, the machining time of WJ process was less than 5 minutes. Compared to the MEDM time, which took about 12 hours for manufacturing of the tool (Figure 5), the machining time required for making the electrode is really negligible. It is worth noticing that the MEDM machining time is related to the relatively low MRR (0.051 mm^3/min) and to the relatively large removed volume (36 mm^3).

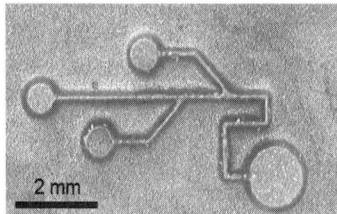

Figure 5. Steel tool for hot embossing Lab-on-chip application.

4. Property analysis of machined parts

As the difficulty of small parts is the measurement of characteristic dimension, the following properties were measured in this study: (1) the dimensions measured from the top profile of the electrode and the tool (Figure 6), (2) the slot profile of the electrode and the tool (Figure 7), and (3) the surface roughness of the tool.

4.1. The top profile

The various measurement points of the top profile of the electrode before and after MEDM machining and of the finished embossing tool were measured by a digital camera (Imaging Source DMK 21F04, 640x480) at positions shown in Figure 6. The results of measurements are collected in Table 2. In the first row the desired dimensions of the micro-channels are given. The specific dimensions were chosen in order to evaluate the discrepancy between expected and obtained values.

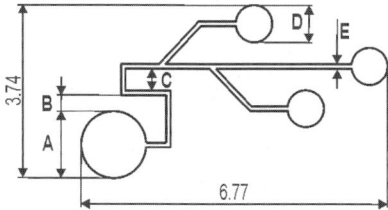

Figure 6. Drawing of a Lab-on-chip feature measurement points.

Table 2. Measurement of the top profile (in mm).

Dimension	A	B	C	D	E
Drawing	1.44	0.36	0.47	0.8	0.1
Tool	1.26	0.44	0.52	0.61	0.04
New electrode	1.45	0.34	0.43	0.77	0.18
Worn electrode	1.61	0.28	0.28	0.95	0.34

The side gap between the electrode and the tool during MEDM was approximately 0.1 mm, which was considered in the phase of electrode machining. The tip of the rib of the tool has the width of only 0.04 mm, which can be considered as a consequence of the poor flushing conditions in the gap.

4.2. The channel profile

In order to achieve the desired dimensions of the finished part, the channel profile has to be checked more thoroughly, since the shape of the embossed channel is exactly the negative shape of the tool.

The new and worn electrodes, as well as the tool were cut in half and the profile was measured by digital camera. The measurements are given in Figure 7, where the measurement points are also marked. The line at 0.7 mm indicates the depth of machining in order to obtain the 0.7 mm high feature of the tool.

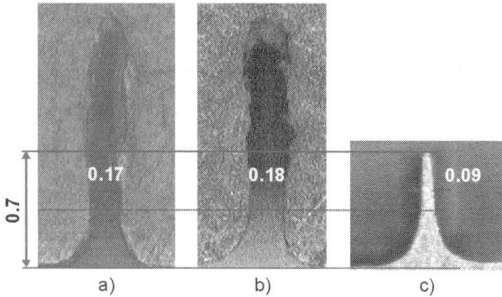

Figure 7. Measurement of slot profile: (a) new electrode, (b) worn electrode and (c) tool rib. The measurements are in millimeters.

The channel on the electrode was found to have very little taper at 0.7 mm, which is the depth of MEDM machining. The chamfer is typical for WJ machining and could be removed by consequent machining before the MEDM phase.

In contrast to the electrode, quite a big taper is noticed on the tool, which is related to the poor flushing conditions during MEDM. Thus, the profiles of the new and worn electrode as well as the tool were measured. Results are given in Figure 7.

4.3. The surface roughness

The surface roughness was measured only on the tool, as its surface profile will be transferred on the final product. The measurements were performed by a Hommel Tester T1000.

The roughness was first measured on the base, where the stylus measurements is applicable. The measuring distance L_t was 1.5 mm. The measured R_a was 0.67 μm and R_z was 3.99 μm. Although this surface has no influence on the tool, it gives us a good estimation of the EDM process performance.

Since it is impossible to measure the surface roughness of the sidewalls, the additional vertical surface was produced by MEDM using the same machining parameters. It should be taken into account that the flushing conditions were not the same as for the machined part, but the measured R_a was 0.65 μm and R_z was 3.99 μm, the same as the surface roughness of the base surface.

388

5. Discussion

It can be observed that WJ performed very well as the deviation from the desired dimension is not so large. There is a slight problem with the width of the main channel, which should be 0.1 mm wide (the diameter of the WJ) but was wider by 80 %. The reason is the back flow of the water jet that is widening the main channel. But this is balanced by the side gap between the electrode and the workpiece (tool). By applying the orbital motion of the electrode during the machining, the size of the channels could be reduced, and the shape of the channels could be improved. Additionally, the flushing of the gap is also improved.

Higher concentration of the debris in the gap results in a higher electrode wear. In the given case, the flushing was performed only by nozzles supplying fresh dielectric in the gap. The flushing could be improved by using additional holes in the electrode for supplying fresh dielectric.

The conical shapes of the tool ribs are due to the poor flushing of the gap and due to the slight taper of the sidewalls made by WJ machining of the electrode. The slope of the rib has to be efficiently controlled through orbital motion of the electrode to better integrate the constraints related to further use of the tool.

The achieved surface roughness is quite good, however, we still need to better analyze the surface roughness to evaluate the impact on the product made by embossing. Only functional surfaces should be considered, e.g. sidewalls and bottom of the channels.

Two prospects have to be developed. First, CNC controlled MEDM that can perform orbital (also called planetary) machining. Second, performing roughing and finishing MEDM. The major job of removing the large part of the material is done by roughing conditions, while the smooth surface and accurate tool rib is done by finishing MEDM parameters. This approach requires fine-tuning of the size of the channels on the electrode, MEDM rough and finishing parameters and orbital motion of the electrode.

6. Conclusions

The results of the case study applying the proposed tooling strategy are promising for the future research. The work will be focused on enhancing the MEDM process by improving the flushing of the gap by orbital motion of the electrode.

Considering WJ machining the orifice diameter has to be further reduced in order to allow the production of smaller features and improvement of their accuracy. Also other WJ machining strategies such as multi-pass cutting should be considered, since early investigations showed an improvement in surface roughness [8].

Further on, different electrode materials, such as graphite, will be used in experiments in order to determine the best conditions for rough and finish MEDM.

Research will also concentrate on the relationship between the expected EDM tool dimensions and the final tool dimensions by taking into account the wear of the electrode, which strongly depends on the machining parameters. Using multiple electrodes in various conditions will surely help to master the key parameters. In general, efforts will be made to characterize the processes and to model them. Existing strong interactions should be highlighted to better master the process.

Finally, a more detailed study of the surface roughness should be held in order to refine the process parameter. Micro-fluid flowing behavior is very sensible to the surface quality and the ability of the process to obtain the desired roughness will make, with no doubt, a decisive advantage.

Acknowledgement

This work is supported by the "Multi-Material Micro Manufacture: Technology and Applications (4M)" Network of Excellence, Contract Number NMP2-CT-2004-500274 and by the "Virtual Research Lab for a Knowledge Community in Production (VRL-KCiP)" Network of Excellence, Contract Number NMP2-CT-2004-507487, both within the EU 6th Framework Program and by the bilateral PROTEUS Project between the Republic of Slovenia and the Republic of France, Contract Number BI-FR//05-06-017.

References

[1] S. Colin: Microfluidique, Lavoisier Paris, 2004.

[2] A.W. Momber, R. Kovacevic: 1998 Principles of abrasive water jet machining, Springer, Berlin.

[3] C. Ojmertz: A Study on Abrasive Waterjet Milling, Ph.D. Thesis, Chalmers University of Technology, Goteborg, Sweden, 1997.

[4] B. Jurisevic, D. Kramar, K.C. Heiniger and M. Junkar: New Perspectives in 3D Abrasive Water Jet Precision Manufacturing. In Bley H., editor, Proceedings of the 36th CIRP-ISMS: Progress in Virtual Manufacturing Systems, pp. 507-512, Saarbruecken, Germany, 03-05 June 2003.

[5] J. Valentincic, M. Junkar: On-line selection of rough machining parameters. Journal of Material Processing Technology. Vol. 149, Issue 1/3, 2004, Pages 256-262.

[6] S. Bigot, A. Ivanov, K. Popov: A study of the micro EDM electrode wear, First Multi-Material Micro Manufacture (4M) Conference, Karlsruhe, Germany 2004, Pages 355-358.

[7] J. Valentincic, M. Junkar: Detection of the eroding surface in the EDM process based on the current signal in the gap. International Journal of Advanced Manufacturing Technology, Vol. 26, Issue 3/4, 2006, Pages 294-301.

[8] B. Jurisevic, K.C. Heiniger, A. Schuetz, and M. Junkar: Feasibilities of Abrasive Water Jet Multipass Cutting Technique. In Summers D., editor, Proceedings of the 2003 WJTA American WaterJet Conference, paper 2-A, Houston, Texas, USA, 16-19 August 2003.

Multi-Material Micro Manufacture
W. Menz, S. Dimov and B. Fillon (Eds.)

Sensor-less measurement of rotational speed for piezoelectric micro-motors

A. Venturi and S. A. Wilson

Materials Department, School of Industrial and Manufacturing Science, Cranfield University, Cranfield Bedfordshire, MK43 0AL United Kingdom

Abstract:

Accurate measurement of speed, acceleration, blocking force and torque for actuators on the micro-scale is problematic by conventional means. Forces tend to be very low and time-scales relatively short. This paper describes experimental research on a non-contact technique that can be used to measure the performance of an ultrasonic piezoelectric micro-motor. A type of sensor-less speed measurement has been achieved by studying the frequency spectrum of the real power absorbed by two different motors. This shows a characteristic spike at low frequencies corresponding to the equilibrium rotational speed. The origin of this low frequency modulation is dependent on the impedance of the motors and the applied mechanical loads.

Keywords: Speed measurement; PZT; micro-motor; real power, flextensional

Introduction

Non-contact systems which do not cause a disturbance in the motion of the rotor appear to be the only means of accurately measuring the performance of motors that operate in the nNm torque range [1]. Rotary encoders are the most versatile and economical non-contact methods of measuring the position and speed of a fast-moving rotor. They avoid some of the problems that afflict the other transducers such as non linearity, drift, low resolution, low accuracy and a relatively poor performance at low velocities [2]. They are very commonly used in a whole range of automatic control systems

For the case of electromagnetic motors, techniques have been developed that can replace conventional speed transducers (electromechanical or electro-optical) by measuring the speed directly from the electrical signal applied to the motors. The basis of these techniques is a frequency analysis of the electric signal that drives the motor and the goal is to identify the harmonic that is linked to the rotational motor speed. Different methods have been studied, firstly based on analogue filters [3]. More recently numerical solutions have been derived through digital evaluation of the electrical harmonic spectrum [4] [5]. In this paper we now introduce a technique to measure the average speed of piezoelectric ultrasonic motors by observation of the fluctuations induced in the power supply by electromechanical coupling to the rotor.

Description of the experiment

The real electrical power absorbed by a piezoelectric micro-motor has been measured

and its fluctuations correlated to the output from a magnetic position signal, in order to see if those fluctuations are regular and repeatable at every revolution. To demonstrate the validity of the method a square wave signal is triggered from the power signal and it is used as a comparison to indicate the speed of the motor.

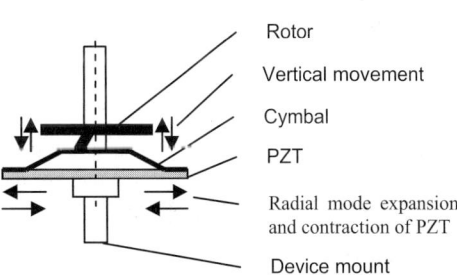

Figure 1: Schematic diagram of the flextensional motor construction

Two different piezoelectric flextensional micro-motors (Flexmotor) fabricated at Cranfield University have been subject of this study. These are standing-wave type motors, working with a single phase electrical supply. A PZT plate, driven by an electrical sinusoidal wave, produces horizontal vibrations that are amplified and transformed in vertical displacement by the cymbal; the cymbal is a beryllium copper flextensional amplifier. The motor converts the vibrations, produced by the PZT plate and amplified by the cymbal, to rotational movement of the rotor by employing elastic fins (Figure 1). Figure 2 illustrates how the friction drive operates. When the vibrator displaces upward, the fins bend under preload and force the rotor to move. When the vibrator then displaces downward, the fins slip across the vibrator surface and the cycle recommences.

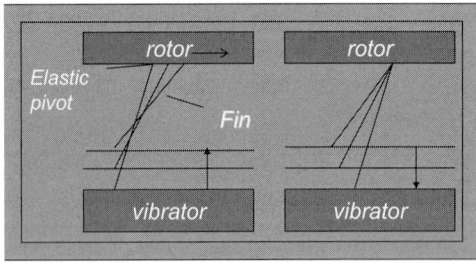

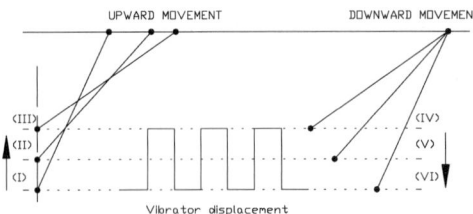

Figure 2: A schematic diagram showing an elastic fin as a rigid strut and the sequence of movements with the motor in operation.

The advantage of these motors lies in their simplicity, both of fabrication and of the electrical supply [6]. In this study two different motors have been tested with different diameters: 25mm and 10mm. The performances of these motors in terms of speed and stall torque are: 0.2 mNm and 240 rpm with 32 V applied for the 10 mm motor, 0.42 mNm and 520 rpm with 60V applied for the 25 mm motor.

Figure 3: Automated test system for the motors

A magnetic sensor system was set up to observe the movement of the rotors. A piece of magnetic material was attached to each of the rotors and a Hall Effect magnetic sensor has been positioned close by in order to measure the rotating magnetic field. Figure 3 shows the

testing rig set up to perform measurements system on the two motors. It incorporates many features used to simultaneously measure torque, speed and power for the motors [7].

To display the real power absorbed by a load driven by a periodic signal it is sufficient to multiply the value of the current passing through the load and the value of the voltage drop on it and integrate over each cycle. [2]. The electrical current passing through the motors was calculated from the voltage drop on a small resistor (1Ohm) placed in series with the motor. The voltage signal was measured using a voltage divider via a signal multiplier. The voltage and the current signals were sent to an Analogue Devices signal multiplier AD633 and its output was low pass filtered in the first order with the pole around 1 Hz. The output of the filter and of the Hall sensor was collected by a National Instruments DAC board (NI 6023E A/D) connected to a personal computer. Three dynamometers by Omega Engineering (LCL-113G for torque and LCL-816G for bias force) measure the torque and the bias force acting on the motors. These are used to build up a full characterisation of motor performance. A program, written in the LabView environment, permits data storage and calculations of the relevant parameters. These are the frequency, the amplitudes of the current and voltage, the phase shift between them, the speed of the motor and the value of the forces acting.

Speed measurements were made using a fiber optic probe by Keyence; with this instrument it is possible to generate a digital signal with a frequency proportional to the speed of the motor. The NI 5102 high speed digitizer counts the digital impulses form the fibre-optic probe, picked up by the NI 6023E, to give a measurement of the speed.

In order to generate a square wave signal from the probe output of the low pass a Schmitt trigger has been designed using an operational amplifier CA3140 by Harris Semiconductors. In this circuit the operational amplifier is configured in positive feedback. This permits two different thresholds to be set, separated by a differential. If the signal is smaller than the first threshold, the operating point, then the output is high. When the operating point is reached, the output passes to the low state and remains at this state until the input reduces to the second threshold called the release point. This technique prevents the circuit from responding to a sudden burst of fast noise. Moreover the transitions are both very fast and independent from the speed of the signal because are driven by positive feedback [8].

Results

Working with the circuit to measure the real power taken by the 10 mm and 25 mm piezomotors it has been determined that the amplitude of signal from the power supply is modulated with a frequency that is identical to the frequency of rotation of the motor. Figures 4 and 5 show the two signals of the Hall sensor and of the real power taken with the 10 and the 25 mm motors. The magnetic signal gives a precise reference according to the position of the rotor.

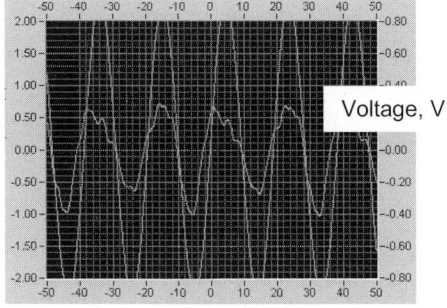

Time, ms

Figure 4: The Hall sensor sinusoidal signal and the power signal of the 25 mm motor.

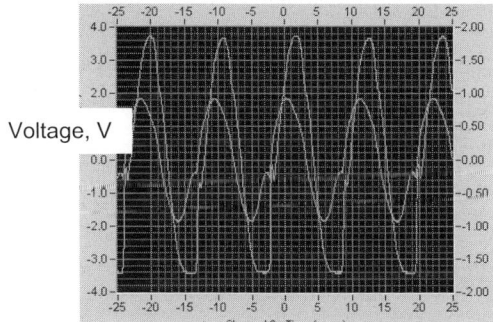

Time - (tenths of a second)

Figure 5: The Hall sensor sinusoidal signal and the power signal of the 10 mm motor.

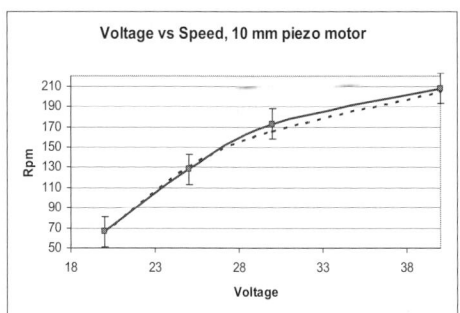

Figure 6: Speed measurement derived from the real power signal compared with measurements taken using a fibre-optic probe

The precise time-correlation of the signals in this case is more important than their magnitude and it is suggested, therefore, that it is possible to use information from the real power signal as a sensor-less speed measurement system.

Figure 6 shows the characteristic voltage/speed curve of the 10 mm motor measured by the two different techniques, the sensor-less (continuous line) and with the fibre-optic probe (dotted line). The root mean square deviation is 15% of the average at low voltage when the movement of the motor is more irregular. This decreases to 9% at high voltage.

Conclusions

The power signal results to be a reliable way to check the position of the rotor during the running of the motor. It is not possible for the moment to use this technique to measure the acceleration of the motor, because when the motor is starting and stopping, the power signal becomes irregular. The reason why of this behaviour is that the origin of the modulation is the movement of the fins on the top of the metal cap. The fins by moving change the distribution of the load and change so the electrical characteristic of the motor; this changing is repeated every revolution, but before that it becomes regular it is necessary wait for some revolutions.

References

[1] Brenner W, F. Suemecz, Evaluation of rotating Microsystems The 10th of International symposium on transport phenomena and dynamics of rotating machinery, Honolulu Hawaii, 2004

[2] Doebelin O Measurement Systems 1990, McGraw-Hill Publishing Company

[3] Ferrah A., K.J. Bradley, G.M. Asher An FFT based Novel Approach to non-invasive Speed Measurement in Induction Motor Drives. IEEE Transactions on Instrumentation and Measurement 1992

[4] Hammerli B., R.Tanner, R. Zwicky A rotor speed detector for induction machines utilizing rotor slot harmonics and an active three phase injection. 2nd European Conference on Power Electronics Applications (EPE), 1997

[5] [5] Hurst K.D., T.G. Habetler Sensor-less speed measurement using current harmonic spectral estimation in induction machine drives 0-7803-1859-5/94 IEEE 1994

[6] Leinvuo J.T. Flextensional piezoelectric motor, PhD Thesis, 2004 Cranfield University

[7] P.J. Rayner, S.A. Wilson, R.W. Whatmore and M. Cain. An automated performance testing system for piezoelectric micro-

motors. IEE Seminar on MEMS Sensors and Actuators, London April 27-28th 2006

[8] Schauder C. Adaptive Speed Identification for vector control of induction motors without a rotational transducer IEEE-IAS Annual Meeting Conference Record, 1989

Multi-Material Micro Manufacture
W. Menz, S. Dimov and B. Fillon (Eds.)

Flow control for high-pressure micro hydraulics

A.J.M. Moers, D. Reynaerts

Department of Mechanical Engineering, Faculty of Engineering, Katholieke Universiteit Leuven
Celestijnenlaan 300B B-3001 Heverlee, Belgium

Abstract

In order to achieve a high force density in microsystems, hydraulics can be used. In our current research, these actuators are used to control surgical instruments inside the human body. For this kind of hydraulic microrobots, flow rates of 100 mm^3.s^{-1} at a supply pressure of 10 bar need to be controlled. Hydraulic valves will be incorporated into the system near the actuators. Existing valves do not meet the specifications. Conventional hydraulic systems are too large because the dimensions are not a critical design factor. They are developed for much higher flow rates and much higher supply pressure. Existing valves for micro hydraulic systems are designed to control low flow rates at low pressure. They are for instance suitable for lab on chip technology [1]. This paper describes the ongoing research on micro valves which is a part of the micro-hydraulics and micro-robotics research at the Micro and Precision Group of K.U.Leuven.

Keywords: micro hydraulics, high force and high power density micro-systems, system integration

1. Introduction and project background

To reduce the trauma to the patient minimally invasive surgery is gaining importance since the eighties. At present, several minimally invasive therapies (MIS) have totally replaced the conventional surgical procedures, or will be if the hospitals are rightly equipped with skilled surgeons and instruments. However, existing MIS methods, including robot-assisted methods (Fig. 1a) have several drawbacks:
- The surgeons lack kinaesthetic and tactile feedback.
- Reduced visual perception.
- Reduced positioning capabilities.

Research focuses on the improvement of the positioning capabilities of the instruments. Existing robotic assisted systems consist of a slave and a master side. At the master side, the surgeon controls the position of the instruments. At the slave side, a robotic manipulator actuates the instrument.

Current manipulators are based on conventional minimally invasive instruments. The instruments are placed at the end of long tubes that are inserted through trocars into the body. Four degrees of freedom are controlled at the outside and advanced systems have two additional degrees of freedom at the instrument tip by a cable transmission (Fig. 1b). To give the surgeon the possibility to operate within a larger volume without trocar replacement, to let him or her work behind organs and to increase the number of operations, instruments with more degrees of freedom inside the human body need to be actuated.

The control and actuation of up to ten degrees of freedom will not be possible with cable transmission systems. Alternative actuation methods are being investigated: shape memory alloy based actuators, polymer actuators and microhydraulics. At present, the research at KULeuven focuses on microhydraulics with physiological water salt solution as transmission fluid [2]. Valves to control the flow, sensors and drive electronics will be integrated into the mechanical structure.

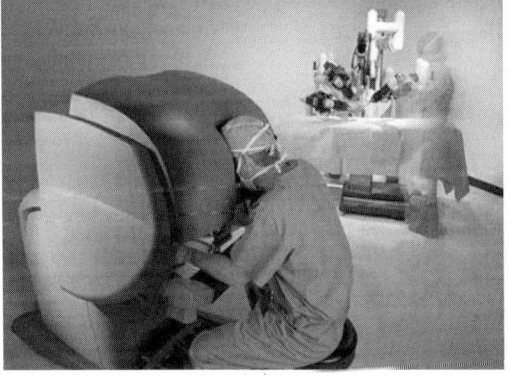

(a)

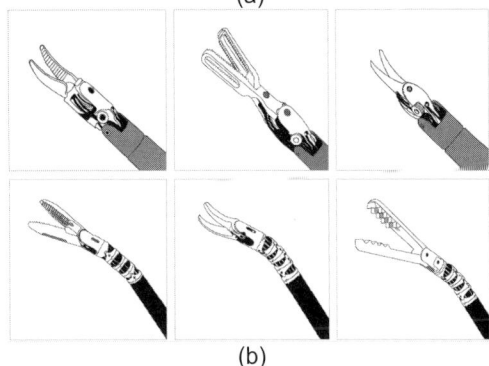

(b)

Fig. 1. (a) Robotic surgery using 5 DOF mechanically actuated instruments (b) Endowrist Cable operated instruments (Intuitive Surgical)

1.1. Specification valve design

In order to achieve the required forces within the small dimensions available within surgical instruments, a supply pressure of 10 bar is necessary.

The maximal flow rate needs to be at least 100

mm^3.s^{-1} to achieve the required actuation speed. For the given medical application, the valve body may not exceed a length of 15 mm and the maximal diameter of the final design is 2 mm.

2. Concept

To avoid a large actuator to control the valve, the required control forces need to be minimised. A small size of the moving part results in low inertia forces. Friction between the moving parts and disturbance forces due to non-symmetric pressure distribution also require special attention.

2.1. Micro hydrostatic bearing

In order to reduce friction, the moving parts are centered by a hydrostatic miniature bearing that has been developed [3]. This bearing also acts as a restriction seal. The leakage that is still present is within the tolerance of the application. Because of the small size, a bearing principle with a simplified configuration to manufacture has been chosen. From the supply side the fluid flows first through a gap of 10 μm and secondly trough a gap of 5 μm [Fig. 2]. If a disturbance moves the cylinder out of the centre, the pressure distribution over the bearing will change and a counteracting force results.

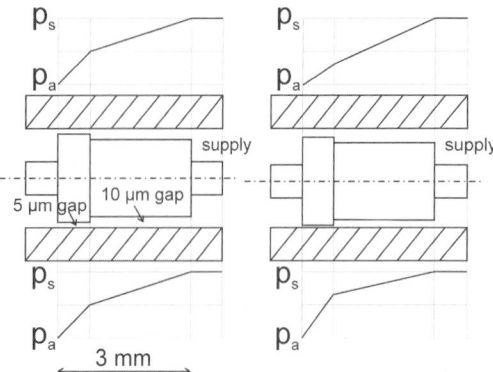

Fig. 2 Bearing principle

2.2. Symmetrical valve design

The geometry of the flow controlling part of the valve assures a symmetrical pressure distribution in all directions. As a result, disturbance forces due to the pressure distribution over the control lands will counteract each other. In figure 3, a schematic explanation of the working principle is shown. By rotating the valve over 90° the flow to the double working hydraulic cylinder can be adjusted.

The connections to each side of the cylinder are implemented twice at 180° to assure the symmetry. A drawback of this design choice is the increased complexity of the bearing housing. In figure 5, a schematic of the unfold valve explains the working. The supply of the system is at position 1. The flow is distributed over channels 2 and 3. At channel 1 and 3 the bearings are supplied at a constant supply pressure. The flow out or in the cylinders is over the shoulders. The gap size of the current prototype is 5 μm, which explains the narrow manufacturing tolerances.

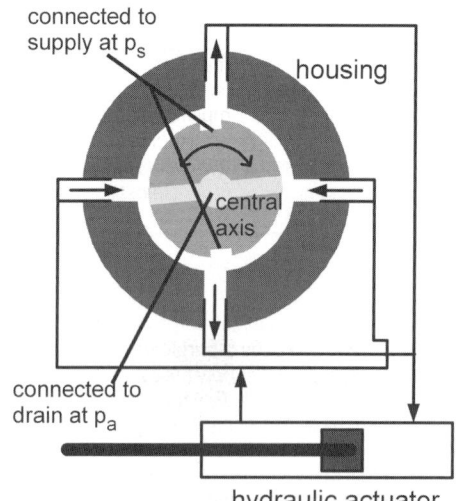

Fig. 3. Operating principle valve.

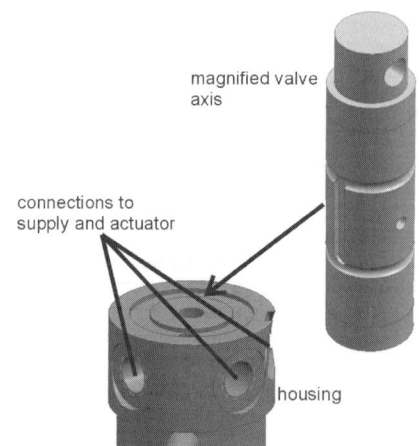

Fig. 4. Housing of the valve (left) and magnified valve axis (right)

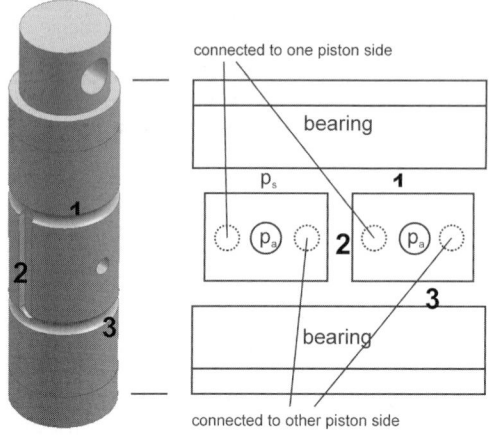

Fig. 5. Unfolded valve axis

3. Simulation

To speed up the simulations, finite element calculations have been performed over the two dimensional surface. This simplified method is possible because of the low height (maximally 20 µm) of the channels in ratio to the width of the channels.

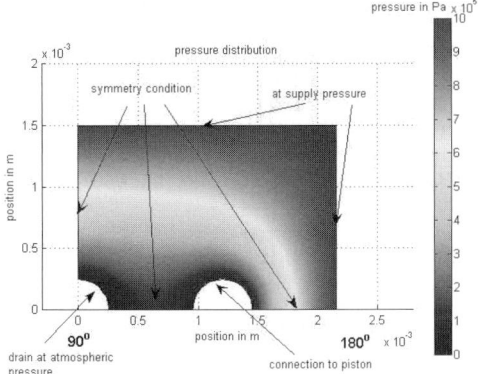

Fig. 6. Simulated pressure distribution

For non-idealised geometries the use of exact analytical models will become too complicated or will even be impossible. Analytical equations have only been used during the research in order to get a rough estimation of the behaviour.

The Reynolds equation [4] is implemented into the FEM model. Because of the low velocities and the fact that the bearings are hydrostatic the Reynolds equation can be rewritten in form [equation 1]:

$$\nabla\left[h^3 \nabla p\right] = 0 \qquad (1)$$

This equation is implemented in FEMLab as a subdomain condition. h is the gap size over the surface and p is the pressure. The subdomain is the surface in the two dimensional plane. At the boundaries of the bearing surface, boundary conditions need to be implemented. Constant pressure and symmetry conditions are used as boundary conditions in the program. Control calculations using control geometries that can be calculated analytically proved that the calculation method is valid.

4. Production

The hydraulic bearing has been produced using conventional micro production technologies. An overview of the parts is given in figure 8. First the housing has been produced on a KERN MMP micro-milling machine. The hole for the bearings and valve has been produced by drilling and reaming. This method does not assure repeatable dimensions. Therefore it was necessary to measure the dimensions after production.

The next step was the production of a matched valve axis (figure 7). The dimensions are thus according to the measured values of the housing dimensions. A precision lathe (Hembrug) with hydrostatic bearings has been used for the turning of

the axis. This method results in a relatively cheap production process with micrometer tolerances.

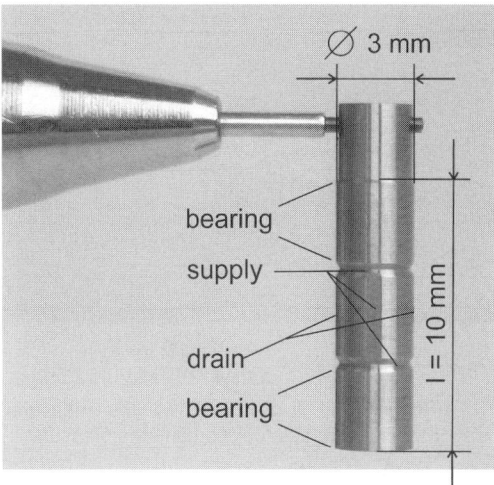

Fig. 7. Valve central axis (distributor)

The drains and grooves in the axis have been produced by micro EDM. Precision milling has been investigated as an alternative. As burs could in this case not be avoided EDM was finally selected.

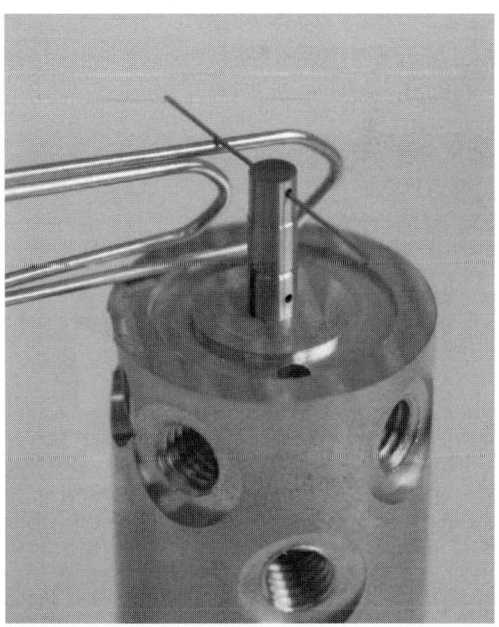

Fig. 8. Assembled valve

5. Experimental results

Three experiments have been performed to verify the behaviour of the valve.

One side of the valve has been closed and the other side of the valve connected to a pressure sensor (figure 9). The pressure has been measured over the full range of the valve.

396

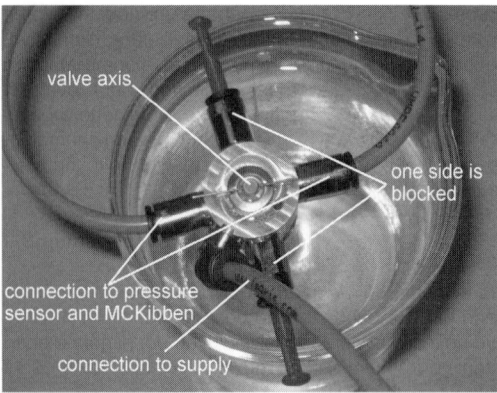

Fig. 9. Connected valve

In figure 10, a graph shows the measured relationship between the position and the pressure.

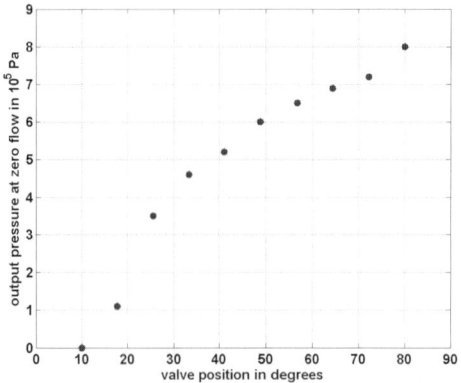

Fig. 10. Measured pressure

The flow rate has been verified at one position at atmospheric pressure at the outflow side. The simulated value is within 20% of the measured value of $100 \text{ mm}^3.\text{s}^{-1}$.

6. Conclusions and future research

In the near future, more measurements will be performed to verify the pressure/flow behaviour and to determine the necessary actuation force. This will be done in a set-up where the valve is connected to a two-way micro hydraulic piston of diameter 1 mm [5].

The rotating part of the present valve has a diameter of 3 mm and a length of 10 mm. In a next step a diameter of 2 mm will be envisaged.

The dimensions of the housing which were of no concern during this design and will be optimised when the remainder parts of the surgical system have been designed

Acknowledgements

The authors would like to thank the Institute for the Promotion of Innovation through Science and Technology in Flanders (IWT-Vlaanderen) and the EU FP6 NoE Multi Material Micro Manufacture (4M) for their support to this research project.

References

[1] Laser D.J. Santiago J.G. A review of micropumps. J. Micromech. Microeng. 14 (2004) R35-R64

[2] Peirs J. Reynaerts D. Van Brussel H. Design of miniature parallel manipulators for integration in self-propelling endoscope. Sensors and Actuators A, 85, (2000), pp409-417

[3] Findeisen D. & F. Öl-Hydraulik, Handbuch für hydrostatische Leistungsübertragung in der Fluidtechnik. Springer-Verlag, 1994

[4] Bassani R. Piccigallo B. Hydrostatic Lubrication. Elsevier, ISBN 0-444-88498-x, 1992.

[5] De Volder M. Peirs J. Reynaerts D. Coosemans J. Puers R. Smal O. Raucent B. Comparison of hydraulic microactuator configurations. Multi-Material Micro Manufacture conference 2005

AUTHOR INDEX

CARDIFF UNIVERSITY
SUSTAINABILITY THROUGH CHANGE, LOGISTICS & TECHNOLOGY

"Re-vitalising the UK manufacturing industry by developing and delivering sustainable solutions"

Cardiff University Innovative Manufacturing Research Centre is an exciting and novel multidisciplinary grouping of three of the UK's leading manufacturing and operations research teams:

- The Lean Enterprise Research Centre
- The Logistics Systems Dynamics Group
- The Manufacturing Engineering Centre

CUIMRC was established in 2004 under funding from the UK Government (Engineering and Physical Sciences Research Council) together with industry to assist UK manufacturing firms develop an *economically sustainable* future using lean, rapid and agile management and engineering approaches.

Why CUIMRC? because the UK manufacturing economy is changing.....

- increased competition from low labour cost economies such as China and East Europe;
- the long term shrinkage of the manufacturing sector as a percentage of GDP;
- the globalisation of supply chains and the increasing mobility of capital;
- the realisation that single tool solutions and consultant quick-fixes are not working.

Our holistic approach helps us tackle issues faced today by manufacturers, whilst ensuring that the solutions developed are implemented in a forward looking long term economically sustainable manner.

For more details and commercial and academic partnering opportunities please call or email us.

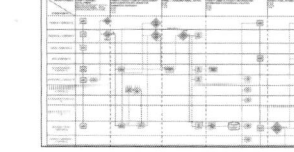

CUIMRC is part of Cardiff University and is funded by EPSRC

Cardiff University IMRC, Cardiff University, Aberconway Building, Colum Drive, CF10 3EU, Wales, UK
W www.cuimrc.cf.ac.uk **E** cuimrc@cardiff.ac.uk **T** 02920 879611 **F** 02920 879633

402

The special lens created by the MEC for Spire, the IR sensor for the Herschel Space Observatory. Looking ten billion light years into space. The School of Physics, Cardiff University using technology from the MEC

Some Results gained from the Welsh Assembly Government, DTI & ERDF funded programmes

Helping victims of Alzheimers disease using Daffodils grown in Wales. Alzeim Ltd. Talgarth, Powys, using expertise at the MEC

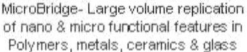

MicroBridge- Large volume replication of nano & micro functional features in Polymers, metals, ceramics & glass

Technically, the most advanced fly fishing line in the world. BVG Airflo, Brecon, Powys, using technology from the MEC

Manufacturing Engineering Centre at Cardiff University, winners of the Queen's Anniversary Prize and the DTI Prize for their work with industry.

MEC, Cardiff University, Queen's Building, The Parade, Cardiff CF24 3AA. Contact: Frank Marsh, Marketing Director. Tel: 029 20 874641 Email: manufacturing@cf.ac.uk, web: www.mec.cf.ac.uk.

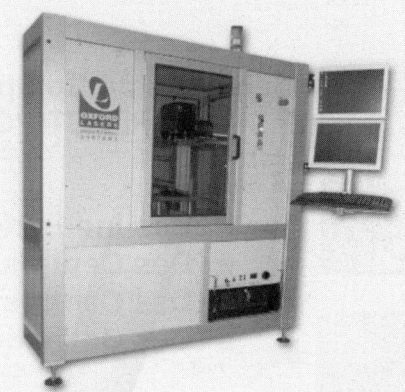

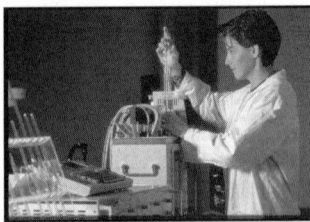

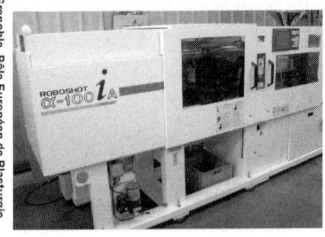

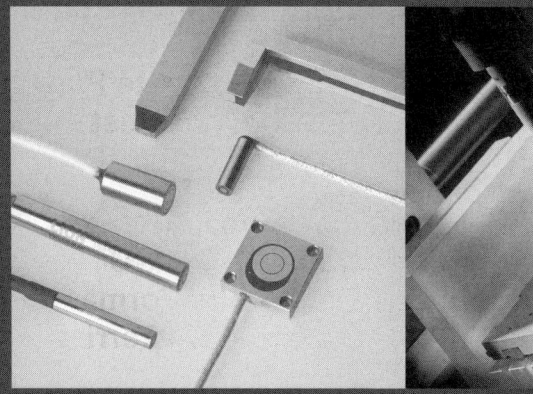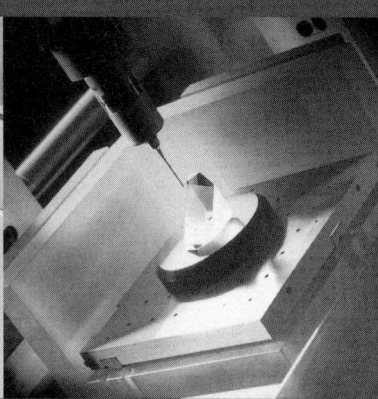

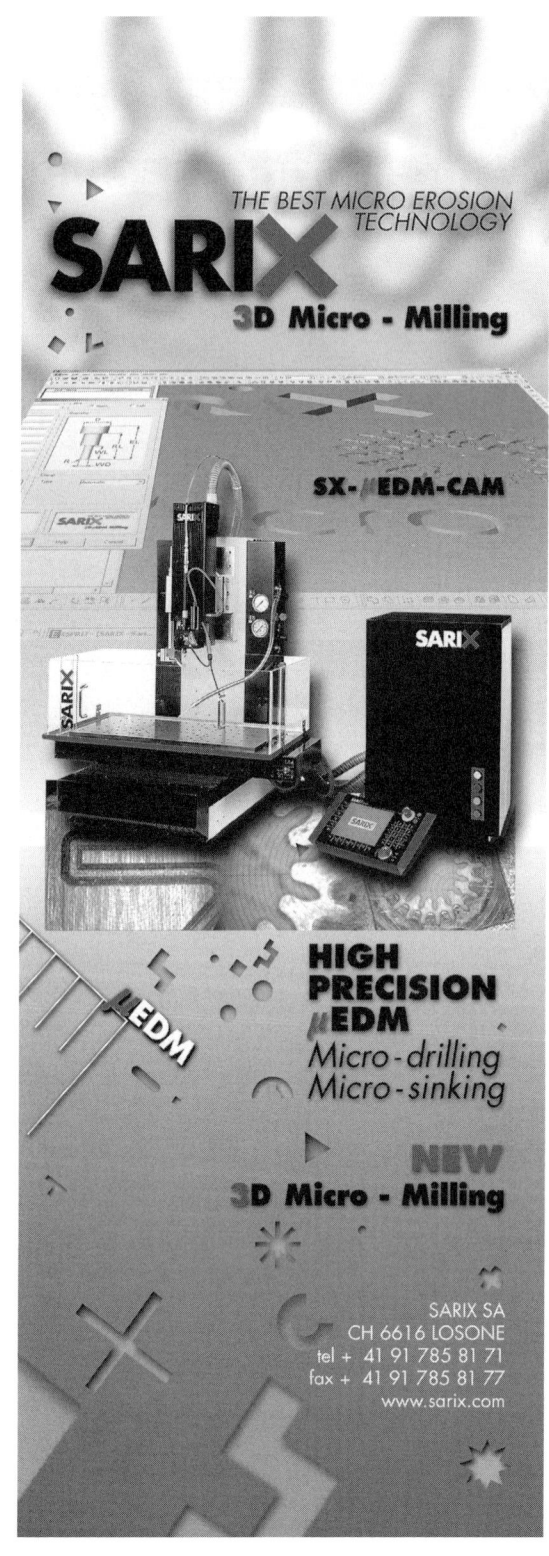

Llywodraeth Cynulliad Cymru
Welsh Assembly Government

The Welsh Assembly Government are proud to sponsor the 2006 4M Network of Excellence on Multi-Material Micro Manufacture.

Our Technology and Innovation Team support businesses and academia and encourage and recognise innovation to embrace the technological change needed to meet the competitive challenges of manufacturing in the 21st Century. The assistance we provide includes:

- Professional and impartial advice from specialist advisers
- Networking opportunities and strategic sector support
- Support for new product and process development
- Access to specialist equipment, facilities and expertise
- Links to internationally respected academic resources

SUBJECT INDEX